ANALYSIS

PART II: INTEGRATION, DISTRIBUTIONS, HOLOMORPHIC FUNCTIONS TENSOR AND HARMONIC ANALYSIS

KRZYSZTOF MAURIN
University of Warsaw

ANALYSIS

PART II

Integration, Distributions, Holomorphic Functions, Tensor and Harmonic Analysis

D. REIDEL PUBLISHING COMPANY
DORDRECHT : HOLLAND/BOSTON : U.S.A./LONDON : ENGLAND

PWN—POLISH SCIENTIFIC PUBLISHERS
WARSAW

Library of Congress Cataloging in Publication Data (Revised)

CIP

Maurin, Krzysztof.
Analysis.
Rev. translation of the author's Analiza.
Includes indexes.
CONTENTS: pt. 1. Elements. — pt. 2. Integration, distributions, holomorphic functions, tensor and harmonic analysis.
1. Mathematical analysis. I. Title.
QA300.M35613 515 74-80525
ISBN 90-277-0484-8 (v. 1)
ISBN 90-277-0865-7 (v. 2)

Revised and enlarged edition based on the original Polish
Analiza, część I–II, PWN–Państwowe Wydawnictwo Naukowe,
Warszawa 1971–1973

Chapters XII, XIII, Sections 3 and 4 of Chapter XV, Sections 1–6
of Chapter XVI, and Chapters XVII–XIX translated from the Polish
by *Eugene Lepa*

Chapter XIV, Sections 1, 2 and 5–7 of Chapter XV, and Sections 7–20
of Chapter XVI written in English by Krzysztof Maurin

Distributors for the U.S.A., Canada and Mexico
D. REIDEL PUBLISHING COMPANY, INC.
Lincoln Building, 160 Old Derby Street, Hingham, Mass. 02043, U.S.A.

Distributors for Albania, Bulgaria, Chinese People's Republic, Cuba, Czechoslovakia, German Democratic Republic, Hungary, Korean People's Democratic Republic, Mongolia, Poland, Rumania, the U.S.S.R, Vietnam and Yugoslavia
ARS POLONA
Krakowskie Przedmieście 7, 00-068 Warszawa 1, Poland

Distributors for all remaining countries
D. REIDEL PUBLISHING COMPANY
P.O. Box 17, Dordrecht, Holland

Printed in Poland by D.R.P.

TO MY PUPILS

PREFACE TO PART II OF THE ENGLISH EDITION

Avant touts, je pense que, si l'on veut faire des progrès en Mathématiques, il faut étudier de maîtres et non les élèves.

N. H. Abel

The mathematical universe is inhabited not only by important species but also by interesting individuals

C. L. Siegel

Die Mathematik ist ein Organ der Erkenntnis und eine unendliche Verfeinerung der Sprache. Sie erhebt sich aus der gewöhnlicher Sprache und Vorstellungswelt wie eine Pflanze aus dem Erdreich und ihre Wurzeln sind Zahlen und einfache räumliche Vorstellungen. Wir wissen nicht, welcher Inhalt die Mathematik als die ihm allein angemessene Sprache verlangt, wir können nicht ahnen in welche Ferne und Tiefe dieses geistige Auge Mathematik den Menschen noch blicken lässt.

E. Kähler

A truly realistic mathematics should be conceived, in line with physics, as a branch of the theoretical construction of the one real world, and should adopt the same sober and cautious attitude toward hypothetic extensions of its foundations as is exhibited by physics.

H. Weyl

Every word, just as every name, carries with it a specific content of its own, and gives rise to various associations, depending on the personal experience of its user. Hence, the word "analysis" means something different to every mathematician. For some, it encompasses not much more that the "differential and integral calculus", whereas for others it is associated with the Riemann–Roch theorem or with harmonic forms.

The position of analysis in mathematics is a special one, quite different from that of algebra, the theory of numbers, or the theory of sets. It is not an independent discipline: it is based on topology and algebra. It came into being at a relatively late date, its original aim being "to satisfy the needs" of mechanics and geometry. The problems which have cropped up within it have led to the formation of such vast "independent" disciplines as set theory, topology, and functional analysis.

The extraordinarily rapid advances made in mathematics since World War II have resulted in analysis becoming an enormous organism spreading in all directions. Gone for good surely are the days of the great French "courses of analysis" which embodied the whole of the "analytical" knowledge of the times in three volumes—as the classical work of Camille Jordan. Perhaps that is why present-day textbooks of analysis are disproportionately modest relative to the present state of the art. More: they have "retreated" to the state before Jordan and Goursat. In recent years the scene has been changing rapidly: Jean Dieudonné is offering us his monumentel *Elements d'Analyse* (10 volumes) written in the spirit of the great French *Course d'Analyse.*

To the best of my knowledge, the present book is the only one of its size: starting from scratch—from rational numbers, to be precise—it goes on to the theory of distributions, direct integrals, analysis on complex manifolds, Kähler manifolds, the theory of sheaves and vector bundles, etc.

My objective has been to show the young reader the beauty and wealth of the unsual world of modern mathematical analysis and to show that it has its roots in the great mathematics of the 19th century and mathematical physics. I do know that the young mind eagerly drinks in beautiful and difficult things, rejoicing in the fact that the world is great and teeming with adventure.

In addition to expounding completed theories, the principal motive for writing this book was that of showing the horizons of analysis and encouraging the reader to embark on an independent journey through the original works of great masters (cf. first Motto).

Whereas the dominant concept of Part I was the derivative (differential) and its various applications, the central concept of Part II is the integral (and its natural extension: the distribution of Laurent Schwartz).

Modern analysis requires general topological spaces: metric spaces no longer suffice and accordingly Part II begins with a systematic exposition of general topology. I have drawn upon the excellent concise treatment of G. Köthe, one of the fathers of the modern theory of locally convex spaces. Since I did not want to bore the reader with such an extensive and abstract presentation of topological structures, I put paracompact spaces and inductive and projective limits where I needed them, e.g. in the discussion of the theory of distributions. The theory of covering manifolds (fundamental group) is presented in Chapter XVI on Riemann surfaces: H. A. Schwarz and H. Poincaré conceived these notions in order to master the uniformization.

Chapter XIII deals with the general theory of the integral as treated by Daniell–Stone. As we know, there are two paths: "from measure to integral" and "from integral to measure". I took the second for the following reasons: (i) It is somewhat more general. (ii) It leads more rapidly to theorems of fundamental importance for applications on the passing to the limit under the integral sign and on changing the order of integration (Fubini, Tonelli). (iii) The most important type of integral in analysis, the Radon integral, is a special case of the Daniell–Stone integral. (iv) The transition to the distribution is very natural: the distribution is a natural extension of the concept of the Radon integral. Of course, not to show the young reader the path from "measure to integral" would be to commit the sin of being fanatic; this path has been shown in Section 30. The excellent mimeographed lecture notes "Integral" by Professor Mikusiński proved to be of much assistance to me in writing the chapter on the integral; it gives me great pleasure to express my wholehearted thanks to my former superior. A natural generalization of the integral of a vector function is "the direct integral," a concept introduced by von Neumann and independently somewhat later (but much more elegantly) by R. Godement for the needs of the theory of representations of concompact groups and the quantum theory. Accordingly, direct integrals and their connection with quantum mechanics end the chapter on integration. Interesting applications of the concept will be found in Part III and, in greater depth, in my previous monographs: *Methods of Hilbert Spaces* and *General Eigenfunction Expansions and Unitary Representations of Topological Groups.*

Classical harmonic analysis, Fourier transformation, Fourier series was crowned by "abstract harmonic analysis" and the theory of representations of topological groups. In order to show these perspectives to the reader a number of fundamental and now classical theorems of this theory have been given, without proof, in the final sections of this part. They are afforded a systematic exposition in Part III. The principal tool in general harmonic analysis is the Haar integral. Therefore, this part of *Analysis* ends with the proof of the existence and uniqueness of the Haar integral.

PREFACE

With his theory of distributions, Laurent Schwartz revolutionized modern analysis: today one can scarcely think of equations with partial derivatives or of harmonic analysis without the distribution concept. In teaching the theory of distributions several years ago, I had at my disposal the splendid lectures of L. Hörmander, which that great mathematician later incorporated into the beginning of his now classical monograph on partial differential equations. The later sections of *Analysis* also owe much to various, mainly mimeographed, lecture notes by Hörmander: when writing the chapters on complex analysis I often drew on the Hörmander treatment.

The central block of the book consists of three chapters XIV–XVI and constitutes what has always been regarded as the core of analysis. Chapter XIV comprises what was one called "surface integrals and their applications" and what we now call the theory of differential forms and their integrals on manifolds with a boundary. The main theorem is, of course, the Poincaré–Stokes theorem; what remains are its applications, culminating in de Rahm's theorems and the theories of harmonic forms of Hodge and Kodaira and of invariant forms in the Lie group. Since the origin of those theories can be found in electrodynamics—what is emphasized by de Rahm's notion of "current"—I could not resist writing a short section on electrodynamics.

The second source of the theory of harmonic forms is the Riemannian geometric theory of functions. It is dealt with in the next two chapters. For Riemann, complex analysis was inseparably connected with physics (potential theory). That is why Chapters XV and XVI are mainly devoted to the potential theory. The theory of Riemann surfaces is undoubtedly the core of mathematics; the requirements of that theory led to the creation of general and algebraic topology, the theory of discontinuous groups of transformations and the theory of function-fields. The theory of Riemann surfaces shows much more clearly than any other theory that the "mathematical universe is inhabited not only by important species but also by interesting individuals" (Siegel). Such are undoubtedly the Weierstrass $\wp$-function, the elliptic modular function, and the $SL_2(\boldsymbol{Z})$ group (Chapter XVI). An interesting phenomenon can now be observed: after a period of constructing very general "abstract" theories, mathematicians are now returning to the sources. It can easily be seen that the great masters of the 19th century knew very well what they were doing (S. Lang). We realize that great mathematicians, like great artists, must not be corrected.

It would be very difficult for me to enumerate all the magnificent works which I consulted while writing this volume. I will only mention the textbooks of Henri Cartan, S. Lang, H. Nickerson – D. Spencer – N. Steenrod and the monographs of L. Ahlfors, R. Nevanlinna, A. Pfluger, O. Forster, and G. Shimura.

As I was reading those brilliant books and trying to make use of them in the preparation of my own, I was haunted by the question on whether it was sensible to distort such supreme edifices by dislodging their fragments. Surely they are perfect. But my textbook sets itself much more modest aims...

This volume owes even more than Part I does to those who were formerly my pupils and are now my colleagues. I lack the words to express my gratitude for their painstaking perusal, on their own initiative, of the thousands of pages of my manuscript and bringing it to a state fit for publication. As regards Chapter XIV, K. Gawędzki wrote large parts of it afresh. I will only mention the names, *sapienti sat*, of the others: J. Komorowski and A. Strasburger (chapter on topology), J. Napiórkowski (theory of the integral), A. Wawrzyńczyk and P. Urbański (theory of distributions and harmonic analysis), A. Strasburger (one-dimensional complex analysis); W. Pusz, J. Kijowski (numerous invaluable critical remarks).

I dedicate this volume to them all.

K. MAURIN

Zakopane, August 1978

TABLE OF CONTENTS

XII. TOPOLOGY
UNIFORM STRUCTURES
FUNCTION SPACES

The present chapter differs in its spirit from the others. Its aim is both to systematize the facts presented in the previous chapters and to generalize them for needs of the modern analysis. In particular, this will concern so-called "uniform notions": uniform continuity, equicontinuity, uniform convergence, completion of (uniform) spaces, etc.

Obviously, this chapter cannot replace such extensive topology courses as those of Bourbaki or Engelking, and several theorems are given without proofs. However, three of the most important theorems, i.e. those of Tychonoff, Baire and Ascoli are proved in details. The reader being in a hurry to get to the "real analysis" may skip this chapter during the first reading and, if necessary, may go back to study those "three theorems".

1. TOPOLOGICAL SPACES

In Chapter II we saw that an enormous role is played by the family $\{\mathcal{O}_\alpha\}$ of all open sets of a metric space. Many fundamental concepts, such as continuity and connectedness, were defined in terms of open sets. An idea that thus suggests itself is that we construct a theory of spaces in which it is meaningful to speak of open sets, or to be more precise, to consider a class of spaces in which a family of sets having certain properties of open sets in metric space is distinguished axiomatically. Such a theory is due to the great German mathematician Hausdorff (Kuratowski introduced other axioms by dealing with the closure operation; cf. Exercise, Chapter II, p. 40). The present chapter treats these general spaces and their mappings.

If (X, d) is a metric space, then the family $\mathcal{T} = \{\mathcal{O}_\alpha\}_{\alpha \in A}$ of open sets has the following properties:

(O.1) The entire space X and the null set $\emptyset$ are open.

(O.2) The union $\bigcup_{\alpha \in I} \mathcal{O}_\alpha$ (of any cardinality) of open sets is an open set.

(O.3) The intersection of a finite number of open sets is an open set.

(O.1) follows directly from the definition of open set in a metric space, while (O.2) and (O.3) were proved as Theorem II.4.1.

A definition of topological space may now be given.

DEFINITION. A *topological space* is a pair $(X, \mathcal{T})$, where $\mathcal{T} = \{\mathcal{O}_\alpha\}_{\alpha \in A}$ is a family of subsets of a nonempty set X, which satisfy axioms (O.1)–(O.3). The family $\mathcal{T}$ is called a *topology* (*topological structure*), and its elements $\mathcal{O}_\alpha$ are the *open sets* of the topological space $(X, \mathcal{T})$. If we are not interested in denoting the topology of a topological space, we shall write simply X instead of $(X, \mathcal{T})$.

Here is a glossary of basic topological terms—all are straightforward generalizations of those already encountered in the context of metric spaces.

A set $F \subset X$ is called *closed* if its complement is open. It is easily verified that the intersection of any family of closed sets is closed and the union of any finite family of closed sets is closed. Also $\emptyset$ and X are closed.

Given a point $x \in X$ or a set $Z \subset X$, any set $U(x)$ ($U(Z)$) such that there exists an open set $\mathcal{O}$ with $x \in \mathcal{O} \subset U(x)$ ($Z \subset \mathcal{O} \subset U(Z)$), is said to be a *neighbourhood of the point* x (*the set* Z).

The *closure of a set* A denoted $\bar{A}$, is the intersection of all closed sets containing A. It is obviously the smallest closed set containing A.

Dually, the *interior of a set* $A \subset X$, denoted $\operatorname{Int} A$ or $\mathring{A}$, is the greatest open set contained in A. Let us remark that for a nonempty set its closure is always nonvoid, whereas its interior may be, and often is, empty. A subset of a topological space is called *dense* if its closure is the whole space.

Another frequently encountered term, a *boundary of a subset* A of a topological space X, is defined as the intersection of the closures of A and that of its complement, i.e.

$$\partial A := \bar{A} \cap \overline{(X - A)}.$$

So general a concept of a space is much too broad for applications. Accordingly, Hausdorff introduced an additional axiom which is plainly satisfied in the case of a metric space:

(O.4) For any two distinct points $x_1, x_2 \in X$ there exist open sets $\mathcal{O}_1, \mathcal{O}_2$ such that $x_i \in \mathcal{O}_i$ and $\mathcal{O}_1 \cap \mathcal{O}_2 = \emptyset$.

A topological space satisfying the axiom (O.4) is called a *Hausdorff space* or a space of type T_2.

Henceforth, we shall be dealing mostly with Hausdorff spaces. Later we shall sometimes bring in other, stronger axioms of separation than (O.4); this will take us to important classes of topological spaces (e.g. regular, normal, and others).

A topological space $(X, \mathcal{T})$ is said to be *metrizable* if a distance d compatible with the topology $\mathcal{T}$ is introducible in it, i.e. if the open sets of the space (X, d) form the topology $\mathcal{T}$. Many function spaces endowed with interesting topological structures are not metrizable. A distance was introduced in the set of bounded functions $B(X, \boldsymbol{R})$ and convergence in this metric space was called a *uniform convergence*. But such simple concept as that of pointwise (simple) convergence of functions $\mathcal{F}(X, \boldsymbol{R})$, leads to a topology which is (in general) nonmetrizable.

Other Ways of Introducing Topology. In a metric space (X, d) topology was defined by the family of balls, $\mathcal{B} = \{K(x, r)\colon x \in X, r \in \boldsymbol{R}\}$; every open set $\mathcal{O}$ is the union of a certain number of balls; in other words, balls constitute a "basis" of the topology of a metric space. Hence the following definition is adopted:

DEFINITION. A family $\mathcal{B} \subset \mathcal{T}$ is called a *basis of a topology* $\mathcal{T}$ when each set of $\mathcal{T}$ is the union of sets of the family $\mathcal{B}$. On the other hand, a family $\mathcal{P} \subset \mathcal{T}$ is called a *subbasis of the topology* $\mathcal{S}$ when the family of all finite products $\mathcal{O}_1 \cap \mathcal{O}_2 \cap \ldots \cap \mathcal{O}_k$, where $\mathcal{O}_i \in \mathcal{P}$, $i = 1, \ldots, k$, constitutes a basis of the topology $\mathcal{T}$.

Thus, if a (sub)basis is known, the topology is determined. The reader can easily verify that $\mathcal{B}$ is a basis of the topology if and only if

$$\bigwedge_{x \in X} \ \bigwedge_{U\text{—neighbourhood of } x} \ \bigvee_{\mathcal{O} \in \mathcal{B}} x \in \mathcal{O} \subset U.$$

It is easily shown that every basis $\mathcal{B}$ is characterized by the following two properties (which can thus be taken for the axioms of the basis):

(B.1) $$\bigwedge_{\mathcal{O}_1, \mathcal{O}_2 \in \mathcal{B}} \ \bigwedge_{x \in \mathcal{O}_1 \cap \mathcal{O}_2} \ \bigvee_{\mathcal{O} \in \mathcal{B}} x \in \mathcal{O} \subset \mathcal{O}_1 \cap \mathcal{O}_2,$$

(B.2) $$\bigwedge_{x \in X} \ \bigvee_{\mathcal{O} \in \mathcal{B}} x \in \mathcal{O}.$$

PROOF. (B.1) follows from the fact that $\mathcal{O}_1 \cap \mathcal{O}_2$ is an open set; (B.2) follows from the fact that X is an open set.

Now, suppose that, conversely, the family $\mathcal{B}$ satisfies conditions (B.1) and (B.2); we construct a topology $\mathcal{T}$, for which $\mathcal{B}$ is a basis, as follows:

$$(\mathcal{O} \in \mathcal{T}) \Leftrightarrow (\mathcal{O} = \bigcup_{s \in S} \mathcal{O}_s \text{ for a subfamily } \{\mathcal{O}_s\}_{z \in S} \text{ of } \mathcal{B}).$$

If for $\{\mathcal{O}_s\}_{s \in S}$ we take the empty family and whole $\mathcal{B}$ then we see that $\emptyset$ and the entire space X (cf. (B.2)) belong to $\mathcal{T}$. Likewise, (O.2) occurs. It is somewhat more difficult to verify (O.3): Suppose that $\mathcal{O}, \mathcal{O}' \in \mathcal{T}$; we thus have

$$\mathcal{O} = \bigcup_{s \in S} \mathcal{O}_s, \quad \mathcal{O}' = \bigcup_{t \in T} \mathcal{O}_t, \quad \text{where } \mathcal{O}_s, \mathcal{O}_t \in \mathcal{B}$$

for $s \in S, t \in T$.

Since

$$\mathcal{O} \cap \mathcal{O}' = \bigcup_{(s,t) \in S \times T} \mathcal{O}_s \cap \mathcal{O}_t,$$

it is thus sufficient to show that each $\mathcal{O}_s \cap \mathcal{O}_t \neq \emptyset$ is the union of sets of the family $\mathcal{B}$. Suppose that $x \in \mathcal{O}_s \cap \mathcal{O}_t$; by condition (B.1), there is a set $\mathcal{O}(x) \in \mathcal{B}$ such that $x \in \mathcal{O}(x) \subset \mathcal{O}_s \cap \mathcal{O}_t$. But

$$\mathcal{O}_s \cap \mathcal{O}_t = \bigcup_{x \in \mathcal{O}_s \cap \mathcal{O}_t} \mathcal{O}(x),$$

and hence axiom (O.3) is satisfied. □

2. A BASIS OF NEIGHBOURHOODS. AXIOMS OF COUNTABILITY

A topology is most often introduced by giving a *fundamental system of neighbourhoods*, or a basis of neighbourhoods of every point $x \in X$.

DEFINITION. If $(X, \mathcal{T})$ is a topological space then a *basis of the topology* $\mathcal{T}$ *at a point* $x \in X$ (a *basis of neighbourhoods of a point* x), is a family $\mathcal{B}(x)$ of neighbourhoods of the point x with the property that for every open set $\mathcal{O} \ni x$ there exists a neighbourhood $U \in \mathcal{B}(x)$ such that $x \in U \subset \mathcal{O}$. It is easily seen that if $\mathcal{B}$ is a basis of the topology $\mathcal{T}$, the family composed of all its elements containing x is a basis (of $\mathcal{T}$) at x.

It is also not difficult to show that if we take for every $x \in X$ a basis $\mathcal{B}(x)$ of neighbourhoods for x, then the family $\{\mathcal{B}(x)\}_{x \in X}$ satisfies the following three conditions (also due to Hausdorff):

(H.1) $$\bigwedge_{x \in X} \mathcal{B}(x) \neq \emptyset, \quad \bigwedge_{x \in X} \bigwedge_{U \in \mathcal{B}(x)} x \in U,$$

i.e. every point x has at least one neighbourhood from $\mathcal{B}(x)$ and lies in each of its neighbourhoods.

(H.2) $$\bigwedge_{x \in X} \bigwedge_{U_1, U_2 \in \mathcal{B}(x)} \bigvee_{U \in \mathcal{B}(x)} U \subset U_1 \cap U_2,$$

(H.3) $$\bigwedge_{x \in X} \bigwedge_{U \in \mathcal{B}(x)} \bigvee_{V \in \mathcal{B}(x)} \bigwedge_{y \in V} \bigvee_{W \in \mathcal{B}(y)} W \subset U.$$

Definition. A family $\{\mathcal{B}(x)\}_{x \in X}$, where each $\mathcal{B}(x)$ is a basis (of the topology $\mathcal{T}$) of a space $(X, \mathcal{T})$ at x, is called a *fundamental neighbourhood system of the space* $(X, \mathcal{T})$ or *of the topology* $\mathcal{T}$ (at times briefly: a *basis of neighbourhoods of the space* $(X, \mathcal{T})$).

As above, it is shown that a basis of neighbourhoods of a space determines its topology:

Given a set X and a family $\{\mathcal{B}(x)\}_{x \in X}$ satisfying conditions (H.1)–(H.3) by $\mathcal{T}$ we denote the family of all sets which are unions of subfamilies of the family $\bigcup_{x \in X} \mathcal{B}(x)$. Then the family $\mathcal{T}$ has properties (O.1)–(O.3), that is, it is a topology. The class $\{\mathcal{B}(x)\}_{x \in X}$ constitutes a fundamental neighbourhood system of the space $(X, \mathcal{T})$. Moreover, if the initial family $\{\mathcal{B}(x)\}_{x \in X}$ is a basis of neighbourhoods of some topology $\mathcal{T}'$, then the above-constructed topology $\mathcal{T} = \mathcal{T}'$; this is why we say that a basis of neighbourhoods of a topological space determines its topology.

Hausdorff also introduced the following important axioms.

Axioms of Countability. I. Every point has a countable basis of neighbourhoods.

II. A space has a countable basis of topology.

If it contains a dense countable subset, a space $(X, \mathcal{T})$ is said to be *separable*. It is not difficult to prove that $\big((X, \mathcal{T})$ satisfies the second axiom of countability$\big) \Rightarrow \big((X, \mathcal{T})$ is separable$\big)$, for upon taking one point from each element of the countable basis we obtain a dense countable subset. The converse does not hold in general; it is true, however, for metric spaces (cf. Theorem XII.6.1).

Examples. 1. Every metric space (X, d) satisfies the first axiom of countability: the sequence of balls $\{K(x, 1/n), n = 1, 2, \ldots\}$ may be taken for $\mathscr{B}(x)$.

2. $\boldsymbol{R}$, or more generally $\boldsymbol{R}^k$, with the ordinary Pythagorean metric satisfies the second axiom of countability. Indeed, the balls $\{K(p, 1/n)\}$, where p runs over rational points (i.e., with rational coordinates), $n = 1, 2, \ldots$, constitute a countable basis of the topology. Let $x \in X$ and let $\mathcal{O}$ be an arbitrary neighbourhood of the point x; a ball $K(x, 1/k) \subset \mathcal{O}$ thus exists: suppose that $d(p, x) < 1/2k$, then $x \in K(p, 1/2k) \subset \mathcal{O}$.

The role of convergent sequences in metric spaces is played, in the case of (general) topological spaces, by convergent nets. A basis $\mathscr{B}(x)$ of neighbourhoods of a point x is (canonically) directed by inclusion: $U_1(x) \prec U_2(x)$ when $U_2(x) \subset U_1(x)$; cf. Chapter III, Example 3, p. 65. Choosing a point $x_U \in U$ for every $U \in \mathscr{B}(x)$, we obtain a net which is convergent in the sense of the following

DEFINITION. Let $(X, \mathscr{T})$ be a topological space. A net $(x_\pi)_{\pi \in \Pi}$ of points of the space X *converges* to x when

$$\bigwedge_{U(x)} \bigvee_{\pi_0} \bigwedge_{\pi_0 < \pi} x_\pi \in U(x).$$

We denote by $\lim\limits_{\pi \in \Pi} x_\pi$ the set of points to which $(x_\pi)_{\pi \in \Pi}$ converges. In the case of a not convergent net $(x_\pi)_{\pi \in \Pi}$ we write $\lim\limits_{\pi \in \Pi} x_\pi = \emptyset$.

It is not difficult to prove the following

THEOREM XII.2.1. Let X be a topological space. (Every convergent net of points of X has only one limit) $\Leftrightarrow$ (X is a Hausdorff space).

Most of the theorems of Chapter II can now be carried over if "sequence" is replaced by "net"; for instance, we then have:

THEOREM XII.2.2. (A subset M of topological space $(X, \mathscr{T})$ is closed) $\Leftrightarrow$ (M contains the limits of all convergent nets with terms belonging to M).

PROOF. $\Rightarrow$: We need to prove the implications: (Let $x_\pi \in M$ and $x_\pi \to x$) $\Rightarrow$ ($x \in M$).

(a.a.) Suppose that $x \in X - M$ which is open. Thus there exists a neighbourhood $U(x) \subset X - M$. Since $x_\pi \to x$, $\bigvee\limits_{\pi_0} \bigwedge\limits_{\pi > \pi_0} x_\pi \in U(x)$; so all these $x_\pi \notin M$ (a contradiction).

$\Leftarrow$ (a.a.) Suppose that M is not closed, i.e. suppose that $X-M$ is not open. Thus there exists a point $x \in X-M$ such that every neighbourhood U from the basis $\mathscr{B}(x)$ has at least one point $x_U \in M$, but, as we know, the net $(x_U)_{U \in \mathscr{B}(x)}$ converges to x, and hence from the assumption it follows that $x \in M$, which is a contradiction! □

When topological spaces are introduced it is natural to distinguish the class of so-called continuous mappings. Let $T: X \to Y$ be a mapping between topological spaces $(X, \mathscr{T}_1)$ and $(Y, \mathscr{T}_2)$; we write also $T: (X, \mathscr{T}_1) \to (Y, \mathscr{T}_2)$.

DEFINITION. A mapping T is *continuous at a point* x_0 when for every neighbourhood $U(y_0)$ of the point $y_0 := T(x_0)$ there is a neighbourhood $\mathcal{O}$ of x_0 such that $T(\mathcal{O}) \subset U(y_0)$. A mapping T is *continuous* when it is continuous at every point of its domain.

The reader will easily prove

PROPOSITION XII.2.3. T: $(X, \mathscr{T}_1) \to (Y, T_2)$, ($T$ is continuous) $\Leftrightarrow$ $\big(T^{-1}(\mathcal{O}) \in \mathscr{T}_1$ for every $\mathcal{O} \in \mathscr{T}_2\big)$.

It can also be shown that the "sequential" definition (in the sense of nets) of the continuity of the mapping $T: X \to Y$ is equivalent to the above neighbourhood definition, i.e.

THEOREM XII.2.4. ($T: X \to Y$ is continuous at x_0) $\Leftrightarrow$ $\big(T(x_\pi) \to T(x_0)$ for every net $x_\pi \to x_0\big)$.

PROOF. $\Leftarrow$: Suppose that T is not continuous at x_0. Then there exists a neighbourhood V of the point $T(x_0)$ such that for every $U \in \{\mathscr{B}(x_0)\}$ there is a point $x_U \in U$ such that $T(x_U) \notin V$. However, the net $x_U \to x_0$, whereas $\big(T(x_U)\big)$ does not converge to $T(x_0)$.

$\Rightarrow$: Follows from the definitions of the continuity and the convergence of nets. □

We thus arrive immediately at the following

COROLLARY XII.2.5. Given topological spaces X and Y. (T: $X \to Y$ is continuous) $\Leftrightarrow$ $\big(T(\lim_{\pi \in \Pi} x_\pi) \subset \lim_{\pi \in \Pi} T(x_\pi)$ for every net $(x_\pi)_{\pi \in \Pi}$ in $X\big)$.

Comparison of Topologies. It is often necessary to be able to compare two topologies $\mathscr{T}_1, \mathscr{T}_2$ defined on the same set X. One obvious way to do this is the following. If the identity mapping from the topological space $(X, \mathscr{T}_2)$ to the topological space $(X, \mathscr{T}_1)$ is continuous, we shall say that $\mathscr{T}_1$ is *weaker* (*coarser*) than $\mathscr{T}_2$ and $\mathscr{T}_2$ is *stronger* (*finer*) than $\mathscr{T}_1$. The reader may convince himself that this happens precisely when every $\mathscr{T}_1$-open set is $\mathscr{T}_2$-open, i.e. $\mathscr{T}_1 \subset \mathscr{T}_2$.

Let us notice that the collection of all topologies on a given set becomes partially ordered set with respect to the relation "to be weaker than". This partial order has the maximal element which is the so-called *discrete topology* consisting of all subsets of a given set, and also minimal element (sometimes called *antidiscrete* or the *coarsest topology*) consisting only of the empty set and the whole space.

Topological Subspace. It is easily seen that if M is a subset of a topological space X with a topology $\mathscr{T} = \{\mathcal{O}_\alpha\}_{\alpha \in A}$ then the family $\mathscr{T}' := \{M \cap \mathcal{O}_\alpha\}_{\alpha \in A}$ is a topology on M. It is called the *induced topology*; more precisely, the topology induced on M by $\mathscr{T}$. The topological space $(M, \mathscr{T}')$ is a (*topological*) *subspace* of the space $(X, \mathscr{T})$. In this situation, the canonical injection $M \to X$ is a continuous mapping. Let us observe that the sets of the form $M \cap Z$, where Z is closed in X, are all closed sets in the subspace $(M, \mathscr{T}')$.

3. FILTERS

The concept of a filter is closely related to the concept of net. It was introduced by H. Cartan and has been used systematically by Bourbaki. Since it is frequently employed in functional analysis, we give in brief a definition of filters, their relation to nets, and several straightforward theorems. How the concept of filter is applied will be demonstrated in the proof of Tychonoff Theorem.

Note that the family $\mathscr{F}$ of all neighbourhoods of a point x of a topological space X satisfies the conditions:

(F.1) If $F \in \mathscr{F}$ and $F \subset U \subset X$, then $U \in \mathscr{F}$.

(F.2) If $F_1, F_2 \in \mathscr{F}$, then $F_1 \cap F_2 \in \mathscr{F}$.

(F.3) $\emptyset \notin \mathscr{F}$.

In general, the following definition is adopted:

DEFINITION. A nonvoid family $\mathscr{F}$ of subsets of a set X is called a *filter* on X, when conditions (F.1)–(F.3) are met.

Examples. 1. The family of all neighbourhoods of a point x constitutes a filter, the *filter of neighbourhoods of the point x.*

2. *The Fréchet Filter.* Let $X = N$ (the set of natural numbers > 0). The family of complements $F_\alpha := \complement A_\alpha$ of all finite parts A_α of the set N constitutes a filter known as the *Fréchet filter*.

3. The family of all neighbourhoods of a set Z in a topological space X.

4. The *filter associated with a net* $(x_\pi)_{\pi \in \Pi}$, $x_\pi \in X$. Take $F_\pi := \{x_\varrho : \varrho \succ \pi\}$ and all supersets[1] in X, of the sets F_π, $\pi \in \Pi$; the family of these sets, as is readily seen, is a filter.

Conversely, a net can be assigned to every filter.

5. A *net associated with a filter* $\mathscr{F}$ on a set X. The set $\Pi := \{(x, F): x \in F \in \mathscr{F}\}$ is directed by inclusions of F's, i.e., $(x_1, F_1) \prec (x_2, F_2)$ when $F_2 \subset F_1$. Thus the mapping $\Pi \ni (x, F) \to x \in X$ defines a net in X. It is worth-while to mention that if we apply to this net the procedure described in Exercise 4 then we get $\mathscr{F}$ as the resulting filter.

It is often more convenient to use a concept which is somewhat simpler than that of filter:

DEFINITION. A *filter basis* on X is a nonempty family $\mathscr{B}$ with the properties:

(B.1) $$\bigwedge_{A_1, A_2 \in \mathscr{B}} \bigvee_{A_3 \in \mathscr{B}} A_3 \subset A_1 \cap A_2.$$

(B.2) $$\emptyset \notin \mathscr{B}.$$

A filter basis $\mathscr{B}$ generates a filter $\mathscr{F}$ as follows: $\mathscr{F}$ consists of those subsets of X which contain at least one set from the family $\mathscr{B}$. This is why the filter basis $\mathscr{B}$ is often called a *basis of the filter* $\mathscr{F}$.

A *filter subbasis* is a family of subsets of X such that all finite intersections of sets of the family are not empty. On adding these intersections together, we obtain a family which constitutes a filter basis. It determines a filter. The initial filter subbasis is called a *subbasis of this filter*.

Examples. 1. Given $(X, \mathscr{T})$; a basis of neighbourhoods $\mathscr{B}(x)$ of a point $x \in X$ is a basis of the filter of neighbourhoods of that point.

2. Let $a_i \in X$, $i = 1, 2, \ldots$ Omitting a finite number of terms of the

---

[1] A set V is a *superset* of a set U when $U \subset V$.

sequence (a_i) we obtain sets which constitute a filter basis $\mathscr{B}$. The filter generated by the filter basis $\mathscr{B}$ consists of all subsets (of X) containing all except a finite number of the elements of our sequence. This is called the *elementary filter associated with the sequence* (a_i).

Two filter bases are said to be *equivalent* when they generate the same filter. As is easily seen, two bases $\mathscr{B}$ and $\mathscr{B}'$ of a filter are equivalent if and only if every element belonging to $\mathscr{B}'$ contains an element belonging to $\mathscr{B}$ and conversely. When $\mathscr{F}$ and $\mathscr{F}'$ are filters on the same set X and when $\mathscr{F} \subset \mathscr{F}'$ (i.e. $\mathscr{F}$ is a subfamily of $\mathscr{F}'$), then $\mathscr{F}'$ is said to be *finer* than $\mathscr{F}$.

A Limit of a Filter. A filter (or filter basis) $\mathscr{F}$ on a topological space X *converges* to x (we write $\mathscr{F} \to x$) if

$$\bigwedge_{U(x)} \bigvee_{F \in \mathscr{F}} F \subset U(x).$$

Accordingly,

$$(\mathscr{F} \to x) \Leftrightarrow (\mathscr{F} \text{ is finer than the filter of neighbourhoods of } x).$$

The set of all limits of a filter (or filter basis) $\mathscr{F} = \{F_\alpha\}_{\alpha \in A}$ is denoted by $\lim \mathscr{F}$ or $\lim_{\alpha \in A} F_\alpha$. If $\mathscr{F}$ is not convergent then we write $\lim \mathscr{F} = \emptyset$.

Filters associated with convergent nets may be expected to be convergent. Indeed, the following theorem holds.

THEOREM XII.3.1. In the case of any of the following two situations:

(i) $(x_\pi)_{\pi \in \Pi}$ is the net associated with a filter $\mathscr{F}$,

(ii) $\mathscr{F}$ is the filter associated with a net $(x_\pi)_{\pi \in \Pi}$,

we have $\lim_{\pi \in \Pi} x_\pi = \lim \mathscr{F}$.

The proof, which is not difficult, is left to the reader.

As a corollary we get the following analogue of Theorem XII.2.1.

THEOREM XII.3.2. Let X be a topological space. (Every convergent filter on X has only one limit) $\Leftrightarrow$ (X is a Hausdorff space).

The continuity of the mapping $T: X \to Y$ can now be characterized in terms of filters as follows: If $\mathscr{F} = \{F_\alpha\}_{\alpha \in A}$ is a filter on X, then the family of images $\{T(F_\alpha)\}_{\alpha \in A}$ constitutes a filter basis on Y; indeed $T(F_\alpha) \neq \emptyset$, $T(F_\alpha \cap F_\beta) \subset T(F_\alpha) \cap T(F_\beta)$, hence it determines a filter on Y, which is called the *image* of $\mathscr{F}$ and is denoted by $T(\mathscr{F})$. This brings us immediately to

THEOREM XII.3.3. (A mapping $T: X \to Y$ is continuous at a point $x \in X$) $\Leftrightarrow$ (For every filter $\mathscr{F} \to x$ the filter $T(\mathscr{F})$ converges to $T(x)$) $\Leftrightarrow$ (For every filter $\{F_\alpha\}_{\alpha \in A} \to a$ the filter basis $\{T(F_\alpha)\}_{\alpha \in A}$ converges to $T(a)$).

Ultrafilters. With the aid of the Kuratowski–Zorn Lemma it can be shown that for every filter on X there exists the finest filter which contains $\mathscr{F}$ and is known as an *ultrafilter*. Ultrafilters are characterized by the following theorem.

THEOREM XII.3.4. (A filter $\mathscr{F}$ is an ultrafilter) $\Leftrightarrow$ (For any subsets $A, B \subset X$, if $A \cup B \in \mathscr{F}$, then $\mathscr{F}$ contains at least one of the sets, A or B).

PROOF. $\Rightarrow$ (a.a): Let $\mathscr{F}$ contains neither A nor B. All subsets $N \subset X$ such that $N \cup A \in \mathscr{F}$ form a filter $\mathscr{G} \supset \mathscr{F}$. Indeed (F.3) is plainly satisfied; when $N_1, N_2 \in \mathscr{G}$ then

$$(N_1 \cup A) \cap (N_2 \cap A) = (N_1 \cap N_2) \cup A \in \mathscr{F}, \quad \text{and thus}$$
$$N_1 \cap N_2 \in \mathscr{G}.$$

If $M \supset N \in \mathscr{G}$, then $M \cup A \supset N \cup A$, and hence $M \cup A \in \mathscr{F}$, or $M \in \mathscr{G}$) Note that $\mathscr{G}$ is finer than $\mathscr{F}$ because it contains the set B which does not belong to $\mathscr{F}$. But $\mathscr{F}$ was an ultrafilter, whereby a contradiction.

$\Leftarrow$ (a.a.): The filter $\mathscr{F}$ contains at least one of the sets A or $X - A$, because it contains the union $X = A \cup (X - A)$. If a filter, strictly finer than $\mathscr{F}$, were to exist, it would have to contain some set A and $X - A$ at once, and hence, being a filter, it would also have to contain $A \cap (X - A.$ $= \emptyset$, which is a contradiction. □

COROLLARY XII.3.5. If $\mathscr{F}$ is an ultrafilter on X, and $A \subset X$, then just one of the sets, A and $X - A$, belongs to $\mathscr{F}$.

Remark. Refining a filter is analogous to extracting a subsequence: Suppose that $\mathscr{F}$, $\mathscr{G}$ are elementary filters associated with the sequences (x_n) and (y_n). If (x_n) is a subsequence of (y_n), then $\mathscr{F} \supset \mathscr{G}$. However, a filter finer than an elementary filter associated with a sequence need not be associated with any subsequence.

Points of Accumulation of Filters and Nets. A point x is called a *point of accumulation of a filter* $\mathscr{F} = \{F_\alpha\}_{\alpha \in A}$ if $x \in \overline{F}_\alpha$ for every F_α.

If a filter $\mathscr{F} \to a$, then a is a point of accumulation of that filter by property (F.2).

The converse fact does not hold, as is shown by the following

Example. $X = \{0, 1, 2, ...\}$ with the discrete topology. Suppose that a filter $\mathscr{F}$ has a basis $F_n = \{0\} \cup [n, \infty[\, , n = 1, 2, ...$ Only $\{0\}$ is a point of accumulation, whereas the filter does not converge to 0.

LEMMA XII.3.6. If x is a point of accumulation of a filter $\mathscr{F} = \{F_\alpha\}_{\alpha \in A}$ and $\{U_\beta\}_{\beta \in \mathscr{B}}$ is a filter of the neighbourhoods of x, then the sets $F_\alpha \cap U_\beta$ constitute a filter basis convergent to x and finer than $\mathscr{F}$.

This fact could also be reformulated as follows.

LEMMA XII.3.7. (A filter $\mathscr{F}$ has a point of accumulation x) $\Leftrightarrow$ (There exists an (ultra)filter finer than $\mathscr{F}$ and convergent to x).

DEFINITION. A point of accumulation of the filter associated with a net $(x_\pi)_{\pi \in \Pi}$ is a *point of accumulation of the net* $(x_\pi)_{\pi \in \Pi}$.

If we recall what was mentioned in Example 5 on p. 11, then we get the following

THEOREM XII.3.8. The points of accumulation of the net associated with a filter $\mathscr{F}$ coincide with those of $\mathscr{F}$ itself.

To complete the analogy between filters and nets we need to introduce finer nets (subnets). A net $(x_\varrho)_{\varrho \in P}$ is called a *finer net* than (*subnet* of) a net $(x_\pi)_{\pi \in \Pi}$ if there is a mapping $f \colon P \to \Pi$ such that $x_\varrho = x_{f(\varrho)}$ for every $\varrho \in P$, and for every $\pi \in \Pi$ there is an element $\varrho \in P$ such that $\sigma \succ \varrho$ implies $f(\sigma) \succ \pi$.

An example which unfortunately is not sufficiently general is the following.

Recall that a subset P of a directed set $(\Pi, \prec)$ is *cofinal* when $\bigwedge_{\pi \in \Pi} \bigvee_{\varrho \in P} \pi \prec \varrho$; then P is directed by the restriction of $\prec$ and the net $(x_\varrho)_{\varrho \in P}$ is then a finer net than $(x_\pi)_{\pi \in \Pi}$.

THEOREM XII.3.9. (x is a point of accumulation of a net $(x_\pi)_{\pi \in \Pi}$) $\Leftrightarrow$ (x is a limit of some finer net than the net $(x_\pi)_{\pi \in \Pi}$).

Ultrafilters have the following properties:

THEOREM XII.3.10. In a topological space, every ultrafilter $\mathscr{F}$ is convergent or has no point of accumulation.

PROOF. Let x be a point of accumulation of the filter $\mathscr{F}$; then (cf. Lemma XII.3.7) there exists a filter $\mathscr{G} \supset \mathscr{F}$ such that $\mathscr{G} \to x$. Since $\mathscr{F}$ is an ultrafilter, therefore $\mathscr{F} = \mathscr{G}$. □

THEOREM XII.3.11. The image of an ultrafilter under a mapping (of sets) is itself an ultrafilter.

PROOF. Suppose that $\mathscr{F} = \{F_\alpha\}_{\alpha \in A}$ is an ultrafilter on X and suppose that $T: X \to Y$.

If the filter generated by $(T(F_\alpha))_{\alpha \in A}$ were not an ultrafilter on Y, there would be a finer filter $\mathscr{G} = \{G_\beta\}_{\beta \in \mathscr{B}}$ on Y such that there would exist a set G_{β_0} not containing any $T(F_\alpha)$. A filter on X, generated by $\{T^{-1}(G_{\beta_0})\} \cup \mathscr{F}$, would indeed be finer than $\mathscr{F}$ since $T^{-1}(G_{\beta_0})$ does not contain any F_α, that is $\mathscr{F}$ would not be an ultrafilter. □

We shall now proceed to prove the counterpart of Theorem XII.2.2.

THEOREM XII.3.12. (A subset M of a topological space X is closed) $\Leftrightarrow$ (M contains the limits of all filters on M which are convergent in the space X).

PROOF. $\Leftarrow$: Let $x \in \overline{M}$, and let $\{U_\alpha\}_{\alpha \in A}$ be the filter of neighbourhoods in X, of the point x. Every $M \cap U_\alpha \neq \emptyset$ and $\{M \cap U_\alpha\}_{\alpha \in A}$ is a filter on M. Obviously, it converges to x. Thus, by hypothesis, $x \in M$, and hence $M = \overline{M}$.

$\Rightarrow$: follows from Theorem XII.3.1. □

Riemann Filter Basis, Riemann Nets and Riemann Integral. The definition of the Riemann integral had us led to the notion of net and the general theory of limits.

Let us recall some notions. Let $[a, b] \subset \boldsymbol{R}$; Π is the set of (all) partitions $\pi = \{P_k\}$ of $[a, b]$; $\delta(\pi) = \max_k |P_k|$ is the diameter of the partition π. On the set Π we have introduced the directing relation: $\pi_1 \prec \pi_2$ if π_2 is a subdivision of π_1. But Riemann considered another directing relation: $\pi_1 \leqslant \pi_2$ if $\delta(\pi_1) \leqslant \delta(\pi_2)$. Plainly $(\pi_1 \prec \pi_2) \Rightarrow (\delta(\pi_1) \leqslant \delta(\pi_2))$, but the opposite implication is false.

DEFINITION. The family $\mathscr{B} := \{\Pi(r): r > 0\}$, where $\Pi(r) := \{\pi' \in \Pi: \delta(\pi') \leqslant r\}$, is a filter basis on Π and it is called a *Riemann filter basis.* This $\mathscr{B}$ generates the filter Π_R called a *Riemann filter*. Let $\Pi[\pi] := \{\pi' \in \Pi: \pi' \succ \pi\}$, i.e. $\Pi[\pi]$ is the set of all subpartitions of π. It is seen that the family $\{\Pi[\pi]: \pi \in \Pi\}$ is a filter basis on Π; it generates the filter Π_D, called a *Darboux filter*.

We see that the Darboux filter is finer than the Riemann filter: to any $\Pi(r)$ take $\pi \in \Pi(r)$ (that means $\delta(\pi) \leqslant r$), then $\Pi[\pi] \subset \Pi(r)$, but the opposite inclusion is false!

On the set Π we have two directing relations:

1° the Riemann directing relation $\leqslant$,

2° the Darboux directing relation $\prec$.

In the theory of the Riemann integral (cf. Part I) we have introduced upper and lower (Darboux) sums:

$$B([a, b]) \times \Pi \ni (f, \pi) \to \overline{S}(f, \pi) := \sum_k \sup f(P_k) \cdot |P_k|,$$

$$B([a, b]) \times \Pi \ni (\pi, f) \to \underline{S}(f, \pi) := \sum_k \inf f(P_k) \cdot |P_k|,$$

where $\pi = \{P_k\}$ and $B([a, b])$ is the space of bounded functions on $[a, b]$. We know that

$$(*) \qquad \overline{\int} f := \inf_{\pi \in \Pi} \overline{S}(f, \pi) = \lim_{\pi \in (\Pi, \prec)} \overline{S}(f, \pi)$$

and

$$\underline{\int} f := \sup_{\pi \in \Pi} \underline{S}(f, \pi) = \lim_{\pi \in (\Pi, \prec)} \underline{S}(f, \pi),$$

cf. Corollary III.4.2.

Whence the upper (lower) Riemann integral of f is the limit of the Darboux net $(\pi, \prec) \ni \pi \to \overline{S}(f, \pi)$ $((\Pi, \prec) \ni \pi \to \underline{S}(f, \pi))$. But the relation $\leqslant$ in Π allows to consider the corresponding Riemann nets:

$$(\Pi, \leqslant) \ni \pi \to \overline{S}(f, \pi) \quad \text{and} \quad (\Pi, \leqslant) \ni \pi \to \underline{S}(f, \pi).$$

Plainly the Darboux nets are finer than the corresponding Riemann nets, therefore the following theorem is not trivial.

THEOREM XII.3.13 (Darboux). 1° For any $f \in B([a, b])$ we have the equalities

$$\overline{\int} f = \lim_{\pi \in (\Pi, \leqslant)} \overline{S}(f, \pi), \qquad \underline{\int} f = \lim_{\pi \in (\Pi, \leqslant)} \underline{S}(f, \pi).$$

2° Let $f \in B([a, b])$. $\big(f$ is Riemann-integrable, i.e. $\underline{\int} f = \overline{\int} f\big) \Leftrightarrow \big(\lim_{\pi \in (\Pi, \leqslant)} \overline{S}(f, \pi) = \lim_{\pi \in (\Pi, \leqslant)} \underline{S}(f, \pi)\big)$.

Whence for the integrability of f it is sufficient and neccessary that upper and lower Riemann nets have the same limit.

PROOF. 2° is an immediate consequence of 1°. We shall prove 1°. It suffices to check the equality for the upper integral. We know that

$$\bigwedge_{\varepsilon > 0} \bigvee_{\pi_\varepsilon} \bigwedge_{\pi \succ \pi_\varepsilon} \left| \overline{S}(f, \pi) - \overline{\int} f \right| < \tfrac{1}{2}\varepsilon.$$

Let $\pi_\varepsilon = \{P_k^\varepsilon\}_1^n$. Let us take a partition $\pi = \{P_j\}$ with diameter $\delta(\pi) \leqslant r := \varepsilon/2n(M-m)$, where $M = \sup f([a, b])$, $m = \inf f([a, b])$. Take $\pi' := \{P_{kj}\}$, where $P_{kj} := P_k^\varepsilon \cap P_j$, provided that $P_k^\varepsilon \cap P_j \neq \emptyset$. Then $\pi' \succ \pi_\varepsilon$ and therefore

$$\overline{S}(f, \pi') < \overline{\int} f + \tfrac{1}{2}\varepsilon.$$

But $\overline{S}(f, \pi) - \underline{S}(f, \pi') \leqslant n(M-m)r = \varepsilon/2$.

Hence, using also (*),

$$\overline{\int} f \leqslant \overline{S}(f, \pi) < \overline{\int} f + \varepsilon.$$

Thus we have proved that

$$\bigwedge_{\varepsilon > 0} \bigvee_{r > 0} \bigwedge_{\pi \in \Pi(r)} \left| \overline{S}(f, \pi) - \overline{\int} f \right| < \varepsilon,$$

which is precisely the thesis of 1°. □

One may consider the set Π_* of all partitions with distinguished points, i.e. $\Pi_* \ni \pi_* = \{(P_k, \xi_k)\}_1^n$, where $\pi := \{P_k\} \in \Pi$ and $\xi_k \in P_k$, $k = 1, \ldots, n$. Π_* is endowed with the Riemann directing relation: $\pi_* \leqslant \pi'_*$ if $\pi \leqslant \pi'$. Then the *Riemann sums* are defined as

$$S(f, \pi_*) := \sum_k f(\xi_k) |P_k|.$$

Plainly

$$\underline{S}(f, \pi) \leqslant S(f, \pi_*) \leqslant \overline{S}(f, \pi).$$

Therefore we have the following

COROLLARY XII.3.14. If f is Riemann-integrable then

$$\lim_{\pi_* \in \Pi_*} S(f, \pi_*) = \int f.$$

4. COMPACT SPACES

In this section $(X, \mathscr{T})$ will denote a Hausdorff space.

In an arbitrary Hausdorff space we cannot define compactness in terms of sequences, but this can be done with the aid of nets and their points of accumulation. A definition operating with coverings is natural; cf. Theorem II.8.6 (Borel, Lebesgue).

DEFINITION. A family $\{A_i\}_{i \in I}$ of subsets of a set M is said to be a *covering* of M when $M = \bigcup_{i \in I} A_i$. An *open* (*closed*) *covering* of a topological space is a covering by open (closed) subsets.

DEFINITION. A Hausdorff space $(X, \mathscr{T})$ is said to be *compact* when a finite covering is contained in each of its open coverings. A set $Z \subset X$ is *compact* if Z endowed with the induced topology is a compact topological space.

The theorem below gives a number of equivalent conditions for compactness: condition (v) generalizes the sequential definition of compact set, given in Chapter II.

THEOREM XII.4.1 (Fundamental Theorem). Let X be a topological space; then the following conditions are equivalent:

(i) The space X is compact,

(ii) Every family $\{B_i\}_{i \in I}$ of closed subsets of X, for which $\bigcap_{i \in I} B_i = \emptyset$, contains a finite subfamily with the empty intersection.

(iii) Every filter on X has at least one point of accumulation (in X).

(iv) Every ultrafilter on X converges (to a limit belonging to X).

(v) Every net in X has at least one point of accumulation (in X).

A family of subsets possesses the *finite intersection property* if each of its finite subfamilies has a nonempty intersection. We can now formulate a condition equivalent to (ii):

(vi) Every family of closed subsets of X, possessing the finite intersection property has a nonempty intersection.

The proof will be given at the end of this section.

COROLLARY XII.4.2. Let $(A_i)_{i\in I}$ be a chain (i.e. of each two elements of this family, one is a subset of the other) of closed nonempty subsets of a compact space X; then $\bigcap_{i\in I} A_i \neq \emptyset$.

This corollary is a consequence of the condition (vi) of the preceding theorem since the intersection of every finite subfamily of the chain is equal to the smallest, and hence nonempty, set.

We shall see that for a space satisfying the second axiom of countability (i.e. possessing a countable basis) the concept of compactness is identical with the concept of sequential compactness (every sequence contains a convergent subsequence).

First, we prove a straightforward lemma:

LEMMA XII.4.3. If a space X satisfies the second axiom of countability, then every open covering $\{\mathcal{U}_\alpha\}_{\alpha\in A}$ contains a countable covering.

PROOF. Let $\mathcal{O}_i$, $i = 1, 2, \ldots$, be a basis of X. Let $\mathcal{O}_{i_n}$, $n = 1, 2, \ldots$, be those sets out of which all sets $\mathcal{U}_\alpha$ can be formed by taking suitable unions, hence $\bigwedge_{\mathcal{O}_{i_n}} \bigvee_{\mathcal{U}_{\alpha_n}} \mathcal{O}_{i_n} \subset \mathcal{U}_{\alpha_n}$. But $X = \bigcup_{\alpha\in A} \mathcal{U}_\alpha = \bigcup_{n=1}^{\omega} \mathcal{O}_{i_n} \subset \bigcup_{n=1}^{\infty} U_{\alpha_n}$. $\sqcup$

This gives the first part of the following

THEOREM XII.4.4. Let X satisfy the second axiom of countability then (X is compact) $\Leftrightarrow$ (A finite covering can be selected from every countable, open covering $\{\mathcal{O}_i\}_1^\infty$ of X)[1] $\Leftrightarrow$ (X is sequentially compact).

PROOF. First note that the requirement of (vi) for every countable family is equivalent to countable compactness. So if a family $\{F_n\}_1^\infty$ of closed sets has the finite intersection property then we may choose a point, say x_n, from every $H_n := \bigcap_{i=1}^{n} F_i$. By sequential compactness

[1] This property is called *countable compactness*.

there is a limit x of some subsequence of the sequence (x_n). But $x_n \in F_i$ for all $n \geqslant i$, and so x must belong to F_i since F_i is closed. Thus $x \in \bigcap_{i=1}^{\infty} F_i$. Conversely, assume X is countably compact. Let (x_n) be a sequence in X and put $B_n := \{x_n, x_{n+1}, \dots\}$. The family $\{\overline{B}_n\}_1^\infty$ has the finite intersection property and so there is a point x common for all $\overline{B}_n$. By the assumption of the second axiom of countability there is a basis $\{U_n(x)\}_1^\infty$ of neighbourhoods of x. But each $U_k(x)$ intersects every B_n and so we may choose a subsequence converging to x. □

THEOREM XII.4.5. A compact set $K \subset X$ is closed.

PROOF. Suppose that $x \in \overline{K}$ and suppose that $\{U_\alpha\}_{\alpha \in A}$ is the filter of neighbourhoods of the point x; then the family $\{U_\alpha \cap K\}_{\alpha \in A}$ is a filter on K, having a point of accumulation $y \in K$; cf. Theorem XII.4.1 (iii). $\{U_\alpha \cap K\}_{\alpha \in A}$ is a filter basis on X, which obviously converges to x. By Lemma XII.3.6, it converges to y. Since X is a Hausdorff space, $x = y \in K$; cf. Theorem XII.3.2. □

THEOREM XII.4.6. Every closed subset M of a compact space X is compact.

PROOF. By the condition (ii) of the fundamental Theorem XII.4.1 a subset M of a Hausdorff space X is compact if and only if:

(ii′) Every family $\{B_i\}_{i \in I}$ of closed subsets of X, for which $M \cap \left(\bigcap_{i \in I} B_i\right) = \emptyset$, contains a finite subfamily possessing the same property.

In our case, X being compact and M closed, for every family $\mathscr{B} = \{B_i\}_{i \in I}$ such that $M \cap \left(\bigcap_{i \in I} B_i\right) = \emptyset$, the family $\{M\} \cup \mathscr{B}$ contains a finite subfamily with the empty intersection. But then the members of $\mathscr{B}$ which belong to this finite subfamily have their intersection disjoint with M as desired. □

Theorem XII.4.7. A continuous image of a compact space (compact set) X is compact.

PROOF. Suppose that $\bigcup_{\alpha \in A} \mathcal{O}_\alpha = T(X)$, where the sets $\mathcal{O}_\alpha$ are open. Since the mapping T is continuous, the sets $T^{-1}(\mathcal{O}_\alpha)$ are open and constitute a covering of the compact space X, and hence there exists a finite family

$T^{-1}(\mathcal{O}_{\alpha_j})$ $j = 1, \ldots, p$, which is a covering of X, whereby $\bigcup_{j=1}^{p} \mathcal{O}_\alpha = T(X)$, that is, the set $T(X)$ is compact. □

The theorems of Chapter II thus may be generalized:

THEOREM XII.4.8. The continuous bijection $T: X \to Y$ of a compact space X in a Hausdorff space Y is a homeomorphism.

PROOF. Note first of all that when a set $Z \subset X$ is closed, the set $T(Z)$ is closed in Y, because it is compact, being an image of a compact set (cf. the two preceding theorems). The continuity of the inverse mapping T^{-1} must be proved, but that has already been done, just above: $(T^{-1})^{-1}(Z) = T(Z)$ is closed for every closed $Z \subset K$. □

Comparison of Topologies on a Compact Space. Let us recall that when we have two topologies $\mathcal{T}_1$ and $\mathcal{T}_2$ on a set X and when $\mathcal{T}_1 \subset \mathcal{T}_2$, we say that $\mathcal{T}_2$ is *finer* (*stronger*) than $\mathcal{T}_1$.

An interesting fact follows from this last theorem:

THEOREM XII.4.9. Suppose that $(X, \mathcal{T}_1)$, $(X, \mathcal{T}_2)$ are Hausdorff spaces and that $(X, \mathcal{T}_2)$ is compact and $\mathcal{T}_1 \subset \mathcal{T}_2$. Then $\mathcal{T}_1 = \mathcal{T}_2$.

PROOF. Consider the identity mapping $I: (X, \mathcal{T}_2) \to (X, \mathcal{T}_1)$. Since $\mathcal{T}_1 \subset \mathcal{T}_2$, then I is continuous, and hence by Theorem XII.4.8 I is a homeomorphism, that is, $\mathcal{T}_1 = \mathcal{T}_2$. □

THEOREM XII.4.10. The union of a finite number of compact sets is compact.

The proof is left to the reader.

Proof of the Fundamental Theorem XII.4.1. We have:

(i) ⇔ (ii) by going over to complements:

(iii) ⇒ (iv) because an ultrafilter is convergent, if it has a point of accumulation;

(iv) ⇒ (iii) because if $\mathcal{F}$ is a filter on X, then there exists an ultrafilter $\mathcal{U}$ which is finer than $\mathcal{F}$ and $\mathcal{U}$ is convergent by assumption; hence it follows from Lemma XII.3.7 that $\mathcal{F}$ has a point of accumulation;

(iii) ⇒ (v) immediately from the definition of a point of accumulation of a net;

(v) ⇒ (iii) for, let $\mathcal{F}$ be a filter of X; there is a net associated with it. By Example 5 on p. 11, $\mathcal{F}$ is the filter associated with a net (namely, the net associated with $\mathscr{F}$). By hypothesis, this net has a point of accumulation which, by definition, is a point of accumulation of $\mathscr{F}$;

(iii) $\Rightarrow$ (ii) for, let $\{B_i\}_{i\in I}$ be a family of closed subsets whose intersection is empty. If the intersection of every finite subfamily were non-empty, $\{B_i\}_{i\in I}$ would constitute a subbasis of a filter on X which could not have a point of accumulation x_0 because, inasmuch as the sets B_i are closed, the point x_0 would belong to every B_i, that is, $\bigcap_{i\in I} B_i \neq \emptyset$;

(ii) $\Rightarrow$ (iii) for, suppose that this is not so and that a filter $\{F_\alpha\}_{\alpha\in A}$ on X has no point of accumulation; then $\bigcap_{\alpha\in A} \overline{F}_\alpha = \emptyset$, and, by hypothesis, there exist $\overline{F}_{\alpha_1}, \ldots, \overline{F}_{\alpha_n}$ with an empty intersection; hence $\bigcap_{i=1}^{n} F_{\alpha_i} = \emptyset$ (contradition with the definition of a filter);

(ii) $\Leftrightarrow$ (vi), obvious.

The equivalence of conditions (i)–(vi) has thus been proved. □

5. THE CARTESIAN PRODUCT OF TOPOLOGICAL SPACES

Suppose that $\{(X_\alpha, \mathscr{T}_\alpha)\}_{\alpha\in A}$ is a family of topological spaces. The *Cartesian product* $\bigtimes_{\alpha\in A} X_\alpha$, or $\prod_{\alpha\in A} X_\alpha$, is the set of all mappings $x\colon A \to \bigcup_{\alpha\in A} X_\alpha$ such that $x(\alpha) =: x_\alpha \in X_\alpha$; the point (mapping) x will be also denoted $(x_\alpha)_{\alpha\in A}$ or simply (x_α). When for every α we have $X_\alpha = Y$, we write Y^A instead of $\prod_{\alpha\in A} X_\alpha$. In the space $\prod_{\alpha\in A} X_\alpha$ we introduce the topology $\prod_{\alpha\in A} \mathscr{T}_\alpha$ by taking as a basis of neighbourhoods of the point $x = (x_\alpha)_{\alpha\in A}$ all the sets of the form $U = \prod_{\alpha\in A} W_\alpha$, where $W_\alpha = X_\alpha$ for all except a finite number of indices α, and $W_\beta = U_\beta(x_\beta)$ for all other indices; here $U_\beta(x_\beta)$ is an arbitrary neighbourhood of the point x_β in X_β. $\prod_{\alpha\in A} X_\alpha$ endowed with this topology is the *topological product of topological spaces* $(X_\alpha, \mathscr{T}_\alpha)$.

The mapping $P_\alpha\big((x_\beta)_{\beta\in A}\big) := x_\alpha$ is called the *projection* of $\prod_{\alpha\in A} X_\alpha$ onto X_α.

We now prove one of the most useful theorems of general topology.

THEOREM XII.5.1 (Tychonoff). The product $\prod_{\alpha \in A} X_\alpha$ of an arbitrary number of topological spaces is a compact space if and only if every X_α is compact.

Remark. This theorem is a far-reaching generalization of Theorem II.8.3 on the finite product of compact sets.

PROOF. $\Rightarrow$: Since a continuous image of a compact set is compact, each $X_\beta = P_\beta\left(\prod_{\alpha \in A} X_\alpha\right)$ is compact, $\beta \in A$.

$\Leftarrow$: By virtue of point (iv) of Theorem XII.4.1 it is sufficient to show that every ultrafilter $\mathscr{F}$ on the product $\prod_{\alpha \in A} X_\alpha$ is convergent. Now, as we know, $\mathscr{F}_\alpha := P_\alpha(\mathscr{F})$ is an ultrafilter on X_α (see Theorem XII.3.11). Since X_α is compact, $\mathscr{F}_\alpha$ converges to some $x_\alpha \in X_\alpha$. We shall prove that $\mathscr{F} \to x := (x_\alpha)$. This follows from the next lemma:

LEMMA XII.5.2. Let $X = \prod_{\beta \in B} X_\beta$ be a topological product, $(x^{(\pi)})_{\pi \in \Pi}$ be a net on X, $x^{(\pi)} = (x_\beta^{(\pi)}) \in X$ and let $\mathscr{F}$ be a filter on X; then

(a) (The net $(x^{(\pi)})_{\pi \in \Pi}$ converges to $x = (x_\beta) \in X) \Leftrightarrow \left(\bigwedge_{\beta \in B} x_\beta^{(\pi)} \to x_\beta\right)$,

(b) (The filter $\mathscr{F}$ on X converges to $x = (x_\beta) \in X) \Leftrightarrow \left(\bigwedge_{\beta \in B} P_\beta \mathscr{F} \to x_\beta\right)$.

PROOF. (a) $\Rightarrow$: follows from the continuity of the projections P_β.

$\Leftarrow$: If $\prod_{\beta \in B} W_\beta$ is a neighbourhood of the point x then only a finite number of $W_\beta \neq X_\beta$; thus there is a common index γ such that $x_\beta^{(\delta)} \in W_\beta(x_\beta)$ for all β and all $\delta \succ \gamma$, that is $(x_\beta^{(\delta)}) \in \prod_{\beta \in B} W_\beta$ or $x^{(\pi)} \to x$.

The equivalence (b) follows immediately from (a) and the theorem on equivalence of the convergence of filters and the nets associated with them (Theorem XII.2.4). □

The topology of the Cartesian product is often called the *Tychonoff topology*; this is the first example of inducing of a topology by a family of mappings, in this case, by the projections. When X and Y are topological spaces, then on strengthening the topology of the space X and weakening the topology of Y, more mappings become continuous. Conversely: by weakening the topology of X and strengthening the topology in Y (i.e. by appending more open sets), we spoil the continuity of many mappings; "in this new situation it is more difficult for a mapping to be continuous".

After this remark, the following theorem may be expected to hold.

THEOREM XII.5.3. Given a set X, a family of topological spaces $X_\alpha = (X_\alpha, \mathcal{T}_\alpha)$, $\alpha \in A$, and a family of mappings $T_\alpha: X \to X_\alpha$. In the set of topologies on X under which all T_α are continuous, there exists the weakest topology; this is the *topology induced by the family of mappings* $(T_\alpha)_{\alpha \in A}$. A basis of this topology consists of all sets of the form $\bigcap_{i=1}^{k} T_{\alpha_i}^{-1}(\mathcal{O}_i)$, where $\alpha_i, \ldots, \alpha_k \in A$, each set $\mathcal{O}_i$ is open in X_{α_i} for $i = 1, \ldots, k$ and $k \in N$.

PROOF. Sets of the above form satisfy axioms (B.1) and (B.2) and thus constitute a basis of some topology $\mathcal{T}$ on X. Clearly, all mappings T_α are continuous under this topology. On the other hand, every topology for which all T_α are continuous is stronger (finer) than $\mathcal{T}$. □

Now, let us consider the Tychonoff Theorem from the point of view of the above theorem, taking $X = \prod_{\alpha \in A} X_\alpha$, $T_\alpha := P_\alpha$ being a projection onto X_α, $\alpha \in A$. Now, the basis of the topology on X induced by the projections P_α, $\alpha \in A$, consists of all sets of the form

$$\bigcap_{i=1}^{k} P_{\alpha_i}^{-1}(\mathcal{O}_i), \qquad \alpha_1, \ldots, \alpha_k \in A, \ \mathcal{O}_i \text{ is an open set in } X.$$

Note that $P_{\alpha_0}^{-1}(\mathcal{O}_{\alpha_0}) = \prod_{\alpha \in A} W_\alpha$, where $W_{\alpha_0} := \mathcal{O}_{\alpha_0}$ and $W_\alpha := X_\alpha$ for $\alpha \neq \alpha_0$. Besides, $\left(\prod_{\alpha \in A} W_\alpha\right) \cap \left(\prod_{\alpha \in A} U_\alpha\right) = \prod_{\alpha \in A} W_\alpha \cap U_\alpha$. Hence the Tychonoff topology is precisely the topology induced by the projections P_α. Thus, we could define the topological product $\prod_{\alpha \in A} X_\alpha$ as follows:

PROPOSITION XII.5.4. The Tychonoff topology on $\prod_{\alpha \in A} X_\alpha$ is the weakest topology such that all projections P_α, $\alpha \in A$, are continuous.

PROPOSITION XII.5.5. If M_α is a subset of $(X_\alpha, \mathcal{T}_\alpha)$, $\alpha \in A$, then the closure $\overline{\left(\prod_{\alpha \in A} M_\alpha\right)} = \prod_{\alpha \in A} \overline{M}_\alpha$, where $\overline{M}_\alpha$ is the closure of M_α in X_α.

PROOF. If $x \in \prod_{\alpha \in A} \overline{M}_\alpha$, then every neighbourhood of x contains at least one element of the set $\prod_{\alpha \in A} M_\alpha$. On the other hand (Lemma XII.5.2), only elements from $\prod_{\alpha \in A} M_\alpha$ come under consideration as points of accumulation of the set $\prod_{\alpha \in A} M_\alpha$. □

DEFINITION. A subset Q of a Hausdorff space is *conditionally compact* when its closure $\overline{Q}$ is a compact set.

COROLLARY XII.5.6. Let M_α be a subset of a Hausdorff space X_α, $\alpha \in A$; $\left(\prod_{\alpha \in A} M_\alpha \text{ is conditionally compact}\right) \Leftrightarrow \bigwedge_{\alpha \in A} M_\alpha$ is conditionally compact).

PROOF. $\Leftarrow$: Every $\overline{M}_\alpha$ is compact, and hence it follows from the Tychonoff theorem that $\prod_{\alpha \in A} \overline{M}_\alpha$ is compact. Proposition XII.5.5, however, yields $\prod_{\alpha \in A} \overline{M}_\alpha = \overline{\prod_{\alpha \in A} M_\alpha}$, and hence $\overline{\prod_{\alpha \in A} M_\alpha}$ is compact.

$\Rightarrow$: Suppose that $\overline{\prod_{\alpha \in A} M_\alpha} = \prod_{\alpha \in A} \overline{M}_\alpha$ is compact. However, the continuous image of a compact set is a compact set, whereby $P_\alpha\left(\prod_{\alpha \in A} \overline{M}_\alpha\right) = \overline{M}_\alpha$ is compact, $\alpha \in A$, and therefore M_α is conditionally compact. □

Remark. As we have shown, when all the sets M_α are closed, then $\prod_{\alpha \in A} M_\alpha$ is also closed, but when the sets $\mathcal{O}_\alpha$ are open, then $\prod_{\alpha \in A} \mathcal{O}_\alpha$ need not be open; this happens when the set A is infinite.

6. METRIC SPACES. BAIRE SPACES

In this section we give one of the central theorems of the theory of metric spaces, the Baire theorem. It plays an enormously important role in functional analysis.

First of all, we shall demonstrate that compact metric spaces, *compacta*, possess a countable basis and hence are sequentially compact. From this we obtain another proof of the fact that the definition of compactness, which we examined in Chapter II, is equivalent to the definition in terms of coverings.

THEOREM XII.6.1 (A metric space (X, d) is separable) $\Leftrightarrow$ ((X, d) has a countable basis of its topology).

Proof. $\Leftarrow$: This part of the theorem holds for arbitrary topological spaces: let $(\mathcal{O}_i)_1^\infty$ be a basis, and let $x_i \in \mathcal{O}_i$ then the set $\{x_i\}_1^\infty$ is dense in X.

$\Rightarrow$: Let $\{x_i\}_1^\infty$ be a dense set in X; we shall show that the balls $K(x_i, r)$, where $i = 1, 2, \ldots$, and r runs over the rational numbers, constitute a basis of the topology. To this end it is sufficient to prove that every point $x \in X$ has a basis of neighbourhoods, consisting of balls of the form $K(x_i, r)$. But, if we have a ball $K(x, r)$, then the ball $K(x_i, \frac{1}{2}r)$, where $d(x, x_i) < \frac{1}{2}r$, is contained in $K(x, r)$. □

Theorem XII.6.2. Every compact metric space X—just as every relatively compact subset of a metric space—(i) is separable, (ii) has a countable basis, and (iii) has a finite diameter.

Proof. For every n there exists a finite covering of X with the balls $K(x_i^n, 1/n)$, $i = 1, 2, \ldots, N(n)$; $x_i^n \in X$. Thus, (iii) follows from this. But the countable set $\{x_i^n;\ i, n = 1, 2, \ldots\}$ is plainly dense in X. □

By Theorem XII.4.4 we obtain

Corollary XII.6.3. In a metric space, the concepts of compactness, sequential compactness, and countable compactness coincide.

Theorem XII.6.4 (Cantor). In a complete metric space, the decreasing sequence $B_1 \supset B_2 \supset \ldots$ of closed sets, whose diameters tend to zero, has a one-point intersection, i.e. there exists a unique point x such that $\bigcap_{i=1}^\infty B_i = \{x\}$.

Proof. Choose a sequence of points $x_i \in B_i$, $i = 1, 2, \ldots$; it is a Cauchy sequence and, by completeness, has a limit x. Since B_i's are closed, x belongs to each of them; thus $\bigcap_{i=1}^\infty B_i \neq \emptyset$. If $x, y \in \bigcap_{i=1}^\infty B_i$, then by the fact the diameters of B_i's tend to 0, $x = y$. □

Before we proceed to formulate the Baire Theorem, we give a definition for a certain class of sets.

Definition. Let X be a topological space. A subset $B \subset X$ is a *boundary set in* X if its interior $\mathring{A}$ (or $\operatorname{Int} A$) $= \emptyset$. A subset $A \subset X$ is said to be *nowhere dense* in X if $\bar{A}$ is a boundary set, i.e. if $\operatorname{Int}(\bar{A}) = \emptyset$. Clearly, a nowhere dense set A is a boundary set. A subset $C \subset X$ is of the *first category* (*meagre*) in X if

$$C = \bigcup_{i=1}^{\infty} A_i, \quad \text{where each } A_i \text{ is nowhere dense in } X.$$

A set $D \subset X$, which is not of the first category in X, is said to be of the *second category* (*nonmeagre*) in X.

A moment consideration shows that every subset of a first category set (boundary; nowhere dense) is itself a set of the first category (boundary; nowhere dense).

The following lemma is straightforward:

LEMMA XII.6.5 Let X be a topological space. (i) (B is a boundary set in X) $\Leftrightarrow$ (for every open set $\mathcal{O} \neq \emptyset$, $(X-B)\cap\mathcal{O} \neq \emptyset$, i.e. $X-B$ is dense).

(ii) (A is nowhere dense in X) $\Leftrightarrow$ ($\mathcal{O}-\overline{A} \neq \emptyset$ for every nonempty open set $\mathcal{O}$, i.e. the interior of $\overline{A}$ is empty) $\Leftrightarrow$ (for every nonempty open set $\mathcal{O}$ there exists an open nonempty set $\mathcal{U} \subset \mathcal{O}$ such that $\mathcal{U} \cap A = \emptyset$).

(iii) Let A be a closed subset of X, then (A is nowhere dense) $\Leftrightarrow$ (A is a boundary set).

DEFINITION. A topological space X is called a *Baire space* if each of its nonempty open subsets is of the second category in X.

Other characterizations of Baire spaces are frequently used.

THEOREM XII.6.6. For a topological space $(X, \mathcal{T})$ the following conditions are equivalent:

(a) $(X, \mathcal{T})$ is a Baire space.

(b) For every open subset $\mathcal{O}$ of X, if $\mathcal{O} = \bigcup_{i=1}^{\infty} M_i$, then for at least one index i the set $\overline{M}_i \cap \mathcal{O}$ contains an open nonempty subset of X.

(c) For every countable family $\{A_i\}_{i=1}^{\infty}$ of nowhere dense subsets of X $\bigcup_{i=1}^{\infty} A_i$ is a boundary set in X. In other words, every set of the first category in X is a boundary set.

(d) For every countable family $\{A_i\}_{i=1}^{\infty}$ of closed nowhere dense subsets of X $\bigcup_{i=1}^{\infty} A_i$ is a boundary set.

(e) The intersection of every countable family of open dense subsets of X is dense in X.

PROOF. (a) $\Rightarrow$ (b). If $\mathcal{O} = \bigcup_{i=1}^{\infty} M_i$, then for some i $\operatorname{Int}(\overline{M}_i) \neq \emptyset$. Hence $\emptyset \neq \operatorname{Int}(\overline{M}_i) \cap M_i \subset \operatorname{Int}(\overline{M}_i) \cap \mathcal{O}$ and the latter set is open in X.

(b) $\Rightarrow$ (c). Let $A = \bigcup_{i=1}^{\infty} A_i$ and assume conversely that $\operatorname{Int} A \neq \emptyset$. Then $\operatorname{Int} A = \bigcup_{i=1}^{\infty} (A_i \cap \operatorname{Int} A)$ and the sets $A_i \cap \operatorname{Int} A$ are nowhere dense as subsets of nowhere dense sets A_i. Hence contradiction.

(c) $\Rightarrow$ (d) is clear.

(d) $\Leftrightarrow$ (e) is clear if we go over to complements.

(d) $\Rightarrow$ (a). If $\mathcal{O} = \bigcup_{i=1}^{\infty} M_i$, where $\mathcal{O}$ is open and nonempty, and M_i are nowhere dense, then $\mathcal{O}$ is a subset of $\bigcup_{i=1}^{\infty} \overline{M}_i$ which is a boundary set by assumption. □

Now we give the Baire theorem.

THEOREM XII.6.7 (Baire). Every complete metric space is a Baire space.

PROOF. We shall prove that a complete metric space satisfies the condition (c) of the preceding theorem. By virtue of Lemma XII.6.5(i) we need to show that given a family $\{A_i\}_{i=1}^{\infty}$ of nowhere dense subsets of the space, for every open nonempty set $\mathcal{O}$, $\mathcal{O} - \bigcup_{i=1}^{\infty} A_i \neq \emptyset$.

Thus, let $\mathcal{O} \neq \emptyset$ be an open set, let $x_0 \in \mathcal{O}$, and let $r_0 > 0$ be such that $\overline{K(x_0, r_0)} \subset \mathcal{O}$. By Lemma XII.6.5 (ii), there exists an $x_1 \in K(x_0, r_0) - \overline{A}_1$. Since the set $K(x_0, r_0) - \overline{A}_1$ is open, there exists an r_1 such that $0 < r_1 < 1$ and $\overline{K(x_1, r_1)} \subset K(x_0, r_0) - \overline{A}_1$. Reasoning as above, we see that there exists an $x_2 \in K(x_1, r_1) - \overline{A}_2$. Now, let $0 < r_2 < \frac{1}{2}$ and let $\overline{K(x_2, r_2)} \subset K(x_1, r_1) - \overline{A}_2$. Repeating this procedure, we arrive at an infinite sequence of balls $K(x_n, r_n)$ such that:

$$\overline{K(x_n, r_n)} \subset K(x_{n-1}, r_{n-1}) - \overline{A}_n \subset K(x_0, r_0),$$

$$0 < r_n < \frac{1}{n}.$$

By the Cantor theorem $\bigcap_{n=0}^{\infty} \overline{K(x_n, r_n)} = \{x\}$. But the point x is clearly in $\mathcal{O}$ and since $\overline{K(x_n, r_n)} \cap A_n = \emptyset$, it is not in $\bigcup_{n=1}^{\infty} A_n$. □

COROLLARY XII.6.8. An open subset $\mathcal{O}$ of a Baire space X is a Baire space.

PROOF. Suppose that $(\mathcal{O}_n)_1^\infty$ is a sequence of dense open subsets of $\mathcal{O}$; $\mathcal{O}_n$ are open in X. Let $T_n := \mathcal{O}_n \cup (X - \overline{\mathcal{O}})$; then T_n are open and dense in X. Hence $\bigcap_n T_n = (X - \overline{\mathcal{O}}) \cup \bigcap_n \mathcal{O}_n$ is dense in X, that is $\mathcal{O}$ is a Baire space. □

Remark. As will be seen further on the locally compact spaces[1] are Baire spaces (cf. Complements and Exercises, p. 58).

The Category Method. The "category method" which Banach and other Polish mathematicians developed for proving the existence of objects endowed with a particular property P (e.g. a continuous function not having a derivative at any point) is as follows: In order to show that an object x with the property P exists in a set X, the set is metrized, the resultant space is shown to be complete and the set of those elements of X which do not possess the property P is shown to be of the first category (meagre) in X. Since it is known from the Baire Theorem that the space X is not of the first category, there are "many" objects with the property P, and objects not possessing it "are the exception."

Example. A great sensation was caused at one time when Weierstrass gave a construction of a continuous function having no derivative at any point. For years it was believed that such functions constituted a rare and pathological case. The category method, however, shows forcefully that what had been regarded as something normal, i.e. that a (piecewise) continuous function "should have" a derivative at least at one point, is in actual fact an exception in the world of continuous functions; the set of such functions is meagre in the metric space $C([a, b])$

[1] A topological space X is *locally compact* if every its point possesses a neighbourhood, the closure of which is compact. If the space X is Hausdorff, then the local compactness means that every point possesses a compact neighbourhood.

equipped with the metric $d(f, g) := \sup_{x \in [a,b]} |f(x)-g(x)|$. The proof (of a slightly stronger fact) runs along the following lines. Denoting by X the set $C([a, b])$ one observes that for every positive integer i the set

$$A_i := \left\{ f \in X: \bigvee_{t \in [a,b]} \bigwedge_{0 \neq h \in R^+} \left| \frac{f(t+h)-f(t)}{h} \right| \leqslant i \right\}$$

is closed. The reader may attempt to prove this shoving that the limit of every convergent sequence of functions in A_i is again in A_i (the proof uses compactness of the interval $[a, b]$). Now, the set $A := \bigcup_{i=1}^{\infty} A_i$ contains all functions which have one-side derivative at least at one point of $[a, b]$. If we show that every A_i is a boundary set, then, as closed, they are nowhere dense sets and therefore, by the Baire Theorem (c) A is of the first category (meagre). Thus, the set $X-A$ consisting of "strange" nowhere differentiable functions which the Weierstrass function is "plentiful" in our metric space $C([a, b])$. Hence we have to prove that every A_i is a boundary set. It is sufficient to notice that $X-A_i$ contains a dense set and, therefore, is dense itself; cf. Lemma XII.6.5(i). As we know (the Stone–Weierstrass theorem), the set of polynomials dense in X and thus it is sufficient to prove that for every polynomial w there is a function at an arbitrarily small distance from w and not belonging to A_i. Such a function is not difficult to construct; it may, for instance, be a piecewise linear function whose sides make an angle α with the x-axis and $|\tan \alpha| > i$. Another example is given in Complements and Exercises (6, p. 59).

7. THE TOPOLOGICAL PRODUCT OF METRIC SPACES

In Chapter II we considered the product of a finite number of metric spaces (X_i, d_i), $i = 1, \ldots, n$. There we proved that $(\prod_{i=1}^{n} X_i, d)$, where $d(x, y) := \sum_{i=1}^{n} d_i(x_i, y_i)$, is a metric in $\prod_{i=1}^{n} X_i$. In Section 5 of the present chapter we considered a general situation: an arbitrary number of arbitrary topological spaces. A natural question arises: in the case of a countable number of metric spaces can such a metric be introduced into the product $\prod_{i=1}^{\infty} X_i$, which determines the Tychonoff topology? It is clear that one should confine oneself to a countable number of spaces inasmuch as a metric space, by the nature of things, satisfies the first axiom of countability: Every point has a countable basis of neighbourhoods. The answer to our question is given by the following theorem.

THEOREM XII.7.1. Let (X_i, d_i), $i = 1, 2, \ldots$, be metric spaces; let $x = (x_i)$, $y = (y_i) \in \prod_{i=1}^{\infty} X_i$ and

$$(*) \qquad d(x, y) := \sum_{i=1}^{\infty} \frac{1}{2^i} \frac{d_i(x_i, y_i)}{1+d_i(x_i, y_i)}.$$

Then

(i) d is a metric on $\prod_{i=1}^{\infty} X_i$ which defines the Tychonoff topology,

(ii) the metric space $(\prod_{i=1}^{\infty} X_i, d)$ is complete if and only if all the spaces (X_i, d_i) are complete.

Remark. Obviously, any $c_i > 0$ such that $\sum_{i=1}^{\infty} c_i < \infty$ could be taken equally well instead of 2^{-i} in the definition $(*)$.

PROOF. (i) The verification that d is a metric is left to the reader. It is easily seen that the metric convergence $\overset{n}{x} \to \overset{0}{x}$ in the space $\prod_{i=1}^{\infty} X_i$ is equivalent to the metric convergence $\overset{n}{x}_i \to \overset{0}{x}_i$ of coordinates in every set X_i, $i = 1, 2, \ldots$ But by Lemma XII.5.2, convergence in the sense of Tychonoff topology possesses this property, and thus the two concepts of convergence are identical. Since every space (X_i, d_i) satisfies the first axiom of countability, then so does $\prod_{i=1}^{\infty} X_i$, and consequently, every point of accumulation of a set is the limit of some convergent sequence in this set, whereby it follows that closed sets in the sense of the metric $(*)$ and the Tychonoff topology are the same, i.e. the two topologies are identical.

(ii) The proof is identical with that in the case of a finite number of metric spaces. □

Example. A cube $\mathscr{D}^\omega = [0, 1]^\omega$ (the product of a countable number of copies of [0, 1], also called a *Hilbert cube* because it is homeomorphic to the product $\prod_{i=1}^{\infty} X_i$, where $X_i = [0, 1/2^i]$.

8. SEMICONTINUOUS FUNCTIONS

It is desirable to extend the concept of real-valued function to that of the numerical function $f: X \to \bar{\boldsymbol{R}}$, where $\bar{\boldsymbol{R}}$ comes into being through the adjunction of two points $-\infty, +\infty$. The linear order in $\boldsymbol{R}$ can be prolonged to $\bar{\boldsymbol{R}}$ as follows: for every $a \in \boldsymbol{R}$ we have $-\infty \leqslant a \leqslant +\infty$. If the intervals $[-\infty, -n]$, $[n, +\infty]$, $n \in \boldsymbol{N}$, are taken as a basis of neighbourhoods of the points $-\infty, +\infty$ (respectively), $\bar{\boldsymbol{R}}$ becomes a topological space.[1] With this natural topology on $\bar{\boldsymbol{R}}$ the mapping

$$[-\tfrac{1}{2}\pi, +\tfrac{1}{2}\pi] \ni x \to \tan x \in \bar{\boldsymbol{R}}$$

is a homeomorphism, and therefore $\bar{\boldsymbol{R}}$ is a compact space. A generalization of such a compactification procedure is described in Chapter XVII, Section 4.

In this section, when speaking of functions we shall have numerical functions in mind.

In the first section of the present chapter (p. 8) we have seen how a basis $\mathscr{B}(x_0)$ of neighbourhoods (of a point x_0 belonging to a space $(X, \mathscr{T})$) is endowed with the canonical directing relation (i.e. by inclusions). Thus $\mathscr{B}(x_0)$ is a filtering family in X; see Chapter III, p. 66. One such a basis $\mathscr{B}(x_0)$ is, of course, the set of all neighbourhoods of x_0, i.e. the filter of neighbourhoods of x_0.

It is easy to verify—as it was done in Chapter III—that if $f: X \to \bar{\boldsymbol{R}}$, then both $\liminf_{U \in \mathscr{B}(x_0)} f(U)$ and $\limsup_{U \in \mathscr{B}(x_0)} f(U)$, belonging to $\bar{\boldsymbol{R}}$, does not depend on a choice of a basis $\mathscr{B}(x_0)$ of neighbourhoods of the point

[1] We define $[-\infty, a] := \{-\infty\} \cup]-\infty, a]$, $[a, +\infty] := [a, \infty[\cup \{+\infty\}$ (cf. Chapter I, p. 25) and $[-\infty, a[,]a, +\infty]$ analogously.

x_0. So we define

$$\liminf_{x\to x_0} f(x) := \liminf_{U\in\mathscr{B}(x_0)} f(U)$$

and

$$\limsup_{x\to x_0} f(x) := \limsup_{U\in\mathscr{B}(x_0)} f(U).$$

Back in Chapter IV (p. 91) we referred to the semicontinuity of a function defined on metric space. Now we give a general definition:

DEFINITION. A function $f\colon X\to\bar{R}$ on a topological space X is said to be *lower (upper) semicontinuous at a point* $x_0\in X$ whenever

$$f(x_0) = \liminf_{x\to x_0} f(x) \quad (\limsup_{x\to x_0} f(x)).$$

Since $\liminf_{x\to x_0} f(x) \leqslant f(x_0)$, the inequality

$$f(x_0) \leqslant \liminf_{x\to x_0} f(x)$$

can be used in the definition instead of the equality.

Obviously (f is upper semicontinuous at x_0) $\Leftrightarrow$ ($-f$ is lower semicontinuous at x_0). Thus, without loss of generality, we may restrict our interest to the lower semicontinuity.

A function is *lower (upper) semicontinuous* if it is lower (upper) semicontinuous at each point of its domain.

We immediately have the following necessary and sufficient conditions for the lower semicontinuity:

THEOREM XII.8.1. A function f is lower semicontinuous at a point x_0 if and only if

$$(*) \qquad \bigwedge_{R\ni m<f(x_0)} \bigvee_{U(x_0)} \bigwedge_{x\in U(x_0)} m < f(x).$$

PROOF. $\Leftarrow$: We have $m \leqslant \inf f(U(x_0))$ for the relevant neighbourhood $U(x_0)$ and thus $f(x_0) \leqslant \sup_{U\in\mathscr{B}(x_0)}\inf f(U) = \liminf_{x\to x_0} f(x_0)$.

$\Rightarrow$ is obvious. $\square$

THEOREM XII.8.2. Let $f\colon X\to\bar{R}$; then (f is lower semicontinuous) $\Leftrightarrow$ ($\bigwedge_{m\in R}\{x\in X\colon f(x)>m\}$ is open) $\Leftrightarrow$ ($\bigwedge_{m\in R}\{x\in X\colon f(x)\leqslant m\}$ is closed).

PROOF. It is sufficient to prove the first equivalence, inasmuch as the second is obtained from the first by going over to complements. The first equivalence, however, follows immediately from Theorem XII.8.1 because the condition (*) states that together with a point x_0, an entire neighbourhood $U(x_0)$ satisfies the inequality $f(x) > m$. □

Examples of Semicontinuous Functions. Taking at every point the limes inferior of a (numerical) function g, we obtain a (numerical) function which we denote by liminf$g$ or $\underset{\sim}{g}$:

$$X \ni x \to \underset{\sim}{g}(x) := \liminf_{y \to x} g(y).$$

PROPOSITION XII.8.3. The function $\underset{\sim}{g}$ is lower semicontinuous.

PROOF. It suffices to show that given $m < \underset{\sim}{g}(x)$ there exists a neighbourhood $U(x)$ with the property that $\inf_{y \in U(x)} \underset{\sim}{g}(y) \geqslant m$. But $m < \underset{\sim}{g}(x)$ means that for some open neighbrourhood $V(x)$ we have $g(y) > m$ for all $y \in V(x)$. But then $\underset{\sim}{g}(y) \geqslant m$ for $y \in V(x)$. □

As follows from the definition, a function is *continuous* if and only if it is lower and upper semicontinuous at the same time.

Characteristic functions of open sets are further examples of semicontinuous functions.

DEFINITION. Let A be a subset of X. The *characteristic function* χ_A of the set A is defined as follows

$$\chi_A(x) := \begin{cases} 1, & \text{when } x \in A, \\ 0, & \text{when } x \notin A. \end{cases}$$

Theorem XII.8.2 immediately leads to the following corollary.

COROLLARY XII.8.4. $\big(A$ is open (closed) in $X\big) \Leftrightarrow \big(\chi_A$ is lower (upper) semicontinuous$\big)$.

In Hausdorff space, lower semicontinuous functions separate points: when $x \neq y$ and when A is an open set such that $x \in A$, $y \notin A$, then $\chi_A(x) = 1 \neq 0 = \chi_A(y)$.

The Upper Envelope of a Family of Functions. Suppose that $\{f_i\}_{i \in I}$ is a family of (numerical) functions on a set X; then the function $X \ni x \to \sup_{i \in I} f_i(x) \in \overline{\boldsymbol{R}}$ is called the *upper envelope of the family* $\{f_i\}_{i \in I}$. The lower envelope is defined analogously. The following theorem is easily proved.

THEOREM XII.8.5. (i) The upper envelope of a family $\{f_i\}_{i \in I}$ of functions which are lower semicontinuous at a point x_0 is also lower semicontinuous at x_0. In particular, the upper envelope of a family of continuous functions is lower semicontinuous.

PROOF. If $m < \sup_{i \in I} f_i(x_0)$, then $m < f_i(x_0)$ for some i; but f_i is lower semicontinuous at x_0, and hence by Theorem XII.8.1 there exists a neighbourhood $U(x_0)$ such that

$$m < f_i(x) \leqslant \sup_{i \in I} f_i(x) \quad \text{for } x \in U(x_0). \ \square$$

The following useful theorem is obtained from the Baire theorem.

THEOREM XII.8.6. Let X be a Baire space (e.g. a complete metric space or a locally compact space). Then:

(i) If f is real-valued, lower semicontinuous on X, the set of all such points that each of them possesses a neighbourhood U on which f is bounded from above, is open and dense in X.

(ii) If the upper envelope of a family of lower semicontinuous functions on X is a real-valued function (i.e. assumes only finite values), then the set of all such points, in whose neighbourhood this family is uniformly bounded from above, is open and dense in X.

PROOF. Since (ii) is an immediate conclusion from (i) and from the preceding theorem, we confine ourselves to proving (i).

Since f is lower semicontinuous, the set $M_n = \{x: f(x) \leqslant n\}$ is closed. For every open subset $\mathcal{O}$ we have by the assumption, $\mathcal{O} = \bigcup_{n=1}^{\infty} M_n \cap \mathcal{O}$. However, X (hence $\mathcal{O}$ as well) is a Baire space, and thus at least one of the sets $\overline{M_n \cap \mathcal{O}} = M_n \cap \mathcal{O}$ (the bar denotes closure in $\mathcal{O}$) contains an open subset. Obviously, f is bounded at this subset by n. $\square$

Now we shall prove the Generalized Weierstrass Theorem which appeared in Chapter IV, p. 91. A motivation of its importance may be found there.

THEOREM XII.8.7 (Generalized Weierstrass Theorem). Let X be a compact space, and let f be a numerical function on X which is lower semicontinuous. Then:

(i) f attains its minimum.

(ii) If f is finite, i.e. $f: X \to \boldsymbol{R}$, then f is bounded from below.

(iii) If f is finite and continuous, then f is bounded and attains its supremum and infimum.

It is sufficient to prove (i); (ii) follows immediately from (i); (iii) follows from (ii) if we consider f and $-f$.

As we know, the sets $M_t = \{x \in X: f(x) \leqslant t\}$ are closed; since $M_t \subset M_s$, whenever $t < s$, then the nonempty sets of this family constitute a basis of a filter on the compact space X and this filter has a point of accumulation a by Theorem XII.4.1. Since the sets M_t are closed, then $a \in M_t$, that is, $f(a) \leqslant f(x)$ for any $x \in X$, or $f(a)$ is a minimum of the function. □

An important and very extensive class of topological spaces is that of regular spaces.

9. REGULAR SPACES

These spaces satisfy the following axiom of separation given by Vietoris:

DEFINITION. A topological space $(X, \mathscr{T})$ is *regular* (or *of the type* T_3) if the following conditions are satisfied:

(R.1) $(X, \mathscr{T})$ is a *space of the type* T_1, i.e. for each pair $x_1, x_2 \in X$ of distinct points there exists $\mathcal{O} \in \mathscr{T}$ such that $x_1 \in \mathcal{O} \not\ni x_2$,

(R.2) disjoint neighbourhoods $U(M) \cap U(x_0) = \varnothing$ exist for every closed $M \subset X$ and $x_0 \notin M$.

Plainly, a regular space is a Hausdorff space (it is sufficient to take $M := \{x_1\}$), but not conversely. The condition (R.2) is equivalent to the following:

(R.2)′ the closed neighbourhoods of every point constitute a basis of neighbourhoods.

A straightforward corollary follows.

COROLLARY XII.9.1. A filter $\{F_i\}_{i \in I}$ in a regular space is convergent to x_0 if and only if x_0 is the only point of accumulation of the filter $\{F_i\}_{i \in I}$.

PROPOSITION XII.9.2. Every subspace S of regular space X is regular.

PROOF. The subspace S is of the type T_1. If $x \in S$ and $\{V_i(x)\}_{i \in I}$ is a basis of neighbourhoods of the point x, which are closed in X, then $\{V_i(x) \cap S\}_{i \in I}$ constitutes a basis of neighbourhood of x, closed in S. □

PROPOSITION XII.9.3. The topological product of regular spaces is a regular space.

PROOF. If V_i is closed in X_i, $i \in I$, then $\prod_{i \in I} V_i$ is closed in $\prod_{i \in I} X_i$, whence, via (R.2)′, the conclusion. □

Extension of Continuous Mapping by Closure (Continuity). In Theorem III.7.2 we gave a necessary and sufficient condition for a continuous mapping of a dense subset of a metric space X into a space (Y, d) to be uniquely extendible to a continuous mapping over the entire space. This is a property which precisely regular spaces possess.

THEOREM XII.9.4. Let A be a dense subset of a topological space X, let Y be a regular space, and let T: $A \to Y$ be a continuous mapping. If

(*) for every $x \in X$ and the filter $\{U_i\}_{i \in I}$ of neighbourhoods of the point x, the filter $\{T(U_i \cap A)\}_{i \in I}$ is convergent,

then the naturally defined mapping $\overline{T}$,

$$\overline{T}(x) := \lim_{i \in I} T(U_i \cap A) \quad \text{for } x \in X,$$

is continuous.

Condition (*) is obviously necessary for the existence of a continuous extension $\overline{T}$: $X \to Y$ of the mapping T. It may be reformulated in the language of nets as follows:

(*) for every $x \in X$ and every net $(x_\pi)_{\pi \in \Pi}$ in A, convergent to x, the net $(T(x_\pi))_{\pi \in \Pi}$ has the limit which does not depend on a choice of a net in A convergent to x.

PROOF. Suppose that V is a neighbourhood of a point $\overline{T}(x)$. By the regularity of the space Y, we may assume that V is closed. Since T is continuous, there is an open (in X) set $\mathcal{O} \ni x$ such that $T(\mathcal{O} \cap A) \subset V$.

Now, suppose that $\bar{x}$ is an arbitrary point in $\mathcal{O}$. Since A is dense, there exists a net $(x_\pi)_{\pi \in \Pi}$ in $\mathcal{O} \cap A$, convergent to $\bar{x}$. But $\overline{T}$ was so defined that $\overline{T}(\bar{x}) = \lim_{\pi \in \Pi} T(x_\pi)$. Since V is closed, $\overline{T}(\bar{x}) \in V$. Whence $\overline{T}(\mathcal{O}) \subset V$. □

Examples of Regular Spaces. 1. A metric space is regular: the closed balls $\overline{K(x, r)}$ may be taken as a basis of (closed) neighbourhoods of a point x.

2. A compact Hausdorff space $(X, \mathcal{T})$ is regular.

PROOF. Let $F \subset X$ be closed. Given a point $x \notin F$, then for every $y \in F$ we choose open disjoint neighbourhoods $\mathcal{O}_y(x) \ni x$ and $\mathcal{O}_x(y) \ni y$. Thus we get an open covering $\{\mathcal{O}_x(y)\}_{y \in F}$ of the set F and since F is compact, there are points $y_1, \ldots, y_k \in F$ such that $\{\mathcal{O}_x(y_i)\}_{i=1}^k$ is also a covering of F. Let $\mathcal{O}(F) := \bigcup_{i=1}^{k} \mathcal{O}_x(y_i)$ and $\mathcal{O}(x) := \bigcap_{i=1}^{k} \mathcal{O}_{y_i}(x)$; these sets are disjoint, open neighbourhoods of F and x respectively. □

3. Every locally compact Hausdorff space X is regular; moreover, compact neighbourhoods constitute a basis of neighbourhoods of a point.

PROOF. Let $x \in X$; then x has a compact neighbourhood V; when U is an arbitrary neighbourhood of the point x, then $V \cap U$ is a neighbourhood of x in the compact Hausdorff space V. However, by 2 we know that V is regular, and thus there exists $W \subset V \cap U$ such that W is a closed neighbourhood in the space V of the point x. But V was a neighbourhood in X of x, and therefore W is also a neighbourhood in X of x. Moreover the above constructed W is compact.

10. UNIFORM SPACES. THE COMPLETENESS OF A SPACE

The very important concept of uniform continuity, in addition to continuity, of a mapping was defined in the class of metric spaces. The fundamental theorem held that both of these concepts coincide on a compact metric space. A question arises as to whether one can also speak of uniform convergence on general topological spaces. This problem was studied by A. Weil when (in 1937) he introduced the concept of uniform structure. It is meaningful to speak of uniform continuity for so-called uniform spaces. Can a uniform structure consistent with the (original) topology be introduced in every topological space? Is that structure unique? These questions will be taken up in conjunction with topological groups which constitute the most important example of spaces with uniform structures. The concept of uniform continuity is easily defined in the case of topological vector spaces (these are commutative topological groups).

Suppose that X and Y are topological vector spaces (over the same field). A mapping $T: X \to Y$ is uniformly continuous when

$$\bigwedge_{U(O_Y)} \bigvee_{U(O_X)} \bigwedge_{x_1, x_2 \in X} (x_1 - x_2) \in U(0_X) \Rightarrow \big(T(x_1) - T(x_2)\big) \in U(0_Y).$$

Here neighbourhoods replace balls. A basis of neighbourhoods of a point x is obtained by taking the sets $x + U = \{x + y\colon y \in U\}$, where U runs over a basis of neighbourhoods of the vector 0. In a metric space as in a topological vector space, it is meaningful to say that two points $x_1, x_2 \in X$ are in some "entourage" prescribed "uniformly" for the entire space: they lie in a ball of radius ε what corresponds to the fact that the difference $(x_1 - x_2) \in U$. The "entourage" thus is a relation $\mathcal{N}$, that is, a subset $\mathcal{N} \subset X \times X$. Clearly, $\mathcal{N}$ should contain the diagonal $\Delta := \{(x, x): x \in X\}$. Some kind of symmetry of the relation $\mathcal{N}$ will also be desirable (in that respect cf. also Theorem XII.10.2).

Now we can give a definition of a uniform structure:

DEFINITION. A *uniform structure* (or *uniformity*) on a (nonempty) set X is a filter $\mathfrak{U}$ on $X \times X$, with the following properties

(U.1) $$\bigwedge_{\mathcal{N} \in \mathfrak{U}} \Delta \subset \mathcal{N},$$

(U.2) $$\bigwedge_{\mathcal{N} \in \mathfrak{U}} \mathcal{N}^{-1} \in \mathfrak{U},$$

(U.3) $$\bigwedge_{\mathcal{N} \in \mathfrak{U}} \bigvee_{\mathcal{M} \in \mathfrak{U}} \mathcal{M}^2 \subset \mathcal{N}.$$

We recall that the composition $\mathcal{N}_1 \mathcal{N}_2 := \mathcal{N}_1 \circ \mathcal{N}_2$ of relations $\mathcal{N}_1$ and $\mathcal{N}_2$ is the relation

$$\Big\{(x_1, x_3): \bigvee_{x_2 \in X} (x_1, x_2) \in \mathcal{N}_1 \quad \text{and} \quad (x_2, x_3) \in \mathcal{N}_2\Big\}.$$

The relation $\mathcal{M}\mathcal{M}$ is denoted by $\mathcal{M}^2$.

A pair $(X, \mathfrak{U})$, where $\mathfrak{U}$ is a uniform structure, is called a *uniform space*. A set (relation) $\mathcal{N} \in \mathfrak{U}$ is an *entourage of the uniform structure* $\mathfrak{U}$. If $(x_1, x_2) \in \mathcal{N} \in \mathfrak{U}$, the points x_1, x_2 are said to be *in an entourage* (*vicinity*) *of order* $\mathcal{N}$, similarly a subset $Y \subset X$ is said to be of *order* $\mathcal{N}$ if $Y \times Y \subset \mathcal{N}$.

Two uniform spaces $(X, \mathfrak{U})$, $(Y, \mathfrak{V})$ are *isomorphic* when there exists a bijection $I: X \to Y$ (hence also a bijection $I \times I: X \times X \to Y \times Y$) for which the family $\mathfrak{U}$ is mapped onto $\mathfrak{V}$. The reader will have no doubt as to what it means to say that a uniform structure $\mathfrak{U}$ is finer than a uniform structure $\mathfrak{V}$ (both on the same set), and that a subset $X_1 \subset X$, where $X = (X, \mathfrak{U})$, i.e. when X is endowed with an uniformity $\mathfrak{U}$, has a uniform structure $\mathfrak{U}|X_1$.

DEFINITION. A *uniformity basis* on a set X is a filter basis $\mathfrak{B}$ on $X \times X$, with the following properties:

(UB.1) $\bigwedge_{\mathcal{N} \in \mathfrak{B}} \Delta \subset \mathcal{N}$,

(UB.2) $\bigwedge_{\mathcal{N} \in \mathfrak{B}} \bigvee_{\mathcal{M} \in \mathfrak{B}} \mathcal{M} \subset \mathcal{N}^{-1}$,

(UB.3) $\bigwedge_{\mathcal{N} \in \mathfrak{B}} \bigvee_{\mathcal{M} \in \mathfrak{B}} \mathcal{M}^2 \subset \mathcal{N}$.

It is easily seen that the filter $\mathfrak{U}$ on $X \times X$, generated by a uniformity basis $\mathfrak{B}$ on X is a uniform structure on X. In this case $\mathfrak{B}$ is called a *uniformity basis of the uniform structure* $\mathfrak{U}$ (or, *of the uniform space* $(X, \mathfrak{U})$).

One can see that if $\mathfrak{U}$ is a uniform structure on X, then each (filter) basis of the filter $\mathfrak{U}$ is a uniformity basis of $\mathfrak{U}$. In other words, every uniform structure (space) has a uniformity basis. Something more will be shown in Theorem XII.10.2.

Examples.1. A metric space (X, d) is uniform if the sequence of relations

$$\mathcal{N}_{1/n} := \{(x_1, x_2) \colon d(x_1, x_2) < 1/n\}$$

is taken as its uniformity basis.

2. A topological vector space $(X, \mathscr{B}(0))$, where $\mathscr{B}(0)$ is a basis of neighbourhoods of zero and

$$\mathcal{N}_U := \{(x_1, x_2) \colon (x_1 - x_2) \in U\}, \qquad U \in \mathscr{B}(0)$$

is taken as its uniformity basis.

By iteration of (UB.3) and the fact that $\mathcal{M}^k \subset \mathcal{M}^{k+1}$, where $\mathcal{M}^k := \underbrace{\mathcal{M} \circ \ldots \circ \mathcal{M}}_{k \text{ times}}$, we get

PROPOSITION XII.10.1. Let $\mathfrak{B}$ be a uniformity basis and n a natural number, then

$$\bigwedge_{\mathcal{N} \in \mathfrak{B}} \bigvee_{\mathcal{M} \in \mathfrak{B}} \mathcal{M}^n \subset \mathcal{N}.$$

The Topology of a Uniform Space. If $(X, \mathfrak{U})$ is a uniform space, then a set $\mathcal{O}_{\mathcal{M}}(x) = \{y \in X \colon (x, y) \in \mathcal{M}\}$ is assigned to every $x \in X$ and $\mathcal{M} \in \mathfrak{U}$. In a moment we shall verify that the family $\{\mathcal{O}_{\mathcal{M}}(x) \colon x \in X,$

$\mathscr{M} \in \mathfrak{U}\}$ satisfies the neighbourhood axioms. The topology so defined is called the *topology of the uniform space* $(X, \mathfrak{U})$.

Let us verify that the family $\{\mathcal{O}_{\mathscr{M}}(x): x \in X, \mathscr{M} \in \mathfrak{U}\}$ satisfies the axioms for a basis of neighbourhoods of a topological space. Indeed,

$$\text{(H.1)} \quad \bigwedge_{x \in X} \mathscr{B}(x) := \{\mathcal{O}_{\mathscr{M}}(x): \mathscr{M} \in \mathfrak{U}\} \neq \varnothing, \qquad \bigwedge_{x \in X} \bigwedge_{\mathscr{M} \in \mathfrak{U}} x \in \mathcal{O}_{\mathscr{M}}(x),$$

$$\text{(H.2)} \quad \bigwedge_{x \in X} \bigwedge_{\mathscr{M}, \mathscr{N} \in \mathfrak{U}} \mathcal{O}_{\mathscr{M}}(x) \cap \mathcal{O}_{\mathscr{N}}(x) \in \mathscr{B}(x)$$

are a direct consequence of the fact that $\mathfrak{U}$ is filter;

$$\text{(H.3)} \quad \bigwedge_{x \in X} \bigwedge_{\mathscr{M} \in \mathfrak{U}} \bigvee_{\mathscr{N} \in \mathfrak{U}} \bigwedge_{y \in \mathcal{O}_{\mathscr{N}}(x)} \bigvee_{\mathscr{P} \in \mathfrak{U}} \mathcal{O}_{\mathscr{P}}(y) \subset \mathcal{O}_{\mathscr{M}}(x)$$

(here the condition (U.3) of the definition of a uniform structure is needed). Thus the axioms are satisfied. Moreover,

$$\bigwedge_{x \in X} \bigwedge_{\mathscr{M} \in \mathfrak{U}} \bigwedge_{A \subset X} (\mathcal{O}_{\mathscr{M}}(x) \subset A) \Rightarrow (A \in \mathscr{B}(x))$$

because $\mathfrak{U}$ is a filter; this shows that every $\mathscr{B}(x)$ is the set of *all* neighbourhoods (in the uniform topology on X) of the point x.

It will now be shown that the discussion could be confined to symmetric entourages (i.e. those $\mathscr{N}$ for which $\mathscr{N}^{-1} = \mathscr{N}$), or, to be more precise, that the following theorem holds.

THEOREM XII.10.2. Let $(X, \mathfrak{U})$ be a uniform space and consider $X \times X$ with the product topology, where X has a uniform topology. Then:

(a) the set $\mathfrak{S}$ of all symmetric entourages of $\mathfrak{U}$ constitutes a basis of the uniform structure $\mathfrak{U}$ and

$$\bigwedge_{\mathscr{N} \in \mathfrak{S}} \bigvee_{\mathscr{M} \in \mathfrak{S}} \overline{\mathscr{M}} \subset \operatorname{Int} \mathscr{N};$$

(b) the set $\mathfrak{S}_o$ ($\mathfrak{S}_c$) of all open (closed) elements of $\mathfrak{S}$ constitutes a basis of the uniform structure $\mathfrak{U}$.

PROOF. (a) By the definition of a uniform structure, if $\mathscr{N} \in \mathfrak{U}$, then $\mathscr{N}^{-1}$ and $\mathscr{N} \cap \mathscr{N}^{-1}$ belong to $\mathfrak{U}$. Moreover, the symmetric entourage $\mathscr{N} \cap \mathscr{N}^{-1} \subset \mathscr{N}$; thus $\mathfrak{S}$ is a basis of $\mathfrak{U}$.

Now, let $\mathcal{N} \in \mathfrak{S}$; by Proposition XII.10.1 there exists $\mathcal{M} \in \mathfrak{S}$, such that $\mathcal{M}^5 \subset \mathcal{N}$. Let us take $(x, y) \in \overline{\mathcal{M}}$. Since (x, y) is an accumulation point of $\mathcal{M}$, the neighbourhood $U := \mathcal{O}_{\mathcal{M}}(x) \times \mathcal{O}_{\mathcal{M}}(y)$ of the point (x, y) intersects $\mathcal{M}$. Let $(x_0, y_0) \in U \cap \mathcal{M}$. So we have $(x, x_0) \in \mathcal{M}$ and $(y, y_0) \in \mathcal{M}$. If we take an arbitrary point $(\bar{x}, \bar{y}) \in U$, then $(x, \bar{x}), (y, \bar{y}) \in \mathcal{M}$. Because of the symmetry of $\mathcal{M}$ we have $(\bar{x}, x), (x, x_0), (x_0, y_0), (y_0, y), (y, \bar{y}) \in \mathcal{M}$; hence $(\bar{x}, \bar{y}) \in \mathcal{M}^5$. So, we have shown that $\mathcal{M}^5$ is a neighbourhood of $\overline{\mathcal{M}}$. Thus $\overline{\mathcal{M}} \subset \operatorname{Int}(\mathcal{M}^5) \subset \operatorname{Int} \mathcal{N}$.

(b) Let $\mathcal{M} \in \mathfrak{S}$. $\overline{\mathcal{M}} \subset X \times X$ is a symmetric relation. But $\mathcal{M} \subset \overline{\mathcal{M}}$ and, by the filter properties, $\overline{\mathcal{M}} \in \mathfrak{U}$. Hence $\overline{\mathcal{M}} \in \mathfrak{S}_c$.

Let $\mathcal{N} \in \mathfrak{S}$. We know from (a) that there exists $\mathcal{M} \in \mathfrak{S}_c$ such that $\overline{\mathcal{M}} \subset \operatorname{Int}\mathcal{N} \subset \mathcal{N}$. Therefore $\mathfrak{S}_c$ is a basis of $\mathfrak{U}$.

Let $\mathcal{N} \in \mathfrak{S}$. Using (a) we see that there exists $\overline{\mathcal{M}} \in \mathfrak{U}$ such that $\overline{\mathcal{M}} \subset \operatorname{Int}\mathcal{N}$. But $\mathfrak{U}$ is a filter and therefore $\operatorname{Int}\mathcal{N} \in \mathfrak{U}$. Moreover, $\operatorname{Int}\mathcal{N}$ is a symmetric relation. Thus $\operatorname{Int}\mathcal{N} \in \mathfrak{S}_o$. At last, it follows from the inclusion $\operatorname{Int}\mathcal{N} \subset \mathcal{N}$ that $\mathfrak{S}_o$ is a basis of $\mathfrak{U}$. □

The following proposition immediately emerges from (b).

Proposition XII.10.3. A uniform Hausdorff space $(X, \mathfrak{U})$ is regular.

Proof. For a given $x_0 \in X$ the injection $X \ni y \to i_{x_0}(y) := (x_0, y) \in X \times X$ is continuous. Let $\mathfrak{S}_c$ be as above. If we take any (closed) $\mathcal{M} \in \mathfrak{S}_c$, then $i_{x_0}^{-1}(\mathcal{M}) = \mathcal{O}_{\mathcal{M}}(x_0)$ is a closed neighbourhood of x_0. But all $\mathcal{O}_{\mathcal{M}}(x_0)$, $\mathcal{M} \in \mathfrak{S}_c$, constitute a basis of neighbourhoods of x_0. By the condition (R.2)′, p. 36, we get the result. □

Corollary XII.10.4. A topological vector space is regular.

An even more general corollary also holds.

Corollary XII.10.5. A topological (Hausdorff) group is regular.

The above corollary is proved if we show how a topological group can be endowed with a uniformity in such a way that the uniform topology on the group coincides with the original one. To being with, let us recall the definition of a topological group.

Definition. Let $(X, \mathcal{T})$ be a Hausdorff space, X being a group with the group operation $X \times X \ni (x, y) \to xy \in X$ and the neutral element (unit element) e. The space $(X, \mathcal{T})$ is called a *topological group* when the mapping

(G) $\quad X \times X \ni (x, y) \to xy^{-1} \in X$

is continuous. The condition (G) is clearly equivalent to the two which follow:

The mappings

$$\text{(G}'\text{)}\qquad \begin{aligned} &X \times X \ni (x, y) \to xy \in X,\\ &X \ni x \to x^{-1} \in X \end{aligned}$$

are continuous.

Whenever $\mathscr{B}$ is a neighbourhood basis of the element e, we can introduce two uniformities: the *left uniformity* determined by the uniformity basis consisting of sets

$$\mathscr{N}_V^l := \{(x, y) : yx^{-1} \in V\}, \qquad V \in \mathscr{B},$$

and the *right uniformity* determined by the uniformity basis consisting of sets

$$\mathscr{N}_V^r := \{(x, y) : x^{-1}y \in V\}, \qquad V \in \mathscr{B}.$$

These uniformities are in general different. They are plainly equal in the case of an Abelian group (e.g. vector space), for then $yx^{-1} = x^{-1}y$. The reader can verify that both of these uniformities determine the initial topology on X.

PROPOSITION XII.10.6. A uniform space $(X, \mathfrak{U})$ is a Hausdorff if and only if

$$\text{(U.4)}\qquad \bigcap_{\mathscr{M} \in \mathfrak{U}} \mathscr{M} = \Delta,$$

i.e. the diagonal Δ is the intersection of all entourages.

PROOF. $\Rightarrow$: Let X be a Hausdorff space and let $x \neq y$; thus there exists an $\mathcal{O}_{\mathscr{M}}(x) \not\ni y$, which means that $(x, y) \notin \mathscr{M}$, and hence (U.4) holds.

$\Leftarrow$: If (U.4) holds, the intersection of all closed entourages is a closed set, and thus Δ is a closed set (in the topology of the product $X \times X$, determined by neighbourhoods of the form $\mathcal{O}_{\mathscr{M}_1}(x) \times \mathcal{O}_{\mathscr{M}_2}(y)$). Therefore, when $x \neq y$, there exists a neighbourhood $\mathcal{O}_{\mathscr{M}_1}(x) \times \mathcal{O}_{\mathscr{M}_2}(y)$ of the point (x, y) which has no points in common with the diagonal. Hence $\mathcal{O}_{\mathscr{M}_1}(x) \cap \mathcal{O}_{\mathscr{M}_2}(y) = \emptyset$. □

If a uniform structure contains the diagonal, $\Delta \in \mathfrak{U}$, the uniformity is said to be *discrete*, for it determines the discrete topology on the space.

The uniform continuity can now be defined naturally.

DEFINITION. Let $(X, \mathfrak{U})$, $(Y, \mathfrak{V})$ be uniform spaces. A mapping T: $X \to Y$ is *uniformly continuous* when

$$\bigwedge_{\mathcal{M}\in\mathfrak{V}} \bigvee_{\mathcal{N}\in\mathfrak{U}} \bigwedge_{x_1, x_2\in X} (x_1, x_2) \in \mathcal{N} \Rightarrow ((T(x_1), T(x_2)) \in \mathcal{M}.$$

Obviously, a uniformly continuous mapping is continuous, since

$$T(\mathcal{O}_{\mathcal{N}}(x)) \subset \mathcal{O}_{\mathcal{M}}(T(x)).$$

Cauchy Nets and Cauchy Filters. Completeness. If $(X, \mathfrak{U})$ is a uniform space, a net $(x_\pi)_{\pi\in\Pi}$ in X is called a *Cauchy net* when

$$\bigwedge_{N\in\mathfrak{U}} \bigvee_{\pi\in\Pi} \bigwedge_{\pi_1,\pi_2 \succ \pi_0} (x_{\pi_1}, x_{\pi_2}) \in \mathcal{N}.$$

Correspondingly, a filter $\{F_\alpha\}_{\alpha\in A}$ is a *Cauchy filter* when

$$\bigwedge_{N\in\mathfrak{U}} \bigvee_{F_\alpha} F_\alpha \times F_\alpha \subset \mathcal{N}.$$

DEFINITION. A uniform space $(X, \mathfrak{U})$ is *complete* if every Cauchy net is convergent, i.e. has a limit in X.

The relation between filters and nets immediately leads to the following corollary.

COROLLARY XII.10.7. A uniform space is complete if and only if every Cauchy filter is convergent.

A uniform Hausdorff space $(X, \mathfrak{U})$ can be completed essentially in the same way as a metric space is, the sole difference being that use must be made of Cauchy nets indexed by a fixed basis $\mathfrak{B}$ of the uniform structure $\mathfrak{U}$, consisting of symmetric entourages. $\mathfrak{B}$ is directed by inclusion, i.e. if $\mathcal{N}, \mathcal{M} \in \mathfrak{B}$, then $\mathcal{N} \prec \mathcal{M}$ means that $\mathcal{N} \supset \mathcal{M}$.

Let $\mathscr{C}$ be the set of all Cauchy nets in X, indexed by $\mathfrak{B}$. For simplicity we shall write $(x_{\mathcal{N}})$ instead of $(x_{\mathcal{N}})_{\mathcal{N}\in\mathfrak{B}}$. We introduce in $\mathscr{C}$ an equivalence relation: $(x_{\mathcal{N}}) \sim (y_{\mathcal{N}})$ if and only if

$$\bigwedge_{\mathcal{M}\in\mathfrak{S}} \bigvee_{\mathcal{N}\in\mathfrak{B}} \bigwedge_{\mathscr{P}\in\mathfrak{B}} (\mathcal{N} \prec \mathscr{P}) \Rightarrow ((x_{\mathscr{P}}, y_{\mathscr{P}}) \in \mathcal{M}).$$

We define $\tilde{X} := \mathscr{C}/\sim$; the equivalence class of $(x_{\mathcal{N}}) \in \mathscr{C}$ will be denoted by $\tilde{x}$.

We are going to construct a uniform structure on $\tilde{X}$ and to prove that it is a complete space containing X in a canonical way as a dense set and therefore it may be considered as a "completion" of X.

The uniform structure $\tilde{\mathfrak{U}}$ on $\tilde{X}$ is defined as follows. Let $\mathcal{M}$ be a symmetric entourage of $\mathfrak{U}$ and let us denote by $\tilde{\mathcal{M}}$ the set consiting of all pairs $(\tilde{x}, \tilde{y})$ of equivalence classes such that for every symmetric $\mathcal{N} \in \mathfrak{U}$ there is $\mathcal{P} \in \mathfrak{B}$ such that for all $\mathcal{Q} \in \mathfrak{B}$, $\mathcal{P} \prec \mathcal{Q}$, $(x_{\mathcal{Q}}, y_{\mathcal{Q}}) \in \mathcal{N}\mathcal{M}\mathcal{N}$. That this is independent of the chosen representatives of $\tilde{x}$ and $\tilde{y}$ is readily seen. In fact, let $(y_{\mathcal{N}}) \sim (z_{\mathcal{N}})$. Take symmetric $\mathcal{N}_1 \in \mathfrak{U}$ such that $\mathcal{N}_1^2 \subset \mathcal{N}$; then there exists $\mathcal{P}_1 \in \mathfrak{B}$ for all $\mathcal{P}_1 \prec \mathcal{Q}$ $(y_{\mathcal{Q}}, z_{\mathcal{Q}}) \in \mathcal{N}_1$. If $\mathcal{P}$ is chosen so that $(x_{\mathcal{Q}}, y_{\mathcal{Q}}) \subset \mathcal{N}_1\mathcal{M}\mathcal{N}_1$ for all $\mathcal{P}_1 \prec \mathcal{Q}$, then $(x_{\mathcal{Q}}, z_{\mathcal{Q}}) \in \mathcal{N}_1\mathcal{M}\mathcal{N}_1^2 \subset \mathcal{N}\mathcal{M}\mathcal{N}$ for all $\mathcal{Q}$ such that $\mathcal{P} \prec \mathcal{Q}$, $\mathcal{P}_1 \prec \mathcal{Q}$. An easy verification of the fact that the family of all $\tilde{\mathcal{M}}$ so defined is indeed a base of uniformity on $\tilde{X}$ can be left to the reader.

Of course $\tilde{X}$ (and $\tilde{\mathfrak{U}}$) depends on a choice of $\mathfrak{B}$. However, it can be shown that starting with another $\mathfrak{B}$ the resulting uniform spaces will be isomorphic (in the sense of uniform spaces isomorphism). Thus the above construction and the theorem below justify that the uniform space $(\tilde{X}, \tilde{\mathfrak{U}})$ is called the *Hausdorff completion of the uniform space* $(X, \mathfrak{U})$.

THEOREM XII.10.8. Let $(X, \mathfrak{U})$ be a Hausdorff uniform space and let $(\tilde{X}, \tilde{\mathfrak{U}})$ be constructed as above (i.e. it is the Hausdorff completion of $(X, \mathfrak{U})$). Then

(a) $\tilde{X}$ is a complete uniform space,

(b) $\tilde{X}$ is a Hausdorff space.

Moreover, let us identify X with a subset of $\tilde{X}$, the identification being that of a point $x \in X$ with the class of constant net determined by x, i.e. the net $(x_{\mathcal{N}})$ where $x_{\mathcal{N}} = x$ for all $\mathcal{N} \in \mathfrak{B}$. Then

(c) X is a dense subset of $\tilde{X}$,

(d) $(X, \mathfrak{U})$ is a subspace of $(\tilde{X}, \tilde{\mathfrak{U}})$ (i.e. the uniform structure $\mathfrak{U}$ is induced by $\tilde{\mathfrak{U}}$).

PROOF. (c) is proved exactly as in the case of a metric space (cf. Lemma I.7.4 and the remark following Theorem II.6.1).

(d) is immediate in virtue of the observation that $\tilde{\mathcal{M}} \cap (X \times X) = \bar{\mathcal{M}}$, where $\tilde{\mathcal{M}}$ is an entourage in $\tilde{X}$ constructed above.

(a) The completeness of $(\tilde{X}, \tilde{\mathfrak{U}})$ will follow easily from a lemma below (we defer the proof until the proof of the theorem is finished).

LEMMA XII.10.9. Let $\mathfrak{B}$ be a uniformity basis of a uniform space $(X, \mathfrak{U})$. Then (X is complete) $\Leftrightarrow$ (every Cauchy net indexed by $\mathfrak{B}$ has a limit in X.

We take as a uniformity basis in $\tilde{X}$ the family of all the sets $\tilde{\mathcal{N}}$ constructed above, with $\mathcal{N}$ ranging over all entourages of the chosen basis $\mathfrak{B}$. Let $(\tilde{x}_{\tilde{\mathcal{N}}})$ be a Cauchy net in $\tilde{X}$ and let us choose for every $\mathcal{N}$, in virtue of (c), a point $z_{\mathcal{N}} \in X$ such that $(\tilde{x}_{\tilde{\mathcal{N}}}, \tilde{z}_{\tilde{\mathcal{N}}}) \in \tilde{\mathcal{N}}$. It is easily seen that $(z_{\mathcal{N}})$ is a Cauchy net in X and its equivalence class $\tilde{z}$ satisfies $\tilde{z} = \lim \tilde{x}_{\tilde{\mathcal{N}}}$.

(b) $\tilde{X}$ is Hausdorff, since if $(\tilde{x}, \tilde{y}) \in \bigcap \tilde{\mathcal{M}}$, then there exists a $\mathcal{P}_{\mathcal{M}} \in \mathfrak{B}$ such that for all $\mathcal{Q} \in \mathfrak{B}$ with $\mathcal{P}_{\mathcal{M}} \prec \mathcal{Q}$ $(x_{\mathcal{Q}}, y_{\mathcal{Q}}) \in \mathcal{M}$; thus $\tilde{x} = \tilde{y}$. □

PROOF of of the Lemma XII.10.9. Since $\Rightarrow$ is obvious we prove $\Leftarrow$: Let $\mathcal{F}$ be a Cauchy filter. For every $\mathcal{N} \in \mathfrak{B}$ we choose $F_{\mathcal{N}} \in \mathcal{F}$ which is of order $\mathcal{N}$ and a point $x_{\mathcal{N}} \in F_{\mathcal{N}}$. Given $\mathcal{P} \in \mathfrak{B}$ let $\mathcal{N} \in \mathfrak{B}$ be such that $\mathcal{N}^2 \subset \mathcal{P}$. If $\mathcal{N} \prec \mathcal{M}$, $\mathcal{N} \prec \mathcal{O}$ and $z \in F_{\mathcal{M}} \cap F_{\mathcal{O}}$, then $(x_{\mathcal{M}}, z) \in \mathcal{M} \subset \mathcal{N}$ and $(x_{\mathcal{O}}, z) \in \mathcal{O} \subset \mathcal{N}$, hence $(x_{\mathcal{M}}, x_{\mathcal{O}}) \in \mathcal{N}^2 \subset \mathcal{P}$. The net $(x_{\mathcal{N}})$ is thus a Cauchy net indexed by $\mathfrak{B}$, and by assumption it has a limit $x_0 \in X$. To show that $\mathcal{F} \to x_0$ let $\mathcal{P} \in \mathfrak{B}$ and let $\mathcal{Q} \in \mathfrak{U}$ be such that $\mathcal{Q}^3 \subset \mathcal{P}$; there exists a point $x_{\mathcal{S}}$ with $\mathcal{S} \subset \mathcal{Q}$, $\mathcal{S} \in \mathfrak{B}$, such that $(x_0, y_{\mathcal{S}}) \in \mathcal{Q}$. If $z \in F_{\mathcal{S}} \cap F_{\mathcal{Q}}$ then $(x_{\mathcal{S}}, z) \in \mathcal{Q}$; if $y \in F_{\mathcal{Q}}$, then $(z, y) \in \mathcal{Q}$. Therefore $(x_0, y) \in \mathcal{Q}^3 \subset \mathcal{P}$ for all $y \in F_{\mathcal{Q}}$, but this means that $\mathcal{F} \to x_0$. □

The reader will have no difficulty in proving the following

COROLLARY XII.10.10. The sets $\tilde{\mathcal{M}}$ are precisely closures in $\tilde{X} \times \tilde{X}$ of sets $\mathcal{M}$, where, as above, $\mathcal{M}$ is an arbitrary symmetric entourage of the uniformity $\mathfrak{U}$.

Example. Let $(X, \mathcal{T})$ be a topological vector space and let $\{U_\alpha\}_{\alpha \in A}$ be a basis of neighbourhoods of zero in X. Then the sets $\bigcup_{x \in X} [(x+U_\alpha) \times \times (x+U_\alpha)]$, $\alpha \in A$, form a basis of the left (and right) uniformity on X. A net $(x_\alpha)_{\alpha \in A}$ is a Cauchy net if and only if

$$\bigwedge_{\gamma \in A} \bigvee_{\alpha(\gamma)} \bigwedge_{\beta_1, \beta_2 \succ \alpha(\gamma)} (x'_{\beta_1} - x_{\beta_2}) \in U_\gamma .$$

The Product of Uniform Spaces. If $(E_i, \mathfrak{U}_i)$, $i \in I$, are uniform spaces, their product $E = \bigtimes_{i \in I} E_i$ becomes a uniform space in a natural manner whenever on the set $E \times E = \bigtimes_{i \in I} (E_i \times E_i)$ we introduce a uniformity basis consisting of all sets $\bigtimes_{i \in I} \mathscr{M}_i$, where $\mathscr{M}_i := \mathscr{N}_i$ for a finite number of indices, and $\mathscr{M}_i := E_i \times E_i$ for the others; here $\mathscr{N}_i$ is an arbitrary entourage from $\mathfrak{U}_i$. This uniformity, called the *product uniformity*, will be denoted $\bigtimes_{i \in I} \mathfrak{U}_i$.

It is immediate that the topology on $\bigtimes_{i \in I} E_i$ determined by the uniformity $\bigtimes_{i \in I} \mathfrak{U}_i$ is the Tychonoff topology. The reader will have no difficulty in proving the following facts:

THEOREM XII.10.11. (i) The uniform product $(\bigtimes_{i \in I} E_i, \bigtimes_{i \in I} \mathfrak{U}_i)$ of uniform spaces $(E_i, \mathfrak{U}_i)$ is complete if and only if all the spaces $(E_i, \mathfrak{U}_i)$ are complete.

(ii) A net (filter) in $\bigtimes_{i \in I} E_i$ is a Cauchy net (filter) if and only if all its projections on E_i's are Cauchy nets (filters).

(iii) The completion $\tilde{E}$ of the product $E := \bigtimes_{i \in I} E_i$ is the topological product $\bigtimes_{i \in I} \tilde{E}_i$ of the completions of the uniform spaces $(E_i, \mathfrak{U}_i)$.

Introduction of a Uniform Structure by a Family of Mappings (Projective Uniformity). With the aid of a family of mappings (T_i), T_i: $X \to X_i$, where X_i were topological spaces, it was possible to introduce a (projective) topology on X as the weakest topology under which all T_i were continuous. In the same way, having the uniform spaces $(X_i, \mathfrak{U}_i)_{i \in I}$ and the mappings T_i: $X \to X_i$, we can introduce on X the weakest uniformity so that all mappings T_i will (still) be uniformly continuous. As its basis we take entourages $\mathscr{N} \subset X \times X$ consisting of all points (x, y) such that $(T_{i_j}(x), T_{i_j}(y)) \in \mathscr{N}_{i_j}$, $j = 1, \dots, n$, where $\mathscr{N}_{i_j}$ is an arbitrary entourage $\in \mathfrak{U}_i$.

The simplest and most important example of this method is the case $X_i = \boldsymbol{R}$, $i \in I$, that is, the introduction of a "projective" uniformity by means of numerical functions f_i: $X \to \boldsymbol{R}$. Then the sets $\mathscr{N}_{i_j} = \{(x, y)$: $|f_{i_j}(x) - f_{i_j}(y)| < \varepsilon$, $\varepsilon > 0$, $j = 1, \dots, n\}$ are entourages. Two questions arise: (i) Can every topological Hausdorff space be uniformized, i.e.

can we introduce in it a uniformity $\mathfrak{U}$ such that the topology it determines is identical with the original one? (ii) When can this be done is exactly one way?

An exhaustive answer to (i) is given by the following theorem to be proved in Chapter XVII.

THEOREM XII.10.12 (A. Weil). A Hausdorff space $(X, \mathscr{T})$ can be uniformized if and only if $(X, \mathscr{T})$ is a *Tychonoff space*.[1]

An answer to question (ii) will be given in Corollary XII.11.5.

11. PRECOMPACT AND COMPACT UNIFORM SPACES

DEFINITION. A subset K of a uniform space $(X, \mathfrak{U})$ is *totally bounded* if for every entourage $\mathscr{N} \in \mathfrak{U}$ there exists a finite family of subsets $M_1, \ldots, M_n \subset X$, of order $\mathscr{N}$, such that $K \subset \bigcup_{j=1}^{n} M_j$.

Let us take a point $x_i \subset M_i$ for every $i = 1, \ldots, n$, then

$$(*) \qquad \bigwedge_{x \in K} \bigvee_{i} (x_i, x) \in \mathscr{N}.$$

A set $\{x_i\}_{i \in I} \subset X$ satisfying (*) is called an $\mathscr{N}$*-net* on the set K. It is easy to check that

PROPOSITION XII.11.1. If K is a subset of a uniform space $(X, \mathfrak{U})$, then (K is totally bounded) $\Leftrightarrow$ ($\bigwedge_{\mathscr{N} \in \mathfrak{U}}$ there exists a finite $\mathscr{N}$-net on K).

Example. A compact set $K \subset X$ is totally bounded. Given $\mathscr{N} \in \mathfrak{U}$; let $\mathscr{M} \in \mathfrak{S}_0$ be such that $\mathscr{M}^2 \subset \mathscr{N}$; then each $\mathcal{O}_{\mathscr{M}}(x)$ is of order $\mathscr{N}$ and it is open (cf. the proof of Proposition XII.10.3). By compactness, a finite covering can be selected from the open covering $\{\mathcal{O}_{\mathscr{M}}(x)\}_{x \in K}$ of K.

Analogously we obtain that a relatively compact subset is totally bounded.

DEFINITION. A uniform space P (or a subset of P) is *precompact* when its completion $\tilde{P}$ is a compact space.

---

[1] $(X, \mathscr{T})$ is a Tychonoff space if 1° it is of type T_1, 2° for every point $x_0 \in X$ and for any open neighbourhood $\mathcal{O}(x_0)$ for that point there exists a continuous function $f: X \to [0, 1]$ such that $f(x_0) = 1$ and $f(x) = 0$ for $x \notin \mathcal{O}(x_0)$. Tychonoff spaces are sometimes referred to as completely regular spaces or $T_{3\frac{1}{2}}$.

PROPOSITION XII.11.2. Let X be a uniform Hausdorff space. Then: (X is precompact) $\Leftrightarrow$ (Every net in X has a Cauchy finer net) $\Leftrightarrow$ (Every ultrafiltei on X is a Cauchy filter).

PROPOSITION XII.11.3. The concepts of precompactness and total boundedness coincide.

PROOF. Let the space E be precompact; in that event its completion $\tilde{E}$ is compact, whereby it is totally bounded and has a finite covering $\{M_j\}_1^n$ with sets of a prescribed order $\tilde{\mathcal{N}}$, and therefore $\{M_j \cap E\}$ is a finite covering of order $\mathcal{N}$.

If the space E is totally bounded and $\{M_j\}_1^n$ is its covering of order $\mathcal{N}$, the sets $\tilde{M}_j$ which are closures in $\tilde{E}$ of the sets M_j constitute a finite covering of order $\tilde{\mathcal{N}}$ of the space $\tilde{E}$. If $\mathcal{F}$ is an ultrafilter on $\tilde{E}$, then since $\bigcup \tilde{M}_j = \tilde{E}$, at least one $\tilde{M}_j \in \mathcal{F}$ (cf. Lemma XII.3.5). Hence the filter $\mathcal{F}$ contains arbitrary small sets and thus it is a Cauchy filter which is convergent owing to the completeness of the space $\tilde{E}$, whereby $\tilde{E}$ is compact. □

It follows from the Urysohn Lemma (Theorem XVII.2.3) that normal spaces are Tychonoff spaces. Compact Hausdorff spaces are normal (Lemma XII.3.6), hence are Tychonoff spaces, and therefore are uniformizable (Theorem XII.10.12). Taking into account

THEOREM XII.11.4. In a uniform compact space the uniformity $\mathfrak{U}$ consists of all the neighbourhoods of the diagonal,

we obtain an important

COROLLARY XII.11.5. Every compact Hausdorff space X can be uniformized in exactly one way (e.g. by means of the family $C(X)$ of continuous functions; cf. § 9). It then becomes a complete uniform space.

PROOF OF THEOREM XII.11.4. Every entourage $\mathcal{N}$ is a neighbourhood of the diagonal in $X \times X$. If the diagonal Δ were to have an open neighbourhood $\mathcal{N}_0$ which did not belong to $\mathfrak{U}$, then for every $\mathcal{N} \in \mathfrak{U}$, since $\mathcal{N} \not\subset \mathcal{N}_0$, we would always have $(X \times X - \mathcal{N}_0) \cap \mathcal{N} \neq \emptyset$. The sets $\{(X \times X - \mathcal{N}_0) \cap \mathcal{N}\}$ would constitute a basis of a filter $\mathcal{F}$ finer than $\mathfrak{U}$; i.e. $\mathcal{F} \supset \mathfrak{U}$. Since $X \times X$ is compact, the filter $\mathcal{F}$ has a point of accumulation $(a, b) \notin \Delta$, and hence the filter $\mathfrak{U}$, being not as fine as $\mathcal{F}$, also has (a, b) as a point of accumulation. Since the topology defined by $\mathfrak{U}$ is by hypothesis a Hausdorff topology $\bigcap_{\mathscr{M}\in\mathfrak{U}} \bar{\mathscr{M}} = \Delta$ (cf. Proposition XII.10.6 and Theorem XII.10.8(b)), which is a contradiction. □

Hence, on considering the compact space as having been uniformized, we can formulate a generalization of Theorem II.9.4.

THEOREM XII.11.6. Let $(Y, \mathfrak{U})$ be a uniform space, and let X be a compact Hausdorff space (uniformized in the only possible way). Then every continuous mapping $T\colon X \to Y$ is uniformly continuous.

PROOF. If $T\colon X \to Y$ is continuous, then the mapping $T\times T\colon X\times X \to Y\times Y$ (defined by the formula $(T\times T)(x, y) := (T(x), T(y))$) is continuous. Thus, if $\mathscr{N} \in \mathfrak{U}$ then $(T\times \bar{T})^{-1}(\mathscr{N}) \supset \Delta_X$ and hence (by Theorem XII.11.4) there exists an entourage $\mathscr{M} \supset \Delta_X$ such that $(T(x), T(y)) \in \mathscr{N}$ when $(x, y) \in \mathscr{M}$. □

12. UNIFORM STRUCTURES ON SPACES OF MAPPINGS

Let X be a set, and let $(Y, \mathfrak{U})$ be a uniform space. It will be shown how various uniform structures can be introduced on the set $\mathscr{F}(X, Y) = Y^X$. This method is a natural generalization of the metrization of the space $B(X, (Y, d))$ of bounded mappings which was introduced in Chapter V, p. 110.

Suppose that $A \subset X$, $\mathscr{N} \in \mathfrak{U}$, and let define the relations in $Y^X \times Y^X$

$$W(A, \mathscr{N}) := \{(f, g) \in Y^X \times Y^X\colon (f(x), g(x)) \in \mathscr{N} \text{ if } x \in A\}.$$

Since formulae

$$W(A, \mathscr{N}_1) \cap W(A, \mathscr{N}_2) = W(A, \mathscr{N}_1 \cap \mathscr{N}_2),$$
$$W(A, N_1) \circ W(A, \mathscr{N}_2) \subset W(A, \mathscr{N}_1 \circ \mathscr{N}_2)$$

hold, we see that the family $\{W(A, \mathscr{N})\colon N \in \mathfrak{U}\}$ constitutes a uniformity basis $\mathfrak{B}_A$ on $\mathscr{F}(X, Y)$; it determines the uniform structure $\mathfrak{U}_A$ on $\mathscr{F}(X, Y)$. The uniform space $(\mathscr{F}(X, Y), \mathfrak{U}_A)$ will be denoted by $\mathscr{F}_A(X, Y)$.

DEFINITION. The uniform structure $\mathfrak{U}_A$ on $\mathscr{F}(X, Y)$ is called a *structure of uniform convergence* (*of the mappings* $X \to Y$) *on the set* A. By a *topology of uniform convergence* (*of the mappings* $X \to Y$) *on the set* A we mean the uniform topology of $\mathscr{F}_A(X, Y)$. If a filter or net converges to $f_0 \in \mathscr{F}(X, Y)$ in the sense of this topology, it is said to be *uniformly convergent to* f_0 *on* A.

Remark. Since $W(A, \mathscr{N}^{-1}) = W(A, \mathscr{N})^{-1}$, so if $\mathscr{N}$ is symmetric, then the entourage $W(A, \mathscr{N})$ is also symmetric.

Now, let Σ be a nonempty family of subsets of X.

DEFINITION. For every $A \in \Sigma$ we have, as above, the uniformity basis $\mathfrak{B}_A$ on $\mathscr{F}(X, Y)$. Obviously, $\bigcup_{A \in \Sigma} \mathfrak{B}_A$ is a filter subbasis; it determines a filter basis $\mathfrak{B}_\Sigma$ which, in addition, is a uniformity basis. Let $\mathfrak{U}_\Sigma$ be the uniform structure determined by $\mathfrak{B}_\Sigma$. We shall denote by $\mathscr{F}_\Sigma(X, Y)$ the uniform space $(\mathscr{F}(X, Y), \mathfrak{U}_\Sigma)$. The uniform structure $\mathfrak{U}_\Sigma$ is called a *structure of uniform convergence* (*of the mappings* $X \to Y$) *on the sets of the family* Σ. The uniform topology of this space is called the *topology of uniform convergence on the sets of the family* Σ.

We now give the most important types of convergence obtained in this way.

Examples. 1. *Pointwise convergence* (also called *simple convergence*); $\mathscr{F}_s(X, Y)$. For the Σ we take a family of one-point sets, or, equivalently, the family of all finite subsets of the space X. The uniform space $\mathscr{F}_\Sigma(X, Y)$ is denoted by $\mathscr{F}_s(X, Y)$. Let us recall that $\mathscr{F}(X, Y) = Y^X = \prod_{x \in X} Y_x$, where $Y_x := Y$ for every $x \in X$. It is easy to see the uniformity $\mathfrak{U}_s := \mathfrak{U}_\Sigma$ is the product uniformity on Y^X; cf. Section 10 of this chapter. Thus the resulting uniform topology on Y^X is the Tychonoff topology; it is the weakest topology in $\mathscr{F}(X, Y)$ for which all evaluation mappings (projections) $\mathscr{F}(X, Y) \ni f \to f(x) \in Y$, $x \in X$, are continuous.

As the second extreme example we obtain the following:

2. *Uniform convergence*; $\mathscr{F}_u(X, Y)$. For Σ we take the one-element family: the entire space X (or, which amounts to be the same thing: Σ is a family of all subsets X); now we denote $\mathfrak{U}_r := \mathfrak{U}_\Sigma = \mathfrak{U}_X$. In classical analysis, this structure is called simply the *uniform convergence*.

3. *Compact* (*almost uniform*) *convergence*; $\mathscr{F}_c(X, Y)$. Let X be a topological space. If the family of all compact subsets is taken for Σ, we obtain the concept of *compact convergence* (in the previous terminology, now obsolete, *almost uniform convergence*).

4. If X is also a uniform space, one may speak of *precompact convergence*: that is, Σ is a family of precompact sets. If X is a topological vector space, convergence on (pre)compact convex sets is then particularly important.

13. FAMILIES OF EQUICONTINUOUS MAPPINGS. GENERAL ASCOLI THEOREM

A family $\mathscr{H} \subset \mathscr{F}(X, Y)$, where X is a topological space, and $Y = (Y, \mathfrak{U})$ is a uniform space, is said to be *equicontinuous* when

$$\bigwedge_{\mathscr{N} \in \mathfrak{U}} \bigwedge_{x \in X} \bigvee_{U(x)} \bigwedge_{h \in \mathscr{H}} \bigwedge_{x' \in U(x)} (h(x), h(x')) \in \mathscr{N},$$

where $U(x)$ is a neighbourhood of the point x; clearly, then $\mathscr{H} \subset \mathscr{C} \times \times (X, Y)$. If X is also a uniform space $X = (X, \mathfrak{B})$, then the family $\mathscr{H}$ is said to be *uniformly equicontinuous* if

$$\bigwedge_{\mathscr{N} \in \mathfrak{U}} \bigvee_{\mathscr{M} \in \mathfrak{B}} \bigwedge_{h \in \mathscr{H}} \bigwedge_{(x, x') \in \mathscr{M}} (h(x), h(x')) \in \mathscr{N}.$$

We can now proceed with the proof of a theorem that is important in functional analysis.

THEOREM XII.13.1. Let X be a compact space, let $(Y, \mathfrak{U})$ be a uniform space, and let $\mathscr{H}$ be an equicontinuous family in $\mathscr{C}(X, Y)$. Then the uniform structures of pointwise convergence and uniform convergence are identical on $\mathscr{H}$, i.e. their intersections with $\mathscr{H} \times \mathscr{H}$ coincide.

PROOF. Since $\mathfrak{U}_u \subset \mathfrak{U}_s$, it is sufficient to show that on $\mathscr{H}$ the filter $\mathfrak{U}_s$ is finer than $\mathfrak{U}_u$. We shall use the notation of the preceding section. Let us take one of sets $W(X, \mathscr{N})$, $\mathscr{N} \in \mathfrak{U}$, constituting a basis of the filter $\mathfrak{U}_u$. There exists $\mathscr{M} \in \mathfrak{U}$ such that $\mathscr{M}^3 \in \mathscr{N}$; cf. Proposition XII.10.1. Without loss of generality we may assume that $\mathscr{M} = \mathscr{M}^{-1}$; cf. Theorem XII.10.8(a). By the equicontinuity of $\mathscr{H}$, to every $x \in X$ we can assign an open neighbourhood $U(x)$ of x, such that

$$(*) \qquad \bigwedge_{h \in \mathscr{H}} \bigwedge_{z \in U(x)} (h(x), h(z)) \in \mathscr{M}.$$

Since X is compact, it is possible to select from its covering $\{U(x)\}_{x \in X}$ a finite one, say $\{U(x_i)\}_1^n$. We prove that $W(\{x_i\}_1^n, \mathscr{M}) \subset W(X, \mathscr{N})$; indeed, $(f, g) \in W(\{x_i\}_1^n, \mathscr{M})$ means that $(f(x_i), g(x_i)) \in \mathscr{M}$ for $i = 1,$

$\ldots, n$; combining $(*)$ with the symmetry of $\mathcal{M}$ we get $(f(z), g(z)) \in \mathcal{M}^3 \subset \mathcal{N}$ for every $z \in X$, i.e. $(f, g) \in W(X, \mathcal{N})$. But $W(\{x_i\}_1, \ldots) \in \mathfrak{U}_s$ which completes the proof. □

COROLLARY XII.13.2. Let X be a topological space, let Y be a uniform space, and let $\mathcal{H}$ be an equicontinuous family in $\mathscr{C}(X, Y)$. Then the uniformities of simple convergence and compact convergence are identical on $\mathcal{H}$.

Example. If U is a neighbourhood of zero in a locally convex complex vector space E, the polar U° of the set U, defined as

$$U^\circ = \{e' \in E' : |\langle x, e' \rangle| \leqslant 1, x \in U\}$$

is clearly an equicontinuous family in $E' = L(E, C)$—the space of complex linear continuous functionals on E.

The Alaoglu–Bourbaki Theorem asserts that it is a compact set in E_s', i.e. in the topology of simple convergence on E' (also called weak topology), and hence also compact in the topology E_0' because of Corollary XII.13.2.

THEOREM XII.13.3 (Ascoli). Let $(X, \mathfrak{B})$, $(Y, \mathfrak{U})$ be uniform spaces and $\mathcal{H} \subset \mathscr{C}(X, Y)$.

(a) Assume X is precompact and $\mathcal{H}$ satisfies

(i) $\mathcal{H} \subset \mathscr{C}(X, Y)$ is a uniformly equicontinuous family,

(ii) the set $\mathcal{H}(x) := \{h(x) : h \in \mathcal{H}\}$ is precompact for every $x \in X$.

Then $\mathcal{H}$ is a precompact subset in $\mathscr{C}(X, Y)$[1].

(b) If the space X is compact, then $\mathcal{H}$ is precompact in $\mathscr{C}_u(X, Y)$ if and only if $\mathcal{H}$ is equicontinuous and satisfies condition (ii). (In the case of compact X the equicontinuity and the uniform equicontinuity of a family $\mathcal{H} \subset \mathscr{C}(X, Y)$ coincide.)

PROOF. (a) Being aware of Propositions XII.11.1 and XII.11.3, and of the fact that $\{W(X, \mathcal{N}) : \mathcal{N} \in \mathfrak{U}\}$ is a basis of the uniformity on $\mathscr{F}_u(X, Y)$, we see that it is sufficient to construct, for a given $\mathcal{N} \in \mathfrak{U}$, a finite $W(X, \mathcal{N})$-net on $\mathcal{H} \subset \mathscr{F}_u(X, Y)$. The elements of this net will be certain "step" mappings, i.e. mappings with a finite range.

Let $\mathcal{M}$ be a symmetric entourage in Y such that $\mathcal{M}^2 \subset \mathcal{N}$. It follows from (i) that for our $\mathcal{M}$ there exists $\mathscr{P} \in \mathfrak{B}$ such that

$$(*) \qquad \bigwedge_{h \in \mathcal{H}} \bigwedge_{x, x' \in \mathscr{P}} (h(x), h(x')) \in \mathcal{M}.$$

---

[1] $\mathscr{C}_u(X, Y)$ denotes $\mathscr{C}(X, Y) \subset \mathscr{F}(X, Y)$ endowed with the uniformity (topology induced from $\mathscr{F}_u(X, Y)$).

Since X is precompact it possesses a covering $\{O_i: i = 1, ..., n\}$ of order $\mathscr{P}$. Now let us construct a covering $\{Z_j\}$ of order $\mathscr{P}$, with disjoint sets:

$$Z_1 := O_1, \quad Z_2 := O_2 - O_2 \cap Z_1, \quad ...,$$

$$Z_k := O_k - O_k \cap \left(\bigcup_{j=1}^{k-1} Z_j\right).$$

As a subset of O_j, clearly Z_j is of order $\mathscr{P}$. We take all nonempty sets $Z'_1, ..., Z'_l$ and, of course $\bigcup_{j=1}^{l} Z_j = X$; we also take $z_j \in Z'_j$ for each $i = 1, ..., l$. The set $P = \bigcup_{j=1}^{l} \mathscr{H}(z_j)$ is precompact as it is a finite union of precompact sets; let $y_1, ..., y_m$ be an $\mathscr{M}$-net on the set P.

Let $T := \mathscr{F}(\{1, ..., l\}, \{1, ..., m\})$; it consists of m^l mappings. For each $t \in T$ we define a piecewise constant mapping $r_t: X \to Y$ as follows: $r_t(Z'_j) := \{y_{t(j)}\}$; we recall that the sets Z'_j are disjoint. If we check that the (finite) set $\{r_t\}_{t \in T}$ is an $W(X, \mathscr{N})$-net on $\mathscr{H}$, the proof will be completed.

Let us take $h \in \mathscr{H}$; then there exists such $t \in T$ that $(y_{t(j)}, h(z_j)) \in \mathscr{M}$, i.e.

$$(**) \qquad (r_t(z_j), h(z_j)) \in \mathscr{M} \quad \text{for every } j = 1, ..., l.$$

If we take any $x \in X$, then there exists (the unique) $Z'_j \ni x$. Thus

$$x \in Z'_j \Rightarrow (z_j, x) \in \mathscr{P} \Rightarrow (h(z_j), h(x)) \in \mathscr{M}, \quad \text{cf. } (*);$$

$$x \in Z'_j \Rightarrow r_t(x) = r_t(z_j).$$

Now, by (**) and by the symmetry of $\mathscr{M}$ we obtain that $(r_t(x), h(x)) \in \mathscr{M}^2 \subset \mathscr{N}$. So, we have shown that $(r_t, h) \in W(X, \mathscr{N})$.

(b) Suppose $\mathscr{H}$ is precompact, let $\mathscr{N}$ be an arbitrary entourage in $\mathfrak{U}$ and let $f_1, ..., f_n$ be a $W(X, \mathscr{N})$-net on $\mathscr{H}$ in $\mathscr{C}_u(X, Y)$, that is

$$\bigwedge_{h \in \mathscr{H}} \bigvee_{i} \bigwedge_{x \in X} (h(x), f_i(x)) \in \mathscr{N};$$

then $\{f_i(x): i = 1, 2, ..., n\}$ constitutes an $\mathscr{N}$-net on the set $\mathscr{H}(x)$ i.e. (ii) follows.

Again take an entourage $\mathscr{N} \in \mathfrak{U}$; then the above chosen family $\{f_i\}_1^n$ is uniformly (because X is compact) equicontinuous (because it is finite). Let $\mathscr{M}$ be a symmetric entourage in $\mathfrak{U}$ such that $\mathscr{M}^3 \subset \mathscr{N}$. By the uniform equicontinuity of $\{f_i\}_1^n$ there exists an entourage $\mathscr{P}$ in

$\mathfrak{U}$ such that for every $(x, x') \in \mathscr{P}$ we have $(f_i(x), f_i(x')) \in \mathscr{M}$ for $i = 1, \ldots, n$. Therefore, for an arbitrary $h \in \mathscr{H}$ and every pair $(x, x') \in \mathscr{P}$ we have $(h(x), h(x')) \in \mathscr{M}^3 \subset \mathscr{N}$, because there is an f_i such that for every $(x, x') \in \mathscr{P}$

$$(h(x), f_i(x)),\ (f_i(x), f(x')),\ ((f_i(x'), h(x')) \in \mathscr{M}. \ \square$$

As an immediate corollary to Theorem XII.13.3 we obtain the following version of the Ascoli theorem for the case of a locally compact space X:

THEOREM XII.13.4 (Ascoli). Let X be a locally compact space, and let $(Y, \mathfrak{V})$ be a uniform space. The family $\mathscr{H} \subset \mathscr{C}_c(X, Y)$ is precompact (the space $\mathscr{C}(X, Y)$ is furnished with a topology of compact convergence) if and only if the following two conditions are satisfied:

(i) $\mathscr{H}$ is equicontinuous,

(ii) the set $\mathscr{H}(x)$ is precompact in Y for every $x \in X$.

A different (nonequivalent) version of the Ascoli theorem, giving a test of the compactness of the family $\mathscr{H} \subset \mathscr{C}_c(X, Y)$, is often presented. To this end we recall that a uniform space is compact if and only if it is precompact and complete (cf. Proposition XII.11.3). On the other hand, we know that for an equicontinuous family $\mathscr{H} \subset \mathscr{C}(K, Y)$, where K is compact, the topologies of $\mathscr{H}_s$, $\mathscr{H}_u$ and $\mathscr{H}_c$[1] are identical. Accordingly,

$$(\mathscr{H}_s \text{ is compact}) \Leftrightarrow (\mathscr{H}_c \text{ is compact}).$$

The preceding two theorems gave the necessary and sufficient conditions for precompactness, and thus if the family is to be compact, its completeness in that topology must be guaranteed. The topology of pointwise convergence obviously suggests itself. However, we shall give a proof based on the properties of the product (Tychonoff theorem). For this purpose we make several remarks.

LEMMA XII.13.5. Let X be a topological space, and let $(Y, \mathfrak{U})$ be a uniform space. If the family $\mathscr{H} \subset \mathscr{F}_s(X, Y)$ is equicontinuous, then its closure $\overline{\mathscr{H}}^s$ (i.e. closure in the space $\mathscr{F}_s(X, Y)$) is also equicontinuous. Moreover, $\overline{\mathscr{H}}^s = \overline{\mathscr{H}}^c$ (i.e. closure in $\mathscr{F}_c(X, Y)$).

---

[1] $\mathscr{H}_s$, $\mathscr{H}_u$, $\mathscr{H}_c$ denote the set $\mathscr{H}$ endowed with the (uniform,topological) structure of simple, uniform, and compact convergence, respectively.

PROOF. As we know, we can take closed entourages $\mathcal{N}$ (because closed entourages constitute a basis of the uniform structure $\mathfrak{U}$ on Y). Since the family $\mathscr{H}$ is equicontinuous in x, then for $\mathcal{N} = \overline{\mathcal{N}}$ there exists a neighbourhood $U(x) \subset X$ such that

$$\bigwedge_{x' \in U(x)} \big(h(x), h(x')\big) \in \mathcal{N} \quad \text{for every } h \in \mathscr{H}.$$

Let us notice that for given $x, x' \in X$ the mapping $\mathscr{F}_s(X, Y) \ni f \to \big(f(x), f(x')\big) \in Y \times Y$ is continuous. Suppose that $\overline{h} \in \overline{\mathscr{H}^s}$; then, inasmuch as the set $\mathcal{N} \subset Y \times Y$ is closed, we also have

$$\bigwedge_{x' \in U(x)} \big(\overline{h}(x), \overline{h}(x')\big) \in \mathcal{N}.$$

In order to prove $\overline{\mathscr{H}^s} = \overline{\mathscr{H}^c}$ it is enough to show that $\overline{\mathscr{H}^s} \subset \overline{\mathscr{H}^c}$. Combining Corollary XII.13.2 with the already proved equicontinuity of $\overline{\mathscr{H}^s}$ we obtain that topologies of the simple and of the compact convergences coincide on $\overline{\mathscr{H}^s}$. Hence, $\mathscr{H}$ is dense in $\overline{\mathscr{H}^s}$ with respect to the topology of compact convergence. Thus $\overline{\mathscr{H}^c} \cap \overline{\mathscr{H}^s} = \overline{\mathscr{H}^s}$ what completes the proof. □

Now we can give the following ("compact") version of the Ascoli Theorem.

THEOREM XII.13.6 (Ascoli). Let X be a topological space, Y be a uniform space, and let $\mathscr{H}$ be a closed subset of $\mathscr{F}_c(X, Y)$.

(a) $\mathscr{H}$ is compact if the following conditions are satisfied:

(i) the family $\mathscr{H}$ is equicontinuous,

(ii) the set $\mathscr{H}(x)$ is a compact subset of Y for every $x \in X$.

(b) If the space X is locally compact and $\mathscr{H} \subset \mathscr{C}_c(X, Y)$ then the conditions (i) and (ii) are also necessary for $\mathscr{H}$ to be compact.

PROOF. (a) In virtue of the above lemma $\mathscr{H}$ is closed in $\mathscr{F}_s(X, Y)$. Since $\prod_{x \in X} \mathscr{H}(x)$ is a compact subset of $\mathscr{F}_s(X, Y) = \prod_{x \in X} Y_x$, here $Y_x = Y$, and $\mathscr{H} \subset \prod_{x \in X} \mathscr{H}(x)$, the set $\mathscr{H}$ is compact.

(b) The conditions (i) and (ii) are necessary: (ii) As we know from the definition of the Tychonoff topology, the projection $p_x\colon F_s(X, Y) \to Y$; $p_x(f) = f(x)$ is continuous. Since a continuous image of a compact set is compact, then $\mathscr{H}(x) = p_x(\mathscr{H})$ is compact in Y.

(i) By Theorem XII.13.4, the precompactness of $\mathscr{H} \subset \mathscr{C}_c(X, Y)$ assures the equicontinuity of $\mathscr{H}$. □

14. COMPLEMENTS AND EXERCISES

Tychonoff Theorem

Because of its paramount importance we give another proof of Tychonoff Theorem XII.5.1, which is more elementary. It does not deal with filters; however, a "filter proof" was given also by Bourbaki. We reproduce the stylization of L. H. Loomis (from his excellent *An Introduction to Abstract Harmonic Analysis*).

PROOF. $\Leftarrow$: We recall that (Theorem XII.4.1(vi)) a Hausdorff space K is compact if and only if any family $\mathscr{F}$ of closed subsets of K which has the finite intersection property (f.i.p.) has a non-void intersection. Let $K := \bigtimes_{\alpha \in A} K_\alpha$, where K_α are compact. Denote by $P_\alpha : K \to K_\alpha$ the projection onto α-th coordinate and let $\mathscr{F}$ be a family of closed subsets of K with f.i.p. By Kuratowski–Zorn Lemma we extend $\mathscr{F}$ to a family $\mathscr{F}^0$ (of not necessarily closed) subsets of K which is maximal with respect to f.i.p. Plainly each family $P_\alpha \mathscr{F}^0 =: \mathscr{F}^0_\alpha$ has f.i.p. and since K_α is compact there is a point x_α which is in the *closure* of every set of $\mathscr{F}^0_\alpha$. Let $x \in K$ with $P_\alpha x = x_\alpha$, $\alpha \in A$. It is sufficient to show that x is in the closure of every set of $\mathscr{F}^0$, and therefore in every set of $\mathscr{F}$. Let $\mathscr{O}$ be any open set in K containing x, then, by the definion of the Tychonoff topology in K, there exists a *finite* set $\alpha_1, \ldots, \alpha_n \in A$ and open sets $\mathscr{O}_{\alpha_i} \subset K_{\alpha_i}$, $i = 1, \ldots, n$, such that

$$x \in \bigcap_{i=1}^{n} P_{\alpha_i}^{-1}(\mathscr{O}_{\alpha_i}) \subset \mathscr{O}.$$

Therefore $x_{\alpha_i} \in \mathscr{O}_{\alpha_i}$ and $\mathscr{O}_{\alpha_i}$ intersects every set in $\mathscr{F}^0_{\alpha_i}$, thus $P_{\alpha_i}^{-1}(\mathscr{O}_{\alpha_i})$ intersects every set of $\mathscr{F}^0$ and so belongs to $\mathscr{F}^0$ because $\mathscr{F}^0$ is maximal with respect to f.i.p. But then, also by the maximality of $\mathscr{F}^0$

$$\left(\bigcap_{i=1}^{n} P_{\alpha_i}^{-1}(\mathscr{O}_{\alpha_i}) \in \mathscr{F}^0\right) \Rightarrow (\mathscr{O} \in \mathscr{F}^0).$$

We have proved that $\mathscr{O}$ intersects every set of $\mathscr{F}^0$, but $\mathscr{O}$ was an arbitrary (open) neighbourhood of x, thus x is in the closure of every set of $\mathscr{F}^0$, hence x is in the intersection of $\mathscr{F}$.

$\Rightarrow$ is trivial; see the proof of the Tychonoff Theorem XII.5.1. $\square$

Baire Spaces

1. A locally compact space X is a Baire space.

PROOF. Let $x \in X$; then x has a compact neighbourhood K. Let $(\mathcal{O}_n)_1^\infty$ be a sequence of open, dense subsets of X. We shall show that $\bigcap_{n=1}^{\infty} \mathcal{O}_n$ is dense in X; to this end it is sufficient to prove that $(\bigcap_{n=1}^{\infty} \mathcal{O}_n) \cap K \neq \emptyset$. Since the set $\mathcal{O}_1$ is dense, there exists an $x_1 \in K^\circ \cap \mathcal{O}_1$ (K° here stands for Int K, the interior of the set K). Let X_1 be a compact neighbourhood of x_1 such that $X_1 \subset \mathcal{O}_1 \cap K^\circ$. Proceeding by induction, we may assume that there exist compact sets $X_1 \supset X_2 \supset \ldots \supset X_{n-1}$, where $X_{n-1} \subset \mathcal{O}_{n-1} \cap \cap X_{n-2}$ and $X_{n-1}^\circ \neq \emptyset$. Since $\mathcal{O}_n$ is dense in X, there exists an $x_n \in X_{n-1}^\circ \cap \mathcal{O}_n$. Let X_n be a compact neighbourhood of the point x_n such that $X_n \subset X_{n-1}^\circ \cap \mathcal{O}_n$. Accordingly, $X_n^\circ \neq \emptyset$, $X_1 \supset X_2 \supset \ldots \supset X_{n-1} \supset X_n$. Thus we have constructed a chain of nonempty compact sets, and hence (cf. Corollary XII.4.2) $\emptyset \neq \bigcap_{n=1}^{\infty} X_n \subset K^\circ \cap (\bigcap_{n=1}^{\infty} \mathcal{O}_n)$, that is, $\bigcap_{n=1}^{\infty} \mathcal{O}_n$ is dense in X.

2. Prove the following theorem: Let X_i, $i \in I$, be complete metric spaces; then $\prod_{i \in I} X_i$ is a Baire space.

3. Corollary: $\boldsymbol{R}^I$ is a Baire space.

4. A countable product of Baire spaces is a Baire space.

5. Give an example to illustrate that an uncountable product of Baire spaces need not be a Baire space.

6. The set of analytic functions is of first category in $C^\infty([a, b])$ provided $C^\infty([a, b])$ is equipped with the metric

$$d(f, g) := \sum \frac{1}{2^p} \frac{||f-g||_p}{1+||f-g||_p},$$

$$\text{where } ||h||_p := \sup |h^{(p)}([a, b])|,$$

which turns $C^\infty([a, b])$ into complete metric space.

Hint. For $x \in [a, b]$, $c > 0$ put

$$W(x,c) := \{f \in C^{\infty}([a,b]) : |f^{(n)}(x)| \leqslant n!\, c^n \text{ for all } n \in N\}.$$

Check that each analytic in a neighbourhood of x function $f \in W(x, c)$ for some c; take $c = \sup_k (|f^{(k)}(x)/k|)^{1/2}$. Hence analytic functions in $[a, b]$ form a subset of the set $\bigcup_{(x,c)} W(x, c)$, where $x \in Q \cap [a, b]$, $c \in N$. Prove that 1° $W(x, c)$ is closed, 2° $\operatorname{Int} W(x, c) = \emptyset$.

To prove 2° construct a function $s \in C^{\infty}[(a, b)]$ such that $d(s, 0) < \varepsilon$ and $|s^{(n)}(x)| > n!\, c^n$.

Completeness

1. A complete subspace F of a uniform space E is closed.
2. A closed subset of a complete uniform space is a complete space.

Uniform Continuity

3. There are two uniform structures, left $\mathfrak{U}^l$ and $\mathfrak{U}^r$ right, on a topological group X. Accordingly, for functions $f \in C^X$ one may speak of left and right uniform continuity. Formulate those concepts.

4. Show that if f is left (right)-uniformly continuous then $f^v(x) := f(x^{-1})$ is right(left)-uniformly continuous.

5. Show that if X is a compact group, the two structures $\mathfrak{U}^l$ and $\mathfrak{U}^r$ are identical, and hence it is not necessary to make a distinction between left and right uniform continuity.

Extension of Uniformly Continuous Maps and Completion of Uniform Spaces

As we know (Proposition XII.10.3) a uniform Hausdorff space is regular. In the case of uniform structures the condition (*) in Theorem XII.9.4 (about the extension of maps by closure) can be replaced by the uniform continuity, i.e.

THEOREM XII.14.1. Let A be a dense subset of a uniform space X, let Y be a complete uniform Hausdorff space. If $T: A \to Y$ is uniformly continuous on A, then T can be extended by closure (continuity) onto the whole X. The extended map $\overline{T}$ is uniformly continuous.

COROLLARY XII.14.2. Let X_1, X_2 be two uniform complete Hausdorff spaces, Y_k a dense subset of X_k, $k = 1, 2$. Then any isomorphism between Y_1 and Y_2 can be extended (uniquely) to an isomorphism between X_1 and X_2.

COROLLARY XII.14.3. (X, Y are uniform spaces) $\Rightarrow$ (To any uniformly continuous map $T: X \to Y$ there exists the unique uniformly continuous map $\tilde{T}: \tilde{X} \to \tilde{Y}$ such that $\tilde{T} \circ i = j \circ T$), where $\tilde{X}$, $\tilde{Y}$ are the completions of X, Y and i: $X \to \tilde{X}$, j: $Y \to \tilde{Y}$ are the canonical injections.

COROLLARY XII.14.4. If $T: X \to Y$, $S: Y \to Z$ are uniformly continuous and $U := S \circ T$, then $\tilde{U} = \tilde{S} \circ \tilde{T}$.

Precompact Sets

From the preceding facts follows

PROPOSITION XII.14.5. If X, Y are uniform spaces and $T: X \to Y$ is uniformly continuous, then for any precompact set $K \subset X$, $T(K)$ is a precompact subset of Y.

A precompact version of Tychonoff Theorem

PROPOSITION XII.14.6. Let X be a set, $(Y_i)_{i \in I}$ a family of uniform spaces. Let, for any $i \in I$, be given $T_i: X \to Y_i$. We assume that the family $(T_i)_{i \in I}$ separates the points of X, i.e. $\bigwedge_{x, y \in X} \bigvee_{i \in I} (x \neq y) \Rightarrow \big(T_i(x) \neq T_i(y)\big)$. Let us provide X with the uniform structure defined by $(T_i)_{i \in I}$, i.e. the weakest uniformity on X with respect to which all T_i are uniformly continuous; see p. 47. Then (X is precompact) $\Leftrightarrow$ (For every $i \in I$, $T_i(X)$ is a precompact subspace of Y_i).

PROOF. $\Rightarrow$ follows from the preceding proposition.

$\Leftarrow$ (sketch of the proof): The uniformity of the Hausdorff completion $\tilde{X}$ of X is the weakest one for which all the extensions $\tilde{T}_i: \tilde{X} \to \tilde{Y}_i$, $i \in I$, are uniformly continuous. Moreover, if j_i is the canonical map $j_i: Y_i \to \tilde{Y}_i$, and $S_i := j_i \circ T_i$, then $\tilde{X}$ can be identified with the closure in $\bigtimes_{i \in I} \tilde{Y}_i$ of the image of X by the map $x \to \big(S_i(x)\big)_{i \in I}$. The thesis follows now by Tychonoff Theorem XII.5.1. □

Ascoli Theorem

Because of the importance of the precompact convergence in functional analysis (cf. Part III) we prove now

Ascoli Theorem (relatively compact version). I. Let X be a topological (uniform) space, Y—a uniform Hausdorff space. Let $\mathscr{H}$ be an equicontinuous (uniformly equicontinuous) set in $\mathscr{C}(X, Y)$. If $\mathscr{H}(x)$ is relatively compact in Y for every $x \in X$, then $\mathscr{H}$ is relatively compact in $\mathscr{C}(X, Y)$ provided with the compact (precompact) convergence.

II. If moreover Y is uniform complete Hausdorff space and if $\mathscr{H}(x)$ is relatively compact in Y for every $x \in D$, where D is dense in X, then the assertions of I are true.

Proof. Part I is exactly the same as that of Theorem XII.13.6. Let $\overline{\mathscr{H}}$ be the closure of $\mathscr{H}$ in $\mathscr{F}_s(X, Y)$. Since for every x, the projection $\mathscr{F}_s(X, Y) \ni f \to f(x) \in Y$ is continuous, $\overline{\mathscr{H}}(x) \subset \overline{\mathscr{H}(x)}$, the Tychonoff Theorem implies that $\overline{\mathscr{H}}$ is a compact subset of $\mathscr{F}_s(X, Y)$. In virtue of Lemma XII.13.5 $\overline{\mathscr{H}}$ is equicontinuous and so the topologies of simple and compact convergence are identical on $\overline{\mathscr{H}}$.

Proof of II. It is sufficient in virtue of I, to prove, that $\mathscr{H}(x)$ is relatively compact for *all* $x \in X$. Since in a complete space the concepts of relative compactness and precompactness coincide it suffices to show the latter property of $\mathscr{H}(x)$. By equicontinuity, for given $x \in X$ and every symmetric entourage $\mathscr{N}$ of $\mathfrak{U}$ there exists a neighbourhood $\mathcal{O}$ of x such that $(h(x), h(x')) \in \mathscr{N}$ for all $x' \in \mathcal{O}$ and $h \in \mathscr{H}$. But D is dense in X, there thus exists $x' \in \mathcal{O} \cap D$ and since $\mathscr{H}(x')$ is precompact in Y, there exists a finite subset $\{y_1, \ldots, y_k\} \subset Y$, such that $\mathscr{H}(x') \subset \mathscr{N}(y_1) \cup \cup \ldots \cup \mathscr{N}(y_k)$. Hence $\mathscr{H}(x)$ is in the union of sets $\mathscr{N}^2(y_1), \ldots, \mathscr{N}^2(y_k)$.

The reader should find it easy to carry over the above proof, step by step, to the case of uniform X and uniformly continuous $\mathscr{H}$. □

XIII. THEORY OF THE INTEGRAL

1. COMPACTIFICATION OF THE REAL LINE

Let us define a homeomorphism of the interval $]-1, 1[$ of the number axis (real line) $\boldsymbol{R}$ as follows:

$$\boldsymbol{R} \ni t \to \varphi(t) = \frac{t}{1+|t|} \in]-1, 1[.$$

The inverse mapping is of the form

$$]-1, 1[\ni y \to \varphi^{-1}(y) = \frac{y}{1-|y|}.$$

Now if the mapping φ is extended formally,

$$(+\infty) = \bar{\varphi}^{-1}(+1), \quad (-\infty) = \bar{\varphi}^{-1}(-1),$$

the set $\bar{\boldsymbol{R}} := \boldsymbol{R} \cup \{+\infty\} \cup \{-\infty\}$ with metric $d(t_1, t_2) := |\bar{\varphi}(t_1) - \bar{\varphi}(t_2)|$ will be a compact set as an isometric image of the compact interval $[-1, 1]$.

The order relation existing in $\boldsymbol{R}$ is extendible to the set $\bar{\boldsymbol{R}}$. Namely, we take

$$(t_1 \leqslant t_2) \Leftrightarrow (\bar{\varphi}(t_1) \leqslant \bar{\varphi}(t_2)).$$

We thus have $-\infty \leqslant t \leqslant +\infty$ for any $t \in \bar{\boldsymbol{R}}$.

Even though algebraic operations defined in $\boldsymbol{R}$ cannot be extended to the set $\bar{\boldsymbol{R}}$ so that $\bar{\boldsymbol{R}}$ be a field, it turns out that for some purposes it is useful to adopt the following definitions of some operations:

$$a+(\pm\infty) = \pm\infty, \quad a-(\pm\infty) = \mp\infty \quad \text{for } a \in \boldsymbol{R},$$

$$a\cdot(\pm\infty) = \pm\infty \text{ for } a > 0, \quad a\cdot(\pm\infty) = \mp\infty \text{ for } a < 0,$$

$$(+\infty)+(+\infty) = +\infty, \quad (-\infty)+(-\infty) = -\infty,$$

$$0\cdot(\pm\infty) = (\pm\infty)\cdot 0 = 0,$$

$$(+\infty)\cdot(\pm\infty) = \pm\infty, \quad (-\infty)\cdot(\pm\infty) = \mp\infty.$$

DEFINITION. A mapping of any set X into $\bar{\boldsymbol{R}}$ is called a *numerical function* (in contradistinction to a number function or real function taking X into $\boldsymbol{R}$).

2. THE DANIELL–STONE INTEGRAL

As is known, the Riemann integral $f \to \int_{[a,b]} f$ is a functional which is linear, positive, and continuous on $C([a, b], \boldsymbol{R})$, the set of continuous functions with uniform convergence.

Dini's Theorem states that a sequence f of continuous functions on $[a, b]$, which converges monotonically to a continuous function f is uniformly convergent to f. Therefore, the Riemann integral is a linear, positive, and monotonically continuous functional on the set $C([a, b], \boldsymbol{R})$. The concept of the Riemann integral, however, proves to be too restricted for applications in other areas of mathematics and in theoretical physics. Hence, this has led to the formulation of many theories extending both the domain of integrable functions and the set of functions themselves. The general theory, inaugurated by Daniell and developed by Stone, replaces $C([a, b], \boldsymbol{R})$ by an arbitrary vector lattice E of real-valued functions defined on any arbitrary set X.

Suppose that X is an arbitrary non-empty set and suppose that E is a family of real functions on X.

DEFINITION. A family E is said to be an *elementary family of functions* if:

(i) E is a vector space,

(ii) E is a lattice, that is, for any two functions $f, g \in E$ we also have $f \cap g \in E$ and $f \cup g \in E$.

The last condition, as we know, may be replaced by the requirement that $|f| \in E$ for every function $f \in E$.

Given a linear functional $\mu \in E'$ (i.e. the mapping $E \to \boldsymbol{R}$).

This functional is said to be *positive* if $\mu(f) \geqslant 0$ for every function $E \ni f \geqslant 0$.

We recall that the monotonic convergence of a sequence of functions f_n to a function f, denoted by $f_n \nearrow f$ (or $f_n \searrow f$), is a pointwise convergence, together with fulfilment of the condition $f_n \leqslant f_{n+1}$ (or $f_n \geqslant f_{n+1}$).

DEFINITION. A functional $\mu \in E'$ is said to be *monotonically continuous* if for every sequence (f_n), from the fact that $f_n \nearrow f$ it follows that $\mu(f_n) \to \mu(f)$.

Remark. If $f_n \nearrow$, then $-f_n \searrow$, and hence the linearity of the functional implies that in the definition of monotonic continuity the condition $f_n \nearrow f$ is replaceable by the condition $f_n \searrow f$.

THEOREM XIII.2.1. The following properties of the functional $\mu \in E'$ are equivalent:

(i) μ is monotonically continuous,

(ii) For any $f_n \nearrow 0$, $\mu(f_n) \to 0$.

PROOF. (i) $\Rightarrow$ (ii)—obvious.

(ii) $\Rightarrow$ (i): Let $E \ni f_n \nearrow f \in E$. Then $(f_n - f) \nearrow 0$, and

$$\mu(f_n) = \mu(f_n - f) + \mu(f) \to 0 + \mu(f) = \mu(f). \quad \square$$

Thus we shall use the term "monotonically continuous functional" when point (ii) of Theorem XIII.2.1 is satisfied.

DEFINITION. A functional which is linear, positive, and monotonically continuous on an elementary family of functions E is called a *Daniell–Stone integral (D–S integral)*. Functions from the family E are called *elementary functions* of that integral.

Remark. Instead of $\mu(f)$ we shall frequently write

$$\int f d\mu \quad \text{or} \quad \int_X f(x) d\mu(x).$$

Examples. 1. Let

$$X = [a, b] \subset R, \quad E = C([a, b]), \quad \mu(f) := \int_{[a,b]} f(x) dx.$$

It is easily verified that the family E so defined is an elementary family. The fact that the Riemann integral is linear, positive and monotonically continuous was proved at the beginning of this section.

2. This example will contain Example 1 as a special case.

Let X be a locally compact space. Take $E = C_0(X)$, the set of continuous functions with compact supports on X (the set $\overline{f} \subset X$, which is the closure of the set on which the function f is nonzero, is called the *support of the function* f).

If μ is a positive linear functional continuous with respect to compact (almost uniform) convergence, then by virtue of Dini's Theorem μ will also be monotonically continuous, i.e. will be a D–S integral. An integral of this kind is usually referred to as a *Radon integral* on a locally compact space X.

3. In Example 2 we set $X = \boldsymbol{R}$, $E = C_0(\boldsymbol{R})$; then

$$\mu(f) = \int_{\boldsymbol{R}} f(x)\,dx.$$

4. In Example 2 we set $X = \boldsymbol{N} \subset \boldsymbol{R}$ ($\boldsymbol{N}$ is the set of natural numbers). The space $C_0(\boldsymbol{N})$ will consist of sequences with a finite number of non-zero terms. For $f \in C_0(\boldsymbol{N})$ we take

$$\mu(f) := \sum_{n=0}^{\infty} f(n).$$

5. Let X be an arbitrary nonempty set. Take the set E of all real-valued functions on X. Suppose a point $x_0 \in X$ is given. For $f \in E$ we define $\mu(f) := f(x_0)$ (the usual notation is $\delta_{x_0}(f) := f(x_0)$). It is easily verified that this is a D–S integral.

6. Let $\varphi \geqslant 0$ be a bounded, nondecreasing function on $\boldsymbol{R}$ and let $f \in C([a, b])$. As in the definition of the Riemann integral, the numbers

$$\overline{S}(f, \pi) := \sum \sup f(P_i)[\varphi(x_i) - \varphi(x_{i-1})],$$

$$\underline{S}(f, \pi) := \sum \inf f(P_i)[\varphi(x_i) - \varphi(x_{i-1})]$$

can be associated with each partition $\pi = (P_i)$ of the interval $[a, b]$. Clearly

$$\inf f([a, b])\,\big(\varphi(b) - \varphi(a)\big) \leqslant \underline{S}(f, \pi) \leqslant \overline{S}(f, \pi)$$
$$\leqslant \sup f([a, b])\,\big(\varphi(b) - \varphi(a)\big).$$

Once again we have an upper and a lower integral,

$$\overline{\int} f\,d\varphi := \inf_{\pi \in \Pi} \overline{S}\,(f, \pi), \qquad \underline{\int} f\,d\varphi := \sup_{\pi \in \Pi} \underline{S}(f, \pi).$$

Of course these integrals exist for any bounded function f. When f is continuous, as we have assumed here, then as in the case of the Riemann integral it is proved that

$$\overline{\int} f d\varphi = \underline{\int} f\,d\varphi.$$

Definition. The common value of both these integrals is called the *Riemann–Stieltjes integral* of the function f and is denoted by

$$\int f\,d\varphi.$$

Note that

$$C_0(R) \ni f \to \int f\,d\varphi \in \boldsymbol{R}$$

is a linear nonnegative functional and, hence, a Radon integral. The D–S extension (the definition of which will be given just below) of this integral is called the *Lebesgue–Stieltjes integral.* The famous theorem of Riesz states that every Radon integral on $\boldsymbol{R}$ is a Lebesgue–Stieltjes integral (cf. p. 121).

LEMMA XIII.2.2. Given two sequences: $E \ni f_n \nearrow f$, $E \ni g_n \nearrow g$, and $f \geqslant g$. Then

$$\lim \mu(f_n) \geqslant \lim \mu(g_n).$$

Remark. If $f, g \in E$, the thesis follows from the monotonic continuity of the functional μ. The lemma however concerns the general case when the functions f and g are arbitrary.

PROOF. Note that when $E \ni h \leqslant \lim f_n = f$, then $f_n \geqslant (f_n \cap h) \nearrow h \in E$. The positivity and continuity of the functional imply that $\mu(f_n) \geqslant \mu(f_n \cap h)$ and $\lim \mu(f_n) \geqslant \lim \mu(f_n \cap h) = \mu(h)$.

Taking $h = g_k \leqslant g \leqslant f$, we have $\lim \mu(f_n) \geqslant \mu(g_k)$, whereby

$$\lim \mu(f_n) \geqslant \lim \mu(g_k). \quad \square$$

COROLLARY XIII.2.3. $(\lim f_n = \lim g_n) \Rightarrow \big(\lim \mu(f_n) = \lim \mu(g_n)\big)$.

PROOF. On invoking Lemma XIII.2.2 twice, we obtain $\lim \mu(f_n) \geqslant \lim \mu(g_n)$ and $\lim \mu(f_n) \leqslant \lim \mu(g_n)$, wherefrom the thesis of the corollary follows.

Let us now define the following function space:

$$\mathscr{E} := \{f\colon \bigvee_{(f_n)} f_n \in E, f_n \nearrow f\}.$$

Functions from the space $\mathscr{E}$ may be numerical, i.e. they may assume the value $+\infty$.

We extend the functional μ to the space $\mathscr{E}$, setting

$$\tilde{\mu}(f) := \lim \mu(f_n).$$

By Corollary XIII.2.3, this definition is correct, that is, the value $\tilde{\mu}(f)$ does not depend on the choice of the approximating sequence $f_n \nearrow f$.

If $f \in E$, then taking $f_n = f$, we have $f_n \nearrow f$, that is $f \in \mathscr{F}$, and $\tilde{\mu}(f) = \lim \mu(f) = \mu(f)$. Hence $E \subset \mathscr{E}$ and $\tilde{\mu}|_E = \mu$.

THEOREM XIII.2.4. (i) The space $\mathscr{E}$ is closed under: addition, multiplication by nonnegative numbers, and the operations $\cup$ and $\cap$ (i.e. is a lattice).

(ii) $\mathscr{E}$ is a closed set under the operation $\nearrow$, that is, if $\mathscr{E} \ni f_n \nearrow f$, then $f \in \mathscr{E}$.

(iii) If $\mathscr{E} \ni f_n \nearrow f$, then $\tilde{\mu}(\sup f_n) = \sup \tilde{\mu}(f_n)$.

(iv) $\tilde{\mu}$ is an additive and positive homogeneous functional (that is, $\tilde{\mu}(cf) = c\tilde{\mu}(f)$ for $c \geqslant 0$).

PROOF. (i) If $E \in f_n \nearrow f \in \mathscr{E}$, $E \ni g_n \nearrow g \in \mathscr{E}$, then $E \ni (f_n+g_n) \nearrow f+g$ and we have $E \ni cf_n \nearrow cf$ for $c \geqslant 0$. Hence $(f+g) \in \mathscr{E}$ and $cf \in \mathscr{E}$.

Also $E \ni (f_n \cap g_n) \nearrow f \cap g$ and $E \ni (f_n \cup g_n) \nearrow f \cup g$, and this means that $f \cap g \in \mathscr{E}$, $f \cup g \in \mathscr{E}$.

(ii) Let $\mathscr{E} \ni f_n \nearrow f \in \mathscr{E}$. For every n there exists a sequence of elementary functions $(f_n^k)_{k=1}^{\infty}$ such that $f_n^k \nearrow f_n$. Let us take

$$E \ni h_m := f_1^m \cup f_2^m \cup \ldots \cup f_m^m.$$

Since $f_i^{m+1} \geqslant f_i^m$, therefore $h_{m+1} \geqslant h_m$, that is $h_n \nearrow h$, where $h = \sup h_n \in E$. But for $n \leqslant m$ we have

$$(1) \qquad f_n^m \leqslant h_m \leqslant f_m.$$

Consequently

$$\lim_{m\to\infty} f_n^m \leqslant \lim_{m\to\infty} h_m \leqslant \lim_{m\to\infty} f_m, \quad f_n \leqslant h \leqslant f.$$

Therefore $f = \lim\limits_{n\to\infty} f_n \leqslant h \leqslant f$, and hence $h = f$, that is $f \in \mathscr{E}$.

(iii) On integrating formula (1), by virtue of Lemma XIII.2.2 we find

$$\mu(f_n^m) \leqslant \mu(h_m) \leqslant \mu(f_m),$$

$$\tilde{\mu}(f_n) \leqslant \tilde{\mu}(h) \leqslant \sup \tilde{\mu}(f_m),$$

$$\sup \tilde{\mu}(f_n) \leqslant \tilde{\mu}(h) \leqslant \sup \tilde{\mu}(f_m).$$

But $h = f = \sup f_n$, whereby the thesis follows.

(iv) $$\tilde{\mu}(f+g) = \lim \mu(f_n+g_n) = \lim\big(\mu(f_n)+\mu(g_n)\big)$$
$$= \lim \mu(f_n) + \lim \mu(g_n) = \tilde{\mu}(f) + \tilde{\mu}(g);$$
$$\tilde{\mu}(cf) = \lim \mu(cf_n) = \lim\big(c\mu(f_n)\big) = c \lim \mu(f_n) = c\tilde{\mu}(f). \quad \square$$

3. THE FUNCTIONAL μ^* AND ITS PROPERTIES

Let us now extend the functional $\tilde{\mu}$ to the set of all nonnegative numerical functions.

DEFINITION. Let $0 \leqslant f$ be a numerical function.

$$\mu^*(f) := \begin{cases} \inf\limits_{f \leqslant h \in E} \tilde{\mu}(h), & \text{if there is a function } h \in \mathscr{E} \text{ such that } f \leqslant h, \\ +\infty & \text{if such a function does not exist.} \end{cases}$$

If by $\mathscr{E}^+$ we denote those functions $\mathscr{E}$ which are nonnegative, then

$$\mu^*|_{\mathscr{E}^+} = \mu|_{\mathscr{E}^+}.$$

THEOREM XIII.3.1. Given two nonnegative numerical functions f and g. The

(i) $(f \leqslant g) \Rightarrow (\mu^*(f) \leqslant \mu^*(g))$ (positivity),

(ii) $(\boldsymbol{R} \ni c > 0) \Rightarrow (\mu^*(cf) = c\mu^*(f))$ (positive homogeneity),

(iii) $\mu^*(f+g) \leqslant \mu^*(f) + \mu^*(g)$ (subadditivity),

(iv) μ^* is a convex functional.

PROOF. (i) If $\mu^*(g) = +\infty$, the thesis holds trivially. Otherwise, every function $h \in \mathscr{E}$ such that $h \geqslant g$ satisfies the condition $h \geqslant f$ and hence

$$\inf_{f \leqslant h \in \mathscr{E}} \tilde{\mu}(h) \leqslant \inf_{g \leqslant h \in \mathscr{E}} \mu(h).$$

(ii) If $\mathscr{E} \ni h \geqslant f$, $\mathscr{E} \ni ch \geqslant cf$ and conversely. Therefore

$$\mu^*(f) = \inf_{f \leqslant h \in \mathscr{E}} \tilde{\mu}(h) = \inf_{cf \leqslant h \in \mathscr{E}} \tilde{\mu}\left(\frac{h}{c}\right) = \inf_{cf \leqslant h \in \mathscr{E}} \frac{1}{c}\tilde{\mu}(h)$$

$$= \frac{1}{c} \inf_{cf \leqslant h \in \mathscr{E}} \tilde{\mu}(h) = \frac{1}{c}\mu^*(cf).$$

(iii) We have

$$\mu^*(f+g) = \inf_{(f+g) \leqslant h \in \mathscr{E}} \tilde{\mu}(h) \leqslant \inf_{\substack{f \leqslant h \in \mathscr{E} \\ g \leqslant k \in \mathscr{E}}} \tilde{\mu}(h+k)$$

$$= \inf_{\substack{f \leqslant h \in \mathscr{E} \\ g \leqslant k \in \mathscr{E}}} (\tilde{\mu}(h) + \tilde{\mu}(k)) = \inf_{f \leqslant h \in \mathscr{E}} \tilde{\mu}(h) + \inf_{g \leqslant k \in \mathscr{E}} \tilde{\mu}(k)$$

$$= \mu^*(f) + \mu^*(g).$$

Here we have made use of the fact that if $f \leqslant h$ and $g \leqslant k$, then $(f+g) \leqslant h+k$.

(iv) Let $\alpha, \beta > 0$, $\alpha+\beta = 1$, then

$$\mu^*(\alpha f+\beta g) \leqslant \mu^*(\alpha f)+\mu^*(\beta g) = \alpha\mu^*(f)+\beta\mu^*(g). \ \square$$

Remark. It is not true that we always have $\mu^*(f+g) = \mu^*(f)+\mu^*(g)$.

Example. Let X be a space consisting of two points: $X = \{a, b\}$. For E let us take all real-valued functions on X which satisfy the condition $f(a) = f(b)$. It is easily seen that $\mathscr{E} = E \cup (+\infty)$ (by $(+\infty)$ we denote here a function assuming the value $+\infty$ throughout all space).

The formula

$$\mu^*(g) = \max\big(g(a), g(b)\big),$$

holds for any numerical function $g \geqslant 0$.

Let us take the following functions g_1 and g_2

$$g_1(a) = 1, \quad g_1(b) = 0, \quad g_2(a) = 0, \quad g_2(b) = 1,$$
$$(g_1+g_2)(a) = (g_1+g_2)(b) = 1.$$

Then

$$\mu^*(g_1+g_2) = 1, \quad \mu^*(g_1) = 1, \quad \mu^*(g_2) = 1.$$

Here is the next property of the functional μ^*:

THEOREM XIII.3.2 (Beppo Levi). Let $(f_n)_1^\infty$ be a sequence of non-negative numerical functions; let $f_n \nearrow$. Then

$$\sup\mu^*(f_n) = \mu^*(\sup f_n).$$

PROOF. Theorem XIII.3.1(i), implies that

$$\mu^*(f_n) \leqslant \mu^*(\sup f_n),$$

whence

$$\sup\mu^*(f_n) \leqslant \mu^*(\sup f_n).$$

It thus remains to show that the inverse inequality holds. If $\sup\mu^*(f_n) = \infty$, the equality is satisfied. Otherwise, we have $\sup\mu^*(f_n) = \mu < \infty$. We shall break up the proof into two parts.

(i) We shall prove that

$$\bigwedge_{\varepsilon>0}\ \bigvee_{\mathscr{E}\supset(g_n)}\ \bigwedge_{n} g_{n+1} \geqslant g_n, \quad f_n \leqslant g_n,$$

$$\mu^*(f_n) \leqslant \tilde{\mu}(g_n) \leqslant \mu^*(f_n)+\varepsilon.$$

In order to do this, note that

$$\text{(1)} \qquad \bigwedge_{\varepsilon>0} \bigvee_{\mathscr{E}\supset(h_n)} \bigwedge_{n} f_n \leqslant h_n, \quad \mu^*(f_n) \leqslant \tilde{\mu}(h_n) \leqslant \mu^*(f_n)+\frac{\varepsilon}{2^n}.$$

This fact follows immediately from the definition of the functional μ^*. Let us assume

$$g_n := h_1 \cup h_2 \cup \ldots \cup h_n .$$

From Theorem XIII.2.4 it follows that $g_n \in \mathscr{E}$. Plainly $g_{n+1} \geqslant g_n$, and the following formulae hold:

$$\text{(2)} \qquad g_{n+1} = g_n \cup h_{n+1},$$

$$\text{(3)} \qquad f_n \leqslant g_n .$$

Since $h_{n+1} \geqslant f_{n+1} \geqslant f_n$, then

$$\text{(4)} \qquad g_n \cap g_{n+1} \geqslant f_n .$$

By induction we shall show that

$$\text{(5)} \qquad \tilde{\mu}(g_n) \leqslant \mu^*(f_n)+\varepsilon\left(1-\frac{1}{2^n}\right).$$

The formula holds for $n = 1$. Suppose that it is valid for some $n \geqslant 1$. We apply the general formula $f+g = f\cap g+f\cup g$ to the functions g_n and h_{n+1}. We obtain

$$g_{n+1} = g_n \cup h_{n+1} = g_n+h_{n+1}-g_n \cap h_{n+1}.$$

From the linearity of the functional $\tilde{\mu}$ it follows that

$$\tilde{\mu}(g_{n+1}) = \tilde{\mu}(g_n)+\tilde{\mu}(h_{n+1})-\tilde{\mu}(g_n \cap h_{n+1}).$$

Relations (1) and (4) yield

$$\tilde{\mu}(g_{n+1}) \leqslant \tilde{\mu}(g_n)+\mu^*(f_{n+1})+\frac{\varepsilon}{2^{n+1}}-\mu^*(f_n).$$

Using the induction hypothesis, we obtain

$$\tilde{\mu}(g_{n+1}) \leqslant \mu^*(f_n)+\varepsilon\left(1-\frac{1}{2^n}\right)+\mu^*(f_{n+1})+\frac{\varepsilon}{2^{n+1}}-\mu^*(f_n)$$

$$= \mu^*(f_{n+1})+\varepsilon\left(1-\frac{1}{2^{n+1}}\right).$$

Formula (5) thus holds for $n+1$, and hence is valid for any n.

(ii) The monotonic continuity of the functional $\tilde{\mu}$ implies that

$$\tilde{\mu}(\sup g_n) = \sup \tilde{\mu}(g_n).$$

Let $\mathscr{E} \ni g := \sup g_n$. Thus we have $g \geqslant \sup f_n$, since $g_n \geqslant f_n$, and

$$\mu^*(\sup f_n) \leqslant \tilde{\mu}(g) = \sup \tilde{\mu}(g_n) \leqslant \sup \mu^*(f_n) + \varepsilon.$$

But ε is arbitrary, and therefore $\mu^*(\sup f_n) \leqslant \sup \mu^*(f_n)$. □

Corollary XIII.3.3 $(0 \leqslant f_n$ — numerical functions$) \Rightarrow \left(\mu^*\left(\sum_{n=1}^{\infty} f_n\right) \leqslant \sum_{n=1}^{\infty} \mu^*(f_n)\right)$.

Proof. Let us denote $s_k := \sum_{n=1}^{k} f_n$. Plainly $s_k \nearrow \sum_{n=1}^{\infty} f_n$. Thus

$$\mu^*\left(\sum_{n=1}^{\infty} f_n\right) = \lim_{k\to\infty} \mu^*(s_k) \leqslant \lim_{k\to\infty}\left(\sum_{n=1}^{k} \mu^*(f_n)\right) = \sum_{n=1}^{\infty} \mu^*(f_n). \ \square$$

Lemma XIII.3.4 (Fatou). $(0 \leqslant f_n$ numerical functions$) \Rightarrow \left(\mu^*(\liminf_{n\to\infty} f_n) \leqslant (\liminf_{n\to\infty} \mu^*(f_n))\right)$.

Proof. Let $g_n := \inf_{p\geqslant 0} f_{n+p}$. It is seen that $g_n \nearrow$. It follows from the definition that $\liminf_{n\to\infty} f_n = \sup_n g_n$, and thus

$$\mu^*(\liminf_{n\to\infty} f_n) = \mu^*(\sup g_n) = \sup \mu^*(g_n).$$

Since $g_n \leqslant f_{n+p}$ for $p \geqslant 0$, therefore

$$\mu^*(g_n) \leqslant \mu^*(f_{n+p}) \quad \text{and} \quad \mu^*(g_n) \leqslant \inf_{p\geqslant 0} \mu^*(f_{n+p}).$$

Accordingly

$$\sup_n \mu^*(g_n) \leqslant \sup_n \left(\inf_{p\geqslant 0} \mu^*(f_{n+p})\right) = \liminf_{n\to\infty} \mu^*(f). \ \square$$

Lemma XIII.3.5. Given a sequence (f_n) of nonnegative numerical functions. Let $\mu^*(f_n) = 0$ for every n. Then

(i) $\mu^*\left(\sum_{n=1}^{\infty} f_n\right) = 0$,

(ii) $\mu^*(\sup_n f_n) = 0$.

The proof is immediate from the Beppo–Levi Theorem and Corollary XIII.3.3.

LEMMA XIII.3.6. Given numerical functions f and g such that $0 \leqslant f \leqslant g$. Let $\mu^*(g) = 0$. Then

(i) $\mu^*(f) = 0$,

(ii) The formula $\mu^*(\alpha_g) = 0$ holds for every number $\alpha \geqslant 0$.

The proof is immediate.

4. THE OUTER MEASURE

DEFINITION. The function

$$\chi_A(x) := \begin{cases} 1 & \text{for } x \in A, \\ 0 & \text{for } x \notin A \end{cases}$$

is called the *characteristic function*, or *indicator* of a set $A \subset X$.

DEFINITION. The number $\mu^*(\chi_A)$ is called the *outer* (or *exterior*) *measure* of a set $A \subset Y$. It is denoted also by $\mu^*(A)$.

Given some property P concerning points of the set X. By N_P let us denote the set of those points which do not possess the property P.

DEFINITION. A property P is said to *occur* μ*-almost everywhere on* X if $\mu^*(N_P) = 0$.

It transpires that the following theorem holds:

THEOREM XIII.4.1. Given a nonnegative numerical function f. If $\mu^*(f) < +\infty$, then f is finite μ-almost everywhere (that is, the set $N_\infty = \{x: f(x) = \infty\}$ has a zero outer measure).

PROOF. The inequality $n\chi_{N\infty} \leqslant f$ holds for every natural number n, and hence

$$n\mu^*(\chi_{N_\infty}) = \mu^*(n\chi_{N_\infty}) \leqslant \mu^*(f), \quad \text{that is,}$$

$$\mu^*(\chi_{N_\infty}) \leqslant \frac{1}{n}\mu^*(f).$$

Therefore $\mu^*(\chi_{N\infty}) = \mu^*(N_\infty) = 0$. □

THEOREM XIII.4.2. ($0 \leqslant f$ numerical function, $f(x) = 0$ μ-almost everywhere) $\Leftrightarrow$ ($\mu^*(f) = 0$).

PROOF. $\Rightarrow$: Let us denote $N = \{x: f(x) \neq 0\}$. The inequality $\sup(n\chi_N) \geqslant f$ holds, and hence by the Beppo–Levi Theorem it follows that

$$\sup(n\mu^*(\chi_N)) = \sup \mu^*(n\chi_N) = \mu^*(\sup n\chi_N) \geqslant \mu^*(f) \geqslant 0.$$

But $\mu^*(\chi_N) = 0$, and thus $\sup(n\mu^*(\chi_N)) = 0$, whereby we have $\mu^*(f) = 0$.

$\Leftarrow$: We have $\sup(n \cdot f) \geqslant \chi_N$, and thus

$$0 = \sup n\mu^*(f) = \sup \mu^*(nf) = \mu^*(\sup nf) \geqslant \mu^*(\chi_N) \geqslant 0,$$

whence

$$\mu^*(\chi_N) = 0. \quad \square$$

THEOREM XIII.4.3. (f, g are nonnegative numerical functions equal to each other μ-almost everywhere) $\Rightarrow$ $(\mu^*(f) = \mu^*(g))$.

PROOF. Let us denote $N = \{x\colon f(x) \neq g(x)\}$. By assumption we have

$$\mu^*(N) = \mu^*(\chi_N) = 0.$$

Let us take the function

$$h(x) := (+\infty) \cdot \chi_N(x) = \begin{cases} +\infty & \text{for } x \in N, \\ 0 & \text{for } x \notin N. \end{cases}$$

Of course, $h(x) = 0$ μ-almost everywhere, and thus $\mu^*(h) = 0$. But $f \leqslant g+h$, which means that

$$\mu^*(f) \leqslant \mu^*(f+h) \leqslant \mu^*(g)+\mu^*(h) = \mu^*(g),$$

and

$$g \leqslant f+h, \text{ therefore } \mu^*(g) \leqslant \mu^*(f+h) \leqslant \mu^*(f)+\mu^*(h) = \mu^*(f)$$

and finally $\mu^*(f) = \mu^*(g)$. $\square$

To make it simpler to express ourselves, we shall adopt the following definition:

DEFINITION. A set $N \subset X$ is said to be a *null* (more precisely, a *μ-null*) *set* when $\mu^*(N) = 0$.

Finally, we shall prove important properties of null sets:

THEOREM XIII.4.4. (i) A subset of a null set is itself a null set.

(ii) A countable union of null sets is a null set.

PROOF. (i) Let $A \subset B \subset X$ and $\mu^*(B) = 0$. Since $\chi_A \leqslant \chi_B$, therefore

$$0 \leqslant \mu^*(A) = \mu^*(\chi_A) \leqslant \mu^*(\chi_B) = \mu^*(B) = 0.$$

(ii) Let $\mu^*(A_n) = 0$, $n = 1, 2, \ldots, A_n \subset X$.

We denote $A = \bigcup_{n=1}^{\infty} A_n$. Since $\chi_A \leqslant \sum_{n=1}^{\infty} \chi_{A_n}$, therefore

$$\mu^*(A) = \mu^*(\chi_A) \leqslant \mu^*(\sum \chi_{A_n}) \leqslant \sum \mu^*(\chi_{A_n})$$
$$= \sum \mu^*(A_n) = 0. \quad \square$$

Examples. 1. Let $X = \boldsymbol{R}$, $E = C_0(\boldsymbol{R})$, and let μ be a Riemann integral. We shall show that a one-point set has zero outer measure (i.e. is a null set).

Let us take $x_0 \in X$,

$$\chi_{\{x_0\}}(x) = \begin{cases} 1 & \text{for } x = x_0, \\ 0 & \text{for } x \neq x_0. \end{cases}$$

Since the proof is identical for all x_0, in order to focus attention we put $x_0 = 0$.

Let us define the following sequence of functions:

$$f_n(x) = \begin{cases} 0 & \text{for } |x| \geqslant \dfrac{1}{n}, \\[2ex] 1-nx & \text{for } 0 \leqslant x \leqslant \dfrac{1}{n}, \\[2ex] 1+nx & \text{for } 0 \geqslant x \geqslant -\dfrac{1}{n}. \end{cases}$$

Since $f_n \in C_0(X)$, then $f_n \in E \subset \mathscr{E}$. Clearly $\mu(f_n) = 1/n$ and $f_n \geqslant \chi_{\{0\}}$. Therefore

$$0 \leqslant \mu^*(\{0\}) = \mu \quad = \inf_{\chi_{\{0\}} \leqslant h \in \mathscr{E}} \tilde{\mu}(h) \leqslant \inf \mu(f_n)$$

$$= \inf \frac{1}{n} = 0.$$

From Theorem XIII.4.4 it follows that every countable set is a null set. In particular, the set of all rational numbers is countable, and hence is a null set.

We shall show that the set of irrational numbers from any interval $[a, b]$ has an outer measure equal to that of the whole interval, that is, equal to the number $b-a$. We have

$$\chi_{[a,b]} = \chi_{N'} + \chi_{Q},$$

where $\chi_{N'}$ (χ_Q) is the characteristic function of the set of irrational numbers N' (rational numbers $\boldsymbol{Q}$). Of course, $\chi_{N'} \leqslant \chi_{[a,b]}$ whence $\mu^*(N') \leqslant \mu^*([a, b])$. But

$$\mu^*([a,b]) = \mu^*(\chi_{[a,b]}) \leqslant \mu^*(\chi_{N'}) + \mu^*(\chi_Q) = \mu^*(\chi_{N'}) = \mu^*(N').$$

It is thus seen that from the point of view of the integral, the set of rational numbers is a negligible quantity in comparison with the irrational numbers.

2. Let $X = N \subset R$ (the set of natural numbers), $\mathscr{E} = C_0(N)$ (sequences with a finite number of nonzero terms),

$$\mu^*(f) = \sum_{n=1}^{\infty} f(n).$$

Here, $\mathscr{E}$ is the set of sequences with a finite number of negative terms. Let us take an arbitrary point $\{n\} \in N$, Since $\chi_{\{n\}} \in E \subset \mathscr{E}$, therefore $\mu^*(\chi_{\{n\}}) = \mu(\chi_{\{n\}}) = 1$.

Let us take an arbitrary nonempty set $A \subset N$. We have

$$\chi_A = \sum_{n \in A} \chi_{\{n\}} \in \mathscr{E}, \qquad \text{that is} \qquad \mu^*(A) = \sum_{n=1}^{\infty} \chi_A(n) \geqslant 1.$$

It is thus seen that there are no null sets with the exception of the empty set.

5. SEMINORMS N_p. THE MINKOWSKI AND HÖLDER INEQUALITIES

Let us now define the following functionals on the set of all numerical functions:

$$N_p(f) := [\mu^*(|f|^p)]^{1/p}, \qquad 1 \leqslant p < \infty.$$

It follows immediately from the definition that

$$N_p(cf) = |c| N_p(f)$$

for $c \in \boldsymbol{R}$. The Hölder inequality is the fundamental property of the functions N_p.

THEOREM XIII.5.1.

$$\left(1 < p < \infty, \frac{1}{p'} + \frac{1}{p} = 1\right) \Rightarrow (N_1(f \cdot g) \leqslant N_p(f) \cdot N_{p'}(g)).$$

PROOF. Let $a = N_p(f)$, $b = N_{p'}(g)$.

(i) If $a \cdot b = 0$, then $|f|^p$ or $|g|^{p'}$ are zero μ-almost everywhere, and hence f or g is zero μ-almost everywhere, that is, $f \cdot g$ is also zero μ-almost everywhere. Thus, the inequality holds in this case.

(ii) If $a \cdot b = +\infty$, the inequality holds.

(iii) Thus, let $0 < a, b < \infty$. From the theorem on the geometric and arithmetic mean (Chapter IV) we have

$$(1) \qquad \alpha^{\frac{1}{p}} \cdot \beta^{\frac{1}{p'}} \leqslant \frac{1}{p}\alpha + \frac{1}{p'}\beta$$

for $0 < \alpha, \beta < \infty$. Let $\alpha = \left(\frac{1}{a}|f(x)|\right)^p$, $\beta = \left(\frac{1}{b}|g(x)|\right)^{p'}$. Therefore

$$\frac{1}{a \cdot b} f(x) \cdot g(x) \leqslant \frac{1}{p \cdot a^p}|f(x)|^p + \frac{1}{p' \cdot b^{p'}}|g(x)|^{p'}.$$

Applying the functional μ^* to both sides of this inequality and making use of its convexity, we have

$$\frac{1}{a \cdot b} N_1(f \cdot g) \leqslant \frac{1}{p \cdot a^p}[N_p(f)]^p + \frac{1}{p' \cdot b^{p'}}[N_{p'}(g)]^{p'}$$

$$= \frac{1}{p} + \frac{1}{p'} = 1$$

(the definitions of the numbers a and b have been taken into account). Consequently $N_1(f \cdot g) \leqslant a \cdot b$. □

It turns out that the functions N_p are convex.

THEOREM XIII.5.2 (Minkowski Inequality). Let f and g be numerical functions such that $f+g$ is defined (that it is not necessary to perform the indeterminate operation $(+\infty)+(-\infty)$). Then $N_p(f+g) \leqslant N_p(f) + N_p(g)$ for $1 \leqslant p < \infty$.

PROOF. When $p = 1$, the inequality follows from the fact that $|f+g| \leqslant |f|+|g|$ and

$$\mu^*(|f+g|) \leqslant \mu^*(|f|+|g|) \leqslant \mu^*(|f|)+\mu^*(|g|).$$

When $p > 1$, we take p' such that $1/p+1/p' = 1$. If $N_p(f) = \infty$ or $N_p(g) = \infty$, this inequality is satisfied. Thus, let $N_p(f), N_p(g) < \infty$. Note that

$$|f| \cup |g| \leqslant |f|+|g| \leqslant 2(|f| \cup |g|).$$

Thus it follows that

$$|f+g|^p \leqslant (2(|f| \cup |g|))^p = 2^p(|f|^p \cup |g|^p) \leqslant 2^p(|f|^p+|g|^p).$$

Applying μ^* to both sides, we find:

$$[N_p(f+g)]^p = 2^p \cdot \mu^*(|f|^p+|g|^p) \leqslant 2^p[N_p(f)]^p+2^p[N_p(g)]^p.$$

Hence $N_p(f+g) < \infty$. Moreover,

$$[N_p(f+g)]^p = \mu^*(|f+g|^p)$$
$$\leqslant \mu^*(|f|\,|f+g|^{p-1})+\mu^*(|g|\,|f+g|^{p-1}).$$

Invoking the Hölder inequality, we obtain

$$[N_p(f+g)]^p \leqslant N_p(f)N_{p'}((f+g)^{p-1})+N_p(g)N_{p'}((f+g)^{p-1})$$
$$= [N_p(f)+N_p(g)] \cdot [\mu^*(|f+g|^{(p-1)p'})]^{\frac{1}{p'}}.$$

But $\frac{1}{p}+\frac{1}{p'} = 1, p'+p = pp'$, that is $p = pp'-p' = p' \cdot (p-1)$. Therefore

$$[N_p(f+g)]^p \leqslant [N_p(f)+N_p(g)] \cdot [\mu^*(|f+g|^p)]^{\frac{1}{p}\cdot\frac{p}{p'}}$$
$$= [N_p(f)+N_p(g)] \cdot [N_p(f+g)]^{p/p'}.$$

Since $p/p' = p-1$, then on dividing both sides by $[N_p(f+g)]^{p-1}$, we have finally

$$N_p(f+g) \leqslant N_p(f)+N_p(g). \ \square$$

Remark. The Minkowski inequality, i.e. the convexity of the functions N_p, is directly demonstrable, without use of the Hölder inequality.

For this purpose, as we know (cf. Chapter IV), it is sufficient to prove the convexity of the indicatrix $V = \{f\colon N_p(f) \leqslant 1\}$ of the functional N_p. We recall that

$$N_p(f) = \inf\left\{t\colon t > 0, \frac{1}{t}f \in V\right\}.$$

The convexity of the set V is easily implied from the convexity of the power $x \to x^p$, as follows: Let $0 \leqslant \theta \leqslant 1$; then

$$(*) \qquad |\theta f+(1-\theta)g|^p \leqslant (\theta|f|+(1-\theta)|g|)^p \leqslant \theta|f|^p+(1-\theta)|g|^p,$$

when $f, g \in V$, that is $\mu^*(|f|^p), \mu^*(|g|^p) \leqslant 1$, then from the positivity and convexity of μ^* we have

$$\mu^*(|\theta f+(1-\theta)g|^p) \leqslant \theta\mu^*(|f|^p)+(1-\theta)\mu^*(|g|^p) \leqslant 1.$$

Let f, g now be arbitrary. Putting $a = N_p(f), b = N_p(g)$, we find

$$\frac{1}{a+\varepsilon}f, \frac{1}{b+\varepsilon}g \in V \quad \text{for any } \varepsilon > 0.$$

Setting $\theta = \dfrac{a+\varepsilon}{a+b+2\varepsilon}$, we obtain

$$\frac{1}{a+b+2\varepsilon}(f+g) = \theta\frac{1}{a+\varepsilon}f+(1-\theta)\frac{1}{b+\varepsilon}g \in V,$$

that is

$$N_p(f+g) \leqslant a+b+2\varepsilon = N_p(f)+N_p(g)+2\varepsilon.$$

Since the number $\varepsilon > 0$ was arbitrary, then

$$N_p(f+g) \leqslant N_p(f)+N_p(g).$$

We shall use the foregoing inequality to prove the following lemma:

LEMMA XIII.5.3. Let $(f_n)_1^\infty$ be a sequence of numerical functions. Then

$$N_p\Big(\sum_{n=1}^{\infty}|f_n|\Big) \leqslant \sum_{n=1}^{\infty}N_p(f_n).$$

PROOF. Let us take

$$f(x) := \sum_{n=1}^{\infty}|f_n(x)|, \qquad s_k(x) := \sum_{n=1}^{k}|f_n(x)|;$$

clearly $s_k \nearrow f$ and by virtue of the continuity of the power function, $(s_k)^p \nearrow f^p$. Therefore

$$[N_p(f)]^p = \mu^*\Big(\sup_k\Big(\sum_{n=1}^{k}|f_n|\Big)^p\Big) = \sup_k\mu^*\Big(\Big(\sum_{n=1}^{k}|f_n|\Big)^p\Big)$$

$$= \sup_k\Big[N_p\Big(\sum_{n=1}^{k}|f_n|\Big)\Big]^p = \Big[\sup_k N_p\Big(\sum_{n=1}^{k}|f_n|\Big)\Big]^p.$$

Hence

$$N_p(f) = \sup_k N_p\Big(\sum_{n=1}^{k}|f_n|\Big) \leqslant \sup_k\Big(\sum_{n=1}^{k}N_p(f_n)\Big) = \sum_{n=1}^{\infty}N_p(f_n).$$

DEFINITION. A vector space W with a nonnegative real function N_0, such that

(i) $\quad \bigwedge\limits_{c\in R}\ \bigwedge\limits_{w\in W} N_0(cw) = |c|N_0(w),$

(ii) $\quad \bigwedge\limits_{w,u\in W} N_0(w+u) \leqslant N_0(w)+N_0(u),$

is called a *seminormed space*. The function N_0 is called a *seminorm*.

This definition differs from that of a normed space in that the space W may contain nonzero vectors whose seminorm vanishes.

A *semidistance* can be introduced into a space with a seminorm; this is a nonnegative function d on the set $W \times W$ and is such that:

(i) $(u = w) \Rightarrow (d(u, w) = 0)$,

(ii) $d(u, w) = d(w, u)$,

(iii) $d(u, w) \leqslant d(u, t) + d(t, w)$.

This semidistance is defined in a seminormed space just as distance is in a normed space, that is,

$$d(u, w) := N_0(u - w).$$

The concept of a Cauchy sequence can be introduced in a seminormed (semimetric) space in formally the same manner as in a metric space:

$$(\{w_n\}\text{—Cauchy sequence}) \Leftrightarrow \Big(\bigwedge_{\varepsilon > 0} \bigvee_{N(\varepsilon)} \bigwedge_{m, n > N(\varepsilon)} (d(w_n, w_m) < \varepsilon\Big).$$

In identical manner we introduce the concept of a sequence convergent to $w_0 \in W$.

One can thus speak of complete space, i.e. spaces in which every Cauchy sequence has a limit.

It turns out that the following theorem holds:

THEOREM XIII.5.4. Let W be a seminormed space with seminorm N_0. Then the relation $\mathscr{R}$: $((u, w) \in \mathscr{R}) \Leftrightarrow (N_0(u - w) = 0)$ is an equivalence relation. The function N defined as

$$N([w]) := N_0(w) \quad \text{where } w \in [w]$$

is well-defined (that is, is independent of the choice of a representative $w \in [w]$) by the norm in the space $W/\mathscr{R}$.

Moreover, if W is a complete space, then $W/\mathscr{R}$ is also a complete space in the norm N, that is, it is a Banach space.

PROOF. The reflexivity and symmetry of the relation $\mathscr{R}$ are obvious. The transitivity follows from the triangle inequality:

$$N_0(w - u) = N_0\big((w - t) + (t - u)\big) \leqslant N_0(w - t) + N_0(t - u).$$

Thus, if $w \mathscr{R} t$, $t \mathscr{R} u$, then $0 \leqslant N_0(w - u) \leqslant 0 + 0 = 0$, that is, $w \mathscr{R} u$.

We shall prove that N is well-defined on $W/\mathscr{R}$. Let $[w] \in W/\mathscr{R}$ and $w_1 \in [w] \subset W$, $w_2 \in [w] \subset W$. It thus follows that $w_1 \mathscr{R} w_2$, that is, $N_0(w_1 - w_2) = 0$, and hence

$$N_0(w_1) = N_0(w_1 - w_2 + w_2) \leqslant N_0(w_1 - w_2) + N_0(w_2) = N_0(w_2),$$

Therefore, $N_0(w_1) = N_0(w_2)$, that is, N does not depend on the choice of a representative.

It is seen that N has all the properties of a seminorm and if $N([w]) = 0$, then $N([w]) = N_0(w) = N_0(w-0) = 0$ and thus $w \mathscr{R} 0 \in W$, that is, $[w] = 0 \in W/\mathscr{R}$. Therefore, N is in fact a norm. Given a sequence $([w_n])_1^\infty$, $[w_n] \in W/\mathscr{R}$ such that

$$\bigwedge_{\varepsilon>0} \bigvee_{M(\varepsilon)} \bigwedge_{n,m>M(\varepsilon)} N([w_n] - [w_m]) < \varepsilon.$$

Taking $w_n \in [w_n] \subset W$, we have

$$N([w_n] - [w_m]) = N([w_n - w_m]) = N_0(w_n - w_m) < \varepsilon,$$

which means that the sequence $(w_n)_1^\infty$ is a Cauchy sequence in the space W.

If W is a complete space, there exists a $w_0 \in W$ such that $w_n \to w_0$. But $N([w_n] - [w_0]) = N([w_n - w_0]) = N_0(w_n - w_0) \to 0$, which means that $[w_n] \to [w_0] \in W/\mathscr{R}$, and hence $W/\mathscr{R}$ is a complete space. □

6. THE SPACES $\mathscr{F}^p$

Using the concepts discussed above, we shall define the following vector spaces of functions defined on a set X:

$$\mathscr{F}^p := \{f\colon f \text{ — real function, } N_p(f) < \infty\}.$$

The fact that $\mathscr{F}^p$ is a vector space follows from the Minkowski inequality:

$$N_p(f+g) \leqslant N_p(f) + N_p(g) < \infty \quad \text{for } f, g \in \mathscr{F}^p, \quad \text{that is}$$

$$f+g \in \mathscr{F}^p.$$

The Minkowski inequality also states that N_p is a seminorm in the space $\mathscr{F}^p$. It turns out that the spaces $\mathscr{F}^p$ are complete. In order to prove this, let us first prove the following lemma.

LEMMA XIII.6.1. Let $(f_n)_1^\infty$ be a sequence of elements from the space $\mathscr{F}^p$ such that $\sum_{n=1}^{\infty} N_p(f_n) < \infty$. Then

(i) The sequence $s_n(x) := \sum_{k=1}^{n} f_k(x)$ is convergent μ-almost everywhere, i.e. the set $M \subset X$ on which s_n do not converge, is a null set.

(ii) If we assume

$$f(x) := \begin{cases} \lim s_n(x) & \text{for } x \notin M, \\ 0 & \text{for } x \in M, \end{cases}$$

the function f is also an element of the space $\mathscr{F}^p$.

(iii) $0 \leqslant N_p(f - s_n) \leqslant \sum_{k=n+1}^{\infty} N_p(f_k) \underset{n\to\infty}{\to} 0$, which means that $s_n \to f$ in the sense of the seminorm N_p.

PROOF. (i) Let $g(x) := \sum_{k=1}^{\infty} |f_k(x)|$. From Lemma XIII.5.3 it follows that $N_p(g) \leqslant \sum_{n=1}^{\infty} N_p(f_n) < \infty$, and hence the function $(g)^p$ is finite μ-almost everywhere. The same may be said of the function g, which means that the series $\sum_{n=1}^{\infty} |f_n(x)|$ is absolutely convergent μ-almost everywhere, and this in turn implies the convergence of the series $\sum_{n=1}^{\infty} f_n(x)$ μ-almost everywhere.

(ii) Since

$$\left|\sum_{k=1}^{\infty} f_k(x)\right| \leqslant \sum_{k=1}^{\infty} |f_k(x)| = g(x),$$

therefore $|f(x)| \leqslant |g(x)|$, whereby $N_p(f) \leqslant N_p(g) < \infty$. Thus $f \in \mathscr{F}^p$.

(iii) $|f(x) - s_n(x)| = \left|\sum_{k=n+1}^{\infty} f_n(x)\right|$ for $x \notin M$.

Thus, if we assume

$$\tilde{f}_n(x) := \begin{cases} f_n(x) & \text{for } x \notin M, \\ 0 & \text{for } x \in M, \end{cases}$$

then

$$|f(x)-s_n(x)| = \left|\sum_{k=n+1}^{\infty} \tilde{f}_n(x)\right| \quad \mu\text{-almost everywhere.}$$

But $N_p(f_n) = N_p(\tilde{f}_n)$ and therefore

$$N_p(f-s_n) = N_p\Big(\sum_{k=n+1}^{\infty} \tilde{f}_k\Big) \leqslant \sum_{k=n-1}^{\infty} N_p(\tilde{f}_k)$$

$$= \sum_{k=n+1}^{\infty} N_p(f_k) \to 0. \ \square$$

We can now go on to the proof of the completeness of the space $\mathscr{F}^p$.

THEOREM XIII.6.2 (Generalized Riesz–Fischer Theorem). The space $\mathscr{F}^p$ is complete, that is, every Cauchy sequence $(f_n)_1^\infty$ of elements of $\mathscr{F}^p$ has a limit $f \in \mathscr{F}^p$

$$N_p(f_n - f) \underset{n\to\infty}{\to} 0.$$

Moreover, there exists a subsequence (f_{n_k}) which is pointwise convergent to f μ-almost everywhere.

PROOF. Let $(f_n)_1^\infty$ be a Cauchy sequence, that is,

$$\bigwedge_{\varepsilon>0} \bigvee_{M(\varepsilon)} \bigwedge_{n,m>M(\varepsilon)} N_p(f_n - f_m) < \varepsilon.$$

We select a sequence of natural numbers (n_k) so that

$$N_p(f_{n_{k+1}} - f_{n_k}) < \frac{1}{2^k}.$$

Then $\sum_{k=1}^{\infty} N_p(f_{n_{k+1}} - f_{n_k}) < \infty$ and hence it follows from Lemma XIII.6.1 that the series $s_m(x) = \sum_{k=1}^{m} (f_{n_{k+1}} - f_{n_k}) = f_{n_{m+1}} - f_{n_1}$ is pointwise convergent μ-almost everywhere to a function $g \in \mathscr{F}^p$ in the sense of the seminorm N_p.

Therefore $f_{n_k} \underset{k\to\infty}{\to} f := g + f_{n_1} \in \mathscr{F}^p$ pointwise μ-almost everywhere and in the sense of the norm N_p. But

$$N_p(f-f_n) \leqslant N_p(f-f_{n_k}) + N_p(f_{n_k} - f_n) \to 0$$

for $k \to \infty$ and $n \to \infty$. $\square$

COROLLARY XIII.6.3. If $(f_n)_1^\infty$ is a Cauchy sequence in $\mathscr{F}^p$ and $f_n(x)\to f(x)$ μ-almost everywhere, then $f\in\mathscr{F}^p$ and $f_n\to f$ in the sense of N_p.

PROOF. From the Riesz–Fischer Theorem it follows that there exists an $\tilde{f}\in\mathscr{F}^p$ such that $\tilde{f}_{n_k}\to\tilde{f}$ μ-almost everywhere and in the sense of the seminorm N_p. Thus, if N (resp. M) denotes the set on which $f_n(\tilde{f}_{n_k})$ do not converge

$$\lim f_{n_k}(x) = \begin{cases} f(x) & \text{for } x\notin N,\\ \tilde{f}(x) & \text{for } x\notin M\end{cases}$$

and $\mu^*(N)=\mu^*(M)=0$, then $f(x)=\tilde{f}(x)$, with the exception of at most the set $N\cup M$, and $\mu^*(N\cup M)=0$.

Accordingly, $\mu^*(f)=\mu^*(\tilde{f})<\infty$, that is $f\in\mathscr{F}^p$ and $f_n\to f$ in the sense of the seminorm N_p. □

7. THE SPACES $\mathscr{L}^p$

Now we define the following spaces:

$$L^p(\mu) = \overline{\mathscr{F}^p\cap E}.$$

The set $\mathscr{F}^p\cap E$ is a vector space as the intersection of two vector spaces. Its closure in the sense of the seminorm N_p is, in view of the completeness of the space $\mathscr{F}^p$, a complete seminormed vector space.

By virtue of Theorem XIII.5.4, the spaces

$$L^p(\mu) := \mathscr{L}^p(\mu)/R,$$

where $\mathscr{R} := \{(f,g)\colon N_p(f-g)=0\}$, are Banach spaces.

We shall occupy ourselves with these spaces further on. For the moment, we confine ourselves to an examination of the space $\mathscr{L}^p$.

Note only that according to Theorem XIII.5.4, if $f\in\mathscr{L}^p$, and $[f]\in L^p$, then

$$\|[f]\|_{L^p} = N_p(f).$$

Next we give the corollary to the Riesz–Fischer Theorem concerning the space $\mathscr{L}^p$:

COROLLARY XIII.7.1. If $(f_n)_1^\infty$ is a Cauchy sequence in $\mathscr{L}^p$ and if the condition $f_n(x)\to f(x)$ is satisfied μ-almost everywhere, then $f\in\mathscr{L}^p$, $f_n\to f$ in the sense of the seminorm N_p.

PROOF. Corollary XIII.6.3 implies that f belongs to $\mathscr{F}^p$ but, being the limit of the sequence of elements of a closed set of $\mathscr{L}^p$, it also belongs to $\mathscr{L}^p$. □

It turns out that the spaces $\mathscr{L}^p$ are lattices:

LEMMA XIII.7.2:

(i) $(E \ni f_n \overset{N_p}{\rightarrow} f) \Rightarrow (E \ni |f_n| \overset{N_p}{\rightarrow} |f|)$.

(ii) If $f \in \mathscr{L}^p$, then $|f| \in \mathscr{L}^p$.

PROOF. Let $N_p(f_n - f) \rightarrow 0$. But $\big||f_n| - |f|\big| \leqslant |f_n - f|$, that is

$$\big||f_n| - |f|\big|^p \leqslant |f_n - f|^p,$$

$$\text{whence } \mu^*(|f_n|^p - |f|^p) \leqslant \mu^*(|f_n - f|^p).$$

Raising both sides to the power $1/p$, we have:

$$0 \leqslant N_p(|f_n| - |f|) \leqslant N_p(f_n - f) \rightarrow 0,$$

which means that $|f|$ is the limit of the elementary functions $|f_n| \in E$, that is, belongs to $\mathscr{L}^p$. □

COROLLARY XIII.7.3. If $f, g \in \mathscr{L}^p$, the following functions also belong to $\mathscr{L}^p$:

$$f \cup g, \quad f \cap g, \quad f^+ := f \cup 0, \quad f^- := -f \cup 0.$$

Remark. The theorem converse to Lemma XIII.7.2 does not hold.

Example. Let $X = \{a, b\}$; $E = \{f\colon f(a) = f(b)\}$. We already know that in this case $\mu^*(g) = \max(g(a), g(b))$. Thus, if $f \in \mathscr{L}^p$, there exists a sequence $(f_n) \in E$ such that

$$\mu^*(|f_n - f|^p) = \max(|f_n(a) - f(a)|^p, |f_n(b) - f(b)|^p) \underset{n \to \infty}{\rightarrow} 0.$$

Accordingly, $f(a) = \lim f_n(a) = \lim f_n(b) = f(b)$.

Thus we have obtained the equality $\mathscr{L}^p = E$.

Let us take the function $h := \{h(a) = 1, h(b) = -1\}$. Plainly, $h \notin \mathscr{L}^p$, but $|h| \in \mathscr{L}^p$.

Interesting situations arise when E is a family closed under exponentiation, i.e. when the proposition

$$(f \in E) \Rightarrow (|f|^\alpha \in E) \quad \text{for each } \alpha > 0$$

is true. Then $E \subset \mathscr{F}^p$ and $\mathscr{L}^p$ is the closure of E in $\mathscr{F}^p$.

We shall show what is the sufficient condition for a function to belong to the space $\mathscr{L}^p$ in this case:

THEOREM XIII.7.4. Let the space E be closed under exponentiation. Let $f \geqslant 0$. If $f^p \in \mathscr{L}^1$, then $f \in \mathscr{L}^p$.

PROOF. Suppose that $f^p \in \mathscr{L}^1$. Hence there exists a sequence of functions $f_n \in E$ such that $\mu^*(|f_n - f^p|) \to 0$. By Lemma XIII.7.2 we have $E \ni |f_n| \to |f^p| = f^p$, in the sense of N_1. Let us assume $E \ni g_n := |f_n|^{1/p}$. Then

$$|f-g_n|^p \leqslant |f^p - g_n^p| = |f^p - |f_n||.$$

Therefore $\mu^*(|f-g_n|^p) < \mu^*(|f^p - |f_n||) \to 0$, which means that $E \ni g_n \to f \in \mathscr{L}^p$. □

LEMMA XIII.7.5. If $f \in \mathscr{L}^p$, and $f^* = f$ μ-almost everywhere, then $f^* \in \mathscr{L}^p$.

PROOF. If $E \ni f_n \to f$ in the sense of N_p, and $f^* - f_n = f - f_n$ μ-almost everywhere, that is, $|f^* - f_n|^p = |f - f_n|^p$ μ-almost everywhere, then $N_p(f^* - f_n) = N_p(f - f_n) \underset{n\to\infty}{\to} 0$, and thus $E \ni f_n \to f^* \in \mathscr{L}^p$ in the sense of N_p. □

8. THE SPACE $\mathscr{L}^1$ OF INTEGRABLE FUNCTIONS. THE INTEGRAL

Since $E \subset \mathscr{F}^1$, then $E \cap \mathscr{F}^1 = E$, that is, $\mathscr{L}^1 = \overline{E}$ (this means that the space $\mathscr{L}^1$ is the completion of the space E in the sense of the norm N_1).

Given a linear functional μ in the set E, dense in $\mathscr{L}^1$. Note that this functional is continuous with respect to the seminorm N_1 which determines the topology in $\mathscr{L}^1$:

$$(1) \qquad |\mu(f)| \leqslant \mu(|f|) = \mu^*(|f|) = N_1(f).$$

Moreover, if f is decomposed into a positive and a negative part, $f = f^+ - f^-$, then

$$(2) \qquad \mu(f) = \mu(f^+) - \mu(f^-) = \mu^*(|f^+|) - \mu^*(|f^-|) = N_1(f^+) - N_1(f^-).$$

As is known, a bounded linear functional, defined on a dense subset of a Banach space, is extendible to a continuous functional on the whole space. A similar fact occurs for seminormed spaces:

THEOREM XIII.8.1. The functional $\int f := N_1(f^+) - N_1(f^-)$, defined for $f \in \mathscr{L}^1$, is a continuous linear extension of the functional μ. The inequality

$$\left|\int f\right| \leqslant \int |f|$$

holds.

PROOF.

$$f^+ = \frac{1}{2}(f+|f|), \quad f^- = \frac{1}{2}(|f|-f).$$

Let $E \ni f_n \to f$ in the sense of the seminorm N_1. From Lemma XIII.7.2 it follows that $|f_n| \to |f|$, wherefrom immediately we have $f_n^+ \to f^+$, $f_n^- \to f^-$, in the sense of the seminorm N_1. Hence

$$\int f = N_1(f^+) - N_1(f^-) = \lim_{n\to\infty}(N_1(f_n^+) - N_1(f_n^-))$$
$$= \lim_{n\to\infty} \mu(f_n).$$

Let $f, g \in \mathscr{L}^1$, $E \ni g_n \to g$ in the sense of the seminorm N_1. Let $a, b \in \boldsymbol{R}$. Then

$$a\int f + b\int g = a\lim \mu(f_n) + b\lim \mu(g_n) = \lim(a\mu(f_n) + b\mu(g_n))$$
$$= \lim \mu(af_n + bg_n) = \int \lim(af_n + bg_n).$$

But

$$N_1(af_n + bg_n - af - bg) \leqslant |a|N_1(f_n - f) + |b|N_1(g_n - g) \to 0,$$

that is, $\lim(af_n + bg_n) = af + bg$. Therefore

$$a\int f + b\int g = \int (af + bg).$$

Moreover

$$\left|\int f\right| = |N_1(f^+) - N_1(f^-)| \leqslant N_1(f^+ - f^-)$$
$$= N_1(f) = \int |f|. \quad \square$$

We shall further prove an important fact about monotonic convergence in the space $\mathscr{L}^1$.

LEMMA XIII.8.2 (Levi Lemma). Let $\mathscr{L}^1 \ni f_n \nearrow f$, $\sup N_1(f_n) < \infty$. Then $f \in \mathscr{L}^1$ and $f_n \to f$ in the sense of the seminorm N_1.

PROOF. Suppose first of all that $f_1 \geqslant 0$. Let us denote:

$$0 \leqslant g_n := f_n - f_{n-1}, \quad g_1 = f_1.$$

Then

$$f_n = f_1 + (f_2 - f_1) + (f_3 - f_2) + \ldots + (f_n - f_{n-1}) = \sum_{k=1}^{n} g_k,$$

which means that $f = \sum_{k=1}^{\infty} g_k$. But

$$N_1(g) = \int g_n = \int (f_n - f_{n-1}) = \int f_n - \int f_{n-1}$$
$$= N_1(f_n) - N_1(f_{n-1}).$$

Therefore

$$\sum_{k=1}^{n} N_1(g_k) = N_1(f_1) + \big(N_1(f_2) - N_1(f_1)\big) + \ldots$$
$$\ldots + \big(N_1(f_n) - N_1(f_{n-1})\big) = N_1(f_n),$$

which means that

$$\sum_{k=1}^{\infty} N_1(g_k) = \sup_n N_1(f_n) < \infty.$$

It follows from Lemma XIII.6.1 that $f_n \to f$ in the sense of the seminorm N_1, and hence the completeness of the space $\mathscr{L}^1$ implies that $f \in \mathscr{L}^1$.

If f_1 is not a nonnegative function, then on putting $\tilde{f}_n := f_n - f_1 \geqslant 0$, we obtain $f_n \to (\lim \tilde{f}_n) + f_1 \in \mathscr{L}^1$.

COROLLARY XIII.8.3 If

$$0 \leqslant f_n \searrow, \quad f_n \in \mathscr{L}^1, \quad \text{then} \quad \lim f_n = \inf f_n \in \mathscr{L}^1.$$

PROOF. Assuming $\mathscr{L}^1 \ni g_n := -f_n$, $N_1(g_n) \leqslant N_1(f_1) < \infty$, we have $g_n \to g \in \mathscr{L}^1$. But $\lim f_n = -\lim g_n = -g \in \mathscr{L}^1$. □

LEMMA XIII.8.4. Let $(f_n)_1^\infty$ be a sequence of functions from $\mathscr{L}^1$. If there exists a function $0 \leqslant g \in \mathscr{L}^1$ such that for every n $|f_n| \leqslant g$, then

$$\sup_n f_n \in \mathscr{L}^1.$$

PROOF. Let us denote $g_n := \sup_{k \leqslant n} f_k$. Since $\mathscr{L}^1$ is a lattice, then $g_n \in \mathscr{L}^1$. Clearly, $|g_n| \leqslant g$, and, therefore, $N_1(g_n) \leqslant N_1(g) < \infty$.

But $g_n \geqslant g_m$, if $n \geqslant m$. Accordingly, $\mathscr{L}^1 \ni g_n \nearrow \sup_k f_k$. The thesis follows immediately from the Levi Lemma. □

We can now proceed with the proof of the following important fact:

THEOREM XIII.8.5 (Fatou; cf. the Fatou Lemma XIII.3.4). Suppose that the hypotheses of Lemma XIII.8.4 are satisfied. Then

(i) $\limsup f_n \in \mathscr{L}^1$,

(ii) $\limsup N_1(f_n) \leqslant N_1(\limsup |f_n|)$, $N_1(\liminf |f_n|) \leqslant \liminf N_1(f_n)$.

PROOF. (i) We have

$$\limsup f_n = \lim_{k \to \infty} (\sup_{n \geqslant k} f_n).$$

From Lemma XIII.8.4 we know that $g_k := \sup_{n \geqslant k} f_n \in \mathscr{L}^1$. But $g_n \searrow$, and hence from Corollary XIII.8.3 we find that

$$\limsup f_n = \lim g_n \in \mathscr{L}^1.$$

(ii) For $n \geqslant k$: $|f_n| \leqslant \sup_{n \geqslant k} |f_n| =: h_k \in \mathscr{L}^1$. Then

$$N_1(f_n) \leqslant N_1(h_k), \qquad \sup_{n \geqslant k} N_1(f_n) \leqslant N_1(h_k).$$

But $0 \leqslant h_k \searrow$. Thus, going to the limit with k, we arrive at (Corollary XIII.8.3)

$$\limsup N_1(f_n) = \lim_{k \to \infty}(\sup_{n \geqslant k} N_1(f_n)) \leqslant \lim_{k \to \infty} N_1(h_k) = N_1(\lim_{k \to \infty} h_k)$$
$$= N_1(\limsup |f_n|). \ \square$$

The most important result of this section will be the following theorem on "majorized convergence."

THEOREM XIII.8.6 (Lebesgue). Let $(f_n)_1^\infty$ be a sequence of functions from the space $\mathscr{L}^1$, and let this sequence be pointwise convergent μ-almost everywhere (i.e. with the exception of the null set Z) to the real-valued function f. If there exists a function $0 \leqslant g \in \mathscr{L}^1$ such that $|f_n| \leqslant g$ μ-almost everywhere (i.e. with the exception of the set Z_1) for every n, then $f \in \mathscr{L}^1$ and $f_n \to f$ in the sense of the seminorm N_1.[1]

---

[1] It is sufficient to assume that $g \in \mathscr{F}^1$, that is, $\mu^*(g) < \infty$.

PROOF. Let us take

$$f_n^*(x) := \begin{cases} f_n(x) & \text{for } x \notin Z \cup Z_1, \\ 0 & \text{for } x \in Z \cup Z_1. \end{cases}$$

Plainly, $f_n = f_n^*$ μ-almost everywhere, and thus $f_n^* \in \mathscr{L}^1$. But $|f_n^*| \leqslant g$. From the Fatou Theorem we find that $f^* := \lim f_n^* = \limsup f_n^* \in \mathscr{L}^1$. But $f^* = f$ μ-almost everywhere. Hence $f \in \mathscr{L}^1$.

Furthermore, since $f-f_n = f^*-f_n^*$ μ-almost everywhere, then

$$\begin{aligned} 0 = \limsup N_1(f-f_n) &= \limsup N_1(f^*-f_n^*) \\ &\leqslant N_1(\limsup|f^*-f_n^*|) = N(0) = 0, \end{aligned}$$

because $|f^*-f_n^*| \to 0$. It thus follows that $\lim N_1(f-f_n) = 0$, that is $f_n \to f$ in the sense of the seminorm N_1.

COROLLARY XIII.8.7. If the hypotheses of Theorem XIII.8.6 are satisfied, then

$$\int f_n \to \int f.$$

PROOF. By virtue of the Lebesgue Theorem we have $N_1(f_n^+) \to N_1(f^+)$ and $N_1(f_n^-) \to N_1(f^-)$, whence

$$\int f_n = N_1(f_n^+) - N_1(f_n^-) \to N_1(f^+) - N_1(f^-) = \int f. \;\square$$

It turns out that a similar theorem also holds for other spaces $\mathscr{L}^p$. This will be discussed in a subsequent section (Theorem XIII.13.7).

9. THE SET $\mathscr{E}$ FOR THE RADON INTEGRAL. SEMICONTINUITY

Let us recall the definition of an upper (lower) semicontinuous function:

(i) (the function f is *upper semicontinuous*) $\Leftrightarrow \big(\limsup\limits_{x \to x_0} f(x) = f(x_0)\big)$;

(ii) (the function f is *lower semicontinuous*) $\Leftrightarrow \big(\liminf\limits_{x \to x_0} f(x) = f(x_0)\big)$.

The following lemma holds:

LEMMA XIII.9.1. (i) (The function f is lower semicontinuous) $\Leftrightarrow \big(\bigwedge\limits_{a \in R} {}^a f := \{x \in X\colon f(x) \leqslant a\}$ is closed$\big)$,

(ii) (the function f is upper semicontinuous) $\Leftrightarrow \big(\bigwedge\limits_{a \in R} {}_a f := \{x \in X\colon f(x) \geqslant a\}$ is closed$\big)$.

PROOF. We give the proof for the case when X is a metric space. The proof of the general case is analogous, the difference being that nets must be taken instead of sequences, and a neighbourhood basis of a point, instead of a sequence of balls of decreasing radii.

(i) $\Rightarrow$: Let ${}^a f \ni x_n \to x_0$. Points of the sequence (x_n) lie in every ball. Since $f(x_n) \leqslant a$, then $\inf_{x \in K(x_0, r)} f(x) \leqslant a$. Hence $\liminf_{x \to x_0} f(x) \leqslant a$, whereby $f(x_0) \leqslant a$, and thus $x_0 \in {}^a f$.

$\Leftarrow$: Let $r_n \to 0$; then

$$\liminf_{x \to x_0} f(x) = \lim_{n \to \infty} \big(\inf f(K(x_0, r_n))\big) \leqslant f(x_0),$$

since $\inf f\big(K(x_0, r_n)\big) \leqslant f(x_0)$. It is thus sufficient to demonstrate the converse inequality.

Let $\liminf_{x \to x_0} f(x) = a = f(x_0) - \varepsilon$, $\varepsilon > 0$. Then $\inf f\big(K(x_0, r_n)\big) \leqslant a + \varepsilon/4$ for $n > N$. But then the ball $K(x_0, r_n)$ contains a point x_n such that $f(x_n) \leqslant a + \varepsilon/2$. Thus we have the sequence $x_n \to x$ and $x_n \in {}^{a+\varepsilon/2} f$. Since ${}^{a+\varepsilon/2} f$ is a closed set, then $f(x_0) \leqslant a + \varepsilon/2 = f(x_0) - \varepsilon/2$. Consequently, $\varepsilon = 0$.

The proof of point (ii) is analogous. □

It turns out that lower semicontinuous functions constitute a family which is closed with respect to attainment of an upper envelope (of the least upper bound).

LEMMA XIII.9.2. Let $\{f_i\}_{i \in I}$ be an arbitrary family of lower semicontinuous functions. Then the function $f(x) := \sup_{i \in I} f_i(x)$ is also lower semicontinuous.

PROOF. Let $a \in \boldsymbol{R}$. Let $x \in {}^a f$, that is, let $\sup_{i \in I} f_i(x) \leqslant a$. Then $f_i(x) \leqslant a$, that is, $x \in {}^a f_i$ for all $i \in I$.

Conversely: if $f_i(x) \leqslant a$ for all $i \in I$, then $\sup_{i \in I} f_i(x) \leqslant a$. This can be written as follows:

$${}^a f = \bigcap_{i \in I} {}^a f_i .$$

But all ${}^a f_i$ were closed, and the intersection of any family of closed sets is a closed set, hence ${}^a f$ is closed. By Lemma XIII.9.1, f is lower semicontinuous. □

As we know, a continuous function is lower semicontinuous; accordingly, the upper limit of continuous functions is a lower semicontinuous function.

Since in the definition of the Radon integral we take $E = C_0(X)$ on a locally compact space X, functions from $\mathscr{E}$ are lower semicontinuous.

LEMMA XIII.9.3. For a Radon integral, functions from the space $\mathscr{E}$ are lower semicontinuous functions and have compact supports of the negative part f^-.

PROOF. Only the second part of the thesis remains to be proved. Let $E \ni f_n \nearrow f, f_n \cap 0 \leqslant f \cap 0$. Since $f_1 \cap 0 \leqslant f_n \cap 0 \leqslant f \cap 0 \leqslant 0$, for $n = 1, 2, \ldots$, and $f_1 \cap 0 \in C_0(X)$, the $f \cap 0$ may differ from zero at most wherever $f_1 \cap 0$ differs from zero. This means that the support of f^- is contained in the compact support f_1^-, that is, it is compact. □

For a broad class of spaces which are important in applications, e.g. metric spaces having a dense countable set ($\boldsymbol{R}^n$ may serve as an example of such a space), the converse theorem holds:

THEOREM XIII.9.4. In a separable metric space (i.e. having a dense countable set) $\mathscr{E}$ is the set of those lower semicontinuous functions whose negative parts have a compact support.

The proof will not be given here.

Examples. 1. Let us take $X = [a, b] \subset \boldsymbol{R}, E = C([a, b])$ and the Riemann integral μ. The functional obtained by extending this integral to the whole space $\mathscr{L}^1$ is called the *Lebesgue integral over* $[a, b]$.

We shall show that if f is Riemann-integrable, then $f \in \mathscr{L}^1$ and $N_1(f) = \int_{[a,b]} |f(x)| dx$.

PROOF. Let us form the upper sum

$$\bar{S}(f, \varkappa) = \sum_{i=1}^{n} (x_{\varkappa_i} - x_{\varkappa_{i-1}}) \cdot \sup_{x \in [x_{\varkappa_{i-1}}, x_{\varkappa_i}]} f(x),$$

where $\varkappa$ is a partition of the interval $[a, b]$ by the points $\{x_{\varkappa_1}, x_{\varkappa_2}, \ldots, x_{\varkappa_n}\}$. The lower sum is of the form

$$\underline{S}(f, \varkappa) = \sum_{i=1}^{n} (x_{\varkappa_i} - x_{\varkappa_{i-1}}) \cdot \inf_{x \in [x_{\varkappa_{i-1}}, x_{\varkappa_i}]} f(x).$$

If we define the "step" functions

$$f_\varkappa^+ := \begin{cases} \sup\limits_{x\in[x_{\varkappa_{i-1}}, x_{\varkappa_i}]} f(x) & \text{for } x \in \,]x_{\varkappa_{i-1}}, x_{\varkappa_i}[, \\ \min(\lim\limits_{x\searrow x_{\varkappa_i}} f_\varkappa^+, \lim\limits_{x\nearrow x_{\varkappa_i}} f_\varkappa^+) & \text{for } x = x_{\varkappa_i}, \end{cases}$$

$$f_\varkappa^- := \begin{cases} \inf\limits_{x\in[x_{\varkappa_{i-1}}, x_{\varkappa_i}]} f(x) & \text{for } x \in \,]x_{\varkappa_{i-1}}, x_{\varkappa_i}[, \\ \max(\lim\limits_{x\searrow x_{\varkappa_i}} f_\varkappa^-, \lim\limits_{x\searrow x_\varkappa} f_\varkappa^-) & \text{for } x = x_{\varkappa_i}, \end{cases}$$

these functions will be semicontinuous; $f_\varkappa^+$ is lower semicontinuous and $f_\varkappa^-$ is upper semicontinuous, that is, by Theorem XIII.9.4 $f_\varkappa^+ \in \mathscr{E}$, $-(f_\varkappa^-) \in \mathscr{E}$ and $\overline{S}(f, \varkappa) = \tilde{\mu}(f_\varkappa^+)$, $-\underline{S}(f, \varkappa) = \tilde{\mu}(-f_\varkappa^-)$. But

$$\mu^*(f_\varkappa^+ - f_\varkappa^-) = \tilde{\mu}(f_\varkappa^+ - f_\varkappa^-) = \overline{S}(f, \varkappa) - \underline{S}(f, \varkappa) \to 0.$$

Since $f \leqslant f_\varkappa^+$, therefore $f - f_\varkappa^- \leqslant f_\varkappa^+ - f_\varkappa^-$, that is $\mu^*(f - f_\varkappa^-) \underset{\varkappa}{\to} 0$. Moreover

$$N_1(f) = \mu^*(f) = \lim_\varkappa \mu^*(f_\varkappa^-) = \lim \underline{S}(f, \varkappa) = \int_{[a,b]} f$$

for a nonnegative f. If f changes sign, then on splitting it up into positive and negative parts, we obtain the same result.

But $-f_\varkappa^- \in \mathscr{E}$, that is, there exists a sequence $f_n \in \mathscr{E}$ such that $f_n \nearrow f_\varkappa^-$ and $\mu(f_n) \underset{n}{\to} \tilde{\mu}(-f_\varkappa^-)$. Therefore

$$\mu^*(-f_\varkappa^- - f_n) = \tilde{\mu}(-f_\varkappa^- - f_n) = \tilde{\mu}(-f_\varkappa^-) + \tilde{\mu}(-f_n)$$
$$= \mu(-f_\varkappa^-) - \mu(f_n) \underset{n}{\to} 0,$$

which means that $-f_\varkappa^- \in \mathscr{L}^1$, that is, $f_\varkappa^- \in \mathscr{L}^1$. The function f thus is the limit in the sense of the seminorm N_1 of the functions $f_\varkappa^- \in \mathscr{L}^1$, that is, $f \in \mathscr{L}^1$. □

2. Let $X = N$ (the set of integers) $\mathscr{E} = C_0(X)$, $\mu(f) = \sum\limits_{n=1}^{\infty} f(n)$. As we know, $\mathscr{E}$ is the set of all sequences with a finite number of negative terms (with compact support of the negative part). We shall show that

$$\left(\sum_{i=1}^{\infty} |f(n)|^p < \infty\right) \Leftrightarrow (f \in \mathscr{L}^p).$$

PROOF. Since $|f|^p \in \mathscr{E}$, and the sequence

$$f_n(k) := \begin{cases} f(k) & \text{for } k \leqslant n, \\ 0 & \text{for } k > n \end{cases}$$

satisfies the condition $E \ni |f_k| \nearrow |f|$ and $E \ni |f_k|^p \nearrow |f|^p$, therefore

$$\mu^*(|f|^p) = \tilde{\mu}(|f|^p) = \lim_{k\to\infty} \mu(|f_k|^p)$$

$$= \lim_{k\to\infty} \sum_{n=1}^{k} |f(n)|^p = \sum_{n=1}^{\infty} |f(n)|^p.$$

Thus, if $f \in \mathscr{L}^p \subset \mathscr{F}^p$, then $\sum_{n=1}^{\infty} |f(n)|^p < \infty$. Conversely, if the series converges, then

$$\mu^*(|f-f_k|^p) = \sum_{n=k+1}^{\infty} |f(n)|^p \underset{k\to\infty}{\to} 0,$$

that is, $f_k \to f \in \mathscr{L}^p$ in the sense of the seminorm N_p. □

Thus we have obtained the interesting result that $\mathscr{F}^p = \mathscr{L}^p$ since every function from $\mathscr{F}^p$ (i.e. one such that $\mu^*(|f|^p) < \infty$ for it) belongs to $\mathscr{L}^p$.

10. APPLICATION OF THE LEBESGUE THEOREM. INTEGRALS WITH A PARAMETER. INTEGRATION OF SERIES

Given a set X, a family of elementary functions E, a Daniell–Stone integral μ, and a metric space $A = (A, d)$. Consider the function f: $A \times X \to \boldsymbol{R}$ such that for every $a \in A$ the function $f(a, \cdot)$ is integrable:

$$\bigwedge_{a\in A} f(a, \cdot) \in \mathscr{L}^1(X, \mu).$$

Let us define the following function on the space A:

$$F(a) := \int f(a, x)\, d\mu(x).$$

The function F is called an *integral with a parameter*.

We shall prove the following property of integrals with a parameter:

THEOREM XIII.10.1. Let the point $a_0 \in A$ have the property that there exists a number $\delta > 0$ and a function $0 \leqslant g \in \mathscr{L}^1(\mu)$ for which the following inequality holds:

$$\bigwedge_{a \in K(a_0, \delta)} |f(a, x)| \leqslant g(x) \quad \mu\text{-almost everywhere.}$$

Furthermore, let the function $f(\cdot\,, x)$ be continuous for a fixed x at the point a_0, μ-almost everywhere in X.

Then $F(a)$ is a continuous function at the point a_0.

PROOF. Take an arbitrary sequence $(a_n)_1^\infty$ of points $a_n \in A$ such that $a_n \to a_0$. Since $\bigvee_N \bigwedge_{n>N} a_n \subset K(a_0, \delta)$, then the sequence of functions

$$f_m(x) := f(a_{N+m}, x)$$

satisfies the hypotheses of the Lebesgue Theorem inasmuch as

$$f_m(x) \to f(a_0, x) \quad \mu\text{-almost everywhere,}$$

$$|f_m| < g \quad \mu\text{-almost everywhere.}$$

It follows from Corollary XIII.8.7 that

$$F(a_{N+m}) = \int f(a_{N+m}, x) \to \int f(a_0, x) = F(a_0),$$

which means that the function F is continuous in a_0. □

Interesting cases arise when instead of the general metric space A we take an open subset of some Banach space. We can then examine the differentiability of integrals with a parameter:

THEOREM XIII.10.2. Let A be an open subset of a Banach space B. Let $a_0 \in A$. Let the function f: $A \times X \to \boldsymbol{R}$ have, for a fixed $x \in X$, a directional derivative in the direction $e \in B$ at the point a_0 μ-almost everywhere with respect to $x \in X$. Let $f(a, \cdot) \in \mathscr{L}^1$ for a fixed $a \in A$.

If there exist a number $\varepsilon > 0$ and a function $0 \leqslant g \in \mathscr{L}^1$, such that

$$\left|\frac{f(a_0+te, x)-f(a_0, x)}{t}\right| \leqslant g(x) \quad \text{for } t \in [-\varepsilon, \varepsilon],$$

then

(i) $\nabla_e f(a_0, \cdot) \in \mathscr{L}^1$.

(ii) An integral with a parameter has a directional derivative in the direction $e \in B$ at the point a_0:

$$\nabla_e F(a_0) = \int \nabla_e f(a_0, x) d\mu(x).$$

PROOF. Given an arbitrary sequence $\boldsymbol{R} \ni t_n \to 0$. Then the functions

$$f_n(x) := \frac{f(a_0+t_n e, x)-f(a_0, x)}{t_n}$$

satisfy the hypotheses of the Lebesgue Theorem for n so large that $t_n \in [-\varepsilon, \varepsilon]$: $f_n \to \nabla_e f(a_0, \cdot)$ μ-almost everywhere and $|f_n| < g$ μ-almost everywhere. Thus, from the Lebesgue Theorem and Corollary XIII.8.7 it follows that $\nabla_e f(a_0, \cdot) \in \mathscr{L}^1$ and

$$\int f_n(x)\, d\mu(x) \to \int \nabla_\varepsilon f(a_0, x)\, d\mu(x).$$

But

$$\lim_{n\to\infty} \int f_n = \lim_{n\to\infty} \frac{F(a_0+t_n e)-F(a_0)}{t_n}.$$

Since this limit exists and is independent of the choice of the sequence $t_n \to 0$, the function F is differentiable:

$$\nabla_e F(a_0) = \int \nabla_e f(a_0, x)\, d\mu(x). \quad \square$$

In practice, we very often have to deal with functions $f(a, x)$ which are continuous in the parameter a and have continuous derivatives. Then there exists a number $0 \leqslant \theta \leqslant 1$, such that

$$\frac{f(a+te, x)-f(a, x)}{t} = \nabla_e f(a+\theta te, x).$$

Hence, thanks to the mean-value theorem, the usually difficult problem of majorizing a difference quotient reduces to one of majorizing a derivative:

THEOREM XIII.10.3. Let A be an open set of a Banach space B. Let the function $f\colon A \times X \to \boldsymbol{R}$ have a continuous directional derivative in the direction $e \in B$, over the whole set A. Suppose that $f(a, \cdot) \in \mathscr{L}^1$ for every $a \in A$.

If there exists a function $0 \leqslant g \in \mathscr{L}^1$ such that for every $a \in A$ we have $|\nabla_e f(a, \cdot)| \leqslant g$ μ-almost everywhere with respect to x, then:

(i) $\nabla_e f(a, \cdot) \in \mathscr{L}^1$ for every $a \in A$,

(ii) the integral with a parameter has a continuous derivative in the direction e:

$$\nabla_e F(a) = \int \nabla_e f(a, x)\, d\mu(x) \quad \text{in the whole set } A.$$

PROOF. Let $a_0 \in A$. Since A is open, there exists an $r > 0$ such that $K(a_0, r) \in A$. Let $\varepsilon = r/\|e\|$. Then $(a_0 + \theta te) \in A$ for $t \in [-\varepsilon, \varepsilon]$ and for $\theta \in [0, 1]$.

Therefore we have

$$\left|\frac{f(a_0+te, x)-f(a_0, x)}{t}\right| = |\nabla_e f(a_0+\theta te, x)| \leqslant g(x)$$

μ-almost everywhere.

The hypotheses of the preceding theorem are thus satisfied, that is

$$\nabla_e F(a) = \int \nabla_e f(a, x) d\mu(x) \quad \text{for every } a_0 \in A.$$

However, the integrand $\nabla_e f$ satisfies the hypotheses of the theorem on the continuity of integrals with a parameter. Hence it follows that $\nabla_e F$ is continuous in A. □

This theorem immediately implies the most useful test for the differentiability of integrals with a parameter:

THEOREM XIII.10.4. Let $f: A \times X \to \boldsymbol{R}$ have a continuous directional derivative in the direction e over the whole open set $A \subset B$ (B being a Banach space). Suppose that $f(a, \cdot) \in \mathscr{L}^1$ for every $a \in A$.

If for every $a \in A$ there exist a neighbourhood $U_a \subset A$ of the point a and a function $g_a \in \mathscr{L}^1$ such that $|\nabla_e f(a', \cdot)| \leqslant g_a$ for $a' \in U_a$, then

$$\nabla_e \int f(a, x) d\mu(x) = \int \nabla_e f(a, x) d\mu(x).$$

PROOF. By Theorem XIII.10.3 the thesis is true in every set U_a. But $\bigcup_{a \in A} U_a = A$, and thus the thesis holds over the whole set A. □

Example. Let $X = [0, \infty[$, and let μ be a Riemann integral on $F = C_0(X)$. Let $f \geqslant 0$ be a continuous function with a continuous derivative f' such that $\lim_{x \to 0} f'(x) = c$ and $f(x) \searrow 0$ for $x \to \infty$.

Consider the function

$$F(a, b) := \int_0^\infty \frac{f(ax)-f(bx)}{x} dx, \quad \text{where } a, b \in]\varepsilon, \infty[.$$

For $x \to 0$ we find by the l'Hospital rule that

$$\lim_{x \to 0} \frac{f(ax)-f(bx)}{x} = \lim_{x \to 0} \frac{af'(ax)-bf'(bx)}{1} = (a-b)c < \infty.$$

Hence the integrand is continuous at $x = 0$.

If we assume that f has an improper integral, then the integral of our function also exists since for $x \geqslant 1$ we have

$$\left|\frac{f(ax)-f(bx)}{x}\right| \leqslant |f(ax)-f(bx)|.$$

Furthermore for $0 < \varepsilon < a, b$

$$\left|\frac{\partial}{\partial a}\left(\frac{f(ax)-f(bx)}{x}\right)\right| = |f'(ax)| \leqslant |f'(\varepsilon x)|,$$

$$\left|\frac{\partial}{\partial b}\left(\frac{f(ax)-f(bx)}{x}\right)\right| = |-f'(bx)| \leqslant |f'(\varepsilon x)|,$$

since $|f'|$ is a monotonically decreasing function.

But

$$\int_0^\infty |f'(\varepsilon x)|\,dt = -\int_0^\infty f'(\varepsilon x)\,dx = -\frac{1}{\varepsilon}\int_0^\infty f'(t)\,dt$$

$$= \frac{1}{\varepsilon} f(0) < \infty,$$

and consequently the function $|f'(\varepsilon x)|$ is integrable.

The integrand $\dfrac{f(ax)-f(bx)}{x}$ thus satisfies the hypotheses of Theorem XIII.10.3. Hence

$$\frac{\partial F(a,b)}{\partial a} = \int_0^\infty \frac{\partial}{\partial a}\left(\frac{f(ax)-f(bx)}{x}\right) dx = \int_0^\infty f'(ax)\,dx$$

$$= \frac{1}{a} f\Big|_0^\infty = -\frac{f(0)}{a}.$$

Identically:

$$\frac{\partial F(a,b)}{\partial b} = \frac{f(0)}{b}.$$

Suppose that $F(a, b) = F_1(a)+F_2(b)$; then

$$\frac{\partial F(a,b)}{\partial a} = \frac{dF_1(a)}{da} = -\frac{f(0)}{a},$$

whence $F_1(a) = -f(0)\ln a + c_1$,

$$\frac{\partial F(a,b)}{\partial b} = \frac{dF_2(b)}{db} = \frac{f(0)}{b},$$

whence $F_2(b) = f(0)\ln b + c_2$.

Therefore

$$F(a,b) = f(0)\ln\frac{b}{a} + c_3, \quad \text{where } c_3 = c_1 + c_2.$$

The function F thus found does indeed satisfy the differential equations it was supposed to satisfy. Accordingly we infer that

$$\int_0^\infty \frac{f(ax)-f(bx)}{x}\,dx = f(0)\ln\frac{b}{a} + c_3.$$

When we put $a = b$, the left-hand side vanishes and $\ln 1 = 0$, whereby $c_3 = 0$. Finally

$$\int_0^\infty \frac{f(ax)-f(bx)}{x}\,dx = f(0)\ln\frac{b}{a}$$

for $a, b \geqslant \varepsilon > 0$. Since ε was arbitrary, on taking the sequence $\varepsilon_n \to 0$, we find that this relation holds for all $a, b > 0$.

The Lebesgue Theorem has one more important practical application:

THEOREM XIII.10.5. Given a sequence of functions $(f_n)_1^\infty, f_n \in \mathscr{L}^1$, on the space X. Suppose that the series $\sum_{n=1}^{\infty} f_n(x)$ converges μ-almost everywhere.

If there exists a function $0 \leqslant g \in \mathscr{L}^1$ such that $\left|\sum_{n=1}^{k} f_n(x)\right| \leqslant g(x)$ μ-almost everywhere for every k, then

(i) $\sum_{n=1}^{\infty} f_n \in \mathscr{L}^1$,

(ii) $\int\left(\sum_{n=1}^{\infty} f_n(x)\right) d\mu(x) = \sum_{n=1}^{\infty}\int f_n(x)\,d\mu(x)$.

PROOF. Applying the Lebesgue Theorem to the function $s_k(x) := \sum_{n=1}^{k} f_n(x)$, we immediately obtain point (i) of the thesis.

But

$$\int s_k = \int \sum_{n=1}^{k} f_n = \sum_{n=1}^{k} \int f_n,$$

whereby we get point (ii) of the thesis. □

Remark. The theorems of Fubini and Tonelli imply a stronger result of similar character.

Example. Let us evaluate the sum of the series $\sum_{n=1}^{\infty} \frac{1}{n \cdot 2^n}$. Note that

$$\frac{1}{n \cdot 2^n} = \int_2^{\infty} \frac{dx}{x^{n+1}}.$$

Thus we take $X = [2, \infty[$ and the Lebesgue integral on X. Let $f_n(x) = 1/x^{n+1}$. Then

$$s_k(x) = \sum_{n=1}^{k} f_n = \sum_{n=1}^{k} \frac{1}{x^{n+1}} = \frac{1}{x^2} \cdot \frac{1-1/x^k}{1-1/x}$$

and

$$\lim_{k\to\infty} s_k(x) = \frac{1}{x^2} \cdot \frac{1}{1-1/x} = \frac{1}{x(x-1)}.$$

However $f_n > 0$, and thus

$$s_k(x) \leqslant \lim_{k\to\infty} s_k(x) = -\frac{1}{x^2\left(1-\frac{1}{x}\right)} \leqslant \frac{1}{x^2 \cdot \frac{1}{2}}$$

for $x \in [2, \infty[$,

and this latter function is integrable. From Theorem XIII.10.5 we find that

$$\sum_{n=1}^{\infty} \frac{1}{n \cdot 2^n} = \sum_{n=1}^{\infty} \int_2^{\infty} f_n = \int_2^{\infty} \sum_{n=1}^{\infty} f_n = \int_2^{\infty} \frac{dx}{x(x-1)}$$

$$= \int_2^{\infty} \left(\frac{1}{x-1} - \frac{1}{x}\right) dx = \ln 2.$$

11. MEASURABLE FUNCTIONS

We now introduce the following notation:

If f, g are numerical functions on X, then

$$(\operatorname{sgn} f)(x) := \begin{cases} +1\,(-1), & \text{when } f(x) = +\infty\,(-\infty), \\ \dfrac{f(x)}{|f(x)|}, & \text{when } 0 < |f(x)| < \infty, \\ 0, & \text{when } f(x) = 0 \end{cases}$$

and

$$f \curlywedge g := \operatorname{sgn}(f \cdot g)\,(|f| \cap |g|).$$

DEFINITION. A numerical function f is said to be *μ-measurable* if

$$f \curlywedge g \in \mathscr{L}^1(\mu)$$

for every function $0 \leqslant g \in \mathscr{L}^1(\mu)$.

COROLLARY XIII.11.1. (A function f is μ-measurable; $\tilde{f} = f$ μ-almost everywhere) $\Rightarrow$ ($\tilde{f}$ is μ-measurable).

PROOF. For every function $0 \leqslant g \in \mathscr{L}^1$ we have $\mathscr{L}^1 \ni f \curlywedge g = \tilde{f} \curlywedge g$ μ-almost everywhere, and therefore $\tilde{f} \curlywedge g \in \mathscr{L}^1$ by Lemma XIII.7.5. □

Since in the sequel we operate with a fixed integral, we shall write simply "measurable function" instead of μ-measurable function.

The lemma which we now prove elucidates the concept of measurability somewhat.

LEMMA XIII.11.2. If $f \in \mathscr{L}^p$, then f is measurable.

PROOF. Let $f \in \mathscr{L}^p$. There exists a sequence $E \ni f_n \to f$ in the sense of the seminorm N_p. It follows from Lemma XIII.7.2 that $E \ni f_n^+ \to f^+$, $E \ni f_n^- \to f^-$ in the sense of the seminorm N_p. Suppose that $0 \leqslant g \in \mathscr{L}^1$. Then

$$f \curlywedge g = (f^+ - f^-) \curlywedge g = f^+ \curlywedge g - f^- \curlywedge g = f^+ \cap g - f^- \cap g.$$

The Riesz–Fischer Theorem enables us to select from the sequence f_n^+ a subsequence $f_{n_k}^+$ such that $E \ni f_{n_k}^+ \to f^+$ in the sense of the seminorm N_p and pointwise μ-almost everywhere. Consequently

$$f_{n_k}^+ \cap g \to f^+ \cap g \quad \mu\text{-almost everywhere and}$$
$$f_{n_k}^+ \cap g \leqslant g \in \mathscr{L}^1.$$

Hence, it follows from the Lebesgue Theorem that $f^+ \cap g \in \mathscr{L}^1$. In identical fashion we prove that $f^- \cap g \in \mathscr{L}^1$. Therefore $f \curlywedge g \in \mathscr{L}^1$. □

LEMMA XIII.11.3. Functions from the space $\mathscr{E}$ are measurable.

PROOF. Let $E \ni f_n \nearrow f$, $0 \leqslant g \in \mathscr{L}^1$. Then $E \ni f_n^+ \nearrow f^+$, $E \ni f_n^- \searrow f^-$. Moreover

$$f \curlywedge g = f^+ \curlywedge g - f^- \curlywedge g = f^+ \cap g - f^- \cap g.$$

But $\mathscr{L}^1 \ni f_n^+ \cap g \to f^+ \cap g$ pointwise and $f_n^+ \cap g \leqslant g \in \mathscr{L}^1$, and hence $f^+ \cap g \in \mathscr{L}^1$ by the Lebesgue Theorem. Identically $f^- \cap g \in \mathscr{L}^1$, and hence $f \curlywedge g \in \mathscr{L}^1$. □

Example. Let X be a two-point set, let $X = \{a, b\}$, and let E be a family of functions which do not separate a and b (that is $f(a) = f(b)$), $\mu(f) = f(a)$. As we know, in this case $\mathscr{L}^p = E$.

Take any function $f \notin \mathscr{L}^p$. Therefore $f(a) \neq f(b)$. Let $c := \max(f(a), f(b))$. Take a function $0 \leqslant g \in \mathscr{L}^1$ defined as $g(a) = g(b) = c$. Then $f \curlywedge g = f \notin \mathscr{L}^1$, and thereby f is not measurable.

We have thus proved that in this case the set of measurable functions coincides with $\mathscr{L}^p$.

In the sequel we prove an important test for measurability.

LEMMA XIII.11.4. A numerical function f is measurable if and only if $f \curlywedge h \in \mathscr{L}^1$ for every function $0 \leqslant h \in E$.

PROOF. The necessity of this condition is obvious, for $E \subset \mathscr{L}^1$.

Let us prove the sufficiency. Suppose that $0 \leqslant g \in \mathscr{L}^1$. Take the sequence $E \ni \tilde{h}_n \to g$ in the sense of the seminorm N_1. From Lemma XIII.7.2 it follows that $0 \leqslant h_n := |\tilde{h}_n| \to g$ in the sense of the seminorm N_1.

But

$$\begin{aligned} f \curlywedge h_n &= \operatorname{sgn}(f)\,(|f| \cap h_n) \in \mathscr{L}^1, \\ f \curlywedge g &= \operatorname{sgn}(f)\,(|f| \cap g), \\ N_1(f \curlywedge h_n - f \curlywedge g) &= N_1(\operatorname{sgn}(f)\,(|f| \cap h_n - |f| \cap g)) \\ &= N_1(|f| \cap h_n - |f| \cap g) \leqslant N_1(g - h_n) \underset{n\to\infty}{\to} 0, \end{aligned}$$

since

$$||f| \cap h_n - |f| \cap g| \leqslant |g - h_n|.$$

Therefore $\mathscr{L}^1 \ni f \curlywedge h_n \to f \curlywedge g$. The completeness of the space $\mathscr{L}^1$ implies that $f \curlywedge g \in \mathscr{L}^1$. Since this holds for every $g \in \mathscr{L}^1$, then f is a measurable function. □

Example. Let X be a locally compact space and let $E = C_0(X)$. Then every continuous function is measurable. Indeed, if f is continuous, $h \in C_0(X)$, then $f \curlywedge h$ is also continuous and has a compact support: $f \curlywedge h \in E \subset \mathscr{L}^1$. □

THEOREM XIII.11.5. If a sequence of measurable functions $(f_n)_1^\infty$ is pointwise convergent μ-almost everywhere, its limit is a measurable function.

PROOF. We have

$$f_n(x) \to f(x) \quad \mu\text{-almost everywhere.}$$

Suppose that $0 \leqslant g \in \mathscr{L}^1$. Then $\mathscr{L}^1 \ni f_n \curlywedge g \to f \curlywedge g$ μ-almost everywhere and $|f_n \curlywedge g| = |f_n| \cap g \leqslant g$. By the Lebesgue Theorem it follows that $f \curlywedge g \in \mathscr{L}^1$. □

Let $\overline{M}$ denote the set of all measurable functions and let M denote the set of all real-valued measurable functions.

The following theorem then holds.

THEOREM XIII.11.6. The space M is a vector lattice.

PROOF. (i) Let $f \in M$, $a \in R$, $0 \leqslant g \in \mathscr{L}^1$. For $a = 0$, we have $af = 0 \in M$. Thus, let $a \neq 0$. Plainly $f \curlywedge g \in \mathscr{L}^1$. We have

$$(af) \curlywedge g = \operatorname{sgn}(a)\operatorname{sgn}(f)\,(|a| \cdot |f| \cap g) = |a| \operatorname{sgn}(a)\operatorname{sgn}(f) \times$$

$$\times \left(|f| \cap \frac{g}{|a|}\right) = a\operatorname{sgn}(af)\left(|f| \cap \frac{g}{|a|}\right) = a\left(f \curlywedge \frac{g}{|a|}\right).$$

But $0 \leqslant g/|a| \in \mathscr{L}^1$, and therefore $f \curlywedge g/|a| \in \mathscr{L}^1$, from which it follows that $(af) \curlywedge g \subset \mathscr{L}^1$. Finally, $a \cdot f \subset M$.

(ii) Let $f, g \in M$, and let $0 \leqslant h \in \mathscr{L}^1$. We write

$$h_n := \big(f \curlywedge (nh) + g \curlywedge (nh)\big) \curlywedge h.$$

Note that $h_n \in \mathscr{L}^1$ and $|h_n| \leqslant h$. If $h(x) = 0$, then $nh(x) = 0$ and $f(x) \curlywedge nh(x) = g(x) \curlywedge nh(x) = 0$, and thus for such x we have

$$\big(f(x) \curlywedge nh(x) + g(x) \curlywedge nh(x)\big) \curlywedge h(x) = \big(f(x) + g(x)\big) \curlywedge h(x).$$

If $h(x) \neq 0$, then $f(x) \curlywedge nh(x) \underset{n\to\infty}{\to} f(x)$, $g(x) \curlywedge nh(x) \underset{n\to\infty}{\to} g(x)$ and therefore for such x: $\lim\limits_{n\to\infty} h_n(x) = \big(f(x)+g(x)\big) \curlywedge h(x)$.

Hence, finally $h_n \to (f+g) \curlywedge h$ pointwise. From the Lebesgue Theorem it follows that $(f+g)\curlywedge h \in \mathscr{L}^1$. Since the function h was arbitrary, $f+g \in M$.

(iii) Let $f \in M$, and let $0 \leqslant g \in \mathscr{L}^1$. In that case

$$|f| \curlywedge g = \operatorname{sgn}(|f|)\cdot(|f| \cap g) = |\operatorname{sgn}(f)\,(|f| \cap g)| = |f \curlywedge g|.$$

Since $f \curlywedge g \in \mathscr{L}^1$, therefore $|f \curlywedge g| \in \mathscr{L}^1$, that is $|f| \curlywedge g \in \mathscr{L}^1$. Consequently $|f| \in M$. □

COROLLARY XIII.11.7. If f is a measurable function, the functions $f^+, f^-, f \cup g, f \cap g$ are measurable functions (for measurable g).

We shall give several more properties of measurable functions.

THEOREM XIII.11.8. If f and g are measurable functions, the function $f \curlywedge g$ is also measurable.

PROOF. Let $0 \leqslant h \in \mathscr{L}^1$; then

$$\begin{aligned}(f \curlywedge g)\curlywedge h &= (f \curlywedge (g^+ - g^-))\curlywedge h = (f \curlywedge g^+ - f \curlywedge g^-)\curlywedge h\\ &= (f \curlywedge g^+)\curlywedge h - (f \curlywedge g^-)\curlywedge h\\ &= \operatorname{sgn}(f)\,(|f| \cap g^+ \cap h) - \operatorname{sgn}(f)\,(|f| \cap g^- \cap h)\\ &= f \curlywedge (g^+ \curlywedge h) - f \curlywedge (g^- \curlywedge h).\end{aligned}$$

But $0 \leqslant g^+ \curlywedge h \in \mathscr{L}'$, $0 \leqslant g^- \curlywedge h \in \mathscr{L}'$, and therefore $f \curlywedge (g^+ \curlywedge h) \in \mathscr{L}^1$ and $f \curlywedge (g^- \curlywedge h) \in \mathscr{L}'$, that is $(f \curlywedge g)\curlywedge h \in \mathscr{L}^1$.

In the first part of the proof we made use of the fact that addition is distributive with respect to the operation $\curlywedge$, if the functions being added (in our case g^+ and g^-) are nonzero on disjoint sets. □

LEMMA XIII.11.9. If f is a measurable function and g is a measurable (or integrable) function, the function $\operatorname{sgn}(f)\cdot g$ is measurable (or integrable).

PROOF. By Theorems XIII.11.6 and XIII.11.8 $nf \curlywedge g \in M$. But

$$nf \curlywedge g = (n|f| \cap |g|)\cdot \operatorname{sgn}(f)\cdot \operatorname{sgn}(g).$$

If $f(x) = 0$, then $n|f(x)| \cap |g(x)| = 0$. If $f(x) \neq 0$, then $(n|f(x)|) \cap |g(x)| \nearrow |g(x)|$. Thus, $nf \curlywedge g \to \operatorname{sgn}(f)\operatorname{sgn}(g)|g| = \operatorname{sgn}(f)\cdot g$ pointwise. The thesis that $\operatorname{sgn}(f)\cdot g$ is measurable follows from Theorem XIII.11.5. On the other hand, if $g \in \mathscr{L}^1$, then

$$|nf \curlywedge g| = (n|f|) \cap |g| \leqslant |g| \quad \text{and}$$

$$nf \curlywedge g = [nf\cdot \operatorname{sgn}(g)] \curlywedge |g| \in \mathscr{L}^1,$$

and hence the thesis follows from the Lebesgue Theorem. □

Now we prove the fundamental relationship between measurability and integrability.

THEOREM XIII.11.10 A measurable function with finite norm $N_1(f)$ is integrable: $\mathscr{F}^1 \cap M = \mathscr{L}^1$.

PROOF. (a) As we know, $\mathscr{L}^1 \subset \mathscr{F}^1$, $\mathscr{L}^1 \subset M$, and hence $\mathscr{L}^1 \subset \mathscr{F}^1 \cap$ $\cap M$.

(b) We shall prove that $\mathscr{F}^1 \cap M \subset \mathscr{L}^1$.

Let $f \in M$ and $N_1(f) = \mu^*(f) < \infty$. Hence there exists a function $h \in \mathscr{E}$ such that $|f| \leqslant h$ and $\mu(h) < \infty$. Therefore, a sequence $\mathscr{E} \ni \tilde{h}_n \nearrow h \geqslant f$ exists. Since $h \geqslant 0$, then also $\mathscr{E} \ni h_n := \tilde{h}_n^+ = \tilde{h}_n \cup 0 \nearrow h \geqslant |f|$. Obviously, $h \in \mathscr{L}^1$ since the functions h_n converge to h in the sense of the seminorm N_1. Accordingly,

$$\mathscr{L}^1 \ni f \curlywedge h_n = \operatorname{sgn}(f) \cdot (|f| \cap h_n) \underset{n\to\infty}{\to} \operatorname{sgn}(f) \cdot |f| = f$$

pointwise and

$$|f \curlywedge h_n| = |f| \cap h_n \leqslant h_n \leqslant h \in \mathscr{L}^1.$$

Thus, it follows from the Lebesgue Theorem that $f \in \mathscr{L}^1$. □

12. MEASURE. INTEGRABLE SETS

Now we introduce the important concept of a measurable set.

DEFINITION. A set $A \subset X$ is said to be *measurable* if its characteristic function χ_A is measurable.

Here are the fundamental properties of measurable sets:

LEMMA XIII.12.1. Let A_i denote sets. Then

(i) (A_1, A_2 measurable) $\Rightarrow$ ($A_1 \cap A_2$ measurable),

(ii) (A_1, A_2 measurable) $\Rightarrow$ ($A_1 - A_2$ measurable),

(iii) (A_i measurable, $i = 1, 2, \ldots$) $\Rightarrow$ ($\bigcup_{n=1}^{\infty} A_i$ measurable).

PROOF. (i) We have $\chi_{A_1 \cap A_2} = \chi_{A_1} \cdot \chi_{A_2} = \operatorname{sgn}(\chi_{A_1}) \cdot \chi_{A_2} \in M$ by virtue of Lemma XIII.11.9.

(ii) $A_1 - A_2 = A_1 - A_1 \cap A_2$. Therefore $\chi_{A_1 - A_2} = \chi_{A_1} - \chi_{A_1 \cap A_2} \in M$, being the sum of measurable functions.

(iii) Let

$$A := \bigcup_{i=1}^{\infty} A_i, \qquad S_n := \bigcup_{i=1}^{n} A_i, \qquad \chi_A = \lim_{n\to\infty} \chi_{S_n}.$$

If S_n were measurable sets, the function χ_A, being the pointwise limit of measurable functions, would also be measurable.

The measurability of S_n will be proved by induction: Clearly the set $S_1 = A_1$ is measurable. Let S_n be measurable. Then $S_{n+1} = S_n \cup A_{n+1}$. But $\chi_{S_{n+1}} = \chi_{S_n} + \chi_{A_{n+1}} - \chi_{S_n \cap A_{n+1}}$, and hence $\chi_{S_{n+1}}$ is measurable as a linear combination of measurable functions.

Remark. The empty set $\emptyset$ is measurable since a function identically equal to zero is measurable.

The following numerical function, called a *measure*, can be defined on a family of measurable sets $\mathscr{M}$:

$$\mathscr{M} \ni A \to \mu(A) := \mu^*(\chi_A) \in \overline{R}_+ .$$

The number $\mu(A)$ is called the *measure of the measurable set A*.

Thus, a measure is a nonnegative numerical function on the family of measurable sets.

Note that the function μ is the restriction of the "outer measure of sets" to the family $\mathscr{M}$.

We shall use the notation $\mu(A) = \int \chi_A$ even when $\chi_A \notin \mathscr{L}^1$.

Below we give the properties of the measure:

LEMMA XIII.12.2.

(a) $\mu(\emptyset) = 0$;

(b) $(A_i \cap A_j = \emptyset) \Rightarrow \left(\mu(\bigcup_{i=1}^{\infty} A_i) = \sum_{i=1}^{\infty} \mu(A_i)\right)$.

PROOF. (a) The proof is trivial.

(b) Let $A \cap B = \emptyset$. Then $\chi_{A\cup B} = \chi_A + \chi_B$. If $\mu^*(\chi_{A\cup B}) < \infty$, then $\chi_{A\cup B} \subset \mathscr{F}^1$, that is, by Theorem XIII.11.10 $\chi_{A\cup B} \in \mathscr{L}^1$. But $|\chi_A| \leqslant \chi_{A\cup B}$ and $|\chi_B| \leqslant \chi_{A\cup B}$, and therefore by virtue of the same theorem, $\chi_A \in \mathscr{L}^1$and $\chi_B \in \mathscr{L}^1$. Hence

$$\begin{aligned}\mu(A\cup B) &= \mu^*(\chi_A + \chi_B) = \int \chi_A + \chi_B \\ &= \int \chi_A + \int \chi_B = \mu(A) + \mu(B).\end{aligned}$$

If $\chi_{A\cup B} \notin \mathscr{L}^1$, then at least one of the functions, χ_A or χ_B, does not belong to $\mathscr{L}^1$, that is, $\mu(A)$ or $\mu(B)$ is infinite. In this case the equality $\mu(A)+\mu(B) = \mu(A \cup B)$ is satisfied trivially.

On using this equality n times, we get

$$\mu\Big(\bigcup_{i=1}^{n} A_i\Big) = \sum_{i=1}^{n} \mu(A_i).$$

Therefore

$$\mu\Big(\bigcup_{i=1}^{\infty} A_i\Big) = \mu^*\Big(\sup_n \sum_{i=1}^{n} \chi_{A_i}\Big) = \sup_n \mu^*\Big(\sum_{i=1}^{n} \chi_{A_i}\Big)$$

$$= \sup_n \mu\Big(\bigcup_{i=1}^{n} A_i\Big) = \sup_n \Big(\sum_{i=1}^{n} \mu(A_i)\Big) = \sum_{i=1}^{\infty} \mu(A_i). \ \square$$

Remark. When a constant function is measurable, $X \in \mathscr{M}$ and then the measurable sets have an additional property:

(iv) $(A \in \mathscr{M}) \Rightarrow ((X-A) \in \mathscr{M})$.

DEFINITION. A family of subsets which is a σ-lattice and which satisfies conditions (i)–(iv) is called a *σ-algebra of sets* (*countably additive algebra*).

In the abstract theory of measure it is assumed that a σ-algebra $\mathscr{M}$ of subsets of the space X is given and that so is a nonnegative σ-additive numerical function with properties (a) and (b), called a measure. As is seen, the theory given here is more general.

We now define an important class of functions which are "integrable on some measurable subsets of the set X."

DEFINITION. Let $Z \in \mathscr{M}$. We say that a *numerical function is integrable on the set* Z if $f \cdot \chi_Z \in \mathscr{L}^1$.

We then write

$$\int_Z f = \int_Z f(x)\, d\mu(x) := \int f \cdot \chi_Z.$$

COROLLARY XIII.12.3. A function $f \in \mathscr{L}^1$ is integrable on every $Z \in \mathscr{M}$.

PROOF. From Lemma XIII.11.9 it follows that

$$f \cdot \chi_Z = f \cdot \operatorname{sgn}(\chi_Z) \in \mathscr{L}^1. \ \square$$

Among measurable sets we distinguish a narrower class:

DEFINITION. A set $Z \subset X$ is said to be *integrable* when $\chi_Z \in \mathscr{L}^1$.

COROLLARY XII.12.4. (Set $Z \in \mathscr{M}$, $\mu(Z) < \infty$) $\Leftrightarrow$ (Z is integrable).

It is very easy to give an example of a non-measurable function (or set) in the case of some "pathological" integrals (see e.g. the example on p. 446). It turns out that there are no simple examples of non-measurable functions for the Lebesgue integral.

Example. Let $X = [0, 1[$. We define an equivalence relation on $\boldsymbol{R}$ by $x \sim y$ whenever $x-y$ is rational. Let A be a subset of $[0, 1[$ which has exactly one common element with each of the equivalence classes of the relation $\sim$. Existence of such a set follows from Axiom of Choice (Part I, p.19). We are going to show that χ_A is not measurable. Let (q_n) denote a sequence which gives a bijection between $\boldsymbol{N}$ and the set $\boldsymbol{Q}$ of rational numbers of the interval $[0, 1[$. By A_n we denote the set $A \dot{+} q_n$, where $\dot{+}$ stands for addition modulo 1. From the definition of A we have $A_n \cap A_m = \emptyset$ for $n \neq m$ since A contains only one element from each class and $\bigcup_n A_n = [0, 1[$ since A has non-empty intersection with each class. If A is measurable then $A_n = ((A+q_n) \cap [0,1[) \cup ((A+q_n- -1) \cap [0, 1[)$ is also measurable and $\mu(A_n) = \mu(A)$ because the Lebesgue measure is invariant under translations. Therefore, we get $1 = \mu([0, 1[) = \mu(\bigcup_n A_n) = \sum_{n=1}^{\infty} \mu(A_n) = \sum_{n=1}^{\infty} \mu(A)$. This is a contradiction because the last term is equal $+\infty$ for $\mu(A) > 0$ and 0 for $\mu(A) = 0$.

13. THE STONE AXIOM AND ITS CONSEQUENCES

Of all the models of integral theory the preferred ones are those which satisfy the *Stone axiom*, i.e. those in which the constant function $f(x) \equiv 1$ belongs to the space of measurable functions. In that case, the set X is measurable and the family of measurable sets is a σ-algebra.

This axiom may be written as follows:

For every function $0 \leqslant h \in \mathscr{L}^1$ we have $1 \wedge h = 1 \cap h \in \mathscr{L}^1$.

It is easily seen that the condition

$$\bigwedge_{0 \leqslant f \in E} 1 \cap f \in \mathscr{L}^1 \tag{1}$$

is equivalent to the Stone axiom. This follows immediately from

Lemma XIII.11.4. Thus, if the condition

(2) $$\bigwedge_{0 \leqslant f \in E} 1 \cap f \in E$$

is satisfied, then so is the Stone axiom. Condition (2) will henceforth be called the *strong Stone axiom.*

Example. Given a locally compact space X and a Radon measure μ. Then

$$\bigwedge_{f \in C_0(X)} 1 \cap f \in C_0(X),$$

that is, the strong Stone axiom is satisfied.

One of the most important consequences of the Stone axiom is the following theorem:

THEOREM XIII.13.1. If the Stone axiom is satisfied, then (Function f is measurable) $\Leftrightarrow$ ($\bigwedge_{a \in R} A_a := \{x \colon f(x) > a\}$ is measurable).

PROOF. $\Rightarrow$: We have

$$g_n := 1 \cap n(f-a)^+.$$

It follows from the Stone axiom that $g(x) \equiv a$ is measurable. Since measurable functions constitute a lattice, then $g_n \in M$.

But $g_n \nearrow \chi_{A_a}$. Hence by Theorem XIII.11.5 it follows that $\chi_{A_a} \in M$.

$\Leftarrow$: Let $a < b$. Then $\{x \colon a < f(x) \leqslant b\} = A_a - A_b$ is a measurable set by virtue of Lemma XIII.12.1. For $x \in A_a - A_b$,

(3) $$0 < f(x) - a(\chi_{A_a}(x) - \chi_{A_b}(x)) \leqslant b - a.$$

Given a number $\varepsilon > 0$. Take the function

$$M \ni f_\varepsilon := \sum_{m=0}^{\infty} m \cdot \varepsilon(\chi_{A_{\varepsilon m}} - \chi_{A_{\varepsilon(m+1)}}) + \sum_{m=0}^{-\infty} m \cdot \varepsilon(\chi_{A_{\varepsilon(m-1)}} - \chi_{A_{\varepsilon m}}).$$

Since the sets $]\varepsilon m, \varepsilon(m+1)]$ are disjoint, then for every x only one term in the whole sum is nonzero. Moreover, that only nonzero term satisfies inequality (3), which means that

(4) $$0 \leqslant |f - f_\varepsilon| < \varepsilon \quad \text{on the entire space } X.$$

Taking the sequence f_{ε_i} for $\varepsilon_i \to 0$, we have

$$M \ni f_{\varepsilon_i} \to f,$$

and hence by Theorem XIII.11.5 it follows that $f \in M$. □

Note that the following relations hold: $|f_\varepsilon| \leqslant |f|$ and

$$f_\varepsilon^+ = \sum_{m=0}^{\infty} m\varepsilon(\chi_{A_{\varepsilon m}} - \chi_{A_{\varepsilon(m+1)}}), \quad 0 \leqslant f^+ - f_\varepsilon^+ \leqslant \varepsilon,$$

$$-f_\varepsilon^- = \sum_{m=0}^{-\infty} m\varepsilon(\chi_{A_{\varepsilon(m-1)}} - \chi_{A_{\varepsilon m}}), \quad 0 \leqslant f^- - f_\varepsilon^- \leqslant \varepsilon.$$

Since $f_\varepsilon^+, f_\varepsilon^- \in M$, then for $f \in \mathscr{L}^1$ (since $|f_\varepsilon| \leqslant |f| \in \mathscr{L}^1$), it follows from Theorem XIII.11.10 that $f_\varepsilon \in \mathscr{L}^1$.

Since $f_{\varepsilon_i} \to f$ pointwise, the Lebesgue theorem implies that $f_{\varepsilon_i} \to f$ in the sense of the seminorm N_1. Furthermore, we have

$$\int f_{\varepsilon_i} = N_1(f_{\varepsilon_i}^+) - N_1(f_{\varepsilon_i}^-) \to N_1(f^+) - N_1(f^-) = \int f.$$

Thus we have proved the following theorem:

THEOREM XIII.13.2. If a function $f \in \mathscr{L}^1$ and

$$f_\varepsilon = \sum_{m=0}^{\infty} m\varepsilon(\chi_{A_{\varepsilon m}} - \chi_{A_{\varepsilon(m+1)}}) + \sum_{m=0}^{-\infty} m\varepsilon(\chi_{A_{\varepsilon(m+1)}} - \chi_{A_{\varepsilon m}}),$$

then $f_\varepsilon \underset{\varepsilon \to 0}{\to} f$ in the sense of the seminorm N_1 and

$$\int f = \lim \int f_\varepsilon.$$

Remark. We shall also use the notation

$$S(f, \mu, \varepsilon) := \int f_\varepsilon d\mu.$$

Such an expression is called a *Lebesgue sum* (by analogy with a Riemann sum). In this notation Theorem XIII.13.2 may be rewritten as:

$$\int f = \lim_{\varepsilon \to 0} S(f, \mu, \varepsilon).$$

We shall prove one more important property of measurable functions in the case when the Stone axiom is satisfied.

THEOREM XIII.13.3. Let φ be a continuous numerical function on $\overline{\boldsymbol{R}}$. Let f be a measurable function on the space X with measure μ. Then $\varphi \circ f$ is also a measurable function.

PROOF. Let

$$f_\varepsilon^k = \sum_{m=0}^{k} m\varepsilon(\chi_{A_{\varepsilon m}} - \chi_{A_{\varepsilon(m+1)}}) + \sum_{m=0}^{-k} m\varepsilon(\chi_{A_{\varepsilon(m-1)}} - \chi_{A_{\varepsilon m}}).$$

Then

$$\varphi \circ f_\varepsilon^k = \sum_{m=0}^{k} \varphi(m\varepsilon)(\chi_{A_{\varepsilon m}} - \chi_{A_{\varepsilon(m+1)}}) + \sum_{m=0}^{-k} \varphi(m\varepsilon)(\chi_{A_{\varepsilon(m-1)}} - \chi_{A_{\varepsilon m}}).$$

The function $\varphi \circ f_\varepsilon^k$ thus is a finite linear combination of measurable functions, that is, $\varphi \circ f_\varepsilon^k \in M$. But $f_\varepsilon^k \underset{k\to\omega}{\to} f_\varepsilon$ pointwise, and consequently by the continuity of the function φ we have $\varphi \circ f_\varepsilon^k \underset{k\to\infty}{\to} \varphi \circ f_\varepsilon$. By virtue of Theorem XIII.11.5 it follows that $\varphi \circ f_\varepsilon \in M$. Now, on going to the limit $f_\varepsilon \underset{\varepsilon\to 0}{\to} f$, by these same arguments we find that $\varphi \circ f \in M$. □

COROLLARY XIII.13.4. For given functions f, g

(i) $(0 \leqslant f \in M, a \in R) \Rightarrow (f^a \in M)$,

(ii) $(f, g \in M) \Rightarrow (f \cdot g \in M)$.

PROOF. (i) Since the function $\varphi(x) = x^a$ is continuous on $\bar{R}$, then $(\varphi \circ f) = f^a$ is measurable.

(ii) $f \cdot g = \frac{1}{4}[(f+g)^2 - (f-g)^2]$. Since $(f+g) \in M$, $(f-g) \in M$, then $(f+g)^2 \in M$ and $(f-g)^2 \in M$, that is $f \cdot g \in M$. □

Next we prove the mean-value theorem of integral calculus.

THEOREM XIII.13.5. Let f be a measurable bounded function

$$m \leqslant f(x) \leqslant M.$$

Let g be an integrable function on a measurable set $Z \subset M$. Then

(i) the function $f \cdot g$ is integrable on Z,

(ii) if $g \geqslant 0$, then there exists a number a: $m \leqslant a \leqslant M$, such that

$$\int_Z f \cdot g = a \int_Z g.$$

PROOF. (i) From Corollary XIII.13.4 it follows that $f \cdot g$ is measurable. But

$$|\chi_Z \cdot f \cdot g| \leqslant \max(|m|, |M|) \cdot |\chi_Z \cdot g| \in \mathscr{L}^1,$$

and therefore $\chi_Z \cdot f \cdot g \in \mathscr{L}^1$.

(ii) Since $mg \leqslant f \cdot g \leqslant Mg$, then from the positivity of the integral we find that

$$m \int_Z g \leqslant \int_Z f \cdot g \leqslant M \int_Z g,$$

which implies the thesis. □

The Stone axiom also yields an interesting relation between the spaces $\mathscr{L}^p$ and $\mathscr{L}^1$, the inverse of Theorem XIII.7.4

THEOREM XIII.13.6. Let $p \geqslant 1$. Then $(0 \leqslant f \in \mathscr{L}^p) \Rightarrow (f^p \in \mathscr{L}^1)$.

PROOF. By Corollary XIII.3.3 we know that $f^p \in M$ since $f \in M$ (cf. Lemma XIII.11.2).

But $N_1(f^p) = [N_p(f)]^p < \infty$, whence $f^p \in \mathscr{F}^1$. It immediately follows from Theorem XIII.11.10 that $f^p \in \mathscr{F}^1 \cap M = \mathscr{L}^1$. □

A consequence of this theorem is a theorem about the majorized convergence in any of the spaces $\mathscr{L}^p$:

THEOREM XIII.13.7. Let $(f_n)_1^\infty$ be a sequence of functions $f_n \in \mathscr{L}^p$ which is pointwise convergent μ-almost everywhere to the function f. If there exists a function $0 \leqslant g \in \mathscr{L}^p$ such that $|f_n| \leqslant g$ μ-almost everywhere for all n, then

$$N_p(f_n - f) \to 0, \quad \text{that is} \quad f \in \mathscr{L}^p.$$

PROOF. Suppose that $f \geqslant 0$. It immediately follows from Theorem XIII.13.6 that $\mathscr{L}^1 \ni f_n^p \to f^p$ μ-almost everywhere and $|f_n^p| \leqslant g^p \in \mathscr{L}^1$ μ-almost everywhere.

The hypotheses of the theorem are thus satisfied for the space $\mathscr{L}^1$. Consequently $N_1(f_n^p - f^p) \to 0$. The convexity of the power function for an exponent $p \geqslant 1$ implies that $|f_n - f|^p \leqslant |f_n^p - f^p|$. Therefore

$$N_p(f_n - f) = [N_1(|f_n - f|^p)]^{1/p} \leqslant [N_1(f_n^p - f^p)]^{1/p} \to 0.$$

In the general case, the thesis is obtained by splitting the function f into $f^+ \geqslant 0$ and $f^- \geqslant 0$. □

14. THE SPACES L^p

Elements of Banach spaces L^p (XIII.7) will be denoted by $\dot{f}, \dot{g}$, etc.

$$\dot{f} := [f], \quad \text{where } f \in \mathscr{L}^p,$$
$$\dot{g} := [g], \quad \text{where } g \in \mathscr{L}^p.$$

We now prove the fundamental theorem about the spaces L^p in the case when the Stone axiom is satisfied:

THEOREM XIII.14.1. Let $\dot{f} \in L^p$, $\dot{g} \in L^{p'}$ (that is $f \in \mathscr{L}^p$, $g \in \mathscr{L}^{p'}$), $p > 1$. Let $\frac{1}{p} + \frac{1}{p'} = 1$. Then

(i) $f \cdot g \in \mathscr{L}^1$,

(ii) $\langle \dot{f} | \dot{g} \rangle := \int f \cdot g \, d\mu$ is a continuous bilinear function.

PROOF. (i) Since $f \in M$ and $g \in M$, by Theorem XIII.13.3 (Corollary XIII.13.4) $f \cdot g \in M$.

But $N_1(f \cdot g) \leqslant N_p(f) N_{p'}(g) < \infty$ and thus $f \cdot g \in \mathscr{F}^1$. It follows from Theorem XIII.11.10 that $f \cdot g \in \mathscr{L}^1$.

(ii) It must be proved that the definition of the form $\langle \dot{f} | \dot{g} \rangle$ is good, i.e. that it does not depend on the choice of representatives. Let

$$f_1 \in [f] \ni f_2, \quad g_1 \in [g] \ni g_2.$$

This means that $f_1 = f_2$ μ-almost everywhere and $g_1 = g_2$ μ-almost everywhere. Therefore $f_1 \cdot g_1 = f_2 \cdot g_2$ μ-almost everywhere, or

$$\int f_1 \cdot g_1 \, d\mu = \int f_2 \cdot g_2 \, d\mu.$$

The bilinearity of the form $\langle \, | \, \rangle$ is trivial:

$$|\langle \dot{f} | \dot{g} \rangle| = \left| \int f \cdot g \, d\mu \right| \leqslant \int |g \cdot f| \, d\mu = N_1(f \cdot g) \leqslant N_p(f) N_{p'}(g)$$
$$= ||\dot{f}||_{L^p} \cdot ||\dot{g}||_{L^{p'}},$$

which means that the form defined above is continuous in both arguments. □

Thus, if $\dot{g} \in L^{p'}$, then $\dot{g}$ defines a continuous linear functional on L^p by the formula: $f \to \langle \dot{f} | \dot{g} \rangle$.

The mapping associating elements from $(L^p)'$ (continuous linear functionals on L^p) with elements from $L^{p'}$, naturally defined as

$$L^{p'} \ni \dot{g} \to T\dot{g} := \langle \dot{g} | \cdot \rangle \in (L^p)'$$

is plainly one-to-one and linear. The latter fact follows from the linearity of the form $\langle \, | \, \rangle$ in the first argument.

It turns out that the mapping T is also isometric, i.e. that the equality

$$||\dot{g}||_{L^{p'}} = ||T\dot{g}||_{(L^p)'} = \sup_{||\dot{f}||=1} |\langle \dot{g} | \dot{f} \rangle|$$

holds. Accordingly, T is the isometric isomorphism of $L^{p'}$ onto $T(L^{p'}) \subset (L^p)'$.

Isomorphic objects are often identified with each other in mathematics. This is what we do in this case, writing

$$L^{p'} \subset (L^p)'.$$

It develops that more can be proven, viz. that $L^{p'} = (L^p)'$. This theorem may be formulated in the following form:

THEOREM XIII.14.2. Let F be a continuous linear functional on a Banach space L^p. Then there exists an element $\dot{g} \in L^{p'}$ such that

$$F(f) = \langle \dot{g} \,|\, \dot{f} \rangle$$

for every $\dot{f} \in L^p$.

The foregoing discussion does not, of course, apply to the case when $p = 1$. For then only $p' = \infty$ satisfies formally the condition $\dfrac{1}{p} + \dfrac{1}{p'} = 1$. It is, however, possible to perform an interesting construction: We introduce the seminorm

$$N_\infty(f) := \operatorname{ess\,sup} |f(x)| = \inf_{\substack{\{f^* : f^* = f \ \mu\text{-almost} \\ \text{everywhere}\}}} \{\sup |f^*(x)|\}$$

and the space $\mathscr{F}^\infty := \{f \text{ real functions}: N_\infty(f) < \infty\}$ ($\mathscr{F}^\infty$ thus is the set of functions bounded μ-almost everywhere). We have

$\mathscr{L}^\infty := \mathscr{F}^\infty \cap M$ (measurable functions, bounded μ-almost everywhere),

$L^\infty := \mathscr{L}^\infty / \mathscr{R}$.

From Theorem XIII.13.5 it follows that for $\dot{f} \in L^1$, $\dot{g} \in L^\infty$ the form

$$\langle \dot{g} \,|\, \dot{f} \rangle = \int g \cdot f d\mu$$

is a continuous linear functional on L^1. It turns out that in this case as well, $L^\infty = (L^1)'$. Examples can, however, be given when $(L^\infty)' \neq L^1$.

Note that $(L^p)'' = (L^{p'})' = L^p$ for $p > 1$.

Banach spaces for which $B'' = B$ are said to be *reflexive*. Hence the following theorem:

THEOREM XIII.14.3. Spaces L^p for $\infty \neq p > 1$ are reflexive.

The case $p = 2$ is especially interesting since then $p' = 2$. This means that

$$(L^p)' = L^2.$$

Note that for elements of the space L^2 the form $\langle \dot{f}|\dot{g}\rangle$ is symmetric, linear in both arguments, and positive definite, since

$$\langle \dot{f}|\dot{f}\rangle = \int f \cdot f \, d\mu = [N_2(f)]^2 = ||\dot{f}||^2_{L^2} \geqslant 0.$$

Moreover

$$(\langle \dot{f}|\dot{f}\rangle = ||\dot{f}||^2 = 0) \Leftrightarrow (\dot{f} = 0).$$

The vector space over the field of real numbers, in which a form satisfying the aforementioned conditions is defined and which is complete, is called the *real Hilbert space*.

The form $\langle \cdot | \cdot \rangle$ is called a *scalar product* and is usually denoted by parentheses:

$$(\dot{f}|\dot{g}) = \langle \dot{f}|\dot{g}\rangle.$$

When $p = 2$, Theorem XIII.14.2 is a corollary to a theorem known from the theory of Hilbert spaces, i.e. the theorem on the form of the linear functional:

THEOREM XIII.14.4 (Fréchet–Riesz Theorem). Let H be a Hilbert space, and let F be a linear functional continuous on H. Then there exists a vector $v \in H$ such that

$$F(u) = (v|u)$$

for every $u \in H$.

The proof of this theorem is given in Section 16.

For the space L^2 the Hölder inequality reduces to the *Schwarz inequality* which holds for any Hilbert space:

$$|(\dot{f}|\dot{g})| \leqslant ||\dot{f}|| \cdot ||\dot{g}||.$$

15. THE HAHN–BANACH THEOREM

In analysis we often come across the problem of extending a linear form l (e.g. in integral theory), defined on some linear subspace $M \subset \boldsymbol{E}$, to the entire space $\boldsymbol{E}$. The problem is interesting when l is continuous for then we would like its extension L also to be continuous. This problem was solved by Hahn and Banach.

THEOREM XIII.15.1 (Hahn–Banach Theorem). Let E be a vector space over the field $K(\mathbf{R}$ or $\mathbf{C})$ and let a seminorm p be defined on E. Let $M \subset E$ be a K-linear subspace; $l\colon M \to K$, where

$$|l(e)| \leqslant p(e) \quad \text{for } e \in M.$$

Then there is a linear form $L\colon E \to K$ such that

$$|L(e)| \leqslant p(e), \quad e \in E \quad \text{and} \quad L|M = l.$$

Remark. The field K is fixed (the theorem would not be true if, in the complex case, M were only a real space).

Before proceeding with the proof, we shall present important corollaries to the Hahn–Banach Theorem.

COROLLARY XIII.15.2. If the seminorm p is continuous, the form L is continuous.

COROLLARY XIII.15.3. Let the space $(E, (p_\alpha)_{\alpha \in A})$ be locally convex (that is, $(p_\alpha)_{\alpha \in A}$ is the family of seminorms which define the topology). Let $p \in (p_\alpha)_{\alpha \in A}$, $0 \neq e_0 \in E$. Then there exists a point $e' \in E'$ such that

$$\langle e_0 | e' \rangle = p(e_0), \quad |e'(e)| \leqslant p(e) \quad \text{for all } e \in E.$$

Thus we see that nontrivial continuous linear forms exist on a locally convex space.

COROLLARY XIII.15.4. (A locally convex space E is a Hausdorff space) $\Leftrightarrow \left(\bigwedge_{0 \neq e_0 \in E} \bigvee_{e' \in E'} e'(e_0) \neq 0 \right)$.

PROOF. Points 0 and e_0 can be separated by the neighbourhoods

$$\{x\colon |e'(x)| < \tfrac{1}{2} |e'(e_0)|\} \quad \text{and}$$
$$\{x\colon |e'(e_0) - e'(x)| < \tfrac{1}{2} e'(e_0)\}. \ \square$$

COROLLARY XIII.15.5. Let the space $(E, \| \ \|)$ be normed; $M \subset E$, $l \in M'$. Then there exists a form $L \in E'$ such that

$$L|M = l, \quad \|L\| = \|l\|_{M'}.$$

PROOF. $|l(x)| \leqslant \|l\|_{M'} \cdot \|x\| := p(x)$. From the Hahn–Banach Theorem it follows that there exists a functional $L \in E'$ since

$$|L(x)| \leqslant p(x) = \|l\|_{M'} \|x\| \quad \text{or} \quad \|L\| \leqslant \|l\|_{M'}.$$

Since it is always the case that $\|L\| \geqslant \|l\|_{M'}$, therefore $\|L\| = \|l\|_{M'}$. $\square$

Remark. Usually in speaking of a locally convex space, one has in mind a Hausdorff space.

THEOREM XIII.15.6 (Mazur Theorem (the geometric form of the Hahn–Banach Theorem, separation theorem)). *Let E be a locally convex space and let D be an absolutely convex closed subset of E. Then for every $x_0 \notin D$ there exists a form $L \in E'$ such that*

$$L(x_0) > 1 \quad \text{and} \quad |L(x)| \leqslant 1 \quad \text{for } x \in D.$$

In other words, every hyperplane $\{x \in E: L(x) = 1+\delta\}$, where $0 < \delta < L(x_0)-1$, separates the point x_0 from D.

We recall that a subset D of a (complex) vector space is called *absolutely convex* if for each pair $x_1, x_2 \in D$ and $\lambda_1, \lambda_2 \in \boldsymbol{R}(\boldsymbol{C})$ such that $\lambda_1+\lambda_2 = 1$, $|\lambda_1| \leqslant 1$, $|\lambda_2| \leqslant 1$ we have $\lambda_1 x_1+\lambda_2 x_2 \in D$.

PROOF. Since the set D is closed, and the space E is regular (Chapter XII deals with regular spaces) there exists an (absolutely) convex neighbourhood O of zero such that

$$D \cap (x_0+O) = \varnothing, \quad \text{thus} \quad \left(D+\tfrac{1}{2}O\right) \cap \left(x_0+\tfrac{1}{2}O\right) = \varnothing.$$

Now, $U := \overline{\left(D+\frac{1}{2}O\right)}$ is an (absolutely) convex closed neighbourhood of zero, because D is (absolutely) convex. Let $p_U(\cdot)$ be the seminorm (so-called *Minkowski functional*) corresponding to the convex set U, i.e.

$$p_U(x) := \inf\{a > 0: x \in aU\}.$$

It is immediately evident that $p_U(x) > 1$ for $x \notin U$, $p_U(x) \leqslant 1$, where $x \in U$. Since $D \subset U$, then for $x \in D$ we have $p_U(x) \leqslant 1$; but $x_0 \notin U$, hence $p_U(x_0) > 1$. Corollary XIII.15.3 thus yields the thesis. □

The proof of the Hahn–Banach Theorem should first be carried out for $K = \boldsymbol{R}$. It turns out that the finite-dimensional case is not at all simpler. The proof proceeds in three steps:

(i) Let $x_0 \notin M$; we shall show that the functional l is extendible to a direct sum $M \oplus [x_0]$ where $[x_0]$ denotes an (l-dimensional) space spanned by the vector x_0. This part of the proof is the most difficult.

(ii) Next, we invoke the Kuratowski–Zorn Lemma in the following manner: We extend the domain of the functional step by step, forming direct sums: adding 1-dimensional spaces, we constantly obtain $|l_\alpha(e)|$

$\leqslant p(e)$. The family of all possible extensions l_α of the form l, such that $|l_\alpha(e)| \leqslant p(e)$, is directed by the relation of being extension. Every linearly ordered subset of this family (chain) contains a maximal element, viz. the extension to the set-theoretical sum of their domains. Hence, the entire family also contains a maximal element; this is the extension to the entire space E, for otherwise, on effecting extension as in (i), we would obtain a true extension (contrary to the maximality).

The entire difficulty comes down to that of correctly determining the value c of the form L at the point x_0, for then $L(x+ax_0) := l(x)+ac$ defines L on $M \oplus [x_0]$. The number $c \in \boldsymbol{R}$ must be so selected that $|l(x)+ac| \leqslant p(x+ax_0)$ for every $0 \neq a \in \boldsymbol{R}$. Hence, on dividing by a, we obtain

$$-l(x_1)-p(x_1-x_0) \leqslant c \leqslant -l(x_1)+p(x_1+x_0),$$
$$\text{for } x_1 \in M.$$

But

$$l(x_2)-l(x_1) = l(x_2-x_1) \leqslant p(x_2-x_1)$$
$$\leqslant p(x_2+x_0)+p(x_1+x_0),$$

and

$$\sup_{x_1 \in M} (-l(x_1)-p(x_1+x_0)) \leqslant \inf_{x_2 \in M} (-l(x_2)+p(x_2+x_0)).$$

It is thus sufficient to take any number from the interval above for c.

It remains to prove the Hahn–Banach Theorem for $K = \boldsymbol{C}$. This transition was carried out independently by Sukhomlinov and Bohnenblust and Sobczyk.

Let $l(x) = l_1(x)+il_2(x)$, where $l_k(x) \in \boldsymbol{R}$ and l_k are $\boldsymbol{R}$-linear. Since $l(ix) = il(x)$, then $l_1(ix) = -l_2(x), l_1(x) = l_2(ix)$. But E may also be considered to be the space over $\boldsymbol{R}$. Let us apply the real case of the Hahn–Banach Theorem to l_1; we have

$$|l_1(x)| \leqslant |l(x)| \leqslant p(x), \quad x \in M.$$

By L_1 we denote the (real) extension of l_1 onto E. Thus

$$|L_1(x)| \leqslant p(x), \quad x \in E, \quad L_1|M = l_1.$$

Obviously we adopt $L_2(x) := -L_1(ix)$ and $L := L_1+iL_2$, and thus $L|M = l$. The additivity of the function L is plain; the $\boldsymbol{C}$-homogeneity is verified immediately. It remains to check whether $|L(x)| \leqslant p(x)$.

Indeed, setting $t = \arg L(x)$, we obtain

$$|L(x)| = e^{-ti}L(x) = L(e^{-ti}x) = L_1(e^{-ti}x),$$

since $\operatorname{Im} L(e^{-ti}x) = 0$, but

$$L_1(e^{-ti}x) \leqslant p(e^{-ti}x) = p(x), \quad \text{that is } |L(x)| \leqslant p(x). \ \square$$

The applications of the Hahn–Banach Theorem are far-flung: the theory of generalized limits, theory of the integral, and recently even quantum field theory. Chapters XVI and XIX give two important applications: the Krein–Milman Theorem and an elegant proof of the Runge Theorem. Here, we give a general form of a linear functional on $C([a, b])$. This is the famous Riesz Theorem which has played such a capital role in the development of theory of integral.

DEFINITION. The *variation of a function* $m\colon [a, b] \to \boldsymbol{R}$ is

$$\overset{b}{\underset{a}{V}} m := \sup \sum_{i=1}^{n} |m(x_i) - m(x_{i-1})|,$$

where sup is taken over all partitions $a = x_0 < x_1 < \ldots < x_n = b$ of the interval $[a, b]$. Clearly $\overset{b}{\underset{a}{V}} m$ could be $+\infty$. When the variation is finite, m is said to have a *bounded variation.*

It can be shown that every function with a bounded variation is the difference of two bounded isotonic functions, $m = m_1 - m_2$, where m_i are nondecreasing functions. As we know, the functions m_i define the Radon integral on $[a, b]$. The Stieltjes integral can be defined analogously,

$$C([a, b]) \ni f \to m(f) = \int_a^b f\,dm \in \boldsymbol{R},$$

for any function m of bounded variation. Of course, $m(f) = m_1(f) - m_2(f)$, and thus $f \to m(f)$ is a continuous linear functional on $C([a, b])$. The Riesz Theorem states that conversely: a linear functional continuous on $C([a, b])$ is of this form.

THEOREM XIII.15.7. (Riesz Theorem). Every functional $l \in C([a, b])'$ is representable in the form of a Stieltjes integral

$$l(f) = \int_a^b f(x)\,dm(x),$$

where

(*) $$\|l\| = \overset{b}{\underset{a}{V}}(m).$$

Thus the Banach space $C([a, b])'$ is the space of functions of bounded variation with the norm $\overset{b}{\underset{a}{V}}$.

PROOF. To simplify the notation we set $a = 0$, $b = 1$. By E let us denote the space of all bounded functions f: $[0, 1] \to K$, normed by

(1) $$\|f\| = \sup|f|([0, 1]).$$

Clearly, the space $C([0, 1])$ equipped with the norm (1) is a Banach space. We have $M = C([0, 1]) \subset E$. We extend the functional l to L while preserving the norm (cf. Corollary XIII.15.5) on the entire space E. Let $m(x) := L(g_x)$, where

$$g_x(y) = \begin{cases} 0 & \text{for } x < y \leqslant 1, \\ 1 & \text{for } 0 \leqslant y \leqslant x. \end{cases}$$

We shall show that $\overset{1}{\underset{0}{V}} m \leqslant \|L\|$. Let $0 = x_0 < x_1 < \ldots < x_n = 1$, $\varepsilon_i = \pm 1$.

We have

$$\sum_{i=1}^{n} |m(x_i) - m(x_i - 1)| = \sum_{i=1}^{n} \varepsilon_i \cdot (m(x_i) - m(x_{i-1}))$$

$$= \sum_{i=1}^{n} \varepsilon_i \cdot (L(g_{x_i}) - L(g_{x_{i-1}})) = L \sum_{i=1}^{n} \varepsilon_i \cdot (g_{x_i} - g_{x_{i-1}})$$

$$\leqslant \|L\| \left\| \sum_{i=1}^{n} \varepsilon_i \cdot (g_{x_i} - g_{x_{i-1}}) \right\| \leqslant \|L\|,$$

that is,

(2) $$\overset{1}{\underset{0}{V}}(m) \leqslant \|L\| = \|l\|.$$

Every function $f \in C([0, 1])$ can be uniformly approximated by step functions:

$$f = \lim_{n \to \infty} \sum_{k=1}^{n} f\left(\frac{k}{n}\right)(g_{k/n} - g_{(k-1)/n}).$$

Thus, by virtue of the linearity and continuity of L (on $\mathscr{E}$) we have

$$l(f) = L(f) = \lim_{n\to\infty} \sum_{k=1}^{n} f\left(\frac{k}{n}\right)(L(g_{k/n}) - L(g_{(k-1)/n}))$$

$$= \lim_{n\to\infty} \sum_{k=1}^{n} f\left(\frac{k}{n}\right)\left(m\left(\frac{k}{n}\right) - m\left(\frac{k-1}{n}\right)\right) = \int_0^1 f dm.$$

Moreover

(3) $$|l(f)| = \left|\int_0^1 |f|\, dm\right| \leqslant \|f\| \overset{1}{\underset{0}{V}}(m), \quad \text{that is, } \|l\| \leqslant \overset{1}{\underset{0}{V}}(m).$$

Associating (3) with (2), we obtain (*). □

16. HILBERT SPACES. THEOREM ON ORTHOGONAL DECOMPOSITION. THE FORM OF A LINEAR FUNCTIONAL

Hilbert spaces constitute undoubtedly the most important category of topological vector spaces. As we know, they comprise the Euclidean spaces $\boldsymbol{R}^n$ and the space $L^2(\mu)$. They also play a fundamental role in physics: it would be difficult to imagine quantum mechanics without the framework of Hilbert spaces. Since a somewhat more general concept than scalar product occurs in many cases, we shall first take up Hermitian forms.

DEFINITION. Let X be a vector space over $\boldsymbol{C}$. The function

$$X \times X \ni (x, y) \mapsto h(x, y) \in \boldsymbol{C}$$

is called a *Hermitian form* when

(i) $\bigwedge_{y} h(\cdot, y)$ is a linear form,

(ii) $\bigwedge_{x,y} h(x, y) = \overline{h(y, x)}$.

Remark. Conditions (i) and (ii) say that $h(x, \cdot)$ is an antilinear form, i.e.

$$h(x, \lambda y) = \bar{\lambda} \cdot h(x, y), \qquad \lambda \in \boldsymbol{C},\ x, y \in X.$$

DEFINITION. A Hermitian form is *nonnegative* when

(iii) $\bigwedge_{x} h(x, x) \geqslant 0$.

A Hermitian form is called a *scalar product* when it is nonnegative and

$$(h(x, x) = 0) \Rightarrow (x = 0).$$

THE SCHWARZ INEQUALITY. Let h be a Hermitian form that is nonnegative on X. Then

$$(1) \qquad |h(x, y)|^2 \leqslant h(x, x)h(y, y), \qquad x, y \in X.$$

PROOF. If $h(x, y) = 0$, then inequality (1) is proven by (iii). Thus, suppose that $h(x, y) \neq 0$. By properties (i)–(iii) we have

$$\begin{aligned} 0 &\leqslant h(x-ay, x-ay) \\ &= h(x, x) - \bar{a}h(x, y) - ah(y, x) + |a|^2 h(y, y) \end{aligned}$$

for any $a \in \mathbf{C}$. Setting $a = h(x, x)/h(y, x)$, we obtain

$$0 \leqslant h(x, x) - 2h(x, x) + (h(x, x))^2 h(y, y)/|h(y, x)|^2,$$

whereby we get the thesis. □

THE MINKOWSKI INEQUALITY. Let h be a nonnegative Hermitian form on X. Then

$$x \to p(x) := h(x, x)^{1/2} \quad \text{is a seminorm.}$$

PROOF. It follows from the Schwarz inequality that

$$\begin{aligned} p(x+y)^2 &= h(x+y, x+y) \\ &= h(x, x) + h(x, y) + h(y, x) + h(y, y) \leqslant (p(x) + p(y))^2. \end{aligned}$$

THE POLARIZATION FORMULA. The formula

$$\begin{aligned} 4h(x, y) &= h(x+y, x+y) - h(x-y, x-y) + \\ &+ ih(x+iy, x+iy) - ih(x-iy, x-iy) \end{aligned}$$

holds for any Hermitian form.

COROLLARY XIII.16.1. Let h be a Hermitian form on X. Let $N_0 = \{x \in X: h(x, x) = 0\}$ and let N be a linear subspace closed in N_0. Then $h(x, y) = 0$ for $x, y \in N$.

COROLLARY XIII.16.2. Let h be a nonnegative Hermitian form, $N = \{x \in X: h(x, x) = 0\}$. Then $N = \{x \in X: \bigwedge_{y \in X} h(x, y) = 0\}$.

THE CANONICAL CONSTRUCTION OF A HILBERT SPACE. Let h be a nonnegative Hermitian form on X. Then the quotient space X/N is in natural fashion a space with a scalar product (not only a normed space). Let $[x], [y] \in X/N$, that is $[x] = x+N$, $[y] = y+N$. Then

$$([x]|[y]) := h(x, y), \quad ||[x]|| := h(x, x)^{1/2} \text{ is a norm.}$$

Corollary XIII.16.2 states that the definition is correct and $(\cdot|\cdot)$ is a nonnegative Hermitian form. It is also a scalar product: for when $[x] \neq 0_{X/N}$, that is, $[x] = x+N$, where $x \notin N$, then $0 < h(x, x) = ([x]|[x])$. Now, completing the space $(X/N, (\cdot|\cdot))$ with respect to the norm $([x]|[x]) =: ||[x]||^2$, we obtain a complete space with scalar product, that is, a Hilbert space. This procedure, that of dividing by a seminorm p and completing the space, is called the *canonical construction of a Hilbert space*. We come across it in many areas of mathematics.

We shall now prove an important extremal property of convex sets in Hilbert space.

THEOREM XIII.16.3. Let W be a closed convex set in a Hilbert space $(H, (\cdot|\cdot))$. Then W contains exactly one element x_0 with the smallest norm.

To be more precise, there exists a single element $x_0 \in W$ such that $||x_0|| = \inf_{x \in W} ||x||$.

PROOF. *Existence*: Let $m = \inf||W||$ and let $W \ni x_n$, $||x_n|| \underset{n\to\infty}{\to} m$. The convexity of the set W implies that

$$\tfrac{1}{2}(x_n + x_k) \in W, \quad \text{and thus} \quad ||x_n + x_k|| \geqslant 2m.$$

From the parallelogram equality it follows that

$$0 \leqslant ||x_n - x_k||^2 = 2(||x_n||^2 + ||x_k||^2) - ||x_n + x_k||^2 \to 0,$$
$$\text{as } n, k \to \infty$$

for the minuend tends to $4m^2$, and the subtrahend $\geqslant 4m^2$. Thus, the sequence (x_n) is convergent. Since the set W is closed, $x_n \to x_0 \in W$. Owing to the continuity of the norm, we have

$$||x_0|| = \lim||x_n|| = m.$$

Uniqueness: Let x be any element of the set W such that $||x|| = m$. Again from the convexity of W we have: $\frac{1}{2}(x_0 + x) \in W$, that is, $||x_0 + x||$

$\geqslant 2m$. But $2m = ||x_0|| + ||x|| \geqslant ||x_0 + x||$, that is, $||x_0|| + ||x|| = ||x_0 + x||$. As we know, however, this is the case only when $x = ax_0$, where $a \geqslant 0$. Thus $a = 1$, and hence $x = x_0$. □

The Topology of the Quotient Space E/M

Definition. Let E be a topological vector space, and let M be a linear subspace of E. Let k: $E \to E/M$ be a canonical mapping. Let $(\mathcal{O}_\alpha)$ be a basis of the neighbourhoods of zero of the space E, then $(k(\mathcal{O}_\alpha))$ is, by definition, a *neighbourhood basis of zero of* E/M. This is clearly the strongest topology for which k is continuous.

Proposition. (E/M is a Hausdorff space) $\Leftrightarrow$ (M is closed).

Proof. $\Rightarrow$: When E/M is a Hausdorff space, the set $\{0_{E/M}\}$ is closed, and thus $k^{-1}(0_{E/M}) = M$ is closed, for k is continuous.

$\Leftarrow$: This is left for the reader to prove. □

Theorem XIII.16.4. Let E and F be topological vector spaces, let M be a vector space $\subset E$, and let the mapping L: $E \to F$ be linear, where $L^{-1}(0_F) \supset M$. Let k be the canonical mapping $E \to E/M$ and let T: $E/M \to F$, where $T(e+M) := L(e)$. Thus $L = T \circ k$, that is

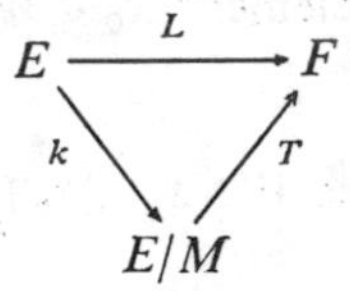

Then:

(i) (the mapping L is continuous) $\Leftrightarrow$ (the mapping T is continuous).

(ii) Let $M := L^{-1}(0_F)$. Then T is a bijection which is continuous if and only if L is continuous.

Proof. Clearly, it is sufficient to prove (i). Let U be an (open) neighbourhood in F; thus the mapping L is continuous if and only if

$$L^{-1}(U) = (T \circ k)^{-1}(U) = k^{-1}(T^{-1}(U))$$

is a neighbourhood in E, that is, if and only if $T^{-1}(U)$ is open in E/M (cf. the definition of the topology E/M), that is, if T is continuous. □

Corollary XIII.16.5. Let $l \in E'$, then $\dim E/l^{-1}(0) = 1$.

This is because $E/l^{-1}(0)$ is isomorphic with $\mathbf{C}$.

Now we shall prove two fundamental theorems of Hilbert space theory.

THEOREM XIII.16.6. (On Orthogonal Projection). Let M be a closed subspace of the Hilbert space $(H, (\cdot|\cdot))$. Then

(i) Every vector $x \in H$ has a single decomposition $x = x_M + x^{\perp}$ where $x_M \in M$, $x^{\perp}$ is orthogonal to M (that is, $(x^{\perp}|h) \equiv 0$ for $h \in M$).

(ii) Of the vectors belonging to M, the vector x_M best approximates x.

(iii) Denoting $M^{\perp} = \{x \in H: (x|m) = 0, m \in M\}$, we have $H = M \oplus \oplus M^{\perp}$, that is, $H/M \cong M^{\perp}$.

PROOF. Since the set $M+x$ is convex and closed, there exists a vector $x^{\perp} \in M+x$ such that $\|x^{\perp}\| = \inf_{v \in M} \|y+x\|$. We shall prove that $x^{\perp}$ is orthogonal to M. Since $\|x^{\perp}+av\| \geqslant \|x^{\perp}\|$ for any $a \in C, v \in M$, therefore

$$\|x^{\perp}\|^2 \leqslant \|x^{\perp}\|^2 + \bar{a}(x^{\perp}|v) + a(v|x^{\perp}) + |a|^2\|v\|^2.$$

Setting $-a = (x^{\perp}|v)/\|v\|^2$, we obtain $-|(x^{\perp}|v)|^2 \geqslant 0$, that is, $x^{\perp} \perp M$. Hence the conclusion that $H/M \cong M^{\perp}$. □

As follows from the Schwarz inequality, every vector $y \in H$ determines a continuous linear form $l_y(x) := (x|y)$. The famous Fréchet–Riesz Theorem states that every linear functional $l \in H'$ is of such a form.

THEOREM XIII.16.7. There exists a unique vector $y \in H$ such that $l(x) = (x|y)$ identically for $x \in H$.

PROOF. When $l = 0$, we take $y = 0$. Let $l \neq 0$; we set $M = l^{-1}(0)$. Since l is continuous, M is a closed subspace. Since $\dim(H/M) = 1$, we have $\dim M^{\perp} = 1$, that is, $H = l^{-1}(0) \oplus \{ah: a \in C\}$, where h is a fixed vector orthogonal to M, that is

$$x = x_M + ah, \quad \text{where } 0 \neq h \perp M.$$

Taking the scalar product with h, we obtain

$$(x|h) = a\|h\|^2, \quad \text{and thus} \quad x = x_M + \frac{(x|h)}{\|h\|^2} h.$$

Therefore,

$$l(x) = \frac{(x|h)}{\|h\|^2} l(h).$$

Let $y := \dfrac{\overline{l(h)}}{\|h\|^2} h$, then $l(x) = (x|y)$. □

Hermitian Operators and Hermitian Forms on Hilbert Space
Let $(H, (\cdot\,|\,\cdot))$ be a Hilbert space and let h be a bounded sesquilinear form on H, that is, let there exist a $c > 0$ such that

$$|h(x, y)| \leqslant c||x|| \cdot ||y||.$$

This form determines mappings $A, B \in L(H, H)$, such that

$$h(x, y) = (Ax|y) = (x|By) \quad \text{identically for } x, y \in H.$$

Indeed, we have $h(x, y) = (x|l_y)$, since the mapping $B\colon y \to l_y \in H$ is linear, we can write $h(x, y) = (x|By)$. Since every antilinear functional $y \to f(y)$ is of the form $(x|y)$, then, reasoning as a moment ago, we obtain the first representation of the form h: $h(x, y) = (Ax|y)$. Clearly, $||A||$, $||B|| \leqslant C$.

The operators A and B are said to be *adjoint* or *conjugate to each other* and this is denoted by $B = A^*$ or $A = B^*$. When the form h is Hermitian,

$$(Ax|y) = h(x, y) = \overline{h(x, y)} = \overline{(y, A^*x)} = (A^*x|y),$$
$$x, y \in H,$$

and therefore $Ax = A^*x$ identically for $x \in H$; hence $A = A^*$.

DEFINITION. An operator $A \in L(H)$ for which $A = A^*$ is called *Hermitian.*

Thus we have obtained an important corollary:

COROLLARY XIII.6.8. Every bounded Hermitian form h determines a Hermitian operator A such that $h(x, y) = (Ax|y)$ and, conversely, every Hermitian operator determines a bounded Hermitian form.

17. THE STRONG STONE AXIOM AND ITS CONSEQUENCES

In this section we shall assume that the strong Stone axiom is satisfied (cf. Section 13).

Let S denote the vector space of functions which are finite linear combinations of characteristic functions of integrable sets, that is,

$$(f \in S(\mu)) \Leftrightarrow \Big(f = \sum_{i=1}^{n} a_i \chi_{Z_i}, \text{ where } \chi_{Z_i} \in \mathscr{L}^1(\mu) \text{ and}$$
$$Z_i \cap Z_j = \emptyset \text{ for } i \neq j\Big).$$

It is seen that S is a lattice. Plainly $S \in \mathscr{L}^1$.

An important property of S is expressed by the following lemma.

LEMMA XIII.17.1. We have

$$\bar{S} = \mathscr{L}^1 \text{ (closure in the sense of the seminorm } N_1).$$

PROOF. It follows from Theorem XIII.13.2 that every function $f \in \mathscr{L}^1$ can be approximated in the N_1 sense by functions f_ε, where

$$f_\varepsilon = \sum_{m=0}^{\infty} m\varepsilon(\chi_{A\varepsilon m} - \chi_{A\varepsilon(m+1)}) + \sum_{m=0}^{-\infty} m\varepsilon(\chi_{A\varepsilon(m-1)} - \chi_{A\varepsilon m}).$$

We define

$$f_\varepsilon^k = \sum_{m=0}^{k} m\varepsilon(\chi_{A\varepsilon m} - \chi_{A\varepsilon(m+1)}) + \sum_{m=0}^{-k} m\varepsilon(\chi_{A\varepsilon(m-1)} - \chi_{A\varepsilon m}).$$

Note that $f_\varepsilon^k \underset{k\to\infty}{\to} f_\varepsilon$ pointwise and that $|f_\varepsilon^k| \leqslant f_\varepsilon$. Thus (by the Lebesgue Theorem) $f_\varepsilon^k \underset{k\to\infty}{\to} f_\varepsilon$ in thesense of the seminorm N_1. Since $f_\varepsilon^k \in S$, hence $f_\varepsilon \in \bar{S}$ for every ε. Therefore $f = \lim_{\varepsilon\to 0} f_\varepsilon \in \bar{\bar{S}} = \bar{S}$. □

Since S is a vector lattice, the question arises: if S were taken instead of E as the family of elementary functions, would the integral theory obtained differ from that obtained from E? The answer is provided by the following theorem.

THEOREM XIII.17.2 (Fundamental Theorem). (i) $S \subset \mathscr{F}^p$ for every $p \geqslant 1$, (ii) $\bar{S} = \mathscr{L}^p$ (closure in the sense of N_p).

PROOF. (i) Let $f \in S, f = \sum_{i=1}^{n} a_i \chi_{Z_i}$. But $|\chi_{Z_i}|^p = \chi_{Z_i}$, and therefore

$$[N_p(\chi_{Z_i})]^p = N_1(\chi_{Z_i}) < \infty, \quad \text{that is,} \quad \chi_{Z_i} \in \mathscr{F}^p.$$

Therefore $f \in \mathscr{F}^p$ as a linear combination of functions from $\mathscr{F}^p$.

(ii) First of all we prove that $S \in \mathscr{L}^p$. Take an arbitrary function χ_Z where Z is an integrable set. Since $\chi_Z \in \mathscr{L}^1$, there exists a sequence $E \ni |h_n| \to \chi_Z$ in the N_1 sense.

From the strong Stone axiom we have $g_n := (|h_n| \cap 1) \in E$. But

$$|g_n - \chi_Z|^p \leqslant |g_n - \chi_Z| \leqslant ||h_n| - \chi_Z|$$

(the first inequality follows from the fact that $|g_n(x) - \chi_Z(x)| \leqslant 1$, whereas $p \geqslant 1$). Hence

$$[N_p(g_n - \chi_Z)]^p = N_1(|g_n - \chi_Z|^p) \leqslant N_1(|h_n| - \chi_Z) \underset{n\to\infty}{\to} 0.$$

Finally, therefore, $E \ni g_n \to \chi_Z$ in the N_p sense, that is, $\chi_Z \in \mathscr{L}^p$.

Thus, if all χ_Z for integrable Z belong to $\mathscr{L}^p$, the functions from S, being linear combinations of them, also belong to $\mathscr{L}^p$.

But if $S \subset \mathscr{L}^p$, then also $\bar{S} \subset \mathscr{L}^p$. It is, therefore, sufficient to prove that $\mathscr{L}^p \subset \bar{S}$.

Suppose that $f \in \mathscr{L}^p$. By virtue of Lemma XIII.11.2 we find that $f \in M$. Thus, by Theorem XIII.13.1 the sets A_a are measurable. They are also integrable since

$$N_1(\chi_{A_a}) = [N_p(\chi_{A_a})]^p = \left[\frac{1}{|a|} N_p(|a| \cdot \chi_{A_a})\right]^p$$

$$\leqslant \left[\frac{1}{|a|} N_p(|f|)\right]^p < \infty,$$

and hence it follows from Theorem XIII.11.10 that $\chi_{A_a} \in \mathscr{F}^1 \cap M = \mathscr{L}^1$.

Accordingly, the functions

$$f_\varepsilon^k = \sum_{m=0}^{k} m\varepsilon(\chi_{A\varepsilon m} - \chi_{A\varepsilon(m+1)}) + \sum_{m=0}^{-k} m\varepsilon(\chi_{A\varepsilon(m-1)} - \chi_{A\varepsilon m})$$

are functions from the space S. By the first part of the proof we find that $f_\varepsilon^k \in \mathscr{L}^p$. But $f_\varepsilon^k \underset{k \to \infty}{\to} f_\varepsilon$ pointwise and $|f_\varepsilon^k| \leqslant |f| \in \mathscr{L}^p$. Therefore, it follows from the Lebesgue Theorem XIII.13.7 that $f_\varepsilon \in \mathscr{L}^p$ and $f_\varepsilon \in \bar{S}$. But $f_\varepsilon \underset{\varepsilon \to 0}{\to} f$ pointwise and $|f_\varepsilon| \leqslant |f| \in \mathscr{L}^p$. Hence, by virtue of the Lebesgue Theorem, $f_\varepsilon \to f$ in the N_p sense, that is, $f \in \bar{S}$. □

It is thus seen that in the case of a theory with the strong Stone axiom satisfied, functions from the space S may be taken as the elementary functions and the spaces $\mathscr{L}^p$ (as well as M) will not suffer any change. Of course, the functional $\mu|_S$ is the Daniell–Stone integral. Its monotonic continuity follows immediately from the Levi Lemma (Lemma XIII.8.2).

The structure of the space $\mathscr{L}^p$ thus is not related to the arbitrariness in the choice of elementary functions, from which we begin the extension of the functional μ, but depends only on the properties of μ itself. Hence, if a σ-algebra of measurable sets and a measure of μ were given *a priori* in the space X, the entire theory of integrals could be reproduced through the space S. The integral would, of course, be given by the Lebesgue sums:

$$\int f = \lim_{\varepsilon \to 0} S(f, \varepsilon, \mu).$$

Note further how lucid the relationship between the spaces $\mathscr{L}^p$ and $\mathscr{F}^p$ becomes in the case when the strong Stone axiom is satisfied:

THEOREM XIII.17.3. We have

$$\mathscr{F}^p \cap M = \mathscr{L}^p.$$

(For $p = 1$ the same result was proved without the Stone axiom (Theorem XIII.11.10).)

PROOF. (i) $\mathscr{L}^p \subset \mathscr{F}^p$. By virtue of Lemma XIII.11.2, $\mathscr{L}^p \subset M$ and hence $\mathscr{L}^p \subset \mathscr{F}^p \cap M$.

(ii) Let $f \in \mathscr{F}^p \cap M$. It follows from Theorem XIII.13.1 that the sets A_a are measurable. They are also integrable since

$$N_1(\chi_{A_a}) = [N_p(\chi_{A_a})]^p = \left[\frac{1}{|a|} N_p(|a| \cdot \chi_{A_a})\right]^p$$

$$\leqslant \left[\frac{1}{|a|} N_p(f)\right]^p < \infty.$$

Further on the proof is identical with that in Theorem XIII.17.2, point (ii), viz.:

$$\mathscr{L}^p \ni f_\varepsilon^k \underset{k\to\infty}{\to} f_\varepsilon \in \mathscr{L}^p, \quad f_\varepsilon \underset{\varepsilon\to 0}{\to} f \in \mathscr{L}^p. \ \square$$

Still one more important property of the space $\mathscr{L}^p$ follows from the strong Stone axiom:

THEOREM XIII.17.4. Given a function $f \geqslant 0$. Then

$$(f \in \mathscr{L}^p) \Leftrightarrow (f^p \in \mathscr{L}^1).$$

PROOF. By Theorem XIII.13.6 the theorem $\Rightarrow$ is true.

$\Leftarrow$: Since S, being the family of elementary functions, is closed with respect to exponentiation, the thesis follows immediately from Theorem XIII.7.4 $\square$

The theorem above justifies the name frequently used for the space $\mathscr{L}^p$: "the space of functions with integrable p-th power".

18. THE TENSOR PRODUCT OF INTEGRALS

Let (X_1, μ_1, E_1), (X_2, μ_2, E_2) be two models of integral theory on two spaces X_1 and X_2. Suppose that the Stone axiom is satisfied. Take $E_1 = S(\mu_1)$, $E_2 = S(\mu_2)$ as the elementary function (the definition of the space S was given in the preceding section). Now we give the construction of a certain integral on the space $X_1 \times X_2$.

Let

$$E = E_1 \otimes E_2 := \Big\{f: f(x_1, x_2) = \sum_{i=1}^{n} f_i^1(x_1)\cdot f_i^2(x_2),$$

$$f_i^k \in E_k,\ k=1,2\Big\}.$$

We shall also use the notation

$$(f^1 \otimes f^2)(x_1, x_2) := f^1(x_1)\cdot f^2(x_2).$$

The space E will be called the *tensor product* of the spaces E_1 and E_2. Note that if $f \in E$, then

$$f = \sum_{i=1}^{n}\Big(\sum_{k=1}^{s} a_k^i \chi_{A_k^i}\Big)\cdot\Big(\sum_{p=1}^{t} b_p^i \chi_{B_p^i}\Big),$$

where the integrable sets: $A_k^i \subset X_1$, $B_p^i \subset X_2$ satisfy the conditions

$$A_k^i \cap A_h^i = \emptyset \text{ for } k \neq h, \quad B_r^i \cap B_p^i = \emptyset \text{ for } r \neq p.$$

Suppose that $A_k^i \cap A_h^j = \emptyset$ for $h \neq k$ and $B_p^i \cap B_r^j = \emptyset$ for $p \neq r$. If that were not the case, the set A_k^i would be decomposable into two sets $(A_k^i \cap A_h^j)$ and $A_k^i - (A_k^i \cap A_h^j)$ and our requirement would be met.

Thus we may write

$$f = \sum_{i=1}^{n}\sum_{k=1}^{s}\sum_{p=1}^{t} a_k^i b_p^i \chi_{A_k^i}\chi_{B_p^i}.$$

Since $\chi_A(x_1)\cdot\chi_B(x_2) = \chi_{A\times B}(x_1, x_2)$, then

$$f = \sum_{\alpha=1}^{N} c_\alpha \chi_{C_\alpha},$$

where $c_\alpha = a_\alpha \cdot b_\alpha \in \boldsymbol{R}$, $C_\alpha = A_\alpha \times B_\alpha \subset X_1 \times X_2$ and $C_\alpha \cap C_\beta = \emptyset$ for $\alpha \neq \beta$.

Hence, we have found that functions from E are finite linear combinations of the characteristic functions of the sets in $X_1 \times X_2$, being Cartesian products of integrable sets in X_1 and in X_2.

Clearly, E is a lattice: $|f| = \sum_{\alpha=1}^{N} |c_\alpha|\chi_{C_\alpha} \in E$.

Now we define the functional μ as follows:

$$\mu\Big(\sum_{i=1}^{n} f_i^1 \otimes f_i^2\Big) := \sum_{i=1}^{n} \mu_1(f_i^1)\mu_2(f_i^2).$$

Note that

$$\mu_1\Big(\sum_{i=1}^{n} f_i^1(\cdot) f_i^2(x_2)\Big) = \sum_{i=1}^{n} \mu_1(f_i^1)\cdot f_i^2(x_2) \in E_2$$

and similarly

$$\mu_2\Big(\sum_{i=1}^{n} f_i^1(x_1)\cdot f_i^2(\cdot)\Big) = \sum_{i=1}^{n} f_i^1(x_1)\cdot \mu_2(f_i^2) \in E_1.$$

The expressions $\mu_1(\mu_2(f))$, $\mu_2(\mu_1(f))$ thus are meaningful. Plainly, the linearity of the two integrals μ_1 and μ_2 implies that

$$\mu(f) = \mu_1(\mu_2(f)) = \mu_2(\mu_1(f)).$$

The linearity and positivity of the functional μ are obvious. Monotonic continuity follows from the fact that when $f_n \searrow 0$, then also $\mu_2(f_n) \searrow 0$, which entails the convergence of $\mu_1(\mu_2(f_n))$ to zero.

The functional μ thus is a Daniell–Stone integral with the family of elementary functions $E_1 \otimes E_2$. We denote

$$\mu = \mu_1 \otimes \mu_2$$

(μ is the tensor product of the integrals μ_1 and μ_2). Hence we have the formula

$$(\mu_1 \otimes \mu_2)(f_1 \otimes f_2) = \mu_1(f_1)\cdot \mu_2(f_2).$$

Having μ and E, we can further develop the theory of integrals onto $X_1 \times X_2$. Note that the procedure which we have described is equivalent to the following:

In the set $X_1 \times X_2$ subsets of the form $A_1 \times A_2$ (where $A_k \subset X_k$, $k = 1, 2$; A_k are integrable sets) are called *integrable product sets*. We define the measure μ:

$$\mu(A_1 \times A_2) = \mu_1(A_1)\cdot \mu_2(A_2).$$

We take the σ-field of sets generated by integrable product sets (i.e. all sets formed from product sets by taking intersections, countable unions, and differences). Since the measure μ is σ-additive, we extend it to the σ-field, taking

$$\mu\left(\bigcup_{i=1}^{\infty} A_i\right) := \sum_{i=1}^{\infty} \mu(A_i),$$

when $A_i \cap A_j = \emptyset$ for $i \neq j$.

Now that we have a measure μ, we can develop the integral theory in the manner described in the preceding section.

Everything that has been said thus far concerning the measure $\mu_1 \otimes \mu_2$ will be assembled in the following theorem.

THEOREM XIII.18.1 (Small Fubini Theorem). If (X_i, μ_i, E_i), $E = S(\mu_i)$, $i = 1, 2$, are two Daniell–Stone integrals, there exists an integral $(X_1 \times X_2, \mu, E)$ such that

1. $\bigwedge_{f \in E} \bigwedge_{x_1 \in X_1} f(x, \cdot) \in E_2$,

1′. $\bigwedge_{f \in E} \bigwedge_{x_2 \in X_2} f(\cdot, x_2) \in E_1$,

2. $\bigwedge_{f \in E} \big(g(x_1) := \mu_2(f(x_1, \cdot))\big) \Rightarrow (g \in E_1)$,

2′. $\bigwedge_{f \in E} \big(h(x_2) := \mu_1(f(\cdot, x_2))\big) \Rightarrow (h \in E_2)$,

3. $\bigwedge_{f \in E} \mu(f) = \mu_1\big(\mu_2(f)\big) = \mu_2\big(\mu_1(f)\big)$.[1]

It transpires that also of interest are theories wherein the condition $E_i = S(\mu_i)$ is not satisfied, but the thesis of the small Fubini Theorem is. In such cases we shall write "the s.F. holds".

LEMMA XIII.18.2. Suppose that the s.F. holds. Then the inequalities

(i) $\mu_1^*\big(\mu_2^*(g)\big) \leqslant \mu^*(g)$,

(ii) $\mu_2^*\big(\mu_1^*(g)\big) \leqslant \mu^*(g)$

hold for every numerical function $g \geqslant 0$ on the space $X_1 \times X_2$.

PROOF. (i) If $\mu^*(g) = +\infty$, the inequality is satisfied trivially. Let $\mu^*(g) < +\infty$. Hence, for any $\varepsilon > 0$ there exists a function $f_\varepsilon \in \mathscr{E}(\mu)$ such that $f_\varepsilon \geqslant g$ and $\mu^*(g) \leqslant \tilde{\mu}(f_\varepsilon) \leqslant \mu^*(g)+\varepsilon$.

But

$$f_\varepsilon = \sup_n f_\varepsilon^n, \qquad \text{where } f_\varepsilon^n \in E \text{ and } \tilde{\mu}(f_\varepsilon) = \sup_n \mu(f_\varepsilon^n).$$

[1] Points 2 and 2′ explain how the iterated integrals $\mu_1(\mu_2(f))$ and $\mu_2(\mu_1(f))$ should be understood.

Therefore

$$\tilde{\mu}(f_\varepsilon) = \sup_n \mu_1(\mu_2(f_\varepsilon^n)) = \tilde{\mu}_1(\sup_n \mu_2(f_\varepsilon^n)) = \tilde{\mu}_1(\tilde{\mu}_2(\sup f_\varepsilon^n))$$
$$= \tilde{\mu}_1(\tilde{\mu}_2(f_\varepsilon)) \leqslant \mu^*(g)+\varepsilon.$$

We have thus proved that $f_\varepsilon(x_1, \cdot) \in \mathscr{E}(\mu_2)$ and $\tilde{\mu}_2(f_\varepsilon) \in \mathscr{E}(\mu_1)$ as the monotonic limit of the elementary functions. Therefore

$$\mu_1^*(\mu_2^*(g)) \leqslant \inf \tilde{\mu}_1(\tilde{\mu}_2(f_\varepsilon)) \leqslant \mu^*(g).$$

(ii) Identically as (i). □

By $\mathscr{L}(\mu_1, \mu_2)$ we now denote the set of functions defined on the space $X_1 \times X_2$ and possessing the following properties:

For every function $f \in \mathscr{L}(\mu_1, \mu_2)$

(i) $\bigvee_{A \subset X_1} \mu_1^*(A) = 0, \ \bigwedge_{x_1 \notin A} f(x_1, \cdot) \in \mathscr{L}^1(\mu_2);$

(i)′ $\bigvee_{B \subset X_2} \mu_2^*(B) = 0, \ \bigwedge_{x_2 \notin B} f(\cdot, x_2) \in \mathscr{L}^1(\mu_1);$

(ii) $g := \mu_2(f) \in \mathscr{L}^1(\mu_1);$

(ii)′ $h := \mu_1(f) \in \mathscr{L}^1(\mu_2).$

A theorem of fundamental significance thus holds:

THEOREM XIII.18.3 (Fubini, Stone). Suppose that the s.F. holds for the measures (X_1, μ_1, E_1), (X_2, μ_2, E_2), $(X_1 \times X_2, \mu, E)$ (e.g. when $\mu = \mu_1 \otimes \mu_2$). Then:

(i) $\mathscr{L}^1(\mu) \subset \mathscr{L}(\mu_1, \mu_2)$.

(ii) The equality

$$\mu(f) = \mu_1(\mu_2(f)) = \mu_2(\mu_1(f))$$

holds for every function $f \in \mathscr{L}^1(\mu)$.

PROOF. (i) Let $f \in \mathscr{L}^1(\mu)$. Then, f can be approximated by the elementary functions $f_n \in E$: $\mu^*(|f-f_n|) \to 0$.

We select a subsequence f_{n_k} such that

$$\mu^*(|f-f_{n_k}|) < \frac{1}{2^k}.$$

Now we introduce the following function:

$$h(x_1) := \sum_{k=1}^{\infty} \mu_2^*(|(f-f_{n_k})(x_1, \cdot)|).$$

As μ_1 is countably subadditive (Corollary XIII.3.3), we have

$$\mu_1^*(h) \leqslant \sum_{k=1}^{\infty} \mu_1^*\big(\mu_2^*(|f-f_{n_k}|)\big) \leqslant \sum_{k=1}^{\infty} \mu^*(|f-f_{n_k}|)$$

$$\leqslant \sum_{k=1}^{\infty} \frac{1}{2^k} = 1.$$

The second of these inequalities is implied by Lemma XIII.18.2.

Hence, it follows from Theorem XIII.4.1 that h is finite μ_1-almost everywhere, which means that the series

$$\sum_{k=1}^{\infty} \mu_2^*\big(|(f-f_{n_k})(x_1, \cdot)|\big)$$

is convergent for all $x_1 \notin A$, where $\mu_1^*(A) = 0$. Thus, for $x_1 \notin A$ we have the convergence

$$\mu_2^*\big(|(f-f_{n_k})(x_1, \cdot)|\big) \to 0.$$

But $f_n(x_1, \cdot) \in E_2$, and thus $f(x_1, \cdot) \in \mathscr{L}^1(\mu_2)$ for $x_1 \notin A$, as the limit of elementary functions. Since this is so, we can define the function

$$g(x_1) := \mu_2\big(f(x_1, \cdot)\big).$$

Since the s.F. holds, $\mu_2\big(f_{n_k}(x_1, \cdot)\big) \in E_1$. But

$$\mu_1^*\big(|g-\mu_2(f_{n_k})|\big) = \mu_1^*\big(|\mu_2(f)-\mu_2(f_{n_k})|\big) = \mu_1^*\big(|\mu_2(f-f_{n_k})|\big)$$
$$\leqslant \mu_1^*\big(\mu_2(|f-f_{n_k}|)\big)$$
$$= \mu_1^*\big(\mu_2^*(|f-f_{n_k}|)\big) \leqslant \mu^*(|f-f_{n_k}|) < \frac{1}{2^k} \to 0,$$

and hence g is the μ_1^*-limit of the elementary functions, that is, $g \in \mathscr{L}^1(\mu_1)$.

We have already proved points (i) and (ii) of the definition of the space $\mathscr{L}(\mu_1, \mu_2)$. The other points will be proved in identical manner, interchanging μ_1 and μ_2.

(ii) We have

$$\mu_1\big(\mu_2(f)\big) = \mu_1(g) = \lim_{k\to\infty} \mu_1\big(\mu_2(f_{n_k})\big) = \lim_{k\to\infty} \mu(f_{n_k}) = \mu(f).$$

The last equality but one follows from the fact that $f_n \in E$ and from the fact that s.F. holds.

In the same way we demonstrate that $\mu_2\big(\mu_1(f)\big) = \mu(f)$. □

As it turns out, the converse theorem is not true. If $f \in \mathscr{L}(\mu_1, \mu_2)$, this function need not be integrable in the sense of the measure μ. For it may happen that

$$\mu_1(\mu_2(f)) \neq \mu_2(\mu_1(f)).$$

Example. Take two Lebesgue integrals μ_1, μ_2 over the interval $]0, 1[$. Let $\mu = \mu_1 \otimes \mu_2$. Then the function

$$f(x, y) := \frac{x^2 - y^2}{(x^2 + y^2)^2} \quad \text{belongs to } \mathscr{L}(\mu_1, \mu_2).$$

1. PROOF. For every $x \in]0, 1[$ we have $f(x, \cdot) \subset C([0, 1])$, and thus this function is μ_2-integrable.

1′. For every $y \in]0, 1[$ we have $f(\cdot, y) \subset C([0, 1])$ and this function is μ_1-integrable.

2. We have

$$\mu_1(f) = \int_0^1 \frac{x^2 - y^2}{(x^2 + y^2)^2}\, dx = \frac{1}{y}\left[\int_0^{1/y} \frac{t^2 dt}{(1+t^2)^2} - \right.$$

$$\left. - \int_0^{1/y} \frac{dt}{(1+t^2)^2}\right] = \frac{1}{y}\left[2\int_0^{1/y} \frac{t^2 dt}{(1+t^2)^2} - \int_0^{1/y} \frac{dt}{1+t^2}\right]$$

$$= \frac{1}{y}\left[-\frac{t}{1+t^2}\Big|_0^{1/y} + \int_0^{1/y} \frac{dt}{1+t^2} - \int_0^{1/y} \frac{dt}{1+t^2}\right]$$

$$= -\frac{1}{y} \cdot \frac{1/y}{1+1/y^2} = -\frac{1}{1+y^2} \in \mathscr{L}^1(\mu_2).$$

2′. Identically we have

$$\mu_2(f) = \int_0^1 \frac{x^2 - y^2}{(x^2 + y^2)^2}\, dy = \frac{1}{1+x^2} \in \mathscr{L}^1(\mu_2).$$

Meanwhile

$$\mu_2(\mu_1(f)) = -\int_0^1 \frac{1}{1+y^2}\, dy = -\arctan 1 = -\frac{1}{4}\pi,$$

$$\mu_1(\mu_2(f)) = \int_0^1 \frac{1}{1+x^2}\, dx = \arctan 1 = \frac{1}{4}\pi.$$

Thus $\mu_2(\mu_1(f)) \neq \mu_1(\mu_2(f))$. Consequently $f \notin \mathscr{L}^1(\mu)$, for if it were that $f \in \mathscr{L}^1$, then the Fubini Theorem would hold.

It might seem that the equality of iterated integrals,

$$\mu_2(\mu_1(f)) = \mu_1(\mu_2(f))$$

is ensured just by the fact that $f \in \mathscr{L}^1(\mu)$. It turns out that this, too, is not true:

Example. Let us take two integrals:

$$\mu_1\text{—Lebesgue integral on } X_1 =]-1, 0[\cup]0, 1[\subset \boldsymbol{R}^1,$$
$$\mu_2\text{—Lebesgue integral on } X_2 =]-1, 0[\cup]0, 1[\subset \boldsymbol{R}^1.$$

Let

$$\mu = \mu_1 \otimes \mu_2 .$$

Plainly, the sets $A_1 =]-1, 0[$ and $A_2 =]0, 1[$ are integrable. Thus, the sets $A_1 \times A_1 \subset X_1 \times X_2$ and $A_2 \times A_2 \subset X_1 \times X_2$ are integrable (product sets).

Let us take the function

$$f(x, y) := \begin{cases} \dfrac{x^2-y^2}{(x^2+y^2)^2} & \text{for } (x, y) \subset A_1 \times A_1, \\ \dfrac{y^2-x^2}{(x^2+y^2)^2} & \text{for } (x, y) \subset A_2 \times A_2, \\ 0 & \text{elsewhere.} \end{cases}$$

In the same fashion as before, it may be proved that $f \in \mathscr{L}(\mu_1, \mu_2)$.

But

$$\mu_1(\mu_2(f)) = \int_0^1 \left\{ \int_0^1 \frac{x^2-y^2}{(x^2+y^2)^2}\, dy \right\} dx +$$

$$+ \int_{-1}^0 \left\{ \int_{-1}^0 \frac{y^2-x^2}{(x^2+y^2)^2}\, dy \right\} dx = \frac{1}{4}\pi - \frac{1}{4}\pi = 0,$$

$$\mu_2(\mu_1(f)) = \int_0^1 \left\{ \int_0^1 \frac{x^2-y^2}{(x^2+y^2)^2}\, dx \right\} dy +$$

$$+ \int_{-1}^0 \left\{ \int_{-1}^0 \frac{y^2-x^2}{(x^2+y^2)^2}\, dx \right\} dy = \frac{1}{4}\pi - \frac{1}{4}\pi = 0.$$

Meanwhile the function f is not integrable, for if it were, then $f\cdot\chi_{A_1\times A_1} = \frac{x^2-y^2}{(x^2+y^2)^2}$ would be integrable (by virtue of Corollary XIII.12.3 and, as we know, this is not true.

It turns out that in the general case even the fact that the function f is measurable and that the iterated integrals $\mu_1(\mu_2(|f|))$ and $\mu_2(\mu_1(|f|))$ exist does not ensure the integrability of the function f (i.e. it need not be that $\mu^*(f) < \infty$).

Such a theorem is valid, however, in certain cases, viz. when both measures are so-called semifinite.

DEFINITION. Let (X, μ) be a Daniell–Stone integral. The set X is said to be of *semifinite measure* if there exists a sequence of integrable sets $\{\Omega_n\}_1^\infty$, $\Omega_n \subset X$, $\chi_{\Omega_n} \subset \mathscr{L}^1(\mu)$ such that $X = \bigcup_{n=1}^{\infty} \Omega_n$.

THEOREM XIII.18.4 (Tonelli). Let (X_1, μ_1, E_1), (X_2, μ_2, E_2), $(X_1\times X_2, \mu, E)$ satisfy the s.F. (e.g. let $\mu = \mu_1 \otimes \mu_2$). Let $X_1\times X_2$ be of semifinite μ-measure. Then every function μ-measurable on $X_1\times X_2$, for which at least one of the integrals $\mu_1^*(\mu_2^*(|f|))$ for $\mu_2^*(\mu_1^*(|f|))$ exists and is finite, belongs to $\mathscr{L}^1(\mu)$, that is, all three of its integrals, $\mu_1(\mu_2(f))$, $\mu_2(\mu_1(f))$, $\mu(f)$, are defined and are equal.

PROOF. Let $\mu_1^*(\mu_2^*(|f|)) = c < \infty$. Let $\bigcup_{n=1}^{\infty} \Omega_n = X_1\times X_2$, where Ω_n is integrable. Let $\theta_k := \bigcup_{k=1}^{k} \Omega_n$. Clearly θ_k is an integrable set and $\theta_k \nearrow X$.

Let $g_k := \chi_{\theta_k} \subset \mathscr{L}^1(\mu)$. Since $f \in M(\mu)$, we have $|f| \in M(\mu)$ which means that

$$h_n := ng_n \curlywedge |f| = ng_n \cap |f| \in \mathscr{L}^1.$$

Plainly, $h_n \nearrow f$. By virtue of the Fubini–Stone Theorem we may write

$$\mu^*(h_n) = \mu(h_n) = \mu_1(\mu_2(h_n)) = \mu_1^*(\mu_2^*(h_n))$$
$$\leqslant \mu_1^*(\mu_2^*(|f|) = c.$$

The Beppo–Levi Theorem yields

$$N_1(f) = \mu^*(|f|) = \mu^*(\sup h_n) = \sup \mu^*(h_n) \leqslant c < \infty.$$

Therefore $f \in \mathscr{F}^1 \cap M = \mathscr{L}^1$. □

Once again, in the light of the results of the Tonelli Theorem let us consider the example of a function for which $\mu_1(\mu_2(f)) \neq \mu_2(\mu_1(f))$.

The set $]0,1[\times]0,1[$ is obviously of semifinite measure; in fact, it is even of finite measure equal to 1.

As we know, a continuous function on a locally compact space is measurable with respect to the Radon measure (cf. Section 11).

Since the function $\dfrac{x^2-y^2}{(x^2+y^2)^2}$ is continuous on $]0,1[\times]0,1[$ and the thesis of the Tonelli Theorem is not satisfied, one may expect the second hypothesis of the theorem not to be satisfied, that is, that the integrals $\mu_1^*(\mu_2^*(|f|))$ and $\mu_2^*(\mu_1^*(|f|))$ do not exist. And indeed

$$\mu_1^*(|f|) = \int_0^1 \frac{|x^2-y^2|}{(x^2+y^2)^2}\,dx = \int_0^y \frac{y^2-x^2}{(x^2+y^2)^2}\,dx +$$

$$+\int_y^1 \frac{x^2-y^2}{(x^2+y^2)^2}\,dx = \frac{x}{x^2+y^2}\Bigg|_{x=0}^{x=y} - \frac{x}{x^2+y^2}\Bigg|_{x=y}^{x=1}$$

$$= \frac{y}{2y^2} - \left(\frac{1}{1+y^2} - \frac{y}{2y^2}\right) = \frac{1}{y} - \frac{1}{1+y^2};$$

$$\mu_2^*(\mu_1^*(|f|)) = \int_0^1 \left(\frac{1}{y} - \frac{1}{1+y^2}\right) dy = \infty .$$

In identical manner, we obtain $\mu_1^*(\mu_2^*(|f|)) = \infty$.

Now we give an important application of the Tonelli Theorem to the integration of series:

THEOREM XIII.18.5 (On Integration of Series). Let X_1 be a space with a semifinite μ_1-measure. Given a sequence $(f_n)_1^\infty$ of μ_1-measurable functions. If

$$\int_{X_1} \sum_{n=1}^{\infty} |f_n|\,d\mu_1 \quad \text{or} \quad \sum_{n=1}^{\infty} \int_{X_1} |f_n|\,d\mu_1$$

exists, then the expressions

$$\int_{X_1} \sum_{n=1}^{\infty} f_n\,d\mu_1 = \sum_{n=1}^{\infty} \int_{X_1} f_n\,d\mu_1$$

exist and are equal to each other.

PROOF. Suppose that $X_2 := N$ (the set of natural numbers), and let μ_2 be the aforementioned Radon integral on N ($\mu_2(f) = \sum_{n=1}^{\infty} f(n)$ for $f \in C_0(N)$).

If g is a function on $X_1 \times N$, then we denote

$$\tilde{g}_n(x) := g(x, n).$$

On $X_1 \times N$ we define the elementary family

$$E := \{g \colon \tilde{g}_n \in E_1(\mu_1), \bigvee_{M(g)} \bigwedge_{n > M(g)} \tilde{g}_m \equiv 0\}$$

and the measure

$$\mu(g) := \sum_{n=1}^{M(g)} \mu_1(\tilde{g}_n) = \sum_{n=1}^{\infty} \mu_1(\tilde{g}_n) = \mu_2(\mu_1(g)).$$

The linearity of the integral μ_1 implies that

$$\mu_2(\mu_1(g)) = \sum_{n=1}^{M(g)} \mu_1(\tilde{g}_n) = \mu_1\Big(\sum_{n=1}^{M(g)} (\tilde{g}_n)\Big) = \mu_1(\mu_2(g)).$$

Therefore, s.F. holds.

The reader should verify that every function which has a finite number of functions $\tilde{g}_n$, not vanishing identically, and for which all $\tilde{g}_n$ are μ_1-integrable, is μ-integrable.

Thus, it follows that every function g, all of whose functions $\tilde{g}_n$ are μ_1-measurable, is μ-measurable, because for any function $0 \leqslant h \in E$ the function $h \curlywedge g$ has only a finite number of nonzero integrable terms $(\widetilde{h \curlywedge g})_n$.

Thus, taking $g(x, n) := f_n(x)$ in our case, we find that g is μ-measurable on a set $X_1 \times N$ of semifinite measure.

The thesis follows immediately from the Tonelli Theorem. □

The *Cavalieri principle* is a direct application of all theorems of the Fubini Theorem type.

Cavalieri (1589–1647) formulated this principle for three-dimensional solids whose volume he was trying to evaluate. It read more or less as follows: If two solids have the property that their sections with all planes parallel to a prescribed plane have the same area, the solids have equal volumes.

In the language of modern integral theory, this principle could be reformulated as follows:

THEOREM XIII.18.6 (Cavalieri Principle). Suppose that the s.F. holds for (X, μ_1, E_1), (Y, μ_2, E_2), $(X\times Y, \mu, E)$. Let $\Omega \subset X\times Y$ be a μ-integrable set. If we denote

$$\Omega_x := \{y \in Y: (x, y) \in \Omega\},$$

$$\Omega_y := \{x \in X: (x, y) \in \Omega\},$$

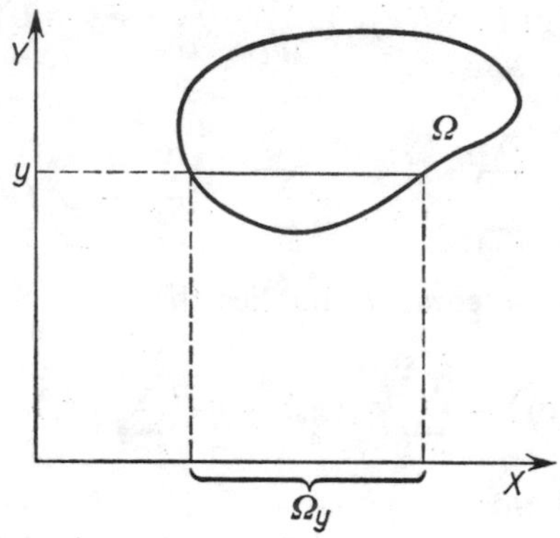

Fig. 1

$$|\Omega_x| := \mu_2(\Omega_x) = \int \chi_{\Omega_x}(y)\,d\mu_2(y),$$

$$|\Omega_y| := \mu_1(\Omega_y) = \int \chi_{\Omega_y}(x)\,d\mu_1(x),$$

then

$$|\Omega| := \mu(\Omega) = \int \chi_\Omega\,d\mu = \int |\Omega_x|\,d\mu_1(x) = \int |\Omega_y|\,d\mu_2(y).$$

PROOF. The Fubini Theorem leads to

$$\int_{X\times Y} \chi_\Omega\,d\mu = \int_Y\Big\{\int_X \chi_\Omega(x, y)\,d\mu_1(x)\Big\}\,d\mu_2(y)$$

$$= \int_Y\Big\{\int_X \chi_{\Omega_y}(x)\,d\mu_1(x)\Big\}\,d\mu_2(y) = \int |\Omega_y|\,d\mu_2(y).$$

The second equality is proved in identical fashion. □

Examples. 1. Let $X = \boldsymbol{R}^2$, $Z = \boldsymbol{R}$, $X\times Z = \boldsymbol{R}^3$ and let μ be the tensor product of Lebesgue integrals.

Take the following two solids:

$$\overset{(1)}{\Omega} = \overline{K(0,1)} \subset \boldsymbol{R}^3,$$

$$\overset{(2)}{\Omega} := \{(x, y, z): x^2+y^2 \leqslant 1; |z| \leqslant 1; x^2+y^2 \geqslant z^2\} \subset \boldsymbol{R}^3.$$

It is thus seen that the set $\overset{(2)}{\Omega}$ was formed from a cylinder of unit radius (and height from $z = -1$ to $z = 1$) by cutting out two cones (the set $x^2+y^2 < z^2$) with bases equal to those of the cylinder and with vertex at the point $(0, 0, 0)$.

Let us calculate $|\overset{(1)}{\Omega}_z|$:

$\overset{(1)}{\Omega}_z$ is a circle:

$$x^2+y^2+z^2 \leqslant 1, \quad x^2+y^2 \leqslant 1-z^2, \quad \sqrt{x^2+y^2} \leqslant \sqrt{1-z^2}$$

and thus $\overset{(1)}{\Omega}_z = K(0, \sqrt{1-z^2}) \subset X$. Therefore

$$|\overset{(1)}{\Omega}_z| = \pi\left(\sqrt{1-z^2}\right)^2 = \pi(1-z^2).$$

$\overset{(2)}{\Omega}_z$ is the set: $z^2 \leqslant x^2+y^2 \leqslant 1$, $|z| \leqslant \sqrt{x^2+y^2} \leqslant 1$.

This is thus an annulus formed by cutting a circle of radius $|z|$ out of the unit circle; accordingly

$$|\overset{(2)}{\Omega}_z| = \pi 1^2 - \pi|z|^2 = \pi(1-z^2).$$

By virtue of the Cavalieri principle

$$|\overset{(1)}{\Omega}| = \int_{-1}^{1} |\overset{(1)}{\Omega}_z|\,dz = \int_{-1}^{1} |\overset{(2)}{\Omega}_z|\,dz = |\overset{(2)}{\Omega}|.$$

We have thus reduced the problem of finding the volume of a sphere to one of evaluating the volume of a cylinder and a cone, i.e. a problem whose solution we know. We obtain

$$|\overset{(1)}{\Omega}| = |\overset{(2)}{\Omega}| = \pi \cdot 1^2 \cdot 2 - 2\left(\tfrac{1}{3}\pi \cdot 1^2 \cdot 1\right) = \tfrac{4}{3}\pi.$$

2. Let us calculate in general the volume of a sphere in $\boldsymbol{R}^n$, where

$$\boldsymbol{R}^n = \boldsymbol{R}^1 \times \boldsymbol{R}^1 \times \ldots \times \boldsymbol{R}^1.$$

We have

$\mu = \mu_1 \otimes \mu_1 \otimes \ldots \otimes \mu_1$, where μ_1 is the Lebesgue integral

Let us introduce the notation

$$K^n := \overline{K(0, 1)} = \{x \in \boldsymbol{R}^n: x_1^2+x_2^2+ \dots +x_n^2 \leqslant 1\}.$$

Employing the Cavalieri principle, we have

$$|K^n| = \int_{-1}^{1} |(K^n)|_{x_n} dx_n,$$

where $(K^n)_{x_n}$ is the section of the sphere K^n by the plane $x_n = \text{const}$. Therefore

$$(K^n)_{x_n} = \{x_1^2+x_2^2+ \dots +x_{n-1}^2 \leqslant 1-x_n^2\}$$
$$= K(0, \sqrt{1-x_n^2}) \subset \boldsymbol{R}^{n-1}.$$

Since, as is seen, $(K^n)_{x_n}$ is an $(n-1)$-dimensional sphere reduced $\sqrt{\dfrac{1-x_n^2}{1}}$ times in relation to the sphere K^{n-1}, the volume of this set will be $(\sqrt{1-x_n^2})^{n-1}$ times smaller than that of the sphere K^{n-1}. Hence

$$|K^n| = \int (\sqrt{1-x_n^2})^{n-1} \cdot |K^{n-1}| dx_n$$
$$= |K^{n-1}| \int (\sqrt{1-x_n^2})^{n-1} dx_n.$$

Substituting $x_n = \sin t$, $\sqrt{1-x_n^2} = \cos t$, we obtain

$$|K^n| = |K^{n-1}| \, 2 \int_0^{\pi/2} \cos^n t \, dt = |K^{n-1}| 2c(n)$$

where

$$c(n) = \begin{cases} \dfrac{\pi}{2} \cdot \dfrac{(n-1)!!}{n!!} & \text{for } n \text{ even,} \\[2ex] \dfrac{(n-1)!!}{n!!} & \text{for } n \text{ odd.} \end{cases}$$

Finally, for $n = 2k$

$$|K^{2k}| = 2c(2k)|K^{2k-1}| = 2^2 c(2k) c(2k-1)|K^{2k-2}|$$
$$= 2^{2k} \left(\frac{\pi}{2}\right)^k \cdot \frac{(n-1)!!}{n!!} \cdot \frac{(n-2)!!}{(n-1)!!} \dots \frac{3!!}{4!!} \cdot \frac{2!!}{3!!} \cdot \frac{1!!}{2!!}$$
$$= \left(\frac{\pi}{2}\right)^k \cdot \frac{2^{2k}}{2k(2k-2)(2k-4) \dots 4 \cdot 2} = \frac{2^{2k}\pi^k}{2^k \cdot 2^k \cdot k!} = \frac{\pi^k}{k!}.$$

For $n = 2k+1$ we similarly find

$$|K^{2k+1}| = \frac{\pi^k}{(2k+1)!!}\, 2^{k+1}.$$

Spheres of radius r will have their volume changed in the ratio of r^n with respect to unit spheres (n is the dimension of the space). Thus

$$|K^{2k}(0, r)| = \frac{\pi^k}{k!} \cdot r^{2k},$$

$$|K^{2k+1}(0,r)| = \frac{\pi^k}{(2k+1)!!}\, 2^{k+1} r^{2k+1}.$$

Applying the formula to the first several natural numbers, we obtain:

$|K^1(0, r)| = 2r$ (length of the interval $[-r, r]$),

$|K^2(0, r)| = \pi r^2$ (area of circle),

$|K^3(0, r)| = \frac{4}{3}\pi r^3$ (volume of three-dimensional sphere),

$|K^4(0, r)| = \frac{1}{2}\pi^2 r^4$ (volume of four-dimensional sphere).

19. THE RADON INTEGRAL. STONE'S SECOND PROCEDURE

In this section we shall collect together the facts we already know from the theory of the Radon measure (integral) and add some new ones.

As we know, the Radon integral on a locally compact space X is a linear, positive and monotonically continuous functional on $E = C_0(X)$.

We now give without proof a fact which is a direct conclusion from the Urysohn Lemma familiar from topology:

LEMMA XIII.19.1. If X is a locally compact space, and a set $K \subset X$ is compact, there exists a function $f_K \in C_0(X)$, such that $0 \leqslant f_K \leqslant 1$ and $f_K(x) = 1$ for $x \in K$.

This fact leads to the following theorem.

THEOREM XIII.19.2. Every positive linear functional on $C_0(X)$ is also monotonically continuous, i.e. is a Daniell–Stone integral.

In other words: The Radon integral is a positive linear functional on $C_0(X)$.

PROOF. Let $f_n \in C_0(X)$. Denote the support of the function f_n by $\underline{f_n} \subset X$, that is,

$$\underline{f_n} := \overline{\{x \in X \colon f(x) \neq 0\}}.$$

Let $f_n \searrow 0$. Thus, for any n $\underline{f_n} \subset \underline{f_1} =: K$; K is a compact set. From the Dini Theorem we infer that $f_n \to 0$ uniformly, that is, $\sup f_n(K) \to 0$. Now, invoking Lemma XIII.19.1, we obtain

$$\sup f_n(K) \cdot f_K \geqslant f_n.$$

Therefore

$$0 \leqslant \mu(f_n) \leqslant \mu\big(f_K \cdot \sup f_n(K)\big) = [\sup f_n(K)] \cdot \mu(f_K) \underset{n\to\infty}{\to} 0. \quad \square$$

Here is the next consequence of Lemma XIII.19.1:

COROLLARY XIII.19.3. A precompact set has a μ^*-finite measure.

PROOF. Let A be precompact. Therefore $\bar{A} = K$ is compact. By virtue of the lemma we have

$$\mu^*(A) \leqslant \mu^*(K) = \mu^*(\chi_K) \leqslant \mu^*(f_K) = \mu(f_K) < \infty,$$

since $f_K \in C_0(X)$. $\square$

In the case when X is a separable metric space, the set $\mathscr{E}$ is the space of all lower-semicontinuous functions, with compact support of the negative part (cf. Section 9, Theorem XIII.9.4). Since the characteristic functions of open sets are lower-semicontinuous, it follows from Lemma XIII.11.3 that in this case all open sets are measurable.

As we have already emphasized, the Radon integral satisfies the strong Stone axiom. Therefore, X is a measurable set and the results of Section 17 are applicable.

THEOREM XIII.19.4. If X is a separable metric space, then:

(i) Open and closed sets are measurable.

(ii) Compact sets are integrable.

(iii) Open sets which are precompact are integrable.

PROOF. (i) We already know that open sets are measurable. Let $D \subset X$ be a closed set, $D = X-(X-D)$. The set $X-D$ is open and, hence, measurable. Thus, as it is the difference of measurable sets, D is measurable.

(ii) If the set K is compact, then it is also precompact, and thus has a finite measure. Being closed, it is also measurable. Therefore, $\chi_K \in \mathscr{F}^1 \cap M = \mathscr{L}^1$.

(iii) The proof is as for (ii). □

By $\mathscr{K}$ let us denote the family of all compact sets in the space X. Let $\mathscr{B} := \mathscr{A}(\mathscr{K})$ denote the least σ-algebra of sets in the space X, containing K. Sets from the family $\mathscr{B}$ thus are formed by taking countable unions, differences, and intersections of compact sets. This family will be called the *family of Borel sets.*

If X is a separable metric space, Borel sets are measurable. Then the following fact is true:

THEOREM XIII.19.5. Let S_1 be the space of functions which are finite combinations of the characteristic functions of integrable Borel sets

$$S_1 \ni f = \sum_{i=1}^{n} a_i \chi_{Z_i}, \qquad Z_i \in \mathscr{B},\ \mu^*(Z_i) < \infty .$$

If X is a separable metric space, then

(i) $S_1 \subset \mathscr{F}^p$,

(ii) $\overline{S}_1 = \mathscr{L}^p$ (closure in the sense of N_p).

PROOF. (i) $S_1 \subset S \subset \mathscr{F}^p$ (cf. Theorem XIII.16.6).

(ii) $\overline{S}_1 \subset \overline{S} = \mathscr{L}^p$. It is to be proved that $\mathscr{L}^p \subset \overline{S}_1$.

Let $f \in \mathscr{L}^p$. Thus there exists a sequence $(f_n)_1^\infty$, $f_n \in E = C_0(X)$ such that $f_n \to f$ in the N_p sense. Let us form the following functions for $\varepsilon > 0$:

$$f_n^\varepsilon := \sum_{m=0}^{\infty} m\varepsilon(\chi_{A_{\varepsilon m}} - \chi_{A_{\varepsilon(m+1)}}) + \sum_{m=0}^{-\infty} m\varepsilon(\chi_{A_{\varepsilon(m-1)}} - \chi_{A_{\varepsilon m}}),$$

where $A_a = \{x: f_n(x) > a\}$.

The sets A_a are Borel sets, if $f_n \in C_0(X)$, because (denoting the support of the function f by $\underline{f}$)

$$A_a = (\underline{f_n} - \{x: f_n(x) \leqslant a\} \cap \underline{f_n})$$

(A_a hence is the difference of compact sets). Therefore, it follows from standard considerations (cf. the proof of Theorem XIII.16.6) that $f_n^\varepsilon \in \overline{S}_1$. But $f_n = \lim_{\varepsilon \to 0} f_n^\varepsilon$, and thus $f_n \in \overline{\overline{S}}_1$, and $f = \lim_{n\to\infty} f_n$, and consequently $f \in \overline{S}_1$. □

It is thus seen that in order to construct a Radon measure, it is sufficient to know the measure on Borel sets since, upon taking S_1 as the elementary functions, we can reconstruct all the spaces $\mathscr{L}^p$ and all the norms N_p.

Note that the Radon integral satisfies a condition stronger than monotonic continuity:

LEMMA XIII.19.6. If $F \subset E = C_0(X)$ is an arbitrary family of non-negative elementary functions, and if this family is directed downwards by the $\leqslant$ relation (i.e. for any two functions $f_1 \in F$, $f_2 \in F$ there exists an $f_3 \in F$ such that $f_3 \leqslant f_1, f_3 \leqslant f_2$), then

$$\left(\inf_{f \in F} f(x) \equiv 0\right) \Rightarrow \left(\inf_{f \in F} \mu(f) = 0\right).$$

PROOF. Take an arbitrary function $f_0 \in F$. Denote $\underline{f_0} = K$.

If $x \in X$, $\varepsilon > 0$, there exists a function $f_x^\varepsilon \in F$ such that $f_x^\varepsilon \leqslant f_0$ and $f_x^\varepsilon(x) \leqslant \frac{1}{2}\varepsilon$. This follows from the fact that $\inf_{f \in F} f(x) = 0$ and from the fact that the family F is directed downwards.

The set $A_x^\varepsilon := \{x \colon f_x^\varepsilon < \varepsilon\}$ is open, because f_x^ε is continuous. But

$$K \subset \bigcup_{x \in K} A_x^\varepsilon.$$

Since K is compact, it is possible to select a finite covering:

$$K \subset \bigcup_{i=1}^{n(\varepsilon)} A_{x_i}^\varepsilon.$$

Now take a function $f^\varepsilon \in F$ such that $f^\varepsilon \leqslant f_{x_i}^\varepsilon$, $i = 1, 2, \ldots, n(\varepsilon)$. If $x \in X$, then either $x \notin K$ (and then $0 \leqslant f^\varepsilon(x) \leqslant f_0(x) = 0$) or x belongs to one of the sets $A_{x_i}^\varepsilon$ (e.g. labelled j) and then $0 \leqslant f^\varepsilon(x) \leqslant f_{x_j}^\varepsilon(x) < \varepsilon$.

Finally, therefore, $0 \leqslant f^\varepsilon \leqslant \varepsilon$. Taking the sequence $\varepsilon_k \to 0$, we get $F \ni f^{\varepsilon_k} \to 0$ uniformly, and hence (cf. the proof of Theorem XIII.19.2) $\mu(f^{\varepsilon_k}) \to 0$. □

A functional μ for which the thesis of the foregoing lemma is satisfied, will be said to be "monotonically continuous in the stronger sense."

THEOREM XIII.19.7. For every function $f \geqslant 0$, which is lower-semicontinuous, there exists a family $F \subset C_0(X)$ such that

$$f(x) = \sup_{g \in F} g(x).$$

PROOF. If f is lower-semicontinuous, $x_0 \in X$, then as we know the sets

$$B_{x_0}^\varepsilon = \{x \colon f(x) \leqslant f(x_0) - \varepsilon\}$$

are closed. Thus the sets $D^{\varepsilon}_{x_0} = X - B^{\varepsilon}_{x_0} = \{x: f(x) > f(x_0) - \varepsilon\}$ are open and $x_0 \in D^{\varepsilon}_{x_0}$.

By the Urysohn Lemma there exists a continuous function with compact support $0 \leqslant \tilde{f}^{\varepsilon}_{x_0} \leqslant 1$, such that $\tilde{f}^{\varepsilon}_{x_0}(x_0) = 1$, $\underline{\tilde{f}^{\varepsilon}_{x_0}} \subset D^{\varepsilon}_{x_0}$. Setting $f^{\varepsilon}_{x_0} := (f(x_0) - \varepsilon) \cdot \tilde{f}^{\varepsilon}_{x_0}$, we obtain

$$f^{\varepsilon}_{x_0}(x_0) = f(x_0) - \varepsilon, \quad f^{\varepsilon}_{x_0} \leqslant f.$$

Hence $f = \sup_{\substack{x_0 \in X \\ \varepsilon > 0}} f^{\varepsilon}_{x_0}$. □

Taking these two facts into account, Stone carried out the construction of a somewhat different theory of the integral from the one we built. This construction is called the "second Stone procedure". Here are the basic steps in this construction:

Integral is a positive, linear functional, monotonically continuous, in the strong sense, on an elementary family E. The space $\mathscr{E}$ is defined as follows:

$$\mathscr{E} := \{g: \bigvee_{F \in E} g = \sup_{F \in E} f\}.$$

It can be shown that with these definitions almost all theorems of Sections 3–19 remain valid.

When the second Stone procedure is applied to the Radon integral, the space $\mathscr{E}$ is a set of lower-semicontinuous functions with compact support for the negative part (an immediate conclusion from Theorem XIII.19.7). Therefore, Theorem XIII.19.4 holds without any assumptions as to the locally compact space X.

If X is a metric separable space, the two procedures coincide.

Remark. We can give a more general version of the preceding theorem:

THEOREM XIII.19.8. Let X be a Tychonoff (i.e. completely regular or uniformisable) space. Then

1° Every lower semi-continuous function $f: X \to \boldsymbol{R}$ is an upper envelope of a family F of continuous functions.

2° If X is compact or second countable then f is a limit of increasing sequence $(f_n)_1^{\infty}$ of continuous functions.

PROOF. The same as of the preceding theorem since we have used only the complete regularity of X.

20. FINITE RADON MEASURES. TOUGH MEASURES

The Radon measure is by definition a positive linear functional on $C_0(X)$, where X is a locally compact space. Often in applications there is need of "integrals" of an arbitrary sign, i.e. some functionals μ linear on $C_0(X)$. Then we break up the functional μ into a positive and a negative part, $\mu = \mu_+ - \mu_-$, where

$$C_0(X) \ni f \to \mu_+(f) := \sup_{0 \leqslant g \leqslant f} \mu(g) \quad \text{and}$$

$$g \in C_0(X), \quad \mu_- := (-\mu)_+.$$

If μ_+ and μ_- assume only finite values on $C_0(X)$ then they are positive functionals and hence we adopt the following definition:

DEFINITION. A linear functional μ on $C_0(X)$ is a *Radon measure* of arbitrary sign, when μ_+, μ_- are positive linear functionals on $C_0(X)$, such that $\mu = \mu_+ - \mu_-$.

Let K be an arbitrary compact subset. On denoting $||\varphi||_K = \sup |\varphi(K)|$, we see that in the space $C(X)$ we can introduce a family of seminorms $|| \; ||_K$, $K \in \{K\}$, where $\{K\}$ is the family of (all) compact subsets. On introducing the notation $C_K(X) = \{f \in C_0(X) \colon \underline{f} \subset K\}$, we see that

$$C_0(X) = \bigcup_{K \in \{K\}} C_K(X).$$

As we have proved, if μ is a positive Radon measure on X, when $\mu_K := \mu | C_K(X)$, then $|\mu_K(f)| \leqslant a(K) ||f||_K$. We can thus adopt a general definition of complex Radon measure:

DEFINITION. A (*complex*) *Radon measure* is a linear functional on $C_0(X)$, continuous in the sense that the inequality

$$|\mu_K(f)| \leqslant a(K) ||f||_K, \quad f \in C_0(X)$$

holds for every compact subset $K \subset X$.

Remark. As will be seen subsequently, this is a "good" definition of continuity:

The family $(|| \; ||_K, K \in \{K\})$ of seminorms determines a natural topology in the space $C_0(X)$ which is the inductive limit of the Banach space $(C_K(X), || \; ||_K)$. This is the strongest (locally convex) topology, such that every injection $(C_K(X), || \; ||_K) \to C_0(X)$ is continuous.

Let μ be a *finite Radon measure* on X, that is, $||\mu|| := \int d|\mu| < \infty$, where $|\mu| := \mu_+ + \mu_-$. Then, applying the Daniell–Stone procedure to $|\mu|$, we can extend μ from $C_0(X)$ to all continuous bounded functions on X. As we know, the inequality

$$(1) \qquad \left|\int f\,d\mu\right| \leqslant \|f\|\,\|\mu\|, \quad f \in B(X),$$

$$\text{where } \|f\| = \|f\|_\infty = \sup|f(X)|$$

holds. This inequality states that we have extended a linear functional to a linear functional continuous over the entire Banach space $B(X)$.

A continuous function f on a locally compact space X is said to *vanish at infinity* when

$$\bigwedge_{\varepsilon>0} \bigvee_{K\in\{K\}} \bigwedge_{x\notin K} |f(x)| \leqslant \varepsilon.$$

(The justification for this terminology is given in Chapter XVII.) The set of such functions will be denoted by $C_\infty(X)$. Plainly $C_\infty(X) \subset B(X)$ and $C_0(X)$ is a subspace dense in $C_\infty(X)$ in the sense of the norm $\|\ \|_\infty$. Thus every finite Radon measure $\mu \in C_0'(X)$, $\|\mu\| < \infty$, is uniquely extendible to the entire space $(C_\infty(X), \|\ \|_\infty)$. But in general (when X is not compact) $C_\infty(X)$ has an infinite codimension in $B(X)$, and thus an infinite number of continuous extensions of μ from $C_\infty(X)$ to $B(X)$ exist (Hahn–Banach Theorem). Hence, the Daniell–Stone extension is a very special extension of the functional $\mu \in C_0(X)'$, $\|\mu\| < \infty$. The problem thus arises of characterizing the Daniell–Stone extensions of finite Radon measures. This is taken up in the following theorem.

THEOREM XIII.20.1. ($m \in B'(X)$ is the Daniell-Stone extension of the restriction $m|C_0(X)$) $\Leftrightarrow$ ($\bigwedge_{\varepsilon>0} \bigvee_{K\in\{K\}}$ [($\|f\|_\infty \leqslant 1$, $f = 0$ on K) $\Rightarrow$ the condition (ε, K) is satisfied: $|m(f)| \leqslant \varepsilon$]).

PROOF. $\Rightarrow$: Let μ be the Daniell–Stone extension of its restriction. Suppose that $\mu \geqslant 0$. Then $\bigwedge_{\varepsilon>0} \bigvee_{K\in\{K\}} \mu(X-K) \leqslant \varepsilon$. Thus when $\|f\|_\infty \leqslant 1$ and $f(k) = 0$ for $k \in K$, then $|\mu(f)| = 1 \cdot \mu\{X-K\} \leqslant \varepsilon$. The case of measures of arbitrary sign (and complex measures) reduces to the case of positive measures, considering the measure $|\mu|$.

$\Leftarrow$: Suppose that m satisfies the condition (ε, K). The restriction $m|C_0(X)$ is a bounded Radon measure; let μ be the Daniell–Stone extension of $m|C_0(X)$. Since both measures m and μ satisfy it, the condition (ε, K) is also satisfied by the measure $l := m-\mu$. The theorem will have been proved when we show that every form $l \in B'(X)$, vanishing on $C_0(X)$ and satisfying the condition (ε, K), vanishes. Suppose that $0 \neq f \in B(X)$, $\varepsilon > 0$, let K be a compact set chosen to match $\varepsilon/\|f\|_\infty$ for the condition (ε, K), and let $\varphi \in C_0(X)$, where $0 \leqslant \varphi \leqslant 1$, and $\varphi(k) = 1$, $k \in K$. Clearly, we have $l(f) = l(\varphi f)+l(f-\varphi f)$, but since $\varphi f \in C_0(X)$, we have

$$(*) \qquad l(\varphi f) = 0.$$

However, $f-\varphi f$ vanishes on K and $\|f-\varphi f\|_\infty \leqslant \|f\|_\infty$, so that consequently $|l(f-\varphi f)| \leqslant \varepsilon$. Associating this inequality with $(*)$, we obtain $|l(f)| \leqslant \varepsilon$ for every $\varepsilon > 0$, and hence $l(f) = 0$ for any $f \in B(X)$, that is, $l = 0$. □

Now let us generalize the concept of finite Radon measure to an arbitrary Hausdorff space X. Let X be a Hausdorff space, and let $B(X)$ be the space of bounded continuous functions with the norm

$$\|f\|_\infty := \sup |f(X)|.$$

DEFINITION. A continuous linear functional m on the Banach space $(B(X), \|\ \|_\infty)$ is called a *tough measure* if it satisfies the (ε, K) condition:

$$\bigwedge_{\varepsilon>0} \ \bigvee_{K\text{—compact}} (\|f\|_\infty \leqslant 1,\ f(k) = 0 \text{ for } k \in K) \Rightarrow |m(f)| \leqslant \varepsilon.$$

This condition states, roughly speaking, that the measure m "condenses to within ε on the set K". Radon measures are tough measures on a locally compact space. Tough measures play a great role in the theory of measures on infinite-dimensional topological vector spaces. We shall be dealing with them in Chapter XVIII (Prokhorov Theorem) and in Chapter XIX (Minlos Theorem).

21. THE TENSOR PRODUCT OF RADON INTEGRALS

As we know, a Radon integral is a nonnegative linear functional on a space of continuous functions with compact supports. Moreover, this functional is bounded (and, hence, continuous). More precisely:

Let X be a locally compact space, and let $C_0(X)$ be a space of continuous functions with compact supports. Usually the space $C_0(X)$ is provided with a topology of compact convergence. By $\|\ \|_K$ let us denote the seminorm $f \to \sup |f(K)|$, where K is a compact subset of X. By

$C_K(X)$ let us denote the set of continuous functions on X whose supports are compact in K. This is the space and seminorm $\| \|_K$.

Reasoning as in the proof of Theorem XIII.19.2, we shall prove

LEMMA XIII.21.1. Let μ be a Radon integral on X; then, for every compact subset $K \subset Y$ there exists a constant $m_K > 0$, such that

(1) $$|\mu(f)| \leqslant m_K \|f\|_K, \quad f \in C_K(X).$$

Remark. The greatest lower bound of such constants m_K is denoted by $\|\mu\|_K$.

LEMMA XIII.21.2. Let X_1, X_2 be locally compact spaces, and let μ_i be a Radon integral on X_i, that is, $E_i = C_0(X_i)$, $i = 1, 2$. Let $E = C_0(X_1 \times \times X_2)$. Then:

(i) On X there exists a unique Radon integral μ such that

(2) $$\mu(f_1 \otimes f_2) = \mu_1(f_1)\mu_2(f_2), \quad f_i \in E_i, \ i = 1, 2.$$

(ii) The thesis of s.F. holds for the triad μ_1, μ_2, μ.

(iii) $E_1 \otimes E_2$ is dense in E, and accordingly the integral μ may be denoted by $\mu_1 \otimes \mu_2$ and called the tensor product of the integrals μ_1 and μ_2.

PROOF. First of all, we shall prove (iii), and then (ii) guarantees uniqueness.

Let $f \in E = C_0(X_1 \times X_2)$; then there exist compact sets $K_i \subset X_i$, $i = 1, 2$, such that $f \subset K_1 \times K_2$.

We shall prove that

(3) $$\bigwedge_{\varepsilon > 0} \bigvee_{f_i \in C_{K_i}(X_i)} \Big\| f - \sum_{j=1}^{n} f_1^j \otimes f_2^j \Big\|_{K_1 \times K_2} < \varepsilon.$$

This follows immediately from the Stone–Weierstrass Theorem, since functions of the form $f_1 \otimes f_2$ separate the points of the set $K_1 \times K_2$, and $E_1 \otimes E_2$ is an algebra spanned by functions of the form $f_1^j \otimes f_2^j$.

(i), (ii). The functional μ defined on $E_1 \otimes E_2$ by the formula

$$\mu\Big(\sum f_1^j \otimes f_2^j\Big) := \sum \mu_1(f_1^j)\mu_2(f_2^j)$$

is extendible to all of E. However, we carry out the construction directly

(4) $$\Big|\mu_2(f(x_1, \cdot)) - \sum f_1^j(x_1)\mu_2(f_2^j)\Big| \leqslant \mu_2\Big(\Big|f(x_1, \cdot) - \sum f_1^j \otimes f_2^j(x_1, \cdot)\Big|\Big) \leqslant \varepsilon \|\mu_2\|_{K_2},$$

and thus $\mu_2(f)$, being the uniform limit of the functions from $C_{K_1}(X_1)$, is a continuous function. Of course, $\underline{\mu_2(f)} \subset K_1$.

On integrating, from inequalities (1) and (4) we obtain

$$\left|\mu_1(\mu_2(f)) - \sum \mu_1(f_1^j)\mu_2(f_2^j)\right| \leqslant \varepsilon \|\mu_2\|_{K_2} \cdot \|\mu_1\|_{K_1}.$$

An analogous evaluation is obtained for $\mu_2(\mu_1(f))$ and hence

$$|\mu_1(\mu_2(f)) - \mu_2(\mu_1(f))| \leqslant 2\varepsilon \|\mu_1\|_{K_1} \cdot \|\mu_2\|_{K_2},$$

since ε was arbitrary, then $\mu_1(\mu_2(f)) = \mu_2(\mu_1(f))$; therefore the s.F. holds.

Thus, the following integral may be defined:

$$\mu(f) = (\mu_1 \otimes \mu_2)(f) := \mu_1(\mu_2(f)) = \mu_2(\mu_1(f))$$

for $f \in E$.

This raises the question of whether the notation $\mu = \mu_1 \otimes \mu_2$ is inconsistent with the results of Section 16 where the tensor product of integrals was defined in terms of S spaces? It turns out that the two definitions are equivalent, that is to say, they both yield the same spaces $\mathscr{L}^p$.

To prove this, note that every function $f \in S(\mu_1) \otimes S(\mu_2)$ belongs to $\mathscr{L}^p$ in the sense of the new definition, since if

$$f = \sum_{i=1}^{n} f_i^1 \otimes f_i^2, \quad f_i^k \in S(\mu_k) \subset \mathscr{L}^p(\mu_k),$$

then, taking the sequences

$$E_1 \ni g_i^m \to f_i^1 \quad \text{in the sense of } N_p(\mu_1),$$
$$E_2 \ni h_i^m \to f_i^2 \quad \text{in the sense of } N_p(\mu_2),$$

we can construct the functions

$$f^m := \sum_{i=1}^{n} g_i^m \otimes h_i^m, \quad f^m \in E.$$

It is easily verified that $f^m \to f$ in the sense of $N_p(\mu)$.

Similarly, every function from $E_1 \otimes E_2$ can be approximated in the N_p sense by functions from $S(\mu_1) \otimes S(\mu_2)$ since functions from E_i can be approximated by functions from $S(\mu_i)$. Now, $E_1 \otimes E_2$ is dense in E

in the topology of uniform convergence, and hence also in the sense of N_p; therefore

$$\overline{S(\mu_1)\otimes S(\mu_2)} = \bar{E}$$

(closure in the sense of every seminorm N_p).

Since the s.F. holds for the measures $\mu_1, \mu_2, \mu_1\otimes\mu_2$, the Fubini–Stone and Tonelli theorems are valid. □

22. THE LEBESGUE INTEGRAL ON R^n. CHANGE OF VARIABLES

One of the most often used examples of a Radon integral is the Lebesgue integral on R, that is, the functional formed by extension of the Riemann integral. The tensor product of n Lebesgue integrals is called the *Lebesgue integral on* R^n.

In the space R^n a ball is compact, which means it is integrable, and $R^n = \bigcup_{k=1}^{\infty} K(0, k)$; hence, R^n is of semifinite measure. The Tonelli Theorem thus holds.

Let us denote the Lebesgue integral on R^n by the symbol λ^n.

As is known, for the integral λ^1 the following change-of-variables formula holds:

$$\text{(1)} \qquad \int f(x)\,dx = \int (f\circ\Phi)(x)\cdot|\Phi'(x)|\,dx,$$

where $f\in C_0(R)$, Φ: $R\to R$, Φ is a topological mapping of class C^1 (i.e. continuous, invertible, with continuous derivative, whose inverse is also continuous and has a continuous derivative). Since the integral is a continuous functional on the space $\mathscr{L}^1$, this equation extends immediately from the set $C_0(X)$ dense in $\mathscr{L}^1$ to the entire space $\mathscr{L}^1$.

It turns out that a similar formula holds for the integrals λ^n.

THEOREM XIII.22.1 (On Change of Variables). Let U and U' be open subsets in R^n (in particular, it may be that $U = U' = R^n$). Let Φ be a topological mapping of class C^1 of the set U onto U'.

If a function $f\in\mathscr{L}^1$ has the property that $\underline{f}\subset U'$, then:

$$\text{(2)} \qquad \int_{R^n} f(y)\,d\lambda^n(y) = \int_{U'} f(y)\,d\lambda^n(y) = \int_U f(\Phi(x))\cdot|\det\Phi'(x)|\,d\lambda^n(x)$$

$$= \int_{R^n} f(\Phi(x))\cdot|\det\Phi'(x)|\,d\lambda^n(x).$$

Remark. The sets U and U' could be replaced by any sets D and D' differing from the others by null sets.

PROOF. (i) We introduce the standard notation:

$$R^n \ni e_i = (0, \dots, 0, 1, 0, \dots, 0) \text{ (unity in the } i\text{-th place)},$$

and the scalar product:

$$(x|y) = \sum_{i=1}^{n} x^i y^i,$$

where $x = (x^1, x^2, \dots, x^n) \in R^n$, $y = (y^1, y^2, \dots, y^n) \in R^n$.

Clearly, $x^i = (x|e_i)$.

The mapping $G\colon U \to R^n$ $(U \subset R^n)$ will be called *primitive* if there exists j: $1 \leqslant j \leqslant n$ such that

$$\big(G(x)|e_i\big) = (x|e_i) \quad \text{for } i \neq j$$

(this mapping changes only the j-th coordinate).

Let Φ satisfy the hypothesis of the change-of-variables theorem and let $0 = (0, \dots, 0) \in U$ and $\Phi(0) = 0 \subset R^n$; we shall show that there exists a neighbourhood $V \subset U$ of the point $0 \in R^n$ such that for $x \in V$

$$\Phi(x) = (G_n \circ B_n \circ G_{n-1} \circ B_{n-1} \circ \dots \circ G_1 \circ B_1)(x),$$

where $B_i(0) = G_i(0) = 0$, G_i are primitive mappings, and B_i are linear mappings which change at most two coordinates.

The proof will be carried out by induction.

Let Φ_m be a mapping satisfying the same hypotheses as Φ and let

$$\big(\Phi_m(x)|e_i\big) = (x|e_i) \quad \text{for } 1 \leqslant i < m$$

(Φ_m does not change the first $m-1$ coordinates).

We denote $A_m := \Phi'_m(0)$. Obviously

(3) $$(A_m x|e_i) = (x|e_i) \quad \text{for } 1 \leqslant i < m$$

($A_m x$ denotes the matrix product, i.e. the operation of the linear mapping A_m on the vector x). Let $a_{ij} := (A_m e_i|e_j)$. By formula (3) it follows that when $j < m$, and $i \geqslant m$, then

$$a_{ij} = (A_m e_i|e_j) = (e_i|e_j) = 0.$$

If it were that $a_{ij} \equiv 0$ for $j = m$, $i \geqslant m$, then for $i \geqslant m$ it would be that

$$A_m e_i = \sum_{j=1}^{n} a_{ij} e_j = \sum_{j=m+1}^{n} a_{ij} e_j.$$

Since the matrix A_m is nonsingular by assumption, the vectors $A_m e_m, A_m e_{m+1}, A_m e_{m+2}, \ldots, A_m e_n$ are linearly independent. But there are $n+m+1$ of these vectors and they lie in $(n-m)$-dimensional space spanned by the vectors $e_{m+1}, e_{m+2}, \ldots, e_n$. Therefore the vectors $A_m e_m, \ldots, A_m e_n$ cannot be linearly independent.

If this contradiction is not to occur there must exist a $k \geqslant m$ such that $a_{km} \neq 0$.

We define the following linear operators on R^n:

$$P_m(e_i) = \begin{cases} e_i & \text{for } i \neq m, \\ 0 & \text{for } i = m; \end{cases}$$

$$B_m(e_i) = \begin{cases} e_i & \text{for } i \neq m, i \neq k, \\ e_k & \text{for } i = m, \\ e_m & \text{for } i = k. \end{cases}$$

Let us define the mapping

$$G_m(x) := P_m(x) + \big(e_m | (\Phi_m \circ B_m)(x)\big) e_m.$$

Of course, for $i \neq m$ we have $\big(G_m(x)|e_i\big) = \big(P_m(x)|e_i\big) = (x|e_i)$, that is, G_m is a primitive mapping, changing only the m-th coordinate.

We shall show that G_m is invertible in the neighbourhood of zero. Since it is of class C^1, it is sufficient to prove that $\det G'_m(0) \neq 0$, or, which amounts to the same thing, that

$$\big(G'_m(0)h = 0\big) \Rightarrow (h = 0).$$

By the inverse mapping theorem (cf. Chapter VIII) this will ensure the invertibility of G_m in the neighbourhood of zero. But $G'_m(0)h = P_m h + (e_m | A_m B_m h) e_m$, because $(\Phi_m B_m)'(0) = A_m B_m$.

If $G'_m(0)h = 0$, then $P_m h = 0$ and $(e_m | A_m B_m h) = 0$. The first of these conditions implies that $h = ce_m$. Accordingly

$$0 = \big(e_m | A_m B_m(ce_m)\big) = c(e_m | A_m e_k) = c(A_m e_k | e_m) = ca_{km}.$$

Now, $a_{km} \neq 0$, and hence $c = 0$, that is $h = 0 \cdot e_m = 0$.

Thus, there exists an open neighbourhood of zero U_m such that G_m is an invertible mapping of class C^1 on U_m.

Let us denote the open set $G_m(U_m) = V_m \subset R^n$.

For $y \in V_m$ we define the mapping

$$\Phi_{m+1}(y) := \Phi_m\big(B_m(G_m^{-1}(y))\big).$$

Since $y = G_m(x)$, $x = G_m^{-1}(y)$, $x \in U_m \subset \boldsymbol{R}^n$, from the definition of the operator G_m it follows that

$$(e_m|y) = (e_m|G_m(x)) = (e_m|(\Phi_m \circ B_m)(x))$$

and

$$(e_i|y) = (e_i|G_m(x)) = (e_i|P_m(x)) = (e_i|x) \quad \text{for } i \neq m.$$

Therefore

$$(e_m|\Phi_{m+1}(y)) = (e_m|(\Phi_m \circ B_m)(x)) = (e_m|y).$$

Since Φ_m does not alter x^i for $i < m$, then

$$(e_i|\Phi_{m+1}(y)) = (e_i|\Phi_m(B_m(x)) = (e_i|B_m(x)) = (e_i|x) = (e_i|y).$$

We have thus found that Φ_{m+1} satisfies the same conditions as Φ_m, except that it does not alter the first m coordinates. If we write $z = B_m(G_m^{-1}(y))$, then $y = G_m(B_m^{-1}(z)) = (G_m \cdot B_m)(z)$, since $B_m^{-1} = B_m$. Hence

$$\Phi_m(z) = [\Phi_{m+1} \circ (G_m \circ B_m)](z).$$

Since $\Phi_1 := \Phi$ satisfies the inductive hypotheses for $m = 1$, then

$$\Phi_1 = \Phi_2 \circ (G_1 \circ B_1) = \Phi_3 \circ (G_2 \circ B_2) \circ (G_1 \circ B_1) = \ldots$$
$$= \Phi_{n+1} \circ (G_n \circ B_n \circ G_{n-1} \circ B_{n-1} \circ \ldots \circ G_1 \circ B_1).$$

But $\Phi_{n+1} = I$ (identity), for this is a mapping which does not alter n coordinates. Our thesis follows from this.

(ii) If now Φ is an arbitrary mapping satisfying the hypothesis of the change-of-variables theorem, then for any point $x_0 \in U$ we have

$$\Phi(x) = \Phi(x_0) + (\Phi(x) - \Phi(x_0)).$$

The mapping $\Psi(y) := \Phi(x_0+y) - \Phi(x_0)$, however, satisfies the condition $\Psi(0) = 0$ and is invertible of class C^1, and hence is locally representable as the composition of the mappings G_m and B_m Finally, therefore, there exists a neighbourhood $O_{x_0} \subset \boldsymbol{R}^n$ of the point x_0 such that for $x \in O_{x_0}$ we have

$$\Phi(x) = \Phi(x_0) + (G_n \circ B_n \circ \ldots \circ G_1 \circ B_1)(x - x_0). \tag{4}$$

(iii) If two mappings satisfy the hypotheses of the change-of-variables theorem, their composition also satisfies them; we find that

$$\int f(y)\,dy = \int f(\Phi(x)) \cdot |\det \Phi'(x)|\,dx$$
$$= \int f(\Phi(\Psi(z)))|\det \Phi'(\Psi(z))| \cdot |\det \Psi'(z)|\,dz$$
$$= \int f(\Phi \circ \Psi)(z)|\det(\Phi \circ \Psi)'(z)|\,dz,$$

for by the theorem on the derivative of the composition and on the product of determinants we have

$$\det(\Phi \circ \Psi)'(z) = \det\{\Phi'(\Psi(z)) \circ \Psi'(z)\}$$
$$= [\det \Phi'(\Psi(z))] \cdot \det \Psi'(z).$$

(iv) If B changes only two coordinates, then by the Fubini–Stone Theorem the change-of-variables formula is satisfied.

If the mapping is primitive,

$$G_m(x^1, \ldots, x^n) = (x^1, \ldots, g(x^m), \ldots, x^n),$$

$\det G'_m = g'$ and thus, on decomposing the n-dimensional integral into n one-dimensional integrals (by virtue of the Fubini–Stone Theorem), we prove the validity of the change-of-variables formula for G_m by means of the formula for a one-dimensional integral. Therefore, it follows from (iii) that the formula holds for mappings of the form (4).

(v) Let $f \in C_0(X)$, $\underline{f} = D$—compact. Let O_x be a neighbourhood of the point x for which formula (4) holds. Then there exists an $r(x) > 0$ such that $K(x, r(x)) \subset O_x$. But $D = \bigcup_{x \in D} K(x, \frac{1}{2} r(x))$.

From this covering we select a finite covering:

$$D = \bigcup_{i=1}^{p} K(x_i, \tfrac{1}{2} r(x_i)).$$

Let $g_i \in C_0(R^n)$ be a function with the following properties:

(i) $\underline{g_i} \subset K(x_i, r(x_i))$,

(ii) for $x \in K(x_i, \frac{1}{2} r(x_i))$ we must have $g_i(x) = 1$.

Let

$$h_1 = g_1, \quad h_j = (1-g_1) \cdot (1-g_2) \ldots (1-g_{j-1}) g_j.$$

It can be verified that $h_1 + h_2 + \ldots + h_p = 1 - (1-g_1)(1-g_2) \ldots (1-g_p)$.

If $x \in D$, then x belongs to one of the balls $K(x_i, \frac{1}{2} r(x_i))$. Hence, one of the factors $(1-g_i(x))$ is equal to zero, that is

$$h_1 + h_2 + \ldots + h_p = 1 \quad \text{on } D.$$

Therefore

$$f = f \cdot \Big(\sum_{i=1}^{p} h_i\Big) = \sum_{i=1}^{p} f \cdot h_i, \quad \underline{(f \cdot h_i)} \subset O_{x_i}.$$

Since formula (4) holds in O_{x_i}, the change-of-variables theorem is valid for each of the functions $f \cdot h_i$, that is to say, is valid for $f \in C_0(\boldsymbol{R}^n)$ by the linearity of the integral.

(vi) Since $C_0(\boldsymbol{R}^n)$ is dense in $\mathscr{L}^1$ and the integral is continuous on $\mathscr{L}^1$, the theorem holds for any integrable function f. □

COROLLARY XIII.22.2. The Lebesgue integral is invariant under group of Euclidean motions, i.e. under the transformations

$$X \ni x \to \Phi(x) := Ax + a,$$

where $a \in \boldsymbol{R}^n$ and A is an orthogonal matrix.

PROOF. We have $\Phi'(x) = A$, and hence $|\det \Phi'(x)| = |\det A| = 1$. □

As is seen, the Lebesgue integral is invariant under diffeomorphisms of class C^1 with Jacobian $\det \Phi' = \pm 1$.

The existence of an invariant Radon integral on $\boldsymbol{R}^n$, the Lebesgue integral, is a special case of an important fact: the existence of an invariant Radon integral on every locally compact group.

DEFINITION. Let X be a group in which operations are denoted by xy and x^{-1} denotes the element inverse to x.

If X is a topological space, and the mapping given by the group operation

$$X \times X \ni (x, y) \to xy^{-1} \in X$$

is continuous, X is called a *topological group*.

Let X be a locally compact topological group. A Radon integral μ on X will be said to be a *left-invariant Haar integral* if the relation

$$\int f(x)\, d\mu(x) = \int f(yx)\, d\mu(x)$$

holds for every function $f \in C_0(X)$ and for every element $y \in X$.

Similarly, *right invariance* means that

$$\int f(x)\, d\mu(x) = \int f(xy)\, d\mu(x).$$

THEOREM XIII.22.3 (Haar). At least one left-invariant and one right-invariant integral exists on every locally compact group.

This theorem is proved in Chapter XIX; cf. Theorem XIX.20.2.

It turns out that the Haar integral is the only invariant integral on the group:

THEOREM XIII.22.4 (von Neumann, A. Weil). Two left- (or right-) invariant integrals μ and ν on a locally compact group are proportional, i.e. there exists a number $a \geqslant 0$ such that

$$\int f d\mu = a \int f d\nu$$

for every function $f \in C_0(X)$.

A proof is given in the last chapter, Theorem XIX.20.4.

It follows from this theorem that the Lebesgue integral is the only (up to a multiplicative constant) invariant Radon integral on $\boldsymbol{R}^n$.

As an example of the application of the change-of-variables theorem we give the *spherical coordinates* on $\boldsymbol{R}^n$.

(i) For $\boldsymbol{R}^2$ the name "polar coordinates" is used.

DEFINITION. The mapping Φ of the set

$$\Omega_2 := \{0 < r < \infty, 0 < \varphi < 2\pi\} \subset \boldsymbol{R}^2$$

into the space $\boldsymbol{R}^2$, given by the formulae $x_1 = r\cos\varphi$, $x_2 = r\sin\varphi$, is called the *system of polar coordinates* in $\boldsymbol{R}^2$.

This is an arbitrarily differentiable, invertible mapping of the set Ω_2 onto $\boldsymbol{R}^2 - R_+$, where $R_+ = \{(x_1, x_2)\colon x_2 = 0, x_1 \geqslant 0\}$.

Since $(\det\Phi')(r, \varphi) = r > 0$, by the inverse mapping theorem it follows that the inverse mapping Φ^{-1} is continuously differentiable. Thus, Φ satisfies the hypotheses of the theorem on the change of variables.

Note that the set R_+ has a zero two-dimensional Lebesgue measure: $\lambda^2(R_+) = 0$. Therefore, for $f \in \mathscr{L}^1$ we have

$$\int_{\boldsymbol{R}^2} f d\lambda^2 = \int_{\Phi(\Omega_2)} f d\lambda^2 + \int_{R_+} f d\lambda^2 = \int_{\Phi(\Omega_2)} f d\lambda^2$$

$$= \int_0^\infty \int_0^{2\pi} f(r\cos\varphi, r\sin\varphi) r\, dr\, d\varphi.$$

Using this formula, we evaluate the integral $J = \int_{-\infty}^{\infty} e^{-ax^2} dx$, where $a > 0$. Take the function $f(x, y) = e^{-a(x^2+y^2)}$. When we invoke the Fubini Theorem we find

$$\iint f(x, y)\, dx\, dy = \int e^{-ax^2} dx \cdot \int e^{-ay^2} dy = J^2.$$

On the other hand, in polar coordinates:

$$J^2 = \iint f(x, y)\,dx\,dy = \int_0^{\infty}\int_0^{2\pi} e^{-r^2} r\,dr\,d\varphi = 2\pi \cdot \int_0^{\infty} e^{-ar^2} r\,dr.$$

Setting $r^2 = t$, we get

$$J^2 = \pi \int_0^{+\infty} e^{-at}dt = \pi/a, \quad \text{whence} \quad J = \sqrt{\pi/a}\,.$$

We have thus obtained an important result:

$$\int_{-\infty}^{\infty} e^{-ax^2}dx = \sqrt{\pi/a}\,.$$

(ii) In the space $\boldsymbol{R}^n$, spherical coordinates are introduced as follows:

$$x_1 = r\cos\theta_1,$$

$$x_i = r\cos\theta_i \cdot \prod_{j=1}^{i-1} \sin\theta_j \quad \text{for } 2 \leqslant i \leqslant n-1,$$

$$x_n = r \prod_{j=1}^{n-1} \sin\theta_j,$$

where $0 < r < \infty$, $0 < \theta_i < \pi$ for $1 \leqslant i \leqslant n-2$, $0 < \theta_{n-1} < 2\pi$. It can be verified that

$$\frac{\partial(x_1, x_2, \ldots, x_n)}{\partial(r, \theta_1, \ldots, \theta_{n-1})} = r^{n-1} \prod_{j=1}^{n-1} (\sin\theta_j)^{n-1-j}.$$

Now let us consider an example of the application of the change-of-variables theorem to evaluation of integrals which appear in probability theory.

Example. Let A be a positive operator in $\boldsymbol{R}^n$, that is, we have $(Ax|x) \geqslant 0$ for $x \in \boldsymbol{R}^n$. If $A = (a_{ij})$, where the matrix elements a_{ij} are evaluated as $a_{ij} = (Ae_i|e_j)$, then

$$(Ax|x) = \sum_{i,j=1}^{n} a_{ij}x^i x^j, \quad x^i = (e_i|x).$$

From algebra we know that there exists an orthogonal matrix $B = (b_{ij})$ such that

$$B^*AB = C = \begin{bmatrix} c_1^2 & & & 0 \\ & c_2^2 & & \\ & & \ddots & \\ 0 & & & c_n^2 \end{bmatrix}.$$

We say that the *matrix B reduces the matrix A to diagonal form.* Only nonnegative terms appear on the diagonals of the matrix C, for A was positive.

As we know, the matrix B^{-1} is also orthogonal.

If now we take the matrix

$$D := \begin{bmatrix} c_1 & & & 0 \\ & c_2 & & \\ & & \ddots & \\ 0 & & & c_n \end{bmatrix},$$

then $D = D^*$ and $D^*D = C$. Since $A = (B^{-1})^*CB^{-1}$, therefore

$$(Ax|x) = \big((B^{-1})^*D^*DB^{-1}x|x\big) = (DB^{-1}x|DB^{-1}x).$$

Writing $DB^{-1} = E$, we have

$$\det E = (\det D)\,(\det B^{-1}) = \det D = \sqrt{\det C}$$
$$= \sqrt{(\det B^*)\,(\det A)\,(\det B)} = \sqrt{\det A},$$

since the determinant of an orthogonal matrix is equal to unity.

Next we calculate the following integral:

$$J_n = \int_{R^n} e^{-(Ax|x)}d\lambda^n(x) = \int_{R^n} e^{-(Ex|Ex)}d\lambda^n(x).$$

Invoking the change-of-variables theorem, we have

$$J_n = \frac{1}{|\det E|}\cdot \int_{R^n} e^{-(Ex|Ex)}\cdot |\det E|\,d\lambda^n(x)$$

$$= \frac{1}{|\det E|}\cdot \int_{R^n} e^{-(y|y)}d\lambda^n(y) = \frac{1}{|\det E|}\cdot \left\{\int_{R^1} e^{-y^2}dy\right\}^n$$

$$= \frac{1}{\sqrt{\det A}}\cdot (\sqrt{\pi})^n.$$

Consequently

$$\int_{R^n} e^{-(Ax|x)} d\lambda^n(x) = \frac{(\pi)^{n/2}}{\sqrt{\det A}}.$$

23. MAPPING OF RADON INTEGRALS

Let X_1 and X_2 be locally compact spaces. Let a Radon integral μ be defined on X_1. Given, moreover, a mapping $T: X_1 \to X_2$. If T has the property that the relation

$$f \circ T \in \mathcal{L}^1(\mu)$$

holds for every function $f \in C_0(X_2)$, then the following Radon integral can be defined on X_2:

$$C_0(X_2) \ni f \to \nu(f) := \mu(f \circ T)$$

(this functional is linear and positive, and thus, by Theorem XIII.19.2, is a Daniell–Stone integral).

We set $\nu = T\mu$. Thus we have the formula

$$(T\mu)(f) = \mu(f \circ T) \quad \text{or} \quad \int_{X_2} f d(T\mu) = \int_{X_1} (f \circ T) d\mu.$$

If T is a homeomorphism, for example, then it preserves the continuity of a function and the compactness of its support, that is, for $f \in C_0(X_2)$ we have $f \circ T \in C_0(X_1) \subset \mathcal{L}^1(\mu)$.

Hence, homeomorphisms of locally compact spaces determine mappings of integrals.

24. INTEGRALS WITH DENSITY. THE RADON–NIKODYM THEOREM

Let μ be a Daniell–Stone integral on the space X and let E be its family of elementary functions. Given a μ-measurable function $0 < g$. We introduce the following notation:

$$E(g) := \{f \in \mathcal{L}^1(\mu) : f \cdot g \in \mathcal{L}^1(\mu)\}.$$

LEMMA XIII.24.1. A set $E(g)$ is a vector lattice.

PROOF. Follows immediately from the properties of measurable and integrable functions. In particular

$$|f| \cdot g = |f \cdot g| \in \mathcal{L}^1, \quad \text{if } f \cdot g \in \mathcal{L}^1, \text{ and hence } |f| \in E(g). \ \square$$

On the space $E(g)$ we define the functional

$$(g\mu)(f) := \int f \cdot g d\mu \quad \text{for } f \in E(g).$$

The functional $g\mu$ is clearly Linear and positive. Its monotonic continuity follows from the Levi Lemma: if $E(g) \ni f_n \searrow 0$, then $\mathscr{L}^1 \ni f_n \cdot g \searrow 0$, hence $\int f_n \cdot g d\mu \to 0$.

The functional $\nu = g\mu$ thus is a Daniell–Stone integral on the space X. The function g is called the *density of the integral* ν with respect to the integral μ and is denoted by $g = d\nu/d\mu$.

Definition. Given two integrals μ and ν on the space X. The integral ν is said to be *absolutely continuous with respect to* μ and we write $\nu \ll \mu$ when the following conditions are met:

(i) The set $\mathscr{L}^1(\mu) \cap \mathscr{L}^1(\nu)$ is dense in $\mathscr{L}^1(\nu)$, that is, every function $f \in \mathscr{L}^1(\nu)$ can be obtained as the limit of the function $f_n \in \mathscr{L}^1(\mu) \cap \mathscr{L}^1(\nu)$ in the sense

$$\nu(|f_n - f|) \to 0.$$

(ii) If $Z \subset X$ and $\mu^*(Z) = 0$, then $\nu^*(Z) = 0$.

Lemma XIII.24.2. The relation

$$g\mu \ll \mu$$

is satisfied.

Proof. The definition of $E(g)$ implies property (i).

(ii) If $\mu^*(Z) = 0$, then $\chi_Z \cdot g = 0$ μ-almost everywhere. Hence $\chi_Z \cdot g \in \mathscr{L}^1(\mu)$ and $(g\mu)(Z) = \int \chi_Z \cdot g d\mu = 0$. □

It turns out that on a space of semifinite measure, the absolute continuity of the measure ν with respect to μ is sufficient for $\nu = g\mu$.

Theorem XIII.24.3 (Radon, Nikodym). Let ν and μ be semifinite integrals on X, satisfying the strong Stone axiom.

If $\nu \ll \mu$, then there exists a μ-measurable function h such that $\nu = h\mu$.

Remark. It is usually assumed that $E(\mu) = E(\nu)$, which simplifies the proof.

Proof. (i) We write $S(\nu + \mu) := S(\nu) \cap S(\mu)$. We recall that

$$S(\nu) = \left\{ f = \sum_{i=1}^{n} a_i \chi_{Z_i}, \chi_{Z_i} \in \mathscr{L}^1(\nu),\ Z_i \cap Z_j = \emptyset \text{ for } i \neq j \right\}.$$

Similarly, we define the space $S(\mu)$.

We shall prove that $S(\nu+\mu)$ is dense in $\mathscr{L}^1(\nu)$.

Let $f \in \mathscr{L}^1(\nu)$. From the absolute continuity of the measure ν with respect to μ we infer that there exists a sequence $f_n \in \mathscr{L}^1(\nu) \cap \mathscr{L}^1(\mu)$ such that $\nu(|f_n - f|) \to 0$. As we know

$$f_n = \lim_{\varepsilon\to 0} f_n^\varepsilon = \lim_{\varepsilon\to 0} (\lim_{k\to\infty} {}^k f_n^\varepsilon),$$

where

$${}^k f_n^\varepsilon = \sum_{m=0}^{k} m\varepsilon(\chi_{A_{\varepsilon m}} - \chi_{A_{\varepsilon(m+1)}}) + \sum_{m=0}^{-k} m\varepsilon(\chi_{A_{\varepsilon(m-1)}} - \chi_{A_{\varepsilon m}}),$$

and $A_a = \{x\colon f_n(x) < a\}$.

Now, $f_n \in \mathscr{L}^1(\nu) \cap \mathscr{L}^1(\mu)$; accordingly $\chi_{A_i} \in \mathscr{L}^1(\nu) \cap \mathscr{L}^1(\mu)$, and thus ${}^k f_n^\varepsilon \in S(\nu+\mu)$. But ${}^k f_n^\varepsilon \underset{\substack{k\to\infty\\ \varepsilon\to 0}}{\to} f_n$ pointwise and $|{}^k f_n^\varepsilon| \leqslant |f_n| \in \mathscr{L}^1(\nu)$ and, therefore, by the Lebesgue Theorem, we have

$$\nu(|{}^k f_n^\varepsilon - f_n|) \underset{\substack{k\to\infty\\ \varepsilon\to 0}}{\to} 0.$$

Since $f_n \in \overline{S(\mu+\nu)}$ (closure in the ν-sense), $f = \lim f_n \in \overline{S(\mu+\nu)}$, that is, our thesis has been proven.

Note that $S(\mu+\nu)$ is a vector lattice satisfying the strong Stone axiom. Thus, on this set we define the Daniell–Stone integral

$$(\mu+\nu)(f) := \mu(f) + \nu(f) \quad \text{for } f \in S(\mu+\nu).$$

(ii) First of all, we prove the Radon–Nikodym Theorem in the case when X has a finite measure: $\mu(X) < \infty$, $\nu(X) < \infty$.

Let $f \in \mathscr{L}^2(\mu+\nu)$. Thus there exists a sequence $f_k \in S(\mu+\nu)$ such that $f_k \underset{k\to\infty}{\to} f$ in the sense of $\mathscr{L}^2(\mu+\nu)$. But

$$\nu^*(|f_k - f|) \leqslant (\mu+\nu)^*(|f_k - f|) \leqslant [(\mu+\nu)^*(|f_k - f|^2)]^{1/2} \times$$
$$\times \Big\{\int_X 1^2 d(\mu+\nu)\Big\}^{1/2} = \|f_k - f\|_{L^2(\mu+\nu)} \cdot (\mu(X) + \nu(X))^{1/2} \underset{k\to\infty}{\to} 0,$$

and hence $f_k \to f$ in the sense of $\mathscr{L}^1(\nu)$, that is, $f \in \mathscr{L}^1(\nu)$.

Identically we prove that

$$\nu(f) \leqslant \nu(|f|) \leqslant \|f\|_{L^2(\mu+\nu)} \cdot (\mu(X) + \nu(X))^{1/2}.$$

Thus the linear functional ν is continuous over the space $\mathscr{L}^2(\mu+\nu)$. Clearly, if $f=f^*$ $(\mu+\nu)$-almost everywhere, then $\nu(f)=\nu(f^*)$, and hence this functional is linear and continuous on $\mathscr{L}^2(\mu+\nu)$.

From the Fréchet–Riesz Theorem we infer that there exists a function $g\in\mathscr{L}^2(\mu+\nu)$ such that

$$\nu(f)=(f|\dot{g})=\int_X f\cdot g\,d(\mu+\nu)=(\mu+\nu)(f\cdot g).$$

Now, suppose we are given an arbitrary function $f\in\mathscr{L}^1(\nu)$. Since $S(\mu+\nu)$ is dense in $\mathscr{L}^1(\nu)$, there exists a sequence

$$L^2(\mu+\nu)\supset S(\mu+\nu)\ni f_n\to f \quad \text{in the } \nu\text{-sense},$$

and $f_{n_l}\to f$ ν-almost everywhere. Since

$$(\mu+\nu)(|f_{n_i}\cdot g-f_{n_j}\cdot g|)=\nu(|f_{n_i}-f_{n_j}|)\to 0,$$

we have $f\cdot g\in\mathscr{L}^1(\mu+\nu)$, because $f_{n_l}\in\mathscr{L}^1(\mu+\nu)$ (cf. Corollary XIII.7.1).

But $\mathscr{L}^1(\mu+\nu)\subset\mathscr{L}^1(\mu)\cap\mathscr{L}^1(\nu)$ (since the convergence of the sequence $f_{n_l}\in S(\mu+\nu)$ in the sense $(\mu+\nu)$ implies ν- and μ-convergence), and hence $f\cdot g\in\mathscr{L}^1(\mu)\cap\mathscr{L}^1(\nu)$.

Thus, since $f\cdot g\in\mathscr{L}^1(\nu)$, then also

$$f\cdot g^n\cdot g\in\mathscr{L}^1(\mu+\nu)\subset\mathscr{L}^1(\mu)\cap\mathscr{L}^1(\nu).$$

Proceeding repeatedly in this manner, we show that if $f\in\mathscr{L}^1(\nu)$ then

$$f\cdot g^n\in\mathscr{L}^1(\mu+\nu)\subset\mathscr{L}^1(\mu)\cap\mathscr{L}^1(\nu)\quad\text{and}$$
$$\nu(f)=(\mu+\nu)(f\cdot g).$$

In particular for $1\equiv f\in\mathscr{L}^1(\nu)$ we have $g^n\in\mathscr{L}^1(\mu+\nu)$.

The following rearrangements are thus meaningful:

$$\begin{aligned}\nu(f)&=(\mu+\nu)(f\cdot g)=\nu(f\cdot g)+\mu(f\cdot g)\\&=(\mu+\nu)(f\cdot g^2)+\mu(f\cdot g)=\nu(f\cdot g^2)+\mu(f\cdot g^2)+\mu(f\cdot g).\end{aligned}$$

Applying this procedure k times, we obtain

$$\nu(f)=\nu(f\cdot g^k)+\sum_{n=1}^{k}\mu(f\cdot g^n)=\nu(f\cdot g^k)+\mu\Big(f\cdot\sum_{n=1}^{k}g^n\Big).$$

We shall demonstrate later that $0\leqslant g<1$ $(\mu+\nu)$-almost everywhere. Thus $\mathscr{L}^1(\nu)\ni f\cdot g^k\to 0$ pointwise ν-almost everywhere and

$$|f\cdot g^k|\leqslant|f|\in\mathscr{L}^1(\nu),\quad\text{and therefore}\quad \nu(f\cdot g^k)\underset{k\to\infty}{\to}0.$$

If we write

$$h_k := \sum_{n=1}^{k} g^n,$$

then $\mathscr{L}^1(\mu) \cap \mathscr{L}^1(\nu) \supset \mathscr{L}^1(\mu+\nu) \ni h_k \nearrow h$, and hence h is measurable under all three measures. But

$$\mathscr{L}^1(\mu) \ni f \cdot h_k = f^+ \cdot h_k - f^- \cdot h_k,$$

whereas

$$\mathscr{L}^1(\mu) \ni f^+ \cdot h_k \nearrow f^+ \cdot h, \quad \mathscr{L}^1(\mu) \ni f^- \cdot h_k \nearrow f^- \cdot h.$$

Since $\mu(f^+ \cdot h_k) = \nu(f^+) - \nu(f^+ \cdot g^k) \leqslant \nu(f^+)$, by the Levi Lemma $f^+ \cdot h \in \mathscr{L}^1(\mu)$ (similarly $f^- \cdot h \in \mathscr{L}^1(\mu)$) and $\mu(f \cdot h) = \lim \mu(f \cdot h_k)$.

Therefore, for $f \in \mathscr{L}^1(\nu)$ we have $f \cdot h \in \mathscr{L}^1(\mu)$ and

$$\nu(f) = \mu(f \cdot h), \quad \text{that is,} \quad \nu = h \cdot \mu.$$

(iii) We shall show that indeed $0 \leqslant g < 1$ $(\mu+\nu)$-almost everywhere.

Let $W := \{x\colon g \geqslant 1\}$. Since $g \in \mathscr{L}^2(\mu+\nu)$, the set W is measurable. Being a subset of the integrable set X, it is itself integrable: $\chi_W \in \mathscr{L}^1(\mu+\nu) \subset \mathscr{L}^1(\mu) \cap \mathscr{L}^1(\nu)$.

Now, $\chi_W \leqslant \chi_W \cdot g$, and thus

$$\nu(W) = \nu(\chi_W) = (\mu+\nu)(\chi_W \cdot g) \geqslant (\mu+\nu)(\chi_W)$$
$$= \mu(W) + \nu(W)$$

whereby $\mu(W) = 0$. From the fact that $\nu \ll \mu$ it follows that $\nu(W) = 0$.

Let $Z := \{x\colon g < 0\}$. By the same arguments as before,

$$\chi_Z \in \mathscr{L}^1(\mu) \cap \mathscr{L}^1(\nu), \quad \text{that is} \quad \chi_Z \in S(\mu+\nu).$$

Since $\chi_Z \cdot g \leqslant 0$ we have

$$0 \leqslant \nu(\chi_Z) = (\mu+\nu)(\chi_Z \cdot g) \leqslant 0,$$

and thus $\chi_Z \cdot g = 0$ $(\mu+\nu)$-almost everywhere.

This function differs from zero on the set Z, whence $(\mu+\nu)(Z) = 0$.

(iv) This theorem was proved for a space X with finite measure. We shall show that in the general case a space with semifinite measures μ and ν is representable as the countable sum of disjoint μ- and ν-integrable sets, on which the theorem holds.

If $X = \bigcup_{i=1}^{\infty} \Omega_i$ and h^i denotes the density of the measure ν with respect to μ on Ω_i, then, taking $h := \sum_{i=1}^{\infty} h^i$, we obtain the sought-after density of ν with respect to μ over all X.

But if X is of semifinite measure ν, then

$$X = \bigcup_{i=1}^{\infty} A_i, \quad \chi_{A_i} \in \mathscr{L}^1(\nu).$$

Since $S(\mu+\nu)$ is dense in $\mathscr{L}^1(\nu)$, there exists a sequence $S(\mu+\nu) \ni f_n = |f_n| \to \chi_{A_i}$. Now, there is a subsequence (f_{n_k}) such that $f_{n_k} \to \chi_{A_i}$ ν-almost everywhere; accordingly the set

$$Z_i := \{x \in A_i \colon \text{for every } k, f_{n_k}(x) = 0\}$$

must have the measure ν equal to zero, for on this set the sequence f_{n_k} does not tend to χ_{A_i}.

Clearly the sets $B_k^i := \{x \in X \colon f_{n_k}(x) > 0\}$ satisfy the condition $\chi_{B_k^i} \in \mathscr{L}^1(\mu) \cap \mathscr{L}^1(\nu)$.

Therefore

$$A_i \subset \left(\bigcup_{k=1}^{\infty} (B_k^i \cup Z_i)\right),$$

and thus

$$X = \left(\bigcup_{i=1}^{\infty} \bigcup_{k=1}^{\infty} B_k^i\right) \cup \left(\bigcup_{i=1}^{\infty} Z_i\right),$$

where the sets B_k^i are μ- and ν-integrable, and Z_i have a zero ν-measure. Clearly the set

$$Z := \bigcup_{i=1}^{\infty} Z_i$$

has a zero measure ν. Also the set

$$Z_0 := X - \left(\bigcup_{i=1}^{\infty} \bigcup_{k=1}^{\infty} B_k^i\right),$$

being a subset of the set Z, has a zero measure ν.

The set Z_0 is a μ-measurable set, as it is the difference of μ-measurable sets.

Since X is of semifinite μ-measure,

$$X = \bigcup_{j=1}^{\infty} D_j, \quad \text{where the sets } D_j \text{ are } \mu\text{-integrable.}$$

The sets $Z_0 \cap D_j$ are, therefore, μ-integrable (as intersections of measurable sets with integrable ones) and ν-measurable (as ν-zero sets).

Thus $X = (\bigcup_{i=1}^{\infty} \bigcup_{k=1}^{\infty} B_k) \cup (\bigcup_{j=1}^{\infty} Z_0 \cap D_j)$, where the sets B_k^i and $Z_0 \cap D_j$ are μ- and ν-integrable.

This can be written as

$$X = \bigcup_{i=1}^{\infty} C_i,$$

where the sets C_i are μ- and ν-integrable. Suppose that $\Omega_n := \{\bigcup_{i=1}^{n} C_i\} - \{\bigcup_{i=1}^{n-1} C_i\}$, we obtain $X = \bigcup_{n=1}^{\infty} \Omega_n$, where the sets Ω_n are disjoint and μ- and ν-integrable. □

Examples. 1. The Radon integral ν is absolutely continuous with respect to the Radon integral μ if and only if every μ-zero set is also ν-zero.

PROOF. The first point of the definition is satisfied automatically, since $E = C_0(X) \subset \mathscr{L}^1(\mu) \cap \mathscr{L}^1(\nu)$ and is dense in both of these spaces.

2. Let T be a mapping $R^n \to R^n$ satisfying the hypotheses of the change-of-variables theorem. It is not difficult to prove that the integral $T\lambda^n$ (cf. Section 23) satisfies the condition written in Example 1. Thus $T\lambda^n \ll \lambda^n$. Accordingly, there must exist a function $g \in M$ such that

$$\int f d(T\lambda^n) = \int (f \circ T) d\lambda^n = \int g \cdot f d\lambda^n.$$

The change-of-variables theorem says that $g = 1/|\det T'|$.

Thus $T\lambda^n = \dfrac{1}{|\det T'|} \lambda^n$ or

$$\frac{d(T\lambda^n)}{d(\lambda^n)} = \frac{1}{|\det T'|}.$$

25. THE WIENER INTEGRAL

Given a family of sets X_i labelled by an arbitrary parameter $i \in J$. Let $X := \bigcup_{i \in J} X_i$.

DEFINITION. The family of all mappings of the set J into X such that $L \ni i \to f(i) \in X_i$ is called the *Cartesian product of the sets* X_i, $i \in J$, and is denoted by $\bigtimes_{i \in J} X_i$ or $\prod_{i \in J} X_i$.

Example. Let $J = \{1, 2\}$. Then $f = (f(1), f(2))$. We denote

$$x^1 := f(1) \in X_1, \quad x^2 := f(2) \in X_2.$$

Every element of the space $\bigtimes_{i \in J} X_i$ is thus representable in the form (x^1, x^2). This definition hence coincides with the familiar definition of the Cartesian product $X_1 \times X_2$.

Let $x \in (\bigtimes_{i \in J} X_i)$, that is, $x = (x_i)_{i \in J}$, $x_i = f(i) \in X_i$.

We introduce the functions $p_i(x) := x_i$. The function p_i will be called a *projection onto the space* X_i.

Now suppose that (X_i, τ_i), $i \in J$, are topological spaces. In the set $\bigtimes_{i \in J} X_i$ we introduce the weakest topology under which all mappings p_i are continuous. We denote this topology by the letter τ. Then the topological space

$$(\bigtimes_{i \in J} X_i, \tau)$$

is called the *Cartesian product of the topological spaces* X_i, $i \in J$.

THEOREM XIII.25.1 (Tychonoff Theorem). If the spaces X_i are compact, then $(\bigtimes_{i \in J} X_i, \tau)$ is a compact space.

This theorem is given here without proof (the proof is presented in Chapter XII).

Now, let $J = R^1$, $X_t = \overline{R^n}$. We write

$$\Omega := \bigtimes_{t \in R} \overline{R^n_t}, \quad \overline{R^n_t} = \overline{R^n} \quad \text{for all } t \in R.$$

Every element $\omega \in \Omega$ thus is a function

$$R \ni t \to \omega(t) \in \overline{R^n}.$$

This function is interpretable as the trajectory of a particle in $\overline{R^n}$.

Suppose that we are given a finite sequence of points in R,

$$t_1 < t_2 < \ldots < t_s,$$

and suppose that $F \in C(\overline{R}^{ns})$. Then the function

$$\varphi(\omega) := F\big(p_{t_1}(\omega), p_{t_2}(\omega), \ldots, p_{t_s}(\omega)\big)$$

is continuous over Ω as it is a composition of the continuous functions p_{t_i} and F. Now let us consider the family $C_{\text{fin}}(\Omega)$ of all functions of this form. This family satisfies the hypotheses of the Stone–Weierstrass Theorem since:

(i) The Tychonoff Theorem implies that Ω is a compact space.

(ii) This family separates the points of the set Ω. Here is the proof: Let $\omega_1 \neq \omega_2$. This means that there exists a $t \in R$ for which $\omega_1(t) \neq \omega_2(t)$. Take the function

$$\overline{R^n} \ni x \to F(x) = \frac{x}{1+|x|}.$$

Plainly $F \in C(\overline{R^n})$. Let $C_{\text{fin}}(\Omega) \ni \varphi := F \circ p_t$. Then $\varphi(\omega_1) = \omega_1(t) \neq \omega_2(t) = \varphi(\omega_2)$, and hence the function φ separates the points ω_1 and ω_2. Therefore, any continuous function over Ω can be uniformly approximated by the polynomials of functions from $C_{\text{fin}}(\Omega)$. But the set $C_{\text{fin}}(\Omega)$ is an algebra, since the product and sum of the functions from $C_{\text{fin}}(\Omega)$ is also a function from $C_{\text{fin}}(\Omega)$.

The Stone–Weierstrass Theorem thus states that $\overline{C_{\text{fin}}(\Omega)} = C(\Omega)$ (closure in the sense of uniform convergence).

Continuing, we define an integral on the set $C_{\text{fin}}(\Omega)$, dense in $C(\Omega)$. To this end, we introduce the notation

$$P^t(x, dy) := \frac{1}{(4\pi Dt)^{n/2}} e^{\frac{-\|x-y\|^2}{4Dt}} d\lambda^n(y).$$

Let $f \in \mathscr{L}^1(\lambda^n)$. Then we define

$$(P^t f)(x) := \int_{R^n} f(y) P^t(x, dy).$$

Remark. If Z is a measurable set, then $(P^t \chi_Z)(x)$ may be interpreted as the probability that a particle performing Brownian motion, with a diffusion coefficient D and starting from the point $\omega(0) = x$ at the instant $t_0 = 0$, will be in the set Z at the instant t.

Now, suppose that $\varphi \in C_{\text{fin}}(\Omega)$

$$\varphi(\omega) = F\big(\omega(t_1), \ldots, \omega(t_s)\big).$$

Take the functional

$$\mu_{t,x}(\varphi) := \int\limits_{R^n} \dots \int\limits_{R^n} P^{t_1-t}(x, dx_1) P^{t_2-t_1}(x_1, dx_2) \dots$$

$$\dots P^{t_s-t_{s-1}}(x_{s-1}, dx_s) F(x_1, \dots, x_s).$$

As calculated in Section 22, the equality

$$\mu_{t,x}(1) = 1 \cdot 1 \cdot \dots \cdot 1 = 1$$

holds.

Since $\mu_{t,x}$ is a positive linear functional,

$$|\mu_{t,x}(\varphi)| \leqslant \mu_{t,x}(\sup|\varphi(\Omega)|) = \sup|\varphi(\Omega)|.$$

The functional $\mu_{t,x}$ thus is continuous in the topology of uniform convergence and hence is extendible from the set $C_{\text{fin}}(\Omega)$ to a positive linear functional over the entire space $C(\Omega)$. The Radon integral so obtained is called a *Wiener integral*. Now, treating $C(\Omega)$ as the space of elementary functions, we can construct the spaces $\mathscr{L}^p(\mu_{t,x})$.

Remark. If Z_i are measurable sets, then the number

$$\int\limits_{Z_1} \dots \int\limits_{Z_s} P_{t_1-t}(x, dx_1) \dots P^{t_s-t_{s-1}}(x_{s-1}, dx_s)$$

expresses the probability that the Brownian particle, starting from the point x at the instant t, will pass through the set Z_i at every instant t_i.

The Wiener integral was the first example of an integral over an infinite-dimensional set. The foregoing elegant treatment of the Wiener integral, which clearly differs very much from the Wiener treatment, is due to E. Nelson.

Studies on "integration in function spaces," which also play a major role in the mathematical foundations of modern quantum field theory, began with the Wiener integral and have continued to this day.

26. THE KOLMOGOROV THEOREM

The construction of Wiener integral given above immediately leads to a generalization.

Generalization. Let X_t be a compact space, let T be an arbitrary set, and let $\Omega := \bigtimes_{t \in T} X_t$. Since the space Ω is compact, we can once again invoke the Stone–Weierstrass Theorem. For this purpose we introduce the following notations: When $S \subset T$, then $\mathscr{C}_S$ is the algebra of all continuous real-valued functions f on Ω such that

$$(*) \qquad \Big(\bigwedge_{s \in S} \omega(s) = \omega_1(s)\Big) \Rightarrow \big(f(\omega) = f(\omega_1)\big).$$

Hence $f(\omega) = f(\omega_1)$, when $p_S\omega_1 = p_S\omega$, where p_S is the projection $x \to p_S(x) = (x_i)_{i \in S}$.

$\mathscr{C}_H := \bigcup \mathscr{C}_S$, when S runs over all finite subsets $S \subset T$.

LEMMA XIII.26.1. $\mathscr{C}_H$ is dense in $\mathscr{C} := \mathscr{C}_T = C(\Omega)$.

PROOF. This is an immediate corollary to the Stone–Weierstrass Theorem since $\mathscr{C}_H$ contains constants and separates the points of Ω. □

Let K be a compact space. $\mathscr{B}(K)$ denotes the σ-algebra of Borel subsets, i.e. the smallest σ-algebra containing all the compact (hence closed, hence open) sets.

A *regular measure* μ is a Borel measure μ: $B \to \mu(B) \in \boldsymbol{R}_+$, $B \in \mathscr{B}(K)$, such that $\mu(B) = \sup_{F \subset B} \mu(F)$, where F runs over the closed subsets $F \subset B$. Regular measures are Radon measures: the one-to-one mapping is given by the formula

$$C(K) \ni \varphi \to \mu(\varphi) = \int \varphi(k)\, d\mu(k).$$

E. Nelson has given the following generalization of the Kolmogorov Theorem XVIII.4.2:

THEOREM XIII.26.2 (Nelson). Suppose that for every $S \subset T$, μ_S denotes a linear positive continuous and normed functional on $\mathscr{C}_S$: $\mu_S(1) = 1$, $\mathscr{C}_S \ni f \to \mu_S(f)$ and let the family $\{\mu_S\}$ be consistent, i.e.

$$\text{(C)} \qquad (S_1 \subset S) \Rightarrow (\mu_S | \mathscr{C}_{S_1} = \mu_{S_1}).$$

Then there exists on Ω exactly one Radon integral μ (or, if one prefers, regular measure μ) such that

$$\mu_S(f) = \int f(\omega)\, d\mu(\omega), \qquad f \in \mathscr{C}_S.$$

In other words, $\mu | \mathscr{C}_S = \mu_S$.

PROOF. For every $f \in \mathscr{C}_H$ there exists a finite set $S \subset T$ such that $f \in \mathscr{C}_S$. Let $\mu(f) := \mu_S(f)$. This definition is correct owing to the consistency condition (C). We verify that when $f \geqslant 0$, then $\mu(f) \geqslant 0$, $\mu(1) = 1$. Thus, μ is a continuous, positive linear functional on $\mathscr{C}_H$.

Since C_H is dense in $C(\Omega)$, there exists a unique continuous extension $\tilde{\mu}$ of the functional μ to the whole of $C(\Omega)$. Plainly, $\tilde{\mu}$ has the desired property: $\tilde{\mu}|\mathscr{C}_S = \mu_S$ by definition. Any other regular measure (Radon integral) with the aforementioned property is equal to the functional μ on $\mathscr{C}_H$.

Remarks. Let us look at the construction above from the point of view taken by Kolmogorov, i.e. given the normed measures μ_S on X^S satisfying the consistency condition (K) given below, construct a measure on Ω. Note that when we write $X^S := \bigtimes_{t\in S} X_t$ we have the bijection

$$\mathscr{C}_S \leftrightarrow C(X^S) \quad \text{for every finite } S \subset T,$$

given by the formulae

$$C(X^S) \ni f^S \to f_S := f^S \circ p_S \in C_S,$$

where p_S is the projection onto X^S,

$$(**) \qquad \mathscr{C}_S \ni f_S \to f^S \in C(X^S) \quad \text{where } f^S(x_S) := f_S(x),$$

where $x_S = p_S(x)$, $x \in \Omega$, that is,

$$f^S(x_S) = f^S \circ p_S(x) = f_S(x).$$

Definition $(**)$ is correct on account of condition $(*)$, which after all can be rewritten as

$$\Big(\bigwedge_{x,y\in\Omega} p_S(x) = p_S(y)\Big) \Rightarrow (f(x) = f(y)).$$

Thus we have a one-to-one mapping of the measures μ^S on X^S and μ_S on Ω:

$$\mu^S(f^S) := \mu_S(f_S).$$

The measures μ^S satisfy the Kolmogorov consistency condition

$$\text{(K)} \qquad \text{if } S_1 \subset S, \text{ then } \mu^{S_1} = p_S^{S_1}\mu^S,$$

where $p_S^{S_1}: X^S \to X^{S_1}$ is the canonical projection.

Condition (K) corresponds to the condition

$$\text{(C')} \qquad \mu^{S_1} = p_{S_1}\mu_S.$$

Theorem XIII.26.2 may thus be reformulated in the following form: Let $\Omega = \bigtimes_{t\in T} X_t$ and suppose that a normed measure μ^S is given on $X^S = \bigtimes_{t\in S} X_t$ for every finite set $S \subset T$, the consistency condition (K) being satisfied. Then there exists on Ω a unique regular measure μ such that $p_S\mu = \mu^S$. This measure is denoted by $\bigotimes_{t\in T} \mu_t$.

Note that in the case of finite Cartesian products, the present construction reduces to the tensor product of Radon measures. In the general case, the construction consists in building cylinders, i.e. subsets of the Cartesian product $\bigtimes_{t\in S} X_t$ of the form $Z_S = p_S^{-1}A_S$, where $A_S \in X^S$, and assigning it the measure

$$\mu(Z_S) = \mu(p_S^{-1}A_S) := \mu^S(A_S).$$

Cylinders do not constitute a σ-algebra and, hence, in the traditional approach it is necessary to build a σ-algebra $\mathscr{A}$ of subsets of the space X_t with the aid of the cylinders and then to demonstrate that a function μ (defined on the cylinders) is uniquely extendible to the (regular) measure on $\mathscr{A}$. This construction is difficult and refined. It is seen that the functional approach is, in this case, simpler: the proof is only a few lines long. This construction is somewhat less general, for we operate with compact sets, whereas the abstract construction is valid for any normed measures.

Of course, in the case of regular measures on compact spaces the two constructions are identical. In conclusion, we note that the Nelson approach fully suffices for applications. The Kolmogorov Theorem will be taken up again in Chapter XVIII, Section 4.

27. INTEGRATION OF VECTOR FIELDS

Let $(B, \|\ \|)$ be a Banach space, and let (X, E, μ) be a space with an integral satisfying the strong Stone axiom.

We shall deal with vector fields on a space X with values in B, that is, the mappings

$$X \ni x \to \vec{f}(x) \in B.$$

In the set of all vector fields, which we denote by $F(X, B)$, we define the functions

$$N_p(\vec{f}) := [\mu^*(\|\vec{f}(\cdot)\|^p)]^{1/p}, \quad p \geqslant 1.$$

It is an obvious fact that N_p has the properties of a seminorm.

Just as in integral theory, we define the spaces $\mathscr{F}^p$ for the functions

$$\mathscr{F}^p(\mu, B) := \{\vec{f} \in F(X, B) \colon N_p(\vec{f}) < \infty\}.$$

In the same manner as in the scalar case, we prove the completeness of the space $\mathscr{F}^p$ (making use of, among other things, the completeness of the space B).

The role of the space of elementary functions E is now taken over by the family $E(B)$ of vector fields, of the form

$$\vec{f}(\cdot) = \sum_{i=1}^{n} b_i \cdot f_i(\cdot), \quad \text{where } b_i \in B, f_i \in E.$$

The spaces $\mathscr{L}^p(\mu, B)$ are once again defined as the closure of the set $E(B) \cap \mathscr{F}^p(\mu, B)$ in $\mathscr{F}^p(\mu, B)$. A number of properties of these spaces may be proved without resorting to scalar theory. Since in the present exposition, the integration of vector fields is treated merely informatively and since a scalar theory has already been built up, we shall proceed "inelegantly", basing ourselves on the scalar-theoretical results.

As we know, it is sometimes more convenient, in scalar theory to replace the family E by the family of simple functions S.

Similarly, we introduce the space:

$$S(\mu,B) := \Big\{F(X, B) \ni \vec{f} = \sum_{i=1}^{n} b_i \cdot \chi_{A_i};\ b_i \in B,\ \chi_{A_i} \in \mathscr{L}^1(\mu)$$
$$\text{or } b_i = 0,\ A_i \cap A_j = \emptyset \text{ for } i \neq j\Big\}.$$

Note that

$$N_p\Big(\sum_{i=1}^{n} b_i \chi_{A_i}\Big) = \Big[\mu^*\Big(\sum_{i=1}^{n} \|b_i\|^p \cdot \chi_{A_i}\Big)\Big]^{1/p}$$
$$= \Big[\sum_{i=1}^{n} \|b_i\|^p \mu(A_i)\Big]^{1/p} < \infty.$$

Therefore, $S(\mu, B) \subset \mathscr{F}^p(\mu, B)$ for every $p \geqslant 1$.

DEFINITION. $\mathscr{L}^p(\mu, B) := \overline{S(\mu, B)}$ (closure in the sense of N_p).

A linear operator, the integral $\mu \colon S(\mu, B) \to B$, can be defined on the space $S(\mu, B)$ as follows

$$\mu\Big(\sum_{i=1}^{n} b_i \chi_{A_i}\Big) := \sum_{i=1}^{n} b_i \cdot \mu(A_i) \in B.$$

It is readily seen that this operator is linear and, moreover,

$$\left\|\mu\left(\sum_{i=1}^{n} b_i\chi_{A_i}\right)\right\| = \left\|\sum_{i=1}^{n} b_i\cdot\mu(A_i)\right\| \leqslant \sum_{i=1}^{n} \|b_i\|\cdot\mu(A_i)$$

$$= N_1\left(\sum_{i=1}^{n} b_i\chi_{A_i}\right),$$

or

$$(1) \qquad \|\mu(\vec{f})\| \leqslant N_1(\vec{f}).$$

This operator thus is continuous on $S(\mu, B)$ and hence is extendible to the entire space $\mathscr{L}^1(\mu, B)$. Finally, therefore, we have an integral on $\mathscr{L}^1(\mu, B)$

$$\mathscr{L}^1(\mu, B) \ni \vec{f} \to \mu(\vec{f}) \in B.$$

Remark. Notation similar to that for scalar functions is often used:

$$\int \vec{f}\, d\mu = \int_X \vec{f}(x)\, d\mu(x) := \mu(\vec{f}).$$

Remark. The space $\mathscr{L}^1(\mu, B)$ was first introduced by S. Bochner. Accordingly, the integral above is frequently called the *Bochner integral.*

The following lemma holds, just as in the scalar case:

LEMMA XIII.27.1 $(\vec{f} \in \mathscr{L}^p(\mu, B)) \Rightarrow (\|\vec{f}\| \in \mathscr{L}^p(\mu, \mathbf{R}))$.

PROOF. If $S(\mu, B) \ni \vec{f}_n \to \vec{f}$ in the N_p sense, then $S(\mu) \ni \|f_n\| \to \|f\|$, since

$$N_p(|\,\|\vec{f}_n\| - \|\vec{f}\|\,|) \leqslant N_p(\|\vec{f}_n - \vec{f}\|) \underset{n\to\infty}{\to} 0. \; \square$$

Therefore, inequality (1), which transposes on extension of the integral μ to the entire space $\mathscr{L}^1(\mu, B)$, can be written as follows:

$$\left\|\int \vec{f}\, d\mu\right\| \leqslant N_1(\vec{f}) = N_1(\|\vec{f}\|) = \int \|\vec{f}\|\, d\mu.$$

Consequently we have:

THEOREM. XIII.27.2. The inequality

$$\left\|\int \vec{f}\, d\mu\right\| \leqslant \int \|\vec{f}\|\, d\mu$$

holds.

As in the scalar case, the Lebesgue Theorem on majorized convergence holds here.

It is easy to see that continuous linear operators can be taken off the Bochner integral. We shall prove here a slightly more general theorem concerning a larger class of operators.

DEFINITION. Let B_1, B_2 be Banach spaces, A a linear operator defined on a dense linear subset $D(A) \subset B_1$, i.e. $A: D(A) \to B_2$. A is called a *closed linear operator* whenever the following condition holds:

$$(\vec{b}_n \in D(A), \vec{b}_n \to \vec{b}_0, A\vec{b}_n \to \vec{c}_0) \Rightarrow (\vec{b}_0 \in D(A), A\vec{b}_0 = \vec{c}_0).$$

Each continuous linear operator is obviously closed, but the converse is not true.

THEOREM XIII.27.3. Let A be a closed linear operator, $D(A)$ its domain. If a vector-valued function $\vec{f} \in \mathscr{L}^1(\mu, B_1)$ satisfies $\vec{f}(x) \in D(A)$ μ-a.e. and $A\vec{f} \in \mathscr{L}^1(\mu, B_2)$ then $\int \vec{f}\, d\mu \in D(A)$ and

$$A\int \vec{f}\, d\mu = \int A\vec{f}\, d\mu.$$

PROOF. Since $\vec{f} \in \mathscr{L}^1(\mu, B_1)$ and $A\vec{f} \in \mathscr{L}^1(\mu, B_2)$, there exist two sequences $\vec{f}_n \in S(\mu, B_1)$ and $\vec{g}_n \in S(\mu, B_2)$ such that $\vec{f}_n \to \vec{f}$, $\vec{g}_n \to A\vec{f}$ μ-a.e. and $\int ||\vec{f}_n - \vec{f}||\, d\mu \to 0$, $||\vec{f}_n(x)|| \leqslant 2||\vec{f}(x)||$, $||\vec{g}_n(x)|| \leqslant 2||A\vec{f}(x)||$ for μ-almost all $x \in X$. For each $(b_1, b_2) \in B_1 \times B_2$ we choose an element $(\tilde{b}_1, A\tilde{b}_1)$, $\tilde{b}_1 \in D(A)$ satisfying

$$||b_1 - \tilde{b}_1|| + ||b_2 - A\tilde{b}_1|| \leqslant 2 \inf \{||b_1 - b'|| + ||b_2 - Ab'||, \; b' \in D(A)\}.$$

From this condition it follows for each $b' \in D(A)$

$$\begin{aligned} &||\tilde{b}_1 - b'|| + ||A\tilde{b}_1 - Ab'|| \\ &\leqslant ||\tilde{b}_1 - b_1|| + ||b_1 - b'|| + ||A\tilde{b}_1 - b_2|| + ||b_2 - Ab'|| \\ &\leqslant 3(||b_1 - b'|| + ||b_2 - Ab'||). \end{aligned}$$

Therefore, the functions $x \to \tilde{f}_n(x)$ and $x \to A\tilde{f}_n(x)$ belong to $S(\mu, B_1)$ (resp. $S(\mu, B_2)$) and satisfy

$$\begin{aligned} &||\tilde{f}_n(x) - \vec{f}(x)|| + ||A\tilde{f}(x) - A\vec{f}(x)|| \\ &\leqslant 3||\vec{f}_n(x) - \vec{f}(x)|| + 3||\vec{g}_n(x) - A\vec{f}(x)||. \end{aligned}$$

The right-hand side of this inequality can be majorized by $6||\vec{f}(x)|| + 6||A\vec{f}(x)||$ which is, by assumption, an integrable function. Moreover, from the inequality it follows that $\tilde{f}_n(x) \to \vec{f}(x)$ and $A\tilde{f}_n(x) \to A\vec{f}(x)$

μ-a.e. Since the functions $\tilde{f}_n$ assume only a finite set of values, we have

$$A\int \tilde{f}_n(x)\,d\mu = \int A\tilde{f}_n(x)\,d\mu.$$

Now we use the assumption that A is closed:

$$\int \tilde{f}_n(x)\,d\mu \to \int \vec{f}(x)\,d\mu \quad \text{and} \quad A\int \tilde{f}(x)\,d\mu \to \int A\vec{f}(x)\,d\mu.$$

Therefore $\int \vec{f}(x)\,d\mu \in D(A)$ and $A\int \vec{f}(x)\,d\mu = \int A\vec{f}(x)\,d\mu$. □

DEFINITION. The vector field $\vec{f} \in F(X, B)$ is said to be *measurable* if for every real-valued function $0 \leqslant h \in \mathscr{L}^1(\mu)$ we have

$$\vec{f} \curlywedge h := \operatorname{dir}(h\cdot\vec{f})\cdot \|\vec{f}\| \cap |h| \in \mathscr{L}^1(\mu, B),$$

where

$$\operatorname{dir}(b) := \begin{cases} \dfrac{b}{\|b\|} & \text{for } 0 \neq b \in B, \\ 0 & \text{for } 0 = b \in B. \end{cases}$$

It follows immediately from this definition that the proofs of all theorems thus far about measurable functions remain valid, except that when a product of functions is involved, one of them must be real-valued.

In particular:

THEOREM XIII.27.4. Let $M(\mu, B)$ denote a space of measurable fields. Then

(i) $(\vec{f} \in M(\mu, B)) \Rightarrow (\|\vec{f}\| \in M(\mu, \boldsymbol{R}))$;

(ii) $\mathscr{L}^p(\mu, B) = \mathscr{F}^p(\mu, B) \cap M(\mu, B)$;

(iii) $(\vec{f} \in \mathscr{L}^p(\mu, B)) \Leftrightarrow (\|f\|^{p-1}\cdot\vec{f} \in \mathscr{L}^1(\mu, B))$.

PROOF. (i) Identically as in the scalar case.

(ii) Clearly $\mathscr{L}^p(\mu, B) \subset \mathscr{F}^p(\mu, B)$. The proof that $\mathscr{L}^p(\mu, B) \subset M(\mu, B)$ is identical with that in the scalar case.

Conversely: Let $\vec{f} \in \mathscr{F}^p(\mu, B) \cap M(\mu, B)$. Then

$$0 \leqslant \|\vec{f}\| \in \mathscr{F}^p(\mu, \boldsymbol{R}) \cap M(\mu, \boldsymbol{R}) = \mathscr{L}^p(\mu, \boldsymbol{R}).$$

Thus, there exists a sequence $0 \leqslant h_n \in S(\mu, \boldsymbol{R})$ such that $N_p(h_n - \|\vec{f}\|)$ tends to zero. Let us take the function

$$\vec{g}_n := h_n \curlywedge \vec{f} = \operatorname{dir}(h_n\cdot\vec{f})(|h_n| \cap \|\vec{f}\|) \in \mathscr{L}^1(\mu, B).$$

Then

$$\|\vec{f} - \vec{g}_n\| = \|\vec{f} - \operatorname{dir}(h_n\cdot\vec{f})(|h_n| \cap \|\vec{f}\|)\|$$

$$= ||\mathrm{dir}(\vec{f})\cdot(||\vec{f}|| - |h_n| \cap ||\vec{f}||)|| = \big|\, ||\vec{f}|| - |h_n| \cap ||\vec{f}||\, \big|$$
$$\leqslant \big| h_n - ||\vec{f}||\, \big|.$$

We have here made use of the fact that $\vec{f} = \mathrm{dir}(\vec{f}) \cdot ||\vec{f}||$.

Hence $N_p(\vec{f}-\vec{g}_n) \leqslant N_p(h_n - ||\vec{f}||) \underset{n\to\infty}{\to} 0$. But $||\vec{g}_n|| \leqslant ||\vec{f}||$, therefore $\vec{g}_n \in \mathscr{F}^p(\mu, B)$. If we could show that $\mathscr{L}^1(\mu, B) \cap \mathscr{F}^p(\mu, B) \subset \mathscr{L}^p(\mu, B)$, we would have $\mathscr{L}^p(\mu, B) \ni \vec{g}_n \to \vec{f}$ in the sense of N_p, and thus we would have $\vec{f} \in \mathscr{L}^p(\mu, B)$.

Thus, suppose we are given a function $g \in \mathscr{L}^1(\mu, B) \cap \mathscr{F}^p(\mu, B)$. We shall prove that $\vec{g} \in \mathscr{L}^p(\mu, B)$.

There exists a sequence of functions $\vec{p}_n \in S(\mu, B)$ such that $\vec{p}_n \to \vec{g}$ in the sence of N_1. From the Riesz–Fischer Theorem we infer that it is possible to choose a subsequence $\vec{p}_{n_k}$ such that $\vec{p}_{n_k} \to \vec{g}$ μ-almost everywhere, i.e. everywhere with the exception of the set Z of measure zero. But $||\vec{g}|| \in \mathscr{L}^1(\mu, \boldsymbol{R})$. As we know, the sequence of functions

$$||\vec{g}||_\varepsilon^k := \sum_{m=0}^{k} m\varepsilon(\chi_{A_{\varepsilon m}} - \chi_{A_{\varepsilon(m+1)}}) + \sum_{m=0}^{-k} m\varepsilon(\chi_{A_{\varepsilon(m-1)}} - \chi_{A_{\varepsilon m}})$$

satisfies the conditions $||\vec{g}||_\varepsilon^k \in S(\mu, \boldsymbol{R})$ and

$$||\vec{g}|| \geqslant ||\vec{g}||_\varepsilon^k \underset{\substack{k\to\infty\\ \varepsilon\to 0}}{\to} ||\vec{g}|| \quad \text{pointwise.}$$

Let us take the functions: $S(\mu, B) \ni \vec{p}_{n_s} \curlywedge ||g||_\varepsilon^k \underset{\substack{k\to\infty,\, \varepsilon\to 0\\ s\to\infty}}{\to} g$ μ-almost everywhere.

Now, if we take $k_i \to \infty$, $\varepsilon_i \to 0$, $s_i \to \infty$, then

$$S(\mu, B) \ni \vec{p}_{n_{s_i}} \curlywedge ||\vec{g}||_{\varepsilon_i}^{k_i} =: \vec{r}_i \to \vec{g} \quad \mu\text{-almost everywhere}$$
$$\text{and } ||\vec{r}_i|| \leqslant ||\vec{g}||,$$

and therefore $||\vec{r}_i - \vec{g}|| \leqslant 2||\vec{g}||$. But

$$||\vec{g}|| \in \mathscr{L}^1(\mu) \cap \mathscr{F}^p(\mu) \subset M(\mu) \cap \mathscr{F}^p(\mu) = \mathscr{L}^p(\mu) = \mathscr{L}^p(\mu, \boldsymbol{R}).$$

Identically

$$||\vec{r}_i - \vec{g}|| \in \mathscr{L}^p(\mu, \boldsymbol{R}).$$

Since $||\vec{r}_i - \vec{g}|| \to 0$ μ-almost everywhere, then by the Lebesgue Theorem (for scalar functions) $N_p(||\vec{r}_i - \vec{g}||) = N_p(\vec{r}_i - \vec{g}) \to 0$ which means that $\vec{g} \in \mathscr{L}^p(\mu, B)$.

(iii) Let $\vec{f} \in \mathscr{L}^p(\mu, B) \subset M(\mu, B)$. Then

$$\|\vec{f}\|^{p-1} \cdot \vec{f} \in M(\mu, B)$$

as the product of measurable functions. Moreover,

$$N_1(\|\vec{f}\|^{p-1} \cdot \vec{f}) = N_1(\|\vec{f}\|^p) = [N_p(\vec{f})]^p < \infty.$$

Accordingly

$$\|\vec{f}\|^{p-1} \cdot \vec{f} \in M(\mu, B) \cap \mathscr{F}^1(\mu, B) = \mathscr{L}^1(\mu, B).$$

Conversely: let $\|\vec{f}\|^{p-1} \cdot \vec{f} \in \mathscr{L}^p(\mu, B)$. Repeating the identical construction as in the proof of (ii), we can select a sequence $\vec{r}_i \in S(\mu, B)$ such that

$$\|\vec{r}_i\| \leqslant \|\vec{f}\|^{p-1} \cdot \|\vec{f}\| = \|\vec{f}\|^p, \quad \vec{r}_i \to \|\vec{f}\|^{p-1} \cdot \vec{f}$$

μ-almost everywhere.

Let us take the sequence of functions: $\vec{g}_i := \operatorname{dir}(\vec{r}_i) \cdot \|\vec{r}_i\|^{1/p}$. Plainly

$$\|\vec{g}_i\| = \|\vec{r}_i\|^{1/p} \leqslant \|\vec{f}\| \quad \text{and} \quad \vec{g}_i \in S(\mu, B), \quad \vec{g}_i \to \vec{f}$$

μ-almost everywhere.

Note however, that $\|\vec{f}\|^p \in \mathscr{L}^1(\mu, \boldsymbol{R})$ and thus $\|\vec{f}\| \in \mathscr{L}^p(\mu, \boldsymbol{R})$ (cf. Section 15). By the Lebesgue majorized-convergence theorem for vector fields it follows that $S(\mu, B) \ni \vec{g}_i \to f$ in the sense of N_p, which means that $\vec{f} \in \mathscr{L}^p(\mu, B)$. □

Below we append an interesting relation for the space $\mathscr{L}^p$ for different p in cases when X has a finite measure.

LEMMA XIII.27.5. If X has a finite measure, then $\mathscr{L}^{p_1}(\mu, B) \subset \mathscr{L}^{p_2}(\mu, B)$ for $p_1 \geqslant p_2$.

PROOF. If $\vec{f} \in \mathscr{L}^{p_1}(\mu, B)$, then $\vec{f}$ is measurable. It is thus sufficient to prove that $N_{p_2}(\vec{f}) < \infty$. But a constant function is integrable. Accordingly, when we write $A = \{x: \|\vec{f}\| > 1\}$, $B = \{x: \|\vec{f}\| \leqslant 1\}$, then

$$\|\vec{f}\|^{p_2} = \|\vec{f}\|^{p_2}\chi_A + \|\vec{f}\|^{p_2}\chi_B.$$

But $\|\vec{f}\|^{p_2}\chi_B \leqslant 1$, and hence

$$\mu^*(\|\vec{f}\|^{p_2}\chi_B) \leqslant \mu^*(X) < \infty$$

and

$$\|\vec{f}\|^{p_2}\chi_A = \|\vec{f}\|^{p_1 \cdot p_2/p_1}\chi_A \leqslant \|\vec{f}\|^{p_1}\chi_A, \quad \text{since} \quad p_2/p_1 \leqslant 1.$$

Therefore

$$\mu^*(\|\vec{f}\|^{p_2}\chi_A) \leqslant \mu^*(\|\vec{f}\|^{p_1}\chi_A) \leqslant N_{p_1}(\vec{f}) < \infty,$$

that is $\mu^*(\|\vec{f}\|^{p_2}) < \infty$. □

A particularly interesting space is $\mathscr{L}^2(\mu, B)$ in the case when B is a Hilbert space $(H, (\ |\ ))$.

LEMMA XIII.27.6. Given a field $\vec{f} \in \mathscr{L}^p(\mu, H)$, $\vec{g} \in \mathscr{L}^{p'}(\mu, H)$ and suppose that $1/p + 1/p' = 1$. Then the scalar function

$$h(x) := (\vec{f}(x)|\vec{g}(x))_H$$

belongs to $\mathscr{L}^1(\mu, \boldsymbol{R})$.

PROOF. Using the inequalities of Schwarz and Hölder, we find

$$\begin{aligned} &N_1\big((\vec{f}(\cdot)|\vec{g}(\cdot))\big) \\ &\leqslant N_1\big(\|\vec{f}(\cdot)\| \cdot \|\vec{g}(\cdot)\|\big) \leqslant N_p(\vec{f}) \cdot N_{p'}(\vec{g}) < \infty. \end{aligned}$$

It thus remains to prove that h is a measurable function. But

$$\begin{aligned} &(\vec{f}(x)|\vec{g}(x)) \\ &= \tfrac{1}{4}\big(\|\vec{f}(x)+\vec{g}(x)\|^2 - \|\vec{f}(x)-\vec{g}(x)\|^2 \|\vec{f}(x)+i\vec{g}(x)\|^2 - \\ &- \|\vec{f}(x)-i\vec{g}(x)\|^2\big), \end{aligned}$$

and hence this is a measurable function, being as it is a linear combination of measurable functions. □

A similar fact occurs for any Banach space:

LEMMA XIII.27.7. Let B be a Banach space, and let B' be its conjugate. Let $\langle b, b'\rangle$ denote the value of the functional $b' \in B'$ at the point $b \in B$.

If $\vec{f} \in \mathscr{L}^p(\mu, B)$, $\vec{g} \in \mathscr{L}^{p'}(\mu, B')$, $1/p + 1/p' = 1$, then the real-valued function $h(x) := \langle \vec{f}(x), \vec{g}(x)\rangle$ is integrable.

PROOF. We have

$$\begin{aligned} &N_1(\langle \vec{f}(\cdot), \vec{g}(\cdot)\rangle) \leqslant N_1\big(\|\vec{f}(\cdot)\|_B \cdot \|\vec{g}(\cdot)\|_{B'}\big) \\ &\leqslant N_p(\vec{f}) N_{p'}(\vec{g}) < \infty. \end{aligned}$$

The function h is measurable as it is the limit of the measurable functions $\langle \vec{f}_n(\cdot), \vec{g}_n(\cdot)\rangle$, where $\vec{f}_n$ and $\vec{g}_n$ belong to $S(\mu, B)$ and $S(\mu, B')$. □

Thus, let $\vec{f} \in \mathscr{L}^p(\mu, B), \vec{g} \in \mathscr{L}^{p'}(\mu, B')$. We take

$$\langle \dot{\vec{f}}, \dot{\vec{g}} \rangle := \int \langle \vec{f}(x), \vec{g}(x) \rangle d\mu(x) \leqslant N_p(\vec{f}) N_{p'}(\vec{g}).$$

The expression $\langle \dot{\vec{f}}, \dot{\vec{g}} \rangle$ is linear and continuous in both factors, and as such may be interpreted by $\dot{\vec{f}}(\dot{\vec{g}})$. One may expect that $(\mathscr{L}^p(\mu, B))' = \mathscr{L}^{p'}(\mu, B')$, just as for real-valued functions.

It turns out that this relation is true (cf. Bourbaki, *Integration*).

The theory expounded in this section enables us to integrate complex functions on X. Knowing how to integrate complex functions, let us take an arbitrary Hilbert space over the field of complex numbers, i.e. a vector space in which a scalar product is defined, which is a form, of the following properties:

(i) $(\alpha h + \beta g | f) = \alpha(h|f) + \beta(g|f)$,
(ii) $(h|f) = \overline{(f|h)}$,
(iii) $(h|h) \geqslant 0$,
(iv) $(h|h) = 0 \Leftrightarrow h = 0$.

Note that

$$(h|\alpha f + \beta g) = \overline{(\alpha f + \beta g|h)} = \overline{\alpha(f|h)} + \overline{\beta(g|h)}$$
$$= \bar{\alpha}(h|f) + \bar{\beta}(h|g).$$

This property is frequently referred to as the *antilinearity* of the scalar product in the second factor.

LEMMA XIII.27.8. If $\vec{f} \in \mathscr{L}^p(\mu, H), \vec{g} \in \mathscr{L}^p(\mu, H), 1/p + 1/p' = 1$, then

$$(\vec{f}(\cdot)|\vec{g}(\cdot)) \in \mathscr{L}^1(\mu, C^1).$$

PROOF — identical with that in the real case. Thus, if we define

$$\langle \dot{\vec{f}}, \dot{\vec{g}} \rangle := \int (\vec{f}(x)|\vec{g}(x)) d\mu(x),$$

the expression $\langle \dot{\vec{f}}, \dot{\vec{g}} \rangle$ is linear in the first factor, and antilinear in the second.

In particular, let us take the space $\mathscr{L}^2(\mu, H)$. This is the Hilbert space over the complex field with the scalar product

$$(\dot{\vec{f}}|\dot{\vec{g}}) := \int (\vec{f}(x)|\vec{g}(x)) d\mu(x).$$

28. DIRECT INTEGRALS OF HILBERT SPACES

In many areas of mathematical analysis and theoretical physics a need arose for generalizing the direct (orthogonal) sum of Hilbert spaces. The expression

$$\bigoplus_{x=1}^{n} H_x, \quad \text{where } H_x = \big(H_x, (\cdot|\cdot)_x\big) \text{ is a Hilbert space,}$$

is the *direct sum* (or *Cartesian product* $\bigtimes_{x=1}^{n} H_x$) in which the scalar product is defined as follows: let $u, v \in \bigtimes_{x=1}^{n} H_x$; then

$$(u|v) := \sum_{x=1}^{n} (u_x|v_x)_x.$$

In this way we once again obtain a Hilbert space. This concept may be generalized to a countable set of Hilbert spaces: Let us consider elements $u \in \bigtimes_{x=1}^{\infty} H(x)$, such that $\sum_{x=1}^{\infty} ||u(x)||_x^2 < \infty$. Clearly, this is a vector subspace of the product $\bigtimes H(x)$; we define the scalar product in it as

$$(u|v) := \sum_{x=1}^{\infty} \big(u(x)|v(x)\big)_x.$$

This space will be denoted by $\bigoplus_{x=1}^{\infty} H(x)$. Note that this is a completely different set than the direct sum defined in identical manner in algebra: there, the sequences considered had only a finite number of nonzero terms. Guided by the needs of quantum mechanics and the theory of group representation, J. von Neumann presented the concept of *direct integral of Hilbert space*. This concept is a natural generalization of both $\bigoplus_{x=1}^{\infty} H(x)$ and the space $\mathscr{L}^2(\mu, H)$.

Let us go on to the definition of this concept.

1. Suppose that (X, μ) is a space with a measure (integral). Consider a family of Hilbert spaces $\big(H(x)\big)_{x\in X}$ and form the product $\prod_{x\in X} H(x)$.

The elements $u, h \in \prod_{x \in X} H(x)$ will be called *vector fields*. A *fundamental family* of μ-measurable fields is a family $\Gamma = (h^\alpha)_{\alpha \in A}$ such that

(i) for all $\alpha, \beta \in A$ the function $X \ni x \to \big(h^\alpha(x)|h^\beta(x)\big)_x \in \mathbf{C}$ is μ-measurable,

(ii) for every $x \in X$ the vectors $\big(h^\alpha(x)\big)_{\alpha \in A}$ span the space $H(x)$.

A field $h \in \prod_{x \in X} H(x)$ is said to be *measurable* when all the functions $x \to \big(h(x)|h^\alpha(x)\big)_x$ are μ-measurable. The pair $\big((H(x))_{x \in X}, \Gamma\big)$ is called a *measurable family of Hilbert spaces*. The following obvious corollary holds.

COROLLARY XIII.28.1. μ-measurable fields constitute a vector subspace of the product $\prod_{x \in X} H(x)$.

Since further on we shall be dealing only with a fixed measure μ, we shall omit the letter μ and write simply measurable, integrable, etc.

Remark. We emphasize that it is meaningful to speak of the measurability of a vector field only when a fundamental family of fields has been given.

In the sequence we shall consider families of Hilbert spaces $H(x)$, $x \in X$, for which there exists a countable fundamental family h^i, $i = 1, 2, \ldots$, of measurable fields.

When $H(x) = H$ for every $x \in X$, we speak of a constant family of Hilbert spaces; then $\prod_{x \in X} H(x) = H^X$ are simply mappings in an established Hilbert space. Precisely such families, where $\dim H < \infty$, are considered in elementary vector analysis. On the other hand, fields of vectors (tensors) tangent to the differentiable manifold X (they will be treated more rigorously in Chapter XIV) are an example of fields in the sense considered here.

2. *Square-integrable Fields. The Direct Integral*

DEFINITION. Let $\mathscr{H} = (H(x)_{x \in X}, \Gamma)$ be a measurable family of Hilbert spaces. A (measurable) field h is *square-integrable* if

$$\int ||h(x)||_x^2 \, d\mu(x) < \infty .$$

The set of square-integrable fields is denoted by

$$\mathscr{L}^2(\mu, \mathscr{H}) = \mathscr{L}^2\big(\mu, (H(x))_{x \in X}, \Gamma\big).$$

LEMMA XIII.28.2. $\mathscr{L}^2(\mu, \mathscr{H})$ is a vector space with scalar product

$$(u, v) := \int_X (u(x)|v(x))_x d\mu(x).$$

Hence, $\mathscr{L}^2(\mu, \mathscr{H})$ is a pre-Hilbert space.

PROOF follows immediately from the polarization formula

$$4(u(x)|v(x))_x = (\|u(x)+v(x)\|_x^2 - \|u(x)-v(x)\|_x^2) + \\ + i\|u(x)+iv(x)\|_x^2 - i\|u(x)-iv(x)\|_x^2. \ \square$$

It is to be expected that the Riesz–Fischer Theorem undergoes the following generalization:

THEOREM XIII.28.3. (i) The space $\mathscr{L}^2(\mu, \mathscr{H})$ is complete.

(ii) If the sequence $h_n \in \mathscr{L}^2(\mu, \mathscr{H})$ is convergent to $h \in \mathscr{L}^2(\mu, \mathscr{H})$ in the sense of the seminorm $\| \, \|$, there exists a subsequence which converges to h almost everywhere.

The proof does not differ from the classical proof of the Riesz–Fischer Theorem.

Clearly, $\mathscr{L}^2(\mu, \mathscr{H})$ is not a Hausdorff space. Dividing it by the relation $\mathscr{R} = \{(h, u): h(x) = u(x) \text{ for almost all } x\}$, we obtain the Hilbert space

$$L^2(\mu, \mathscr{H}) = \mathscr{L}^2(\mu, \mathscr{H})/\mathscr{R}.$$

DEFINITION. The space $L^2(\mu, \mathscr{H})$ is called the *direct integral of the Hilbert spaces* $H(x)$, $x \in X$, and is denoted by

$$\int H(x) d\mu(x) \quad \text{or} \quad \int \oplus H(x) d\mu.$$

Examples. 1. Let $X = N$, $\mu(\{n\}) = 1$, then

$$\int_X H(x) d\mu(x) = \bigoplus_{x \in N} H(x).$$

2. Let $H(x) = C$ for every $x \in X$; then

$$\int_X H(x) d\mu(x) = \mathscr{L}^2(\mu).$$

3. When $H(x) \underset{x}{\equiv} H$, where H is an established Hilbert space, then

$$\int H(x) d\mu(x) = \mathscr{L}^2(\mu, H).$$

The most important direct integrals are obtained when X is a locally compact space and μ is the Radon integral on X. In this case we could define continuous fields with compact supports and we would show, for instance, that the space of these fields is dense in $\mathscr{L}^2(\mu, \mathscr{H})$. In other words, $\mathscr{L}^2(\mu, \mathscr{H})$ could be defined as the completion of the pre-Hilbert space of the fields $C_0(X, \mathscr{H})$ with compact supports in the relevant scalar product; however, the concept of a Hilbert bundle is required for this purpose (cf. Part III).

3. *Fields of Linear Mappings*

DEFINITION. If we have two families $\mathscr{H}_i = (\mathscr{H}_i(x)_{x \in H}, \Gamma_i, \mu_i)$, $i = 1, 2$, of Hilbert spaces, the field of linear mappings

$$(T(x))_{x \in X} = T \in \prod_{x \in X} \mathscr{L}(H_1(x), H_2(x)),$$

that is,

$$\bigwedge_{x \in X} T(x) \in \mathscr{L}(H_1(x), H_2(x)),$$

is (μ_1, μ_2)*-measurable,* if for every μ_1-measurable field $h_1 \in H_1$ the field $Th_1 = (T(x)h_1(x))_{x \in X}$ is μ_2-measurable. When $\mathscr{H}_1 = \mathscr{H}_2$, we speak of a *field of operators.*

The operators $T \in \mathscr{L}(\mathscr{H})$ of the above form are said to be *decomposable. Diagonal operators* are those of a particularly simple nature; these are fields of operators of the form $X \ni x \to T(x) = f(x)1(x)$, where $f(x)$ is a measurable function, $1(x)$ is the identity operator in the space $H(x)$; we shall denote them by T_f;

Example. Let A be a measurable subset in X. Then the operator $P_X := T_{\chi_A}$ is an orthogonal projection operator: $P_A^2 = P_A, P_A^* = P_A$; where P^* denotes the adjoint operator in the sense of the scalar product in $\mathscr{L}^2(\mu, \mathscr{H})$. The proof follows from the next theorem:

THEOREM XIII.28.4. Let $(T(x))_{x \in X}$ be a μ-measurable field of operators. If the function $(x \to ||T(x)||_x) \in \mathscr{L}^\infty(\mu)$ (i.e. is essentially bounded), then the operator $(T(\cdot))$ is defined over the entire space and its norm is $= ||T(\cdot)||_\infty < \infty$; we have

$$(Th|u) \underset{u,v \in H}{\equiv} (h|T^*u), \quad \text{where } T^*(x) = (T(x))^* \text{—almost everywhere.}$$

Diagonal operators are commutative and are related by

$$T_f \circ T_g = T_{f \cdot g}, \quad (T_f)^* = T_{\bar{f}}, \quad \text{where } \overline{f(x)} = \bar{f}(x).$$

Plainly, diagonal operators commute with decomposable operators. The inverse theorem (von Neumann) also holds and plays an important role in quantum mechanics, group theory, and harmonic analysis. These topics are treated in a monograph by the present author: *General Eigenfunction Expansions and Unitary Representations of Topological Groups* (Warsaw 1968, second edition in preparation).

With this, we end our excursion into the domain of direct integrals.

4. *Spectral Measures*

As was seen in the example above, an orthogonal projection operator $\mathscr{M} \ni A \to P_A := T_{\chi_A}$ was assigned to each μ-measurable set $A \subset X$. When we have a countable family A_i, $i = 1, 2, \ldots$, of disjoint measurable sets, then since $\chi_{\cup A_i} = \sum_{i=1}^{\infty} \chi_{A_i}$, we have

(i) $P_{\cup A_i} = T_{\chi_{\cup A_i}} = T_{\Sigma \chi_{A_i}} = \sum T_{\chi_{A_i}} = \sum P_{A_i}$,

(ii) $P_X = T_{\chi_X} = T_1 = 1$

The mapping $A \to P_A$ is said to be a *spectral family* (*measure*).

We now adopt the general definition:

DEFINITION. Let H be a (separable) Hilbert space. Let $\mathscr{A}$ be a σ-algebra of subsets of space X. A mapping $\mathscr{A} \ni A \to P_A \in \text{Proj}(H)$, where $\text{Proj}(H)$ denotes the set of projection operators in H, is called a *spectral measure* when the following conditions are satisfied:

(i) For a countable disjoint family $(A_i)^{\infty}$ (that is, $A_i \cap A_k = 0$, $i \neq k$)

$$P_{\cup A_i} = \sum_{i=1}^{\infty} P_{A_i}$$

(ii) $P_X = 1$.

From the foregoing definition it follows immediately that $P_B \cdot P_A = P_A \cdot P_B = P_{A \cap B}$; thus if $A \cap B = \emptyset$, then $P_A \cdot P_B = 0$, that is, the operators P_A, P_B project onto mutually orthogonal subspaces. If $A, B \in \mathscr{A}$ and $A \subset B$, then

$$(*) \qquad P_A \leqslant P_B \quad \text{in the sense that } (P_A h|h) \leqslant (P_B h|h)$$

for every $h \in H$. Since $||P_A h|| = (P_A h|h)$, it is seen that the inequality $(*)$ denotes that $P_A H \subset P_B H$: the operator P_A projects onto a subspace contained in the subspace $P_B H$.

As we shall see later (cf. Part III) every commutative family of Hermitian operators determines a direct integral, and hence every spectral measure $\{P_A\}_{A\in\mathscr{A}}$ determines (up to isomorphism) a direct integral, the diagonal operators T_{χ_A} corresponding to the operators P_A. John von Neumann built up a quantum mechanics on the basis of spectral measures. "Properties" (or statements) could be associated with projection operators: they have a characteristic of "filters"—their spectrum is (at most) a two-point set $\{0\}\cup\{1\}$, and they decompose the space H into an orthogonal sum $H_P \oplus H_{1-P} = H$. The space H_P is an eigenspace corresponding to the eigenvalue 1:

$$Ph = \begin{cases} 1\cdot h & \text{for } h \in H_P = PH, \\ 0 & \text{for } h \in H_{1-P} = H - H_P. \end{cases}$$

Just as the characteristic function (of a set), so the operator takes on only two values: 1, 0. As we have seen, this analogy is not fortuitous; its justification (the Gel'fand–Naimark Theorem) will be given further on. Then, too, we shall give further details of von Neumann's view of quantum mechanics.

5. *The Isomorphism of Direct Integrals*

DEFINITION. When (X_i, μ_i), $i = 1, 2$, are spaces with measures, the mapping $T\colon X_1 \to X_2$ is said to be *(μ_1, μ_2)-measurable* if for every μ_2-measurable set $A_2 \subset X_2$ the set $T^{-1}(A_2)$ is μ_1-measurable. When T is a bijection and when T^{-1} is (μ_2, μ_1)-measurable, then (X_i, μ_i), $i = 1, 2$, are said to be *isomorphic*.

A close analogy is seen here with the continuity of a mapping of topological spaces ("the inverse image of an open set is open").

Let $H_i = \int\limits_X H_i(x)\,d\mu_i(x)$, $i = 1, 2$, be direct integrals on the same space X. Then, if

(i) the identical mapping id: $X \to X$ is an isomorphism,

(ii) $(\mu_1(N) = 0) \Leftrightarrow (\mu_2(N) = 0)$,

(iii) there exists a measurable null set N and a mapping $U(\cdot)\colon (H_1(\cdot)) \to (H_2(\cdot))$, such that for $x \notin N$ the mapping $U(x)\colon H_1(x) \to H_2(x)$ is unitary

(iv) the mapping $\int U(x)\,d\mu_1(x)$ is a unitary mapping of the direct integral $\int H_1\,d\mu_1$ onto $\int H_2\,d\mu_2$,

then these direct integrals are called *isomorphic*.

29. ON THE EQUIVALENCE OF THE STONE AND RADON INTEGRAL THEORIES

The abstract integral theory due to Stone, an exposition of which has been given in this chapter, came into being in 1948 and 1949, that is, at about the same time as Bourbaki was developing his treatment of the Radon integral. It might seem that Stone's second procedure is incomparably more general than is the Radon integral theory, for after all it does treat the integration of functions on an arbitrary set X whereas the Radon integral concerns only a locally compact space.

Since the two theories employed analogous methods, a question arose as to whether the second Stone procedure (with strong Stone axiom) is "truly" more general.

In 1957 Heinz Bauer proved an amazing theorem stating that the two theories are equivalent.

To be more precise: the following theorem holds.

THEOREM XIII.29.1 (Bauer). Let μ be an abstract Daniell–Stone integral on the set X, with a family of elementary functions E. Let $\mathscr{L}^p_E(X,\mu)$, $p \geqslant 1$, be $\mathscr{L}^p$ spaces constructed from E according to the second Stone procedure. Then there exists a locally compact space X' such that

(i) X is a dense subset of X',

(ii) every function $f \in E$ is extendible to a continuous function f' on X', vanishing at infinity,

(iii) $\bigwedge_{x' \in X'} \bigvee_{f \in E} f'(x') \neq 0$,

(iv) $\bigwedge_{x', y' \in X'} \bigvee_{f \in E} f'(x') \neq f'(y')$.

Properties (i)–(iv) determine the space X' up to homeomorphisms not affecting the set X.

Moreover:

(v) The integral μ is uniquely extendible to a Radon integral μ' on X';

(vi) Every function $f \in \mathscr{L}^p_E(X, \mu)$ is extendible to $f' \in \mathscr{L}^p_{C_0(X')}(X, \mu')$, a nonnegative function being extendible to a nonnegative function. The following equation holds:

$$\int_X f d\mu = \int_{X'} f' d\mu'.$$

30. FROM MEASURE TO INTEGRAL

In this book we have endeavoured consistently to represent a point of view of integral theory that is being taken ever more frequently and is fundamental for analysis: an integral is a linear functional on a linear space of "elementary" functions. In this theory, measure is a secondary concept. Small wonder that integrable functions enjoy pride of place in this treatment; measurable functions are defined in terms of integrable functions.

The transition to subsets of the space X is effected with the aid of characteristic functions; definitions are obtained for, respectively, the integrability and measurability of the set Z: this is a set such that its characteristic function χ_Z is integrable, or possibly measurable. Thus, measurable sets are at the very end in the theory and play quite a mediocre role.

I have chosen this approach for various reasons:

(i) This approach leads much more quickly to the fundamental theorems of analysis on the limit under the integral sign, and hence also theorems on integrals with a parameter.

(ii) The integrals most often met with in application are Radon integrals, "*née*" linear functionals. Thus, integration and "measure" on the sheets of surfaces and, more generally, on differentiable manifolds, are obtained immediately. What a tedious procedure it is to determine "measure" integrals on these objects is common knowledge.

(iii) It is a natural transition to go over to the theory of distributions (generalized functions) which are by definition linear functionals on suitable spaces of test functions. Distribution theory is expounded in Chapter XIX. There is, however, a branch of mathematics where the measure is the natural, primary concept. This is probability theory and related theories. Consequently it would seem desirable to mention the more traditional "from measure to integral" approach.

In writing this section, however, I had another consideration in mind; it is always instructive to look at a matter from different points of view, for different aspects then come to our attention and we penetrate more deeply into the nature of things. It would be a barren exercise to enter into a dispute over what came first, the measure or the integral (the chicken or the egg)?

Measure Space. The starting point is a space X in which has been distinguished a certain σ-algebra $\mathscr{A}$ of subsets of the space X. A set A is said to be *measurable* if $A \in \mathscr{A}$.

Measurability of a Mapping. Let $(X_i, \mathscr{A}_i)$, $i = 1, 2$, be measurable spaces. The mapping $T: (X_1, \mathscr{A}_1) \to (X_2, \mathscr{A}_2)$ is called $(\mathscr{A}_1 - \mathscr{A}_2)$-*measurable* when $T^{-1}(A_2) \in \mathscr{A}_1$ for every $A_2 \in \mathscr{A}_2$. Note that measurability is defined by analogy with continuity: the role of the topology $\mathscr{T}_i$ of the space X_i is taken over by the σ-algebra $\mathscr{A}_i$. A question immediately comes to mind: where do the σ-algebras of the subsets come from? Usually, we are given some family $\mathscr{H}$ which in general is not a σ-algebra; e.g., the topology $\mathscr{T}$ of a topological space $(X, \mathscr{T})$. The reader can easily prove the following theorem:

THEOREM XIII.30.1. For every family $\mathscr{H}$ of subsets of the space X there exists a smallest σ-algebra $\mathscr{A}(\mathscr{H})$ containing $\mathscr{H}$.

Examples. 1. $\mathfrak{P}(X)$, the family of all subsets of the space X, is clearly a σ-algebra.

2. When $\mathscr{A}$ is a σ-algebra on X, and $X' \subset X$, then

$$X' \cap \mathscr{A} := \{X' \cap A\colon\ A \in \mathscr{A}\}$$

is a σ-algebra on X'. The alert reader will notice that this procedure is analogous to that of obtaining a relative topology $X' \cap \mathscr{T}$ on the subspace $X' \subset X$.

3. Let $(X_2, \mathscr{A}_2)$ be a space with a σ-algebra, let X_1 be an arbitrary set, and let $T: X_1 \to X_2$. Then the family

$$T^{-1}(\mathscr{A}_2) := \{T^{-1}(A_2)\colon\ A_2 \in \mathscr{A}_2\}$$

is a σ-algebra on X_1.

Borel Sets. Suppose that $(X, \mathscr{T})$ is a topological space. In general $\mathscr{T}$ is not a σ-algebra (give an example when this is the case!). The elements of the family $\mathscr{A}(\mathscr{T})$ of the σ-algebra generated by the topology $\mathscr{T}$ are called *Borel sets*. This is a much richer family than $\mathscr{T}$, for it also contains all closed sets and their countable intersections.

Using $\mathscr{C}$ to denote the family of closed sets of the space $(X, \mathscr{T})$, we have

$$\mathscr{A}(\mathscr{T}) = \mathscr{A}(\mathscr{C}).$$

Borel Sets on $\boldsymbol{R}$. Since the intervals $]a, b[$, $a, b \in \boldsymbol{R}$, constitute the basis of the topology on $\boldsymbol{R}$ and, as readily evident, $[a, b[= \bigcup_{n=1}^{\infty}]a_n, b[$, where $a_n = a - 1/n$, hence, on denoting the intervals $[a, b[$, $a, b \in \boldsymbol{R}$, by $\mathscr{I}$ for the moment, we see that $\mathscr{A}(\mathscr{T}) = \mathscr{A}(\mathscr{I})$. The reader should prove the following exercise.

Exercise. On $\boldsymbol{R}$ we have $\mathscr{A}(\mathscr{T}) = \mathscr{A}(\mathscr{K})$, where $\mathscr{K}$ stands for the family of compact subsets of the space $\boldsymbol{R}$.

Clearly, all of this also holds for $\boldsymbol{R}^p$. Thus we have

THEOREM XIII.30.2. On $\boldsymbol{R}^p$ the family $\mathscr{B}^p$ of Borel sets in $\boldsymbol{R}^p$ is of the form

$$\mathscr{B}^p = \mathscr{A}(\mathscr{T}) = \mathscr{A}(\mathscr{C}) = \mathscr{A}(\mathscr{I}) = \mathscr{A}(\mathscr{K}).$$

Borel Sets on $\overline{\boldsymbol{R}}$. Since we operate with numerical functions in integral theory, it is a good thing to know how Borel sets look on the compactification $\overline{\boldsymbol{R}} = \boldsymbol{R} \cup \{\infty\} \cup \{-\infty\}$ of the real line. These are sets of the form $B_0, B_0 \cup \{\infty\}, B_0 \cup \{-\infty\}, B_0 \cup \{\infty\} \cup \{-\infty\}$, where $B_0 \in \mathscr{B}^1$. These are easily shown to be sets $B \subset \overline{\boldsymbol{R}}$ for which $B \cap \boldsymbol{R} \in \mathscr{B}^1$.

Usually, when speaking of measurable sets on $\boldsymbol{R}$, $\boldsymbol{R}^p$, or $\overline{\boldsymbol{R}}$, we have Borel sets in mind. Accordingly, the following definition is adopted:

DEFINITION. Let $(X, \mathscr{A})$ denote a space with a σ-algebra. A real-valued numerical function $f\colon X \to \overline{\boldsymbol{R}}\ (\boldsymbol{R})$ is said to be *$\mathscr{A}$-measurable* when it is $(\mathscr{A} - \overline{\mathscr{B}})$-measurable.

It is not difficult to prove

THEOREM XIII.30.3. (A numerical function on $(X, \mathscr{A})$ is $\mathscr{A}$-measurable)

(m) $\quad \Leftrightarrow \big(\{x \in X\colon f(x) \geqslant a\} \in \mathscr{A}$ for every $a \in \boldsymbol{R}\big)$.

Remark. The inequality sign $\geqslant$ in condition (m) could, of course, be replaced by any of the signs $>, \leqslant, <$.

It is thus seen that, on confrontation with Theorem XIII.13.1, both definitions of measurability lead to the same family of functions.

COROLLARY XIII.30.4. Let $(X, \mathscr{T})$ be a topological space. Then a continuous function $f\colon X \to \boldsymbol{R}$ is Borel-measurable, i.e., is $A(\mathscr{T})$-measurable.

PROPOSITION. Let (f_n) be a sequence of $\mathscr{A}$-measurable numerical functions. Then each of the following functions is $\mathscr{A}$-measurable:

$$\sup f_n, \quad \inf f_n, \quad \limsup f_n, \quad \liminf f_n.$$

DEFINITION. Let $(X, \mathscr{A})$ be a space with a σ-algebra. A *measure* on $(X, \mathscr{A})$ is a function $\mu: \mathscr{A} \to \overline{\boldsymbol{R}}_+$ satisfying

$$\mu\left(\bigcup_{i=1}^{\infty} A_i\right) = \sum_{i=1}^{\infty} \mu(A_i)$$

for each sequence of mutually disjoint sets $A_i \in \mathscr{A}$.

The triplet $(X, \mathscr{A}, \mu)$ is called a *measure space.*

Now, suppose that a measure space $(X, \mathscr{A}, \mu)$ is given.

Integral. We now pass from measure to integral. To this end, we first define the elementary functions:

DEFINITION. A real-valued function $u: X \to \boldsymbol{R}$ is said to be *$\mathscr{A}$-elementary* when

(i) $u \geqslant 0$,

(ii) u is measurable,

(iii) the set of values $u(X)$ is finite and $\mu(\{x \in X: f(x) \neq 0\}) = 0$.

(We often speak of *nonnegative step functions*).

$E = E(\mathscr{A})$ will stand for the family of $\mathscr{A}$-elementary functions. This definition immediately implies that $\{a_1, \ldots, a_n\} = u(X)$, whence $A_i := u^{-1}(a_i)$, $i = 1, \ldots, n$ are measurable and disjoint. Clearly

$$(1) \qquad u = \sum_{i=1}^{n} a_i \chi_{A_i}, \qquad A_i \in \mathscr{A}, \ A_i \cap A_k = \emptyset, \ i \neq k.$$

Expression (1) is called the *normal form* of the elementary function u.

Every function $u \in E(\mathscr{A})$ can, of course, have different normal forms, but the following lemma shows that normal forms are particularly useful in integral theory:

LEMMA XIII.30.5. For two normal forms of an elementary function

$$u = \sum_{i=1}^{n} a_i \chi_{A_i} = \sum_{j=1}^{m} b_j \chi_{B_j}$$

the equation

$$(*) \qquad \sum_{i=1}^{n} a_i \mu(A_i) = \sum_{j=1}^{m} b_j \mu(B_j)$$

holds.

The proof is left for the reader.

DEFINITION. The number (∗), which is independent of the normal form, is called the *integral of the function u* (*over X*) and is denoted by $\mu(u)$ or $\int u\,d\mu$:

$$\int u\,d\mu := \sum_{i=1}^{n} a_i \mu(A_i).$$

The mapping $E \ni u \to \int u\,d\mu \in \overline{\boldsymbol{R}}_+$ has thus been defined. The reader should verify the following proposition:

PROPOSITION.

$$\int \chi_A d\mu = \mu(A), \quad A \in \mathscr{A},$$

$$\int (au)\,d\mu = a\int u\,d\mu, \quad a \in \boldsymbol{R}_+,\ u \in E,$$

$$\int (u+v)\,d\mu = \int u\,d\mu + \int v\,d\mu, \quad u, v \in E,$$

$$(u \leqslant v) \Rightarrow \left(\int u\,d\mu \leqslant \int v\,d\mu\right), \quad u, v \in E.$$

As in the theory of the Daniell–Stone integral, we extend μ to the functions which are the monotonic limits of the elementary functions. An important theorem follows.

THEOREM XIII.30.6. For every isotonic sequence (u_n) of $\mathscr{A}$-elementary functions and every $v \in E$

$$(v \leqslant \sup_n u_n) \Rightarrow \left(\int v\,d\mu \leqslant \sup \int u_n\,d\mu\right).$$

COROLLARY XIII.30.7. For arbitrary $u_n \nearrow, v_n \nearrow, u_n, v_n \in E$

$$(\sup u_n = \sup v_n) \Rightarrow \left(\sup \int u_n d\mu = \sup \int v_n d\mu\right).$$

Notation. $\mathscr{E}^*$ denotes the set of all nonnegative numerical functions $: X \to \boldsymbol{R}$ which are the limits of isotonic sequences of elementary functions: $f = \sup u_n$, where $u_n \nearrow$. As the following corollary shows, we can adopt the definition of integral given below:

COROLLARY XIII.30.8. $\sup \int u_n d\mu$ depends not on the sequence (u_n) but only on the function f.

DEFINITION. For an arbitrary function $f \in \mathscr{E}^*$ let $f = \sup u_n$;

$$\int f d\mu := \sup \int u_n d\mu,$$

$\int f d\mu$ being called the (μ-)*integral of the function f over the set X*.

Clearly, $E \subset \mathscr{E}^*$. We have the immediate corollary:

COROLLARY XIII.30.9. (i) $f, g \in \mathscr{E}^* \Rightarrow af, f+g, f \cap g, f \cup g \in \mathscr{E}^*, a \in \boldsymbol{R}_+$.

(ii) The integral $\mathscr{E}^* \ni f \to \int f d\mu \in \overline{\boldsymbol{R}}$ has all the properties mentioned in the proposition, i.e. is a positive homogeneous, additive, and monotonic function on $\mathscr{E}^*$.

Now, without any great difficulty, we obtain the following theorem:

THEOREM XIII.30.10 (B. Levi). For every $f_n \nearrow$, where $f_n \in \mathscr{E}^*$, we have

$$\sup f_n \in \mathscr{E}^* \quad \text{and} \quad \int \sup f_n d\mu = \sup \int f_n d\mu .$$

As we know, $\mathscr{E}^*$ is a subset of $\mathscr{A}$-measurable functions. The next theorem states that these are all the nonnegative measurable functions.

THEOREM XIII.30.11. $(f \in \mathscr{E}^*) \Leftrightarrow (0 \leqslant f,\ f$ being $\mathscr{A}$-measurable$)$.

PROOF. $\Leftarrow$: Let $0 \leqslant f$ be measurable. The sets

$$A_{in} = \{x : i2^{-n} \leqslant f(x) < (i+1)2^{-n}\} \in \mathscr{A},$$
$$i = 0, \ldots, 2^n - 1,$$
$$A_{mn} = \{x : n \leqslant f(x), \text{ where } m = 2^n\} \in \mathscr{A}$$

are pairwise disjoint for every $n = 1, 2, \ldots$, and hence the function

$$u_n := \sum_{i=0}^{n2^n} i2^{-n} \chi_{A_{in}}$$

is elementary. Since u_{n+1} assumes on A_{in} only the values $2^{-n-1}2i$, $2^{-n-1}(2i+1)$, $i = 0, \ldots, m-1$ whereas $n \leqslant u_{n+1}(x)$ on A_{mn}, then $u_n \nearrow$. But $f = \sup u_n$, since for any $x \in X$ either $f(x) = \infty$, and then $u_n(x) = n$ for arbitrary n, or $f(x) < \infty$, and then $u_n(x) \leqslant f(x) < u_n(x) + 2^{-n}$ for all $n > f(x)$, that is, $f = \sup u_n$. $\square$

Integrability. We must now, of course, adopt the following definition:

DEFINITION. A function $f : X \to \overline{\boldsymbol{R}}$ is said to be *μ-integrable* when
(i) it is measurable,
(ii) $\int f^{\pm} d\mu < \infty$.
Then

$$\int f d\mu := \int f^+ d\mu - \int f^- d\mu$$

is called the *integral of the function f with respect to the measure μ*. The space $\mathscr{L}^1(\mu)$ is the space of μ-integrable functions.

Further the theory flows along a straight channel: It is shown that the integral is a linear form over $\mathscr{L}^1(\mu)$ and that the inequality

$$\left|\int f d\mu\right| \leqslant \int |f| d\mu, \quad f \in \mathscr{L}^1(\mu)$$

holds.

Next, we introduce the sets of measure zero and the concept of the property μ-almost everywhere, and we verify that integrable functions which are equal μ-almost everywhere have the same integrals; the spaces $\mathscr{L}^p(\mu)$ and spaces $L^p(\mu)$ are introduced, and are proved to be complete (exactly as in Daniell–Stone theory); the Fatou Lemma and the majorized-convergence theorem are proved. It should be remarked that the application of Daniell–Stone procedure to the elementary family of functions $E(\mathscr{A})$ leads, in general, to a larger σ-algebra of measurable subsets than the algebra $\mathscr{A}$ used as the starting point.

The theorems of Fubini and Tonelli are proved with much more difficulty than in Daniell–Stone theory owing to the fact that the tensor product of the measures $\mu_1 \otimes \mu_2$ must be determined and its properties examined. On the other hand, change of variables in the integral appears in a much more natural manner. Hence, the transport and convolution of measures will be taken up in Chapter XVIII.

The reader who wants to become more familiar with this approach to integral theory is referred to the book by H. Bauer, *Wahrscheinlichkeitstheorie und Grundzüge der Masstheorie*, Berlin 1974, upon which I drew in writing this section. (There is already an English translation!)

XIV. TENSOR ANALYSIS
HARMONIC FORMS
COHOMOLOGY
APPLICATIONS TO ELECTRODYNAMICS

IN MEMORIAM HERMANN WEYL

This chapter, which is central to the whole treatise, consists of two parts.

The first part (Sections 1–5) is fairly elementary: it develops the theory of differential forms and their integration (the "Stokes Theorem") for finite domains of R^n. Harmonic forms, Laplace operators and fundamental integral formulas (Green formulas) and elements of the de Rham cohomology are introduced. The most important case of three dimensions ("vector analysis") is derived as a corollary of the general theory.

The second part (Sections 6–16) introduces the most important notions of modern analysis, namely those of a differentiable manifold, a tangent bundle, Riemann manifold, and a Lie group. The Poincaré–Stokes Theorem for manifolds with a boundary is proved.

At this point I cannot refrain from quoting one of the greatest and most interesting mathematicians of our times, René Thom. In his provocative encyclopaedic article *La science malgré tout…* he writes: "*…Le progrès scientifique est avant tout une affaire de cerveaux, et non une question de laboratoires ou d'expérimentation. De ce point de vue, il est bon de signaler un fait dont bien peu de scientifiques sont conscients*: (…) *c'est l'extrême misère théorique de la plupart de sciences. Je ne connais guère qu'une science réellement difficile*: *c'est la mathématique* (…). *De ait, on peut se poser la question*: *quel est le théorème mathématique le plus profond, le plus difficile, dont il existe une interprétation physique concrète et indubitable?* (…) *Pour moi, c'est le théorème de Stokes qui est le candidat numéro un. Et cela témoigne d'un fait*: *la différentielle extérieure est une notion très mystérieuse, dont la véritable nature, je crois, recèle encore bien des énigmes, et cela en dépit de la simplicité de sa définition formelle.*"

Dual objects to differential forms were introduced by Hermann Weyl: contravariant skew-symmetric tensor densities (of weight 1). I call this dual pairing the "Weyl duality" (Section 11). On a Riemann manifold the Weyl duality reduces to the famous Hodge star operator.

Several important dualities of algebraic topology and analysis were unified by Georges de Rham in his theory of currents (Section 13). One of the most beautiful chapters of modern analysis, the theory of harmonic forms (conceived by W. V. D. Hodge and developed by Kunihiko Kodaira, Weyl, de Rham and others), is also treated in Section 13.

Lie groups, invariant differential forms and invariant integrals on a Lie group are introduced in Section 15. Compact connected Lie groups are perhaps the most perfect objects of mathematics. The cohomology of these groups, discovered by the greatest geometrical genius of our century, Élie Cartan, culminates in the following beautiful theorem:

$$\mathscr{H}^k(G) = \Lambda^k_{\text{inv}}(G);$$

bi-invariant forms and harmonic forms are the same entities in the case of a compact, connected Lie group G (Section 15).

It would be unpardonable not to mention the physical theory which was the main source of the theories developed in the present chapter, namely electrodynamics. Section 14 is a short introduction to these important ideas, written (as I hope) in the spirit of Weyl.

I devote this chapter to the memory of Hermann Weyl—the last great universalist.

The present chapter passed through several stages: the first part of it (Sections 1–5) was wholly rewritten by Dr. Krzysztof Gawędzki. But the second part also owes very much to him: He not only made infinitely many "corrections" but also wrote the best parts of Sections 6–14.

1. ALTERNATING MAPS. GRASSMANN ALGEBRA

Let E and F be Banach spaces over $\boldsymbol{R}$ and let $L_p(E, F)$ be the Banach space of continuous p-linear maps $f\colon \underbrace{E\times \ldots \times E}_{p \text{ times}} \to P$ (see Section VIII.7).

DEFINITION. A map $f \in L_p(E, F)$ is *alternating* (*skew-multilinear*) if (i) for any permutation σ of the set $\{1, \ldots, p\}$

(1) $\quad f(e_{\sigma(1)}, \ldots, e_{\sigma(p)}) = \operatorname{sgn}(\sigma) f(e_1, \ldots, e_p),$

where $\operatorname{sgn}(\sigma) = \pm 1$ is the sign of the permutation σ (i.e. $= +1$ for even permutations and -1 for odd ones), or equivalently

(2) $\quad f(e_1, \ldots, e_p) = 0$

whenever $e_i = e_j$ for some $i \neq j$.

Denote by $A_p(E, F)$ the space of alternating p-linear maps. $A_p(E, F)$ is a closed subspace of $L_p(E, F)$. We shall denote $A_p(E, R) \equiv A_p(E)$ and call its elements *alternating forms.*

Exterior Multiplication of Alternating Forms. About 1840 Hermann Grassmann introduced the very important notion of exterior multiplication of alternating forms. This notion was the cornerstone of his *Ausdehnungslehre* and served him in the investigation of linear (in)dependence and afterwards in his fundamental work on the Pfaff problem.

DEFINITION. Let $f \in A_p(E)$, $g \in A_q(E)$. Define $f \wedge g \in A_{p+q}(E)$ by

$$(3) \qquad f\wedge g(e_1, \ldots, e_{p+q}) := \frac{1}{p!q!} \sum_{\sigma \in S_{p+q}} \operatorname{sgn}(\sigma) f(e_{\sigma(1)}, \ldots, e_{\sigma(p)}) g(e_{\sigma(p+1)}, \ldots, e_{\sigma(p+q)}),$$

where S_r denotes the symmetric group of r elements, i.e., the set of all permutations of the set $\{1, \ldots, r\}$ and $\operatorname{sgn}(\sigma)$ is the sign of σ.

Remark. If we allow only such $\sigma \in S_{p+q}$ that

$$(4) \qquad \sigma(1) < \ldots < \sigma(p) \quad \text{and} \quad \sigma(p+1) < \ldots < \sigma(p+q),$$

then we can rewrite (3) as

$$(5) \qquad f\wedge g(e_1, \ldots, e_{p+q}) = \sum_{\sigma}{}' \operatorname{sgn}(\sigma) f(e_{\sigma(1)}, \ldots, e_{\sigma(p)})\, g(e_{\sigma(p+1)}, \ldots, e_{\sigma(q+q)}),$$

where $\sum_{\sigma}'$ denotes the restricted sum over $\sigma \in S_{p+q}$. The exterior multiplication defines a bilinear continuous mapping

$$A_p(E) \times A_q(E) \to A_{p+q}(E).$$

PROPOSITION XIV.1.1. Let $f \in A_p(E)$, $g \in A_q(E)$; then

$$(6) \qquad g\wedge f = (-1)^{pq} f\wedge g.$$

PROOF. The proof is almost obvious since

$$f(e_{\sigma(1)}, \ldots, e_{\sigma(p)}) g(e_{\sigma(p+1)}, \ldots, e_{\sigma(p+q)}) = g(e_{\tau(1)}, \ldots, e_{\tau(q)}) f(e_{\tau(q+1)}, \ldots, e_{\tau(q+p)})$$

if

$$\tau(i) := \begin{cases} \sigma(p+i) & \text{if } 1 \leqslant i \leqslant q, \\ \sigma(i-q) & \text{if } q+1 \leqslant i \leqslant q+p, \end{cases}$$

but $\operatorname{sgn}(\sigma) = (-1)^{pq}\operatorname{sgn}(\sigma)$. □

PROPOSITION XIV.1.2. Exterior multiplication is associative:

(7) $$(f \wedge g) \wedge h = f \wedge (g \wedge h)$$

for any $f \in A_p(E)$, $g \in A_q(E)$, $h \in A_r(E)$.

PROOF.

$$\begin{aligned}
&(f \wedge g) \wedge h(e_1, \ldots, e_{p+q+r}) \\
&= \sum_{\sigma \in S_{(p+q)+r}}' \operatorname{sgn}(\sigma) f \wedge g(e_{\sigma(1)}, \ldots, e_{\sigma(p+q)}) \times \\
&\times h(e_{\sigma(p+q+1)}, \ldots, e_{\sigma(p+q+r)}) \\
&= \sum_{\sigma \in S_{(p+q)+r}}' \operatorname{sgn}(\sigma) \sum_{\tau \in S_{p+q}}' \operatorname{sgn}(\tau) f(e_{\sigma(\tau(1))}, \ldots, e_{\sigma(\tau(p))}) \times \\
&\times g(e_{\sigma(\tau(p+1))}, \ldots, e_{\sigma(\tau(p+q))}) h(e_{\sigma(p+q+1)}, \ldots, e_{\sigma(p+q+r)}) \\
&= \sum_{\substack{\varrho \in S_{p+q+r} \\ \varrho(1) < \ldots < \varrho(p) \\ \varrho(p+1) < \ldots < \varrho(p+q) \\ \varrho(p+q+1) < \ldots < \varrho(p+q+r)}} \operatorname{sgn}(\varrho) f(e_{\varrho(1)}, \ldots, e_{\varrho(p)}) \times \\
&\times g(e_{\varrho(p+1)}, \ldots, e_{\varrho(p+q)}) h(e_{\varrho(p+q+1)}), \ldots, e_{\varrho(p+q+r)}) \\
&= f \wedge (g \wedge h)(e_1, \ldots, e_{p+q+r}),
\end{aligned}$$

where

$$\varrho(i) = \begin{cases} \sigma(\tau(i)), & 1 \leqslant i \leqslant p+q, \\ \sigma(i), & p+q+1 \leqslant i \leqslant p+q+r. \end{cases} \; □$$

The associativity of the exterior multiplication allows any finite products $f^1 \wedge \ldots \wedge f^p$ to be considered as linear forms. Grassmann proved the following

PROPOSITION XIV.1.3. Let $f^1, \ldots, f^p \in L(E, \mathbf{R}) \equiv E'$. Then

(8) $$\begin{aligned} &f^1 \wedge f^2 \wedge \ldots \wedge f^p(e_1, \ldots, e_p) \\ &= \sum_{\sigma} \operatorname{sgn}(\sigma) f^1(e_{\sigma(1)}) \ldots f^p(e_{\sigma(p)}) = \det(f^i(e_j)) \\ &\equiv \det(\langle e_j \mid f^i \rangle). \end{aligned}$$

PROOF. The proof follows by induction with respect to n (the assertion is obvious for $n = 1$ and $n = 2$). □

COROLLARY XIV.1.4. ($f^1, \ldots, f^p \in E'$ are linear dependent) $\Leftrightarrow$ ($f^1 \wedge \ldots \wedge f^p = 0$).

PROOF. $\Rightarrow$: If $f^p = \sum_{i=1}^{p-1} a_i f^i$ then $f^1 \wedge \ldots \wedge f^p = 0$, since $f^k \wedge f^k = -f^k \wedge f^k = 0$.

$\Leftarrow$: If $f_1, \ldots, f^p$ are linearly independent then there exist vectors $e_1, \ldots, e_p \in E$ such that $f^i(e_k) = \delta^i_k$. Thus $1 = \det(f^i(e_k)) = f^1 \wedge \ldots \wedge f^p(e_1, \ldots, e_p)$. Hence $f^1 \wedge \ldots \wedge f^p \neq 0$. □

Case of E with $\dim E = n < \infty$. Let $(f^1, \ldots, f^n)$ be a basis in E'.

THEOREM XIV.1.5. Each p-linear alternating map $f \in A_p(E, F)$ can be uniquely written in the form

$$f = \sum_{1 \leqslant j_1 < \ldots < j_p \leqslant n} C_{j_1 \ldots j_p} f^{j_1} \wedge \ldots \wedge f^{j_p} \text{ with } C_{j_1 \ldots j_p} \in F. \tag{9}$$

PROOF. Let $(v_1, \ldots, v_n)$ be the basis in E dual to $(f^1, \ldots, f^n)$. Let $e_1, \ldots, e_p \in E$ and let

$$e_i = \sum_{j=1}^{n} e_i^j v_j.$$

Then

$$f(e_1, \ldots, e_p) = \sum_{j_1, \ldots, j_p=1} e_1^{j_1} \ldots e_p^{j_p} f(v_{j_1}, \ldots, v_{j_p})$$
$$= \sum_{1 \leqslant j_1 < \ldots < j_p \leqslant n} \Big(\sum_{\sigma \in S_p} \operatorname{sgn}(\sigma) e_1^{j_{\sigma(1)}}, \ldots, e_p^{j_{\sigma(p)}} \Big) f(v_{j_1}, \ldots, v_{j_p})$$
$$= \sum_{1 \leqslant j_1 < \ldots < j_p \leqslant n} f(v_{j_1}, \ldots, v_{j_p}) f^{j_1} \wedge \ldots \wedge f^{j_p}(e_1, \ldots, e_p),$$

which proves (9) with $c_{j_1 \ldots j_p} = f(v_{j_1}, \ldots, v_{j_p})$.

But the latter holds for $c_{j_1 \ldots j_p}$ whenever (9) holds. Thus uniqueness follows. □

COROLLARY XIV.1.6. 1° $A_p(E, F) = 0$ for $p > n$.

2° For $p \leqslant n$ $(f_{j_1} \wedge \ldots \wedge f_{j_p})_{j_1 < \ldots < j_p}$ form a basis of $A_p(E)$.

3° Each element of $A_n(E, F)$ is of the form $c f_1 \wedge \ldots \wedge f_n$ where $c \in F$.

PROOF. 1° There exists no strictly increasing sequence of natural numbers $1 \leqslant j_1 < \ldots < j_p \leqslant n$. 2° and 3° are obvious consequences of Theorem XIV.1.5. □

2. DIFFERENTIAL FORMS

In the sequel E and F will denote Banach spaces over $\boldsymbol{R}$, and U an open subset of E. We remind that $A_p(E, F)$, as a closed subspace of $L_p(E, F)$, is a Banach space.

DEFINITION. A map

$$\omega: U \to A_p(E, F)$$

will be called a *differential form* on U of degree p with values in F.

The space of differential forms on U of degree p with values in F and of class C^r will be denoted by $\Lambda^p_{(r)}(U, F)$. We shall also write $\Lambda^p(U, F)$ instead of $\Lambda^p_{(\infty)}(U, F)$ and $\Lambda^p(U)$ instead $\Lambda^p(U, \boldsymbol{R})$. For $x \in U$ and $e_1, \ldots, e_p \in E$ we shall often denote

$$\omega(x)(e_1, \ldots, e_p) \equiv \omega(x; e_1, \ldots, e_p).$$

Example. Let f: $U \to F$ be a map of class C^r, $r \geqslant 1$. Then the derivative f': $U \to L(E, F)$ is a differential form on U of degree 1 and class C^{r-1}.

The Cartan Algebra. If $F = \boldsymbol{R}$ we can define exterior multiplication of differential forms.

DEFINITION. For $\overset{p}{\omega} \in \Lambda^p_{(r)}(U)$, $\overset{q}{\omega} \in \Lambda^q_{(r)}(U)$, $\overset{p}{\omega} \wedge \overset{q}{\omega} \in \Lambda^{p+q}_{(r)}(U)$ is defined by

$$\text{(1)} \qquad \overset{p}{\omega} \wedge \overset{q}{\omega}(x) := \overset{p}{\omega}(x) \wedge \overset{q}{\omega}(x).$$

Remark. That $\overset{p}{\omega} \wedge \overset{q}{\omega}$ is also of class C^r follows easily as it is a superposition of a C^r map $(\overset{p}{\omega}, \overset{q}{\omega})$: $U \to A_p(E) \times A_q(E)$ and a continuous bilinear map Λ: $A_p(U) \times A_q(U) \to A_{p+q}(E)$ which is of class C^∞.

Example. For $\eta, \omega \in \Lambda^1(U)$

$$\eta \wedge \omega(x; e_1, e_2) = \eta(x; e_1)\omega(x; e_2) - \eta(x; e_2)\omega(x; e_1).$$

2. DIFFERENTIAL FORMS

It is useful to treat maps from $C^r(U, F)$ as 0-forms on U with values in F. Hence we put

$$\Lambda^0_{(r)}(U, F) \equiv C^r(U, F), \qquad \Lambda^0(U, F) = C^\infty(U, F).$$

For $\alpha \in \Lambda^0_{(r)}(U, F)$ and $\omega \in \Lambda^p_{(r)}(U, F), p \geqslant 0$, we define

$$(\alpha \wedge \omega)(x) := (\alpha\omega)(x) = \alpha(x)\omega(x) = (\omega \wedge \alpha)(x).$$

DEFINITION. Denote by $\Lambda(U, \mathbf{R})$ or $\Lambda(U)$ the direct sum

$$\Lambda(U) := \bigoplus_{p \geqslant 0} \Lambda^p(U).$$

This space provided with exterior multiplication

$$\Lambda\colon \Lambda^p(U) \times \Lambda^q(U) \to \Lambda^{p+q}(U)$$

forms a (graded) associative algebra called the *Cartan algebra* over U.

The most important construct of Éli Cartan, however, is the

Exterior Derivative. It is possible to assign to each differential p-form ω of class C^r, $r \geqslant 1$, a $(p+1)$-form $d\omega$ of class C^{r-1} given by the

DEFINITION. Let $\omega \in \Lambda^p_{(r)}(U, F)$. Define $d\omega\, \Lambda^{p+1}_{(r-1)}(U, F)$:

$$d\omega(x; e_0, e_1, \ldots, e_p) := \sum_{i=0}^{p} (-1)^i (\omega'(x) \cdot e_i)(e_0, \ldots, \hat{e}_i, \ldots, e_p) \tag{2}$$

where $\omega'(x)$ is the derivative of the map $\omega\colon U \to A_p(E, F)$ at $x \in U$ and $\ldots, \hat{e}_i, \ldots$ means that we omit e_i.

Example 1. Let $f\colon U \to F$. Then

$$df(x; e) = f'(x) \cdot e.$$

Example 2. For $p = 1$ we have

$$d\omega(x; e_1, e_2) = (\omega'(x) \cdot e_1) \cdot e_2 - (\omega'(x) \cdot e_2) \cdot e_1.$$

Hence we obtain the following

LEMMA XIV.2.1. Let $\omega \in \Lambda^1_{(1)}(U, F)$. Then $(d\omega = 0) \Leftrightarrow$ (for any $x \in U$ the bilinear map

$$E \times E \ni (e_1, e_2) \to (\omega'(x) \cdot e_1) \cdot e_2 \in F$$

is symmetric).

We have an important fact due to Poincaré:

THEOREM XIV.2.2. For any $\omega \in \Lambda^p_{(2)}(U, F)$

(3) $\quad d(d\omega) = 0.$

PROOF. Notice that by virtue of (2)

$$d\omega(x)\,(e_0, \ldots, \hat{e}_i, \ldots, e_{p+1})$$
$$= \sum_{k=0}^{i-1} (-1)^k (\omega'(x) \cdot e_k)(e_0, \ldots, \hat{e}_k, \ldots, \hat{e}_i, \ldots, e_{p+1}) -$$
$$- \sum_{k=i+1}^{p+1} (-1)^k (\omega'(x) \cdot e_k)(e_0, \ldots, \hat{e}_i, \ldots, \hat{e}_k, \ldots, e_{p+1}).$$

Hence

$$\big((d\omega)'(x) e_i\big)\,(e_0, \ldots, \hat{e}_i, \ldots, e_{p+1})$$
$$= \sum_{k=0}^{i-1} (-1)^k \big(\omega''(x)\,(e_i, e_k)\big)(e_0, \ldots, \hat{e}_k, \ldots, \hat{e}_i, \ldots, e_{p+1}) -$$
$$- \sum_{k=i+1}^{p+1} (-1)^k \big(\omega''(x)(e_i, e_k)\big)(e_0, \ldots, \hat{e}_i, \ldots, \hat{e}_k, \ldots, e_{p+1}).$$

But

$$d(d\omega)\,(x)\,(e_0, \ldots, e_{p+1})$$
$$= \sum_{i=0}^{p+1} (-1)^i \big((d\omega)'(x) \cdot e_i\big)(e_0, \ldots, \hat{e}_i, \ldots, e_{p+1})$$
$$= \sum_{i=0}^{p+1} \sum_{k=0}^{i-1} (-1)^{i+k} \big(\omega''(x)(e_i, e_k)\big)(e_0, \ldots, \hat{e}_k, \ldots, \hat{e}_i, \ldots, e_{p+1}) -$$
$$- \sum_{i=0}^{p+1} \sum_{k=i+1}^{p+1} (-1)^{i+k} \big(\omega''(x)\,(e_i, e_k)\big)(e_0, \ldots, \hat{e}_i, \ldots$$
$$\ldots, \hat{e}_k, \ldots, e_{p+1})$$
$$= \sum_{0 \leqslant k < i \leqslant p+1} (-1)^{i+k} \big(\omega''(x)\,(e_i, e_k)\big)(e_0, \ldots, \hat{e}_k, \ldots$$
$$\ldots \hat{e}_i, \ldots, e_{p+1}) -$$
$$- \sum_{0 \leqslant i < k \leqslant p+1} (-1)^{i+k} \big(\omega''(x)\,(e_i, e_k)\big)\,(e_0, \ldots, \hat{e}_i, \ldots$$
$$\ldots, \hat{e}_k, \ldots, e_{p+1})$$

$$= \sum_{0 \leqslant k < i \leqslant p+1} (-1)^{i+k} (\omega''(x)(e_i, e_k)) -$$
$$- (\omega''(x)(e_k, e_i))(e_0, \dots, \hat{e}_k, \dots, \hat{e}_i, \dots, e_{p+1}) = 0,$$

where we have interchanged i and k in the last but one line and used the symmetry of the second derivative. □

Both fundamental operations on scalar differential forms: external derivation and external multiplication are related by

THEOREM XIV.2.3. Let $\overset{p}{\omega} \in \Lambda^p_{(1)}(U)$, $\overset{q}{\eta} \in \Lambda^q_{(1)}(U)$. Then

$$d(\overset{p}{\omega} \wedge \overset{q}{\eta}) = d\overset{p}{\omega} \wedge \overset{q}{\eta} + (-1)^p \overset{p}{\omega} \wedge d\overset{q}{\eta}. \tag{4}$$

PROOF. First we recall a simple formula for the derivative of a bilinear map which is an easy consequence of the theorem on the derivative of a composite map (Theorem VII.4.2.).

LEMMA XIV.2.4. Let $B\colon E_1 \times E_2 \times \dots \times E_n \to F$ be a bilinear continuous map (of Banach spaces).

If the maps $f_k\colon U \to E_k$, $k = 1, 2, \dots, n$, U being an open subset of a Banach space E, are of class C^r $(r \geqslant 1)$ then for the map $w\colon U \to F$, $w(\cdot) := B(f_1(\cdot), \dots, f_n(\cdot))$, which is of class C^r as a superposition of C^r map $(f_1, \dots, f_n)\colon U \to E_1 \times \dots \times E_n$ and B being of class C^∞, we have

$$\begin{aligned} &w'(x) \cdot h \\ &= B(f_1'(x) \cdot h, f_2(x), \dots, f_n(x)) + \dots \\ &\dots + B(f_1(x), \dots, f_{n-1}(x), f_n'(x) \cdot h). \end{aligned} \tag{5}$$

From (5) we have for $x \in U$, $e \in E$

$$\begin{aligned} (\overset{p}{\omega} \wedge \overset{q}{\eta})'(x) \cdot e &= (\overset{p}{\omega}{}'(x) \cdot e) \wedge \overset{q}{\eta}(x) + \overset{p}{\omega}(x) \wedge \overset{q}{\eta}{}'(x) \cdot e \\ &= (\overset{p}{\omega}{}'(x) \cdot e) \wedge \overset{q}{\eta}(x) + (-1)^{pq} (\overset{q}{\eta}{}'(x) \cdot e) \wedge \overset{p}{\omega}(x). \end{aligned} \tag{6}$$

Hence (2) applied to both sides of (6) gives after easy manipulations

$$\begin{aligned} d(\overset{p}{\omega} \wedge \overset{q}{\eta}) &= (d\overset{p}{\omega}) \wedge \overset{q}{\eta} + (-1)^{pq} (d\overset{q}{\eta}) \wedge \overset{p}{\omega} \\ &= (d\overset{p}{\omega}) \wedge \overset{q}{\eta} + (-1)^{pq + p(q+1)} \overset{p}{\omega} \wedge (d\overset{q}{\eta}) \\ &= (d\overset{p}{\omega}) \wedge \overset{q}{\eta} + (-1)^p \overset{p}{\omega} \wedge (d\overset{q}{\eta}). \ \square \end{aligned}$$

Change of Variables in Differential Forms. Let E_1, E_2 be Banach spaces and $U_1 \subset E_1$, $U_2 \subset E_2$ open subsets. Let Φ: $U_1 \to U_2$ be a map of class C^{r+1}.

DEFINITION. Let ω: $U_2 \to A_p(E_2, F)$ be a differential p-form of class C^r. Then $\Phi^*\omega$ will denote a p-form of class C^r on U_1 defined by

$$(\Phi^* \omega)(y; e_1, \ldots, e_p) := \omega\big(\Phi(y); \Phi'(y)e_1, \ldots, \Phi'(y)e_p\big). \tag{7}$$

Remark. That $\Phi^*\omega$ is of class C^r again follows easily as it is a superposition of C^r maps. We leave the details to the reader. $\Phi^*\omega$ is often called the *pull-back* of ω by Φ.

Both fundamental operations in the Cartan algebra are preserved by Φ^* operation. More exactly, there holds

THEOREM XIV.2.5. Let $\Phi \in C^{r+1}(U_1, U_2)$.

1° If $\omega \in \Lambda^p_{(r)}(U_2, F)$, then

$$d(\Phi^*\omega) = \Phi^*(d\omega).$$

2° If $\omega \in \Lambda^p_{(r)}(U)$, $\eta \in \Lambda^q_{(r)}(U)$, then

$$(\Phi^*\omega) \wedge (\Phi^*\eta) = \Phi^*(\omega \wedge \eta).$$

Moreover,

3° If $\Phi \in C^{r+1}(U_1, U_2)$, $\Psi \in C^{r+1}(U_2, U_3)$ (U_i being open subsets of Banach spaces E_i, $i = 1, 2, 3$) and $\omega \in \Lambda^p_{(r)}(U_3, F)$, then

$$\Phi^*\big(\Psi^*(\omega)\big) = (\Psi \circ \Phi)^*(\omega). \tag{8}$$

PROOF. 1° Differentiation (using (7) and (5)) gives

$$\begin{aligned}
&((\Phi^*\omega)'(x)e_0)(e_1, \ldots, e_p) \\
&= \big(\omega'(\Phi(x)) \circ \Phi'(x)e_0\big)\big(\Phi'(x)e_1, \ldots, \Phi'(x)e_p\big) + \\
&+ \sum_{k=1}^{p} \omega\big(\Phi(x); \Phi'(x)e_1, \ldots, \Phi''(x)(e_0, e_k), \ldots, \Phi'(x)e_p\big).
\end{aligned}$$

Hence

$$\begin{aligned}
&d(\Phi^*\omega)(x; e_0, \ldots, e_p) \\
&= \sum_{i=0}^{p} (-1)^i \big((\Phi^*\omega)'(x)e_i\big)(e_0, \ldots, \hat{e}_i, \ldots, e_p) \\
&= \sum_{i=0}^{p} (-1)^i \big(\omega'(\Phi(x)) \circ \Phi'(x)e_i\big)\big(\Phi'(x)e_0, \ldots
\end{aligned}$$

$$\ldots, \widehat{\Phi'(x)e_i}, \ldots, \Phi'(x)e_p\big)+$$
$$+\sum_{0\leqslant k<i\leqslant p}(-1)^i\omega\big(\Phi(x); \Phi'(x)e_0, \ldots, \Phi''(x)(e_i, e_k), \ldots$$
$$\ldots, \widehat{\Phi'(x)e_i}, \ldots, \Phi'(x)e_p\big)+$$
$$+\sum_{0\leqslant i<k\leqslant p}(-1)^i\omega\big(\Phi(x); \Phi'(x)e_0, \ldots, \widehat{\Phi'(x)e_i}, \ldots$$
$$\ldots, \Phi''(x)(e_i, e_k), \ldots, \Phi'(x)e_p\big).$$

The last sum can be rewritten in the form

$$\sum_{0\leqslant k<i\leqslant p}(-1)^k\omega\big(\Phi(x); \Phi'(x)e_0, \ldots, \widehat{\Phi'(x)e_k}, \ldots$$
$$\ldots, \Phi''(x)(e_k, e_i), \ldots, \Phi'(x)e_p\big)$$
$$=-\sum_{0\leqslant k<i\leqslant p}(-1)^i\omega\big(\Phi(x); \Phi'(x)e_0, \ldots$$
$$\ldots, \Phi''(x)(e_k, e_i), \ldots, \widehat{\Phi'(x)e_i}, \ldots, \Phi'(x)e_p\big)$$

and thus cancels the preceding sum. Finally

$$d(\Phi^*\omega)(x; e_0, \ldots, e_p)$$
$$=\sum_{i=0}^{p}(-1)^i\big(\omega'(\Phi(x))\circ\Phi'(x)\circ e_i\big)\big(\Phi'(x)e_0, \ldots$$
$$\ldots, \widehat{\Phi'(x)e_i}, \ldots, \Phi'(x)e_p\big)=\Phi^*(d\omega).$$

2° $$(\Phi^*\omega)\wedge(\Phi^*\eta)(x; e_1, \ldots, e_{p+q})$$
$$=\sum_{\sigma}{}'\operatorname{sgn}(\sigma)\omega\big(\Phi(x); \Phi'(x)e_{\sigma(1)}, \ldots$$
$$\ldots, \Phi'(x)e_{\sigma(p)}\big)\eta\big(\Phi(x); \Phi'(x)e_{\sigma(p+1)}, \ldots, \Phi'(x)e_{\sigma(p+q)}\big)$$
$$=(\omega\wedge\eta)\big(\Phi(x), \Phi'(x)e_1, \ldots, \Phi'(x)e_{p+q}\big)$$
$$=\big(\Phi^*(\omega\wedge\eta)\big)(x; e_1, \ldots, e_{p+q}).$$

3° $$\big(\Phi^*(\Psi^*(\omega)\big)(x; e_1, \ldots, e_p)$$
$$=(\Psi^*\omega)\big(\Phi(x); \Phi'(x)e_1, \ldots, \Phi'(x)e_p\big)$$
$$=\omega\big(\Psi(\Phi(x)); \Psi'(\Phi(x))\circ\Phi'(x)e_1, \ldots, \Psi'(\Phi(x))\circ\Phi'(x)e_p\big)$$
$$=\omega\big(\Psi\circ\Phi(x); (\Psi\circ\Phi)'(x)e_1, \ldots, (\Psi\circ\Phi)'(x)e_p\big)$$
$$=\big((\Psi\circ\Phi)^*\omega\big)(x; e_1, \ldots, e_p). \quad \square$$

Differential Forms on Finite-Dimensional Space. Suppose that $\dim E = n < \infty$, U is an open subset of E. Let $(f^1, \ldots, f^n)$ form a basis of E'. Then df^i is a differential form of degree 1 (and class C^∞) on E (or after restriction on U).

Theorem XIV.1.5 leads immediately to

THEOREM XIV.2.6. Each differential form $\omega \in \Lambda^p_{(r)}(U, F)$ can be uniquely written in the form

$$\omega = \sum_{0 \leqslant j_1 < \ldots < j_p \leqslant n} c_{j_1 \ldots j_p} df^{j_1} \wedge \ldots \wedge df^{j_p},$$

where

$$c_{j_1 \ldots j_p} \in C^r(U, F).$$

Remark. If $F = \boldsymbol{R}$ (scalar differential forms) then $\Lambda^p(U)$ is a module over the ring $C^\infty(U)$ and the elements $df^{j_1} \ldots df^{j_p}$, $1 \leqslant j_1 < \ldots < j_p \leqslant n$, form a basis of this module.

If $E = \boldsymbol{R}^n$ take the basis of the dual space formed by coordinate maps x^i.

COROLLARY XIV.2.7. Each differential form $\omega \in \Lambda^p_{(r)}(U, F)$, where U is an open subset of $\boldsymbol{R}^n$, can be uniquely written in the form

$$\omega = \sum_{0 \leqslant j_1 < \ldots < j_k \leqslant n} c_{j_1 \ldots j_p} dx^{j_1} \wedge \ldots \wedge dx^{j_p}, \tag{9}$$

where

$$c_{j_1 \ldots j_p} \in C^r(U, F).$$

The form (9) is called the *canonical form of a differential form on* $\boldsymbol{R}^n$. It is easy to give the external derivative in canonical form.

COROLLARY XIV.2.8. Let

$$\omega = \sum_{j_1 < \ldots < j_p} c_{j_1 \ldots j_p} dx^{j_1} \wedge \ldots \wedge dx^{j_p}.$$

Then

$$\begin{aligned} d\omega &= \sum_{j_1 < \ldots < j_p} dc_{j_1 \ldots j_p} \wedge dx^{j_1} \wedge \ldots \wedge dx^{j_p} \\ &= \sum_{j_1 < \ldots < j_p} \sum_{j=1}^{n} \frac{\partial c_{j_1 \ldots j_p}}{\partial x_j} dx^j \wedge dx^{j_1} \wedge \ldots \wedge dx^{j_p}. \end{aligned} \tag{10}$$

Example. Let $n = 3$, $p = 1$. Then

$$\omega = c_1\,dx^1 + c_2\,dx^2 + c_3\,dx^3,$$

$$d\omega = \left(\frac{\partial c_3}{\partial x^2} - \frac{\partial c_2}{\partial x^3}\right) dx^2 \wedge dx^3 + \left(\frac{\partial c_1}{\partial x^3} - \frac{\partial c_3}{\partial x^1}\right) dx^3 \wedge dx^1 +$$

$$+ \left(\frac{\partial c_2}{\partial x^1} - \frac{\partial c_1}{\partial x^2}\right) dx^1 \wedge dx^2$$

(we take account of: $dx^i \wedge dx^i = 0$, $dx^i \wedge dx^j = -dx^j \wedge dx^i$, $j \neq i$).

COROLLARY XIV.2.9. Let

$$\omega = \sum_{j_1 < \ldots < j_p} c_{j_1 \ldots j_p} dx^{j_1} \wedge \ldots \wedge dx^{j_p},$$

$$\eta = \sum_{k_1 < \ldots < k_q} e_{k_1 \ldots k_q} dx^{k_1} \wedge \ldots \wedge dx^{k_q}.$$

Then

$$\omega \wedge \eta = \sum_{j_1 < \ldots < j_p} \sum_{k_1 < \ldots < k_q} c_{j_1 \ldots j_p} e_{k_1 \ldots k_q} dx^{j_1} \wedge \ldots \wedge dx^{j_p} \wedge$$

$$\wedge\, dx^{k_1} \wedge \ldots \wedge dx^{k_q}.$$

We end the section with a description of the coordinate form of change of variables.

COROLLARY XIV.2.10. Let Φ: $U_1 \to U_2$ be a map of class C^r, $r \geqslant 1$) U_1 being an open subset of $\boldsymbol{R}^n$ and U_2 of $\boldsymbol{R}^m$. Let $\omega \in \Lambda^p_{(r-1)}(U_2, F,$

$$\omega = \sum_{1 \leqslant j_1 < \ldots < j_p \leqslant m} c_{j_1 \ldots j_p} dy^{j_1} \wedge \ldots \wedge dy^{j_p},$$

where y^j are coordinate maps on U_2. Then

$$\text{(11)} \qquad \Phi^*\omega = \sum_{1 \leqslant j_1 < \ldots < j_p \leqslant m} c_{j_1 \ldots j_p} \circ \Phi \sum_{1 \leqslant k_1 < \ldots < k_p \leqslant n} \frac{\partial(\Phi^{j_1}, \ldots, \Phi^{j_p})}{\partial(x^{k_1}, \ldots, x^{k_p})} \times$$

$$\times\, dx^{k_1} \wedge \ldots \wedge dx^{k_p},$$

where $\Phi^j := y^j \circ \Phi$ and x^k are coordinate maps on U_1.

PROOF.

$$\Phi^*\omega = \sum_{j_1<\ldots<j_p} c_{j_1\ldots j_p} \circ \Phi d\Phi^{j_1} \wedge \ldots \wedge d\Phi^{j_p}$$

$$= \sum_{j_1<\ldots<j_p} c_{j_1\ldots j_p} \circ \Phi \sum_{\substack{1\leqslant k_1,\ldots,k_p\leqslant n\\ k_i\neq k_j, i\neq j}} \frac{\partial\Phi^{j_1}}{\partial x^{k_1}} \cdots \frac{\partial\Phi^{j_p}}{\partial x^{k_p}} dx^{k_1} \wedge \ldots \wedge dx^{k_p}$$

$$= \sum_{j_1<\ldots<j_p} c_{j_1\ldots j_p} \circ \Phi \sum_{k_1<\ldots<k_p} \frac{\partial(\Phi^{j_1}, \ldots, \Phi^{j_p})}{\partial(x^{k_1}, \ldots, x^{k_p})} dx^{k_1} \wedge \ldots \wedge dx^{k_p}. \square$$

3. COHOMOLOGY SPACES. POINCARÉ LEMMA

Let us return to the general situation: E, F are Banach spaces and U is an open subset of E. By definition set $\Lambda^p(U, F) := \{0\}$ for $p < 0$. We shall say that $\omega \in \Lambda^p(U, F)$ is *closed* (is a *cocycle*) if $d\omega = 0$. Denote the vector space of closed p-forms by $Z^p(U, F)$. So $Z^p(U, F) = \ker \overset{p}{d}$. Similarly the image of $\overset{p-1}{d} : \Lambda^{p-1}(U, F) \to \Lambda^p(U, F)$ is denoted by $B^p(U, F)$ and its elements are called *exact p-forms* (*coboundaries*). By virtue of the Poincaré Theorem XIV. 2.2

$$B^p(U, F) \subset Z^p(U, F).$$

Therefore we can form the quotient spaces

$$Z^p(U, F)/B^p(U, F).$$

DEFINITION. $H^p(U, F) := Z^p(U, F)/B^p(U, F)$ is called the *p-th de Rham cohomology space.*

For example, since $Z^0(U, F)$ is composed of maps $f\colon U \to F$ constant on each connected component of U and $B^0(U, F) = \{0\}$, we have

$$H^0(U, F) \simeq F^J,$$

where J is the set of connected components of U.

If $\Phi\colon U_1 \to U_2$ is a C^∞ map, then $\Phi^*\colon \Lambda^p(U_2, F) \to \Lambda^p(U_1, F)$ maps $Z^p(U_2, F)$ $(B^p(U_2, F))$ into $Z^p(U_1, F)$ $(B(U_1, F))$. Indeed, if $\omega \in \Lambda^p(U_2, F)$ is exact ($d\omega = 0$), then $d\Phi^*\omega = \Phi^*d\omega = 0$. If $\omega = d\eta$, then $\Phi^*\omega = d\Phi^*\eta$. Thus Φ^* defines a linear map

$$\Phi^\#\colon H^p(U_2, F) \to H^p(U_1, F).$$

If $U_2 = U_1 = U$ and Φ is the identity map, then $\Phi^{\#} = \mathrm{id}_{H^p(U,F)}$. Moreover if $\Phi: U_1 \to U_2$ and $\Psi: U_2 \to U_3$, then we have

$$\Phi^{\#} \circ \Psi^{\#} = (\Psi \circ \Phi)^{\#}.$$

In this way we have proved the important

THEOREM XIV.3.1. If Φ is a C^∞ diffeomorphism of U_1 onto U_2 then $\Phi^{\#}$ is an isomporphism of $H^p(U_2, F)$ onto $H^p(U_1, F)$ for $p = 0, 1, \ldots$ Moreover, $(\Phi^{\#})^{-1} = (\Phi^{-1})^{\#}$.

This theorem shows that de Rham cohomology spaces are invariants of diffeomorphisms, and thus have a (differential) topological sense. This is one of the chief arguments for investigating them. In Part III we shall develop a general cohomology theory, that of Čech cohomology, and we shall prove that the de Rham cohomology spaces are isomorphic with the Čech cohomology spaces, i.e., we shall prove the famous de Rham Theorem.

For a wide class of Φ, the induced maps $\Phi^{\#}$ coincide.

DEFINITION. Let Φ_k: $U_1 \to U_2, k = 0, 1,$ where U_i is an open subset of a Banach space E_i, $i = 1, 2$. Φ_0 and Φ_1 are *smoothly homotopic* if there exists a C^∞ map (see Section VII.13)

$$H\colon [0, 1] \times U_1 \to U_2$$

such that

$$H(0, \cdot) = \Phi_0(\cdot) \quad \text{and} \quad H(1, \cdot) = \Phi_1(\cdot).$$

H is called a *differential homotopy* between Φ_0 and Φ_1.

Less rigorously but more intuitively we could say that two maps are differentiable homotopic if they can be smoothly deformed into each other.

THEOREM XIV.3.2. If Φ_i: $U_1 \to U_2$, $i = 0, 1$ are differential homotopic, then

$$(1) \qquad \Phi_0^{\#} = \Phi_1^{\#}.$$

PROOF. We start with the following

Remark. If $V \subset E$ is not an open set, one can still define the spaces $\Lambda^q(V, F)$ and the operations of external differentiation and pull-back of forms such that Theorem XIV.2.5 holds as long as V has the property that the derivatives of continuously differentiable mappings defined on

V are determined uniquely (see Section VII.13). This holds if V is comprised in the closure of its interior.

In the sequel we shall often make use of the above Remark. Here it will enable us to consider differential forms on the set $[0, 1] \times U \subset \boldsymbol{R}^1 \times \times E$. Furthermore we shall be able to use them as if they were defined on an open set.

We define the maps

$$j_i\colon U_1 \to [0, 1] \times U_1, \quad i = 0, 1,$$
$$j_0(x) := (0, x), \quad j_1(x) := (1, x).$$

They induce the linear maps

$$j_i^*\colon \Lambda^q([0, 1] \times U_1, F) \to \Lambda^q(U_1, F).$$

We define another linear map

$$k\colon \Lambda^q([0, 1] \times U_1, F) \to \Lambda^{q-1}(U_1, F),$$

$$k\omega(x; e_1, \ldots, e_{q-1}) := \int_0^1 \omega(t, x; \tau, e_1, \ldots, e_{q-1})\,dt,$$

where $\tau = (1, 0) \in \boldsymbol{R}^1 \times E$ and we identify $e \in E$ with $(0, e) \in \boldsymbol{R}^1 \times E$.

Lemma XIV.3.3. Let $\omega \in \Lambda^p([0, 1] \times U_1, F), p \geqslant 0$. Then

(2) $$k\,d\omega + dk\omega = j_1^*\omega - j_0^*\omega.$$

Proof:

(3) $$(k\,d\omega)(x; e_1, \ldots, e_p) = \int_0^t d\omega(t, x; \tau, e_1, \ldots, e_p)\,dt$$

$$= \int_0^t \frac{\partial \omega}{\partial t}(t, x)(e_1, \ldots, e_p)\,dt +$$

$$+ \sum_{i=1}^p (-1)^i \int_0^t (\omega'(t, x) \cdot e_i)\,(\tau, e_1, \ldots, \hat{e}_i, \ldots e_p)\,dt$$

$$= (j_1^*\omega - j_0^*\omega)(x; e_1, \ldots, e_p) +$$

$$+ \sum_{i=1}^p (-1)^i \int_0^t (\omega'(t, x) \cdot e_i)\,(\tau, e_1, \ldots, \hat{e}_i, \ldots, e_p)\,dt,$$

(4) $$(dk\omega)(x; e_1, \ldots, e_p)$$
$$= -\sum_{i=1}^{p} (-1)^i \big((k\omega)'(x)\cdot e_i\big)(e_1, \ldots, \hat{e}_i, \ldots, e_p)$$
$$= -\sum_{i=1}^{p} (-1)^i \int_0^t \big(\omega'(t, x)\cdot e_i\big)(\tau, e_1, \ldots, \hat{e}_i, \ldots, e_p)\,dt.$$

From (3) and (4) we get (2). □

Now let H: $[0, 1]\times U_1 \to U_2$ be a differential homotopy between Φ_1 and Φ_2. Let $\omega \in Z^p(U_2, F)$. Apply Lemma XIV.3.3 to $H^*\omega$. We get

(5) $$dkH^*\omega = j_1^* H^*\omega - j_0^* H^*\omega$$
$$= (H\circ j_1)^*\omega - (H\circ j_0)^*\omega = \Phi_1^*\omega - \Phi_0^*\omega.$$

Thus

(6) $$\Phi_0^*\omega = \Phi_1^*\omega \ \big(\operatorname{mod} B^p(U_1, F)\big).$$ □

Theorem XIV.3.2 is very important.

It provides for, example, a very useful sufficient condition for all $H^p(U, F)$, $p \geqslant 1$, to vanish.

DEFINITION. Open $U \subset E$ is said to be *(smoothly) contractible (to a point)* if its identical mapping is smoothly homotopic to a constant mapping, i.e., if there exists a C^∞ mapping

$$H\colon [0, 1]\times U \to U$$

such that

(7) $$H(0, x) \equiv x \quad \text{and} \quad H(1, x) \equiv a.$$

Example. $U \subset E$ is called *star-shaped with respect to* $a \in U$ if for any $x \in U$

$$[a, x] := \{y \in E\colon y = ta + (1-t)x,\ t \in [0, 1]\} \subset U.$$

If U is star-shaped with respect to a, put

$$H(t, x) := ta + (1-t)x.$$

Thus U is contractable.

In particular any convex subset U is differential contractible.

Theorem 1 implies the famous

COROLLARY XIV.3.4 (Poincaré Lemma). If $U \subset E$ is contractible then

$$H^p(U, F) = 0 \quad \text{for } p \geqslant 0.$$

In other words, if ω is a closed p-form on U of class C^∞, then there exists a $(p-1)$-form α of class C^∞ on U such that

$$\omega = d\alpha. \tag{8}$$

PROOF. Take $\Phi_0 = 1_U$ and $\Phi_1 = \text{const} = a$ (see (7)). Then $\Phi_0^* = 1_{A^p(U,F)}$ and $\Phi_1^* = 0$ for $p \geqslant 1$. Thus

$$1_{H^p(U,F)} = \Phi_0^\# = \Phi_1^\# = 0$$

and consequently $H^p(U, F) = 0$. □

Remark. (5) gives a possible form of α satisfying (8) since in our case it reads

$$\omega = -dkH^*\omega. \tag{9}$$

Any other possible α differs in that it is a closed form.

Example 1. Let $E = R^n$, $p = 1$. In canonical form we have

$$\omega = \sum_{i=1}^{n} c_i dx^i,$$

where

$$c_i \colon R^n \to F.$$

The assumption $d\omega = 0$ (that ω is closed) means that

$$\frac{\partial c_i}{\partial x^k} - \frac{\partial c_k}{\partial x^i} = 0 \quad \text{for } 1 \leqslant i, j \leqslant n.$$

Now take H, $H(t, x) := (1-t)x$, as the contracting homotopy. Then (9) defines a map $\alpha\colon R^n \to F$ (0-form)

$$\alpha(x) = \sum_{i=1}^{p} \int_0^1 c_i((1-t)x)x^i dt = \sum_{i=1}^{p} \int_0^1 c_i(tx)x^i dt.$$

It is sufficient to check that

$$d\alpha = \omega,$$

i.e., that

$$\frac{\partial \alpha}{\partial x^i} = c_i.$$

Example 2. Let $U = K(0, 1) \subset \boldsymbol{R}^3$, $p = 2$. Plainly, $K(0, 1)$ is contractible and we can again set $H(t, x) := (1-t)x$. Let

$$\overset{2}{\omega} = x^1x^2 dx^1 \wedge dx^2 + 2x^1 dx^2 \wedge dx^3 + 2x^2 dx^1 \wedge dx^3.$$

Let us first write down a general formula

$$\omega = \sum_{i<k} c_i dx^i \wedge dx^k,$$

$$\alpha = -kH^*\omega = \sum_{k=1}^{n} \Big(\sum_{i<k} \int_0^1 tx^i c_{ik}(tx)\,dt - $$
$$- \sum_{i>k} \int_0^1 tx^i c_{ki}(tx)\,dt \Big) dx^k.$$

In our case

$$c_{12} = x^1x^2, \qquad c_{23} = 2x^1, \qquad c_{13} = 2x^2.$$

Therefore we have finally:

$$\overset{1}{\alpha} = -\Big(\int_0^1 \big(tx^2 c_{12}(tx) + tx^3 c_{13}(tx)\big)dt \Big) dx^1 +$$
$$+ \Big(\int_0^1 \big(tx^1 c_{12}(tx) - tx^3 c_{23}(tx)\big)dt \Big) dx^2 +$$
$$+ \Big(\int_0^1 \big(tx^1 c_{13}(tx) + tx^2 c_{23}(tx)\big)dt \Big) dx^3$$
$$= -\big(\tfrac{1}{4}x^1(x^2)^2 + \tfrac{2}{3}x^2x^3\big)dx^1 +$$
$$+ \big(\tfrac{1}{4}(x^1)^2x^2 - \tfrac{2}{3}x^1x^3\big)dx^2 + \tfrac{4}{3}x^1x^2\,dx^3.$$

We check immediately that $d\overset{1}{\alpha} = \overset{2}{\omega}$.

4. INTEGRATION OF DIFFERENTIAL FORMS

Now we are ready to begin with the main topic of the present chapter. Since any n-form α in $\boldsymbol{R}^n$ has the canonical form

$$\alpha = a\,dx^1 \wedge \dots \wedge dx^n$$

it is natural to define an integral of α on a compact subset $K \subset \boldsymbol{R}^n$ $\int_K \alpha$ as $\int_K a\,d\lambda^p$. We shall extend this notion in a natural way to p-forms defined on an open subset U of a Banach space E.

DEFINITION. A *singular p-cube of class* C^r $(p > 0, r \geqslant 0)$ in $U \subset E$ is a map Φ: $P^p \to U$ of class C^r, where

$$P^p = [a^1, b^1] \times \dots \times [a^p, b^p], \quad a^i < b^i \in \boldsymbol{R}.$$

It is often useful to write a singular cube as a pair

$$s_p := (P^p, \Phi), \quad \text{where } \Phi\colon P^p \to U.$$

If T: $U \to V$, where V is an open subset in a Banach space G, is of class C^r, and if $s^p = (P^p, \Phi)$ is a singular p-cube of class C^r in U, then $(P^p, T \circ \Phi)$ is a singular p-cube (of class C^r) in V. We shall denote it by Ts_p. The image of P^p under Φ is often called the *support* of $s_p = (P^p, \Phi)$. It is a compact subset of U. We shall denote it by $|s_p|$. Let ω be an F-valued p-form in $U \subset E$ of class C^0 and let $s_p = (P^p, \Phi)$ be a singular p-cube in U of class C^1.

DEFINITION. Define

$$(1) \qquad \langle s_p | \omega \rangle \equiv \int_{s_p} \omega := \int_{P^p} \Phi^* \omega.$$

Remark 1. By virtue of the remark in the proof of Theorem XIV.3.2. the fact that Φ is defined on a closed subset of $\boldsymbol{R}^p$ should not bother us, If $U \subset \boldsymbol{R}^n$, then

$$\omega = \sum_{j_1 < \dots < j_p} c_{j_1 \dots j_p} dx^{j_1} \wedge \dots \wedge dx^{j_p}$$

and by Corollary XIV.2.10

$$(2) \qquad \langle s_p | \omega \rangle = \sum_{j_1 < \dots < j_p} \int_{P^p} c_{j_1 \dots j_p} \circ \Phi(t) \frac{\partial(\Phi^{j_1}, \dots, \Phi^{j_p})}{\partial(t^1, \dots, t^p)} d\lambda^p(t).$$

Remark 2. It is useful to define a singular 0-cube as a map Φ: $P^0 \to U$, where $P^0 = \{0\}$ is any one-point subset of $\boldsymbol{R}$. Then, for ω as a 0-form on U, we write

$$(3) \qquad \langle S_0 | \omega \rangle := \omega(\Phi(0)).$$

Remark 3. Of course we can also integrate a p-form ω defined on a non-open set over p-cube s_p, provided that $|s_p|$ is in the domain of ω. From (1) and Theorem XIV.2.5 we immediately have

PROPOSITION XIV.4.1. Let T: $U \to V$ be a (C^1) map of open subsets in Banach spaces. Let $s_p = (P^p, \Phi)$ be a singular p-cube in U (of class C^1)

and ω an F-valued p-form on V (of class C^0). Then

$$(4) \qquad \langle s_p | T^* \omega \rangle = \langle T s_p | \omega \rangle .$$

The notion of a singular p-cube is a natural generalization of that of the 1-cube, i.e., of the parametrical curve. The p-cubes (P^p, Φ) under discussion are called *singular*, because the map Φ is not necessarily injective.

In the set of singular p-cubes of class C^r we introduce an equivalence relation

$$S_p^1 = (P_1^p, \Phi_1) \sim (P_2^p, \Phi_2) = S_p^2$$

if there exists a diffeomorphism S: $P_1^p \to P_2^p$ of class C^r with positive Jacobian ($\det S^1 > 0$) such that

$$\Phi_1 = \Phi_2 \circ S .$$

DEFINITION. The class $[s_p]$ of equivalent singular p-cubes (of class C^r) is called an *oriented p-cell.*

The importance of the equivalence relation considered above is due to

THEOREM XIV.4.2. Let $\omega \in \Lambda^p_{(0)}(U, F)$, s_p^1, s_p^2 be equivalent singular p-cubes of class C^1 in U. Then

$$\int_{s_p^1} \omega = \int_{s_p^2} \omega .$$

PROOF. The theorem follows from the transformation formula for the Lebesgue integral (Theorem XIII.22.1) and Corollary XIV.2.10

$$\int_{s_p^1} \omega = \int_{P_1^p} \Phi_1^* \omega = \int_{P_1^p} S^*(\Phi_2^* \omega) = \int_{P_2^p} \Phi_2^* \omega = \int_{s_p^2} \omega . \ \square$$

Singular Chains. Formal Linear Combinations. Let S be an arbitrary non-empty set. Denote by $\boldsymbol{R}^S$ the space of all $\boldsymbol{R}$-valued functions on S. We know that $\boldsymbol{R}^S$ is a real vector space: if $f_1, f_2 \in \boldsymbol{R}^S$, then

$$(a_1 f_1 + a_2 f_2)(s) := a_1 f_1(s) + a_2 f_2(s),$$
$$s \in S, \ a_1, a_2 \in \boldsymbol{R} .$$

DEFINITION. The *real vector space generated by* S is the linear subspace of $\boldsymbol{R}^S$ consisting of functions on S which vanish at all but a finite number of elements of S. This space will be denoted by $(\boldsymbol{R}^S)_0$ and its elements will be called *(finite) formal linear combinations of elements of S*.

We have an elementary

PROPOSITION XIV.4.3. Denote by $a \cdot s$ the function which takes the value a at s and 0 at all other elements of S. Then the set $\{1 \cdot s, s \in S\}$ is a basis of $(\boldsymbol{R}^S)_0$. The dimension of $(\boldsymbol{R}^S)_0$ is infinite iff S is infinite. If S is finite then $(\boldsymbol{R}^S)_0 = \boldsymbol{R}^S$.

PROOF. If $f = \sum_{i=1}^{N} a_i \cdot s^i$, where s^i are different, then f has a value a_i at s_i and vanishes on $S - \{s^1, \ldots, s^N\}$. Thus any $f \in (\boldsymbol{R}^S)_0$ can be written uniquely in this form. □

Let $S_p^{(r)}(U)$ denote the set of all C^r singular p-cubes in U.

DEFINITION. Elements of $C_p^{(r)}(U) := (\boldsymbol{R}^{S_p^{(r)}(U)})_0$ are called *singular p-chains* in U. We shall write $C_p^{(\infty)}(U) \equiv C_p(U)$.

If T: $U \to V$ is of class C^r and $c_p \in C_p^{(r)}(U)$, $C_p = \sum a_i s_p^i$, define $Tc_p := \sum a_i T s_p^i$.

If $C_p = \sum a_i s_p^i \in C_p^{(1)}(U)$ and $\omega \in \Lambda_{(0)}^p(U)$ we define the integral of ω over c_p by

$$\langle c_p | \omega \rangle \equiv \int_{c_p} \omega := \sum_i a_i \int_{s_p^i} \omega .$$

Now we are ready to define a boundary of a p-cube, which happens to be a $(p-1)$-chain.

Let $P^p = [a^1, b^1] \times \ldots \times [a^p, b^p]$, $a^i < b^i \in \boldsymbol{R}$. For $p \geqslant 2$ define

$$P_m^{p-1} := [a^1, b^1] \times \ldots \times \widehat{[a^m, b^m]} \times \ldots \times [a^p, b^p]$$

(where $\widehat{[a^m, b^m]}$ means the $[a^m, b^m]$ term is absent) and put $P_1^0 := \{0\}$. By $P_{m,\pm}^{p-1}$ we denote the mappings

$$p_{m,\pm}^{p-1} : P_m^{p-1} \to P^p,$$

$$p_{m,+}^{p-1}(t^1, \ldots, \widehat{t^m}, \ldots, t^p) := (t^1, \ldots, b^m, \ldots, t^p),$$

$$p_{m,-}^{p-1}(t^1, \ldots, \widehat{t^m}, \ldots, t^p) := (t^1, \ldots, a^m, \ldots, t^p)$$

for $p \geqslant 2$, $p_{1,+}^0(0) := b^1$, $p_{1,-}^0(0) := a^1$. Let $s_p = (P^p, \Phi)$ be a singular p-cube.

DEFINITION. The singular $(p-1)$-chain

$$(5) \qquad \partial s_p = \sum_{m=1}^{p} (-1)^m [(P^{p-1} \ \Phi \circ p^{p-1}_{m,-}) - (P^{p-1}_m, \Phi \circ p^{p-1}_{m,+})]$$

is called the *(algebraic) boundary of the singular p-cube* $s_p = (P^p, \Phi)$.

The following figure visualizes this construction at least in the case $p = 2$.

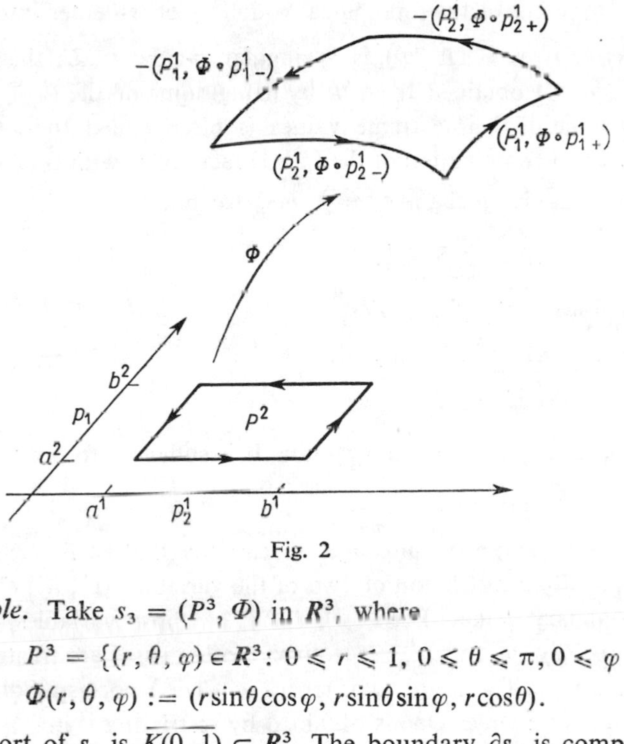

Fig. 2

Example. Take $s_3 = (P^3, \Phi)$ in R^3, where

$$P^3 = \{(r, \theta, \varphi) \in R^3 : 0 \leqslant r \leqslant 1,\ 0 \leqslant \theta \leqslant \pi,\ 0 \leqslant \varphi \leqslant 2\pi\},$$
$$\Phi(r, \theta, \varphi) := (r \sin\theta \cos\varphi, r \sin\theta \sin\varphi, r \cos\theta).$$

The support of s_3 is $\overline{K(0, 1)} \subset R^3$. The boundary ∂s_3 is composed of the following terms

$-(P^2_1, \Phi \circ p^2_{1,-})$, where $(\Phi \circ p^2_{1,-})(\theta, \varphi) = 0$,

$(P^2_1, \Phi \circ p^2_{1,+})$,
where $(\Phi \circ p^2_{1,+})(\theta, \varphi) = (\sin\theta \cos\varphi, \sin\theta \sin\varphi, \cos\theta)$,

$(P^2_2, \Phi \circ p^2_{2,-})$, where $(\Phi \circ p^2_{2,-})(r, \varphi) = (0, 0, r)$,

$-(P^2_2, \Phi \circ p^2_{2,+})$, where $(\Phi \circ p^2_{2,+})(r, \varphi) = (0, 0, -r)$,

$-(P_3^2, \Phi \circ p_{3,-}^2)$,
where $(\Phi \circ p_{3,-}^2)(r, \theta) = (r\sin\theta, 0, r\cos\theta)$,
$(P_3^2, \Phi \circ p_{3,+}^2)$,
where $(\Phi \circ p_{3,+}^2)(r, \theta) = (r\sin\theta, 0, r\cos\theta)$.

It is easy to see that if we compute $\langle \partial s_3, \overset{2}{\omega} \rangle$ only the second term gives nonzero contributions (fifth and sixth terms cancel out). Thus only values of $\overset{2}{\omega}$ at points of the topological boundary of $|s_3|$ enter into $\langle \partial s_3, \overset{2}{\omega} \rangle$.

Remark. If $s_p = (P^p, \Phi)$ is a singular p-cube in U, then a j-cube $(0 \leqslant j \leqslant p-1)$ obtained from Φ by fixing some of the t^j of $(t^1, \ldots, t^p) \in P^p$ at some of their extreme values is often called the j-face of s_p. Thus, ∂s_p is a combination of the $(p-1)$-faces of s_p with coefficients ± 1.

For a singular p-chain $c_p = \sum a_i s_p^i$ we put

$$\partial c_p := \sum a_i \, \partial s_p^i .$$

It is obvious that $\partial T c_p = T \partial c_p$.

THEOREM XIV.4.4. Let $p \geqslant 2$.

(6) $\qquad \partial(\partial c_p) = 0.$

PROOF. Let s_p be a singular p-cube. It is sufficient to prove that

$$\partial(\partial s_p) = 0.$$

With a slight abuse of language we can say that each term in $\partial(\partial s_p)$ corresponds to a restriction of two of the variables $(t^1, \ldots, t^p)$ of P^p to their boundary values. Let t^i, t^j, $i < j$, be those variables, which we restrict to, say, b^i and a^j, respectively (other cases are treated in complete analogy). There arc two terms in $\partial(\partial s_p)$ corresponding to the restriction considered. One is obtained by restricting t^i in

$$(-1)^j (P_j^{p-1}, \Phi \circ p_{j,-}^{p-1})$$

which thus equals

$$-(-1)^{i+j} \sigma_{p-2},$$

where

$$\sigma_{p-2} = (P_{ij}^{p-2}, \Phi \circ p_{ij,+-}^{p-2})$$

with

$$p_{ij,+-}^{p-2}: P_{ij}^{p-2} \to P^p,$$

$$P_{ij}^{p-2} = [a^1, b^1] \times \ldots \times \widehat{[a^i, b^i]} \times \ldots \times \widehat{[a^j, b^j]} \times \ldots \times [a^p, b^p],$$

$$p_{ij,+-}^{p-2}(t^1, \ldots, \hat{t}^i, \ldots, \hat{t}^j, \ldots, t^p) := (t^1, \ldots, b^i, \ldots, a^j, \ldots, t^p)$$

for $p \geqslant 2$ and

$$P_{12}^0 = \{0\}, \quad p_{12,+-}^{p-2}(0) := (b^1, a^2).$$

The second term in $\partial(\partial s_p)$ corresponding to the restriction of t^i to b^i and t^j to a^j is obtained by restricting t^j in $-(-1)^i (P_i^{p-1}, \Phi \circ p_{i,+}^{p-1})$ and equals $(-1)^{i+j}\sigma_{p-2}$. In this way we exhibit complete cancellations in $\partial(\partial s_p)$. □

The Stokes Theorem. In this section we shall prove a version of the Poincaré–Stokes Theorem in the sequel called "the Stokes Theorem", which is perhaps the most important theorem in mathematics. This is the opinion of one of the greatest living mathematicians René Thom. It embraces and generalizes such integral theorems of classical analysis as that of Gauss, Green, Stokes. The following formulation is more or less due to Poincaré and Éli Cartan.

THEOREM XIV.4.5 (Poincaré–Stokes). Let ω be an F-valued $(p-1)$-form in U (of class C^1) and let c_p be a singular p-chain (of class C^1) in $U \subset E$. Then

$$\int_{c_p} d\omega = \int_{\partial c_p} \omega \tag{7}$$

or in another notation

$$\langle c_p | d\omega \rangle = \langle \partial c_p | \omega \rangle. \tag{8}$$

Remark. The last notation stresses the duality between chains and forms. $\langle \cdot | \cdot \rangle$ is a bilinear F-valued form on $C_p(U) \times \Lambda^p(U, F)$ and (8) shows that operations ∂ and d are transposed (adjoint) relative to each other with respect to $\langle \cdot | \cdot \rangle$.

PROOF. Let η be a p-form (of class C^1) on P^p. It is sufficient to prove that

$$\int_{P^p} d\eta = \int_{\partial P^p} \eta, \tag{9}$$

where $P = (P^p, \mathrm{id})$. Indeed if $s_p = (P^p, \Phi)$, then

$$\int_{s_p} d\omega = \int_{P^p} \Phi^* d\omega = \int_{P^p} d\Phi^*\omega,$$

$$\int_{\partial s_p} \omega = \int_{\partial P^p} \Phi^*\omega.$$

Thus (9) for $\eta = \Phi^*\omega$ implies (7). We have

$$\eta = \sum_{i=1}^{p} (-1)^{i+1} c^i dt^1 \wedge \ldots \wedge \widehat{dt^i} \wedge \ldots \wedge dt^p,$$

$$d\eta = \sum_{i=1}^{p} \frac{\partial c^i}{\partial t^i} dt^1 \wedge \ldots \wedge dt^p,$$

$$\int_{P^p} d\eta = \sum_i \int_{P^p} \frac{\partial c^i}{\partial t^i}(t)\, d\lambda^p(t)$$

$$= \sum_i \Big(\int_{P_i^{p-1}} c^i(t^1, \ldots, b^i, \ldots, t^p)\, d\lambda^{p-1}(t) - \int_{P_i^{p-1}} c^i(t^1, \ldots$$

$$\ldots, a^i, \ldots, t^p)\, d\lambda^{p-1} \Big),$$

where we have used Fubini Theorem to change the multiple integral to an iterated one with integration over t^i performed first by applying the fundamental theorem of calculus. Now

$$\int_{\partial P^p} \omega = \sum_{i=1}^{p} (-1)^{i+1} \int_{\partial P^p} c^i dt^1 \wedge \ldots \wedge \widehat{dt^i} \wedge \ldots \wedge dt^p$$

$$= \sum_i \Big(\int_{P_i^{p-1}} c^i(t^1, \ldots, b^i, \ldots, t^p)\, d\lambda^{p-1}(t) -$$

$$- \int_{P_i^{p-1}} c^i(t^1, \ldots, a^i, \ldots, t^p)\, d\lambda^{p-1}(t) \Big),$$

where we have used (2) and (5) and the facts

$$\frac{\partial(t^1, \ldots, \widehat{t^i}, \ldots, t^p)}{\partial(t^1, \ldots, \widehat{t^i}, \ldots, t^p)} = 1,$$

$$\frac{\partial(t^1, \ldots, a^j, \ldots, \widehat{t^i}, \ldots, t^p)}{\partial(t^1, \ldots, \widehat{t^i}, \ldots, t^p)} = \frac{\partial(t^1, \ldots, b^j, \ldots, \widehat{t^i}, \ldots, t^p)}{\partial(t^1, \ldots, \widehat{t^i}, \ldots, t^p)} = 0.$$

Homology Spaces. Now we shall introduce the notion dual to that of cohomology. We have seen that boundary operation $\partial: C_p(U) \to C_{p-1}(U)$ has the property $\partial\partial = 0$ (we put $C_p(U) = \{0\}$ for $p < 0$).

A singular p-cube c_p is called a *p-cycle* if $\partial c_p = 0$, and is called a *p-boundary* if $c_p = \partial c_{p+1}$. By $Z_p(U)$ and $B_p(U)$ we denote the linear spaces of p-cycles and p-boundaries, respectively. We have

$$B_p(U) \subset Z_p(U)$$

by virtue of (6). Therefore, we can form the quotient spaces

$$Z_p(U)/B^p(U).$$

DEFINITION.

$$H_p(U)(\equiv H_p(U, R)) := Z_p(U)/B_p(U)$$

are called (*real*) *homology spaces of* $U (p = 0, 1, 2, \ldots)$.

As mentioned before, the notions of homology and cohomology are dual. We have the following bilinear pairing between the space of cycles $Z_p(U)$ and of closed p-forms $Z^p(U, R) \equiv Z^p(U)$:

$$\text{(10)} \qquad Z_p(U) \times Z^p(U) \ni (c_p, \overset{p}{\omega}) \to \langle c_p | \overset{p}{\omega} \rangle \equiv \int_{c_p} \overset{p}{\omega} \in R^1.$$

$\int_{c_p} \overset{p}{\omega}$ is called a *period of a closed form* $\overset{p}{\omega}$ *on a cycle* c_p (if we took F-valued forms, then the periods would be elements of F).

LEMMA XIV.4.6. Let $\omega \in \Lambda^p(U, F)$. Then

1° $(d\omega) = 0 \Rightarrow \int_{c_p} \omega = 0$ for all $c_p \in B_p(U)$,

2° $(\omega = d\eta, \eta \in \Lambda^{p-1}(U, F)) \Rightarrow \int_{c_p} \omega = 0$ for all $c_p \in Z_p(U)$.

PROOF. 1° If $c_p = \partial c_{p+1}$ then by the Stokes Theorem

$$\int_{c_p} \omega = \int_{c_{p+1}} d\omega = 0.$$

2° Again, by the Stokes Theorem

$$\int_{c_p} \omega = \int_{\partial c_p} \eta = 0. \quad \square$$

COROLLARY XIV.4.7. The pairing (10) defines a bilinear pairing between homology and cohomology spaces

$$H_p(U)\times H^p(U)\ni([c_p],[\overset{p}{\omega}])\to\langle[c_p]|[\overset{p}{\omega}]\rangle :=\int_{c_p}\overset{p}{\omega}.$$

PROOF. We must show that

$$\int_{c_p+\partial c_{p+1}}(\overset{p}{\omega}+d\eta^{p-1})=\int_{c_p}\overset{p}{\omega}.$$

Indeed,

$$\int_{c_p+\partial c_{p+1}}(\overset{p}{\omega}+d\eta^{p-1})=\int_{c_p}\overset{p}{\omega}+\int_{c_p}d\eta^{p-1}+\int_{\partial c_{p+1}}\overset{p}{\omega}+\int_{\partial c_{p+1}}d\eta^{p-1}$$
$$=\int_{c_p}\overset{p}{\omega}$$

by virtue of Lemma XIV.4.6. □

Now suppose that U is an open subset of $\boldsymbol{R}^m$.

The inverse implications to those of LEMMA XIV.4.6 also hold. The first one is rather trivial.

LEMMA XIV.4.8. Let $\omega\in\Lambda^p(U,F)$. Then

$$\Big(\int_{c_p}\omega=0 \text{ for all } c_p\in B_p(U)\Big)\Rightarrow(d\omega=0).$$

PROOF. For any $c_{p+1}\in C_{p+1}(U)$ we have

$$\text{(11)}\qquad \int_{c_{p+1}}d\omega=\int_{\partial c_{p+1}}\omega=0.$$

We have

$$d\omega=\sum_{j_1<\ldots<j_{p+1}}c_{j_1\ldots j_{p+1}}dx^{j_1}\wedge\ldots\wedge dx^{j_{p+1}}.$$

Suppose that $d\omega\neq 0$. Hence there exists $x_0\in U$ and $1\leqslant j_1<\ldots<j_{p+1}\leqslant n$ such that $c_{j_1\ldots j_{p+1}}(x_0)\neq 0$. Take $c_{p+1}=1\cdot s_{p+1}$, where $s_{p+1}=(P_\varepsilon^{p+1},\Phi)$, $P_\varepsilon^{p+1}:=\bigtimes_{p+1}[-\varepsilon,\varepsilon]$,

$$\Phi(t^1,\ldots,t^{p+1})$$
$$=(x_0^1,\ldots,x_0^{j_1-1},x_0^{j_1}+t^1,x_0^{j_1+1},\ldots,x_0^{j_2}+t^2,\ldots,x_0^{j_{p+1}}+$$
$$+t^{p+1},\ldots,x_0^n).$$

Then

$$\int_{c_{p+1}} d\omega = \int_{\underset{p+1}{\times}[-\varepsilon,\varepsilon]} c_{j_1\dots j_{p+1}}(\Phi(t^1, \dots, t^{p+1}))d\lambda^{p+1}(t) \neq 0$$

for sufficiently small ε, which contradicts (11). Thus $d\omega = 0$. □

As we mentioned above, the inverse of the second implication of Lemma XIV.4.6 also holds and is known as the de Rham Theorem. Its proof, however, is highly non-trivial and we are not going to give it here.

THEOREM XIV.4.9 (de Rham Theorem). Let $\omega \in \Lambda^p(U, F)$. Then

$$\left(\int_{c_p} \omega = 0 \text{ for all } c_p \in Z_p(U)\right) \Rightarrow (\omega = d\eta,\ \eta \in \Lambda^{p-1}(U, F)).$$

In another formulation

$$(\langle [c_p] \rangle | [\overset{p}{\omega}] \rangle \underset{[c_p]}{\equiv} 0) \Rightarrow ([\overset{p}{\omega}] = 0).$$

Instead of giving the full proof, we shall prove Theorem XIV.4.9 in the case $p = 1$. As usual let U be an open subset of a Banach space E. A sequence $(s_1, \dots, s_N)$ of singular 1-cubes (of class C^∞) in U will be called a *piecewise smooth path in U* beginning at x_1 and ending at x_2 if $S_i = ([a_i, b_i], \Phi_i)$ and $\Phi_1(a_1) = x_1$, $\Phi_i(b_i) = \Phi_{i+1}(a_{i+1})$, $1 \leqslant i \leqslant N-1$, and $\Phi_N(b_N) = x_2$.

Let $\omega \in \Lambda^1(U, F)$ and let $(s_1, \dots, s_N)$ be a piece-wise smooth path in U and let $(s_1, \dots, s_N)$ define a singular 1-chain $\sum_{i=1}^{N} 1 \cdot s_i$.

$\int_{\sum_{i=1}^{N} 1\cdot s_i} \omega$ will be also called the *integral of* ω *along* $(s_1, \dots, s_N)$. We shall say that the integral of ω is *path-independent* if it assumes equal values on all paths with the same starting points and end-points.

PROPOSITION XIV.4.10. Let $\omega \in \Lambda^1(U, F)$ and suppose that the integral of ω is path-independent. Then there exists a C^∞-function f on U such that $df = \omega$.

PROOF. Without loss of generality we can suppose that U is connected (otherwise we can work with each connected component separately).

Fix an $x_0 \in U$. For any $x \in U$ there exists a path $(s_1, \ldots, s_N)$ beginning at x_0 and ending at x. Indeed, the set of points, which can be connected with x_0 by a path, is open and closed in U and comprises x_0. Hence it coincides with U. Define $f(x) := \int\limits_{\sum_i 1 \cdot s_i} \omega$. By virtue of our assumption $f(x)$ is uniquely defined. For $x \in K(x_1, r) \subset U$ let $s_{x_1, x} = ([0, 1], \Phi_{x_1 x})$ be a 1-cube defined by

$$\Phi_{x_1 x}(t) := (1-t)x_1 + tx.$$

As

$$(12) \qquad f(x) = f(x_1) + \int\limits_{s_{x_1 x}} \omega$$

$$= f(x_1) + \int_0^1 \omega((1-t)x_1 + tx;\, x - x_1)\,dt.$$

The smoothness of f follows by the theorem on differentiation under the integral sign. Setting $x = x_1 + se$, $e \in E$, in (12), we get

$$f(x_1 + se) = f(x_1) + s\int_0^1 \omega(x_1 + ste;\, e)\,dt.$$

Differentiation over s at $s = 0$ gives

$$f'(x_1) \cdot e = \int_0^1 \omega(x_1; e)\,dt = \omega(x_1; e),$$

showing that $df = \omega$.

Of course, f is defined up to a constant (on each connected component). □

Remark. If for any 1-cycle c_1 $\int\limits_{c_1} \omega = 0$, then, of course, ω is path-independent. For, if $(s_1, \ldots, s_N)$ and $(s'_1, \ldots, s'_{N'})$ are two paths with the same starting points and end-points, then

$$\int\limits_{\sum_i 1 \cdot s_i} \omega - \int\limits_{\sum_j 1 \cdot s'_j} \omega = \int\limits_{\sum_i 1 \cdot s_i + \sum_j (-1) s'_j} \omega = 0$$

since $\sum_i 1 \cdot s_i + \sum_j (-1) \cdot s'_j$ is a 1-cycle. Thus, Proposition XIV.4.10 implies the case $p = 1$ of Theorem XIV.4.9.

Now we shall give an example of a 1-cycle c_1 which is not a boundary and of a 1-cocycle $\overset{1}{\omega}$ which is not a coboundary. By virtue of Lemma XIV.4.6 it is sufficient to ensure that of $\langle c_1 | \overset{1}{\omega} \rangle$ does not vanish.

Example. Let $U = \{(x^1, x^2) \in \boldsymbol{R}^2 : \frac{1}{4} < (x^1)^2 + (x^2)^2 < 4\}$ and let

$$\overset{1}{\omega} = -\frac{x^2}{(x^1)^2 + (x^2)^2} dx^1 + \frac{x^1}{(x^1)^2 + (x^2)^2} dx^2.$$

Thus, $d\overset{1}{\omega} = 0$.

Let $c_1 = 1 \cdot s_1$ with $s_1 = ([0, 2\pi], \Phi)$, where $\Phi(t) = (\cos t, \sin t)$. Plainly, $\partial c_1 = 0$. But

$$\int_{c_1} \overset{1}{\omega} = \int_0^{2\pi} dt = 2\pi \neq 0.$$

Thus, $H_1(U) \neq 0$ and $H^1(U) \neq 0$. Hence, U is not contractible.

Integration Over Finite Domains. From the point of view of integration of differential forms the algebraic notions of singular chains are neither very convincing nor absolutely necessary. In practice, e.g., in physics in continuum mechanics, electrodynamics, etc. one has to integrate differential forms over some compact subsets and their topological boundaries, but the subdivision of these objects into supports of (singular) cubes is, to some extent, an arbitrary and technical operation. One would like to get rid of this contingency. This is the idea behind the notion of a finite domain. However, a more satisfactory solution to this problem is offered by the theory of differentiable manifolds which will be our main concern in Sections 6–14.

Let s_p be a p-cube of class C^1 in $U \subset E$. We shall say that $s_p = (P^p, \Phi)$ is regular if

$$\Phi: P^p \ni (t^1, \ldots, t^p) \to \Phi(t^1, \ldots, t^p) \in U$$

is injective and of maximal rank (i.e., vectors $\dfrac{\partial \Phi}{\partial t^j}(t)$, $j = 1, \ldots, p$, are linearly independent at each $t \in P^p$).

Let E be of finite dimension n and let $(e_1, \ldots, e_n)$ and $(f_1, \ldots, f_n)$ be its bases. We shall say that they are *equivalent* if $\det[A_i^j] > 0$, where $[A_i^j]$ is the transfer matrix:

$$f_i = \sum_j A_i^j e_j.$$

Remark. For the notion of equivalence of bases of E not only the set of vectors in each basis is important but also their order. For example, the bases $(e_1, e_2, e_3, \ldots, e_n)$ and $(e_2, e_1, e_3, \ldots, e_n)$ are not equivalent.

DEFINITION. An equivalence class $\vartheta = [(e_1, \ldots, e_n)]$ (one of two distinct classes) of bases of E will be called an *orientation of* E. Together with a fixed orientation, E will be called an *oriented vector space*.

Let $s_n = (P^n, \Phi)$ be a regular n-cube in E ($n = \dim E$). The cube s_n will be said to be *compatible with orientation* ϑ *of* E if at each $t \in P^n$

$$\left[\frac{\partial \Phi}{\partial t^1}(t), \ldots, \frac{\partial \Phi}{\partial t^n}(t)\right] = \vartheta.$$

Let us fix an orientation of E.

DEFINITION (after Nickerson–Spencer–Steenrod). Let M be a compact set in E. The set M will be called a *finite domain if* there exists an n-chain c_n in E with following properties:

1° $c_n = \sum_i 1 \cdot s_n^i$.

2° Each s_n^i is a regular n-cube of class C^1 compatible with the orientation of E.

3° $M = \bigcup_i |s_n^i|$.

4° For $i_1 \neq i_2$ $|s_n^{i_1}| \cap |s_n^{i_2}|$ is either empty or coincides with the common support of k-faces of both $s_n^{i_1}$ and $s_n^{i_2}$, $0 \leqslant k \leqslant n-1$, which, in the case $k = n-1$, cancel out in the expression for ∂c_n.

5° Write $\partial c_n = \sum_j a_j s_{n-1}^j$, where $a_j = \pm 1$ and all s_{n-1}^j are different (so that the $(n-1)$-faces of the n-cubes of c do not occur since they cancel out by virtue of 4). Then

$$\partial M = \bigcup_j |s_{n-1}^j|.$$

Here ∂M denotes the topological boundary of M.

6° For $j_1 \neq j_2$ $|s_{n-1}^{j_1}| \cap |s_{n-1}^{j_2}|$ is either empty or the support of a k-face of both $s_{n-1}^{j_1}$ and $s_{n-1}^{j_2}$.

We shall say that c_n *parametrizes* M.

Remark. We did not pay attention to independence of 1°–6°.

Now we can give the definition of the integral of an n-form over a finite domain.

DEFINITION. Let $\omega \in \Lambda^n_{(0)}(U, F)$, $\eta \in \Lambda^{n-1}_{(1)}(U, F)$. Let $M \subset U$ be a finite domain parametrized by an n-chain c_n. Define

$$\int_M \omega := \int_{c_n} \omega, \qquad \int_{\partial M} \eta := \int_{\partial c_n} \eta.$$

This definition is justified by the following

PROPOSITION XIV.4.11. Suppose that c_n and c_n' are two n-chains parametrizing a finite domain $M \subset U$. Let $\omega \in \Lambda^n_{(0)}(U, F)$ and $\eta \in \Lambda^{n-1}_{(1)}(U, F)$. Then

$$\int_{c_n} \omega = \int_{c_n'} \omega \quad \text{and} \quad \int_{\partial c_n} \eta = \int_{\partial c_n'} \eta.$$

PROOF. Notice that the second equality follows from the first one since by the Stokes Theorem

$$\int_{\partial c_n} \eta = \int_{c_n} d\eta.$$

Without loss of generality we can assume that $E = \boldsymbol{R}^n$ with the canonical orientation given by the canonical basis. Let $\omega = c\, dx^1 \wedge \ldots \wedge dx^n$. Let $c_n = \sum_{i=1}^{N} 1 \cdot s_n^i$, where $s_n^i = (P_i^n, \Phi_i)$. Then

$$\int_{c_n} \omega = \sum_i \int_{P_i^n} c(\Phi_i(t)) \frac{\partial(\Phi_i^1, \ldots, \Phi_i^n)}{\partial(t^1, \ldots, t^n)}\, d\lambda^n(t)$$

$$= \sum_i \int_{\operatorname{Int} P_i^n} c(\Phi_i(t)) \frac{\partial(\Phi_i)}{\partial(t)}\, d\lambda^n(t).$$

We have used the fact that the topological boundary of $P_i^n = P_i^n - \operatorname{Int} P_i^n$ is of Lebesgue measure zero (see lemma below). But Φ_i is a C^1-diffeomorphism of $\operatorname{Int} P_i^n$ onto the image (since it is an injection and is a local diffeomorphism). Moreover, since s_n^i are compatible with the canonical orientation of $\boldsymbol{R}^n$, the Jacobian

$$\frac{\partial(\Phi_i)}{\partial(t)} > 0.$$

Thus, by the theorem on change of variables in the Lebesgue integral we have

$$\int_{\operatorname{Int} P_i^n} c(\Phi_i(t)) \frac{\partial(\Phi_i)}{\partial(t)}\, d\lambda^n(t) = \int_{\Phi(\operatorname{Int} P_i^n)} c(x)\, d\lambda^n(x)$$

and

$$\int_{c_n} \omega = \int_{\bigcup_i \Phi(\mathrm{Int} P_i^n)} c(x)\, d\lambda^n(x).$$

Since $M - \bigcup_i \Phi(\mathrm{Int}\, P_i^n)$ is contained in a finite sum of supports of $(n-1)$-faces of cubes s_n^i, once we know that such supports are of measure zero we get

$$\int_{c_n} \omega = \int_M c(x)\, d\lambda^n(x).$$

Since the right-hand side does not depend on the choice of c_n, we shall have completed the proof.

Hence we have only to prove

LEMMA XIV.4.12. Let $s_{n-1} = (P^{n-1}, \Phi)$ be an $(n-1)$-cube of class C^1 in $\boldsymbol{R}^n$. Then $\Phi(P^{n-1})$ is of Lebesgue measure zero in $\boldsymbol{R}^n$.

PROOF. Let $r = \sup_{t \in P^{n-1}} \|\Phi'(t)\|$. Then for $t, t' \in P^{n-1}$

$$\|\Phi(t) - \Phi(t')\| \leqslant r\|t - t'\|.$$

Divide P^{n-1} into m^{n-1} cubes P_j^{n-1} of equal dimension and diameter of order c/m. We then have

$$\int_{\Phi(P^{n-1})} d\lambda^n \leqslant \sum_{j=1}^{m^{n-1}} \int_{\Phi(P_j^{n-1})} d\lambda^n \leqslant \sum_{j=1}^{m^{n-1}} \int_{K\left(y_j, \frac{c}{m}\right)} d\lambda^n$$

$$= \mathrm{const} \cdot m^{n-1} \cdot \frac{1}{m^n} = \frac{\mathrm{const}}{m},$$

where y_j is a point in $\Phi(P_j^{n-1})$ and const is independent of m. Taking $m \to \infty$ we have completed the proof. □

Remark 1. We have proved much more: $\Phi(P^{n-1})$ has (n-dimensional Jordan) content zero, since (a set A has Lebesgue measure zero: $\lambda^n(A) = 0$) $\Leftrightarrow$ (A is a countable union of sets of content zero).

Remark 2. Lemma XIV.4.12 implies also the easier fact used above that $|\partial P^n|$ is of Lebesgue measure zero in $\boldsymbol{R}^n$.

Remark 3. In the proof of Proposition XIV.4.11 we have used only properties 1°, 2°, 3° and (part of) 4° of a chain parametrizing a finite domain. The other ones, however, guarantee that the $(n-1)$-chain ∂c_n

used in the definition of the integral of an $(n-1)$-form over the topological boundary of M really parametrize ∂M in a decent way.

The following figure illustrates how one could "divide" a ring, a sphere, and a triangle (case $n = 2$) into (supports) of a parametrizing n-chain, proving that these sets are finite domains.

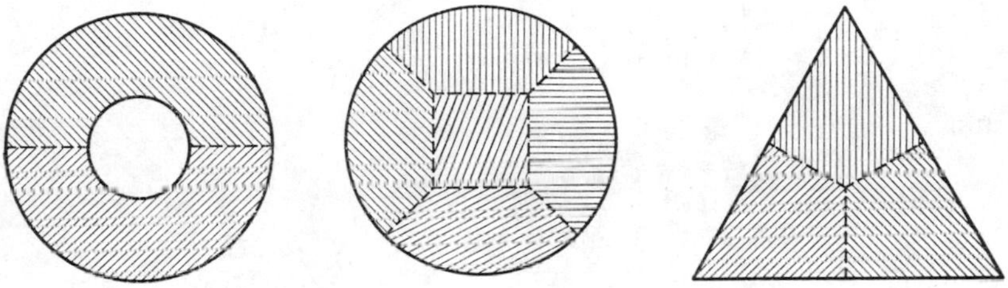

Fig. 3

Now we can reformulate a special case of the Stokes Theorem.

THEOREM XIV.4.13 (Poincaré–Stokes for Finite Domains). Let $\omega \in \Lambda^{n-1}_{(1)}(U, F)$ and let $M \subset U$ be a finite domain. Then

$$\int_{\partial M} \omega = \int_M d\omega .$$

5. ELEMENTS OF VECTOR ANALYSIS

Throughout this section, E will be a Euclidean space, i.e., a real vector space of finite dimension $n < \infty$ endowed with a scalar product $(\cdot\,|\,\cdot)$. The results of the section also hold (with obvious modifications) for a pseudo-Euclidean space, that is, in the case when $(\cdot\,|\,\cdot)$ is a bilinear non-degenerate form.

Scalar Product in $\Lambda^p(E)$. Notice that $(\cdot\,|\,\cdot)$ defines a linear isomorphism $\gamma\colon E \to E'$, defined by

(1) $\qquad \langle f|\gamma(e)\rangle \equiv \gamma(e)(f) := (e|f), \qquad e, f \in E.$

The isomorphism γ transports the scalar product from E to E'. Namely for $u, v \in E'$ we put

(2) $\qquad (u|v) := \big(\gamma^{-1}(u)|\gamma^{-1}(v)\big)$

(we denote the scalar products in E and E' and also other ones introduced below by the same symbol).

Let $u_i^1, \ldots, u_i^p, v_j^1, \ldots, v_j^p \in E'$, $1 \leqslant p \leqslant n$, $1 \leqslant i \leqslant I$, $1 \leqslant j \leqslant J$.

PROPOSITION XIV.5.1 Let c^i, $d^j \in \boldsymbol{R}$, $1 \leqslant i \leqslant I$, $1 \leqslant j \leqslant J$. The number

$$(3) \qquad \sum_{i,j} c^i d_j \det_{k,l}[(u_i^k | v_j^l)]$$

depends only on the elements

$$(4) \qquad u = \sum_{i=1}^{I} c^i u_i^1 \wedge \ldots \wedge u_i^p$$

and

$$(5) \qquad v = \sum_{j=1}^{J} d^j v_j^1 \wedge \ldots \wedge v_j^p$$

of $A_p(E)$.

PROOF. Fix a basis $(f^1, \ldots, f^n)$ of E'. Let

$$(6) \qquad u_i^k = \sum_{r=1}^{n} \alpha_{ir}^k f^r.$$

Then

$$u_i^1 \wedge \ldots \wedge u_i^p = \sum_{1 \leqslant r_1 < \ldots < r_p \leqslant n} \det_{k,s}[\alpha_{ir_s}^k] f^{r_1} \wedge \ldots \wedge f^{r_p},$$

$$(7) \qquad u = \sum_{r_1 < \ldots < r_p} \Big(\sum_{i=1}^{I} c^i \det_{k,s}[\alpha_{ir_s}^k] \Big) f^{r_1} \wedge \ldots \wedge f^{r_p}$$

$$= \sum_{r_1 < \ldots < r_p} \beta_{r_1 \ldots r_p} f^{r_1} \wedge \ldots \wedge f^{r_p}.$$

But

$$(8) \qquad (u_i^k | v_j^l) = \sum_{r=1}^{n} \alpha_{ir}^k (f^r | v_i^l).$$

We shall use the following algebraic

LEMMA XIV.5.2 (Generalized Cauchy Theorem). Let $[A_r^k]_{\substack{1 \leqslant k \leqslant p \\ 1 \leqslant r \leqslant n}}$ and $[B_l^r]_{\substack{1 \leqslant r \leqslant n \\ 1 \leqslant l \leqslant p}}$ be two matrices. Then

$$(9) \qquad \det_{k,l}\Big(\sum_r A_r^k B_l^r\Big)$$

$$= \sum_{1 \leqslant r_1 < \ldots < r_p \leqslant n} \det_{k,s}[A_{r_s}^k] \det_{s,l}[B_l^{r_s}].$$

Setting

$$A_r^k := \alpha_{ir}^k, \quad B_l^r := (f^r|v_j^l)$$

in Lemma XIV.5.2, we get

$$\det_{k,l}[(u_i^k|v_j^l)] = \sum_{r_1<\ldots<r_p} \det_{k,s}[\alpha_{ir_s}^k]\det_{s,l}[(f^{r_s}|v_j^l)]. \tag{10}$$

Hence

$$\begin{aligned}\sum_{i,j} c^i d^j \det_{k,l}[(u_i^k|v_j^l)] &= \sum_{i,j}{}' c^i d^j \sum_{r_1<\ldots<r_p}{}' \det_{k,s}[\alpha_{ir_s}^k]\det_{s,l}[(f^{r_s}|v_j^l)] \\ &= \sum_{\substack{r_1<\ldots<r_p \\ j}} \beta_{r_1\ldots r_p} d^j \det_{s,l}[(f^{r_s}|v_j^l)].\end{aligned} \tag{11}$$

Since the coefficients $\beta_{r_1\ldots r_p}$ in (7) are uniquely determined by u (Theorem XIV.1.5), we see that (3) for fixed d_j, v_j^l depends only on u. However, by symmetry, (3) depends on d_j, v_j^l only through v, which completes the proof. □

Define for u, v as in (4) and (5)

$$(u|v) := \sum_{i,j} c^i d^j \det_{k,l}[(u_i^k|v_j^l)]. \tag{12}$$

PROPOSITION XIV.5.3. $A_p(E)\times A_p(E) \ni (u, v) \to (u|v) \in \boldsymbol{R}$ defines a scalar product on $A^p(E)$.

PROOF. Bilinearity of $(\cdot|\cdot)$ is obvious. Let for an orthonormal basis $(f^1, \ldots, f^n)$ of E'

$$u = \sum_{j_1<\ldots<j_p} \beta_{j_1\ldots j_p} f^{j_1}\wedge \ldots \wedge f^{j_p}.$$

Then

$$(u|u) = \sum_{j_1<\ldots<j_p} (\beta_{j_1\ldots j_p})^2 \geqslant 0$$

and the equality holds only if $u = 0$. □

We shall also define a scalar product in $A_0(E) := \boldsymbol{R}$, taking $(s|t) := st$, $s, t \in \boldsymbol{R}$.

Remark. If $u = u^1 \wedge \ldots \wedge u^p$, $u^i \in E'$, then $||u|| := \sqrt{(u|u)}$ is often called the *volume of the parallelepiped spanned by* $(u^1, \ldots, u^p)$.

Volume Element of E. For the rest of the present section we shall fix an orientation of E. Let $(e_1, \ldots, e_n)$ be an orthonormal basis of E compatible with the orientation. Let $(f^1, \ldots, f^n)$ be the dual basis of E'.

PROPOSITION XIV.5.4. The element

$$\mu := f^1 \wedge \ldots \wedge f^n \in A_n(E) \tag{13}$$

does not depend on the choice of an orthonormal basis $(e_1, \ldots, e_n)$ compatible with the orientation.

PROOF. Take another basis compatible with the orientation $(g_1, \ldots, g_n)$ and its dual $(h^1, \ldots, h^n)$. Then $e_i = \sum_j A_i^j g_j$, where $\det[A_i^j] = 1$, and $h^l = \sum_m {}^tA_m^l f^m$. Hence

$$h^1 \wedge \ldots \wedge h^n = \det[{}^tA] f^1 \wedge \ldots \wedge f^n = f^1 \wedge \ldots \wedge f^n. \; \square$$

DEFINITION. μ is called the *volume element of the oriented Euclidean space E.*

Remark. μ changes sign if we change the orientation of E.

*Hodge * Isomorphism.* Let $\varphi \in A_p(E), 0 \leqslant p \leqslant n$. For any $\eta \in A_{n-p}(E)$ denote by $\dfrac{\eta \wedge \varphi}{\mu}$ the unique number such that $\eta \wedge \varphi = \dfrac{\eta \wedge \varphi}{\mu} \cdot \mu$. Then

$$A_{n-p}(E) \ni \eta \to \frac{\eta \wedge \varphi}{\mu} \in \mathbf{R}$$

is a linear form and, hence, there exists a unique element $*\varphi \in A_{n-p}(E)$ such that

$$\frac{\eta \wedge \varphi}{\mu} = (\eta | *\varphi). \tag{14}$$

Obviously, $*$: $A_p(E) \to A_{n-p}(E)$ is a linear map.

DEFINITION. $*$: $A_p(E) \to A_{n-p}(E)$ is called the *Hodge *operation.*

Remark. We shall not keep any distinction between different *operations obtained for different p in our notation.

PROPOSITION XIV.5.5.

1° $*1 = \mu$.

2° Let $(f^1, \ldots, f^n)$ be an orthonormal basis in E' whose dual is compatible with the orientation of E. Let $\varphi = f^{j_1} \wedge \ldots \wedge f^{j_p}, j_1 < \ldots < j_p$.

Then

(15) $\quad *\varphi = \sigma f^{j_{p+1}} \wedge \dots \wedge f^{j_n}$

if

$$\{j_1, \dots, j_p\} \cup \{j_{p+1}, \dots, j_n\} = \{1, \dots, n\}.$$

Here, σ is the sign of the permutation $(j_{p+1}, \dots, j_n, j_1, \dots, j_p)$.

3° For $\varphi \in A_p(E)$

(16) $\quad **\varphi = (-1)^{p(n-p)}\varphi.$

4° For $\varphi, \eta \in A_p(E)$

(17) $\quad (\eta|\varphi) \cdot \mu = (*\varphi) \wedge \eta = (*\eta) \wedge \psi.$

5° For $\varphi, \eta \in A_p(E)$

(18) $\quad (*\varphi|*\eta) = (\varphi|\eta).$

PROOF. 1° Obvious. 2° Take $\eta = f^{j_{p+1}} \wedge \dots \wedge f^{j_n}$. Then $(\eta|*\varphi) = 0$ unless

$$\{j_1, \dots, j_p\} \cup \{j_{p+1}, \dots, j_n\} = \{1, \dots, n\},$$

in which case

$$(\eta|*\varphi) = \frac{f^{j_{p+1}} \wedge \dots \wedge f^{j_n} \wedge f^{j_1} \wedge \dots \wedge f^{j_p}}{f^1 \wedge \dots \wedge f^n} = \sigma.$$

Hence, (15).

3° $**f^{j_1} \wedge \dots \wedge f^{j_p} = \sigma * f^{j_{p+1}} \wedge \dots \wedge f^{j_n} = \sigma\sigma' f^{j_1} \wedge \dots \wedge f^{j_p}$, where

$$\sigma = \operatorname{sgn}(j_{p+1}, \dots, j_n, j_1, \dots, j_p),$$
$$\sigma' = \operatorname{sgn}(j_1, \dots j_p, j_{p+1}, \dots, j_n).$$

Thus, $\sigma\sigma' = (-1)^{p(n-p)}$.

4° By (14) $(\eta|**\varphi) = \dfrac{\eta \wedge *\varphi}{\mu}$. Hence,

$$(\eta|\varphi) = (-1)^{p(n-p)} \frac{\eta \wedge *\varphi}{\mu} = \frac{(*\varphi) \wedge \eta}{\mu}.$$

The other equality follows from the symmetry of $(\cdot|\cdot)$.

5° $(*\eta|*\varphi) \cdot \mu = (**\varphi) \wedge (*\eta) = (*\eta) \wedge \varphi = (\eta|\varphi) \cdot \mu.$ □

From (16) we get

COROLLARY XIV.5.6. $*: A_p(E) \to A_{n-p}(E)$ is an orthogonal isomorphism,

(19) $$*^{-1} = (-1)^{p(n-p)}*.$$

Surface Measure. We have defined the scalar product and the Hodge $*$operation in spaces $A_p(E)$. Substituting $E' \to E$ we get a scalar product and a Hodge $*$operation in spaces $A_p(E')$ (we take E' with scalar product (2) and with orientation defined by the dual basis to any one of E compatible with the orientation). By virtue of Theorem XIV.1.5 and since there is a canonical isomorphism $E'' \simeq E$, $A_p(E')$ is spanned by elements $e_1 \wedge \ldots \wedge e_p$, $e_i \in E$. This is why $A_p(E')$ is often (in the case of finite-dimensional spaces) called the *p-fold exterior product of E.*

Let $s_p = (P^p, \Phi)$, $1 \leqslant p \leqslant n$ be a regular p-cube (of class C^1) in E. We remind that Φ: $P^p \to E$ is an injective C^1 map of maximal rank $(= p)$. Define ϱ: $P^p \to \boldsymbol{R}$ by

$$\varrho(t) := \left\| \frac{\partial \Phi}{\partial t^1}(t) \wedge \ldots \wedge \frac{\partial \Phi}{\partial t^p}(t) \right\|.$$

Since the vectors

$$\frac{\partial \Phi}{\partial t^1}(t), \ldots, \frac{\partial \Phi}{\partial t^p}(t)$$

are linearly independent, $\varrho(t)$ is (strictly) positive.

Remark. We notice that $\varrho(t)$ is the volume of the parallelepiped spanned by

$$\left(\frac{\partial \Phi}{\partial t^1}(t), \ldots, \frac{\partial \Phi}{\partial t^p}(t) \right).$$

Now, the expression

$$\int_{P^p} f(\Phi(t))\varrho(t)\, d\lambda^p(t)$$

defines a (finite) Radon measure on E. It is easy to see that this measure is supported by $|s_p|$. Denote this measure by $d\sigma_{s_p}$ or by $d\sigma$ for short:

$$\int_{|s_p|} f d\sigma_{s_p} := \int_{P^p} f(\Phi(t))\varrho(t)\, d\lambda^p(t).$$

DEFINITION. $d\sigma_{s_p}$ is called a *surface measure* for s_p. In the case $p = 0$ we shall define $\int_{|s_0|} f d\sigma_{s_0}$ as the value of f at the point of $|s_0|$.

PROPOSITION XIV.5.7. Suppose that $s_p^1 = (P_1^p, \Phi_1)$, $s_p^2 = (P_2^p, \Phi_2)$ are two regular p-cubes and that there exists a C^1-diffeomorphism $S: P_1^p \to P_2^p$, such that $\Phi_1 = \Phi_2 \circ S$. Then $d\sigma_{s_p^1} = d\sigma_{s_p^2}$.

PROOF. Since

$$\frac{\partial \Phi_1}{\partial t^1}(t) = \sum_{j=1}^{p} \frac{\partial S^j}{\partial t^i}(t) \frac{\partial \Phi_2}{\partial t^j}(S(t)),$$

$$\frac{\partial \Phi_1}{\partial t^1}(t) \wedge \ldots \wedge \frac{\partial \Phi_1}{\partial t^p}(t)$$

$$= \det\left[\frac{\partial S^j}{\partial t^i}(t)\right] \frac{\partial \Phi_2}{\partial t^1}(S(t)) \wedge \ldots \wedge \frac{\partial \Phi_2}{\partial t^p}(S(t)).$$

Thus

$$\varrho_1(t) = |\det S'(t)|\, \varrho_2(S(t))$$

Hence, by the theorem on change of variables in the Lebesgue integral

$$\int_{P_1^p} f(\Phi_1(t))\varrho_1(t)\, d\lambda^p(t) = \int_{P_2^p} f(\Phi_2(t))\varrho_2(t)\, d\lambda^p(t). \ \square$$

Remark. We have not used the orientation of E in defining $d\sigma_{s_p}$. Now, given a regular p-cube s_p we have two integration theories related to it: integration of p-forms and integration with surface measure. We shall relate those two theories in the cases $p = 0, 1, n-1, n$.

The case $p = 0$. Obviously if $f \in C(U)$ and s_0 is a 0-cube in U then

$$\int_{s_0} f = \int_{|s_0|} f d\sigma_{s_0}.$$

The case $p = 1$. For a regular 1-cube (curve) $s_1 = (P^1, \Phi)$ in U we define a vector field $\tau: |s_1| \to E$ by

$$\tau(\Phi(t)) := \frac{1}{\varrho(t)} \frac{d\Phi}{dt}(t);$$

τ is called the *unique vector field tangent to* $|s_1|$. It is easy to see that τ does not change if we change s_1 within the equivalence class. For $\omega \in \Lambda^1_{(0)}(U)$ we have

$$\int_{s_1} \omega = \int_{P^1} \omega\left(\Phi(t); \frac{d\Phi}{dt}(t)\right) dt$$

$$= \int_{P^1} \omega\big(\Phi(t); \tau(\Phi(t))\big)\varrho(t)\, dt = \int_{|s_1|} \langle \tau | \omega \rangle d\sigma_{s_1}.$$

Since there exists an isomorphism $\gamma\colon E \to E'$, we can work with vector fields (E-valued maps) instead of covector fields (E'-valued maps). Then

$$\int_{s_1} \omega = \int_{|s_1|} (\gamma^{-1} \circ \omega | \tau) \, d\sigma_{s_1}.$$

In the case $p = 1$, $d\sigma$ is often denoted by ds (length measure).

The case $p = n-1$. Let $s_{n-1} = (P^{n-1}, \Phi)$ be an $(n-1)$-regular cube in E. We define a vector field $n\colon |s_1| \to E$ by

$$n(\Phi(t)) = \frac{1}{\varrho(t)} * \left(\frac{\partial \Phi}{\partial t^1}(t) \wedge \dots \wedge \frac{\partial \Phi}{\partial t^{n-1}}(t) \right).$$

By virtue of Corollary XIV.5.6

$$\left\| * \left(\frac{\partial \Phi}{\partial t^1}(t) \wedge \dots \wedge \frac{\partial \Phi}{\partial t^{n-1}}(t) \right) \right\| = \left\| \frac{\partial \Phi}{\partial t^1}(t) \wedge \dots \wedge \frac{\partial \Phi}{\partial t^{n-1}}(t) \right\|,$$

which shows that $n(\Phi(t))$ is a unique vector. Moreover, if $(e_1, \dots, e_n)$ is a basis in E compatible with the orientation, then by (14) we have

$$\left(n(\Phi(t)) \middle| \frac{\partial \Phi}{\partial t^i}(t) \right)$$
$$= \frac{1}{\varrho(t)} \frac{\partial \Phi}{\partial t^1}(t) \wedge \dots \wedge \frac{\partial \Phi}{\partial t^{n-1}}(t) \wedge \frac{\partial \Phi}{\partial t^i}(t) = 0.$$

This justifies the name given to n: *unique normal vector field for* s_{n-1}. The field n does not change if we change s_{n-1} within the equivalence class. Indeed, if $s_{n-1}^1 = (P_1^{n-1}, \Phi_1)$ and $s_{n-1}^2 = (P_2^{n-1}, \Phi_2)$ are equivalent, i.e., if there exists a C^1-diffeomorphism $S\colon P_1^{n-1} \to P_2^{n-1}$, $\det S' > 0$, such that $\Phi_1 = \Phi_2 \circ S$, then

$$n_1(\Phi_1(t)) = \frac{1}{\varrho_1(t)} * \left(\frac{\partial \Phi_1}{\partial t^1}(t) \wedge \dots \wedge \frac{\partial \Phi_1}{\partial t^{n-1}}(t) \right)$$
$$= \frac{\det S'(t)}{\varrho_1(t)} * \left(\frac{\partial \Phi_2}{\partial t^1}(S(t)) \wedge \dots \wedge \frac{\partial \Phi_2}{\partial t^{n-1}}(S(t)) \right)$$
$$= n_2\big(\Phi_2(S(t))\big) = n_2(\Phi_1(t)).$$

However, since it is defined by means of Hodge $*$operation, n is orientation-dependent: it changes sign if we change the orientation of E. Hence, it is often called a *pseudo-vector field.*

We need still one more formula.

Let $e_1, \ldots, e_{n-1} \in E$, $u \in A_{n-1}(E)$. Then

$$(20) \qquad \langle *(e_1 \wedge \ldots \wedge e_{n-1}) | *U \rangle = U(e_1, \ldots, e_{n-1}).$$

It is sufficient to prove (20) for orthonormal $e_1, \ldots, e_{n-1}$. Thus, suppose that $(e_1, \ldots, e_n)$ is an orthonormal basis. We can also assume without loss of generality that $(e_1, \ldots, e_n)$ is compatible with the orientation of E. Let $(f^1, \ldots, f^n)$ be the dual basis. It is sufficient to take $u = f^1 \wedge \ldots \wedge \widehat{f^i} \wedge \ldots \wedge f^n$. Then, by Propositions XIV.5.5 and XIV.1.3 we have

$$\begin{aligned} &\langle *(e_1 \wedge \ldots \wedge e_{n-1}) | *(f^1 \wedge \ldots \wedge \widehat{f^i} \wedge \ldots \wedge f^n) \rangle \\ &= (-1)^{n-1}(-1)^{i-1}\langle e_n, f^i \rangle = \delta^i_n \\ &= (f^1 \wedge \ldots \wedge \widehat{f^i} \wedge \ldots \wedge f^n)(e_1, \ldots, e_{n-1}). \end{aligned}$$

Thus, (20) is proved. Now, if $\omega \in \Lambda^{n-1}_{(0)}(U)$ and if $s_{n-1} = (P^{n-1}, \Phi)$ is a regular $(n-1)$-cube in U, then (using (20)) we get

$$\begin{aligned} (21) \qquad \int_{s_{n-1}} \omega &= \int_{P^{n-1}} \omega\left(\Phi(t); \frac{\partial \Phi}{\partial t^1}(t), \ldots, \frac{\partial \Phi}{\partial t^{n-1}}(t)\right) d\lambda^{n-1}(t) \\ &= \int_{P^{n-1}} \langle n(\Phi(t)) | *\omega(\Phi(t)) \rangle \varrho(t)\, d\lambda^{n-1}(t) \\ &= \int_{|s_{n-1}|} \langle n | *\omega \rangle \, d\sigma_{s_{n-1}}. \end{aligned}$$

Here, $*\omega$ is a covector field on U. Using γ, we can rewrite (21) as

$$(22) \qquad \int_{s_{n-1}} \omega = \int_{|s_{n-1}|} (\gamma^{-1} \circ * \circ \omega | n) \, d\sigma_{s_{n-1}}.$$

The last integral is often called the *flux of the vector field* $\gamma^{-1} \circ * \circ \omega$ through the $(n-1)$-cube s_{n-1}.

Remark. Since both n and $\gamma^{-1} \circ * \circ \omega$ are pseudo-vector fields (they change sign with change of the orientation of E), $(\gamma^{-1} \circ * \circ \omega | n)$ is orientation independent.

The case $p = n$. If ω is an n-form on $U \subset E$, then $\omega = f\mu$, where f is a function on U and μ is the volume element. Let $s_n = (P^n, \Phi)$ be a regular n-cube in E. Then

$$\int_{s_n} \omega = \int_{P^n} f(\Phi(t)) \cdot \mu\left(\frac{\partial \Phi}{\partial t^1}(t), \ldots, \frac{\partial \Phi}{\partial t^n}(t)\right) d\lambda^n(t).$$

Let $(e_1, \ldots, e_n)$ be an orthonormal basis compatible with the orientation of E and let $(f^1, \ldots, f^n)$ be its dual. Then (Proposition XIV.1.3) we have

$$\mu\left(\frac{\partial \Phi}{\partial t^1}(t), \ldots, \frac{\partial \Phi}{\partial t^n}(t)\right) = \det\left[\left\langle \frac{\partial \Phi}{\partial t^i}(t) \mid f^j \right\rangle\right].$$

But

$$\int_{s_n} f \, d\sigma_{s_n} = \int_{P^n} f(\Phi(t)) \varrho(t) \, d\lambda^n(t).$$

Since

$$\frac{\partial \Phi}{\partial t^i}(t) = \sum_{j=1}^{n} \left\langle \frac{\partial \Phi}{\partial t^i}(t) \mid f^j \right\rangle e_j,$$

we have

$$\frac{\partial \Phi}{\partial t^1}(t) \wedge \ldots \wedge \frac{\partial \Phi}{\partial t^n}(t) = \det\left[\left\langle \frac{\partial \Phi}{\partial t^i}(t) \mid f^j \right\rangle\right] e_1 \wedge \ldots \wedge e_n.$$

Hence,

$$\varrho(t) = \left| \det\left[\left\langle \frac{\partial \Phi}{\partial t^i}(t) \mid f^j \right\rangle\right] \right|$$

and thus

$$\int_{s_n} \omega = \pm \int_{|s_n|} f \, d\sigma_{s_n},$$

where the $+$ sign appears if s_n is compatible with the orientation of E and the $-$ sign appears otherwise.

Let M be a finite domain in $U \subset E$ and fix a chain c_n parametrizing M. We then write

$$\text{(23)} \qquad \partial c_n = \sum_j a_j \cdot s_{n-1}^j,$$

where $a_j = \pm 1$ and all s_{n-1}^j are different. Then $\bigcup_j |s_{n-1}^j| = \partial M$. For $f \in C(U)$ we write

$$\text{(24)} \qquad \int_{\partial M} f \, d\sigma = \sum_j \int_{|s_{n-1}^j|} f \, d\sigma_{s_{n-1}^j}.$$

Equation (24) defines a measure $d\sigma$ support by ∂M. On $|s_{n-1}^j|$ we define a unique normal vector field n as that for s_{n-1}^j if in (23) $a_j = 1$ or as the opposite one if $a_j = -1$. Then n is well defined on ∂M $d\sigma$-almost everywhere. With this definition, for $\omega \in \Lambda_{(0)}^{n-1}(U)$ we have

$$(25) \qquad \int_{\partial M} \omega = \int_{\partial M} (\gamma^{-1} \circ * \circ \omega | n) d\sigma.$$

Remark. It can be shown that $d\sigma$ does not depend on the parametrization c_n. However, as we have seen, parametrization-independence of $\int_{\partial M} (\gamma^{-1} \circ * \circ \omega | \eta) d\sigma$ follows from the Stokes Theorem if ω is of class C^1.

Similarly, if s_2 is a regular 2-cube in U and if $\partial s_2 = \sum_{j=1} b_j \cdot s_1^j$ with $b_j = \pm 1$, we define a measure ds on the topological boundary $\partial |s_2|$ by

$$\int_{\partial |s_2|} f ds = \sum_j \int_{|s_1^j|} f d\sigma_{s_1^j}.$$

On $|s_1^j|$ we define the unique tangent vector field τ as that for s_1^j if $b_j = 1$ or the opposite one if $b_j = -1$. Then τ is well-defined ds-almost everywhere on $\partial |s_2|$. Moreover, for $\omega \in \Lambda_{(0)}^1(U)$ we have

$$\int_{\partial s_2} \omega = \int_{\partial |s_2|} (\gamma^{-1} \circ \omega | \tau) ds.$$

Now we can rewrite the Stokes Theorem in a form which uses surface integrals rather then chain integration for the cases which involve integrals of 0- and 1-forms, of $(n-1)$- and of n-forms and, when $n = 3$, of 1- and 2-forms. Before doing that, we shall introduce some new operations on functions and vector fields on E.

DEFINITION. Let $f \in C^1(U)$. Then the vector field $\operatorname{grad} f$: $U \to E$ (gradient of f) is defined by $\operatorname{grad} f = \gamma^{-1} \circ df$.

Let B: $U \to E$ be a C^1 vector field on U. Then the function $\operatorname{div} B$: $U \to \boldsymbol{R}^1$ (divergence of B) is given by

$$\operatorname{div} B = * \circ d(*^{-1} \circ \gamma \circ B).$$

Let $\dim E = n = 3$. Let A: $U \to E$ be a C^1 vector field on U. Then the vector field $\operatorname{curl} A$: $U \to E$ (rotation of A) is defined by

$$\operatorname{curl} A = \gamma^{-1} \circ * \circ d(\gamma \circ A).$$

Remark. Notice that $\operatorname{curl} \operatorname{grad} f = 0$ and $\operatorname{div} \operatorname{curl} A = 0$.

THEOREM XIV.5.8 (Stokes Theorem in Vector Notation). 1. Let $f \in C^1(U)$ and let s_1 be a regular 1-cube in U with beginning x_1 and end x_2. Then

$$\int_{|s_1|} (\operatorname{grad} f \,|\, \tau)\, ds = f(x_2) - f(x_1).$$

2. Let $A\colon U \to E$ be a C^1 vector field on U. Let M be a finite domain in U. Then

$$\int_{\partial M} (A \,|\, n)\, d\sigma = \int_M \operatorname{div} A \, d\lambda^n.$$

3. Let $\dim E = n = 3$. Let $A\colon U \to E$ be a C^1 vector field on U. Let s_2 be a regular 2-cube in U. Then

$$\int_{\partial|s_2|} (A \,|\, \tau)\, ds = \int_{|s_2|} (\operatorname{curl} A \,|\, n)\, d\sigma.$$

PROOF. Straightforward by virtue of the Stokes Theorem for chains forms, the last definition, and the discussion on the interrelation between chain and surface integrals. It is also useful to give a vector formulation of the Poincaré Lemma (Corollary XIV.3.4).

THEOREM XIV.5.9 (Poincaré Lemma in Vector Formulation). Let $U \subset E$ be (smoothly) contractible and let $\dim E = n = 3$. Let $A\colon U \to E$ be a C^∞ vector field such that $\operatorname{curl} A = 0$. Then there exists an $f \in C^\infty(U)$ such that $A = \operatorname{grad} f$.

Let $B\colon U \to E$ be a C^∞ vector field such that $\operatorname{div} B = 0$. Then there exists a C^∞ vector field $A\colon U \to E$ such that $B = \operatorname{curl} A$.

Remark. A vector field A such that $\operatorname{curl} A = 0$ is called *irrotational* or *curl-free*. A vector field B such that $\operatorname{div} B = 0$ is called *solenoidal* *source-free*. If $A = df$, f is called a *scalar potential* for A. If $B = \operatorname{curl} A$, A is called a *vector potential* for B.

For completeness we give the coordinate formulas in the case when $E = \boldsymbol{R}^n$. Then

$$\operatorname{grad} f = \left(\frac{\partial f}{\partial x^1}, \ldots, \frac{\partial f}{\partial x^n}\right),$$

$$\operatorname{div}(B^1, \ldots, B^n) = \sum_{i=1}^{n} \frac{\partial B^i}{\partial x^i}.$$

Moreover, if $n = 3$, then

$$\operatorname{curl}(A^1, A^2, A^3) = \left(\frac{\partial A^3}{\partial x^2} - \frac{\partial A^2}{\partial x^3}, \frac{\partial A^1}{\partial x^3} - \frac{\partial A^3}{\partial x^1}, \frac{\partial A^2}{\partial x^1} - \frac{\partial A^1}{\partial x^2}\right).$$

These formulas follow easily from definition of operations grad, div, curl as

$$*\left(\sum_{i=1}^{n} (-1)^{i-1} B_i dx^1 \wedge \dots \wedge \widehat{dx^i} \wedge \dots \wedge dx^n\right) = \sum_{i=1}^{n} B_i dx^i.$$

Coderivative and Laplace Operators. Greens Formulas. Let us notice that on a Euclidean space E one can define a Lebesgue measure. Namely take any orthonormal basis $(e_1, \dots, e_n)$ on E. Then the linear isomorphism

$$\boldsymbol{R}^n \ni (x^1, \dots, x^n) \to \sum_{i=1}^{n} x^i e_i \in E$$

can be used to transport the Lebesgue measure from $\boldsymbol{R}^n$ to E. That the resulting measure $d\lambda$ is independent of the choice of the basis $(e_1, \dots, e_n)$ easily follows from the theorem on change of variables in the Lebesgue integral.

Fix an orientation of E. Let $\tilde{\mu}$ be the constant form $\tilde{\mu} \equiv \mu$, where μ is the volume element ($\tilde{\mu}$ is often called the *volume form* on E). Let M be a finite domain in E and let $f \in C_0(E)$. Then, as easily follows from the proof of Proposition XIV.4.11, we have

$$(26) \qquad \int_M f\tilde{\mu} = \int_M f d\lambda.$$

Now the scalar products in $A_p(E)$ allow us to introduce scalar products in spaces $\Lambda_0^p(U)$ of C^∞ p-forms with compact supports in U. Namely, for $\omega, \chi \in \Lambda_0(U)$ we put

$$(27) \qquad \langle \omega | \chi \rangle := \int_U (\omega(x) | \chi(x)) d\lambda(x).$$

We define a linear operator

$$d^*\colon \Lambda^p(U) \to \Lambda^{p-1}(U)$$

which (restricted to forms with compact support) is an adjoint of the external derivative d with respect to the scalar product (27).

Let us write

$$d^*\omega = (-1)^p * \circ d(*^{-1} \circ \omega).$$

DEFINITION. d^* is called the *coderivative operator*.

We have the following

PROPOSITION XIV.5.10. Let $\omega \in \Lambda_0^p(U)$, $\chi \in \Lambda_0^{p+1}(U)$. Then

$$\langle d\omega \mid \chi \rangle = \langle \omega \mid d^*\chi \rangle.$$

PROOF. We start with the technical

LEMMA XIV.5.11. Given a compact subset $K \subset U$ there exists a finite domain M in U such that $K \subset M$.

PROOF OF THE LEMMA. We can assume that $E = \boldsymbol{R}^n$. Take

$$\varepsilon = \tfrac{1}{2}\operatorname{dist}(K, E - U), \qquad \varepsilon > 0.$$

Now, the union of closed squares of the lattice $\varepsilon \boldsymbol{Z}^n \subset \boldsymbol{R}^n$ which intersect K is a finite domain with the required properties. □

Choose M to be a finite domain in U such that $\operatorname{suppt}\omega \cap \operatorname{suppt}\chi \subset M$. Then (see (26), (14))

$$\begin{aligned}
\langle d\omega \mid \chi \rangle &= \int_M (d\omega(x) \mid \chi(x))\, d\lambda(x) \\
&= \int_M (d\omega(\cdot) \mid \chi(\cdot))\tilde{\mu} = \int_M d\omega \wedge *^{-1} \circ \chi \\
&= \int_M d(\omega \wedge *^{-1} \circ \chi) - (-1)^p \int_M \omega \wedge d(*^{-1} \circ \chi) \\
&= \int_{\partial M} \omega \wedge *^{-1} \circ \chi - (-1)^p \int_M (\omega(\cdot) \mid * \circ d(*^{-1} \circ \chi)(\cdot))\tilde{\mu} \\
&= (-1)^{p+1} \int_M (\omega(x) \mid * \circ d(*^{-1} \circ \chi)(x))\, d\lambda(x) + \int_{\partial M} \omega \wedge *^{-1} \circ \chi \\
&= \langle \omega \mid d^*\chi \rangle + \int_{\partial M} \omega \wedge *^{-1} \circ \chi = \langle \omega \mid d^*\chi \rangle,
\end{aligned}$$

since $\int_{\partial M} \omega \wedge *^{-1} \circ \chi = 0$ inasmuch as the form under the integral sign vanishes on ∂M. □

PROPOSITION XIV.5.12. Let $\omega \in \Lambda^p(U)$. Then

1° $d^*d^*\omega = 0$,

2° $d^*(* \circ \omega) = (-1)^{n-p} * \circ d\omega$,

3° $d(* \circ \omega) = (-1)^{n-p+1} * \circ d^*\omega$,

4° $* \circ d^*d\omega = dd^*(* \circ \omega)$,

5° $d^*d(* \circ \omega) = * \circ dd^*\omega$.

PROOF. 1° $d^*d^*\omega = (-1)^p d^*(* \circ d(*^{-1} \circ \omega)) = - * \circ dd(*^{-1} \circ \omega) = 0$.

2° $d^*(* \circ \omega) = (-1)^{n-p} * \circ d(*^{-1} \circ * \circ \omega) = (-1)^{n-p} * \circ d\omega$.

3° Application of 2° to $* \circ \omega$ gives

$$d^*(* \circ * \circ \omega) = (-1)^p * \circ d(* \circ \omega).$$

Now, application of $*$ gives

$$* \circ d^*(* \circ * \circ \omega) = (-1)^p * \circ * \circ d(* \circ \omega),$$

$$(-1)^{p(n-p)} * \circ d^*\omega = (-1)^p (-1)^{(n-p+1)(p-1)} d(* \circ \omega),$$

$$* \circ d^*\omega = (-1)^{n-p+1} d(* \circ \omega),$$

which is 3°.

4° By 3° and 2° we have

$$* \circ d^* d\omega = (-1)^{n-p} d(* \circ d\omega) = d\big(d^*(* \circ \omega)\big) = dd^*(* \circ \omega).$$

5° follows by writing 4° for $* \circ \omega$ and applying $*$. □

PROPOSITION XIV.5.13. If $E = \mathbf{R}^n$ and ω is given in its canonical form

$$\omega = \sum_{1 \leqslant j_1 < \dots < j_p \leqslant n} c_{j_1 \dots j_p} dx^{j_1} \wedge \dots \wedge dx^{j_p},$$

then

$$(28) \qquad d^*\omega = \sum_{j_1 < \dots < j_p} \sum_{l=1}^{p} (-1)^l \frac{\partial c_{j_1 \dots j_p}}{\partial x^{j_l}} dx^{j_1} \wedge \dots \wedge \widehat{dx^{j_l}} \wedge \dots \wedge dx^{j_p}.$$

PROOF. To check (28) it is sufficient to assume that ω has a compact support (d^* is local) and to show that for any $f \in C_0^\infty(U)$ and any $(k_1, \dots, k_{p-1})$, $1 \leqslant k_1 < \dots < k_{p-1} \leqslant n$, we have

$$\langle d(f dx^{k_1} \wedge \dots \wedge dx^{k_{p-1}}) | \omega \rangle = \langle f dx^{k_1} \wedge \dots \wedge dx^{k_{p-1}} | d^*\omega \rangle.$$

We have

$$\langle d(f dx^{k_1}\wedge \dots \wedge dx^{k_{p-1}}) | c_{j_1\dots j_p} dx^{j_1}\wedge \dots \wedge dx^{j_p}\rangle$$

$$= \sum_{k=1}^{n} \Big\langle \frac{\partial f}{\partial x^k} dx^k \wedge dx^{k_1}\wedge \dots \wedge dx^{k_{p-1}} | c_{j_1\dots j_p} dx^{j_1}\wedge \dots \wedge dx^{j_p}\Big\rangle$$

$$= \begin{cases} (-1)^{l+1}\displaystyle\int \frac{\partial f}{\partial x^{j_l}} c_{j_1\dots j_p} d\lambda & \text{if } (k_1, \dots, k_{p-1}) \\ & \quad = (j_1, \dots, \widehat{j_l}, \dots, j_p) \\ & \text{for some } 1 \leqslant l \leqslant p, \\ 0 & \text{otherwise,} \end{cases}$$

$$\langle f dx^{k_1}\wedge \dots \wedge dx^{k_{p-1}} | d^*(c_{j_1\dots j_p} dx^{j_1}\wedge \dots \wedge dx^{j_p})\rangle$$

$$= \sum_{l=1}^{p} (-1)^l \langle f dx^{k_1}\wedge \dots \wedge dx^{k_{p-1}} | \frac{\partial c_{j_1\dots j_p}}{\partial x^{j_l}} dx^{j_1}\wedge \dots$$

$$\dots \wedge \widehat{dx^{j_l}}\wedge \dots \wedge dx^{j_p}$$

$$= \begin{cases} (-1)^l \displaystyle\int f \frac{\partial c_{j_1\dots j_p}}{\partial x^{j_l}} d\lambda & \text{if } (k_1, \dots, k_{p-1}) = (j_1, \dots, \widehat{j_l}, \dots, j_p) \\ & \text{for some } 1 \leqslant l \leqslant p, \\ 0 & \text{otherwise,} \end{cases}$$

$$= \begin{cases} (-1)^{l+1}\displaystyle\int \frac{\partial f}{\partial x^{j_l}} c_{j_1\dots j_p} d\lambda, \\ 0. \ \square \end{cases}$$

We can now define the most important (differential) operator on differential forms.

DEFINITION. For $\omega \in \Lambda^p(U)$ we define $\Delta\omega := -(dd^* + d^*d)\omega$, where Δ is called the *Laplace operator*.

PROPOSITION XIV.5.14.

1° $\Delta\colon \Lambda^p(U) \to \Lambda^p(U)$ is a linear operator,

2° $\Delta\omega = -(d+d^*)(d+d^*)\omega$,

3° $d\Delta\omega = \Delta d\omega$,

4° $* \circ \Delta\omega = \Delta(* \circ \omega)$.

PROOF. 1° is obvious.

2° $-(d+d^*)(d+d^*)\omega = -(dd\omega + dd^*\omega + d^*d\omega + d^*d^*\omega) = -(dd^* + d^*d)\omega$.

3° $d\Delta\omega = -d(dd^*+d^*d)\omega = -dd^*d\omega,$
$\Delta d\omega = -(dd^*+d^*d)d\omega = -dd^*d\omega.$

4° By Proposition XIV.5.12 4° and 5° we have

$$\begin{aligned} * \circ \Delta\omega &= -* \circ dd^*\omega - * \circ d^*d\omega \\ &= -d^*d(* \circ \omega) - dd^*(* \circ \omega) = \Delta(* \circ \omega). \ \square \end{aligned}$$

Propositions XIV.5.10 and XIV.5.14 2° imply immediately

COROLLARY XIV.5.15. The restriction of Δ to $\Lambda_0^p(U)$ is

1° a symmetric operator

$$\langle \Delta\omega \,|\, \chi \rangle = \langle \omega \,|\, \Delta\lambda \rangle, \qquad \omega, \chi \in \Lambda_0^p(U),$$

2° a negative operator

$$\langle \Delta\omega \,|\, \omega \rangle \leqslant 0.$$

The coordinate form of Δ is given in

PROPOSITION XIV.5.16. If $E = \boldsymbol{R}^n$ and

$$\omega = \sum_{1 \leqslant j_1 < \ldots < j_p \leqslant n} c_{j_1 \ldots j_p} dx^{j_1} \wedge \ldots \wedge dx^{j_p},$$

then

$$\Delta\omega = \sum_{j_1 < \ldots < j_p} \sum_{k=1}^{n} \frac{\partial^2 c_{j_1 \ldots j_p}}{\partial (x^k)^2} dx^{j_1} \wedge \ldots \wedge dx^{j_p}. \tag{29}$$

PROOF. Using Propositions XIV.2.8 and XIV.5.13, we get

$$\begin{aligned} &dd^* c_{j_1 \ldots j_p} dx^{j_1} \wedge \ldots \wedge dx^{j_p} \\ &= d \sum_{l=1}^{p} (-1)^l \frac{\partial c_{j_1 \ldots j_p}}{\partial x^{j_l}} dx^{j_1} \wedge \ldots \wedge \widehat{dx^{j_l}} \wedge \ldots \wedge dx^{j_p} \\ &= \sum_{k=1}^{n} \sum_{l=1}^{p} (-1)^l \frac{\partial^2 c_{j_1 \ldots j_p}}{\partial x^k \partial x^{j_l}} dx^k \wedge dx^{j_1} \wedge \ldots \wedge \widehat{dx^{j_l}} \wedge \ldots \wedge dx^{j_p}, \end{aligned} \tag{30}$$

$$\begin{aligned} &d^* d c_{j_1 \ldots j_p} dx^{j_1} \wedge \ldots \wedge dx^{j_p} \\ &= d^* \sum_{k=1}^{n} \frac{\partial c_{j_1 \ldots j_p}}{\partial x^k} dx^k \wedge dx^{j_1} \wedge \ldots \wedge dx^{j_p} \end{aligned} \tag{31}$$

$$= -\sum_{l=1}^{p}\sum_{k=1}^{n}(-1)^l \frac{\partial^2 c_{j_1\dots j_p}}{\partial x^{j_l}\partial x^k} dx^k \wedge dx^{j_1} \wedge \dots$$

$$\dots \wedge \widehat{dx^{j_l}} \wedge \dots \wedge dx^{j_p} - \sum_{k=1}^{n} \frac{\partial^2 c_{j_1\dots j_p}}{\partial (x^k)^2} dx^{j_1} \wedge \dots \wedge dx^{j_p};$$

(30) and (31) produce (29). □

Now we shall prove some fundamental formulas which generalize the famous classical formulas of Gauss, Green, Ostrogradski.

THEOREM XIV.5.17 (Generalized Green's Formulas). Let $\omega, \varphi \in \Lambda^p(U)$, $\chi \in \Lambda^{p+1}(U)$. Let M be a finite domain in U. Then

1° $\int_M (d\omega(x)\,|\,\chi(x))\,d\lambda(x) = \int_M (\omega(x)\,|\,d^*\chi(x))\,d\lambda(x) + \int_{\partial M} \omega \wedge *^{-1} \circ \chi,$

2° $\int_M (\Delta\omega(x)\,|\,\varphi(x))\,d\lambda(x) + \int_M (d\omega(x)\,|\,d\varphi(x))\,d\lambda(x) +$

$$+ \int_M (d^*\omega(x)\,|\,d^*\varphi(x))\,d\lambda(x) = \int_{\partial M} (\varphi \wedge *^{-1} - d\omega \circ d^*\omega \wedge *^{-1} \circ \varphi),$$

3° $\int_M [(\Delta\omega(x)\,|\,\varphi(x)) - (\omega(x)\,|\,\Delta\varphi(x))]\,d\lambda(x)$

$$= \int_{\partial M} \{\varphi \wedge *^{-1} \circ d\omega - d^*\omega \wedge *^{-1} \circ \varphi - \omega \wedge *^{-1} \circ d\varphi + d^*\varphi \wedge *^{-1} \circ \omega\}.$$

PROOF. Proof of 1° follows that of Proposition XIV.5.10.
2° Applying 1° to ω and $\chi = d\varphi$, we get

$$\text{(32)} \qquad \int_M (d\omega(x)\,|\,d\varphi(x))\,d\lambda(x)$$

$$= \int_M (\omega(x)\,|\,d^*d\varphi(x))\,d\lambda(x) + \int_{\partial M} \omega \wedge *^{-1} \circ d\varphi.$$

Applying 1° to $\omega = d^*\varphi$, we get

$$\text{(33)} \qquad \int_M (dd^*\varphi(x)\,|\,\chi(x))\,d\lambda(x)$$

$$= \int_M (d^*\varphi(x)\,|\,d^*\chi(x))\,d\lambda(x) + \int_{\partial M} d^*\varphi \wedge *^{-1} \circ \chi.$$

We can rewrite (32) and (33) as

$$(34)\qquad \int_M (d\omega(x)\,|\,d\varphi(x))\,d\lambda(x) - \int_M (d^*d\omega(x)\,|\,\varphi(x))\,d\lambda(x)$$

$$\int_{\partial M} \varphi \wedge *^{-1} \circ d\omega,$$

$$(35)\qquad \int_M (d^*\omega(x)\,|\,d^*\varphi(x))\,d\lambda(x) - \int_M (dd^*\omega(x)\,|\,\varphi(x))\,d\lambda(x)$$

$$= -\int_{\partial M} d^*\omega \wedge *^{-1} \circ \varphi.$$

Summing yields 2°.

4° follows by subtracting two versions of 3°, one for the pair (ω, φ) another one for (φ, ω). □

For the sake of illustration we shall write down Green's formulas for the case $p = 0$ in vector notation introduced previously.

COROLLARY XIV.5.18 (Green's Formulas in Vector Notation). Let $f, g \in C^\infty(U)$ and let $A\colon U \to E$ be a C^∞ vector field on U. Let M be a finite domain in U. Then

$$1° \quad \int_M (\operatorname{grad} f\,|\,A)\,d\lambda = -\int_M f \operatorname{div} A + \int_{\partial M} f(A\,|\,n)\,d\sigma,$$

$$2° \quad \int_M (\Delta f) g\,d\lambda + \int_M (\operatorname{grad} f\,|\operatorname{grad} g)\,d\lambda = \int_M (\operatorname{grad} f\,|\,n) g\,d\sigma,$$

$$3° \quad \int_M ((\Delta f)g - f(\Delta g))\,d\lambda = \int_{\partial M} \{(\operatorname{grad} f\,|\,n)g - f(\operatorname{grad} g\,|\,n)\}\,d\sigma.$$

PROOF. 1° In Theorem XIV.5.17 1° we put $\omega = f$, $\chi := \gamma \circ A$. We use the formula $d^*\gamma \circ A = -\operatorname{div} A$ and (25).

2° In Theorem XIV.5.17 2° we put $\omega := f$, $\varphi := g$ and proceed as when proving 1°.

3° From Theorem XIV.5.17 3° as above. □

DEFINITION. A form $\omega \in \Lambda^p(U)$ is called *harmonic* if $d\omega = d^*\omega = 0$. In Section 11 we shall devote much time to this notion. Analysis of the space of harmonic forms on general differential manifolds, especially on compact manifolds, is of utmost importance. The most important theorems of Hodge–Kodaira and of de Rham in this field are milestones of global analysis.

6. THE DIFFERENTIABLE MANIFOLD

In this section we shall introduce the fundamental notion of the differentiable manifold (in a Banach space). This notion is perhaps the most important notion of analysis and is indispensable for applications of mathematics to physics, natural sciences, and recently even to economics and the social sciences. This notion was constructed in order to transfer the most important notions of calculus (derivative, differential forms, ...) onto objects which "are parametrized" by the Euclidean space $\boldsymbol{R}^m$ or, more generally, by a (real) Banach space E.

Such objects first occurred in analytic dynamics, statistical mechanics, and differential geometry. The notion of a differentiable manifold developed very slowly (over almost a hundred years): as most fundamental notions, it is a difficult one. In order not to discourage the reader we will first introduce the notion of a differentiable manifold (embedded) in a Banach space F and afterwards outline the general notion of an (abstract) differentiable manifold modelled on a Banach space E. But in order to obtain a nice formulation we have to extend slightly the notion of a differentiable map given in Chapter VII, Section 13.

DEFINITION. Let E, F be real Banach spaces and let $X \subset E$ and $Y \subset F$ be arbitrary subsets of these spaces. A map φ: $X \to Y$ is of class C^k if for each $x \in X$ there exists an open set $U \subset E$ containing x and a differentiable map $\bar{\varphi}$: $U \to F$ of class C^k, i.e., $\bar{\varphi} \in C^k(U, F)$ such that $\bar{\varphi}|U \cap \cap X \equiv \varphi|U \cap X$. If $\bar{\varphi}$ is smooth, i.e., of class C^∞ we say that φ is smooth (of class C^∞).

Remark. This definition seems less restrictive than the one adopted in Chapter VII, Section 13. However, using the concept of a smooth partition of unity which will be introduced later, one can easily show that in many cases (e.g., if E is of finite dimension) the two definitions coincide. In what follows we shall use the extended notion of differentiability just introduced.

Example. The identity map id_X of X is smooth.

Plainly we have the following

LEMMA XIV. 6.1. Let E, F, G be Banach spaces and $X \subset E$, $Y \subset F$, $Z \subset G$. If f: $X \to Y$, g: $Y \to Z$ are of class C^k, then the superposition $g \circ f$: $X \to Z$ is of class C^k.

DEFINITION. A map $f: X \to Y$ is a *diffeomorphism of class* C^k if f is a bijection onto Y and both f and f^{-1} are of class C^k.

After this preparation we can now define a manifold of class C^k.

DEFINITION. Let E, F be Banach spaces (over $\boldsymbol{R}$). A subset $M \subset F$ is called a *differentiable (of class* C^k*) manifold modelled on* E if each $x \in M$ has a neighbourhood $W \cap M$ which is diffeomorphic (of class C^k) to an open subset U of E. Any particular diffeomorphism $\varphi: U \to W \cap M$ is called a *parametrization* of $W \cap M$. The (inverse) diffeomorphism $u := \varphi^{-1}: W \cap M \to U$ is called a *system of coordinates* on $W \cap M$. If the model E is of dimension m, the manifold M is said to be *of dimension m* and we write $\dim M = m$.

We thus have the following situation:

The Hausdorff space M is covered by subsets $(M_i)_{i \in I}$ open in M, which are diffeomorphic (of class C^k) with open subsets $U_i \subset E$; $u_i: M_i \to U_i$, $i \in I$, are diffeomorphisms of class C^k—the systems of coordinates. The family $(u_i, M_i)_{i \in I}$ is called a C^k*-atlas of* M.

Abstract Differentiable Manifold. We can now make one more step in abstraction: in the definition of a C^k-atlas of M the ambient Banach space F does not appear explicitly but does exist in the definition of differentiability of the maps $u_i^{-1}: U_i \to M_i \subset F$. But, as Hermann Weyl remarked in his classic *Idee der Riemannschen Fläche* (in 1912) and, in the general case of m-dimensional manifolds, H. Whithey wrote in his fundamental paper *Differentiable manifolds*, Ann. of Math. 37 (1936), 645–680, one can completely drop the ambient space F.

DEFINITION (Whitney). A *topological manifold modelled on a Banach space* E is a Hausdorff space M such that there exists an open covering $(M_i)_{i \in I}$ of M (i.e. $M = \bigcup_{i \in I} M_i$, M_i open) and corresponding homeomorphisms $u_i: M_i \to U_i \subset E$ onto open subsets of E (called *coordinate systems* or *charts of* M). The family $(u_i, M_i)_{i \in I}$ is also called *topological atlas of* M. An atlas $(u_i, M_i)_{i \in I}$ is of class C^k if, for all $(i, k) \in I \times I$,

$$u_k \circ (u_i | M_i \cap M_k)^{-1}: u_i(M_i \cap M_k) \to u_k(M_i \cap M_k)$$

is a C^k-diffeomorphism.

Two C^k-atlases $(u_i, M_i)_{i \in I}$, $(u_j, M_j)_{j \in J}$ are equivalent if their union is a C^k-atlas.

The abstract differentiable manifold M modelled on a Banach space E is a topological manifold (with model E) together with an equivalence class of C^k-atlases.

Complex (Holomorphic) Manifolds. We have considered differentiable manifolds modelled on real Banach spaces and the notion of differentiability was the corresponding one: a derivative at e of a map was by definition an $\boldsymbol{R}$-linear map. The natural question arises how one should modify the notion of differentiability of maps between complex Banach spaces. This leads to the notion of holomorphic functions, which will be considered in some detail in the next chapter. Here we only want to give some definitions.

DEFINITION. Let E and F be complex Banach spaces; let U be an open subset of E. A map f: $U \to F$ is *holomorphic* if

1° $f \in C^1(U, F)$, i.e. f is continuously differentiable in U;

2° The derivative df_e is a $\boldsymbol{C}$-linear continuous map for every $e \in U$:

$$df_e \in L_{\boldsymbol{C}}(E, F).$$

This definition immediately implies that the composition $f \circ g$ of holomorphic mappings is holomorphic.

DEFINITION. Let E be a complex Banach space. Let M be a topological manifold modelled on M. An atlas $(u_i, M_i)_{i \in I}$ is called a *holomorphic atlas* if, for all $(i, k) \in I \times I$,

$$u_k \circ (u_i | M_i \cap M_k)^{-1}: u_i(M_i \cap M_k) \to u_k(M_k \cap M_i)$$

is a holomorphic map.

The equivalence of holomorphic atlases is defined in the same way as for C^k-atlases (by replacing "of class C^k" by "holomorphic"). A topological manifold M modelled on a complex Banach manifold E together with an equivalence class of holomorphic atlases is called a *complex* (or *holomorphic*) manifold.

Remark 1. We shall see later that holomorphicity is an incomparably stronger property than differentiability and even than smoothness.

Remark 2. If V is an open subset of the manifold M provided with an atlas (u_i, M_i), then $(u_i | V, M_i \cap V)_{i \in I}$ makes V a manifold of the same class as M; it is called an *open submanifold of M*.

Remark 3. One often replaces the equivalence class of C^k-atlases by a maximal C^k-atlas. Plainly, since any C^k-atlas can be extended in a unique way to a maximal C^k-atlas, one often adopts the "simplified" definition of C^k-manifold:

The *C^k-manifold* is a topological manifold M with a C^k-atlas.

The *product $M \times N$ of differentiable manifolds* M and N can now be introduced in a natural way. Let $(u_i, M_i)_{i \in I}$ and $(v_j, N_j)_{j \in J}$ be atlases of class C^k of manifolds M and N modelled on E and F. Then $(u_i \times v_j, M_i \times \times M_j)$, $(i, j) \in I \times J$ is a C^k-atlas of $M \times N$ modelled on $E \times F$. We leave the details to the reader.

We already encountered examples of differentiable manifolds in Part I: the first theorem of Lusternik (Theorem VIII.4.1) gives the construction of a differentiable manifold as a locus (in a Banach space). Let us consider that theorem in the context of differentiable manifolds.

We recall briefly the relevant notions. Let F, Y be Banach spaces and let $h\colon F \to Y$ be a differentiable map. Let $y_0 \in \operatorname{im} h$. Consider the set $M = h^{-1}(y_0)$. The point $x_0 \in M$ is regular if the derivative $h'(x_0)$ is a surjection onto Y: $\operatorname{im} h'(x_0) = Y$. We say that the kernel of $h'(x)$ splits if

$$F = \ker h'(x) \oplus F_2,$$

where F_2 is a closed subspace of F. We can now reformulate the first theorem of Lusternik in a version slightly stronger than that of VIII.4.1, namely using C^k maps.

THEOREM XIV.6.1 (Lusternik). Let $h\colon F \to Y$ be of class C^k and let x be a regular point of $M = h^{-1}(y_0)$. If the kernel E of $h'(x)$ splits, then a neighbourhood M_1 of x is a C^k-manifold in F modelled on $E := \ker h'(x)$.

Remark. The kernel of $h'(x)$ splits in the following important cases:

(a) F is a Hilbert space,

(b) $\dim F < \infty$,

(c) $\dim\big(F/\ker h'(x)\big) < \infty$: $\ker h'(x)$ is of finite codimension.

Combining the preceding remark with the Rank Theorem (Theorem VIII.2.8), we obtain the following very useful

COROLLARY XIV.6.2. Let $h: \boldsymbol{R}^n \to \boldsymbol{R}^l$ be of class C^k and of constant rank p on $\boldsymbol{R}^n$ (i.e. $\dim(h'(x)\boldsymbol{R}^n) = p$). Then for any $y \in \boldsymbol{R}^l$ $h^{-1}(y)$ is a closed C^k manifold of dimension $n-p$.

Examples. 1. The unit sphere M in a Hilbert space $(F, (\cdot|\cdot))$: $M = \{x \in F: (x|x) = 1\}$ is a C^∞-manifold (of codimension 1).

PROOF. Take $h(x) := (x|x)$. Then $h'(x)h = 2(x|h)$, and for $x \neq 0$ the derivative $h'(x)$ is surjective; moreover its kernel splits: it is the orthogonal complement to the subspace $[x] = \boldsymbol{R}x$ generated by x. Hence the codimension of the kernel of $h'(x)$ is one.

2. The units sphere S^n in $\boldsymbol{R}^{n+1}$ is a C^∞-manifold of dimension n.

3. The torus T^2 in $\boldsymbol{R}^3$ is a manifold of dimension 2. We can check this in the following way. Let $h: \boldsymbol{R}^3 \to \boldsymbol{R}$ be given by

$$h(x, y, z) = (\sqrt{x^2+y^2} - a)^2 + z^2.$$

Then h has rank 1 at each point in $G^{-1}(b^2)$, $a > b > 0$. Hence $T^2 := h^{-1}(b^2)$ is a manifold of dimension 2.

4. Very important manifolds are matrix groups: for instance, the orthogonal group $O(n)$: the group of $(n \times n)$-orthogonal matrices. The reader may try to prove that $O(n)$ is a manifold, in $\boldsymbol{R}^{n^2}$, of dimension $n(n-1)/2$.

7. TANGENT SPACES

We shall now define the important notion of a tangent space. Since the notion is not easy, we will introduce it first in the special case of a manifold in a Banach space.

If we have a smooth manifold in a Banach space F, then the tangent space $T_x M$ is a natural generalization of the notion of tangent plane introduced in Section VIII.4. Let $M \subset F$ be a manifold modelled on a Banach space E and let

$$\varphi: U \to M \subset F$$

be a parametrization of a neighbourhood $\varphi(U)$ of x in M, with $\varphi(u) = x$. Think of φ as a map from U to F; therefore the derivative

$$\varphi'(u): E \to F$$

is defined.

DEFINITION. $\{x\}\times\varphi'(u)\cdot E \subset \{x\}\times F$ is called the *tangent space of M at x* and is denoted by $T_x M$ or $T_x(M)$. We shall prove that

1° this construction does not depend on the particular choice of parametrization φ

2° $T_x M$ is isomorphic to the model E.

Ad 1°. Let $\psi\colon V \to M \subset F$ be another parametrization of a neighbourhood $\psi(V)$ of x in M and let $x = \psi(v)$. Then $\psi^{-1}\circ\varphi$ maps a neighbourhood U' of u diffeomorphically onto a neighbourhood V' of v. We have the commutative diagrams:

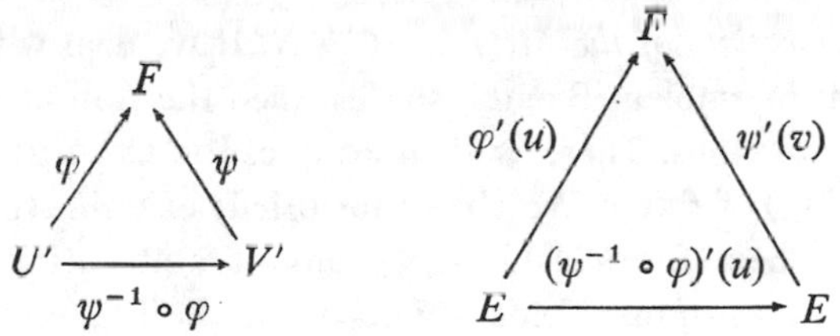

$$\varphi = \psi\circ(\psi^{-1}\circ\varphi), \qquad \varphi'(u) = \psi'(v)\circ(\psi^{-1}\circ\varphi)'(u).$$

$(\psi^{-1}\cdot\varphi)'(u)$ is an isomorphism of E and $\varphi'(u)\cdot E = \psi'(v)\cdot E$, hence 1°.

2° is obvious: $\varphi'(u)\cdot E$ is an isomorphic image of E by $\varphi'(u)$ (or $\psi'(v)$). □

Elements of the tangent space $T_x M$ are called *tangent vectors at $x\in M$*. The vector $(x, \varphi'(u)e)$ can be represented by a triple (x, e, u) where $e\in E$, $u = \varphi^{-1}$ is a chart around x. Two such triples (x, u, e), $(x, \tilde u, \tilde e)$ are equivalent, or define the same tangent vector at x, if

$$\tilde e = (\tilde u\cdot u^{-1})'\big(u(x)\big)\cdot e.$$

The reader will observe that this notion does not use the ambient Banach space: it involves only the atlas (u_i, M_i) of the manifold M and, hence, can be taken as a definition of a tangent vector in the abstract case. In this way we arrive at the following important

DEFINITION. Let M be a differentiable manifold M modelled on a Banach space E. Fix $x\in M$. Consider the set of triples (x, u, e), where u is a chart (a coordinate system) in a neighbourhood of x and $e\subset E$. We shall say that (x, u, e) and $(x, \tilde u, \tilde e)$ are *equivalent* if

(1) $$(\tilde u\circ u^{-1})'\big(u(x)\big)\cdot e = \tilde e.$$

An equivalence class of such triples is called a *vector tangent to M*

at x. The set of tangent vectors is called the *tangent space of* M *at* x and is denoted by $T_x M$ or $T_x(M)$.

Remarks. 1. Plainly (1) defines an equivalence by the chain rule (for derivatives).

2. Each chart (u_i, M_i) defines a bijection of $T_x M$ onto the model E: $[(x, u_i, e)] \to e$.

This bijection allows us to transfer the Banach structure of E to the tangent space $T_x M$. The reader may check that this structure is independent of the chart u_i.

Tangent (Derivative) of the Map f: $M \to N$. If we deal with differentiable manifolds in ambient Banach spaces, then the notion of differentiability is quite obvious. There is then no question as to how to define the derivative $f'(x)$ of f at x. We shall thus briefly examine these notions and obtain for them formulas independent of ambient spaces: since they use only atlases of manifolds M and N, we shall take them as definitions for the notions of the class C^k and of the derivative f' in the general situation of abstract manifolds.

Let $M \subset F$, $N \subset G$ be differentiable manifolds in Banach spaces F, G. Let f: $M \to N$ be a C^k-map. Then the derivative (or rather: the tangent) of f

$$f'(x):\ T_x M \to T_{f(x)} N$$

is defined as follows: since f is of class C^k there exists an open set $W \ni x$ and a mapping of class C^k

$$\tilde{f}:\ W \to G, \quad \tilde{f}|W \cap M = f|W \cap M.$$

We define

$$f'(x) \cdot v := \tilde{f}'(x) \cdot v \qquad \text{for all } v \in T_x M.$$

We have to prove that this definition is independent of the extension $\tilde{f}$ and that $\tilde{f}'(x)v \in T_{f(x)}N$.

Take the charts (u, O), (w, V) around $x \in O$, and $y = f(x) \in V$. We may assume that

$$O \subset W \quad \text{and} \quad f(O) \subset V.$$

Therefore

$$(2) \qquad w \circ f \circ u^{-1}:\ u(O) \to w(V)$$

is of class C^k. We have the commutative diagrams

$$\begin{array}{ccc} F \supset W & \xrightarrow{\tilde{f}} & G \\ \uparrow u^{-1} & & \uparrow w^{-1} \\ E_1 \supset u(O) & \xrightarrow{w \circ f \circ u^{-1}} & w(v) \subset E_2 \end{array}$$

and the corresponding commutative diagram of derivatives

$$\begin{array}{ccc} F & \xrightarrow{\tilde{f}'(x)} & G \\ \uparrow (u^{-1})'(t) & & \uparrow (w^{-1})'(s) \\ E_1 & \xrightarrow{(w \circ f \circ u^{-1})'(t)} & E_2 \end{array}$$

where E_1 is the model of M, and E_2 is the model of N, $t = u(x)$, $s = w(y) = w(f(x))$. We see that

$$\tilde{f}'(x) \cdot T_x M = \tilde{f}'(x) \operatorname{im}(u^{-1})'(x) \subset \operatorname{im}(w^{-1})'(s) = T_{f(x)}N.$$

Moreover, $f'(x)$ does not depend on the choice of the extension $\tilde{f} \supset f$ since we see from the second diagram that

$$\tilde{f}'(x) \cdot (u^{-1})'(t) = (w^{-1})'(s) \circ (w \circ f \circ u^{-1})'(t).$$

These considerations suggest the following

DEFINITION. Let (u_i, M_i), $i \in I$ be an atlas of M and let $(w_j, N_j)_{j \in J}$ be an atlas of N. A continuous mapping $f\colon M \to N$ is of class C^k if, for each $i, j \subset I, J$,

$$f_{w_j u_i} := w_j \circ f \circ u_i^{-1}\colon u_i(M_i \cap f^{-1}(N_j)) \to w_j(N_j)$$

is of class C^k. The derivative (or tangent) of f at x, denoted by $f'(x)$ (or $T_x f$ or $d_x f$) is a mapping

$$f'(x)\colon T_x M \to T_{f(x)}N$$

defined as follows: if (u_i, M_i) is a chart at x and (w_j, N_j) a chart at $f(x)$ such that $f(M_i) \subset N_j$ and e_x is a tangent vector at x represented by the triple (x, u_i, e), then $(T_x f) \cdot e_x$ is a tangent vector at $f(x)$ represented by $(f(x), w_j, \tilde{e})$ where $\tilde{e} = f'_{w_j u_i}(u_i(x)) \cdot e$. Thus

$$(T_x f) \cdot e_x := [f(x), w_j, f'_{w_j u_i}(u_i(x)) \cdot e].$$

PROPOSITION XIV.7.1. 1° The definition of $T_x f$ is independent of the choice of atlases.

2° If $f\colon M \to N$ and $g\colon N \to P$ are of class C^k, then so is $g \circ f$ and

$$T_x(g \circ f) = T_{f(x)}(g) \circ T_x(f)$$

$$T_x(\mathrm{id}) = \mathrm{id}.$$

PROOF. 1° follows from the obvious equation

$$w_{j'} \circ f \circ u_{i'}^{-1} = (w_{j'} \circ w_j^{-1}) \circ (w_j \circ f \circ u_i^{-1}) \circ (u_i \circ u_{i'}^{-1}).$$

2° The chain rule for $T_x(g \circ f)$ follows from the local representation of the derivatives of $f, g, g \circ f$ and from the chain rule for derivatives in (open sets of) Banach spaces. □

Tangent Bundles. Tangent spaces $(T_x M)$ of a manifold M form a family of Banach spaces indexed by points of M. The following natural question arises: is it possible to organize the union

$$TM = \bigcup_{x \in M} T_x M$$

into a differentiable manifold in such a way that the natural projection $\pi\colon TM \to M$

$$\pi | T_x M := x \text{ is differentiable?}$$

The answer is "yes!". In order to make more accessible the notion of a tangent bundle which, although natural, is quite abstract (and, hence, difficult) and to facilitate formulation of the corresponding theorem on the differentiable structure on TM, we first introduce the notion of a

Tangent Bundle of a Subset of a Banach Space. Let e_0 be an element of a Banach space E. Then we can regard the tangent space of E at e_0 as

$$\{e_0\} \times E =: T_{e_0} E$$

The bijection $T_{e_0}E \to E$ given by

$$T_{e_0} E \ni (e_0, e) \to e \in E$$

provides $T_{e_0} E$ with the Banach structure. The structure of a vector space, for example, is given by

$$(e_0, e) + (e_0, \bar{e}) := (e_0, e + \bar{e}), \qquad \lambda(e_0, e) := (e_0, \lambda e).$$

Let O be a subset of E. The tangent bundle TO of O is defined as a union $\bigcup_{e \in O} T_e E$ of the tangent spaces $T_e E$, $e \in O$. There is a canonical bijection of TO onto $O \times E$

$$TO \supset T_{e_0}E \ni (e_0, e) \to (e_0, e) \in O \times E.$$

We see that one can regard TO as a subset $O \times E$ of $E \times E$. If O is open, then $O \times E$ is open in $E \times E$ and the notion of a map of class C^k

$$\Phi\colon TO \to F, \quad F\text{—a Banach space}$$

is well defined and we may accept the following

DEFINITION. Let E, F be Banach spaces (over R) and let O be an open subset of E. Let $\varphi\colon O \to F$ be of class C^k; then for each $e_0 \in O$ the map

(3) $$T_{e_0}\varphi\colon T_{e_0}E \to T_{e_0}F, \quad (e_0, e) \to (\varphi(e_0), \varphi'(e_0)e)$$

is defined. Therefore the mapping

(4) $$T\varphi\colon TO \to TF, \quad (T\varphi)|T_{e_0}E := T_{e_0}\varphi$$

is well defined (and of class C^{k-1}). $T\varphi$ is called the *tangent* of φ.

Remark. If one identifies, in canonical way, $T_{e_0}E$ with E and $T_{e_0}F$ with F, then one (usually) denotes $T_{e_0}\varphi$ also by $\varphi'(x_0)$ or $d\varphi_{x_0}$ or $d\varphi(x_0)$. We shall use all these notations.

Tangent Bundle TM of the Manifold M. We are now ready to introduce the general notion of the tangent bundle of a differentiable manifold. This construction is given in the following

THEOREM XIV.7.2. Let M be a differentiable C^k-manifold modelled on E. Denote by TM the (disjoint) union $\bigcup_{x \in M} T_xM =: TM$ of tangent spaces. Let

$$\pi\colon TM \to M, \quad \pi|T_xM := x.$$

Then

1° TM is a differentiable manifold of class C^{k-1} (modelled on $E \times E$): if $(u_i, M_i)_{i \in I}$ is an atlas of M, then the atlas (Tu_i, TM_i), $i \in I$ of TM is defined by

$$TM_i := \bigcup_{x \in M_i} T_xM, \quad Tu_i\colon TM_i \to TO_i = O_i \times E,$$

$$Tu_i([(x, u_i, e)]) := (u_i(x), e).$$

2° The projection $\pi\colon TM \to M$ is differentiable.

PROOF. 1° As we know (from the definition of u_i and Tu_i), Tu_i: $TM_i \to TO_i$ is a bijection. In this way we transfer the topology of TO_i to TM_i.

Since the bijection $Tu_k \circ (Tu_i)^{-1}$: $TO_i \to TO_k$ has the form $T(u_k \circ u_i^{-1})$, it is a diffeomorphism. Therefore the topology of TM_i is independent of the choice of charts and (Tu_i, TM_i), $i \in I$ is a differentiable atlas (of class C^{k-1}).

2° π: $TM \to M$ is differentiable since its restriction $\pi | TM_i$ is given in the charts by $u_i \circ \pi \circ Tu_i^{-1}$— the projection on the first factor: $TO_i \equiv O_i \times E \to O_i$. □

DEFINITION. The tangent bundle TM of the differentiable manifold M is the differentiable manifold defined in the preceding theorem. More precisely, the tangent bundle TM is the triple (TM, π, M), where π: $TM \to M$ is called the *projection.*

We can now give an elegant definition of a vector field.

DEFINITION. A *vector field* X (of class C^k) on M (a subset V of M) is a map (of class C^k)

(5) $\qquad X\colon V \to TM$ such that $\pi \circ X = \mathrm{id}_V$,

i.e.,

$$X(x) \in T_x(M), \qquad x \in V.$$

Remark. Equation (5) defines a section X (over V) of the tangent bundle (TM, π, M).

We can therefore say that a vector field (over $V \subset M$) is a section (over V) of the tangent bundle.

Let us denote the set of vector fields on M of class C^r by $X_{(r)}(M)$, or by $X(M)$ if $r = \infty$. If $X \in X(M)$ and f is a C^1-function on M, then we denote the function $X(f)(x) := f'(x) \cdot X(x)$ by $X(f)$.

PROPOSITION XIV.7.3. Let X and Y be two vector fields of class C^r on a manifold M of class C^p, $p \geqslant 2$. Let (u_i, M_i) be an atlas for M, and let

$$X(x) = \big[\big(x, u_i, e_i(u_i(x))\big)\big], \qquad Y(x) = \big[\big(x, u_i, f_i(u_i(x))\big)\big]$$

for $x \in M_i$. There exists a unique vector field $[X, Y]$ (of class C^{r-1}) such that for $x \in M$

(6) $$[X, Y](x) = \big[\big(x, u_i, f_i'(u_i(x)) \cdot e_i(u_i(x)) - \\ - e_i'(u_i(x)) \cdot f_i(u_i(x))\big)\big].$$

PROOF. Let $x_0 \in M_i \cap M_j$. We must check that

$$[(x_0, u_i, f_i'(u_i(x_0)) \cdot e_i(u_i(x_0)) - e_i'(u_i(x_0)) \cdot f_i(u_i(x_0)))]$$
$$= [(x_0, u_j, f_j'(u_j(x_0)) \cdot e_j(u_j(x_0)) - e_j'(u_j(x_0)) \cdot f_j(u_j(x_0)))].$$

For $t \in u_j(M_i \cap M_j)$ we have

$$e_j(t) = (u_j \circ u_i^{-1})'(u_i \circ u_j^{-1}(t)) \cdot e_i(u_i \circ u_j^{-1}(t))$$

and

$$f_j(t) = (u_j \circ u_i^{-1})'(u_i \circ u_j^{-1}(t)) \cdot f_i(u_i \circ u_j^{-1}(t)).$$

Hence for $t_0 := u_j(x_0)$, $u_{ij} := u_i \circ u_j^{-1}$:

$$\begin{aligned}
&f_j'(t_0) \cdot e_j(t_0) - e_j'(t_0) \cdot f_j(t_0)\\
&= (u_{ij}^{-1})''(u_{ij}(t_0)) \cdot (u_{ij}'(t_0) \cdot e_j(t_0), f_i(u_{ij}(t_0))) +\\
&+ (u_{ij}^{-1})'(u_{ij}(t_0)) \cdot f_i'(u_{ij}(t_0)) \cdot u_{ij}'(t_0) \cdot e_j(t_0) -\\
&- (u_{ij}^{-1})''(u_{ij}(t_0)) \cdot (u_{ij}'(t_0) \cdot f_j(t_0), e_i(u_{ij}(t_0))) -\\
&- (u_{ij}^{-1})'(u_{ij}(t_0)) \circ e_i'(u_{ij}(t_0)) \cdot u_{ij}'(t_0) \cdot f_j(t_0)\\
&= (u_{ij}^{-1})'(u_{ij}(t_0))(f_i'(u_{ij}(t_0)) \cdot e_i(u_{ij}(t_0)) -\\
&- e_i'(u_{ij}(t_0)) \cdot f_i(u_{ij}(t_0))). \quad \square
\end{aligned}$$

Remark. One can easily check that $X(M)$ with the bracket $[\cdot, \cdot]$ is a Lie algebra.

We can define algebraic operations on the tangent bundle. Let us mention two of them:

the direct sum

$$\underbrace{TM \oplus \dots \oplus TM}_{p \text{ times}} := \bigcup_{x \in M} \underbrace{T_xM \oplus \dots \oplus T_xM}_{p \text{ times}}$$

and

the dual bundle

$$T^*M := \bigcup_{x \in M} T_xM',$$

where T_xM' is the (strong) dual of T_xM.

Both $TM \oplus \dots \oplus TM$ and T^*M can easily be provided with a structure of a manifold of class C^{k-1}. Namely, given an atlas (u_i, M_i) of M, we take $(TM_i \oplus \ldots \oplus TM_i, Tu_i \oplus \ldots \oplus Tu_i)$ and (T^*M_i, T^*u_i) as the corresponding atlas for $TM \oplus \ldots \oplus TM$ and T^*M respectively, where

$$TM_i \oplus \ldots \oplus TM_i := \bigcup_{x \in M_i} T_x M \oplus \ldots \oplus T_x M,$$

$$Tu_i \oplus \ldots \oplus Tu_i([(x, u_i, e_1)], \ldots, [(x, u_i, e_p)])$$

$$:= (u_i(x), e_1 \oplus \ldots \oplus e_p) \in O_i \times E \oplus \ldots \oplus E,$$

$$T^*M_i := \bigcup_{x \in M} T_x M',$$

$$T_x M' \ni \eta_x \xrightarrow{T^*u_i} (u_i(x), \tilde{\eta}_x) \in O_i \times E',$$

$$\langle e | \tilde{\eta}_x \rangle := \langle [(x, u_i, e)] | \eta_x \rangle.$$

8. COVARIANT TENSOR FIELDS. RIEMANNIAN METRIC AND DIFFERENTIAL FORMS ON A MANIFOLD

We can now define covariant tensor fields (and especially differential forms) on the manifold M:

Let M be a manifold of class C^k ($k \geqslant 1$) modelled on E. If (u_i, M_i), $i \in I$, is an atlas of M denote by (φ_i, O_i) the corresponding parametrization of M:

$$\varphi_i = u_i^{-1}, \quad O_i = u_i(M_i), \quad \text{thus } M_i = \varphi_i(O_i),\ i \in I.$$

DEFINITION. *A p-covariant tensor field* Φ (of class C^r, $r < k$) on M is a map (of class C^r)

$$(1) \qquad \Phi: \underbrace{TM \oplus \ldots \oplus TM}_{p \text{ times}} \to R$$

such that for each $x \in M$ its restriction Φ_x to $T_x M \oplus \ldots \oplus T_x M$ is a p-linear form

$$(2) \qquad \Phi_x \in L_p(T_x M, R).$$

If Φ_x is alternating, $\Phi_x \in A_p(T_x M)$, then the tensor field Φ is called a *differential p-form on M* (of class C^r). The space of such forms is denoted by $\Lambda^p_{(r)}(M)$, and the linear space of p-covariant tensor fields on M of class C^r will be denoted by $\mathcal{T}_{p,(r)}(M)$. If $r = \infty$, we omit the index (r) and write simply

$$\Lambda^p(M) \text{ respectively } \mathcal{T}_p(M) \text{ or } \mathcal{T}^0_p(M).$$

Remark 1. We can also consider F-valued p-covariant tensor fields, F being a Banach space.

Remark 2. If we do not want to introduce the differentiable structure (atlas) on $TM \oplus \dots \oplus TM$, we can define the C^r-differentiability of a p-form (tensor field) on M in the following way.

DEFINITION. A local representation of the p-form Φ on M is a map $\varphi_i^* \Phi \equiv \Phi_{O_i}$ from O_i to $A_p(E)$:

$$\varphi_i^* \Phi(t; e_1, \dots, e_p) := \Phi(\varphi_i(t); \varphi_i'(t)e_1, \dots, \varphi_i'(t)e_p).$$

Plainly

$$\varphi_i'(t)e \in T_x M, \qquad x = \varphi_i(t) \qquad \text{for } e \in E.$$

Exercise. Prove that Φ is of class C^r if and only if

$$\Phi_{O_i} \in \Lambda^p_{(r)}(O_i)$$

If $\omega \in \Lambda^p_{(r)}(M)$ and $\eta \in \Lambda^q_{(r)}(M)$ we can define the exterior product $\omega \wedge \eta \in \Lambda^{p+q}_{(r)}(M)$ by

$$(\omega \wedge \eta)_x = \omega_x \wedge \eta_x \in A_{p+q}(T_x M).$$

We can now define the

Pull Back (Inverse Image) of Covariant Tensor Fields. In Section 2 (formula (7)) we introduced the pull back (or "change of variables") of a differential form. A natural generalization of this operation is given in the following

DEFINITION. Let X and Y be differentiable manifolds. Let $f\colon X \to Y$ be a map (of class C^{r+1}) and let Φ be a covariant tensor field on Y of class C^r, $\Phi \in \mathscr{T}_{p,(r)}(Y)$. Then the covariant tensor field $f^*\Phi$ of class C^r on X, defined by

$$f^*\Phi(x; e_x^1, \dots, e_x^p) := \Phi(f(x); T_x f \cdot e_x^1, \dots, T_x f \cdot e_x^p)$$

for any vector $e_x^1, \dots, e_x^p \in T_x X$ is called the *pull back of Φ by the map f*.

We have the following

PROPOSITION XIV.8.1. If $f\colon X \to Y$ and $g\colon Y \to Z$ are C^{r+1}-differentiable maps, then $f^*\colon \mathscr{T}_{p,(r)}(Y) \to \mathscr{T}_{p,(r)}(X)$ is linear. We have

$$(g \circ f)^* = f^* \circ g^*,$$

or more precisely: for any tensor field $\Phi \in \mathscr{T}_p(Z)$

$$(g \circ f)^*(\Phi) = f^*(g^*(\Phi)) \in \mathscr{T}_p(X).$$

The proof is left to the reader.

The exterior derivative of a p-form is defined by the local representation.

DEFINITION. Let ω_{O_i} be the local representation of a p-form $\omega \in \Lambda^p_{(r)}(M)$, $O_i = u_i(M_i)$ and let $e_k(\cdot)$, $k = 1, \dots, p$ be local representations of vector fields $X_1(\cdot), \dots, X_p(\cdot)$ on M. Denote by $\omega'_{O_i}(t)$ the derivative of ω_{O_i} at $t \in O_i$. We know that

$$\omega'_{O_i}(t) \in L(E, A_p(E, \mathbf{R})).$$

Then the exterior derivative $d\omega \in \Lambda^{(p+1)}_{(r-1)}(M)$ is defined by its local representative by means of formula (2.2)

$$(*) \qquad d\omega_{O_i}(t; e_0(t), e_1(t), \dots, e_p(t))$$
$$:= \sum_{j=0}^{p} (-1)^j (\omega'_{O_i}(t) \cdot e_j(t))(e_0(t), \dots, \hat{e}_j(t), \dots, e_p(t)).$$

One can check that this definition of $d\omega$ is independent of the local representation, or prove the following

PROPOSITION XIV.8.2. Let U be an open subset of $\varphi_i(O_i) = M_i$. Let $X_0(\cdot), \dots, X_p(\cdot)$ be any C^1-vector fields on U and let ω be a p-form on U. Then $d\omega$ can be written in the following (obviously) invariant manner:

$$(**) \qquad d\omega(X_0, \dots, X_p) = \sum_{j=0}^{p} (-1)^j X_j(\omega(X_0, \dots, \hat{X}_j, \dots, X_p)) +$$
$$+ \sum_{j<k}^{p} (-1)^{j+k} \omega([X_j, X_k], X_0, \dots, \hat{X}_j, \dots, \hat{X}_k, \dots, X_p).$$

In the case of a 1-form this equation reads

$$d\omega(X_0, X_1) = X_0(\omega(X_1)) - X_1(\omega(X_0)) - \omega([X_0, X_1]),$$
$$X_0, X_1 \in X(M).$$

PROOF (after S. Lang). Denote by S_1 and S_2 the two sums occurring in $(**)$. We have to check that $S_1 + S_2 = L$, the local expression in $(*)$.

We consider S_1 and apply the definition of the vector field X as a derivation $X(f)$ of a function on M $Xf(x) = (T_x f)X(x)$ (cf. Proposition XIV.7.2). We obtain

$$S_1 = \sum_{j=0}^{p} (-1)^j \omega_{O_i}(e_0, \ldots, \hat{e}_j, \ldots, e_p)'(x) e_j(t).$$

Applying this definition and discarding second-order terms, we see that S_1 is equal to

$$\sum_j (-1)^j (\omega'_{O_i}(t) e_j(t);\ e_0(t), \ldots, \hat{e}_j(t), \ldots, e_p(t)) +$$
$$+ \sum_j \sum_{i<j} (\omega_{O_i}(t);\ e_0(t), \ldots, e'_i(t) e_j(t), \ldots, \hat{e}_j, \ldots, e_p(t)) +$$
$$+ \sum_j \sum_{i<j} (\omega_{O_i}(t); e_0(t), \ldots, \hat{e}_j(t), \ldots, e'_i(t) e_j(t), \ldots, e_p(t)).$$

The first sum is equal to L; in the third one we permute i and j and move the term $e'_i(t)e_j(t)$ to the first position, and we see that they give the following expression

$$-\sum_j \sum_{j<i} (-1)^{j+i} \omega(e'_i e_j - e'_j e_i, e_0, \ldots, \hat{e}_j, \ldots, \hat{e}_i, \ldots, e_p)$$

evaluated at t. Using Proposition XIV.7.2, we can see that this sum is equal to $-S_2$, hence $S_1 + S_2 = L$. □

This definition of the exterior derivative ensures that pull back, exterior derivative and exterior multiplication are connected in the same way as in the flat case. We shall recall these facts in the following

THEOREM XIV.8.3. Let M be a differentiable manifold modelled on a Banach space, then

1° $\omega \in \Lambda^p_{(r)}(M), \quad \eta \in \Lambda^q_{(r)}(M) \Rightarrow \omega \wedge \eta \in \Lambda^{p+q}_{(r)}(M)$;

2° If f: $N \to M$ is of class C^{r+1}, then

$$f^*(\omega \wedge \eta) = (f^*\omega) \wedge (f^*\eta);$$

3° $d(f^*\omega) = f^*(d\omega)$.

Hence the pull back is a "natural map": it commutes with the exterior derivative and exterior multiplication.

4° $(g \circ f)^* = f^* \circ g^*$ (cf. Theorem XIV.2.5.3°).

5° If ω is a 0-form on Y, then

$$d(\omega \circ f) = f^*(d\omega).$$

Riemannian Structure (*Metric*). We know that the most important class of infinite-dimensional Banach spaces are Hilbert spaces. A real Hilbert space E is a real vector space provided with a scalar product $(\cdot \mid \cdot)$, i.e., a symmetric positive bilinear form such that with the norm $\|e\| := (e|e)^{1/2}$, the normed space $(E, \|\cdot\|)$, is complete. We know that for any differentiable manifold M modelled on a Banach space E each tangent space $T_x M$ is provided with a Banach structure isomorphic to that of the model E. But this does not mean that there is a function $\|\cdot\|: TM \to \boldsymbol{R}$ which is differentiable. Such a function is called a *Finsler metric* and it is an additional structure on a Banach manifold M.

If the model E is a Hilbert space, then we can provide each tangent space $T_x M$ with such a scalar product $G_x \equiv (\cdot \mid \cdot)_x$ that the pair $(T_x M, (\cdot \mid \cdot)_x)$ is a Hilbert space isomorphic to the model E. This leads us to a notion of paramount importance for the whole of mathematics.

Definition. Let M be a differentiable manifold modelled on a Hilbert space $(E, (\cdot \mid \cdot))$. A *Riemannian structure on* M is a covariant tensor field $G \in \mathscr{T}_{0,(2)}(M)$ such that its restriction G_x to each tangent space $T_x M$ is a scalar product such that the pair $(T_x M, G_x)$ is a Hilbert space isomorphic to the model $(E, (\cdot|\cdot))$. A pair (M, G), where G is a Riemannian structure of M, is called a *Riemannian manifold* or a *Riemannian space*.

Remarks. 1. If M is an m-dimensional manifold (the model $M = \boldsymbol{R}^m$) then the definition of a Riemannian structure is slightly simplified.

A Riemannian structure G on an m-dimensional differentiable manifold is a smooth tensor field which is symmetric and positive.

The point is that any finite-dimensional normed space is complete, and hence each finite-dimensional vector space with a scalar product is complete (i.e., a Hilbert space).

2. The preceding remark is of importance in the theory of infinite-dimensional Hilbert–Riemann manifolds. These objects are very important for many applications, e.g., in the calculus of variations, modern

hydrodynamics, the problem of a closed geodesics and many others (cf. Part III).

3. A construction of a Riemannian metric will be given in Section 9. An invariant Riemannian metric on a Lie group will be considered in Section 15.

Partitions of Unity on Differentiable Manifolds. In 1938 Jean Dieudonné discovered an important method of constructing significant objects on manifolds. This method is now indispensable in analysis and differential geometry. The fundamental notion is that of "partition of unity". Investigations of the existence of this object led Dieudonné to the discovery of a large and important class of topological spaces, namely paracompact spaces. This notion is investigated in Chapter XVII. We only remark here that an m-dimensional manifold with a countable basis is paracompact. This is one of the main reasons why we consider "only" such manifolds in the sequel.

For convenience we introduce the following

DEFINITION. A covering $(U_i)_{i\in I}$ of a Hausdorff space M is *locally finite* if each point of M has an open neighbourhood, which intersects U_i only for finitely many $i \in I$.

DEFINITION. Let M be a differentiable manifold. A *partition of unity* of class C^k on M subordinate to the open covering $(O_j)_{j\in N}$ of M is the family $(f_j)_{j\in N}$ of C^k-functions on M with the following properties:

1° $f_j \geqslant 0$ on M, $\operatorname{suppt} f_j \subset O_j$;

2° The supports $\operatorname{suppt} f_j$ of $f_j, j \in J$, form a locally finite covering of M;

3° $\sum_{j\in J} f_j(x) = 1$ for every $x \in M$.

Note that the sum in 3° is a well-defined function of class C^k since each point has a neighbourhood on which only a finite number of f_j's do not vanish.

The following theorem asserts the existence of partitions of unity on a reasonably large class of manifolds.

THEOREM XIV.8.4. Let M be a manifold of class C^k which is locally compact and whose topology admits a countable base (for instance a manifold in $\boldsymbol{R}^n$). Then M admits partitions of unity of class C^k, i.e., for each open covering of M there exists a smooth partition of unity subordinate to it.

PROOF. Let $(O_j)_{j\in J}$, be an open covering of M. Then there exists an open locally finite covering (U_i), $i \in I$, of M such that each U_i is contained in (at least) one O_j and a compact covering (K_i), $i \in I$, $K_i \subset U_i$ (since M is paracompact, cf. Chapter XVII). Moreover, we can assume that U_i are coordinate neighbourhoods of M of charts $(\varkappa_i, U_i)$. Let $V_i := \varkappa_i(U_i) \subset \boldsymbol{R}^m$, $L_i := \varkappa_i(K_i)$. By virtue of Theorem XIX.2.1 there exists a function $\varphi_i \in C_0^k(\boldsymbol{R}^m)$ such that $\varphi_i \geqslant 0$ suppt$\varphi_i \subset V_i$, $\varphi_i(y) = 1$ for $y \in L_i$. Put

$$g_i(x) = \begin{cases} \varphi_i\big(\varkappa_i(x)\big) & \text{for } x \in U_i, \\ 0 & \text{for } x \notin U_i. \end{cases}$$

It is easy to check that g_i are of class C^k. $\sum_{i\in I} g_i \equiv g$ is a well-defined positive function on M of class C^k. For each $i \in I$ choose $j(i) \in J$ such that $U_i \subset O_{j(i)}$. Set $f_j = g^{-1} \sum_{\{i:j(i)=j\}} g_i$ or 0 if there is no i such that $j = j(i)$. The family $(f_j), j \in J$, is the required partition of unity of class C^k subordinate to the covering (O_j). □

In subsequent sections we shall present important applications of the existence of partitions of unity:

1° existence of a Riemannian metric,

2° existence of a volume form,

3° existence of a smooth measure.

9. ORIENTATION OF MANIFOLDS. EXAMPLES

All manifolds considered in the remainder of this chapter are of finite dimension!

In Section 4 we introduced the notion of orientation of an m-dimensional space. This notion has its analogue in the theory of differentiable manifolds.

DEFINITION. Let M be a manifold of class C^k $(k \geqslant 1)$ modelled on $\boldsymbol{R}^m$. The manifold M is *orientable* if it possesses an atlas (u_i, M_i), $i \in I$ such that for all $i, j \in I$

(1) $\qquad \det\big((u_j \circ u_i^{-1})'\big) > 0.$

An atlas with property (1) is said to be *oriented.* Two atlases define the same orientation, or are *orientation-equivalent*, if their union is oriented. A chart (u, O) is *orientation-compatible* with the oriented atlas (u_i, M_i), $i \in I$, if the union $\{(u, O)\} \cup \{(u_i, M_i): i \in I\}$ is oriented.

A class of orientation equivalent atlases of M is called an *orientation of* M or is said to *define an orientation of* M.

Remarks. 1. If M is connected and orientable, then it has two distinct orientations.

2. Not every manifold is orientable. The best known examples of non-orientable manifold are the real projective plane $P^2(R)$ and the Möbius strip (cf. the examples below).

Examples. 1. The 2-sphere $M = S_r^2(0) = \{(x, y, z) \in R^3: x^2+y^2+z^2 = r\}$, $r > 0$ provided with the relative topology of R^3. We take the atlas (u_i, M_i), $i = 1, 2$:

$$M_1 = M - \{(0, 0, -r)\}, \qquad M_2 = M - \{(0, 0, r)\}.$$

Let u_1 (u_2) be stereographic projections from the south (north) pole:

$$u_1(x, y, z) := \left(\frac{2rx}{r+z}, \frac{2ry}{r+z}\right),$$

$$u_2(x, y, z) := \left(\frac{2rx}{r-z}, \frac{2ry}{r-z}\right) =: (\xi, \eta).$$

Since

$$x = \frac{4\xi r^2}{\xi^2+\eta^2+4r^2}, \qquad y = \frac{4\eta r^2}{\xi^2+\eta^2+4r^2},$$

$$z = \frac{r(\xi^2+\eta^2-4r^2)}{\xi^2+\eta^2+4r^2},$$

we find that

$$u_1 \circ u_2^{-1}(\xi, \eta) = \left(\frac{8r^2\xi}{\xi^2+\eta^2}, \frac{8r^2\eta}{\xi^2+\eta^2}\right)$$

is a diffeomorphism of $u_2(M_1 \cap M_2) = R^2 - \{(0, 0)\}$ onto $u_1(M_1 \cap M_2)$.

The Jacobian of $u_1 \circ u_2^{-1}$ is equal to $-64r^4(\xi^2+\eta^2)^{-2} < 0$. Evidently $S_r^2(0)$ is orientable.

2. The *projective plane* $M = \boldsymbol{P}^2 = \boldsymbol{P}^2(\boldsymbol{R})$ (called also the *elliptic plane*) can be defined in the following way:

$$\boldsymbol{P}^2 = \{\text{non-ordered pairs } \{a, -a\}: a \in S^2 = S_1^2(0)\}.$$

Consider the map $f\colon S^2 \to \boldsymbol{P}^2, f(a) := \{a, -a\}$. If $B \subset S^2$ is in an open hemisphere, then $f|B\colon B \to \boldsymbol{P}^2$ is injective. We take as the basis of open sets of $\boldsymbol{P}^2$ the images $f(B)$ of such $B \subset S^2$, B open. A differentiable atlas $(u_j, S_j^2), j \in J$, of S^2, with S_j^2 in a hemisphere, gives the differentiable atlas $(u_j \circ (f|S_j^2)^{-1}, fS_j^2), j \in J$, of the manifold $\boldsymbol{P}^2$. Subdivide the atlas (u_1, M_1), (u_2, M_2) of S from Example 2 into sufficiently small maps.

We shall show that the projective plane $\boldsymbol{P}^2$ is not orientable. To this end we take the antipodal map $A\colon S^2 \to S^2$, $A(a) := -a$. In the atlas of Example 2 this map has the form

$$u_2 \circ A \circ u_2^{-1}\colon (\xi, \eta) \to \left(\frac{-\xi}{\xi^2+\eta^2}, \frac{-\eta}{\xi^2+\eta^2}\right).$$

It changes the orientation of S^2. Let us suppose that $\boldsymbol{P}^2$ has an oriented atlas $(u_j, \boldsymbol{P}_j^2), j \in J$. Since $f^{-1}\boldsymbol{P}_j^2$ can be decomposed into $S_j^2 \cup A(S_j^2)$ in such a way that

$$f\colon S_j^2 \to \boldsymbol{P}_j^2 \quad \text{and} \quad f\colon A(S_j^2) \to \boldsymbol{P}_j^2$$

are diffeomorphisms, we infer that

$$\left(u_j \circ f, S_j^2; u_j \circ f, A(S_j^2)\right), \quad j \in J$$

is an oriented atlas of S^2. But the map $A\colon S_j^2 \to S_j^2$ has in this atlas the form id: $u_j(\boldsymbol{P}_j^2) \to u_j(\boldsymbol{P}_j^2)$, and hence preserves the orientation—a contradiction! This example suggests the consideration of a general

3. *Projective space* $\boldsymbol{P}(E)$. Apart from Euclidean spaces n-dimensional real and complex projective spaces are perhaps the most important objects in mathematics. They are of paramount importance not only in algebraic geometry but also in complex analysis and differential geometry. In this section we only want to give the definitions and to examine more closely the complex projective space $\boldsymbol{P}(\boldsymbol{C}^1)$ of dimension 1.

Since the projective space $\boldsymbol{P}(H)$, where H is a Hilbert space, is of great importance for quantum mechanics, let us begin with the general

DEFINITION. Let E be a Banach space over $\mathscr{F}$ ($\mathscr{F} = \boldsymbol{R}$ or $\boldsymbol{C}$). By $\boldsymbol{P}(E)$ we denote the space of all lines through the origin 0 of E (in another wording, $\boldsymbol{P}(E)$ is the set of one-dimensional subspaces of E). $\boldsymbol{P}(E)$ is called the *projective space defined by E*. $\boldsymbol{P}(\mathscr{F}^{n-1})$ is also denoted by $\boldsymbol{P}^n(\mathscr{F})$.

We can give an alternative description of $\boldsymbol{P}(E)$.

Let an equivalence relation $\sim$ on $E-\{0\}$ be defined by: $x \sim y$ if there exists a number $t \in \mathscr{F}$ such that $y = tx$. Denote by $[x]$ the equivalence class of x. $\boldsymbol{P}(E) = \{[x]: x \in E-\{0\}\}$; thus $\boldsymbol{P}(E) = (E-\{0\})/\sim$.

By π we denote the natural map (projection)

$$\pi: E-\{0\} \to \boldsymbol{P}(E), \quad \pi(x) := [x].$$

We introduce a topology in $\boldsymbol{P}(E)$ in the standard way: $O \subset \boldsymbol{P}(E)$ is open iff $\pi^{-1}(O)$ is open in E. This is the finest topology on $\boldsymbol{P}(E)$ for which π is continuous. Since π takes each $x \neq 0$ to the line (through 0) which contains it, O is the set of all lines which meet a given open set $\pi^{-1}(O)$.

If $E = \mathscr{F}^{n+1}$ then $x = (x^1, \ldots, x^{n+1})$ are called the *homogeneous coordinates of* $[x] = \pi(x)$.

Since we want to prove that projective spaces are Hausdorff, we recall a simple fact from general topology.

LEMMA XIV.9.1. Let $\sim$ be an equivalence relation on a topological space E fulfilling the following requirement:

(2) if O is open in E, then so is

$$\pi^{-1}(\pi(O)) = \{x \in E: \text{there exists an } y \in O \text{ such that } x \sim y\}.$$

Then

(the relation $\sim$ is a closed subset R of $E \times E$) $\Leftrightarrow$ ($E/\sim$ is Hausdorff).

PROOF. $\Rightarrow$ Suppose that $[x] \neq [y]$, i.e., that $(x, y) \notin R$. There exist open sets $U \ni x$, $V \ni y$ in E such that $U \times V \cap R = \emptyset$. But then also $\pi^{-1}(\pi(U)) \times \pi^{-1}(\pi(V)) \cap R = \emptyset$. Thus $\pi(U) \cap \pi(V) = \emptyset$. But $\pi(U)$ and $\pi(V)$ are open in $E/\sim$ as $\pi^{-1}(\pi(U))$ and $\pi^{-1}(\pi(V))$ are open in E. Moreover, $[x] \in \pi(U)$ and $[y] \in \pi(V)$.

$\Leftarrow$ The proof (along the same lines) is left to the reader. □

We shall now prove the following important

THEOREM XIV.9.2. The real (complex) projective space $\boldsymbol{P}^n(\boldsymbol{R})$ ($\boldsymbol{P}^m(\boldsymbol{C})$) is:

1° Hausdorff;

2° compact and, moreover, $\boldsymbol{P}^m(\boldsymbol{R})$ is a continuous image of the m-sphere S^m ($(2m+1)$-sphere S^{2m+1});

3° $\boldsymbol{P}^m(\boldsymbol{R})$ can be provided with a differentiable (analytic) structure.

The proof is given here for the real case—in the complex case the proof is analogous. On the open subset $(\boldsymbol{R}^{m+1}-\{0\})\times(\boldsymbol{R}^{m+1}-\{0\}) \subset \boldsymbol{R}^{m+1}\times\boldsymbol{R}^{m+1}$ we define a real-valued function

$$\varphi(x;y) = \varphi(x^1, \ldots, x^{m+1}; y^1, \ldots, y^{m+1})$$
$$:= \sum_{k\neq j} |x^k y^j - x^j y^k|^2.$$

Plainly φ is continuous and vanishes iff $y = tx$ for some $t \in \boldsymbol{R}, t \neq 0$; hence

$$\big(\varphi(x;y) = 0\big) \Leftrightarrow (x \sim y).$$

But since φ is continuous, $\varphi^{-1}(0)$ is closed and by virtue of the preceding Lemma $\boldsymbol{P}^m(\boldsymbol{R})$ is Hausdorff, as (2) is easy to verify.

2° $\boldsymbol{P}^m(\boldsymbol{R})$ is compact. Indeed, let

$$S^m = \{x \in \boldsymbol{R}^{m+1}: \; ||x|| = 1\}.$$

For $x \in \boldsymbol{R}^{m+1}-\{0\}$ the point $\tilde{x} = x/||x||$ is in S^m and $\pi(\tilde{x}) = \pi(x)$. Therefore, $\pi|S^m\colon S^m \to \boldsymbol{P}^m(\boldsymbol{R})$ is a continuous surjection. Since $\boldsymbol{P}^m(\boldsymbol{R})$ is Hausdorff and S^m is compact the space $\boldsymbol{P}^m(\boldsymbol{R})$ is compact.

3° Let us define $m+1$ charts (u_k, M_k) as follows:

$$\pi^{-1}(M_k) = \{x \in \boldsymbol{R}^{m+1}: \; x^k \neq 0\}, \quad k = 1, 2, \ldots, m+1,$$

$$u_k\colon M_k \to \boldsymbol{R}^m,$$

$$u_k([x]) := \left(\frac{x^1}{x^k}, \ldots, \frac{x^{k-1}}{x^k}, \frac{x^{k+1}}{x^k}, \ldots, \frac{x^{m+1}}{x^k}\right),$$

where $x = (x^1, \ldots, x^{m+1})$. We verify that

$$\big(u_k([x]) = u_k([y])\big) \Leftrightarrow (x \sim y);$$

hence u_k is properly defined and is a continuous injection. We have to prove that u_k is invertible. But

$$u_k^{-1}(t^1, \ldots, t^m) = \pi(t^1, \ldots, t^{k-1}, 1, t^k, \ldots, t^m);$$

therefore, u_k^{-1} is continuous as a superposition of two continuous maps. Thus $(u_k, M_k), k = 1, \ldots, m+1$ is a topological atlas and $\boldsymbol{P}^m(\boldsymbol{R})$ is modelled on $\boldsymbol{R}^m$. Plainly, $u_k \circ u_j^{-1}$ is smooth for $k, j = 1, 2, \ldots, m+1$. □

Remarks. 1. In Part III we shall consider projective spaces from another point of view and we shall prove that

$\boldsymbol{P}^m(\boldsymbol{R})$ is not orientable for even m.

2. We see that $\boldsymbol{P}^m(\boldsymbol{R})$ can be identified with the set {non-ordered pairs $\{x, -x\}$: $x \in S^m$} since such pairs are in a one-to-one correspondence with the lines passing through the origin of $\boldsymbol{R}^{m+1}$: each such line intersects S^m precisely at two points x and $-x$. Therefore the definition of $\boldsymbol{P}^2(\boldsymbol{R})$ given in Example 2 is consistent with the general definition given here.

3. *Riemann sphere* $\overline{\boldsymbol{C}}$ *and* $\boldsymbol{P}^1(\boldsymbol{C})$. In Chapter XVI an 1-point compactification of $\boldsymbol{C}$ $(= \boldsymbol{R}^2)$, denoted by $\overline{\boldsymbol{C}}$, will be considered. By definition, $\overline{\boldsymbol{C}}$ is the union $\boldsymbol{C} \cup \{\infty\}$. The construction of $\overline{\boldsymbol{C}}$ was suggested by the stereographic projection (from the north pole): Let S be the sphere of radius r in $\boldsymbol{R}^3$ with centre $(r, 0, 0)$. It is tangent to $\boldsymbol{R}^2(=\boldsymbol{C})$ at the origin (the south pole of S) and has the north pole $p = (2r, 0, 0)$. The *stereographic projection* is the map σ: of $S - \{p\}$ onto $\boldsymbol{R}^2$

$$\sigma\colon S - \{p\} \to \boldsymbol{R}^2.$$

In order to extend σ to the whole of S we have to "complete" $\boldsymbol{R}^2$ with the "point at infinity" ∞. In this way $\bar{\sigma}\colon S \to \boldsymbol{R}^2 \cup \{\infty\} =: \overline{\boldsymbol{C}}$. We introduce a topology on $\overline{\boldsymbol{C}}$ in such a way that σ is a homeomorphism.

Open sets are: 1° all open subsets of $\boldsymbol{R}^2$, 2° all sets $(\boldsymbol{C} - K) \cup \{\infty\}$ where K is a compact subset of $\boldsymbol{C}$. Therefore the basis of neighbourhoods (of points) of $\overline{\boldsymbol{C}}$ are open circles of $\boldsymbol{C}$ and the basis of $\{\infty\}$ are sets $\{|z| > \varrho, \varrho > 0\} \cup \{\infty\}$. In Chapter XVI.7 it is shown in a general way that $\overline{\boldsymbol{C}}$ is a compact differentiable manifold. $\overline{\boldsymbol{C}}$ can be provided with an atlas $(v_1, \boldsymbol{C}), (v_2, \overline{\boldsymbol{C}} - \{0\})$, where $v_1(z) := z, v_2(z) := z^{-1}$. Plainly, $v_1 \circ v_2^{-1}\colon \boldsymbol{C} - \{0\} \to \boldsymbol{C} - \{0\}$ and $v_2 \circ v_1^{-1}\colon \boldsymbol{C} - \{0\} \to \boldsymbol{C} - \{0\}$ are smooth (even holomorphic). We shall show that $\overline{\boldsymbol{C}}$ is compact and connected.

Let $V_1 = \{z \in \boldsymbol{C}\colon |z| \leqslant 1\}, V_2 = \{z \in \overline{\boldsymbol{C}}\colon |z| \geqslant 1\}$. Then V_1 is compact. But $v_2 | V_2\colon V_2 \to V_1$ is a homeomorphism. Therefore V_2 is also compact. Since $\overline{\boldsymbol{C}} = V_1 \cup V_2$, we have proved that $\overline{\boldsymbol{C}}$ is compact. But

V_1 and V_2 are connected; since $V_1 \cap V_2 \neq \emptyset$ the Riemann sphere is connected. We can now formulate an interesting result.

THEOREM XIV.9.3. 1° The Riemann sphere $\bar{\boldsymbol{C}} = \boldsymbol{C} \cup \{\infty\}$ is a compact, connected, holomorphic manifold.

2° There exists a biholomorphic mapping of $\bar{\boldsymbol{C}}$ onto the complex projective space $\boldsymbol{P}^1(\boldsymbol{C})$

$$\varphi\colon \bar{\boldsymbol{C}} \to \boldsymbol{P}^1(\boldsymbol{C}).$$

PROOF. We have to prove only 2°. Define φ by

$$\varphi(\infty) := \pi((0, 1)) = [(0, 1)], \qquad \varphi(z) := \pi((1, z)).$$

Plainly is φ a bijection. Denote $S_1 := \boldsymbol{C}$, $S_2 := \bar{\boldsymbol{C}} - \{0\}$. Then

$$(\varphi|S_i) \circ v_j^{-1} = \begin{cases} \pi(1, z) & \text{for } j = 1, \\ \pi(z, 1) & \text{for } j = 2. \end{cases}$$

Let

$$M_1 := \pi(\{(1, z_2) : z_2 \in \boldsymbol{C}\}),$$
$$M_2 := \pi(\{(z_1, 1) : z_1 \in \boldsymbol{C}\}).$$

We see that $\varphi|S_j\colon S_j \to M_j$ is diffeomorphic (even biholomorphic) for $j = 1, 2$. $\square$

Remark. We shall prove in the next chapter that any holomorphic map $\boldsymbol{C}^n \supset O \xrightarrow{f} \boldsymbol{C}^m$ considered as a map between real spaces $\boldsymbol{R}^{2m}$ has a positive Jacobian. Therefore we have

PROPOSITION XIV.9.4. Any holomorphic (complex) manifold modelled on $\boldsymbol{C}^m$ is orientable.

We see here that complex manifolds possess a richer structure than real manifolds.

Construction of a Volume Form on an Oriented Manifold. As the first application of partitions of unity we give the construction of a non-vanishing m-form on a manifold provided with an oriented atlas: in this way we obtain an important criterion of orientability.

PROPOSITION XIV.9.5. Let M be an m-dimensional differentiable manifold. The two statements:

1° On M exists a $\boldsymbol{C}^0$ m-form Ω which does not vanish at any point,

2° M has an oriented atlas (u_i, M_i), $i \in I$,
are equivalent.

Remark. Therefore we can take 1° as a definition of orientability.

PROOF. 2° ⇒ 1°: Let (u_i, M_j) be an oriented atlas of M and let (f_j) be a partition of unity subordinated to it. Let $x_j = (x_j^1, \ldots, x_j^m) := u_j(x)$; define $\Omega \in \Lambda^m(M)$ by

$$\Omega = \sum f_j u_j^* (dx_j^1 \wedge \ldots \wedge dx_j^m). \tag{3}$$

We have to check that $\Omega_x \neq 0$ for an arbitrary $x \in M$. Let (O, φ) be a coherently oriented chart around x and let $y = \varphi(x)$. By the formula of change of coordinates in an m-form we have from (3)

$$\Omega_x = \sum f_j(x) \det\left[\frac{\partial x_j^k}{\partial y^l}\right](\varphi(x))\varphi^*(dy^1 \wedge \ldots \wedge dy^m). \tag{4}$$

But at least one of the $f_j(x) > 0$ and, since *all* Jacobians are positive,

$$\det\left[\frac{\partial x_j^k}{\partial y^l}\right] > 0,$$

we have $\Omega_x \neq 0$. But x is arbitrary and, hence, Ω never vanishes on M.

1° ⇒ 2°: Let $\Omega \in \Lambda^m(M)$; since Ω does not vanish anywhere we can choose an atlas (u_i, M_i) such that Ω has the local representation

$$(u_i^{-1})^*\Omega | M_i = \lambda_i(x) dx_i^1 \wedge \ldots \wedge dx_i^m \quad \text{with } \lambda_i > 0. \tag{5}$$

If $x \in M_i \cap M_j$, $i, j \in I$, then the transformation formula gives:

$$\lambda_i \det\left[\frac{\partial x_i^k}{\partial x_j^l}\right] = \lambda_j. \tag{6}$$

But $\lambda_i, \lambda_j > 0$, hence the Jacobian in (6) is positive, and this means precisely that the atlas (u_i, M_i), $i \in I$ is oriented. □

Volume Element of the Oriented Riemannian Manifold. If M is an orientable manifold then any m-form giving orientation is called a *volume form*. Such a form is defined only modulo a positive factor. In case of a Riemannian manifold there is a distinguished m-form, defined by the metric.

PROPOSITION XIV.9.6. Let (M, G) be an m-dimensional oriented manifold provided with Riemannian metric $G \in T_2^0(M)$. Then, corresponding to the orientation of M, there exists a unique m-form μ which gives the orientation and takes the value $+1$ on every oriented orthonormal system of vectors. The form μ is called the *(Riemannian) volume element.*

PROOF. Let $e_1(x), \ldots, e_m(x)$ be any oriented orthonormal basis of $T_x M$. Put $\mu_x(e_1(x), \ldots, e_m(x)) = 1$. This defines a unique element $\mu_x \in A_m(T_x M)$. Let (u, V) be any oriented chart around x and let $(f_1(x), \ldots$ $\ldots, f_m(x)$ be the corresponding coordinate basis of $T_x M$. We have

$$f_i(x) = \sum_{k=1}^{m} a_i^k(x) e_k(x), \qquad G(e_i(x), e_k(x)) = \delta_{ik}.$$

If we define $g_{ik}(x) := G(f_i(x), f_k(x))$, $i, k = 1, \ldots, m$, then

$$g_{ik}(x) = G\Big(\sum_g a_i^g(x) e_g(x), \sum_\lambda a_k^\lambda(x) e_\lambda(x)\Big)$$

$$= \sum_{g,l} a_i^g(x) a_k^l(x)\, \delta^{gl} = \sum_g a_i^g(x) a_k^g(x).$$

Therefore, taking determinants on both sides, we obtain

$$\tilde{g}(x) := \det[g_{ik}(x)] = (\det[a_i^r(x)])^2.$$

But, from the transformation formula for p-forms we have

$$\mu_x(f_1(x), \ldots, f_m(x))$$
$$= \det[a_i^r(x)]\mu_x(e_1(x), \ldots, e_n(x)) = \det[a_i^r(x)] \cdot 1.$$

From the preceding two equations we obtain the formula

$$\mu(f_1(x), \ldots, f_m(x)) = \sqrt{\tilde{g}(x)} = \sqrt{\det[g_{ik}(x)]} > 0.$$

Since the map $x \to \det[g_{ik}(x)]$ is smooth, we see that the form μ defined by the equation

$$\mu(x; e_1, \ldots, e_m) := \mu_x(e_1, \ldots, e_m)$$

is of the same class as G, giving the orientation. □

Remark. The function $\sqrt{\tilde{g}(\cdot)}$ is called the *density* of the Riemannian measure on (M, G). This terminology will be explained in later sections. The Riemannian volume element μ is often denoted by vol. Denoting $g(t) := \det[g_{ik}(\varphi(t)]$, where $\varphi = u^{-1}$ is a parametrization of M, we have the local representation of the volume element

$$\mathrm{vol}_O = \varphi^*\mu = \mu_O = \sqrt{g(t)}\,dt^1 \wedge \ldots \wedge dt^m .$$

The function $\sqrt{g(\cdot)}$ plays an important role, not only in integration formulas. Its transformation formula is given by

LEMMA XIV.9.7. Let (u_i, M_i), (u_j, M_j) be two oriented charts around $x \in M_i \cap M_j$; then if $t' = (u_i \circ u_j^{-1})(t)$, then

$$\sqrt{g(t')} = \det[d(u_i \circ u_j^{-1})_{(t)}]\sqrt{g(t)} = \left|\frac{\partial(t')}{\partial(t)}\right|\sqrt{g(t)}$$

since $\dfrac{\partial(t')}{\partial(t)} > 0$.

Construction of a Riemannian Metric. We recall the notion of a Riemannian metric on a finite-dimensional manifold.

DEFINITION. A Riemannian metric G on a differentiable manifold M (of finite dimension) is a covariant tensor field $G \in \mathcal{T}_2(M)$ such that, for each $x \in M$, G_x is a bilinear, symmetric, positive definite form on $T_x(M)$.

PROPOSITION XIV.9.8. Let M be a C^∞ differentiable manifold modelled on $E = \mathbf{R}^m$ (or, more generally, a smooth manifold modelled on a Hilbert space $(E, (\cdot\,|\,\cdot))$, admitting smooth partitions of unity. Then M admits a Riemannian metric.

PROOF. Let (u_j, M_j) be an atlas of M and (f_j) a partition of unity subordinate to the covering (M_j). Take on $u_j(M_j) \subset E$ the scalar product $h = (\cdot\,|\,\cdot)$ and pull h back to M_j by the map u_j: $g_i := u_j^* h$. Now define

$$g := \sum f_j \cdot g_j = \sum f_j u_j^* h.$$

Plainly g is a Riemannian metric on M: it is symmetric since each g_i is symmetric, it is positive definite by a similar argument:

$$(7) \qquad g_x(X_x, X_x) = \sum f_j(x) g_{jx}(X_x, X_x) > 0$$

for any tangent vector $0 \neq X_x \in T_x M$. □

Radon Measures Determined by Volume Forms. Integration Over Finite Domains in a Differentiable Manifold. As the next application of the existence of partitions of unity we prove that a volume form ω

on an m-dimensional differential manifold defines a Radon integral (measure) μ_ω on M. We recall that a Radon integral μ on a locally compact space M is a positive linear functional on the space $C^0(M)$ of continuous functions with compact support:

$$\mu(f) \geqslant 0 \quad \text{for } f \geqslant 0,\ f \in C_0(M).$$

THEOREM XIV.9.9. Let M be an oriented m-dimensional differentiable manifold. Let ω be its volume form (of class C^0). Then there exists a unique Radon integral μ_ω on M such that if $(\varkappa, U)$ is a chart of M and if

$$(8) \qquad \omega_O = \varrho_\varkappa dt^1 \wedge \ldots \wedge dt^m$$

is the local representation of ω in this chart ($O = \varkappa(U)$), then

$$(9) \qquad \mu_\omega(f) = \int_O f(\varkappa^{-1}(t))\varrho_\varkappa(t)\, d\lambda^m(t),$$

where λ^m is a Lebesgue integral (on $\boldsymbol{R}^m$).

Since the proof is typical of such situations, we proceed with the following

LEMMA XIV.9.10. Let $(M_j), j \in J$, be an open covering of M. If

1° for each $j \in J$ there is a (positive) linear functional μ_j on $C_0(M_j)$,

2° for $i, j \in J$, $\mu_i = \mu_j$ on $C_0(M_i \cap M_j)$,

then there exists a unique (positive) linear functional μ on $C_0(M)$ such that $\mu \restriction C_0(M_j) = \mu_j$.

Remark. We could say that we construct μ by glueing together its "pieces" μ_j. This is an example of the "principle of glueing together", which works in many domains of mathematics (cf. distribution theory) and which has led to the theory of sheaves (this theory will be developed in Part III; the notion of a sheaf is mentioned in Chapter XIX).

PROOF OF THEOREM XIV.9.9. Plainly (9) defines a positive functional on $C_0(U)$. The change-of-variables formula for a Lebesgue integral shows that if $(\varkappa, U)$ and (φ, V) are two charts and $\operatorname{suppt} f \subset U \cap V$, then the value of (9) is independent of the choice of the chart used for its computation. Now Lemma XIV.9.10 completes the proof. □

We still have to give the

PROOF OF LEMMA XIV.9.10. Let K be the support of $f \in C_0(M)$. Choose a finite subcovering (M_{j_i}), $i = 1, \ldots, k$ of K and a partition of unity (f_i) on $\bigcup_i M_{j_i}$ subordinate to it (this can be always done). Then $f = \sum_i f_i f$ and $\mu_\omega(f) = \sum_i \mu_{j_i}(f_i f)$, which shows the uniqueness of μ_ω. Now if (M_{j_l}), $l = 1, \ldots, n$ is another subcovering of K with a partition of unity (g_l), then

$$\begin{aligned}\sum_i \mu_{j_i}(f_i f) &= \sum_i \mu_{j_i}\Big(\sum_l f_i g_l f\Big) \\ &= \sum_{i,l} \mu_{j_i}(f_i g_l f) = \sum_{i,l} \mu_{j_l}(f_i g_l f) \\ &= \sum_l \mu_{j_l}\Big(\sum_i f_i g_l f\Big) = \sum_l \mu_{j_l}(g_l f).\end{aligned}$$

Thus, $\mu_\omega(f)$ is well defined. ☐

Let η be an m-form on M (of class C^0). Then $\eta = f\omega$, where ω is a volume form for M.

For compact $K \subset M$ define

$$\int_K \eta := \int_K f d\mu_\omega. \tag{10}$$

If η has a compact support, put

$$\int \eta := \int_{\operatorname{suppt} \eta} \eta. \tag{11}$$

One can easily check that (10) and (11) do not depend on the choice of the volume form ω (which is arbitrary up to a positive function).

THEOREM XIV.9.11.

1° $\Lambda^m_{(0)}(M) \ni \eta \to \int_K \eta \in \boldsymbol{R}^1$ is linear.

2° If η is a volume form of M, then

$$\int_K \eta \geqslant 0.$$

3° If we change the orientation of M, then $\int_K \eta$ changes the sign.

4° If φ: $M \to N$ is a diffeomorphism of oriented differentiable manifolds preserving the orientations and if $\eta \in \Lambda^m_{(0)}(N)$ and K is compact in N, then

$$\int_K \eta = \int_{\varphi^{-1}(K)} \varphi^* \eta. \tag{12}$$

PROOF. 1°, 2° and 3° are straightforward. But 4° also follows as

$$\int_K \eta = \int_K f\mu_\omega = \int_{\varphi^{-1}(K)} f \circ \varphi \, \mu_{\varphi^*\omega} = \int_{\varphi^{-1}(K)} \varphi^*\eta$$

if $\eta = f\omega$ and ω is a volume form for N and $\varphi^*\omega$ a volume form for M. □

We can repeat the definitions of a singular p-cube, a regular p-cube, a p-chain, its boundary, and the integral of a p-form over a singular p-cube (p-chain) replacing the underlying Banach space E by a differentiable manifold M. If M is oriented, a definition of a finite domain can also be repeated with obvious changes only. Moreover, as in the proof of Proposition XIV.4.11, it can be shown that if c_M is a chain parametrizing a finite domain $M_0 \subset M$, then

$$\int_{c_M} \eta = \int_{M_0} \eta \quad \text{for } \eta \in \Lambda^m_{(0)}(M).$$

Defining, as in the flat case, $\int_{\partial M_0} \chi := \int_{\partial c_M} \chi$ for $\chi \in \Lambda^{m-1}_{(1)}(M)$, we obtain

THEOREM XIV.9.12 (Poincaré–Stokes Theorem for a Finite Domain). Let M_0 be a finite domain in an m-dimensional oriented differentiable manifold M. Then, for any $(m-1)$-form χ of class C^1 on M,

$$\int_{M_0} d\chi = \int_{\partial M_0} \chi .$$

10. POINCARÉ–STOKES THEOREM FOR A MANIFOLD WITH BOUNDARY

In the preceding section we formulated a Poincaré–Stokes Theorem for a finite domain M_0 in an (oriented) manifold M. The method consisted in subdividing M_0 into a number of cubes, each referred to its own chart. However, the boundary surfaces may cause trouble: it is a difficult task to prove that a differentiable manifold is "triangulable". Therefore, in "modern topology there is a general tendency to replace dissection by coverage" (H. Weyl). We have already practised this by using partitions of unity.

However, for many applications there is a need of the Stokes Theorem for the whole manifold and not just for a finite domain in it (cf. the

Hodge–Kodaira theory, integration in Lie groups, general relativity, hydrodynamics, etc.). The notion of a manifold is still too narrow. Many objects of geometry, such as a segment, a disk, a half-plane, a closed ball, $\boldsymbol{R}^3$ with a ball removed and so on, are not manifolds: they are "*manifolds with boundary*".

Denote by $\boldsymbol{R}^m_-$ the half space $\{x \in \boldsymbol{R}^m : x^1 \leqslant 0\}$ provided with the relative topology of $\boldsymbol{R}^m$. The following definition is quite natural:

Definition. 1° A *topological manifold with boundary of dimension m* is a Hausdorff space M such that there exists an open covering (M_i), $i \in I$, of M and corresponding homeomorphisms $u_i \colon M_i \to U_i \subset \boldsymbol{R}^m_-$ onto open subsets of $\boldsymbol{R}^m_-$ (called *coordinate systems* or *charts* of M).

The family (u_i, M_i) is also called a *topological atlas of M*. An atlas (u_i, M_i), $i \in I$, is *of class* C^k if, for all $(i, k) \in I \times I$,

$$u_k \circ (u_i | M_i \cap M_k)^{-1} \colon u_i(M_i \cap M_k) \to u_k(M_i \cap M_k)$$

is a C^k-diffeomorphism.

Two C^k-atlases are *equivalent* if their union is a C^k-atlas.

2° The *differentiable manifold with boundary* (*of dimension m*) *and of class* C^k is a topological manifold with boundary together with an equivalence class of C^k-atlases (called a *differentiable structure of M*).

3° Suppose (u_i, M_i), $i \in I$, is an atlas for M. With respect to this atlas we define a subset ∂M of M—called the *boundary of M*—as follows: $x \in \partial M$ if there is a chart (u_i, M_i) with $x \in M_i$ and $u_i(x) \in \partial \boldsymbol{R}^m_- := \{x \in \boldsymbol{R}^m : x^1 = 0\}$.

It can be shown that if $u(x) \in \partial \boldsymbol{R}^m_-$ in one chart, then it holds for all charts. The proof is left to the reader (it is easy if (u_i, M_i) is a C^k atlas with $k \geqslant 1$). Thus ∂M is defined intrinsically.

Remark 1. The union of atlases in an equivalence class is the largest atlas in the equivalence class. The maximal atlas is, therefore, sometimes identified with a differentiable structure of M.

Remark 2. If $\partial M = \emptyset$, then M is simply a manifold, sometimes called a *manifold without boundary*. A compact manifold is sometimes called *closed*.

Proposition XIV.10.1. The boundary ∂M of M is (in a natural way) a manifold (without boundary) of the same class and dimension lower by one.

PROOF. Let (u_i, M_i), $i \in I$, be an atlas for M and let $I' \subset I$ be the set of such indices i that $M_i \cap \partial M \neq \emptyset$. Then plainly

$$(u_i | M_i \cap \partial M, M_i \cap \partial M)_{i \in I'}$$

is an atlas for ∂M. □

Differentiable functions, differentiable maps, tangent vectors, tangent maps, and differential forms are now defined in the same way as in the case of manifolds.

The orientation of a differentiable manifold with boundary is also defined in the same way as for manifolds: by a coherently oriented atlas (u_i, M_i), i.e., such that

$$\det[u_k \circ u_i^{-1}]' > 0.$$

This can be proven equivalent to the existence of a nowhere vanishing m-form on M.

In order to integrate over ∂M we must have a volume form on it or, in other words, we must orient this manifold: this is always possible if M is oriented.

THEOREM XIV.10.2. Let M be an oriented manifold with boundary ∂M. Then ∂M is orientable and the orientation of M determines an orientation of ∂M.

PROOF. Let (φ, O) and (ψ, U) be coordinate neighbourhoods of $p \in \partial M$ with local coordinates $(x^1, \ldots, x^m)$ and $(y^1, \ldots, y^m)$, respectively. By the definition of coordinates of boundary points, $x^1 = 0$ or $y^1 = 0$ if the point in O or U is on ∂M and negative otherwise. Writing, for simplicity,

$$y^k = y^k(x^1, \ldots, x^m), \qquad k = 1, \ldots, m$$

for the change of coordinates, we have $0 = y^1(0, x^2, \ldots, x^m)$; therefore,

$$\frac{\partial y^1}{\partial x^2}(\varphi(q)) = 0 = \frac{\partial y^1}{\partial x^m}(\varphi(q)) \quad \text{for each } q \in O \cap \partial M.$$

Hence the derivative of $\psi \circ \varphi^{-1}$ has the form

$$(1) \qquad (\psi \circ \varphi^{-1})'(\varphi(q)) = \begin{bmatrix} \partial_1 y^1 & \partial_1 y^2 & \ldots & \partial_1 y^m \\ 0 & \partial_2 y^2 & \ldots & \partial_2 y^m \\ \vdots & \vdots & & \vdots \\ 0 & \partial_m y^2 & \ldots & \partial_m y^m \end{bmatrix}_{\varphi(q)} .$$

Since the determinant of the left-hand side is positive, $\partial y^1/\partial x^1 \neq 0$ at $\varphi(q)$. We have to verify that it is even positive. Indeed, if $\varphi(q) = (0, c^2, \ldots, c^m)$ and $h(t) := y^1(t, c^2, \ldots, c^m)$, then

$$\frac{dh}{dt}(0) = \frac{\partial y^1}{\partial x^1}(\varphi(q)) \geqslant 0$$

since $h(0) = 0$ and $h(t) < 0$ for $-\varepsilon < t < 0$. Therefore, $\partial_1 y^1(\varphi(q)) > 0$ and the minor determinant of (1) obtained by striking out the first row and column has the same sign: it is positive. We have thus proved that the atlas for ∂M obtained from an oriented atlas for M is oriented. □

In the sequel *we shall only use second countable manifolds with boundary*. For such manifolds with boundary we can use partitions of unity, just as we did in the case without boundary. Now, provided that our m-dimensional manifold with boundary is oriented, we can imitate the definition of the integral of a continuous m-form over compacts in M or, if the m-form has a compact support, over M. Theorem XIV.9.9 also holds in this case. We can now prove the fundamental theorem of this chapter.

THEOREM XIV.10.3 (Poincaré–Stokes Theorem for Manifolds with Boundary). Let M be a C^1-oriented m-dimensional manifold with the boundary ∂M and let ∂M have the induced orientation. Let ω be an $(m-1)$-form on M of class C^1 and assume that ω has a compact support. Then

$$\int_M d\omega = \int_{\partial M} \omega .$$

If M is a compact manifold without boundary (a so-called closed manifold) then

$$\int_M d\omega = 0.$$

PROOF. Let $(f_i)_{i\in I}$ be a partition of unity of class C^2 subordinate to a coordinate covering. Then

$$\omega = \sum_{i\in I} f_i\omega .$$

Since the support of ω is compact, this sum has only a finite number of non-zero terms. By the additivity of the integral and exterior derivative we have

$$\int_M d\omega = \sum_{i\in I} \int_{M_i} d(f_i\omega).$$

But suppt $f_i \subset M_i \subset M$ (M_i—open). If we suppose that we have already proved

$$(2) \qquad \int_{M_i} d(f_i\omega) = \int_{M_i \cap \partial M} f_i\omega,$$

we can write

$$\int_{M_i \cap \partial M} f_i\omega = \int_{\partial M} f_i\omega, \qquad \int_{M_i} d(f_i\omega) = \int_M d(f_i\omega).$$

By the additivity of the integral we obtain

$$\int_M d\omega = \sum_{i\in I} \int_M (f_i\omega) = \sum_{i\in I} \int_{\partial M} f_i\omega = \int_{\partial M} \omega.$$

Therefore, the Poincaré–Stokes Theorem will be proved if we prove (2). But for the purpose we can use local coordinates given by charts (u_i, M_i) of an oriented atlas. Let $O_i := u_i(U_i) \subset \boldsymbol{R}^m_-$ and let η_{O_i} be the local representation of $f_i\omega$:

$$\eta_{O_i} = \sum_{i=1}^{m} (-1)^{i+1} h_i\, dt^1 \wedge \ldots \wedge \widehat{dt^i} \wedge \ldots \wedge dt^m.$$

Then

$$\int_{M_i} d(f_i\omega) = \int_{O_i} \sum_i \frac{\partial h_i}{\partial t^i} dt.$$

h_i have compact supports in O_i, which is an open subset of $\boldsymbol{R}^m_-$. By integration by parts

$$\int_{M_i} d(f_i\omega)$$

$$= \begin{cases} \displaystyle\int_{O_i \cap \partial \boldsymbol{R}^m_-} h_1(0, t^2, \ldots, t^m)\, dt^2 \ldots dt^m & \text{if } O_i \cap \partial \boldsymbol{R}^m_- \neq \emptyset. \\ 0 & \text{if } O_i \cap \partial \boldsymbol{R}^m_- = \emptyset. \end{cases}$$

Hence

$$\int_{M_i} d(f_i\omega) = \begin{cases} \displaystyle\int_{\partial M_i} f_i\omega & \text{if } \partial M \cap M_i \neq \emptyset, \\ 0 & \text{if } \partial M \cap M_i = \emptyset, \end{cases}$$

and (2) follows. □

11. TENSOR DENSITIES. WEYL DUALITY. HOMOLOGY

Let E be an m-dimensional vector space over $\boldsymbol{R}$. We shall identify $(E^*)^*$ with E.

DEFINITION. A tensor of type $\binom{p}{q}$ is an element of E^p_q.

$$E^p_q := L_{p+q}(\underbrace{E, \ldots, E}_{q \text{ times}}, \overbrace{E^*, \ldots, E^*}^{p \text{ times}}; \boldsymbol{R}).$$

This linear space can be identified with the following tensor product: $(\overset{q}{\otimes} E^*) \otimes (\overset{p}{\otimes} E)$. Elements of E^p_q are also called *p-times contravariant* and *q times covariant tensors.*

Example 1. $\Lambda^p E := \Lambda_p(E^*)$ is the space of alternating (or skew-symmetric) contravariant tensors. Such tensors are called *p-vectors.* In this way (cf. Section 1) we obtain the notion of exterior product of a p-vector t and an r-vector s. It is a $(p+r)$-vector

$$t \wedge s \in \Lambda^{p+r} E.$$

All notions and theorems for Grassmann algebra are therefore valid for skew-symmetric contravariant tensors (called also *multivectors*). We have for instance

PROPOSITION XIV.11.1. 1° $\dim \Lambda^p E = \binom{m}{p}$ if $m = \dim E$. Therefore, $\Lambda^m_0 E \cong \boldsymbol{R}$.

2° Let $(e_1, \ldots, e_m)$ be a basis of E and let

$$f_j = \sum_{n=1}^{m} f_j^n e_n, \quad j = 1, 2, \ldots, m.$$

Then

$$f_1 \wedge \ldots \wedge f_m = \det[f_j^k] e_1 \wedge \ldots \wedge e_m, \tag{1}$$

$$t \wedge s = (-1)^{pq} s \wedge t \quad \text{for } t \in \Lambda^p E,\ s \in \Lambda^q E. \tag{2}$$

Multiplication of Tensors.

DEFINITION. Let $v \in E^p_q$ and $w \in E^{p'}_{q'}$; then we can define a tensor $v \otimes w \in E^{p+p}_{q+q}$ of type $\binom{p+p'}{q+q'}$ by

$$\begin{aligned} v \otimes w(e_1, \ldots, e_q, e_{q+1}, \ldots, e_{q+q'}; f^1, \ldots, f^p, f^{p+1}, \ldots, f^{p+p'}) \\ := v(e_1, \ldots, e_q; f^1, \ldots, f^p) w(e_{q+1}, \ldots, e_{q+q'}; f^{p+1}, \ldots, f^{p+p'}). \end{aligned} \tag{3}$$

The tensor $v \otimes w$ is called the *tensor product* of v and w. If $v_1, \ldots, v_p \in E$, $v'_1, \ldots, v'_q \in E^*$, then a tensor of type $\binom{p}{q}$ $v_1 \otimes \ldots \otimes v_p \otimes v'_1 \otimes \ldots \otimes v'_q$ is called a *decomposable* or a *simple tensor*. Each tensor of type $\binom{p}{q}$ is a linear combination of decomposable tensors.

Contraction. Let E^p_q be a tensor space over E with $p > 0$ and $q > 0$. Then for each $1 \leqslant i \leqslant p$ and $0 \leqslant j \leqslant q$ there is a linear map $c^i_j\colon E^p_q \to E^{p-1}_{q-1}$ such that to every decomposable tensor $z = v_1, \ldots, v_p, v'_1, \ldots, v'_q$ there corresponds the tensor

$$c^i_j(z) := \langle v_i, v'_j \rangle\, v_1 \ldots \overbrace{v_{i-1} v_i}^{i} \ldots v_p \cdot v'_1 \ldots \overbrace{v'_{j-1} v_{j+1}}^{j} \ldots v_q . \tag{4}$$

Transvection is a superposition of some multiplications and contractions.

Let $(e_1, \ldots, e_m)$ be a basis of E and let $(f^1, \ldots, f^m)$ be the dual basis, i.e., $\langle e_i, f^k \rangle = \delta^k_i$ for $i, k = 1, \ldots, m$. Let $\xi \in E^p_q$. We denote by $\xi^{i_1, \ldots, i_p}_{j_1, \ldots, j_q}$

$$\xi(e_{j_1}, \ldots, e_{j_q}, f^{i_1}, \ldots, f^{i_p}) =: \xi^{i_1, \ldots, i_p}_{j_1, \ldots, j_q} \tag{5}$$

where $j_1, \ldots, j_q, i_1, \ldots, i_p \in \{1, \ldots, m\}$ and call them *components of* ξ *in the basis* $e_1, \ldots, e_m$.

Evidently they determine ξ uniquely.

Plainly we have the following

PROPOSITION XIV.11.2. Let $\xi \in E^p_q$, $\eta \in E^{p'}_{q'}$, then:

$$(\xi \otimes \eta)^{i_1, \ldots, i_p, k_1, \ldots, k_{p'}}_{j_1, \ldots, j_q, l_1, \ldots, l_{q'}} = \xi^{i_1, \ldots, i_p}_{j_1, \ldots, j_q} \cdot \eta^{k_1, \ldots, k_{p'}}_{l_1, \ldots, l_{q'}}, \tag{6}$$

$$c^r_j(\xi)^{i_1 \ldots \overset{i}{\hat{}} \ldots i_p}_{j_1 \ldots \underset{j}{\hat{}} \ldots j_q} = \sum_{\alpha=1}^{m} \xi^{i_1 \ldots i_{r-1} \alpha i_{r+1} \ldots i_p}_{j_1 \ldots j_{s-1} \alpha j_{s+1} \ldots j_q}.$$

There is a procedure which we tacitly used in defining exterior multiplication; it is

Alternation. If we have a function of several indices, $f_{\alpha_1 \ldots \alpha_n}$ form the sum

$$\frac{1}{k!} \sum_{\sigma \in S_k} \operatorname{sgn} \sigma f_{\alpha_{\sigma(1)} \ldots \alpha_{\sigma(k)} \alpha_{k+1} \ldots \alpha_n} =: f_{[\alpha_1 \ldots \alpha_k] \alpha_{k+1} \ldots \alpha_n}, \tag{7}$$

the brackets [ ] denote those indices which are antisymmetrized (alternated).

Examples 1. $a^{[i}b^{k]} = \frac{1}{2}(a^i b^k - b^k a^i)$.

2. The exterior derivative of a p-form $\omega = \frac{1}{p!}\omega_{i_1 \dots i_p} dx^{i_1} \wedge \dots \wedge dx^{i_p}$ is

$$(d\omega)_{\mu i_1 \dots i_p} = (p+1)\partial_{[\mu}\omega_{i_1 \dots i_p]}.$$

3. The exterior product of linear forms is

$$f^1 \wedge \dots \wedge f^p(e_1, \dots, e_p) := p!f^1(e_{[1})f^2(e_2)\dots f^p(e_{p]}).$$

Symmetrization (mixing) is performed in a similar manner but the term $\operatorname{sgn}\sigma$ is omitted.

We denote symmetrization by taking the symmetrized indices in brackets $\{\ \}$.

It is easy to see that if we perform alternation or mixing on groups of convariant or (separately) contravariant indices of components of a tensor ξ in a basis $(e_1, \dots, e_m)$, then the resulting data are coordinates of a tensor which does not depend on the actual choice of the basis (see (7) below). Thus one can define operations of alternation or mixing of tensors in E^p_q.

Einstein's Convention. Since in tensor calculus one has to perform many summations, Albert Einstein introduced the following summation convention:

"If an index appears twice in the same term, once as a subscript and once as a superscript, the sign Σ will be omitted".

Example. Change of a basis. Let $(e_1, \dots, e_m)$ be a basis in E and let $(e'_1, \dots, e'_m)$ be a new basis. Instead of writing

$$(8) \qquad e'_k = \sum_{l=1}^{m} a^l_k e_l,$$

we write $e'_k = a^l_k e_l$. Let $(f^1, \dots, f^m)$ be the dual basis of E^* and $(f'^1, \dots, f'^m)$ the dual basis of $(e'_1, \dots, e'_m)$. Then we obtain the new basis $(f'^1, \dots, f'^m)$ from the old one $(f^1, \dots, f^m)$ by the $m \times m$-matrix $[b^k_l]$

$$(9) \qquad f'^k = b^k{}_l f^l,$$

where $B = [b^k{}_l] = {}^t[a_n{}^m]^{-1}$ is the matrix contragradient to $A = [a_n{}^m]$.

We have the following

PROPOSITION XIV.11.3. Let $(\xi^{i_1 \dots i_p}_{j_1 \dots j_q})$ be components of a tensor ξ in the basis $(e_1, \dots, e_m)$. Then if the new bases $(e'_1, \dots, e'_m)$ and $(f'^1, \dots, f'^m)$

are obtained by (8) and (9), then the components of ξ in both bases are related by the following formula:

$$(10)\qquad \xi^{i_1 \dots i_p}_{j_1 \dots j_q} = \xi'^{k_1, \dots, k_p}_{l_1, \dots, l_q} a_{k_1}{}^{i_1} \dots a_{k_p}{}^{i_p} b^{l_1}{}_{j_1} \dots b^{l_q}{}_{j_q}$$

(Einstein convention!).

The easy proof is left to the reader.

Raising and Lowering of Indices. Let g be a scalar product in E. Therefore $g \in L_2(E, \boldsymbol{R})$. Let

$$g_{ij} = g(e_i, e_j), \qquad i, j = 1, \dots, m,$$

be components of the metric tensor g in the basis $(e_1, \dots, e_m)$. Plainly the $m \times m$-matrix $[g_{ij}]$ is invertible and symmetric, $g_{ij} = g_{ji}$, for all $i, j = 1, 2, \dots, m$. Denote by $[g^{kl}]$ the inverse (in this case contragradient) matrix to $[g_{ij}]$. Therefore, $g^{kl} = \check{g}(e^k, e^l)$, where $\check{g}$ is the induced scalar product on E^*. Plainly

$$g_{ij} g^{jk} = \delta_i{}^k.$$

Now if we have a tensor ξ, say of type $\binom{0}{3}$, we can obtain by transvection tensors, say of type $\binom{2}{1}$, $\xi^{ij}{}_k := g^{il} g^{jn} \xi_{lnk}$.

Therefore, in an Euclidean space we have a procedure for passing from covariant to contravariant tensors and vice versa. This was the main reason why the difference between covectors and contravariant vectors was discovered so late.

We will now introduce some important quantities, which are very useful for definitions of new objects and interesting dualities.

Generalized Kronecker Deltas. Permutation Symbols.

DEFINITION. We introduce the *generalized Kronecker deltas* by

$$(11)\qquad \delta^{k_1 \dots k_n}_{l_1 \dots l_n} = \begin{cases} +1 & \text{if } k_1, \dots, k_m \text{ are distinct integers} \in \{1, \dots, m\} \text{ and if } l_1, \dots, l_n \text{ is an even permutation of } k_1, \dots, k_n, \\ 0 & \text{if the } \{k_1, \dots, k_n\} \neq \{l_1, \dots, l_n\} \text{ or if any two } k_1, \dots, k_n \text{ are equal or any two } l_1, \dots, l_n \text{ are equal}, \\ -1 & \text{if } k_1, \dots, k_n \text{ are distinct integers} \in \{1, \dots, m\} \text{ and } l_1, \dots, l_n \text{ is an odd permutation of } k_1, \dots, k_n. \end{cases}$$

Permutation symbols (*densities of Levi–Civita*) are now defined in the following way:

$$\varepsilon_m^{k_1,\dots,k_m} := \delta_{1,\dots,m}^{k_1,\dots,k_m}, \quad \varepsilon_{i_1,\dots,i_m}^m := \delta_{i_1,\dots,i_m}^{1,\dots,m}.$$

LEMMA XIV.11.4. We have the following important formulas:

(12) $$\delta_{l_1,\dots,l_n}^{k_1,\dots,k_n} S^{l_1,\dots,l_n} = n!\, S^{k_1,\dots,k_n}$$

for any skew-symmetric (coordinate components of a) tensor of type $\binom{n}{0}$,

(13) $$\delta_{l_1,\dots,l_n}^{k_1,\dots,k_n} = \operatorname{sgn}\binom{k_1,\dots,k_n}{l_1,\dots,l_n}$$

if $\{k_1, \dots, k_n\} = \{l_1, \dots, l_n\}$ is a set of n distinct elements,

(14) $$\varepsilon_m^{\alpha_1,\dots,\alpha_m} \varepsilon_{\alpha_1,\dots,\alpha_m}^m = m!,$$

(15) If $n > m$ then $\delta_{l_1,\dots,l_n}^{k_1,\dots,k_n} = 0$,

(16) $$\varepsilon_m^{k_1,\dots,k_m} \varepsilon_{l_1,\dots,l_m}^m = \delta_{l_1,\dots,l_m}^{k_1,\dots,k_m},$$

(17) $$\varepsilon_m^{k_1,\dots,k_n r_1,\dots,r_{m-n}} \varepsilon_{s_1,\dots,s_n r_1,\dots,r_{m-n}}^m = (m-n)!\, \delta_{s_1,\dots,s_n}^{k_1,\dots,k_n} \text{ for } m > n.$$

The proof of these simple formulas (12)–(17) is left to the reader.

Exercise. Prove that exterior product of $\varphi \in \Lambda^p(E)$ and $\psi \in \Lambda^q(E)$ can be defined by $(\varphi \wedge \psi)_{i_1,\dots,i_p j_1,\dots,j_q} = \dfrac{1}{p!}\dfrac{1}{q!} \delta_{i_1,\dots,i_p, j_1,\dots,j_q}^{\alpha_1,\dots,\alpha_p, \beta_1,\dots,\beta_q} \varphi_{\alpha_1,\dots,\alpha_p} \psi_{\beta_1,\dots,\beta_q}$.

Tensor Fields. Let M be a finite-dimensional C^r-manifold modelled on E. Since each tangent space $T_x M$ is isomorphic to E, we can form the space $(T_x M)_q^p$ of tensors at $x \in M$ p times contravariant and q times covariant. In the same way as we formed the tangent bundle TM of M we can define tensor bundles.

DEFINITION. A *tensor bundle* of type $\binom{p}{q}$ of the differentiable manifold M is the triple

$$(TM_q^p, \pi, M),$$

where

$$(TM)_q^p = \overset{p}{\otimes} TM \otimes \overset{q}{\otimes} T^*M := \bigcup_{x \in M} (T_x M)_q^p.$$

Projection π: $(TM)_q^p \to M$, is defined by $\pi \restriction (T_x M)_q^p := x$. A tensor field S of type $\binom{p}{q}$ on $O \subset M$ is a mapping S: $O \to (TM)_q^p$, such that $S(x) \in (T_x M)_q^p$, for $x \in O$. In another wording S is a cross-section in the tensor bundle.

Remark. A C^l atlas (u_i, M_i), $i \in I$, of M induces in a natural way an atlas of the bundle space $(TM)^p_q$. The reader may construct it in a similar way to that followed for the tangent bundle TM and bundles T^*M, $TM \oplus \ldots \oplus TM$ in Section XIV.7. Since M and $(TM)^p_q$ are now differentiable manifolds, it makes sense to speak of C^k-tensor fields, $0 \leqslant k < l$; especially if M is a smooth (i.e., C^∞) manifold, the notion of a smooth tensor field is well defined.

Each tensor field S of type $\binom{p}{q}$ on $U \subset M$ can be represented by a mapping $S_O\colon O \to \boldsymbol{R}^{m^{p+q}}$, provided U is the domain of a chart (u, U) and $\boldsymbol{R}^m \supset O = u(U)$. Namely $S_O(t)$ is given by the components $S(t)^{i_1 \ldots i_p}_{j_1 \ldots j_q}$ of $S(u^{-1}(t))$ in the basis of $T_{u^{-1}(t)}M$ obtained by the application of $(u^{-1})'(t)$ to the vectors of the canonical basis of $\boldsymbol{R}^m$. Then S is of class C^k if S_O is of class C^k for (u, U) from an atlas of M.

Tensor Densities. Weyl Densities. We have noticed that a tensor density S of type $\binom{p}{q}$ on an m-dimensional differentiable manifold M of class C^r defines, given an atlas (u_i, M_i) of M, a system of mappings $S_{O_i}\colon O_i \to \boldsymbol{R}^{m^{p+q}}$, where $\boldsymbol{R}^m \supset O_i = u_i(M_i)$. Those mappings are related to each other by the following formula which can be easily obtained from the definition of S_{O_i} and (10):

$$(18)\qquad S_{O_i}(t)^{i_1,\ldots,i_p}_{j_1,\ldots,j_q} = S_{O_j}(t')^{k_1,\ldots,k_p}_{l_1,\ldots,l_q} \frac{\partial t^{i_1}}{\partial t'^{k_1}}(t') \ldots \ldots \frac{\partial t^{i_p}}{\partial t'^{k_p}}(t') \frac{\partial t'^{l_1}}{\partial t^{j_1}}(t) \ldots \frac{\partial t'^{l_q}}{\partial t^{j_q}}(t),$$

where $t \in u_i(M_i \cap M_j)$, $t' \in u_j(M_i \cap M_j)$ and $t' = u_j \circ u_i^{-1}(t)$. In fact, it is not hard to see that a system of maps (S_{O_i}) uniquely defines the tensor field S. The field S is of class C^k, $k < r$, iff maps S_{O_i} are of class C^k.

This suggests a possible generalization of the notion of a tensor field.

DEFINITION. Fix $w \in \boldsymbol{R}^1$. Given an atlas $(u_i, M_i)_{i \in I}$ for M with $O_i := u_i(M_i) \subset \boldsymbol{R}^m$, consider a family $(S_{O_i})_{i \in I}$ of mappings $S_{O_i}\colon O_i \to \boldsymbol{R}^{m^{p+q}}$ such that

$$(19)\qquad S_{O_i}(t)^{i_1,\ldots,i_p}_{j_1,\ldots,j_q} = \left|\frac{\partial(t'^1, \ldots, t'^m)}{\partial(t^1, \ldots, t^m)}(t)\right|^w S_{O_j}(t')^{k_1,\ldots,k_p}_{l_1,\ldots,l_q} \frac{\partial t^{i_1}}{\partial t'^{k_1}}(t') \ldots \ldots \frac{\partial t^{i_p}}{\partial t'^{k_p}}(t') \frac{\partial t'^{l_1}}{\partial t^{j_1}}(t) \ldots \frac{\partial t'^{l_q}}{\partial t^{j_q}}(t),$$

for each $(i, j) \in I \times I$ such that $M_i \cap M_j \neq \emptyset$, $t \in u_i(M_i \cap M_j)$, $t' \in u_j(M_i \cap \cap M_j)$, $t' = u_j \circ u_i^{-1}(t)$.

Two such families, corresponding to different atlases, are considered equivalent if they have a common extension to a family corresponding to a larger atlas comprising both the initial ones.

An equivalence class of such families is called a *tensor density of type* $\binom{p}{q}$ *and weight* w on M. It is said to be of class C^k (k smaller than the class of M) if the functions S_{O_i} defining it are of class C^k. We called S_{O_i} the components of the tensor density in the coordinate system (u_i, M_i).

Example. 1° With a slight abuse of language we can call tensor densities of type $\binom{p}{q}$ and weight 0 tensor fields of type $\binom{p}{q}$.

2° *Weyl p-densities* are by definition alternating tensor densities of type $\binom{p}{0}$ and weight $w = 1$. Thus they are given by families (S_{U_i}) S_{U_i}: $O_i \to R^{m^p}$ such that

$$(20) \qquad S_{O_i}(t)^{i_1 \ldots i_n \ldots i_m \ldots i_p} = -S_{O_i}(t)^{i_1 \ldots i_m \ldots i_n \ldots i_p}$$

and that (19) holds with $w = 1$.

Since the notion of a Weyl density is of fundamental importance, we mention here the history of its discovery. In 1919 Hermann Weyl discovered objects dual to differential forms which are necessary to describe in a precise way many physical quantities. They were introduced in his classic *Raum Zeit Materie*. Weyl writes: "...it seems that we have grasped rigorously the difference between *quantity* and *intensity*... tensor densities are the magnitudes of quantity."

We shall denote the space of Weyl p-densities of class k by $W_{p,(k)}(M)$ if $k = \infty$ by $W_p(M)$.

Operations on tensor densities. Tensor densities of type $\binom{p}{q}$ and weight w form a vector space. They can be contracted in the same way as tensor fields: a contraction C_j^i of such a density yields a tensor density of type $\binom{p-1}{q-1}$ and the same weight w.

Also multiplication of tensor densities is performed in the same way as for tensor fields. We immediately have the following:

PROPOSITION XIV.11.5. 1° Let S be a tensor density of type $\binom{p}{q}$ and weight w, and let T be a tensor denstiy of type $\binom{p'}{q'}$ and weight w', then the product $S \otimes T$ is a tensor density of type $\binom{p+p'}{q+q'}$ and weight $w+w'$. $S \otimes T$ is of course defined by

$$(21) \qquad ((S\otimes T)_{Oi})^{j_1,\ldots,j_{p+p'}}_{i_1,\ldots,i_{q+q'}} := (S_{Oi})^{j_1,\ldots,j_p}_{i_1,\ldots,i_q}(T_{Oi})^{j_{p+1}\ldots j_{p+p'}}_{i_{q+1}\ldots i_{q+q'}}.$$

2° Let G be a Riemannian metric on M. Then G defines its local components G_{O_ι}. Let $g_{O_\iota} := |\det[(G_{O_\iota})_{jk}]|$. Then the family of functions $(\sqrt{g_{O_\iota}})_{\iota \in I}$ defines a scalar Weyl density $\left(\text{i.e., of type } \binom{0}{0}\right)$. We shall denote it by $\sqrt{g}$. If R is any tensor field of type $\binom{p}{q}$ on M, then $\sqrt{g}R$ is a tensor density of type $\binom{p}{q}$ and weight $w = 1$.

Proof. 1° is obvious. 2° Since g_{O_ι} define a scalar density of weight $w = 2$, $\sqrt{g}$ is scalar density of weight 1. □

Remark. Item 2° of the preceding proposition is the main source of (Weyl) densities in Riemannian geometry and general relativity. The simplest tensor densities are

Numerical Tensors. Kronecker deltas and permutation symbols can be considered as tensor densities in the following way:

Define a tensor Δ^n of type $\binom{n}{n}$, where $n \leqslant m = \dim M$, by taking its components in the coordinate system $(t^1, \ldots, t^n)$ to be

$$(22) \qquad \Delta^{n k_1\ldots k_n}_{O l_1\ldots l_n} \equiv \delta^{k_1\ldots k_n}_{l_1\ldots l_n}.$$

Its components $\Delta^{k_1\ldots k_n}_{O' l_1\ldots l_n}$ in another coordinate system $(t'^1, \ldots, t'^m)$ are

$$\Delta^{n' k_1,\ldots,k_n}_{O' l_1,\ldots,l_n} \equiv \delta^{i_1,\ldots,i_n}_{j_1,\ldots,j_n} \frac{\partial t^{j_1}}{\partial t'_{l_1}} \cdots \frac{\partial t^{j_n}}{\partial t'_{l_n}} \frac{\partial t'^{k_1}}{\partial t^{i_1}} \cdots \frac{\partial t'^{k_n}}{\partial t^{i_n}}$$

$$= \det \begin{bmatrix} \dfrac{\partial t'^{k_1}}{\partial t'^{l_1}} & \cdots & \dfrac{\partial t'^{k_1}}{\partial t'^{l_n}} \\ \dfrac{\partial t'^{k_n}}{\partial t'^{l_1}} & \cdots & \dfrac{\partial t'^{k_n}}{\partial t'^{l_n}} \end{bmatrix} = \delta^{k_1,\ldots,k_n}_{l_1,\ldots,l_n}.$$

Therefore, there is a unique tensor Δ^n whose components in any coordinate system satisfy the equations

$$\Delta^{n\, k_1\ldots k_n}_{O l_1\ldots l_n} = \delta^{k_1\ldots k_n}_{l_1\ldots l_n}.$$

Now suppose that M is oriented. Then, as can easily be seen, if we take an atlas $(u_i, M_i)_{i\in I}$ defining the orientation and put $(O_i = u_i(M_i))$

$$(23) \qquad (\varepsilon_m)_{O_i}^{i_1\ldots i_m} \equiv \delta_{1\ldots m}^{i_1\ldots i_m} = \varepsilon_m^{i_1\ldots i_m}, \qquad (\varepsilon^m)_{O_i}^{i_1\ldots i_m} \equiv \delta_{i_1\ldots i_m}^{1\ldots m} = \varepsilon_{i_1\ldots i_m}^m$$

then the family $((\varepsilon_m)_{O_i})_{i\in I}$ defines a Weyl m-density and the family $((\varepsilon^m)_{O_i})_{i\in I}$ a tensor density of type $\binom{0}{m}$ and weight -1.

Thus we have defined a numerical tensor field of type $\binom{n}{n}$ Δ^n and, for an oriented manifold, a numerical Weyl m-density ε_m and a numerical tensor density of type $\binom{0}{m}$ and weight -1 ε^m. The importance of Weyl densities stems from the fact that Weyl p-densities are in natural duality with p-forms with compact support. Suppose first that we are given a scalar Weyl density $\bar{R}$ which is positive (i.e., is represented by a positive function in any coordinate system). Such densities always exist since we can set $\bar{R} = \sqrt{g}$ for some Riemannian structure on M. A Radon measure $\mu_{\bar{R}}$ on M can be associated with $\bar{R}$ in a natural way. As shown by Lemma XIV.9.10 it is sufficient to define a compatible system $(\mu_{\bar{R}}^i)_{i\in I}$ of measures on open sets of a covering $(M_i)_{i\in I}$ of M. Take an atlas (u_i, M_i) for M and, for $f \in C_0(M_i)$ and $O_i = u_i(M_i)$, and set

$$\int f d\mu_{\bar{R}}^i := \int_{O_i} (f \circ u_i^{-1})(t)\bar{R}_{O_i}(t)\,d\lambda^m(t).$$

Then, if $f \in C_0(M_i \cap M_j)$, we have

$$\begin{aligned}\int f d\mu_{\bar{R}}^i &= \int_{O_i} (f \circ u_i^{-1})(t)\bar{R}_{O_i}(t)\,d\lambda^m(t)\\ &= \int_{O_i} (f \circ u_i^{-1})(t)\bar{R}_{O_j}(t'(t))\left|\frac{\partial(t')}{\partial(t)}(t)\right| d\lambda^m(t)\\ &= \int_{O_j} (f \circ u_j^{-1})(t')\bar{R}_{O_j}(t')\,d\lambda^m(t') = \int f d\mu_{\bar{R}}^j,\end{aligned}$$

where we have used (19) for scalar densities of weight 1 and the formula for the change of variables in the Lebesgue integral. Let us denote

$$\int_M f\bar{R} := \int_M f d\mu_{\bar{R}} \qquad \text{for } f \in C_0(M).$$

If R is any scalar Weyl density on M, then $R = r\bar{R}$ where r is a function on M and $\bar{R}$ is a positive scalar Weyl density. Put

$$\int_M fR = \int_M fr\bar{R} \quad \text{for } f \in C_0(M).$$

It is easy to see that the left-hand side does not depend on the choice of $\bar{R}$.

Now we are ready to define a (Weyl) pairing $\langle \cdot | \cdot \rangle_p$ between $\Lambda_O^p(M)$ and $W_p(M)$:

$$\text{(24)} \qquad \langle \omega | \iota \rangle_p := \int_M \omega \rfloor \iota$$

where $\omega \rfloor \iota$ is a transvection of ω and ι, given in coordinates by

$$\text{(25)} \qquad (\omega \rfloor \iota)_O = \frac{1}{p!} \omega_{O i_1 \ldots i_p} \iota_O^{i_1 \ldots i_p}$$

($\omega \rfloor \iota$ is a scalar Weyl density on M).

Define also a mapping

$$\overset{p}{\delta} = \delta \colon W_p(M) \to W_{p-1}(M)$$

on coordinate components:

$$\text{(26)} \qquad (\delta \iota)_O^{i_1 \ldots i_{p-1}} := -\partial_i \iota_O^{i i_1 \ldots i_{p-1}}.$$

The reader will check the correctness of this definition, δ is called a *Weyl divergence operator* (*coderivative*). It easily follows from the definition that $\delta\delta = 0$.

LEMMA XIV.11.6. δ is the formal adjoint of the exterior derivative in the following sense:

$$\text{(27)} \qquad \langle d\omega | \iota \rangle_p = \langle \omega | \delta \iota \rangle_{p-1}$$

for $\omega \in \Lambda_O^{p-1}(M)$, $\iota \in W_p(M)$.

PROOF. By a partition of unity argument it is sufficient to assume that ω has a support in a chart domain. Thus we can use local coordinates:

$$\langle d\omega | \iota \rangle_p = \frac{1}{p!} \int (d\omega)_{O i_1 \ldots i_p} \iota_O^{i_1 \ldots i_p} d\lambda^m$$

$$= \frac{1}{p!} p \int (\partial_{i_1} \omega_{O i_2 \ldots i_p}) \iota_O^{i_1 \ldots i_p} d\lambda^m$$

$$= -\frac{1}{(p-1)!} \int \omega_{O i_2 \ldots i_p} \partial_{i_1} \iota_O^{i_1 i_2 \ldots i_p} d\lambda^m = \langle \omega | \delta \iota \rangle_{p-1}. \quad \square$$

We shall examine Weyl pairing and Weyl divergence more closely in the case of an oriented manifold M. For such manifolds there exists an important

Weyl Duality. We have the following

THEOREM XIV.11.7. Let M be a smooth oriented manifold of dimension m.

1° Then, for each $0 \leqslant p \leqslant m$ there exists an isomorphism $\overset{p}{D}$, called a *Weyl duality*

$$\overset{p}{D}\colon \Lambda^p(M) \to W_{m-p}(M),$$

which is defined by

(28) $$\overset{p}{D}\omega := \varepsilon_m \rfloor \omega,$$

where

$$(\varepsilon_m \rfloor \omega)_{o}^{j_{p+1}\dots j_m} := \frac{1}{p!}\,\varepsilon_m^{j_1\dots j_p j_{p+1}\dots j_m}\omega o_{j_1\dots j_p}.$$

The inverse isomorphism $\overset{p}{D}{}^{-1}$

(29) $$\overset{p}{D}{}^{-1}\colon W_{m-p}(M) \to \Lambda^p(M)$$

is given by

(30) $$(\overset{p}{D}{}^{-1}\iota)o_{i_1\dots i_p} := (\varepsilon^m \rfloor \iota)o_{i_1\dots i_p},$$

where

$$(\varepsilon^m \rfloor \iota)o_{i_1\dots i_p} = \frac{1}{(m-p)!}\,\varepsilon^m_{i_1\dots i_p i_{p+1}\dots i_m}\iota_o^{i_{p+1}\dots i_m}.$$

2°

(31) $$\delta \circ \overset{p}{D} = (-1)^{p+1}\overset{p+1}{D} \circ d$$

or, if $q := m-p$,

(32) $$\overset{q}{\delta} = (-1)^{p+1}\overset{p+1}{D} \circ d \circ \overset{p}{D}{}^{-1}.$$

3°

(33) $$\langle \omega | \iota \rangle_p = (-1)^{p(m-p)} \int_M \omega \wedge (\overset{m-p}{D})^{-1}\iota.$$

PROOF. 1° follows immediately from (28), (30) and (12) and (17).

2° Let $q = m-p$ and $\omega \in \Lambda^p(M)$; then in coordinate components

$$(\delta \overset{p}{D}\omega)_O^{i_2\dots i_q} = -\frac{1}{p!}\partial_\mu(\omega_{O_{\alpha_1\dots\alpha_p}}\varepsilon^{\alpha_1\dots\alpha_p\mu i_2\dots i_q})$$

$$= \frac{(-1)^{p+1}}{(p+1)!}(p+1)\,\partial_{[\mu}\omega_{O_{\alpha_1\dots\alpha_p]}}\varepsilon^{\mu\alpha_1\dots\alpha_p i_2\dots i_q}$$

$$= (-1)^{p+1}(\overset{p+1}{D}(d\omega))_O^{i_2\dots i_q}$$

(32) follows from (31).

3° For w localized:

$$(-1)^{p(m-p)}\int_M \omega\wedge(\overset{m-p}{D})^{-1}\iota$$

$$= (-1)^{p(m-p)}\left(\frac{1}{p!}\right)^2\frac{1}{(m-p)!}\int_M \varepsilon^{i_1\dots i_m}\omega_{O_{i_1\dots i_p}}\times$$

$$\times\,\varepsilon_{i_{p+1}\dots i_m j_1\dots j_p}\iota_O^{j_1\dots j_p}d\lambda^m$$

$$= \frac{1}{(p!)^2}\frac{1}{(m-p)!}\int_M \varepsilon^{i_1\dots i_p i_{p+1}\dots i_m}\times$$

$$\times\,\varepsilon_{j_1\dots j_p i_{p+1}\dots i_m}\omega_{O_{i_1\dots i_p}}\iota_O^{j_1\dots j_p}d\lambda^m$$

$$= \frac{1}{(p!)^2}\int_M \delta^{i_1\dots i_p}_{j_1\dots j_p}\omega_{O_{i_1\dots i_p}}\iota_O^{j_1\dots j_p}d\lambda^m$$

$$= \frac{1}{p!}\int \omega_{O_{i_1\dots i_p}}\iota_O^{i_1\dots i_p}d\lambda^m = \langle\omega|\iota\rangle_p.$$

Homology and Cohomology Spaces of de Rham.

DEFINITION. Let

$$(34)\qquad H_q(M,\boldsymbol{R}) := \frac{\ker\delta^q}{\operatorname{im}\delta^{q+1}} \equiv \frac{\{\iota\in W_q(M):\ \delta\iota = 0\}}{\delta W_{q+1}(M)}.$$

We shall call $H_q(M,\boldsymbol{R})$ the *de Rham q-th homology space*. Let

$$(35)\qquad H^p(M,\boldsymbol{R}) = \frac{\ker\overset{p}{d}}{\operatorname{im} d^{p-1}} \equiv \frac{\{\omega\in\Lambda^p(M):\ d\omega = 0\}}{d\Lambda^{p-1}(M)}.$$

We shall call $H^q(M,\boldsymbol{R})$ the *de Rham p-th cohomology space*. Plainly, $\overset{q}{\delta}\circ\overset{q+1}{\delta} = 0$ implies that $\operatorname{im}\overset{q+1}{\delta}\subset\ker\overset{q}{\delta}$, whence $H_q(M,\boldsymbol{R})$ is well defined. Similarly $H^q(M,\boldsymbol{R})$ are also well defined.

We can now prove the following duality theorem.

THEOREM XIV.11.8 (Duality of Poincaré–Weyl). We have the following linear isomorphism of de Rham spaces:

(36) $$H^p(M, \boldsymbol{R}) \cong H_{m-p}(M, \boldsymbol{R}) \quad \text{for } 0 \leqslant p \leqslant m.$$

PROOF. The Weyl isomorphism $\overset{p}{D}$ projects to the quotient spaces (34), (35) by virtue of (32). □

In the next section we shall show that for compact oriented manifolds $H^p(M, \boldsymbol{R}) \simeq H^{m-p}(M, \boldsymbol{R})$. This isomorphism and duality (36) will imply

$$H_q(M, \boldsymbol{R}) \cong H_{m-p}(M, \boldsymbol{R}) \qquad (\textit{Poincaré duality}).$$

Dual of Poincaré Lemma. In Section 3 we have seen the importance of the vanishing of cohomology spaces $H^p(U, \boldsymbol{R})$. The famous Poincaré Lemma (Corollary XIV.3.4) asserts that $H^p(U, \boldsymbol{R}) = 0$ for $p > 0$ if U is any contractible (e.g., convex) open subset of a Banach space; in other words: if $d\overset{p}{\omega} = 0$, then there exists a p-form η such that $d\eta = \overset{p}{\omega}$. By virtue of the Weyl duality we have a corresponding theorem for p-densities.

COROLLARY XIV.11.9. Let M be a contractible open subset of $\boldsymbol{R}^m$. Then, if $p < m$,

(37) $$(\iota \in W_p(M) \text{ and } \delta\iota = 0)$$
$$\Rightarrow (\text{there exists an } \mathfrak{n} \in W_{p+1}(M) \text{ such that } \delta\mathfrak{n} = \iota).$$

In another wording: if M is contractible then

$$H_p(M; \boldsymbol{R}) = 0 \quad \text{for } p < m.$$

We shall use this fact in electrodynamics, in Section 13.

Remark. Special cases of (37) were already known to Volterra (1889) and Brouwer (1906). The operators d and δ were also investigated by the Polish physicist Jan Weysenhoff in 1937 and he called them *Grossrotation* and *Grossdivergenz*.

Example. We illustrate the Weyl duality by a particular but very important example: a differentiable manifold M of dimension 4.

This is the case of relativistic electrodynamics. It is quite certain that the contemplation of the Maxwell–Minkowski electrodynamics led Hermann Weyl to the discovery of his densities and the Weyl duality.

Let $T \in \Lambda^2(M)$, $\dim M = 4$. Then (working in coordinates) we have

$$(\overset{2}{D}T)^{ik} = \tfrac{1}{2}\varepsilon_4^{ik\alpha\beta}\, T_{\alpha\beta}, \qquad T_{jl} = \tfrac{1}{2}(\overset{2}{D}T)^{\mu\nu}\varepsilon^4_{\mu\nu jl}.$$

More explicitly

$$(\overset{2}{D}T)^{23} = T_{14}, \quad (\overset{2}{D}T)^{31} = T_{24}, \quad (\overset{2}{D}T)^{12} = T_{34},$$

$$(\overset{2}{D}T)^{14} = T_{23}, \quad (\overset{2}{D}T)^{24} = T_{31}, \quad (\overset{2}{D}T)^{34} = T_{12}.$$

Exercise. Let $T \in \Lambda^2(M)$, $\dim M = 4$. Then

$$T_{14}T_{23} + T_{24}T_{32} + T_{34}T_{12} = \tfrac{1}{2}(\overset{2}{D}T)^{\mu\nu}T_{\mu\nu}$$

is a scalar density.

In Section 14 we shall write down the (differential) equations of the relativistic electrodynamics of Lorentz–Minkowski.

12. WEYL DUALITY AND HODGE * OPERATOR. GENERALIZED GREEN'S FORMULAS ON RIEMANNIAN MANIFOLDS

We have seen that on a Riemannian manifold (M, G) there is a natural procedure of passing from covariant tensors to contravariant tensors "by raising indices" and to contravariant tensor densities of weight $+1$ by means of multiplication by the scalar density $\sqrt{g}$. In this section we shall, following Kodaira, use this natural procedure to pass from the Weyl duality to the corresponding isomorphism

$$\Lambda^p(M) \to \Lambda^{m-p}(M),$$

which is the Hodge isomorphism considered in Section 5 in the special case of a Euclidean domain.

In this section (M, G) will denote an oriented m-dimensional smooth Riemannian manifold. Let μ denote the Riemannian volume element on M (cf. Proposition XIV.9.6).

As we have already mentioned, there is a one-to-one correspondence

$$\text{(1)} \qquad \Lambda^p(M) \ni c \underset{I^p}{\overset{I_p}{\rightleftarrows}} \iota \in W_p(M)$$

given in coordinate components by the equations

$$\text{(2)} \qquad \iota_O^{i_1\ldots i_p} = \sqrt{g_O}\, c_O^{i_1\ldots i_p}, \qquad \text{where } c_O^{i_1\ldots i_p} = g_O^{i_1j_1}\ldots g_O^{i_pj_p}c_{Oj_1\ldots j_p},$$

(3) $$c_O^{j_1\dots j_p} = \frac{1}{\sqrt{g_O}}\,\iota^{j_1\dots j_p} \qquad c_{O_{i_1\dots i_p}} = g_{O_{i_1k_1}}\cdots g_{O_{i_pk_p}} c^{k_1\dots k_p},$$

where $[g_O^{ij}]$ is the matrix inverse to $[g_{O_{kl}}]$.

(Plainly, we also have similar correspondences for objects which are not skew-symmetric.) Therefore, any p-form can be considered as a representative of the corresponding Weyl p-density and vice versa. In the case of oriented Riemannian manifold, therefore, we could call $c_{O_{i_1\dots i_p}}$ covariant components of the p-density ι.

By the superposition $I^{m-p}\circ D^p$ of the Weyl duality D^p and the correspondence (2) we obtain the so-called *Hodge isomorphism*.

DEFINITION. The Hodge star isomorphism $\Lambda^p(M)\to\Lambda^{m-p}(M)$ is defined by:

(4) $$\overset{p}{*} = * := (D^{p-m})^{-1}\circ I_p$$

or explicitly:

(5) $$(*\varphi)_{O_{j_1\dots j_{m-p}}} = \frac{1}{p!}\sqrt{g_O}\,\varepsilon_{j_1\dots j_{m-p}l_1\dots l_p} g_O^{l_1k_1}\dots \varphi_{O_{k_1\dots k_p}}.$$

From (5) follows

LEMMA XIV.12.1. Let $\varphi,\psi\in\Lambda^p(M)$. Then

(6) $$*\varphi\wedge\psi = *\psi\wedge\varphi = (\varphi|\psi)\mu \qquad (= \text{in coordinate components})$$

$$= \frac{1}{p!}\varphi_{O_{i_1\dots i_p}} g_O^{i_1j_1}\dots g_O^{i_pj_p}\psi_{O_{j_1\dots j_p}}\sqrt{g_O}\,\varepsilon_{1\dots m},$$

where $(\varphi|\eta)$ is a function on M given in local coordinates by

$$\frac{1}{p!}\varphi_{O_{i_1\dots i_p}} g_O^{i_1j_1}\dots g_O^{i_pj_p}\psi_{O_{j_1\dots j_p}}.$$

It is now clear what constitutes the connection between the star operation defined in Section 5 for a Euclidean oriented vector space and the star operation just introduced. Each tangent space T_xM to our oriented Riemannian manifold M is, in a natural way, a Euclidean space. Thus, as in Section 5 we can introduce a scalar product $(\cdot|\cdot)_x$ in $A_p(T_xM)$. It is easy to see that for $x\in M$

$$(\varphi|\psi)(x) = (\varphi(x)|\psi(x))_x.$$

Comparing (6) and (5.17) gives

$$(*\varphi)(x) = *_x \varphi(x)$$

where $*_x$ is the Hodge star operation for the Euclidean space $T_x M$.

In this way we have "returned" to the $*$-operator as it was defined in Section 5. We shall now show that the coderivative δ defined in Section 11 is "the same" as the coderivative d^* defined in Section 5 by means of the Hodge operator. We have a general rule: in order to pass from an operation defined on Weyl densities to the corresponding operation on differential forms one has to use the correspondence I_p (2) to perform the operation on the densities and then to "descend" by correspondence I^q to the forms.

PROPOSITION XIV.12.2. $I^{p-1}\,\delta I_p \varphi = (-1)^{p(m-p+1)} *d*\varphi = (-1)^p *d*^{-1}\varphi =: d^*\varphi$ for $\varphi \in \Lambda^p(M)$. In local coordinates

$$(d^*\varphi)_{O_{i_1\ldots i_{p-1}}} = \frac{-1}{\sqrt{g_O}}\, g_{O_{i_1 j_1}} \cdots$$

$$\cdots g_{O_{i_{p-1} j_{p-1}}} \partial_i \left(\sqrt{g_O}\, g_O^{ik} g_O^{j_1 k_1} \ldots g_O^{j_{p-1} k_{p-1}} \varphi_{O_{k k_1 \ldots k_{p-1}}}\right).$$

The proof is left to the reader as an excercise.

Let us define the Laplace–Beltrami–Hodge operator as in the flat case:

$$(7) \qquad -\Delta = -\Delta_p := d^* \circ d + d \circ d^*.$$

Since the formulas of Section 5 used only the Stokes theorem, $*$ operators, and the Riemannian metric G, all the proofs from that Section 5 apply verbatim to general Riemann manifolds. We will put together the most important facts in the following

THEOREM XIV.12.3. $*: \Lambda^p(M) \to \Lambda^{m-p}(M)$ is an isomorphism;

$$(8) \qquad (\overset{p}{*})^{-1} = (-1)^{p(m-p)} * \quad \text{is the inverse of } 1°;$$

$$(9) \qquad **\varphi = (-1)^{p(m-p)}\varphi \quad \text{for } \varphi \in \Lambda^p(M);$$

$$(10) \qquad (*\varphi | *\psi) = (\varphi|\psi) \quad \text{for } \varphi, \psi \in \Lambda^p(M);$$

$$(11) \qquad d^* \circ d^* = 0;$$

$$(12) \qquad d^*(*\omega) = (-1)^{m-p} * d\omega \quad \text{for } \omega \in \Lambda^p(M);$$

$$(13) \qquad d(*\varphi) = (-1)^{m-p+1} * d^*\varphi \quad \text{for } \varphi \in \Lambda^p(M);$$

(14) $\quad *d^*d\varphi = dd^*(*\varphi) \quad$ for $\varphi \in \Lambda^p(M)$;

(15) $\quad d^*d*\varphi = *dd^*\varphi$;

(16) $\quad$ Operator $-\Delta_p$ is symmetric and positive;

(17) $\quad (\Delta_p\varphi|\psi) = (\varphi|\Delta\varphi) \quad$ for $\varphi \in \Lambda_0^p(M)$;

(18) $\quad (-\Delta_p\varphi|\varphi) \geqslant 0 \quad$ for $\varphi \in \Lambda_0^p(M)$,

operator Δ commutes with $*$

(19) $\quad *\Delta = \Delta *$.

Also all the generalized Green' formulas (Theorem XIV.5.17) are valid:

THEOREM XIV.12.4 (Generalized Green's Formulas). Let M be a compact oriented Riemannian manifold with boundary ∂M; then for ω, $\varphi \in \Lambda^p(M)$, $\chi \in \Lambda^{p+1}(M)$ we have

$$\int_M (d\omega|\chi)\mu = \int_M (\omega|d^*\chi)\mu + \int_{\partial M} \omega \wedge *^{-1}\chi, \tag{20}$$

$$\int_M (\Delta\omega|\varphi)\mu + \int_M (d\omega|d\varphi)\mu + \int_M (d^*\omega|d^*\varphi)\mu = \int_{\partial M} (\varphi \wedge *^{-1}d\omega - d^*\omega \wedge *^{-1}\varphi), \tag{21}$$

$$\int_M [(\Delta\omega|\varphi) - (\omega|\Delta\psi)]\mu = \int_{\partial M} \{\varphi \wedge *^{-1}d\omega - d^*\omega \wedge *^{-1}\varphi - \omega \wedge *^{-1}d\varphi + d^*\varphi \wedge *^{-1}\omega]. \tag{22}$$

Case of Indefinite Metric. In physical applications one encounters also pseudo-Riemannian manifolds (M, G): $M = \boldsymbol{R}^4$ with the Minkowski metric or more generally M being a pseudo-Riemannian manifold with the Einstein metric. All the preceding formulas remain valid with minor modifications. We recall that the signature of a pseudo-Riemannian metric G is defined in the following way: in T_xM we take an orthonormal basis $(e_1, \ldots, e_m)$: $G(e_i, e_k) = \pm\delta_{ik}$, $i, k = 1, \ldots, m$. If we have plus sign r times and minus sign s times then $t := r - s$ is called the *signature* of G (at x). Signature is independent of the basis. The reader may check the following formulas.

PROPOSITIONS XIV.12.5. Let (M, G) be an oriented pseudo–Riemannian manifold with signature t; then for any $\varphi, \psi \in \Lambda^p(M)$ we have

$$(23) \qquad **\overset{p}{\varphi} = (-1)^{p(m-p)+\frac{1}{2}(m-t)}\overset{p}{\varphi},$$

$$*\varphi \wedge \psi = *\psi \wedge \varphi = (-1)^{\frac{1}{2}(m-t)}(\varphi|\psi)\mu.$$

13. HARMONIC FORMS. HODGE–KODAIRA–DE RHAM THEORY

In this section and the next M denotes an oriented compact (smooth) Riemannian manifold without boundary (in older terminology M is a closed oriented Riemannian manifold).

Hence, in generalized Green' formulas, terms with $\int_{\partial M}$ vanish. W. V. D. Hodge in his classic 1941 monograph (revised edition 1951) *The Theory and Applications of Harmonic Integrals* introduced the fundamental notions of harmonic form and harmonic integral. We quote the words of founder of this theory:

"The theory of harmonic integrals has its origin in an attempt to generalize the well-known existence theorem of Riemann for the everywhere finite integrals on Riemann surface..." This theory was afterwards developed independently by the great Japanese mathematician Kunihiko Kodaira in a series of excellent papers. We have used some of his notions and notations. Both authors were strongly influenced by Hermann Weyl's classical *Idee der Riemannschen Fläche*. It is interesting to see that Hodge's original proof of his theorem was faulty and he later replaced it by Weyl's proof. Hodge had denoted the dual form by ω^* and not by $*\omega$. The latter (and far better) notation, now fully accepted, was introduced by Hermann Weyl in his article *On Hodge's Theory of Harmonic Integrals*, Ann. of Math. 44 (1943), 1–6; hence the "Hodge $*$ operator" was invented by Weyl. We will return to this theory and its generalizations in Part III.

DEFINITION. A p-form φ on a Riemannian manifold M is harmonic if

$$d\varphi = 0 = d^*\varphi.$$

The vector space of harmonic p-forms is denoted by $\mathscr{H}^p(M)$. From this definition immediately follows another characterization of harmonic forms.

THEOREM XIV.13.1. ($\varphi \in \Lambda^p(M)$ is harmonic) $\Leftrightarrow$ ($\Delta\varphi = 0$) (M is closed).

PROOF. $\Rightarrow$ is obvious since $\Delta\varphi = dd^*\varphi + d^*d\varphi = 0+0$.

$\Rightarrow$: If $\Delta\varphi = 0$, then from Green's formula (12.21)

$$0 = \int_M (\Delta\varphi\,|\,\varphi)\mu = -\int_M (d\varphi\,|\,d\varphi)\mu - \int_M (d^*\varphi\,|\,d^*\varphi)\mu.$$

But for $\chi \in \Lambda_0^p(M)$ we have $\int_M (\chi\,|\,\chi)\mu \geqslant 0$ and the equality holds if and only if $\chi = 0$. □

For M closed we can introduce a scalar product in $\Lambda^p(M)$ by

$$\langle\omega\,|\,\eta\rangle := \int_M (\omega\,|\,\eta)\mu.$$

LEMMA XIV.13.2 (Hodge–Kodaira). Let $\partial M = 0$.

1° (A p-form is orthogonal to $d\Lambda^{p-1}(M)$) $\Leftrightarrow$ ($d^*\varphi = 0$).

2° (A p-form is orthogonal to $d^*\Lambda^{p+1}(M)$ $\Leftrightarrow$ ($d\varphi = 0$).

3° The subspaces $\operatorname{im}\overset{p-1}{d}$, $\operatorname{im}\overset{p+1}{d^*}$ and $\mathscr{H}^p(M)$ are mutually orthogonal.

PROOF. 1° $\Leftarrow$: If $d^*\varphi = 0$, then for each $\psi \in \Lambda^{p-1}(M)$ we have $\langle\varphi|d\psi\rangle = \langle d^*\varphi|\psi\rangle = 0$; hence $\varphi \perp \operatorname{im} d$.

$\Rightarrow$: If $\varphi \perp d\Lambda^{p-1}(M)$, then

$$0 = \langle\varphi, dd^*\varphi\rangle = \langle d^*\varphi\,|\,d^*\varphi\rangle, \text{ hence } d^*\varphi = 0.$$

The same argument gives 2°. But 3° is a corollary of 1° and 2° and of the relation $dd = 0$, $d^*d^* = 0$. □

COROLLARY XIV.13.3. (A p-form φ is zero) $\Leftrightarrow$ (φ is orthogonal to the three subspaces $d(\Lambda^{p-1}(M))$, $d^*\Lambda^{p+1}(M)$, $\mathscr{H}^p(M)$ of $\Lambda^p(M)$).

PROOF. In fact, if φ is orthogonal to $d(\Lambda^{p-1}(M))$ and to $d^*(\Lambda^{p+1}(M))$, then it is harmonic (by the preceding lemma). Moreover, if φ is also orthogonal to $\mathscr{H}^p(M)$, $\langle\varphi\,|\,\varphi\rangle = 0$ and $\varphi = 0$. □

But Hodge and Kodaira showed a much stronger result, which will be proved in Part III.

THEOREM XIV.13.4 (Hodge–Kodaira). Let M be closed: then

(a) $\Lambda^p(M) = \mathscr{H}^p(M) \oplus \operatorname{im}\overset{p-1}{d} \oplus \operatorname{im}\overset{p+1}{d^*}$.

(b) The space $\mathscr{H}^p(M)$ of harmonic forms has finite dimension.

We cannot prove this profound theorem now because its proof involves a theory of elliptic operators on compact manifolds. This theory will be presented in the first part of Part III. Here we give an important

COROLLARY XIV.13.5 (Theorem of Hodge). Each (de Rham) cohomology class $H^p(M)$ contains one and only one harmonic form. Hence the homomorphism $\mathscr{H}^p(M) \to H^p(M)$ is an isomorphism.

PROOF. Let P be the orthogonal projection $P\colon \Lambda^p(M) \to \mathscr{H}^p(M)$. If φ is a closed p-form, then φ is orthogonal to $d^*\Lambda^{p+1}(M)$ (cf. Lemma XIV.12.2, 2°). Hence, by virtue of the Hodge–Kodaira Theorem, we have

$$\varphi = P\varphi + d\psi \quad \text{for } \psi \in \Lambda^{p-1}(M).$$

Therefore, φ is cohomologous to a harmonic form ($\varphi \sim P\varphi$). Conversely since $\mathscr{H}^p(M)$ is orthogonal to $d\Lambda^{p-1}(M)$ two distinct harmonic forms cannot be cohomologous. □

Poincaré Duality. For the moment, we retain the assumption that M is a closed oriented Riemannian manifold. We can now prove the famous duality of Poincaré, which asserts that the vector spaces $H^p(M)$ and $H^{m-p}(M)$ are isomorphic. The Hodge Theorem allows us to replace these (de Rham) spaces by finite-dimensional vector spaces $\mathscr{H}^p(M)$ and $\mathscr{H}^{m-p}(M)$ of harmonic forms. Hence the Poincaré duality follows from the more precise

THEOREM XIV.13.6. Let M be a closed oriented Riemannian manifold. The Hodge star operator $*\colon \mathscr{H}^p(M) \to \mathscr{H}^{m-p}(M)$ is an isomorphism. Hence it induces the (natural!) isomorphism of cohomology spaces

$$H^p(M) \cong H^{m-p}(M).$$

PROOF. The proof follows immediately from (12.12) and (12.13). □

There is a natural bilinear (Poincaré) pairing $\langle\ ,\ \rangle$ between $H^p(M)$ and $H^{m-p}(M)$. We recall that $H^p(M) = \ker \overset{p}{d}/\operatorname{im} \overset{p-1}{d} = Z^p(M)/B^p(M)$. If $\beta \in Z^{m-p}(M)$, $\alpha \in Z^p(M)$, then for $\gamma \in \Lambda^{p-1}(M)$

$$\int_M (\alpha + d\gamma) \wedge \beta = \int_M \alpha \wedge \beta + \int_M (d\gamma) \wedge \beta = \int_M \alpha \wedge \beta + \int_M d(\gamma \wedge \beta)$$
$$= \int_M \alpha \wedge \beta.$$

In the same way we have for $\psi \in \Lambda^{m-p-1}(M)$

$$\int_M \alpha \wedge (\beta + d\psi) = \int_M \alpha \wedge \beta.$$

Hence we can project the mapping

$$Z^p(M)\times Z^{m-p}(M)\ni(\alpha,\beta)\to\int_M \alpha\wedge\beta\in \boldsymbol{R}$$

to a bilinear form—denoted by $\langle\cdot,\cdot\rangle$—on $H^p(M)\times H^{m-p}(M)$ (it should not be confused with a scalar product on $\Lambda^p(M)\langle\cdot|\cdot\rangle$).

We check that $\langle\cdot,\cdot\rangle$ is a non-degenerate pairing, i.e., that $(\langle[\varphi],[\psi]\rangle = 0$ for all $[\varphi]\in H^p(M))\Rightarrow([\psi]=0)$ and that $(\langle[\varphi],[\psi]\rangle=0$ for all $[\psi]\in H^{m-p}(M))\Rightarrow([\varphi]=0)$.

Indeed, take φ and ψ to be harmonic. Then

$$\langle[\varphi],[*\varphi]\rangle=\int_M\varphi\wedge*\varphi=(-1)^{p(m-p)}\langle\varphi|\varphi\rangle$$

is non-zero if φ is non-zero. Similarly,

$$\langle[*\psi],[\psi]\rangle=\int_M *\psi\wedge\psi=\langle\psi|\psi\rangle\neq 0\qquad\text{for non-zero } \psi.$$

DEFINITION. The *Euler number* of M is $e(M):=\sum_p(-1)^p\dim H^p(M)$.

COROLLARY XIV.13.7. Let M be orientable, closed and $m=\dim M = 2k-1$, $k>0$.

Then the Euler number $e(M)$ is equal to 0.

PROOF. From the Poincaré duality it follows that

$$\dim H^p(M)=\dim H^{m-p},\ \text{thus}\ \sum_p(-1)^p\dim H^p(M)=0.\ \square$$

Remark. The Euler number $e(M)$ of M is often called the *Euler–Poincaré characteristic and* denoted by $\chi(M)$.

Currents of de Rham. The Poincaré Duality Once More. In the preceding sections we have encountered several important pairings:

1° $\Lambda^p_0(M)\ni\varphi\to\langle\varphi,c_p\rangle_1:=\int_{c_p}\varphi\in\boldsymbol{R}$

where a p-chain c_p defines a linear functional on $\Lambda^p(M)$.

But an $(n-p)$-form ψ also induces a linear functional if M is oriented.

2° $\Lambda^p_0(M)\ni\varphi\to\langle\varphi,\psi\rangle_2:=\int_M\varphi\wedge\psi\in\boldsymbol{R}$.

A p-density ι_p induces another dual pairing:

3° $\Lambda^p_0(M)\ni\varphi\to\langle\varphi,\iota_p\rangle_3:=\langle\varphi|\iota_p\rangle_p=\int_M\varphi\perp\iota_p\in\boldsymbol{R}$.

If M is oriented and provided with a Riemannian metric, then we have the "Riemann–Hilbert" pairing: for any p-form ψ

4° $\Lambda_0^p(M) \ni \varphi \to \langle \varphi, \psi \rangle_4 := \langle \varphi | \psi \rangle \equiv \int_M \varphi \wedge * \psi \in \boldsymbol{R}.$

This rich assortment of dualities suggestes a general theory. This was done remarkably early, back in 1938, by Georges de Rham. In his theory of currents, de Rahm was guided by physical intuitions and by Hermann Weyl's *Raum—Zeit—Materie*. But his theory was premature: at that time functional analysis was too young. After the Second World War the theory of distributions of Laurent Schwartz (cf. Chapter XIX) enabled de Rham to give the final shape to his theory.

One should note, that four pairings $\langle \cdot , \cdot \rangle_k$, $k = 1, \ldots, 4$ define continuous functionals and, therefore, we have to provide $\Lambda_0^p(M)$ with a reasonable topology. This was done for $\Lambda_0(M) = C_0^\infty(M)$ by L. Schwartz (1946).

The Schwartz (Pseudo)Topology on $\Lambda_0^p(M)$. Let M be an m-dimensional differentiable manifold of class C^∞. Let us consider the vector space $\Lambda_0^p(M)$ of p-forms with compact supports. A sequence $(\varphi_n)_1^\infty$, $\varphi_n \in \Lambda_0^p(M)$ is said to *converge* to 0 if

(a) The supports suppt $\varphi_n \subset K$—a fixed compact subset of M.

(b) All derivatives $\dfrac{\partial^{|\alpha|}}{\partial t^{\alpha_1} \ldots \partial t^{\alpha_m}} (\varphi_n \circ u^{-1})$ in coordinate representation converge to zero uniformly on compacts.

Remark. If M is compact, then (a) is satisfied in an obvious way. We will use the following notion of continuity:

DEFINITION. A linear functional S on $\Lambda_0^p(M)$ is *continuous* if $T(\varphi_n) \to 0$, $n \to \infty$ for any sequence $\varphi_n \to 0$. Plainly, our four functionals $\langle \cdot , T \rangle_k$, $k = 1, \ldots, 4$, are continuous in this sense.

We are now ready to define the currents of de Rham.

DEFINITION. A *p-current* (or a *current of dimension* $m-p$) T is a linear continuous functional on $\Lambda_0^{m-p}(M)$.

We shall often write $\langle \varphi, T \rangle$ for the value of T on $\varphi \in \Lambda_0^{m-p}(M)$. It can now be seen that p currents of de Rham are defined by $(m-p)$-chains, Weyl $(m-p)$-densities, p-forms if M is oriented, and $(m-p)$-forms if M is oriented and provided with a Riemannian structure.

Operations on Currents. Plainly, p-currents form a linear space. It will be denoted by $\Lambda^p(M)'$. Consider for a p-current T a linear form

(1) $\quad \Lambda_0^{m-p-1}(M) \ni \varphi \to (-1)^{m-p}\langle d\varphi^{m-p-1}, T\rangle \in \boldsymbol{R}.$

This is a linear continuous functional on $\Lambda_0^{m-p-1}(M)$ and hence a $(p+1)$-current. It is denoted by dT and called the *exterior derivative of T.*

Examples. 1. Let $\overset{p}{T} = \psi \in \Lambda^p(M)$:

$$\langle \varphi, \psi\rangle = \int_M \varphi \wedge \psi \qquad (M \text{ is oriented}).$$

Then, from Green's formula it follows that

$$\langle \varphi, d\psi\rangle = (-1)^{m-p}\langle d\varphi, \psi\rangle.$$

Hence the derivative of a p-form ψ, considered as a p-current, coincides with the exterior derivative $d\psi$ of ψ.

2. Let $\overset{p}{T} = c_{m-p}$ be an $(m-p)$-chain:

$$\langle \varphi, c_{m-p}\rangle := \int_{c_{m-p}} \varphi.$$

From the Poincaré–Stokes Theorem we have

$$\langle d\overset{m-p-1}{\varphi}, c_{m-p}\rangle = \langle \varphi, \partial c_{m-p}\rangle = (-1)^{m-p}\langle \varphi, dT\rangle.$$

Thus $dT = (-1)^{m-p}\partial c_{m-p}$: the derivative of an $(m-p)$-chain, considered as a p-current, coincides with the boundary ∂c_{m-p} multiplied by $(-1)^{m-p}$.

This result suggests the following

DEFINITION. The boundary ∂T of a p-current is defined as $(-1)^{m-p}dT$. Thus

$$\langle \varphi, \partial T\rangle = \langle d\varphi, T\rangle.$$

3. Let $\overset{p}{T} = \iota_{m-p} \in W_{m-p}(M)$. Then, by Lemma XIV.11.6,

$$\langle d\overset{m-p-1}{\varphi} \,|\, \iota\rangle_{m-p} = \langle \varphi^{m-p-1} \,|\, \delta\iota\rangle_{m-p-1}.$$

Thus,

$$\partial T = (-1)^{m-p}dT = \delta\iota_{m-p}.$$

The boundary $\partial\iota_{m-p}$ of ι_{m-p}, considered as a p-current, coincides with the coboundary $\delta\iota_{m-p}$.

4. Let $\overset{p}{T} = \psi^{m-p} \in \Lambda^{m-p}(M)$. Then

$$\langle d \overset{m-p-1}{\varphi} | \psi^{m-p} \rangle = \langle \varphi^{m-p-1} | d^* \psi \rangle$$

Thus, $dT = (-1)^{m-p} d^* \psi$. The derivative of $\overset{m-p}{\psi}$, considered as a p-current, coincides with the coderivative $d^*\psi$ of ψ multiplied by $(-1)^{m-p}$.

The Riemann–Hilbert duality and the star operator. If M is oriented and provided with a Riemannian structure, then the Hodge star operator is defined for currents on M as

$$\text{(2)} \qquad \langle \overset{p}{\varphi}, *\overset{p}{T} \rangle := \langle *^{-1}\varphi, T \rangle = (-1)^{p(m-p)} \langle * \varphi, T \rangle$$

($\varphi \in \Lambda_0^p(M)$). Hence, if T is a p-current, then $*T$ is an $(m-p)$-current. We have

$$\text{(3)} \qquad **\overset{p}{T} = (-1)^{p(m-p)} \overset{p}{T}.$$

Now, the Riemann–Hilbert pairing can be extended to $\Lambda^p(M) \times \times \Lambda^{m-p}(M)'$ by setting

$$\text{(4)} \qquad \langle \overset{p}{\varphi} | \overset{p}{T} \rangle := \langle * \varphi, T \rangle = \langle \varphi, *^{-1} T \rangle.$$

The relation

$$\langle \varphi | d^* \psi \rangle = \langle d\varphi | \psi \rangle$$

for $\varphi \in \Lambda_0^{m-p-1}(M)$, $\psi \in \Lambda^{m-p}(M)$ (resulting from Theorem XIV.12.4) suggests the following

DEFINITION. The coderivative $d^*\overset{p}{T}$ of a p-current $\overset{p}{T}$ is a $(p-1)$-current given by

$$\text{(5)} \qquad \langle \overset{m-p-1}{\varphi} | d^* T \rangle = \langle d\varphi | T \rangle = \langle * d\varphi, T \rangle.$$

COROLLARY XIV.13.8. 1° $d^* d^* \overset{p}{T} = 0$.

2° $d^* \overset{p}{T} = (-1)^p * d *^{-1} \overset{p}{T}$.

PROOF. 1° is obvious since $dd = 0$.

$$2° \ \langle \overset{p-1}{\varphi} | * d *^{-1} \overset{p}{T} \rangle = \langle \varphi, d *^{-1} \overset{p}{T} \rangle = (-1)^p \langle d\varphi, *^{-1} \overset{p}{T} \rangle$$
$$= (-1)^p \langle d\varphi \mid T \rangle = (-1)^p \langle \varphi \mid d^* T \rangle. \ \square$$

Examples. 1. Let $\overset{p}{T} = \overset{p}{\psi} \in \Lambda^p(M)$. Then the coderivative d^*T coincides with $d^*\psi$.

2. Let $\overset{p}{T} = \overset{m-p}{\psi} \in \Lambda^{m-p}(M)$. Then,

$$\langle \overset{p-1}{\varphi} \,|\, d^*\overset{p}{T} \rangle = \langle *d\varphi, T \rangle = \langle *d\varphi |\, \overset{m-p}{\psi} \rangle$$

$$= \langle d\varphi \,|\, *^{-1}\psi \rangle = \langle \varphi \,|\, d*\,*^{-1}\psi \rangle = \langle *\varphi \,|\, *d*\,*^{-1} \overset{m-p}{\psi} \rangle$$

$$= (-1)^p \langle *\varphi \,|\, **d*^{-1}*^{-1} \overset{m-p}{\psi} \rangle = (-1)^{m-p+1} \langle *\varphi \,|\, d\psi \rangle .$$

Thus,

$$d^*\overset{p}{T} = (-1)^{m-p+1} d \overset{m-p}{\psi} ,$$

i.e., the coderivative of an $(m-p)$-form, treated as a p-current, coincides with its exterior derivative times $(-1)^{m-p+1}$, treated as a $(p-1)$-current.

The Laplace–Beltrami operator is given by

$$\Delta T := -(dd^* + d^*d)T.$$

Now, de Rham extends the notion of (co)homology in a natural way.

DEFINITION. A p-current T is said to be *closed* (*coclosed*) if $dT = 0$ ($d^*T = 0$). It is called a *coboundary* (*boundary*) if, for a $(p-1)$-current ($(p+1)$-current) S,

$$T = dS \quad (T = d^*S).$$

We can introduce homology and cohomology spaces for currents:

$$H_p(M) := \ker \overset{p}{d^*} / \operatorname{im} \overset{p+1}{d^*}$$

$$H^p(M) := \ker \overset{p}{d} / \operatorname{im} \overset{p-1}{d}.$$

A p-current T is said to be *harmonic* if $\Delta T = 0$. In Part III we shall prove that harmonic currents are regular: they can be represented by harmonic p-forms.

The theory of currents enabled de Rham to unify the homology and cohomology theories for differentiable manifolds: chains, differential forms and Weyl densities are currents.

But, even more important, the Hodge–Kodaira theory can be extended, in a natural way, to currents.

THEOREM XIV.13.9 (de Rham). Let M be oriented, closed Riemannian manifold of dimension m. Then

1° The space of harmonic p-currents is of finite dimension and can be identified with $\mathscr{H}^p(M)$, the space of harmonic p-forms.

2° Let us denote by H the orthogonal projection onto $\mathscr{H}^p(M)$: if $h_1, \ldots, h_n$ is an orthonormal basis of harmonic p-forms

$$HT := \sum_{i=1}^{n} \langle h_i | T \rangle h_i,$$

then

$$d \cdot H = H \cdot d, \ \delta H = H\delta, \ *H = H*,$$

$$\langle \varphi | HT \rangle = \langle H\varphi | T \rangle.$$

3° There is one and only one mapping ("Green's operator")

$\mathscr{G}\colon \Lambda^p(M)' \to \Lambda^p(M)'$ such that

(i) $-\Delta \mathscr{G} T = T - HT$,

(ii) $H\mathscr{G}T = \mathscr{G}HT = 0, \ \mathscr{G}d = d\mathscr{G}, \ \delta\mathscr{G} = \mathscr{G}\delta$.

From this theorem and Theorem XIV.13.4 we easily obtain the important

THEOREM XIV.13.10 (de Rham). For any p-current we have the Hodge decomposition

(a) $T = -\Delta\mathscr{G}T + HT = d(d^*\mathscr{G}T) + d^*(d\mathscr{G}T) + HT$. This decomposition of T into a sum of three currents, one cohomologous to zero, another homologous to zero, and the last equal to a harmonic form, is unique.

(b) A closed (coclosed) current T is cohomologous (homologous) to one and only one harmonic form HT. Thus, we have the following far-reaching generalization of

(c) The *Poincaré duality* $H^p(M) \cong H_{m-p}(M) \cong \mathscr{H}^p(M) \cong \mathscr{H}^{m-p}(M) \cong \mathscr{H}_p(M)$.

Remarks. 1. Poincaré proved the duality $H_p(M) \cong H_{m-p}(M)$ for homology classes of chains.

2. From the de Rham decomposition formula it follows that:

A closed current T is cohomologous to zero if and only if $HT = 0$.

We now obtain another famous theorem of Hodge about the existence of a harmonic form with assigned periods. Let us recall that a *period* of a closed p-form φ on a p-cycle z_p is the value of the integral

$$\int_{z_p} \varphi = \langle \varphi, z_p \rangle .$$

p-cycles z_k are said to be *homologically independent* if they define linearly independent elements in $H_p(M)$.

THEOREM XIV.13.11 (Hodge). There is one and only one harmonic p-form on M having arbitrarily assigned periods on $\dim H_p(M) = \dim H^p(M)$, homologically independent p-cycles z_k.

PROOF. The uniqueness follows from the de Rham decomposition.

Existence: Let h be a harmonic $(m-p)$-form such that $(h|Hz_k) = p_k$, where z_k, $k = 1, 2, \ldots, \dim\mathscr{H}^p(M)$ are the p-cycles and p_k are the prescribed periods. Then $*h$ is a harmonic p-form and, if we consider the cycles as currents, we have

$$\int_{z_k} *h = (h|z_k) = (Hh|z_k) = (h|Hz_k) = p_k . \; \square$$

Remark. In his famous thesis (1931) de Rham proved "only" the existence of a p-form with assigned periods. Currents were introduced by de Rham in the paper *Über mehrfache Integrale*, Abh. Math. Sem. Hansischen Univ. 12 (1938).

We shall return to these problems in Part III. We merely remark here that the notion of period was introduced by Riemann for $\dim M = 2$.

14. APPLICATIONS TO ELECTRODYNAMICS

In this short section we shall write down the fundamental differential equations of relativistic electrodynamics. The language of differential forms and tensor densities allows us to reveal the beautiful symmetry of this physical theory. Special relativity arose as an attempt to understand and describe electrodynamical phenomena in bodies moving with great speed. It is an edifice erected by such giants as Poincaré, Lorentz, Einstein, and Minkowski. The special relativity principle of Einstein, Poincaré and Lorentz was given an adequate mathematical formulation by Hermann Minkowski: we are indebted to him not only for the idea of four-dimensional world-geometry but also for the marvellous discovery of the energy-momentum tensor, which contains the Maxwell stress tensor and all the other relevant quantities of classical electromagnetism.

We will not proceed by first defining the physical notions and then "deducing" the relevant equations, which is perhaps necessary for describing the historical development of the theory of "electricity" and other theories. Instead, we shall follow the advice of Hermann Weyl (whom the Nobel Prize winner F. Dyson called "one of three or four men who have most deeply understood physics in the twentieth century"). In his excellent lecture *Mathematik als symbolische Konstruktion des Menschen*, describing a physical theory on the example of Maxwellian electrodynamics, Weyl says:

"*Man fängt daher besser überhaupt nicht mit einer Definition der Elektromagnetischen Feldstärke an, sondern stellt Gleichungen auf, in denen unerklärte Symbole, wie Raum-Zeit-Koordinaten, die Komponenten der Feldstärke, Ladungsdichte u.s.w. auftreten. Neben den immanenten Feldgesetzen müssen die Gleichungen angegeben werden nach denen sich die auf die erzeugenden Konduktoren wirkenden Kräfte aus den primitiven das Feld kennzeichnenden Größen bestimmen. Erst am Schluß der ganzen Konstruktion wird der Übergang von der physikalischen Welt zur Wahrnehmung vollzogen, indem man zu beschreiben versucht, in welcher Weise die Werte gewisser abgeleiteter Größen in den Wahrnehmungen sich kundgeben.*"

Instead of giving a clumsy English translation of these profound philosophical remarks, I shall give another citation, namely the words of one of the most interesting theoretical physicists, John A. Wheeler, written in the same spirit and probably under Weyl's influence.

"Here and elsewhere in science, as stressed not least by Henry Poincaré, that view is out of date which used to say "Define your terms before you proceed."

"All the laws and theories of physics, including the Lorentz force law, have this deep and subtle character, that they both define the concepts they use (here B and E) and make statements about these concepts. Contrariwise, the absence of some body of the theory, law and principle, deprives us of the means properly to define or even to use concepts. Any forward step in human knowledge is truly creative in this sense: that theory concept, law and method of measurement—forever insepara-ble—are born into the world in union." [Wheeler in *Gravitation* (written together with C. W. Misner and K. S. Thorne), 1973].

Let (M, G) denote a $(4 = 1+3)$-dimensional differentiable manifold where the metric tensor has the signature 2 $(+, -, -, -)$. In the sequel it is $\boldsymbol{R}^4$ provided with an atlas composed of one map in which G has components

(1) $$g_{00} = +1, \quad g_{11} = -1, \quad g_{22} = -1, \quad g_{33} = -1,$$

$$[g_{\alpha\beta}] = \begin{bmatrix} +1 & & & 0 \\ & -1 & & \\ & & -1 & \\ 0 & & & -1 \end{bmatrix}.$$

(But most considerations are valid on any differentiable manifold of dimension 4 and orientable for which $H^2(M, \boldsymbol{R}) = 0$.) Electrody-namics (of special relativity) is dominated by two quantities $F \in \Lambda^2(M)$ and $H \in W_2(M)$ which, in the case of a vacuum ("ether"), are dual to each other, $H = I_2 F$.

All other quantities and fundamental equations are derived from these quantities by the invariant processes of tensor analysis developed in the preceding sections. It is interesting to note that the two fundamen-tal quantities F ("electromagnetic field") and H (magnetic field density) are skew-symmetric. From these tensor densities Minkowski constructs a symmetric tensor density of type $\binom{2}{0}$ S, the famous Minkowski ener-gy-momentum tensor

(2) $$S^{\mu\kappa} = H^{\mu\lambda} F_{\nu\lambda} g^{\nu\kappa} - \tfrac{1}{4} H^{\nu\lambda} F_{\nu\lambda} g^{\mu\kappa}.$$

We can now write two dual equations

(3) $$dF = 0,$$

(4) $$\delta H = s, \quad \text{where } s \text{ is given } (s \in W_1(M)).$$

Equations (3), (4) are called the *Maxwell equations*.

If we assume that $H^2(M, \boldsymbol{R}) = 0$, then (3) implies the existence of a 1-form $A \in \Lambda^1(M)$ such that

(5) $$F = dA.$$

Now, (4) implies the "continuity condition":

(6) $\delta s = 0.$

In a vacuum, where $H = I_2 F$,

(7) $d^* dA = I^1 s.$

One often adds an additional "Lorentz gauge" condition on A:

(8) $d^* A = 0.$

Then (7) reads

(9) $\Box A = I^1 s,$

where $\Box = d^* d + dd^*$ is the d'Alembert (wave) operator. Therefore, (9) is a wave equation for A, which is a system of partial differential equations of hyperbolic type.

To complete the picture, we need equations of motion of charges in electromagnetic fields. To this end, let us introduce the line element

(10) $ds^2 = g_{\alpha\beta} dx^\alpha dx^\beta.$

Let m_0 be the "rest mass" and e the charge of a particle moving in the field F with the velocity $u = (u^\alpha) = \left(\dfrac{dx^\alpha}{ds}\right)$. Then the equations of motion of a charged particle are supplied by the so-called *Lorentz force equations*

(11) $$m_0 \frac{du^\alpha}{ds} = e u^\nu F^\alpha_\nu.$$

In this way we have a very beautiful set of equations (3), (4), (11), but in order to gain contact with "physical reality" we have to connect our tensor densities F and H with classical physical quantities. Since we can identify contravariant and covariant vectors in the Euclidean space $\boldsymbol{R}^3$, we shall not be very pedantic about indices. Let us use the following notation:

$\vec{E} = (E^1, E^2, E^3)$ — electric field (strength),
$\vec{D} = (D^1, D^2, D^3)$ — dielectric displacement,
$\vec{H} = (H^1, H^2, H^3)$ — magnetic field,
$\vec{B} = (B^1, B^2, B^3)$ — magnetic induction,
$\vec{j} = (J^1, J^2, J^3)$ — current density,
ϱ — charge density.

The (2-form) tensor F is composed of $\vec{E}$ and $\vec{B}$, the tensor density H is composed of $\vec{D}$ and $\vec{H}$ and s of ϱ and $\vec{j}$ in the following way:

Tensor $H_{\mu\nu} = -H_{\nu\mu}$

(12)

ν \ μ	0	1	2	3
0	0	$-D^1$	$-D^2$	$-D^3$
1	D^1	0	H^3	$-H^2$
2	D^2	$-H^3$	0	H^1
3	D^3	H^2	$-H^1$	0

Tensor $F_{\alpha\beta} = -F_{\beta\alpha}$

(13)

β \ α	0	1	2	3
0	0	$-E^1$	$-E^2$	$-E^3$
1	E^1	0	B^3	$-B^2$
2	E^2	$-B^3$	0	B^1
3	E^3	B^2	$-B^1$	0

(14) $$H^{\varkappa\lambda} = \sqrt{|g|}\, g^{\varkappa\mu} g^{\lambda\nu} F_{\mu\nu},$$

(15) $$s = (\varrho, J^1, J^2, J^3),$$

(16) $$A = (\varphi, A_1, A_2, A_3).$$

Now (3) and (4) assume the familiar three-dimensional form

(3_3) $$\operatorname{curl} \vec{E} + \frac{\partial \vec{B}}{\partial t} = 0, \quad \operatorname{div} \vec{B} = 0,$$

(4_3) $$\operatorname{curl} \vec{H} - \frac{\partial \vec{D}}{\partial t} = \vec{j}, \quad \operatorname{div} \vec{D} = \varrho.$$

The Lorentz force equations (11) are now

(11_3) $$\frac{d}{dt}(m\vec{v}) = e(\vec{E} + \vec{v} \times \vec{B}), \quad m = \frac{m_0}{\sqrt{1 + \|\vec{v}\|^2}}.$$

We have seen that fundamental equations (3), (4), (9) and the Minkowski (energy-momentum) tensor $(S^{\mu\nu})$ have an invariant form valid not only in the (flat) Minkowski space-time but also in any 4-dimensional manifold. This implies that the equations of the Maxwell–Lorentz–

Poincaré electrodynamics are Lorentz-invariant, i.e., invariant with respect to a linear group of transformations of $\boldsymbol{R}^4$ leaving the fundamental quadratic form G invariant. We will finish this section by a theorem which we owe to Hermann Minkowski and which demonstrates the wonderful symmetry of relativistic electrodynamics in Minkowski's form.

THEOREM XIV.14.1 (Minkowski). 1° In the vacuum the tensor S is symmetric

$$(17) \qquad S^{\alpha\beta} = S^{\beta\alpha}, \qquad \alpha, \beta = 0, 1, 2, 3.$$

2° (S^{ik}) is the Maxwell stress-tensor, $i, k = 1, 2, 3$, $S_i^k = F_{ir}F^{kr} - \frac{1}{4}\delta_i^k F_{\alpha\beta}F^{\alpha\beta}$.

3° S^{00} is the energy density $W = \frac{1}{2}(||\vec{E}||^2 + ||\vec{B}||^2)$.

S^{0i} are components of the "Poynting vector" $\vec{S} := \vec{E}\times\vec{H}$.

4° S^{k0} are components of the momentum density $\vec{D}\times\vec{B}$.

Therefore, we can express 1°–4° in the matrix form

$$[S^{\mu\nu}] = \begin{bmatrix} W & \vec{S} \\ \vec{D}\times\vec{B} & [S^{ik}] \end{bmatrix}.$$

5° The "divergence" of the Minkowski tensor is equal to the density of electromagnetic (Lorentz) force

$$(18) \qquad \partial_\nu S_\mu^\nu = s^\nu F_{\nu\mu}.$$

Remarks. 1. We see that the Theorem on the Conservation of Energy is only one component, the time component, of a law which is invariant under Lorentz transformations, the other components being the space components which express the conservation of momentum,

2. The symmetry of the Minkowski tensor shows that $\vec{S} = \vec{D}\times\vec{B}$: energy flux = momentum density.

3. We see that in special relativity the conservation of energy is indissolubly connected with the principle of the conservation of momentum.

PROOF OF THEOREM. 1° is obvious from the definition of $S^{\alpha\beta}$.

3° follows from the definition of both tensors: the Maxwell stress-tensor and the Minkowski tensor.

2° follows immediatelly from definitions of tensors $(F_{\alpha\beta})$ and $(H^{\mu\nu})$.

4° —ditto.

The only difficulty is to verify 5°, i.e. to prove

$$F_{\alpha\beta}s^{\alpha} = \partial_{\beta} S^{\beta}_{\alpha}.$$

In a vacuum we have $S^{\beta}_{\alpha} = F_{\alpha\nu}F^{\beta\nu} - \frac{1}{2}\delta^{\beta}_{\alpha}|F|^2$, where $|F|^2 := \frac{1}{2}F_{\alpha\beta}F^{\alpha\beta}$. But we have

$$\partial_{\beta} S^{\beta}_{\alpha} = F_{\alpha\nu}\partial_{\beta}F^{\beta\nu} + F^{\beta\nu}\partial_{\beta}F_{\alpha\nu} - \tfrac{1}{2}F^{\beta\nu}\partial_{\alpha}F_{\beta\nu}.$$

The first term on the right-hand side gives

$$+F_{\alpha\nu}s^{\nu}.$$

If we write the coefficient of $F^{\beta\nu}$ skew-symmetrically, we get for the second term

$$\tfrac{1}{2}F^{\beta\nu}(\partial_{\beta}F_{\alpha\nu} + \partial_{\nu}F_{\beta\alpha}),$$

which together with the third term gives

$$\tfrac{1}{2}F^{\beta\nu}(\partial_{\beta}F_{\alpha\nu} + \partial_{\nu}F_{\beta\alpha} + \partial_{\alpha}F_{\nu\beta}) = 3F^{\beta\nu}\partial_{[\alpha\beta}F_{\nu]} = 0$$

by the first Maxwell equation (3). □

Exercise 1. Show that (7) has the form of four wave equations

$$-\frac{\partial^2\varphi}{\partial t^2} + \Delta\varphi = \varrho, \tag{19}$$

$$-\frac{\partial^2 A}{\partial t} + \Delta A^k = J^k, \qquad k = 1, 2, 3. \tag{20}$$

Exercise 2. Check that (19) has a solution

$$4\pi\varphi(y) = \int \frac{\varrho(t - |x-y|)}{|x-y|}\, d\lambda^3(x),$$

the so-called *retarded potential*. □

15. INVARIANT FORMS (HURWITZ INTEGRAL). COHOMOLOGY OF COMPACT LIE GROUPS

Lie groups are perhaps the most beautiful objects of analysis. They are finite-dimensional differentiable manifolds provided with a group structure and these both structures are intertwined in a most harmonious way. As was supposed by Hilbert (his famous fifth problem) a topological manifold which is a topological group must, in a natural way, be a differentiable (and even analytic) manifold and its group operations are differentiable, i.e., it is a Lie group. After fifty years of ingenious work by the best mathematicians, this modified Hilbert hypothesis has found glorious confirmation.

As a by-product of these efforts a new flourishing mathematical discipline arose: the theory of locally compact groups.

Lie Groups. Hurwitz Integral

DEFINITION. A *Lie group* G is a differentiable (finite-dimensional) manifold which is at the same time a group and the group operations are smooth.

If, for $x, y \in G$, we denote their product by xy and the inverse of x by x^{-1}, then the mapping

(1) $$G \times G \ni (x, y) \to xy^{-1} \in G$$

is smooth.

A homomorphism of Lie groups is (by definition) a C^∞-map.

Remarks. 1. If the differentiable manifold G is modelled on a Banach space of infinite dimension, we speak of a *Banach Lie group*.

2. We can replace (1) by two conditions:

(2) $$G \times G \ni (x, y) \to xy \in G,$$

(3) $$G \ni x \to x^{-1} \in G$$

are C^∞-mappings.

3. Sometimes discrete groups are called *Lie groups of dimension* 0.

Examples. 1. $G = \boldsymbol{R}^m$, $x \cdot y := x+y$. Then $\boldsymbol{R}^m$ is a Lie group.

2. $\mathrm{Gl}(n, \boldsymbol{R})$, the set of non-singular $n \times m$ matrices with real entries, is an open submanifold of $M_n(\boldsymbol{R})$, the set of all real $n \times n$ matrices identified with $\boldsymbol{R}^{n^2}$. The map $(A, B) \to AB^{-1}$ is smooth (even analytic) since the elements of AB^{-1} are rational functions (of the entries of A and B) with non-vanishing denominators.

3. More generally: Let V be a Banach space (over $\boldsymbol{R}$ or $\boldsymbol{C}$); then the group $\mathrm{Aut}(V)$, of automorphisms of V, is also denoted by $\mathrm{Gl}(V)$, and is called a "general linear group".

Especially: $\mathrm{Gl}(n, \boldsymbol{R}) \equiv \mathrm{Gl}(\boldsymbol{R}^n)$, $\mathrm{Gl}(n, \boldsymbol{C}) \equiv \mathrm{Gl}(\boldsymbol{C}^n)$.

4. $C^* := C - \{0\}$ is the multiplicative group of non-zero complex numbers.

5. Let G_1, G_2 be Lie groups. Then the direct product $G_1 \times G_2$ is a Lie group, where the group $G_1 \times G_2$ is provided with the differentiable structure of the Cartesian product of differentiable manifolds.

6. $T^n := R^n/Z^n = (R/Z)^n$, the n-dimensional torus.

7. A real orthogonal $O(n)$ group, and, a unitary group $U(n)$, are compact Lie groups of dimension $n(n-1)/2$ and n^2, respectively.

Since in Part III we shall give many other examples of Lie groups and since a major part of Part III is devoted to the study of transformation groups, and Lie groups, and their representations and applications, we shall stop here.

The present section is only an illustration, or an "example" of a differentiable manifold. The preceding examples should convince the young reader that Lie groups are very important objects.

We can look at a Lie group G as a transformation group: each element g defines an automorphism L_g of G:

$$L_g \colon G \to G, \quad L_g x := gx.$$

But G also acts on G from the right:

$$R_g x := xg.$$

Applying the notion of the pull back of a covariant tensor field, therefore, we arrive at the following

DEFINITION. A covariant tensor field Φ on a Lie group G is *left-(right-)invariant* if for each $g \in G$

$$(4) \qquad L_g^* \Phi = \Phi \quad (R_g^* \Phi = \Phi),$$

or, more precisely

$$L_g^* \Phi_{ga} = \Phi_a \quad (R_g^* \Phi_{ag} = \Phi_a).$$

A left-invariant 1-form is called a *Maurer–Cartan form*.

The tensor field Φ is bi-invariant: it is both left- and right-invariant. Left-, right-, and bi-invariant forms are denoted by $\Lambda_l(G)$, $\Lambda_r(G)$ and $\Lambda_{inv}(G)$, respectively. This definition immediately implies

PROPOSITION XIV.15.1. Every Lie group G has a left- (right-)invariant volume element; in particular, every Lie group is orientable. Every Lie group has a left- (right-)invariant Riemannian metric.

PROOF. Let $m = \dim_R G$ and take a skew-symmetric non-vanishing tensor Φ_e of type $\binom{m}{0}$. Denote

$$\Phi_g = L_g^* \Phi_e.$$

Then, $g \to \Phi_g$ is a left-invariant non-vanishing m-form on G.

Taking a scalar product $\gamma_e := (\cdot|\cdot)_e$ in $T_e G$, we obtain a left-invariant Riemannian metric $\gamma_g := L_g^* \gamma_e$ on G.

Remark. A left-invariant volume element is unique up to a constant factor.

COROLLARY XIV.15.2 (Hurwitz). 1° On every Lie group there exists the unique, modulo positive constant factor, left-invariant Radon integral $\mu_1 := \mu_{\omega_1}$, defined by the left-invariant volume element ω_1 (cf. Theorem XIV.9.9). We have

(5) $\mu_1(L_g^* f) = \mu_1(f)$ for any $f \in C_0(G)$ and each $g \in G$.

2° If G is compact, then ω_1 is automatically right-invariant.

PROOF. 1°

$$\mu_1(L_g^* f) = \int_G L_g^* f \omega_1 = \int_G f L_{g^{-1}}^* \omega_1 = \int_G f \omega_1 = \mu_1(f).$$

2° is valid even for the Haar integral on *any* compact group (cf. Section XIX.20 and Theorem XIV.16.2).

Remark. Adolf Hurwitz (1859–1919), the eldest of three great mathematical friends Koenigsberg's (Hurwitz, Minkowski, and Hilbert), discovered his "invariant integral" in 1897 and applied it to the theory of invariants rather than to the theory of groups. The method of invariant integration developed in the hands of Isai Schur and Hermann Weyl into the most powerful method in the theory of group representation. Alfred Haar proved in 1933 that any second countable locally compact group has a left-invariant Radon integral. André Weil and Stefan Banach proved shortly afterwards that the second countability is superfluous. A construction of the Haar integral is given in Chapter XIX, Section 23.

The Lie Algebra of a Lie Group. There is a dual notion to the Maurer–Cartan form: that of a left-invariant vector field:

DEFINITION. A vector field $X \in X(G)$ on a Lie group G is left-invariant if $T(L_g)X = X$; more precisely:

$$TL_g X(a) = X(ga) \quad \text{for every } g, a \in G.$$

Left-invariant vector fields form a Lie algebra under the commutator of the vector fields, which as a vector space is isomorphic to $T_e G$, the tangent space at the neutral element e of G.

DEFINITION. The vector space $T_e G \equiv LG$, denoted also by $L(G)$ (tangent space at e), is called the *Lie algebra* of the Lie group G.

A homomorphism $f\colon G \to H$ defines a homomorphism

$$Lf := T_e f\colon LG \to LH$$

of the corresponding Lie algebras.

DEFINITION. A one-parameter subgroup of a Lie group G is a homomorphism

$$\varphi\colon \mathbf{R} \to G$$

(i.e., $\varphi(a+b) = \varphi(a)\varphi(b)$, $\varphi(0) = e \in G$).

The method of Section IX.17 ("dynamical systems") produces

THEOREM XIV.15.3. The map {one-parameter subgroups of G} $\ni \varphi \to L\varphi(1) \in LG$ defines a bijection between one-parameter subgroups of G and left-invariant vector fields on G. Each vector $X \in T_e G$ defines a one-parameter subgroup φ_X denoted also by $\exp(tX)$.

This notation is a natural generalization of the exponential map $\exp(\cdot)$ in a Banach space E considered in Section IX.10:

$$\exp(tA) := \sum_{k=0}^{\infty} \frac{(tA)^n}{n!}, \qquad A \in L(E, E). \tag{6}$$

If we take $M = G$ and consider only left-invariant vector fields on the Lie group G, we see that the following proposition is valid.

PROPOSITION XIV.15.4. Let $X, Y \in L(G)$ be left-invariant vector fields on G and let $\varphi(t) := \exp(tY)$ be the one-parameter group defined by Y. Then

$$(7) \qquad [Y, X] = \lim_{t \to 0} \frac{1}{t} \big((TR_{\varphi(-t)} X(\varphi(t)) - X(e) \big).$$

The proof is left to the reader.

We can now prove the following

LEMMA XIV.15.5. Let Φ be a bi-invariant tensor field of type $\binom{0}{r}$ on a Lie group G.

Then for any $X_1, \ldots, X_r, Y \in L(G)$ we have

$$(8) \qquad \sum_{k=1}^{r} \Phi(X_1, \ldots, [Y, X_k], \ldots, X_r) = 0.$$

PROOF. Since Φ is bi-invariant, $R^*_{\varphi(-t)}\Phi - \Phi = 0$, for any $X_1, \ldots, X_r \in LG$ we have

$$(9) \qquad \Phi(TR_{\varphi(-t)} X_1, \ldots, TR_{\varphi(-t)} X_r) - \Phi(X_1, \ldots, X_r) = 0.$$

Adding to and subtracting from (9)

$$\Phi(X_1, \ldots, X_{k-1}, TR_{\varphi(-t)} X_k, \ldots, TR_{\varphi(-t)} X_r), \qquad k = 1, \ldots, r,$$

then multiplying by t^{-1} and letting $t \to 0$, we obtain (8). □

We can now prove the following important

PROPOSITION XIV.15.6 (E. Cartan). Every bi-invariant differential form ω on a Lie group G is closed.

PROOF. Let ω be a bi-invariant r-form on G and let $X_0, X_1, \ldots, X_r \in LG$. Then by Proposition XIV.8.2 we have

$$d\omega(X_0, \ldots, X_r) = \sum_{j=0}^{r} (-1)^j X_j \omega(X_0, \ldots, \hat{X}_j, \ldots, X_r) +$$

$$+ \sum_{j<k}^{r} (-1)^{j+k} \omega([X_j, X_k], X_0, \ldots, \hat{X}_j, \ldots, \hat{X}_k, \ldots, X_r) = 0;$$

since the first term on the right-hand side is 0 (by the invariance of ω and $X_0, \ldots, X_r$), the second term vanishes by virtue of (8). □

Cohomology of Compact Lie Groups. In his famous monograph *The Classical Groups,* H. Weyl writes:

"One of the most beautiful applications of the integration methods is E. Cartan's theory of invariant differentials on a compact Lie group."

This theory was probably the starting point of the cohomology theory of de Rham. Cartan's theory establishes a close relationship between bi-invariant differential forms and cohomology spaces:

THEOREM XIV.15.7 (E. Cartan). Let G be a connected compact Lie group. Then

1° Every bi-invariant form on G is closed.

2° Every closed form is cohomologous to a bi-invariant form.

3° A bi-invariant exact form is equal to 0.

Therefore, $H^k(G) \cong \Lambda^k_{\text{inv}}(G)$, $k = 0, 1, \dots, \dim G$, and in each element of $H^k(G)$ there exists only one (closed) bi-invariant form.

The Betti numbers

$$b_p(G) := \dim H^p(G) = \dim \Lambda^p_{\text{inv}}(G).$$

PROOF. 1° was proved for any Lie group. In order to prove 2°, 3° we introduce an important linear map M_l.

LEMMA XIV.15.8 (E. Cartan). Let dg be the bi-invariant volume element for which $\int_G dg = 1$. Denote by M_l (resp. M_r) the operation of the left (resp. right) average over G:

$$(M_l \omega)(X_1, \dots, X_k) := \int_G (L_g^* \omega)(X_1, \dots, X_k)\, dg$$

for any $\omega \in \Lambda^k(G)$, $X_1, \dots, X_k \in X(G)$.

In the same manner we introduce the right mean

$$M_r \omega := \int_G R_g^* \omega \, dg.$$

Then

(a) $M_l \omega$ is left-invariant.

(b) $M_r \omega$ is right-invariant.

(c) (ω is left-invariant) $\Rightarrow$ ($M_l \omega = \omega$); therefore, $M_l \Lambda^k(G) = \Lambda^k_l(G)$.

(d) (ω is right-invariant) $\Rightarrow$ ($M_r \omega = \omega$); hence, $M_r \Lambda^k(G) = \Lambda^k_r(G)$.

(e) $d \circ M_l = M_l \circ d$.

(f) $d \circ M_r = M_r \circ d$.

(g) $M_l M_r \Lambda(G) = \Lambda_{\text{inv}}(G)$.

PROOF. It is sufficient to prove (a), (c), and (e).

(a) $L_a^* M_1 \omega(X_1, \ldots, X_k) = (M_1\omega)(TL_a X_1, \ldots, TL_a X_k)$

$$= \int_G (L_g^* \omega)(TL_a X_1, \ldots, TL_a X_k) dg$$

$$= \int_G L_a^* \circ L_g^* \omega(X_1, \ldots, X_k) dg = \int_G L_{ag}^* \omega(X_1, \ldots, X_k) dg$$

$$= \int_G L_g^* \omega(X_1, \ldots, X_k) dg = M_1 \omega(X_1, \ldots, X_k).$$

(c) If $L_g^* \omega = \omega$ for all $g \in G$, then

$$M_1\omega(X_1, \ldots, X_k) = \int_G (L_g^* \omega)(X_1, \ldots, X_k) dg$$

$$= \omega(X_1, \ldots, X_k) \int_G dg = \omega(X_1, \ldots, X_k).$$

(e) This follows from the theorem on differentiating under the integral sign.

PROOF OF THEOREM XIV.15.7 (continued). Since G is connected, L_g is homotopic to $L_e = \mathrm{id}_G$. Therefore L_g^* induces the identity map of the de Rham cohomology spaces $H^k(G)$, which means that if ω is closed, then $L_g^* \omega \sim \omega$; similarly, $R_g^* \omega \sim \omega$ if $d\omega = 0$. Hence, $\langle c, R_g^* \omega \rangle \equiv \langle c, \omega \rangle$ for all k-cycles and all $g \in G$. By virtue of the Fubini Theorem, integration with respect to dg gives

$$\langle c, M_1\omega \rangle = \langle c, \omega \rangle \quad \text{for all } k\text{-cycles } c.$$

From the de Rham Theorem we obtain

$$\omega \sim M_1\omega \sim M_r M_1 \omega,$$

but $M_r M_1 \omega$ is bi-invariant.

Let $\omega = d\alpha$ and let $L_g^* \omega = \omega = R_{g'}^* \omega$ for all $g, g' \in G$. Therefore, (by Lemma XIV.15.8 and 1°)

$$\omega = M_1 M_r \omega = M_1 M_r d\alpha = d(M_1 M_r \alpha) = 0.$$

Remark. 3° very much resembles the Hodge Theorem $H^p(G) \cong \mathscr{H}^p(G)$. A natural question arises: are bi-invariant forms harmonic? Since they are closed, one has to prove that they are co-closed. This is indeed the case. Moreover, we have

COROLLARY XIV.15.9. Let G be a connected compact Lie group. We provide G with a bi-invariant Riemannian metric and define the corresponding codifferential δ. Then, 1° M_l and M_r are symmetric idempotents projecting on $\Lambda_l^k(G)$ and $\Lambda_r^k(G)$.

2° Each bi-invariant k-form ω is harmonic:

$$d\omega = 0 = \delta\omega:$$

$$\Lambda_{\text{inv}}^k(G) \subset \mathscr{H}^k(G),\ k = 1, 2, \ldots, \dim G.$$

PROOF. Let γ be any Riemannian metric; then $\langle \cdot | \cdot \rangle := M_l M_r \gamma$ is a bi-invariant Riemannian metric on G. This metric induces invariant metrics $\langle \cdot | \cdot \rangle$ on $\Lambda^k T^*(G), k = 0, \ldots, \dim G$. (Plainly, we could take the metric on $\Lambda^k T^*(G)$ induced by γ and obtain an invariant metric on $\Lambda^k T^*(G)$ performing right- and left-averaging over G.) We now define in a standard way the (pre-)Hilbert structure $(\cdot | \cdot)$ on $\Lambda^k(G)$ by

$$(\omega | \varphi) := \int_G \langle \omega(g) | \varphi(g) \rangle (g)\, dg.$$

Plainly, $(L_a^* \omega | L_a^* \varphi) = (\omega | \varphi)$ for all $a \in G$ and $\omega, \varphi \in \Lambda^k(G)$. Therefore, L_a^* extends to a unitary map and $(L_a^*)^* = L_{a^{-1}}^*$. The same is valid for the operators R_a^*.

Now, 1° follows from the bi-invariance of the Hurwitz–Haar integral:

$$M_l = \int_G L_g^*\, dg, \quad \text{i.e.,} \quad M_l \omega = \int_G L_g^* \omega\, dg,$$

$$(M_l \omega | \varphi) = \left(\int_G L_g^* \omega\, dg \,\Big|\, \varphi \right) = \int_G (L_g^* \omega | \varphi)\, dg$$

$$= \int_G (\omega | L_{g^{-1}}^* \varphi)\, dg = \int_G (\omega | L_a^* \varphi)\, da = (\omega | M_l \varphi).$$

Since M_l is bounded and $M_l M_l = M, M_l$ is a self-adjoint idempotent, i.e., an orthogonal projection. But

$$M_l \Lambda^k(G) \subset \Lambda_l^k(G);$$

therefore

$$M_l \Lambda^k(G) = \Lambda_l^k(G), \quad \text{for } k = 0, 1, \ldots, \dim G.$$

2° We recall that $(\delta\varphi|\omega) = (\varphi|d\omega)$; hence by virtue of 1° we obtain

$$(\omega\,|\,\delta M_1 M_r \varphi) = (d\omega\,|\,M_1 M_r \varphi) = (M_1 M_r d\omega\,|\,\varphi)$$
$$= (dM_1 M_r \omega\,|\,\varphi) = 0 \quad \text{for all } \omega \in \Lambda^k(G).$$

Therefore, $\delta(M_1 M_r \varphi) = 0$. But $d(M_1 M_r \varphi) = 0$ for each $\varphi \in \Lambda^k(G)$. Since

$$\Lambda^k_{\text{inv}}(G) = M_1 M_r \Lambda^k(G),$$

we have proved that each bi-invariant form is harmonic.

Corollary XIV.15.9 and the Hodge Theorem imply

THEOREM XIV.15.10. Let G be a connected compact Lie group. Then

1° $\Lambda^k_{\text{inv}}(G) = \mathscr{H}^k(G)$, $k = 0, 1, \ldots, \dim G$; bi-invariant k-forms are identical with harmonic k-forms.

2° $\Lambda^k(G) = \delta\Lambda^{k+1}(G) \oplus d\Lambda^{k-1}(G) \oplus \Lambda^k_{\text{inv}}(G)$.

PROOF. 1° The Cartan Theorem and the Hodge Theorem assert that

$$\Lambda^k_{\text{inv}}(G) \cong H^k(G) \cong \mathscr{H}^k(G),$$

i.e., all these spaces have the same dimension. But

$$\Lambda^k_{\text{inv}}(G) \subset \mathscr{H}^k(G), \quad \text{therefore} \quad \Lambda^k_{\text{inv}}(G) = \mathscr{H}^k(G).$$

2° now follows from 1° and from the Hodge–Kodaira–Weyl orthogonal decomposition of $\Lambda^k(G)$. □

16. COMPLEMENTS. EXERCISES

Lie Groups and Lie Algebras. Denote by $\mathscr{X}_1(G)$ left-invariant vector fields on a Lie group G:

$$X \in \mathscr{X}_1(G) \quad \text{if } (TL_g)X(x) = X(gx), \quad g, a \in G.$$

We have the isomorphism $L(G) \cong \mathscr{X}_1(G)$:

(1) $\quad L(G) \ni h \to X^h \in \mathscr{X}_1(G), \quad X^h(e) := h.$

$L(G) := T_e G$ is endowed with the structure of a Lie algebra

(2) $\quad [h, k] := [X^h, X^k](e).$

If G and H are Lie groups and $f\colon G \to H$ is a homomorphism, then we denote

(3) $\quad Lf\colon LG \to LH, \quad Lf := T_e f.$

DEFINITION. Denote $\alpha(g): G \to G$, $\alpha(g)x := gxg^{-1}$; then

(4) $$\mathrm{Ad}: G \to \mathrm{Aut}(LG), \quad g \to L\alpha(g),$$

is called the *adjoint representation* of G.

ad $:= L\mathrm{Ad} := LG \to L\mathrm{Aut}(LG) = \mathrm{End}(LG)$ is called the *adjoint representation* of the Lie algebra LG.

Exercise 1. Prove that $\mathrm{Ad}: G \to \mathrm{Aut}(LG)$ is smooth.

Exercise 2. A covariant tensor field Φ on G is bi-invariant iff

(5) $$(\mathrm{Ad}\, g)^* \Phi_e = \Phi_e \quad \text{for all } g \in G.$$

Exercise 3. Denote by ω_l a unique left-invariant volume form on G such that $\omega_l(e) = \mathrm{vol}_e$, where vol_e is a fixed orientation of $T_e G$.

Exercise 4. Prove the following equations:

(6) $$R_a^* \omega_l = \det[\mathrm{Ad}\, a^{-1}]\omega_l,$$

(7) $$\omega_r(a) = \det[\mathrm{Ad}\, a]\, \omega_l(a).$$

In this way the following fact is proved.

PROPOSITION XIV.16.1. A Lie group admits a bi-invariant volume form if and only if

(8) $$\det[\mathrm{Ad}(g)] \equiv 1 \quad \text{for all } g \in G.$$

DEFINITION. A Lie group G is called *unimodular* if $|\det[\mathrm{Ad}(\cdot)]| = 1$, i.e., if the left Hurwitz integral μ_l is also right-invariant.

On several occasions we have used the following

THEOREM XIV.16.2. 1° The Lie groups G for which $\mathrm{Ad}(G)$ is compact are unimodular. Therefore,

2° Compact Lie groups are unimodular,

3° Abelian Lie groups are unimodular.

PROOF. 1° The group $\{|\det \mathrm{Ad}(g)|: g \in G\}$ is a compact subgroup of R_+, the multiplicative group of positive real numbers. Such a subgroup consists of one element, $\{1\}$, and so G is unimodular

2° follows from 1°.

3° Obvious either from definitions, since left- and right-invariance is the same in the case of commutative groups, or by 1° since $\alpha(g) = \mathrm{id}_G$, and, therefore, $\mathrm{Ad}\, g = \mathrm{id}_{LG}$. There is a very important class of unimodular groups, the semisimple Lie groups first investigated by Wilhelm Killing and E. Cartan.

DEFINITION. Let E be a Lie algebra (of finite dimension) over $\boldsymbol{R}$ or $\boldsymbol{C}$. The bilinear form

$$B(X, Y) := \operatorname{Trace}(\operatorname{ad} X \operatorname{ad} Y)$$

is called the *Killing form* of E.

Exercise 5. Prove the following invariance of the Killing form:

$$B(\varphi X, \varphi Y) = B(X, Y) \quad \text{for } \varphi \in \operatorname{Aut}(E).$$

DEFINITION. A Lie algebra is semisimple if its Killing form is non-degenerate. A Lie group G is semisimple if its Lie algebra LG is semisimple.

We have the following important

PROPOSITION XIV.16.3. Semisimple Lie groups G are unimodular.

PROOF. Each $\operatorname{Ad}(a)$ leaves invariant a non-degenerate 2-linear form on LG, namely the Killing form B. Therefore,

$$(\det \operatorname{Ad}(a))^2 = 1, \quad a \in G$$

and, by virtue of Theorem XIV.16.2, G is unimodular. □

Exercise 6. Prove that $\operatorname{ad}(X)Y = [X, Y]$.

This fact suggests the following

DEFINITION. Let $E = (E, [\cdot, \cdot])$ be a Lie algebra; then the map $E \to \operatorname{End}(E)$ given by

$$X \to [X, Y]$$

is denoted by $\operatorname{ad}(X)$ and called the *adjoint representation* of the Lie algebra E.

XV. ELEMENTARY PROPERTIES OF HOLOMORPHIC FUNCTIONS OF SEVERAL VARIABLES HARMONIC FUNCTIONS

This chapter consists of two parts. In the first we prove some elementary properties of holomorphic maps and complex-valued differential forms on complex manifolds.

The most important theorem of this part is perhaps the theorem of Montel–Stieltjes–Vitali, which asserts that the space $A(\mathcal{O})$ of holomorphic functions (on a domain $\mathcal{O} \subset \boldsymbol{C}^n$) is a Montel space. More profound theorems of complex analysis will be considered in Chapter XVI and in Part III.

The real and imaginary parts of a holomorphic function are harmonic functions; therefore, harmonic functions have many properties of holomorphic functions: they are analytic, satisfy the maximum principle, possess the mean-value property (on spheres), and the space $\mathscr{H}(\mathcal{O})$ of harmonic functions is a Montel space. Therefore, the second part (Sections 5–7) of the present chapter is devoted to harmonic functions, which are called also potential functions.

We have in the "Poisson Integral Formula "(Section 6) a very powerful instrument for investigating harmonic functions. The maximum principle is valid for a larger class $\underline{\mathscr{H}}(\mathcal{O})$ of functions: subharmonic functions. These functions and the "Harnack Principle" allowed Oskar Perron to develop a beautiful, simple, and general method of solving the Dirichlet problem for quite general domains in $\boldsymbol{R}^m$ (Section 7). This part of the present chapter may be considered as a first introduction to potential theory.

1. HOLOMORPHIC MAPPINGS. CAUCHY–RIEMANN EQUATIONS

Let us recall some elementary notions of linear algebra.

Linear and Antilinear Mappings. Let E, F be complex vector spaces.

DEFINITION. (i) A map $u\colon E \to F$ is $\boldsymbol{C}$*-linear* if

$$u(\alpha e_1+\beta e_2) = \alpha u(e_1)+\beta u(e_2)$$

for any $\alpha, \beta \in \boldsymbol{C}$, $e_1, e_2 \in E$.

(ii) $v\colon E \to F$ is *antilinear* if

$$v(\alpha e_1+\beta e_2) = \bar{\alpha} v(e_1)+\bar{\beta} v(e_2),$$

for $\alpha, \beta \in \boldsymbol{C}$, $e_1, e_2 \in E$.

(iii) $w\colon E \to F$ is ***R*-linear** if

$$w(\alpha e_1+\beta e_2) = \alpha w(e_1)+\beta w(e_2)$$

for any $\alpha, \beta \in \mathbf{R}, e_1, e_2 \in E$.

We shall denote by $L_{\mathbf{C}}(E, F)$ (resp. $L_{\bar{\mathbf{C}}}(E, F), L_{\mathbf{R}}(E, F)$) the space of $\mathbf{C}$-linear, (resp. antilinear, resp. $\mathbf{R}$-linear) maps $E \to F$.

We have the following simple

LEMMA XV.1.1. Let E, F be $\mathbf{C}$-vector spaces. For any $\mathbf{R}$-linear map $u\colon E \to F$ denote

$$(1)\qquad \begin{aligned} u'(e) &:= \tfrac{1}{2}\big(u(e)-iu(ie)\big), \quad e \in E \\ u''(e) &:= \tfrac{1}{2}\big(u(e)+iu(ie)\big), \quad e \in E. \end{aligned}$$

Then u' is $\mathbf{C}$-linear and u'' is antilinear. Moreover, we have a canonical direct sum decomposition

$$(2)\qquad L_{\mathbf{R}}(E, F) = L_{\mathbf{C}}(E, F) \oplus L_{\bar{\mathbf{C}}}(E, F).$$

Remark. If E, F are complex Banach spaces, then by $L_{\mathbf{R}}(E, F)$, $L_{\mathbf{C}}(E, F), L_{\bar{\mathbf{C}}}(E, F)$ we denote the Banach spaces of the corresponding bounded (i.e., continuous) maps and (2) is then a decomposition into a direct sum of Banach spaces.

PROOF. We shall only check that u'' is antilinear. It is sufficient to prove that $u''(ie) = -iu''(e)$; $2u''(ie) = u(ie)+iu(iie) = u(ie)-iu(e) = -i\big(u(e)+iu(ie)\big) = -2u''(e)$. □

Holomorphic Maps. Let E and F be complex Banach spaces. They can be treated as vector spaces over $\mathbf{R}$. Let U be an open subset of E and let $u\colon U \to F$ be differentiable in U. Then for any $x \in U$ the derivative $d_x u$ of u at x is a continuous $\mathbf{R}$-linear map $E \to F$, whence by Lemma XV.1.1 we can decompose d_x into a $\mathbf{C}$-linear part $(d_x u)'$ and an antilinear part $(d_x u)''$

$$d_x u = d_x u' + d_x u'', \qquad \text{for any } x \in U.$$

Usually one writes $du' \equiv d'u, du'' \equiv d''u$. Therefore, for any differentiable map $u\colon U \to F$ we have

$$(3)\qquad d_x u = d'_x u + d''_x u,$$

where $d'_x u \in L_{\mathbf{C}}(E, F), d''_x u \in L_{\bar{\mathbf{C}}}(E, F)$.

Remark. d'' is called the *Cauchy–Riemann operator*, and in American literature it is mostly denoted by $\bar{\partial}$ (and d' is denoted by ∂).

We can now introduce the fundamental notion of holomorphicity:

DEFINITION. Let U be an open subset of a complex Banach space E, and let F also be a complex Banach space; a map $u\colon U \to F$ is *holomorphic* if:

(i) $u \in \boldsymbol{C}^1(U, F)$.

(ii) In any $x \in U$ the derivative $d_x u$ of u is $\boldsymbol{C}$-linear.

This means that $du = d'u$ and, hence, that the antilinear part of du vanishes; therefore, we can replace (ii) by

(ii)′ $d''_x u = 0$ identically for $x \in U$.

Hence we can express this briefly:

COROLLARY XV.1.2. A $\boldsymbol{C}^1$-map $u\colon U \to F$ is holomorphic iff it satisfies the Cauchy–Riemann (C–R) equation

$$d''u = 0 \quad (\text{or } \bar{\partial}u = 0 \text{ in the other notation}). \tag{4}$$

The space of holomorphic maps is denoted by $A(U, F)$ or $\mathrm{Hol}(U, F)$ or $\mathcal{O}(U, F)$. If $F = \boldsymbol{C}$, holomorphic functions are denoted by $A(U)$ or $\mathcal{O}(U)$.

The chain rule and the definition immediately imply

THEOREM XV.1.3. 1° The composition $u \circ v$ of holomorphic maps is holomorphic.

2° The inverse map u^{-1} of a holomorphic diffeomorphism u is holomorphic.

3° If $F = \boldsymbol{C}^m$ and $u = (u_1, \ldots, u_m)$, where $u_1, \ldots, u_m$ are components of u, then $u\colon U \to \boldsymbol{C}^m$ is holomorphic iff all components $u_1, \ldots, u_m$ are holomorphic functions.

We can identify $\boldsymbol{R}^{2n} = \boldsymbol{R}^n \times \boldsymbol{R}^n$ with $\boldsymbol{C}^n$ by the $\boldsymbol{R}$-linear isomorphism $\lambda\colon \boldsymbol{R}^n \times \boldsymbol{R}^n \to \boldsymbol{C}^n$,

$$\lambda(x, y) := x + iy, \qquad \lambda^{-1}(z) := \left(\tfrac{1}{2}(z+\bar{z}), \tfrac{1}{2}i(z-\bar{z})\right)$$

where

$$x = (x^1, \ldots, x^n), \qquad z = (z^1, \ldots, z^n), \qquad \bar{z} = (\overline{z^1}, \ldots, \overline{z^n}),$$

$$i = \sqrt{-1}.$$

We can now define the following differential operators in $\boldsymbol{C}^n$:

$$(5)\qquad \frac{\partial}{\partial z} := \frac{1}{2}\left(\frac{\partial}{\partial x} - i\frac{\partial}{\partial y}\right)$$

in coordinates $\dfrac{\partial}{\partial z^k} = \dfrac{1}{2}\left(\dfrac{\partial}{\partial x^k} - i\dfrac{\partial}{\partial y^k}\right)$,

$$(6)\qquad \frac{\partial}{\partial \bar{z}} := \frac{1}{2}\left(\frac{\partial}{\partial x} + i\frac{\partial}{\partial y}\right)$$

in coordinates $\dfrac{\partial}{\partial \bar{z}^k} = \dfrac{1}{2}\left(\dfrac{\partial}{\partial x^k} + i\dfrac{\partial}{\partial y^k}\right)$.

We immediately check that

$$(7)\qquad d'u = \frac{\partial u}{\partial z}, \qquad d''u = \frac{\partial u}{\partial \bar{z}}.$$

Hence $\operatorname{Re} u_\mu$ and $\operatorname{Im} u_\mu$ are harmonic functions.

We thus have "another" definition of holomorphicity:

PROPOSITION XV.1.4. Let U be an open subset of $\boldsymbol{C}^n$. A C^1-map $u\colon U \to F$ is holomorphic iff u satisfies the system of C–R-equations

$$\text{(C–R)}\qquad \frac{\partial u}{\partial x} = -i\frac{\partial u}{\partial y}$$

or in coordinates

$$(8)\qquad \frac{\partial u}{\partial x^k} = -i\frac{\partial u}{\partial y^k}, \qquad k = 1, \ldots, n.$$

Consider a more special case: Let $F = \boldsymbol{C}^m$. Identifying $\boldsymbol{R}^{2m}$ with $\boldsymbol{C}^m$ by means of the isomorphism λ and writing $u\colon U \to \boldsymbol{C}^m$, $u = u' + iu''$, where $u'_\mu = \operatorname{Re} u_\mu$, $u''_\mu = \operatorname{Im} u_\mu$ we can rewrite (8) as

$$\frac{\partial u}{\partial x} = \frac{\partial u'}{\partial x} + i\frac{\partial u''}{\partial x}, \qquad \frac{\partial u}{\partial y} = \frac{\partial u'}{\partial y} + i\frac{\partial u''}{\partial y},$$

$$\frac{\partial u'}{\partial x} + i\frac{\partial u''}{\partial x} = -i\frac{\partial u'}{\partial y} + \frac{\partial u''}{\partial y},$$

whence, by comparing the real and the imaginary parts, we obtain

COROLLARY XV.1.5 (Real Form of C–R-equations). Let U be an open subset of $\boldsymbol{C}^n$, then C^1-map $u\colon U \to \boldsymbol{C}^m$ is holomorphic iff u' and

u'' satisfy the system of first-order equations (C–R)

$$\frac{\partial \operatorname{Re} u}{\partial x} = \frac{\partial \operatorname{Im} u}{\partial y}, \qquad \frac{\partial \operatorname{Im} u}{\partial x} = -\frac{\partial \operatorname{Re} u}{\partial y} \tag{9}$$

or in coordinates

$$\frac{\partial u'_\mu}{\partial x^k} = \frac{\partial u''_\mu}{\partial y^k}, \qquad \frac{\partial u''_\nu}{\partial x^l} = -\frac{\partial u'_\nu}{\partial y^l}, \tag{9'}$$
$$k, l = 1, \ldots, n, \qquad \mu, \nu = 1, \ldots, m.$$

Therefore

$$\sum \frac{\partial^2 u'_\mu}{(\partial x^k)^2} + \sum \frac{\partial^2 u'_\mu}{(\partial y^k)^2} = 0 = \sum \frac{\partial^2 u''_\mu}{(\partial x^k)^2} + \sum \frac{\partial^2 u''_\mu}{(\partial y^k)^2} = 0. \square \tag{10}$$

Remark. Equations (8) and (9) are also called Cauchy–Riemann equations. For $m = n = 1$ they were already known to d'Alembert and Gauss. (Equations (10) are called *Laplace equations.*)

Let $u \in A(U, C^m)$, $U \subset C^n$. By virtue of the C–R-equations (9′) we can write down on the following form the Jacobian matrix of du:

$$\frac{\partial(u)}{\partial(x)} = d(u) = \begin{bmatrix} \frac{\partial u'_1}{\partial x^1} & \cdots & \frac{\partial u'_1}{\partial x^n} & -\frac{\partial u''_1}{\partial x^1} & \cdots & -\frac{\partial u''_1}{\partial x^n} \\ \cdots & \cdots & \cdots & \cdots & \cdots & \cdots \\ \frac{\partial u'_m}{\partial x^1} & \cdots & \frac{\partial u'_m}{\partial x^n} & -\frac{\partial u''_m}{\partial x^1} & \cdots & -\frac{\partial u''_m}{\partial x^n} \\ \frac{\partial u''_1}{\partial x^1} & \cdots & \frac{\partial u''_1}{\partial x^n} & -\frac{\partial u'_1}{\partial x^1} & \cdots & \frac{\partial u'_1}{\partial x^n} \\ \cdots & \cdots & \cdots & \cdots & \cdots & \cdots \\ \frac{\partial u''_m}{\partial x^1} & \cdots & \frac{\partial u''_m}{\partial x^n} & \frac{\partial u'_m}{\partial x^1} & \cdots & \frac{\partial u'_m}{\partial x^n} \end{bmatrix}.$$

If $m = n$, we obtain for the Jacobian

$$\frac{\partial(u'_1, \ldots, u'_n, \ldots, u''_n)}{\partial(x^1, \ldots, x^n, y^1, \ldots, y^n)}$$
$$= \begin{vmatrix} \frac{\partial u'_1}{\partial x^1} & \cdots & \frac{\partial u'_1}{\partial x^n} \\ \cdots & \cdots & \cdots \\ \frac{\partial u'_n}{\partial x^1} & \cdots & \frac{\partial u'_n}{\partial x^n} \end{vmatrix}^2 + \begin{vmatrix} \frac{\partial u''_1}{\partial x^1} & \cdots & \frac{\partial u''_1}{\partial x^n} \\ \cdots & \cdots & \cdots \\ \frac{\partial u''_n}{\partial x^1} & \cdots & \frac{\partial u''_n}{\partial x^n} \end{vmatrix}^2 \geqslant 0.$$

PROPOSITION XV.1.6. Let u be a biholomorphism (it is a holomorphic diffeomorphism), $U \overset{u}{\rightarrow} V \subset C^n$; then the Jacobian of u is > 0.

We have thus proved the following

COROLLARY XV.1.7. Every complex n-dimensional manifold is orientable.

Holomorphic Maps of Complex Manifolds. We recall that a topological manifold M modelled on a complex Banach space is a complex (holomorphic) manifold if there is a holomorphic atlas $(u_i, M_i)_{i \in I}$ of M, where each u_i: $M_i \to E$ is a homeomorphism, and for all $i, k \in I$

$$u_k \circ (u_i | M_i \cap M_k)^{-1}, \quad u_i(M_i \cap M_k) \to u_k(M_k \cap M_i)$$

are holomorphic. We can now extend the notion of holomorphic maps in the same manner as in case of differentiable maps.

DEFINITION. Let M be a complex manifold modelled on E provided with a holomorphic atlas $\{(u_i, M_i)$: $i \in I\}$, let N be a complex manifold modelled on a complex Banach space F, and let $\{(v_j, N_j)$: $j \in J\}$ be a holomorphic atlas of N. f: $M \to N$ is *holomorphic* if for any two charts u_i, v_j, $(i, j) \in I \times J$

$$f_{u_i, v_j} := v_j \circ f \circ u^{-1} \cdot u_i(M_i) \to v_j(N_j)$$

is a holomorphic map (of the open subset $u_i(M_i)$ of E into an open subset $v_j(N_j)$ of E).

Remark. Plainly, any complex manifold M is of class C^1 by definition. More precisely, it is provided with a real differentiable structure of class C^1. In order to avoid confusion, one denotes by M^R the manifold M endowed with this C^1-real structure.

Therefore, as a corollary of Proposition XV.1.6, we have

THEOREM XV.1.8. Let M be a complex manifold of complex dimension n. Then the manifold M^R is a real manifold of real dimension $2n$; it is orientable and endowed with a natural orientation.

The Implicit Map Theorem. From Theorem VIII.3.1 we obtain the following holomorphic version of the Graves Theorem:

THEOREM XV.1.9. Let E and F be complex Banach spaces and let h be a mapping of an open subset V of the product $E \times F$ into F. Suppose that there exist a point $(e_0, f_0) \in V$ such that $h(e_0, f_0) = 0$ and an open neighbourhood $U \subset V$ of (e_0, f_0). Suppose that $h \in A(U, F)$.

If $\nabla_F \bar{h}(e_0, f_0) \in L_C(F, F)$ has a bounded inverse, then h generates an implicit mapping $G: K(e_0, r_1) \to K(f_0, r_2) \subset F$ for some $r_1, r_2 > 0$. The mapping G is holomorphic and its derivative is given by

$$(*) \qquad \nabla G(e) = -(\nabla_F h(e,f))^{-1} \circ \nabla_E h(e,f).$$

PROOF. From the Graves theorem we know that G is of class C^1 and formula $(*)$ shows that $\nabla G(e) \in L_C(E, F)$ is C-linear, and thus G is holomorphic. □

Remark. The Rank Theorem VIII.2.8 is valid *mutatis mutandis* for holomorphic mappings.

Examples. 1. Let $u: C \to C$, where $u(z) = z^k$, $k \in N$, then u is holomorphic since $h \to \nabla u(z_0) \cdot h = kz_0^{k-1} \cdot h$ is C-linear. Plainly, $\dfrac{\partial z^k}{\partial \bar{z}} = 0$.

2. Let $p(z)$ $(=p(z^1, \ldots, z^n))$ be a polynomial in C^n of order k:

$$p(z) = \sum_{|\alpha| \leq k} c_\alpha z^\alpha, \qquad \alpha \in N^n$$

where $z^\alpha = z_1^{\alpha_1}, \ldots, z_n^{\alpha_n}$,

$$\nabla p(a) = \left(\sum \alpha_1 \ldots \alpha_n c_{\alpha_1 \ldots \alpha_n} a_1^{\alpha_1 - 1} a_2^{\alpha_2 - 1} a_n^{\alpha_{n-1}}\right),$$

is plainly a C-linear map $C^n \to C^n$.

3. The function $f: C \to C$, $f(z) := \bar{z}$ is not holomorphic (it is antiholomorphic) since $\dfrac{\partial f}{\partial \bar{z}} = 1$.

4. The function $\exp(z) := \cos y + i \sin y_2$ $(z := x + iy)$ is holomorphic on the whole of C since $\nabla \exp(z_0) = \exp(z_0)$.

5. Projective spaces $P(C^n)$ are holomorphic manifolds and, therefore, $P(R^{2n}) \cong P(C^n)^R$ are orientable for every $n = 1, 2, \ldots$

Remark. We have seen that $P(R)$ is not orientable. In Part III we shall prove that $P(R^{2k+1})$, $k = 0, 1, \ldots$ is not orientable.

2. DIFFERENTIAL FORMS ON COMPLEX MANIFOLDS. FORMS OF TYPE (p, q). OPERATORS d' AND d''

Let M be a complex manifold. Since the space $L(T_x M, C)$ can be considered as a "complexification" of $T(M)^*$, let us consider the notion of complexification in complete generality.

Complexification. Let V be a real vector space, and assume that there is an $\boldsymbol{R}$-automorphism

$$(1) \qquad J\colon V \to V \quad \text{such that } J^2 = -\mathrm{id}_V;$$

then the definition

$$(2) \qquad (a+ib)v := av+bJ(v)$$

extends scalar multiplication (by real numbers) to $\boldsymbol{C}$ (scalar multiplication by complex numbers). Therefore, (2) converts V into a vector space over $\boldsymbol{C}$.

Conversely, if E is a complex vector space, then $E \ni e \to \sqrt{-1}\, e \in E$ is an automorphism J of E such that $J \circ J = -\mathrm{id}_E$. We have proved

PROPOSITION XV.2.1. If V is a vector space over $\boldsymbol{R}$, then scalar multiplication in V has an extension to $\boldsymbol{C}$ iff there exists an automorphism $J\colon V \to V$ such that $J^2 = -\mathrm{id}_V$.

The pair (V, J) is called a *complex structure on the real space* V.

If V is a vector space over $\boldsymbol{R}$, it is often desirable to imbed V $\boldsymbol{R}$-linearly into a complex vector space W. A "minimal" such W will be called a *complexification* of V.

We shall now give "two" constructions of such embeddings.

1° Consider the product $V\times V$ over $\boldsymbol{R}$ and the mapping

$$(3) \qquad J\colon V\times V \to V\times V, \qquad J(x, y) := (-y, x).$$

Plainly, $J^2 = -\mathrm{id}_{V\times V}$, and thus scalar multiplication (in $V\times V$) can be extended by (2) to a map $\boldsymbol{C}\times(V\times V) \to V\times V$.

Thus $i(y, 0) = (0, y)$ and if we write $(x, 0) \equiv x$ for all $x \in V$, then each $z \in V\times V$ has a unique representation $z = x+iy$, with $x, y \in V$.

DEFINITION. If V is a vector space over $\boldsymbol{R}$, then $V\times V$ with the above constructed multiplication by complex scalars is a complex vector space called a complexification of V and denoted $V^{\boldsymbol{C}}$. Considered as a vector space over $\boldsymbol{R}$ $V^{\boldsymbol{C}}$ is isomorphic to $V \oplus V$.

We can arrive at the same result by a more formal procedure. Let V be a vector space over $\boldsymbol{R}$.

2° Consider the space $V^{\boldsymbol{C}}$ of formal sums

$$z = x+iy, \quad \text{where } x, y \in V.$$

We have

$$(x_1+iy_1)+(x_2+iy_2) := (x_1+x_2)+i(y_1+y_2),$$
$$(a+ib)(x+iy) = (ax-by)+i(bx+ay) \quad \text{for } a, b \in \boldsymbol{R}.$$

In this way $V^C = V \oplus iV$. If $z = x+iy$, then $\bar{z} := x-iy$.

Remarks. 1. We see that the complexification of V has a richer structure than an arbitrary complex vector space.

2. Not every real vector space permits the introduction of a complex structure.

Example. $V = \boldsymbol{R}$. Assume that J is $\boldsymbol{R}$-linear and $J^2 = -\mathrm{id}_{\boldsymbol{R}}$; then $Jx = ax$, $a \in \boldsymbol{R}$. Therefore, $J^2x = a^2x = -x$, and thus $a^2 = -1$, a contradiction.

Exercise. Prove that (more generally) no real vector space of odd dimension allows a complex structure.

Example. $(\boldsymbol{R}^n)^C = \boldsymbol{R}^n \oplus i\boldsymbol{R}^n = \boldsymbol{C}^n$.

Since $\boldsymbol{R}^n$ can be considered as a subspace of $(\boldsymbol{C}^n)^{\boldsymbol{R}}$, the following question arises: is it the case for general complex vector spaces? The answer is "yes":

PROPOSITION XV.2.2. Let E be a vector space over $\boldsymbol{C}$. Call a real subspace F of E properly real if $F \cap iF = \{0\}$. Then there exists a properly real subspace G of F such that $E = G+iG$.

PROOF (by the Kuratowski–Zorn Lemma) is left to the reader.

Let X and V be real Banach spaces and let $E = X \oplus iX = X^C$; $F = V \oplus iV = V^C$.

If $U \subset E$, u: $U \to V \oplus iV$, then

$$u = u' + iu'',$$

where

$$u' := \operatorname{Re} u, \quad u'' := \operatorname{Im} u$$

are well defined. We can now write the real form of C–R-equations.

PROPOSITION XV.2.3 (Real Form of C–R-Equations, General Case). Let U be an open subset of $E := X \oplus iX$ and let $F := V \oplus iV$, where X and V are real Banach spaces. If u: $U \to F$, then we denote $u' := \operatorname{Re} u$,

$u'' := \operatorname{Im} u$. Thus, $u = u' + iu''$. Then a $u \in C^1(U, F)$ is holomorphic ($u \in A(U, F)$) iff the real and the imaginary parts u' and u'' of u satisfy the equations

$$\frac{\partial u'}{\partial x} = \frac{\partial u''}{\partial y}, \qquad \frac{\partial u''}{\partial x} = -\frac{\partial u'}{\partial y},$$

where $\frac{\partial}{\partial x}$ denotes the partial derivative with respect to the real subspace of F and $\frac{\partial}{\partial y}$ is the partial derivative with respect to the imaginary subspace of F.

Alternating Forms of Type (p, q). In Section 1 of Chapter XIV we considered vector-valued alternating forms. We shall now consider $\boldsymbol{C}$-valued forms in more detail.

DEFINITION. Let E be a complex Banach space; an *alternating* $\boldsymbol{C}$*-valued r-form* φ *is of type* (p, q) with $p, q, \in \boldsymbol{N}, p+q = r$, if

$$\varphi(ce_1, \ldots, ce_r) = c^p \bar{c}^q \varphi(e_1, \ldots, e_r) \quad \text{for all } c \in \boldsymbol{C}.$$

LEMMA XV.2.4. If $0 \neq \varphi \in A_r(E, \boldsymbol{C})$ and is of type (p, q), then p and q are uniquely defined.

PROOF. Let φ be of type (p, q) and (p', q'). Then

$$\varphi(ce_1, \ldots, ce_r) = \begin{cases} c^p \bar{c}^q \varphi(e_1, \ldots, e_r), \\ c^{p'} \bar{c}^{q'} \varphi(e_1, \ldots, e_r). \end{cases}$$

Therefore, $c^p c^q = c^{p'} c^{q'}$ for all $c \in \boldsymbol{C}$.

Let $c = \exp(it)$ with arbitrary $t \in \boldsymbol{R}$. Then

$$\exp\big(it(p-q)\big) = \exp\big(it(p'-q')\big) \quad \text{for all } t \in \boldsymbol{R}.$$

Therefore, $p - q = p' - q'$. But $p + q = p' + q' = r$, whence

$$p = p', \quad q = q'. \ \square$$

Let $A_{(p,q)}(E)$ denote forms of type (p, q). Define $\bar{\varphi}$ by $\bar{\varphi}(e_1, \ldots, e_r) := \overline{\varphi(e_1, \ldots, e_r)}$. The reader can verify the following

PROPOSITION XV.2.5. 1° $\big(\varphi \in A_{(p,q)}(E)\big) \Rightarrow \big(\bar{\varphi} \in A_{(q,p)}(E)\big)$,

2° $\big(\varphi \in A_{(p,q)}(E), \psi \in A_{(p',q')}(E)\big) \Rightarrow \big(\varphi \wedge \psi \in A_{(p+p',q+q')}(E)\big)$,

3° $A_{(p,q)}(E)$ is a complex vector space.

We can now define

Differential Forms on Complex Manifolds. Let M be a smooth complex manifold modelled on a complex Banach space E. Denote by $T_x(M)$ or $T_x M$ the (real) tangent space at $x \in M$. Then $A_r(T_x M, \boldsymbol{C})$ is the complexification of the real Banach space $A_r(T_x M, \boldsymbol{R})$.

Example. $A_1(T_x, \boldsymbol{C}) = (T_x M)' \oplus i(T_x M)' = (T_x M)'^{\boldsymbol{C}}$.

DEFINITION. Let U be an open subset of a complex manifold. A continuous mapping

$$\varphi\colon U \to \coprod_{x \in U} A_r(T_x M, \boldsymbol{C})$$

is called a *complex differential r-form on U*.

The linear space of a complex r-form on U is denoted by $\Lambda^r(U, \boldsymbol{C})$ or shortly $\Lambda^r(U)$. The space of differential r-forms of type (p, q) is denoted by $\Lambda^{(p,q)}(U)$. As we have seen, p and q are uniquely defined.

Exterior multiplication and exterior derivative are defined in the same way as for $\boldsymbol{R}$-valued forms and of course they have the same properties:

THEOREM XV.2.6. The exterior derivative has the following properties:

1° $d\colon \Lambda^r(U, \boldsymbol{C}) \to \Lambda^{r+1}(U, \boldsymbol{C})$ and is $\boldsymbol{C}$-linear,

2° $d \circ d = 0$,

3° $d(\overset{r}{\varphi} \wedge \psi) = d\varphi \wedge \psi + (-1)^r \overset{r}{\varphi} \wedge d\psi$,

4° d is a real operator: $d\bar{\varphi} = \overline{d\varphi}$,

5° d is $\boldsymbol{C}$-linear $d(c\varphi) = c\,d\varphi$, $c \in \boldsymbol{C}$.

Differential Forms on n-dimensional Complex Manifolds. Let M be a (smooth) complex manifold of complex dimension n, i.e., let M be modelled on $\boldsymbol{C}^n = \boldsymbol{R}^{2n}$. The tangent space $T_{x_0}(M)$ is a real vector space of real dimension $2n$ and

$$\frac{\partial}{\partial x^1}, \ldots, \frac{\partial}{\partial x^n}, \frac{\partial}{\partial y^1}, \ldots, \frac{\partial}{\partial y^n}$$

form a basis $T_{x_0} M$. We shall now construct a basis of $A_1(T_{x_0} M, \boldsymbol{C}) = T_{x_0}(M)^* \oplus i(T_{x_0} M)^*$.

Let $(z^1, \dots, z^n)$ be a coordinate system in a neighbourhood U of $x_0 \in M$. Let $\xi = \xi_{x_0} \in T_{x_0}M$; then we define $2n$ complex 1-forms

$$(4) \qquad dz^k(\xi) := \xi(z^k), \quad d\bar{z}^k(\xi) := \overline{dz^k}(\xi) = \overline{\xi(z^k)} = \xi(\bar{z}^k), \quad k = 1, \dots, n.$$

LEMMA XV.2.7. $\{dz^1, \dots, dz^n, d\bar{z}^1, \dots, d\bar{z}^n\}$ is a basis of $A_1(T_{x_0}C^n)$. We have

$$(5) \qquad dz^k = dx^k + idy^k, \quad d\bar{z}^k = dx^k - idy^k, \; k = 1, \dots, n,$$
$$dz^k \in \Lambda^{(1,0)}(U), \quad d\bar{z}^l \in \Lambda^{(0,1)}(U).$$

More generally

$$dz^{k_1} \wedge \dots \wedge dz^{k_p} \wedge d\bar{z}^{j_1} \wedge \dots \wedge d\bar{z}^{j_q} \in \Lambda^{(p,q)}(U).$$

We shall show that each differential complex r-form has a canonical (or normal) form (cf. Corollary XIV.2.7).

THEOREM XV.2.8. 1° Let $\varphi \in \Lambda^r_{(l)}(U, C)$; then

$$(6) \qquad \varphi = \sum_{p+q=r} \varphi^{(p,q)}, \text{ where } \varphi^{(p,q)} \text{ are of type } (p, q).$$

This representation is unique. Each $\varphi^{(p,q)}$ has in the chart $(z^1, \dots, z^n)$ the representation

$$(7) \qquad \varphi^{(p,q)} = \sum a_{I,J}\, dz^I \wedge d\bar{z}^J,$$

where $I = (i_1, \dots, i_p)$, $J = (j_1, \dots, j_q)$ are multi-indices

$$0 \leqslant i_1 < \dots < i_p \leqslant n, \quad 1 \leqslant j_1 < \dots < j_q \leqslant n$$

and $a_{I,J} \in C^l(U, C)$.

PROOF. We see that (6), (7) is a representation of the required shape. We only have to prove its uniqueness. Let

$$\varphi = \sum_{p+q=r} \varphi^{(p,q)} = \sum_{p+q=r} \psi^{(p,q)};$$

then

$$\sum_{p+q=r} \eta^{(p,q)} = 0, \quad \text{where } \eta^{(p,q)} := \varphi^{(p,q)} - \psi^{(p,q)}.$$

We have for any $c \in C$

$$(8) \qquad \sum_{p+q=r} \eta^{(p,q)}(c\xi_1, \dots, c\xi_r) = \sum_{p+q=r} c^p \bar{c}^q \eta^{(p,q)}(\xi_1, \dots, \xi_r) = 0,$$

where

$$\xi_1, \ldots, \xi_r \in T_{x_0}M.$$

For fixed $\xi_1, \ldots, \xi_r$ (8) is a polynomial equation in the polynomial ring $\boldsymbol{C}[c, \bar{c}]$.

It is well known that the coefficients $\psi^{(p,q)}(\xi_1, \ldots, \xi_r)$ vanish for all p, q. Since $\xi_1, \ldots, \xi_r$ can be chosen arbitrarily, we obtain $\eta^{(p,q)} = 0 = \varphi^{(p,q)} - \psi^{(p,q)}$. □

Operators d' and d''. We can now prove the important

THEOREM XV.2.9. 1° If $\varphi \in \Lambda^{(p,q)}(U)$ then

$$d\varphi = d'\varphi + d''\varphi, \text{ where } d'\varphi \in \Lambda^{(p+1,q)}(U), d''\varphi \in \Lambda^{(p,q+1)}(U).$$

2° d' and d'' are $\boldsymbol{C}$-linear maps and $d = d' + d''$.

3° $d' \circ d' = 0 = d'' \circ d'' = d' \circ d'' + d'' \circ d'$.

4° d' and d'' are not real operators since we have

$$\overline{d'\varphi} = d''\bar{\varphi}, \qquad \overline{d''\varphi} = d'\bar{\varphi}.$$

5° For any complex r-form φ and arbitrary form ψ we have

$$d'(\varphi \wedge \psi) = d'\varphi \wedge \psi + (-1)^r \varphi \wedge d'\psi,$$
$$d''(\varphi \wedge \psi) = d''\varphi \wedge \psi + (-1)^r \varphi \wedge d''\psi.$$

PROOF. 1° Take the canonical form

$$\varphi^{(p,q)} = \sum a_{I,J} dz^I \wedge d\bar{z}^J;$$

then

$$d\varphi^{(p,q)} = \sum_{I,J} da_{I,J} \wedge dz^I \wedge d\bar{z}^J$$

$$= \sum_{I,J} \sum_{\nu=1}^{n} \frac{\partial a_{I,J}}{\partial z^\nu} dz^\nu \wedge dz^I \wedge d\bar{z}^J + \sum_{I,J} \sum_{\nu=1}^{n} \frac{\partial a_{I,J}}{\partial \bar{z}^\nu} d\bar{z}^\nu \wedge dz^I \wedge d\bar{z}^J.$$

Denoting the summands by $d'\varphi^{(p,q)}$ and $d''\varphi^{(p,q)}$, we see that

$$d'\varphi^{(p,q)} \in \Lambda^{(p+1,q)}(U), \qquad d''\varphi^{(p,q)} \in \Lambda^{(p,q+1)}(U).$$

2° is obvious.

3° $0 = dd\varphi = (d' + d'') \circ (d' + d'')\varphi = d'd'\varphi + d'd''\varphi + d''d'\varphi + d''d''\varphi$.

If φ is of type (p, q), then the summands are of the following types:

$d'd'\varphi$ is of type $(p+2, q)$,

$d''d''\varphi$ is of type $(p, q+2)$,

$(d'd'' + d''d')\varphi$ is of type $(p+1, q+1)$.

Since the decomposition into forms of pure type is unique (Theorem XIV.2.7), each of the three summands must vanish.

4° Since $\overline{d\varphi} = d\overline{\varphi}$, we have

$$(*)\qquad (\overline{d'\varphi}+\overline{d''\varphi} = d'\overline{\varphi}+d''\overline{\varphi}) \Rightarrow ((\overline{d'\varphi}-d''\overline{\varphi})+(\overline{d''\varphi}-d'\overline{\varphi}) = 0).$$

But

$$(\overline{d'\varphi}-d''\overline{\varphi}) \in \Lambda^{(q,p+1)}, \qquad (\overline{d''\varphi}-d'\overline{\varphi}) \in \Lambda^{(q+1,p)};$$

therefore, both summands in $(*)$ vanish.

5° follows from the corresponding formula for d and by comparison of types as in the proof of 3° and 4°. □

Since the operator d'' is a natural generalization of the Cauchy–Riemann operator for functions, the following definition is not shocking.

DEFINITION. 1° A complex r-form is called *holomorphic* if φ is of type $(r, 0)$ and $d''\varphi = 0$. They are also called *forms of the first kind.*

2° A q-form ψ is *antiholomorphic* if ψ is of type $(0, q)$ and $d'\psi = 0$.

3° Holomorphic forms of type $(1, 0)$ are called *Abelian differentials* (or *forms*) *of first kind.*

COROLLARY XV.2.10. (φ is holomorphic) $\Leftrightarrow$ ($\overline{\varphi}$ is antiholomorphic).

Forms of the first kind play a fundamental role in the cohomology of compact complex manifolds. The dimension ($\dim \mathscr{A}^{(1,0)}(M)$) of the space of Abelian differentials of the first kind on a compact Riemann surface M is called the *genus* of M and is the most important topological invariant of the surface M. If was discovered by Riemann. A major part of Part III and Chapter XVI will be devoted to a global analysis of Riemann surfaces and complex manifolds.

3. CAUCHY'S FORMULA AND ITS APPLICATIONS

To begin with, let us give the *Cauchy formula for functions of one variable* for an arbitrary finite domain. By n-fold iteration this formula is transformed into the Cauchy formula for functions of n complex variables for sets which are the Cartesian products of n disks of circles, called *polydisks*:

$$D_n(\mathring{z}; r) = D_n(\mathring{z}; r_1, \ldots, r_n) := \bigtimes_{j=1}^{n} D(\mathring{z}_j, r_j),$$

where $D(\mathring{z}_j, r_j) := \{z \in \boldsymbol{C}\colon |z_j - \mathring{z}_j| < r_j\}$.

The Cauchy formula is the most important formula in the elementary theory of holomorphic functions.

Suppose that Ω is an open subset of $\boldsymbol{C}$ to which Stokes's Theorem is applied, e.g., let Ω be a finite domain in $\boldsymbol{R}^2 = \boldsymbol{C}$. Suppose that $u \in C^1(\overline{\Omega})$. Note that

$$du \wedge dz = \frac{\partial u}{\partial z} dz \wedge dz + \frac{\partial u}{\partial \bar{z}} d\bar{z} \wedge dz = \frac{\partial u}{\partial \bar{z}} d\bar{z} \wedge dz$$

$$= (2i) \frac{\partial u}{\partial \bar{z}} dx \wedge dy.$$

Stokes's formula thus becomes

$$(*) \qquad \int_{\partial\Omega} u\, dz = \int_{\partial\Omega} du \wedge dz = 2i \iint_{\Omega} \frac{\partial u}{\partial \bar{z}} dx \wedge dy = \int_{\Omega} \frac{\partial u}{\partial \bar{z}} d\bar{z} \wedge dz.$$

COROLLARY XV.3.1 (Cauchy Theorem). If $f \in C^1(\overline{\Omega}), f \in A(\Omega)$, then

$$(1) \qquad \int_{\partial\Omega} f\, dz = 0.$$

The formula

$$(2) \qquad f(\zeta) = \frac{1}{2\pi i} \int_{\partial\Omega} \frac{f(z)}{z-\zeta} dz$$

holds for every $\zeta \in \Omega$.

Instead of proving (2), we give a more general formula for any function $f \in C^1(\overline{\Omega})$.

THEOREM XV.3.2. If $f \in C^1(\Omega)$, then the formula

$$(3) \qquad f(\zeta) = \frac{1}{2\pi i} \left\{ \int_{\partial\Omega} \frac{f(z)}{z-\zeta} dz + \iint_{\Omega} \frac{1}{z-\zeta} \cdot \frac{\partial f}{\partial \bar{z}} dz \wedge d\bar{z} \right\}$$

holds for $\zeta \in \Omega$.

Note that if $f \in A(\Omega)$, then $\bar{\partial} f = 0$ and (3) reduces to (2).

PROOF. Suppose that $\varepsilon > 0$ is smaller than the distance from ζ to $\partial\Omega$. Next cut a disk $K(\zeta, \varepsilon)$ out of Ω. Let $\Omega_\varepsilon := \Omega - K(\zeta, \varepsilon)$; then the chain $\partial\Omega_\varepsilon$ is the difference between the chain $\partial\Omega$ and the circle $|z-\zeta| = \varepsilon_1$. We apply formula (*) to the domain Ω_ε, taking $u(z) = f(z)\, \dfrac{1}{z-\zeta}$.

Since $\dfrac{\partial}{\partial \bar{z}}\left(\dfrac{1}{z-\zeta}\right) = 0$ in Ω_ε, on going over to polar coordinates in the integral over the circle we obtain

$$\iint_{\Omega_\varepsilon} \frac{\partial f}{\partial \bar{z}} \cdot \frac{1}{z-\zeta} dz \wedge d\bar{z} = \int_{\partial\Omega} f(z) \frac{dz}{z-\zeta} - \int_0^{2\pi} f(\zeta+\varepsilon e^{i\theta}) i\, d\theta.$$

Since the function f is continuous in ζ, and the function $(z-\zeta)^{-1}$ is integrable (with respect to the Lebesgue measure on $\boldsymbol{R}^2$) over the set Ω, when we make use of the mean-value theorem for an integral over a circle and go to the limit with ε, we arrive at

$$i\int_0^{2\pi} f(\zeta+\varepsilon e^{i\theta})\, d\theta = 2\pi i f(\xi) \underset{\varepsilon\to 0}{\to} 2\pi i f(\zeta),$$

$$\iint_{\Omega_\varepsilon} \frac{\partial f}{\partial \bar{z}} \cdot \frac{1}{z-\zeta} dz \wedge d\bar{z} \underset{\varepsilon\to 0}{\to} \iint_{\Omega} \frac{\partial f}{\partial \bar{z}} \cdot \frac{1}{z-\zeta} dz \wedge d\bar{z}. \ \square$$

Suppose that D_n is a polydisk; $D_n = \bigtimes_{j=1}^{n} D(w_j, r_j)$. The set

$$\partial_0 D := \bigtimes_{j=1}^{n} \partial D(w_j, r_j)$$

is called the *distinguished boundary* of the polydisk D_n.

THEOREM XV.3.3. If $f \in C^1(\overline{\Omega})$ and if f is holomorphic with respect to every variable in Ω, then

(i) the formula

$$f(z) = \frac{1}{(2\pi i)^n} \int_{\partial_0 D} \frac{f(\zeta_1, \ldots, \zeta_n)}{(\zeta_1-z_1)\ldots(\zeta_n-z_n)} d\zeta_1 \ldots d\zeta_n, \qquad z \in D$$

holds for every closed polydisk $D \subset \Omega$;

(ii) the function f belongs to the space $C^\infty(\Omega)$ and is even analytic in Ω;

(iii) the Taylor formula holds:

$$f(z) = \sum_{\alpha} a_\alpha (z-w)^\alpha, \qquad \text{where } a_\alpha = \frac{1}{\alpha!}\partial^\alpha f(w).$$

Remark. Part (ii) is known as the *Osgood Lemma.*

Before we proceed with the proof, let us recall that a complex function f on an open set $\Omega \subset C^n$ is analytic in Ω if every point $w \in \Omega$ has a neighbourhood $O(w)$ such that f has a power-series expansion

$$f(z) = \sum_{\alpha_1, \dots, \alpha_n = 0}^{\infty} a_{\alpha_1 \dots \alpha_n} (z_1 - w_1)^{\alpha_1} \dots (z_n - w_n)^{\alpha_n}.$$

We shall frequently use the multi-index $\alpha := (\alpha_1, \dots, \alpha_n)$; we have

$$(z-w)^\alpha := \prod_{j=1}^{n} (z_j - w_j)^{\alpha_j}, \qquad \alpha! := \alpha_1! \dots \alpha_n!.$$

PROOF. (i) Take an arbitrary point $w \in \Omega$ and a closed polydisk centred at the point w, lying in Ω. Since the function f is holomorphic with respect to each of the variables (the others being fixed), we can use the Cauchy formula for a single variable. Application of this formula n times yields

$$f(z) = \frac{1}{(2\pi i)^n} \int_{|w_1 - \zeta_1| = r_1} \frac{d\zeta_1}{\zeta_1 - z_1} \int_{|w_2 - \zeta_2| = r_2} \frac{d\zeta_2}{\zeta_2 - z_2} \dots$$
$$\dots \int_{|w_n - \zeta_n| = r_n} f(\zeta) \frac{d\zeta_n}{\zeta_n - z_n}$$

for all $z \in D_n(w, r_1, \dots, r_n)$.

For a fixed z, the integrand is continuous on the compact set $\partial_0 D$; accordingly, we can employ Fubini's Theorem and replace the repeated integral by an integral over $\partial_0 D$. This gives us

$$(4) \qquad f(z) = \frac{1}{(2\pi i)^n} \int_{\partial_0 D} \frac{f(\zeta)\, d\zeta_1 \dots d\zeta_n}{(\zeta_1 - z_1) \dots (\zeta_n - z_n)}.$$

(ii) We know that for fixed $z \in D_n(w, r)$, the geometric series

$$(5) \qquad \frac{1}{(\zeta_1 - z_1) \dots (\zeta_n - z_n)} = \sum_{\alpha_1, \dots, \alpha_n = 0}^{\infty} \frac{(z-w)^\alpha}{(\zeta_1 - w_1)^{\alpha_1 + 1} \dots (\zeta_n - w_n)^{\alpha_n + 1}}$$

converges absolutely and uniformly for all $\zeta \in \partial_0 D$. Substituting (5) into (4) and changing the order of summation and integration yields:

$$(6) \qquad f(z) = \sum_{\alpha} a_\alpha (z-w)^\alpha,$$

$$(7) \qquad a_\alpha := \frac{1}{(2\pi i)^n} \int_{\partial_0 D} \frac{f(\zeta)\, d\zeta_1 \dots d\zeta_n}{(\zeta_1 - w_1)^{\alpha_1 + 1} \dots (\zeta_n - w_n)^{\alpha_n + 1}}.$$

(iii) We differentiate formula (4):

$$(8)\qquad \partial^\beta f(z) := \frac{\partial^{\beta_1+\dots+\beta_n}}{z_1^{\beta_1}\dots z_n^{\beta_n}} f$$

$$= \frac{\beta!}{(2\pi i)^n} \int\limits_{\partial_0 D} \frac{f(\zeta)\,d\zeta_1 \dots d\zeta_n}{(\zeta_1 - z_1)^{\beta_1+1} \dots (\zeta_n - z_n)^{\beta_n+1}}.$$

Putting $z = w$ in formula (8) and equating formulae (7) and (8), we see that the coefficients in the power-series expansion (6) are given by the formula

$$(9)\qquad a_\alpha = \frac{1}{\alpha!}\,\partial^\alpha f(w).\ \square$$

COROLLARY XV.3.4. The power-series expansion of the function $f \in A(\Omega)$ is uniquely determined. This is a Taylor expansion convergent in every polydisk $D_n \subset \Omega$.

The Cauchy formula implies

THEOREM XV.3.5 (Cauchy Inequality). Let $f \in A\big(D_n(z, r)\big)$ and let $|f| < M$; then

$$(10)\qquad |\partial^\alpha f(z)| \leqslant \frac{M\alpha!}{r^\alpha}.$$

PROOF. Inequality (10) follows immediately from formula (8), first for small polydisks $D(z, r')$, $r_j' < r_j$, $j = 1, \dots, n$. Then, taking the limit as $r_j' \to r_j$, $j = 1, \dots, n$, we obtain the hypothesis. $\square$

COROLLARY XV.3.6 (Liouville's Theorem). The only function holomorphic on $\boldsymbol{C}^n$ and bounded is constant.

PROOF. Going over to the Cauchy inequalities as $r \to \infty$, we see that all the derivatives $\partial^\alpha f(z)$ equal 0. $\square$

Analytic Continuation. Now we give two versions of the uniqueness theorem for analytic continuation.

THEOREM XV.3.7 (Identity Theorem). Let $\Omega \subset \boldsymbol{C}^n$ be a connected open set. If $f, g \in A(\Omega)$ and if $\partial^\alpha f(z_0) = g(z_0)$, $|\alpha| = 0, 1, \dots$, for some $z_0 \in \Omega$, then $f = g$ on Ω.

PROOF. Suppose that $u = f - g$. Since $\partial^\alpha u$ are continuous functions, the set Z_α: $\{z \in \boldsymbol{C}^n \colon \partial^\alpha u(z) = 0\}$ is:

(i) nonempty, because $z_0 \in Z_\alpha$, and

(ii) closed in Ω.

The set Z, for which $\partial^\alpha f(z) = \partial^\alpha g(z)$, $|\alpha| = 0, 1, \ldots$, is of the form $Z = \bigcup_\alpha Z_\alpha$, and hence is nonempty and closed in Ω. However, it is seen from the Taylor expansion that the set Z is open because it contains polydisks. However, the entire set Ω is the only nonempty set which is simultaneously closed and open in Ω. Thus, $Z = \Omega$. □

THEOREM XV.3.8. If $\Omega \subset C^n$ is open and connected, and O is an open nonempty subset of the set Ω, $f(z) = g(z)$, $z \in O$, then $f = g$ on Ω.

PROOF. At every point of the set O we have $\partial^\alpha f(z) = \partial^\alpha g(z)$, $|\alpha| = 0, 1, \ldots$ The theorem follows from the preceding theorem. □

COROLLARY XV.3.9. Let Ω_1 and Ω_2 be open sets in C^n such that $\Omega_1 \cap \Omega_2$ is a nonempty connected set. If $f_j \in A(\Omega_j)$, $j = 1, 2$, and if $\partial^\alpha f_1(z_0) = \partial^\alpha f_2(z_0)$, $|\alpha| = 0, 1, \ldots$, at some point $z_0 \in \Omega_1 \cap \Omega_2$, then there exists a unique function $f \in A(\Omega_1 \cup \Omega_2)$ which is the continuation of the functions f_1 and f_2.

The function f is called the *analytic continuation* of the functions f_1 and f_2.

LEMMA XV.3.10 (Mean-Value Theorem). Let Ω be an open set, $f \in A(\Omega)$. Then, for every polydisk $D_n(\mathring{w}, r) \subset \Omega$ the mean value of the function f on the polydisk $D_n(\mathring{w}, r)$ is equal to the value of the function at the centre of the polydisk:

$$f(\mathring{w}) = \frac{1}{|D_n|} \int\limits_{D_n(w,r)} f(\zeta)\, d\lambda^{2n}(\zeta),$$

where λ^{2n} is the Lebesgue measure on $R^{2n} = C^n$, and $|D_n|$ is the measure of the polydisk $D_n(\mathring{w}, r)$.

PROOF. The Cauchy formula for n variables gives us

$$f(\mathring{w}) = \frac{1}{(2\pi i)^n} \int\limits_{\partial_0 D_n(w,r)} \frac{f(\zeta_1, \ldots, \zeta_n)\, d\zeta_1 \ldots d\zeta_n}{(\zeta_1 - \mathring{w}_1) \ldots (\zeta_n - \mathring{w}_n)},$$

$\varrho_k \leqslant r_k$, $k = 1, \ldots, n$.

Performing the parametrization $\zeta_k = \mathring{w}_k + \varrho_k e^{i\varphi_k}$, $1 \leqslant k \leqslant n$, we have

$$f(\mathring{w}) = \frac{1}{(2\pi)^n} \int\limits_0^{2\pi} \ldots \int\limits_0^{2\pi} f(\mathring{w}_1 + \varrho_1 e^{i\varphi_1}, \ldots, \mathring{w}_n + \varrho_n e^{i\varphi_n})\, d\varphi_1 \ldots d\varphi_k.$$

Hence

$$\int_0^{r_1} \dots \int_0^{r_n} \varrho_1 \dots \varrho_n f(\mathring{w})\, d\varrho_1 \dots d\varrho_n$$

$$= \frac{1}{(2\pi)^n} \int_0^{r_1} \dots \int_0^{r_n} \int_0^{2\pi} \dots \int_0^{2\pi} f(w_1 + \varrho_1 e^{i\varphi_1}, \dots$$

$$\dots, w_n + \varrho_n e^{i\varphi_n}) \varrho_1 \dots \varrho_n \, d\varphi_1 \dots d\varphi_n \, d\varrho_1 \dots d\varrho_n.$$

Therefore

$$f(\mathring{w}) = \frac{1}{\pi^n r_1^2 \dots r_n^2} \int_{D_n(w,r)} f(\zeta)\, d\lambda^{2n}(\zeta).$$

However, $|D_n| = \pi^n \cdot r_1^2 \dots r_n^2$. □

THEOREM XV.3.11. Suppose that Ω is a region (i.e., an open connected set). Let $f \in A(\Omega)$ and let $|f|$ attain a maximum at the point $w \in \Omega$, that is, $|f(z)| \leqslant f(w)$, $z \in O(w)$.

Then $f =$ const on Ω.

PROOF. Take a polydisk such that $|f(w)| - |f(z)| \geqslant 0$, $z \in D_n(w, r)$. Then

$$(11) \qquad 0 \leqslant \int_{D_n(w,r)} (|f(w)| - |f(z)|)\, d\lambda^{2n}(z)$$

$$= |D_n| |f(w)| - \int_{D_n(w,r)} |f(z)|\, d\lambda^{2n}(z).$$

The mean-value theorem (Lemma XV.3.10), however, yields

$$f(w) = \frac{1}{|D_n|} \int_{D_n(w,r)} f(z)\, d\lambda^{2n}(z).$$

Therefore,

$$(12) \qquad |D_n| \cdot |f(w)| \leqslant \int_{D_n(w,r)} |f(z)|\, d\lambda^{2n}(z).$$

Formulae (11) and (12) imply

$$\int_{D_n(w,r)} (|f(w)| - |f(z)|)\, d\lambda^{2n}(z) = 0,$$

and, hence, $|f(w)| = |f(z)|$ for all $z \in D_n(w, r)$. It thus follows from Theorem XV.3.6 that $f =$ const on $D_n(w, r)$ and the identity theorem implies that $f =$ const on Ω. □

COROLLARY XV.3.12 (Maximum Modulus Principle). Let Ω be an open bounded set, $f \in A(\Omega)$ and $f \in C(\overline{\Omega})$. Then, $\sup|f(\overline{\Omega})|$ is attained at the boundary of Ω.

PROOF. Since $\overline{\Omega}$ is compact and $f \in C(\overline{\Omega})$, there exists a point $w \in \overline{\Omega}$ such that $|f(w)| = \sup|f(\overline{\Omega})|$. If $w \in \Omega$, then it follows from Theorem XV.3.2 that f is constant on the connected component containing the point w and, hence, f assumes the same value at some point of the boundary of the set Ω. □

This section ends with a complex analogue of the Poincaré Lemma, which plays an enormous role in the theory of the homology of complex manifolds. We shall not be using it until the second part of Chapter XXIV.

THEOREM XV.3.11 (Dolbeault–Grothendieck Lemma). Let $D = D_n(O, r)$ be a polydisk in C^n and set $\alpha \in C^\infty_{(p,q)}(D)$, $q \geqslant 1$, whereas $d''\alpha = 0$. Then there exists a form $\beta \in C^\infty_{(p,q-1)}(D')$ and $D' = D_n(O, r')$, $0 < r'_j < r_j, j = 1, \ldots, n$, such that $d''\beta = \alpha$ in D'.

Remark. The theorem remains valid, of course, if one operates with polydisks centred at $w \neq 0$.

PROOF. To begin with, note that the theorem holds for the case $n = 1$, since then $(p, q) = (0, 1)$ (forms of other types vanish!), and on writing $\bar{z}$ instead of $\bar{z}_1$, we have $\alpha = f(z)\,d\bar{z}$, where $f \in C^\infty(D)$. Then the required form, i.e., function, β satisfies the nonhomogeneous Cauchy–Riemann equation

$$d''\beta = \frac{\partial\beta}{\partial\bar{z}} = f(z),$$

which, as we know, has the solution

$$(*) \qquad \beta(z) = \frac{1}{2\pi i}\int_{D'} \frac{f(\zeta)\,d\zeta \wedge d\bar{\zeta}}{\zeta - z} + g(z), \qquad \text{where } g \in A(D').$$

It is seen from formula $(*)$ that if $f(\zeta)$ is a holomorphic function of further complex variables (parameters), $\partial f(\zeta)/\partial\bar{z}_k = 0$; then, since we can differentiate under the integral sign with respect to $\bar{z}_k$, we see that $\beta(z)$ is holomorphic with respect to $\bar{z}_k$.

After this remark, let us consider the general case. To this end we introduce the hypothesis

(H_j) $\quad \alpha$ does not contain $d\bar{z}_{j+1}, \ldots, d\bar{z}_n$.

It will be shown that if the Dolbeault–Grothendieck Lemma is valid with the additional hypothesis (H_{j-1}), then it also holds with the additional hypothesis (H_j). Note that under (H_0) the form α does not depend on $d\bar{z}_1, \ldots, d\bar{z}_n$, but $q \geqslant 1$ and, therefore, $\alpha = 0$, whereby the lemma does hold ($\beta = 0$). Now, in the second limiting case the hypothesis (H_n) is satisfied trivially, i.e., is not in fact an "additional" hypothesis.

Suppose, therefore, that the lemma holds under the restriction (H_{j-1}). Now, α does not contain $d\bar{z}_{j+1}, \ldots, d\bar{z}_n$; we may thus assume that $\alpha = \mu + d\bar{z}_j \wedge \lambda$, where μ is of type (p, q), while λ is of type $(p, q-1)$, and neither form contains $d\bar{z}_j, \ldots, d\bar{z}_n$ any longer. Since $\bar{\partial}\alpha = 0$, the coefficients of the forms λ, μ are holomorphic in $z_{j+1}, \ldots, z_n$. From the remark above we know that there exists a form λ_1 of type $(p, q-1)$ with coefficients holomorphic in $z_{j+1}, \ldots, z_n$ which satisfies in D the equation

$$\frac{\partial \lambda_1}{\partial \bar{z}_j} = \lambda;$$

here, of course, $\partial\lambda_1/\partial\bar{z}_j$ denotes differentiation with respect to $\bar{z}_j$ of all coefficients of the form λ_1; thus, ν: $\bar{\partial}\lambda_1 - d\bar{z}_j \wedge \lambda$ does not contain $d\bar{z}_j, \ldots, d\bar{z}_n$. Hence, $\alpha = d''\lambda_1 + \mu - \nu$. But $(\mu - \nu)$ is of type (p, q) and does not contain $d\bar{z}_j, \ldots, d\bar{z}_n$, and $d''(\mu - \nu) = d'(\alpha - \bar{\partial}\lambda_1) = d''\alpha - - d''^2\lambda_1 = d''\alpha = 0$. Accordingly, the form $(\mu - \nu)$ satisfies the hypotheses of the lemma and (H_{j-1}) and, hence, there exists in D a form π of type $(p, q-1)$, satisfying the equation $\mu - \nu = d''\pi$ in D'. Thus,

$$\alpha = d''\lambda_1 + d''\pi = d''(\lambda_1 + \pi),$$

that is, the lemma holds under the additional hypothesis (H_j). □

As will be seen in Part III, the foregoing theorem makes it possible to prove the *Dolbeault isomorphism*, which is of fundamental impor tance. Suppose that M is an open set in $\boldsymbol{C}^n$ or, more generally, an n-dimensional complex manifold (cf. p. 340). The form $\alpha \in C_{(p,q)}(M)$

is said to be *d''-closed* if $d''\alpha = 0$. We denote

$$Z^{p,q}(M) = \{\alpha \in C^{\infty}_{(p,q)}(M):\ d''\alpha = 0\},$$
$$B^{p,q}(M) = d''C^{\infty}_{(q,p-1)}(M).$$

It can be proved that the quotient space $Z^{p,q}(M)/B^{p,q}(M)$ is isomorphic to the q-th cohomology group of the manifold M with values in the sheaf $C^{\infty}_{(p,0)}(M)$ of forms of type $(p, 0)$ on M. This theorem of Dolbeault, as well as its generalization, will be proved in the second part of Chapter XXI.

4. THE TOPOLOGY OF THE SPACE OF HOLOMORPHIC FUNCTIONS $A(\Omega)$

Since an open subset $\Omega \subset C^n$ is a locally compact space, it is natural to introduce in the set $C(\Omega)$ a topology of compact convergence (in other words, almost uniform convergence). This topology is given by the family of seminorms

$$||f||_K := \sup|f(K)|,$$

where K runs over the family of compact subsets $K \subset \Omega$. The function $C(\Omega) \ni f \to ||f||_K$ is only a seminorm for there are functions which have a support in $\Omega - K$.

The neighbourhood basis of zero in $C(\Omega)$ is given by the sets

$$U(K, \varepsilon) = \{f \in C(\Omega): ||f||_K < \varepsilon\}.$$

It is not difficult to show that the set Ω is exhaustible by a growing sequence of compact sets $K_j \nearrow \Omega$, that is,

$$K_j \subset K_{j+1}, \qquad \bigcup_{j=1}^{\infty} K_j = \Omega.$$

Since in a vector space the topology is defined by the neighbourhood basis of the zero element, the topology of the space $C(\Omega)$ is given by a countable (increasing) family of seminorms $||\ ||_{K_1} \leqslant ||\ ||_{K_2} \leqslant ||\ ||_{K_3} \dots$

Spaces with a countable number of seminorms were studied by the "Lvov school" (Mazur, Orlicz), and after the Second World War mainly by the school of N. Bourbaki (Dieudonné, Schwartz, Grothendieck) They are metrizable spaces: e.g., the expression

$$(1) \qquad d(f, g) := \sum_{j=1}^{\infty} \frac{1}{2^j} \cdot \frac{||f-g||_{K_j}}{1+||f-g||_{K_j}}$$

is a distance consistent with the (locally convex) topology given by the family of seminorms $|| \; ||_{K_j}, j = 1, 2, \ldots$

If a countably seminormable space is complete for a distance (1), it is called a *Fréchet space* (B_0 in the Lvov terminology). Thus, the space $C(\Omega)$ is, in a natural manner, a Fréchet space.

The purpose of this section is to show that the space $A(\Omega)$ of holomorphic functions has a highly interesting topology, that it is a so-called *Montel space*: bounded closed sets are compact and, hence, $A(\Omega)$ has some features of $\boldsymbol{R}^n$. A Banach space is a Montel space only if it is finite-dimensional, and only in such normed spaces are bounded sets precompact. First of all, we must introduce the concept of a bounded set in a locally convex space.

DEFINITION. Let $(E, || \cdot ||_j, j \in J)$ be a topological vector space whose topology is determined by a family (in general, uncountable) of seminorms $|| \; ||_j, j \in J$. A subset $B \subset E$ is bounded if $||B||_j < \infty$ for every seminorm $|| \; ||_j, j \in J$.

An interesting fact follows:

(The locally convex space E contains a bounded neighbourhood of zero) $\Leftrightarrow$ (E is normable).

(The principal facts concerning locally convex spaces can be found in Chapter XX or in Chapter III of the author's monograph, *Methods of Hilbert Spaces*.) The theory of convex spaces developed first on the model of the space of holomorphic functions $A(\Omega)$ and later on models of spaces occurring in the theory of distributions: $\mathscr{E}(\Omega)$, $\mathscr{D}(\Omega)$, $\mathscr{S}(\boldsymbol{R}^n)$, ... These spaces are introduced and examined in subsequent chapters.

Let us consider what bounded sets are like in $A(\Omega)$. This is the subject of the following

LEMMA XV.4.1 (The set $B \subset A(\Omega)$ is bounded) $\Leftrightarrow$ (For every compact set $K \subset \Omega$ we have $||B||_K < \infty$).

The proof is obvious.

THEOREM XV.4.2 (Fundamental Inequality). For every compact $K \subset \Omega$ (Ω—open subset in C^n), every neighbourhood $\mathcal{O}(K)$ of the set K, and every multi-index α there exist constants $C_\alpha = C_\alpha(\alpha, K, \mathcal{O}(K))$ such that

$$(2) \qquad \sup|\partial^\alpha f(K)| \leqslant C_\alpha \|f\|_{L^1(\mathcal{O}(K))}, \qquad f \in A(\Omega),$$

that is,

$$\|\partial^\alpha f\|_K \leqslant C_\alpha \int_{\mathcal{O}(K)} |f| \, d\lambda^{2n}.$$

Before we proceed with the proof of this fundamental inequality, we give several important corollaries.

COROLLARY XV.4.3. The space $A(\Omega)$ is complete (i.e. is a Fréchet space).

PROOF. We have to show that if $f_k \to f$, $k \to \infty$, where $f_k \in A(\Omega)$ are convergent in the sense of the topology of $A(\Omega)$, i.e., compactly convergent, the limit (which, as we know from Chapter V, is a continuous function) is holomorphic on Ω.

Inequality (2) implies that f_k and $\partial^\alpha f_k$ are compactly convergent. Now, $d'' f_k = 0$ and

$$\frac{\partial}{\partial x_j} = \frac{\partial}{\partial z_j} + \frac{\partial}{\partial \bar{z}_j}, \qquad i\frac{\partial}{\partial y_j} = \frac{\partial}{\partial z_j} - \frac{\partial}{\partial \bar{z}_j}.$$

Thus, if $\|\partial f_k\|_K \to 0$, as $k \to \infty$ then

$$\left\|\frac{\partial f_k}{\partial x_j}\right\|_K \to 0, \quad \left\|\frac{\partial f_k}{\partial y_j}\right\|_K \to 0$$

as $k \to \infty$ and, hence, $f \in C^1(\Omega)$ and

$$\frac{\partial f}{\partial \bar{z}_j} = \lim_{k\to\infty} \frac{\partial f_k}{\partial \bar{z}_j} = 0, \qquad \text{whereby } f \in A(\Omega). \ \square$$

COROLLARY XV.4.4 (Montel, Stieltjes, Vitali). If $f_i \in A(\Omega)$, and the set $\{f_j, j = 1, 2, \ldots\}$ is bounded (i.e. the sequence $|f_j|$ is uniformly bounded on every compact subset), there exists a subsequence (f_{k_j}) which is compactly convergent to $f \in A(\Omega)$.

This corollary can be formulated as follows:

THEOREM XV.4.5 (Generalized Theorem of Montel, Stieltjes, and Vitali). The space $A(\Omega)$ is a metrizable Montel space.

PROOF. As in the proof of Corollary XV.4.3, we note that the first-order partial derivatives of the functions $f_j, j = 1, 2, \ldots$, are uniformly bounded on every compact set K; hence, the functions $f_j, j = 1, 2, \ldots$, are equicontinuous and uniformly bounded on every compact subset $K \subset \Omega$, and we can thus apply the Ascoli Theorem. Now, take the sequence $K_p \nearrow \Omega, p \to \infty$. Thus there exists a subsequence $(f_{(1,j)})$ uniformly convergent on K_1. Since the sequence $(f_{(1,j)})$ satisfies the hypotheses of the theorem, there exists a subsequence $(f_{(2j)})$ which converges uniformly on K_2. By repeating this procedure, we obtain for every natural m a sequence $(f_{(m,j)})$ which converges uniformly on K_m.

The "diagonal" sequence $(f_{(j,j)})$ is uniformly convergent on every set K_s, $s = 1, 2, \ldots$ Since a set $K_s \supset K$ exists for any compact set $K \subset \Omega$, the "diagonal" sequence $(f_{(j,j)})$ is uniformly convergent on K. □

Remark. Every precompact set (i.e., such that its closure is compact) is bounded in any complete, locally convex space. Thus, in a Montel space (precompactness) ⇔ (boundedness). The Montel–Stieltjes–Vitali Theorem thus gives a necessary and sufficient condition for the compactness of sets in the space $A(\Omega)$.

Example. The sum of a power series in $\mathbf{C}$, $f(z) = \sum_{j=0}^{\infty} a_j z^j$, is a holomorphic function because the sequence of partial sums of the series is compactly convergent in the circle of convergence.

It remains to prove the fundamental inequality.

PROOF OF THEOREM XV.4.2. Consider a function $g \in C_0^\infty(\mathcal{O}(K))$ such that $g = 1$ in a neighbourhood of the set K (the construction of such a function will be given in Chapter XVII). Since $f \in A(\Omega)$, we have $\bar{\partial} f = 0$ and hence $d''(gf) = f \cdot d''g$; the application of the generalized Cauchy formula to gf yields

$$(3) \qquad g(\zeta)f(\zeta) = \frac{1}{2\pi i} \int f(z) \frac{\partial g}{\partial \bar{z}} \cdot \frac{1}{z - \zeta} dz \wedge d\bar{z}$$

for $\xi \in K$. Now $g = 1$ in a neighbourhood of K, and $|z - \zeta| > m > 0$ if z lies in the support of g and if ζ runs over K (the derivative of a constant is zero and the support is a compact set, hence the greatest lower bound is attained). Formula (3) thus gives

$$\sup |f(K)| \leqslant c_0 \int_{\mathcal{O}(K)} |f| \, d\lambda^{2n}.$$

Differentiating formula (3) and reasoning in an analogous manner, we obtain inequality (2).

Now, let $n > 1$. If $\mathcal{O}(K)$ is a polydisk, the fundamental inequality is obtained by an n-fold application of the inequality for each of the variables and by the Fubini Theorem. Since the set $K \subset \mathcal{O}(K)$ for any $z \in K$, there exist polydisks $D_n(z, r)$, $D_n(z, r')$ such that we have $\overline{D_n(z, r)} \subset D_n(z, r')$ for $r_j < r_j'$. The set K, however, is compact, and thus there exists a finite covering of the set K with the polydisks $D_n(z^s, r^s)$, $s = 1, \ldots, p$. The sets $K_s := K \cap D_n(z^s, r^s)$, $s = 1, \ldots, p$, are compact subsets of the polydisks and $K = \bigcup_{s=1}^{p} K_s$, whereas $K_s \subset D(z^s, r'^s) \subset \mathcal{O}(K)$, $s = 1, \ldots, p$. Accordingly, we have

$$\sup|\partial^\alpha f(K_s)| \leqslant C(\alpha, s, \mathcal{O}(K))\int |f| \, d\lambda^{2n}$$

for $f \in A(\Omega)$. Putting $C_\alpha := \max_{1 \leqslant s \leqslant p} C(\alpha, s, \mathcal{O}(K))$, we obtain the hypothesis. □

More interesting applications of the Montel–Stieltjes–Vitali Theorem will be presented in subsequent chapters.

Remark. Corollary XV.4.3 can be formulated for series: Let $g_j \in A(\Omega)$ and let $f_n := \sum_{j=0}^{n} g_j$ converge compactly in Ω. Then

(i) the function $f := \sum_{j=0}^{\infty} g_j$ is holomorphic on Ω,

(ii) The series $\sum_{j=0}^{\infty} \partial^\alpha g_j$ is compactly convergent to $\partial^\alpha f$ in Ω.

5. ELEMENTARY PROPERTIES OF HARMONIC FUNCTIONS

We have several times encountered harmonic functions, solutions of the Laplace equation $\Delta u = 0$.

In Section 1 of this chapter (Corollary XV.1.5) we saw that the real and imaginary parts of a holomorphic function are harmonic:

$$(f \in A(\Omega)) \Rightarrow (\Delta \operatorname{Re} f = 0 = \Delta \operatorname{Im} f).$$

(But not every harmonic function h is a real part of a holomorphic function f; it must satisfy the Cauchy–Riemann equations. Such functions are called pluriharmonic and wil be investigated and used in Part III). Therefore, we can expect that some fundamental properties of holomorphic functions also have harmonic functions. We shall prove that indeed harmonic functions

1° have the mean-value property,
2° satisfy the maximum and minimum principle,
3° satisfy the identity theorem,
4° are analytic,
5° satisfy the Liouville Theorem,
6° have the following property: the space $\mathscr{H}(\mathcal{O})$ of harmonic functions is a Montel space.

Remark. Harmonic functions are also called *potential functions.* Since the Green's formulas are the main tool, let us recall them in the case of the finite domain $M \subset \boldsymbol{R}^m$.

Green's Formulas. Let M be a finite domain in $\boldsymbol{R}^m$ and let $\partial M \ni x \to n(x) \in \boldsymbol{R}^n$ be the unit normal to the boundary ∂M of M. We shall denote $\dfrac{du}{dn} := (\nabla u | n)$ by u_n, $u'_x = (\partial_1 u, \ldots, \partial_m u) \cdot v'_x$, $u'_x := \sum \dfrac{\partial v}{\partial x^i} \dfrac{\partial u}{\partial x^i}$, $D(u, v) := \int\limits_M u'_x \cdot v'_x$, $D(u) := D(u, u) = \int\limits_M ||u'_x||^2$.

Then the Green's formulas read

$$(1) \qquad \int\limits_M v\Delta u + D(v, u) = \int\limits_{\partial M} v \cdot u_n,$$

$$(2) \qquad \int\limits_M u\Delta u + D(u) = \int\limits_{\partial M} uu_n,$$

$$(3) \qquad \int\limits_M v\Delta u - u\Delta v = \int\limits_{\partial M} (vu_n - uv_n).$$

COROLLARY XV.5.1 (Theorem of Gauss). Let $u \in \mathscr{H}(M)$; then

$$(4) \qquad \int\limits_{\partial M} \frac{du}{dn} = 0.$$

PROOF. Put $v = 1$ into (1); since $\Delta u = 0$, (4) follows. In the same way, (1) implies

THEOREM XV.5.2. Let u be of class C^1 in a neighbourhood of M.
1° ($u \in \mathscr{H}(M)$ and $u | \partial M = 0$) $\Rightarrow$ ($u \equiv 0$),
2° ($u \in \mathscr{H}(M)$ and $u_n = 0$ on ∂M) $\Rightarrow$ ($u =$ const).

PROOF. In both cases $\partial_i u = 0, i = 1, \dots, n,$ and thus $u = \text{const}$. If we assume that $u|\partial M = 0$, we obtain 1°. □

Dirichlet and Neumann Problems. These most famous boundary value problems are formulated as follows.

DEFINITION. Let M be a finite domain in $\boldsymbol{R}^m$. The problem: "To each $b \in C(\partial M)$ find such $u \in C(\overline{M})$ that $u \in \mathscr{H}(M)$ and $u|\partial M = b$" is called the *Dirichlet problem.*

"To each $f \in C(\partial M)$ such that $\int\limits_{\partial M} f = 0$ find a $u \in \mathscr{H}(M)$ such that $u_n|\partial M = f$" is called the *Neumann problem.* From Theorem XV.5.2 follows

COROLLARY XV.5.3. 1° The Dirichlet problem has at most one solution.

2° The Neumann problem has at most one solution modulo additive constants.

PROOF. 1° Let u_1, u_2 be two solutions of the Dirichlet problem such that $u_k|\partial M = b, k = 1, 2$. Then $u := u_1 - u_2 \in \mathscr{H}(M)$ and $u|\partial M = 0$. Therefore, from Corollary XV.5.1.1° follows 1°. Applying Corollary XV.5.1.1° we obtain 2°. □

Fundamental Solutions. Potentials. Let us consider the Laplace operator Δ in $\boldsymbol{R}^m$ in spherical (polar) coordinates; $(r, \vartheta_1, \dots, \vartheta_{m-1})$, $\vartheta = (\vartheta_1, \dots, \vartheta_{m-1})$ are coordinates on the sphere $S^{m-1} = \{x \in \boldsymbol{R}^m\colon |x| = 1\}$, where $|x|^2 = \sum_{j=1}^{m} (x^j)^2$. Let us denote $|x|$ by $r \cdot \vartheta = \vartheta(x/|x|), x \neq 0$.

Our first task is to find all solutions u of $\Delta u = 0$ with radial symmetry, i.e., functions which are dependent only on $r\colon u = u(r)\colon \dfrac{\partial u}{\partial \vartheta_j} = 0$ for $j = 1, \dots, m-1$. Since

$$\frac{\partial r}{\partial x^j} = \frac{x^j}{r}, \qquad \frac{\partial^2 r}{\partial x^i \partial x^i} = \frac{1}{r}\left(1 - \frac{(x^i)^2}{r^2}\right),$$

we have

$$\frac{\partial u}{\partial x^i} = \frac{\partial u}{\partial r}\frac{\partial r}{\partial x^i}, \qquad \frac{\partial^2 u}{\partial x^i \partial x^i} = \frac{\partial^2 u}{(\partial r)^2}\left(\frac{\partial r}{\partial x^i}\right)^2 + \frac{\partial u}{\partial r}\frac{\partial^2 r}{(\partial x^i)^2}.$$

For $u = u(r)$ we get the ordinary differential equation

$$\Delta u = \frac{d^2 u}{dr^2} + \frac{m-1}{r}\frac{du}{dr} = 0. \tag{5}$$

We immediately find the general solutions of (5).

For $m = 1$, $u(r) = ar+b$,

for $m = 2$, $u(r) = a\log r+b$,

for $m \geqslant 3$, $u(r) = \dfrac{a}{r^{m-2}}+b$.

It is interesting to note that in the nontrivial case $m > 1$ the only radial symmetric potential functions are singular at the origin. These solutions are called the *fundamental solutions of the Laplace equation.* The most important case is, of course, $m = 3$

$$\psi(r) = \frac{a}{r}.$$

This function plays a fundamental role in electrostatics and gravitation: its physical interpretation is the potential of a point mass. By integration one forms the very important functions

$$v(x) = \int_D \frac{\varrho(\xi)}{|x-\xi|} d\lambda^3(\xi),$$

or, more generally,

$$\int \frac{d\mu(\xi)}{|x-\xi|},$$

where μ is a Radon measure with compact support.

In $\boldsymbol{R}^m$ we form the corresponding integrals

$$v(x) = \int_D \psi(|x-\xi|)\varphi(\xi)\,d\xi, \quad \text{where } \psi(r) = a\cdot r^{2-m}$$

or, more generally,

$$\int \psi(|x-\xi|)\,d\mu(\xi).$$

These functions are called *potentials* or, more precisely, *Newtonian potentials* in the case $m > 2$ (and *logarithmic potentials* in the case $m = 2$) since they are of great importance in Newton's theory of gravitation (in the case $m = 2$ the fundamental solution has a logarithmic singularity, hence the notion "logarithmic potential").

Represntation of a Harmonic Function by Potentials.

THEOREM XV.5.4. Let $u \in C^2(\overline{M})$ and $u \in C(\overline{M})$. Let us assume that $\Delta u = h$, where $h \in C(\mathcal{O})$ is a given function; then the following theorem is valid:

THE GAUSS–GREEN FORMULA.

$$(6) \qquad (m-2)\omega_m u(y) = \int_M \frac{h(x)}{|y-x|^{m-2}}\,dx + \int_{\partial M} u_n(x)\frac{ds(x)}{|y-x|^{m-2}} - $$
$$- \int_{\partial M} u(x)\frac{d}{dn}\left(\frac{1}{|y-x|^{m-2}}\right)ds(x), \qquad n > 2.$$

Remark. The integrals on the right-hand side are called:

1° a *volume potential* (or *potential of volume charge*),

2° a *surface potential* (or *potential of surface charge*),

3° a *potential of double layer* (or *dipole potential*).

These notions were taken from electrostatics (and gravitation) since in case $m = 3$ we can interprete 1° as a potential originating from the charge density h; similarly 2°, and 3° have an electrostatic interpretation.

The proof of (6) follows from the application of (3) to $M_\varepsilon = M - K(y, \varepsilon)$, where $\varepsilon > 0$ is sufficiently small, $M \ni y$ is fixed, and

$$v(x) = |y-x|^{2-m} \equiv \psi(r), \qquad r = |y-x|.$$

We know that $\Delta_x v = 0$ on M_ε. But $\partial M_\varepsilon = \partial M \cup -S^{m-1}(y, \varepsilon)$ and on $S^{m-1}(y, \varepsilon)$ $v = \varepsilon^{2-m}$ and $v_n = \varepsilon^{1-m}(2-m)$. We have

$$\int_{M_\varepsilon} vh - u0 = \int_{\partial M} vu_n - \int_{\partial M} uv_n + c\varepsilon^{1-m} \int_{S^{m-1}(y,\varepsilon)} u - \varepsilon^{2-m} \int_{S^{m-1}(y,\varepsilon)} u_n .$$

If we let $\varepsilon \to 0$, then by the continuity of u and u_n we see that the last term converges to 0 (since $\int_{S^{m-1}(y,\varepsilon)} u_n = O(\varepsilon^{m-1})$). But $\varepsilon^{1-m} \int_{S^{m-1}(y,\varepsilon)} u \to u(y)\omega_m$ with $\varepsilon \to 0$. Thus the Gauss–Green formula is proved. □

Newtonian Potential. Modern potential theory works with potentials of general mass (charge) distributions

$$U^\mu(x) = \int \frac{d\mu(y)}{|x-y|^{m-2}},$$

where μ is a Radon measure of arbitrary sign with compact support. If $\mu \geqslant 0$ (or $\mu \leqslant 0$), then U^μ is called the *Newtonian potential* of μ. L. Schwartz's theory of distributions has extended this notion, replacing μ by more general distributions. These extensions of the notion of potential were necessary and are dictated by applications. We shall prove later that potentials of measures $d\mu = \varrho d\lambda^m$, where $\varrho \in \boldsymbol{C}^2$, satisfy the Poisson equation:

$$\Delta u = \varrho .$$

Mean-Value Theorems for Harmonic Functions. We shall now prove two mean-value theorems which are characteristic of harmonic functions. Both theorems were already known to Gauss.

THEOREM XV.5.5 (First Mean-Value Theorem). Let $\mathcal{O}$ be a finite domain in $\boldsymbol{R}^m$. Let u be harmonic in $\mathcal{O}$. Then, for each $y \in \mathcal{O}, u(y)$ is equal to the mean-value (average of its values) over any $(m-1)$-sphere $S^{m-1}(y, r)$ such that $\overline{K(y, r)} \subset \mathcal{O}$: Let ω_m denote the measure of the unit $(m-1)$-sphere in $\boldsymbol{R}^m$; then

$$(7) \qquad u(y) = \frac{1}{r^{m-1}\omega_m} \int\limits_{S^{m-1}(y,r)} u = \frac{1}{|S^{m-1}(y, r)|} \int\limits_{S^{m-1}(y,r)} u .$$

PROOF. Let us denote $M = M(r, \varrho) = K(y, r) - K(y, \varrho)$, $0 < \varrho < r$; then $\partial M = S^{m-1}(y, r) \cup S^{m-1}(y, \varrho)$. Apply formula (6) to M, har-

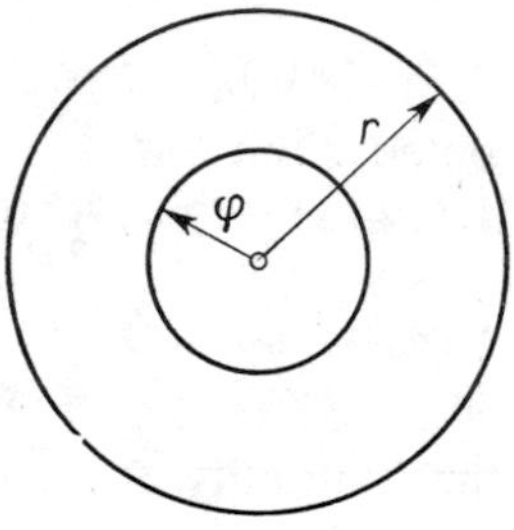

Fig. 4

monic u and $v(x) = |x-y|^{-m+2}$ if $m > 2$ (and $v(x) = \log|x-y|$ for $m = 2$). Since

$$v_n(x) = \begin{cases} -(m+2)|x-y|^{-m+1} & \text{for } x \in S^{m-1}(y, r), \\ \ \ (m+2)|x-y|^{-m+1} & \text{for } x \in S^{m-1}(y, \varrho), \end{cases}$$

from (6) we obtain

$$(8) \qquad \frac{1}{\varrho^{m-1}\omega_m} \int\limits_{S^{m-1}(y,\rho)} u = \frac{1}{r^{m-1}\omega_m} \int\limits_{S^{m-1}(y,r)} u .$$

Since u is continuous, we see from the integral mean-value theorem for $\varrho \downarrow 0$ that the left-hand side of (8) converges to $u(y)$. □

By multiplying both sides of (7) by r and integrating $\int_0^r \dots dr$, we immediately obtain

THEOREM XV.5.6 (Second Mean-Value Theorem). Let $\mathcal{O}$ be a finite domain in R^m and let u be harmonic in $\mathcal{O}$. Then, for each $y \in \mathcal{O}$, $u(y)$ is equal to the mean value (average of its values) over any m-ball $K(y, r)$ such that $\overline{K(y, r)} \subset \mathcal{O}$:

$$(9) \qquad u(y) = \frac{1}{|K(y, r)|} \int_{K(y,r)} u.$$

We shall now show, as a converse of Theorem XV.5.6, that the mean-value property is characteristic of harmonicity and that harmonic functions are analytic.

THEOREM XV.5.7. Let $\mathcal{O}$ be a (finite) domain, let $u \in C(\mathcal{O})$ and let u satisfy the mean-value property. Then

(a) u is harmonic in $\mathcal{O}$,

(b) u is analytic in $\mathcal{O}$.

COROLLARY XV.5.8. Harmonic functions are analytic.

PROOF. (a) We shall first prove that $u \in C^\infty(\mathcal{O})$. Since u has the same average $u(y)$ over any $(m-1)$-sphere, $S^{m-1}(y, r)$, we can obtain $u(y)$ also as a weighted average over a sphere $S^{m-1}(y, r)$ if the weighting function depends only on the radius. Let $0 \leqslant f \in C_0^\infty(R)$, $\operatorname{supp} f \subset]0, \varepsilon[$ with sufficiently small $\varepsilon > 0$, and such that

$$\int_0^\infty f(t)\,dt = 1.$$

(Such functions will be constructed in Chapter XVII). Therefore, by virtue of (6) we have

$$\omega_m u(y) = \int_0^\varepsilon \omega_m u(y) f(t)\,dt = \int_0^\varepsilon \int_{|s-y|=t} u(s) t^{m-1} ds f(t)\,dt$$

$$= \int_{|x-y| \leqslant \varepsilon} u(x) f(|x-y|)\,dx = \int_{R^m} u(x) f(|x-y|)\,dx,$$

since $\operatorname{supp} f \subset \,]0, \varepsilon[$. But the integrand is differentiable with respect to y; therefore, we have

$$(10) \qquad D_y^\alpha u(y) = \frac{1}{\omega_m} \int_{R^m} u(x) D_y^\alpha f(|x-y|)\, dx,$$

and we have proved that $u \in C^\infty(\mathcal{O})$.

Let us assume that u is not harmonic in $\mathcal{O}$. Then there is a ball $K = K(y, r) \subset \mathcal{O}$ such that $\Delta u > 0$ on K; therefore, from the Green's formula (3) applied for $M = K$, $v = 1$ and from the formula (9) we obtain

$$0 < \int_K u_n = \frac{1}{\omega_m r^{m-1}} \int_K u_n = \frac{d}{dr} \frac{1}{\omega_m} \int_{S^{m-1}(0,1)} u(y+r\xi)\, d\xi = 0,$$

which is a contradiction!

(b) We have to prove that u is analytic. Let us apply the Gauss–Green formula (6) for $h = 0$ (u is harmonic!). Then

$$(*) \qquad u(y) = \int_S \big(u(x)\psi_n(|x-y|) - u_n(x)\psi(|x-y|)\big).$$

But only ψ is dependent on y and, moreover, is an analytic function of y, if only y is off a neighbourhood of S. We can now expand the integrand into a power series in y; but since we can interchange the integration and the summation of power series, we see that the right-hand side of $(*)$ is analytic in y. □

From the mean-value theorem we obtain the maximum and minimum principle for harmonic functions in the same way as the maximum modulus principle for holomorphic functions.

THEOREM XV.5.8 (Maximum and Minimum Principle). Let $\mathcal{O}$ be a domain and u a non-constant harmonic function on $\mathcal{O}$. Then u does not attain its maximum (minimum).

PROOF (indirect). Let us assume that there exists a $p \in \mathcal{O}$ such that $u(p) = M := \sup f(\mathcal{O})$; then the set $E = \{x \in \mathcal{O}\colon u(x) = M\}$ is not empty and closed (since u is continuous). The proof is complete if we show that E is also open; then, since $\mathcal{O}$ is connected, $E = \mathcal{O}$, which means that $u = \text{const}$ and we shall arrive at a contradiction.

Let $x_0 \in E$ and then

$$(*) \qquad M = u(x_0) = \frac{1}{|\partial K(x_0, r)|} \int_{\partial K(x_0, r)} u.$$

Since the integrand $\leqslant M$ and $\partial K(x_0, r) \not\subset E$, we have $u(\xi) < M$ for some $\xi \in \partial K(x_0, r)$; then from the continuity of u it follows that

$$M = \frac{1}{|\partial K(x_0, r)|} \int_{\partial K(x_0, r)} u < \frac{1}{|\partial K(x_0, r)|} \int_{\partial K(x_0, r)} M = M,$$

a contradiction. Therefore, for each $0 < \varrho \leqslant r$, $\partial K(x_0, \varrho) \subset E$, whence $K(x, \varrho) \subset E$ and E is open. The same proof is valid for the minimum of u. □

Corollary XV.5.9. 1° Let $\mathcal{O}$ be relatively compact domain and $u_1, u_2 \in \mathscr{H}(\mathcal{O})$ and continuous on $\overline{\mathcal{O}}$. If $u_1|\partial\mathcal{O} < u_2|\partial\mathcal{O}$, then $u_1 < u_2$ on $\mathcal{O}$.

2° If $u \in \mathscr{H}(\mathcal{O})$ and $u|\partial\mathcal{O}$ is const, then $u = $ const. Hence a harmonic function is determined by its boundary values.

Proof. 1° $(u_1 - u_2) \in \mathscr{H}(\mathcal{O})$ and $(u_1 - u_2)|\partial\mathcal{O} < 0$; therefore, either $u_1 - u_2 = $ const and we have finished, or $\max(u_1 - u_2)$ and $\min(u_1 - u_2) < 0$.

2° follows from 1°: If $u_1, u_2 \in \mathscr{H}(\mathcal{O})$ and $u_1|\partial\mathcal{O} = u_2|\partial\mathcal{O}$, then $(u_1 - u_2)|\partial\mathcal{O} = 0$, and hence $u_1 - u_2 = 0$. □

Remark. In the proof of the Maximum (Minimum) Principle we used only two properties of the function u:

1° u is continuous in $\mathcal{O}$,

2° the inequality

$$u(x_0) \underset{(\geqslant)}{\leqslant} \frac{1}{|\partial K(x_0, r)|} \int_{\partial K(x_0, r)} u$$

holds for each $\overline{K(x_0, r)} \subset \mathcal{O}$.

It is reasonable to have a special name for this class of functions.

Definition. A function which satisfies 1° and 2° is called *subharmonic* (*superharmonic*) in $\mathcal{O}$. The space of those functions will be denoted by $\underline{\mathscr{H}}(\mathcal{O})$ ($\overline{\mathscr{H}}(\mathcal{O})$). Plainly $\mathscr{H}(\mathcal{O}) = \underline{\mathscr{H}}(\mathcal{O}) \cap \overline{\mathscr{H}}(\mathcal{O})$. We have thus proved the following general

THEOREM XV.5.10. Let $\mathcal{O}$ be a domain in $\boldsymbol{R}^m$. Then for each subharmonic (superharmonic) function on $\mathcal{O}$ the maximum (minimum) principle is valid.

Remark. In Section 7 we shall give another equivalent definition of subharmonicity.

Exercise. Prove that in 2° the definition of subharmonicity we can replace mean values over spheres by mean values over balls. Hence we have an equivalent

DEFINITION. $u \in C(\mathcal{O})$ is subharmonic in $\mathcal{O}$ if

$$\text{(m.v.)} \qquad u(x) \leqslant \frac{1}{|K(x, r)|} \int_{K(x, r)} u$$

for any ball $K(x, r) \subset \mathcal{O}$.

Exercises. 1. Show that the functions $|x|$, $|z|^\alpha$ $(\alpha \geqslant 0)$, $\log(1+|z|^2)$ are subharmonic.

2. Prove that (f is holomorphic) $\Rightarrow$ ($|f(\cdot)|^\alpha$ $(\alpha \geqslant 0)$ and $\log(1+ +|f(\cdot)|^2)$ are holomorphic).

3. Prove that the compact limit of a sequence of subharmonic functions is subharmonic.

4. Let $u \in C^2(\mathcal{O})$ and show that (u is subharmonic) $\Leftrightarrow$ $(\Delta u \geqslant 0)$.

5. Extend the theorem of Exercise 4 to any continuous (or even locally integrable) functions as follows: Δu is defined in the weak sense: $\int_{\mathcal{O}} \Delta u \varphi \, d\lambda^m := \int_{\mathcal{O}} u \Delta \varphi \, d\lambda^m$ identically for $\varphi \in C_0^\infty(\mathcal{O})$.

Prove that

$$(u \in C(\mathcal{O}) \text{ is subharmonic}) \Leftrightarrow (\Delta u \geqslant 0),$$

where Δu is a positive Radon measure on $\mathcal{O}$.

Remark. In the modern theory of potential one uses a more general class of functions: a numerical function f is subharmonic if

0° $-\infty \leqslant f(x) < \infty$, and $f \not\equiv -\infty$,

1° f is upper semicontinuous,

2° f satisfies the mean-value inequality.

It is not difficult to prove the maximum principle for this wider class of subharmonic functions.

Frederic Riesz has given the following beautiful characterization of subharmonic functions

THEOREM XV.5.11 (Decomposition of F. Riesz). Let u be a (general) subharmonic function in $\mathcal{O}$, then u is a sum of a harmonic function h in $\mathcal{O}$ and a Newtonian potential U^μ (of negative mass): $u = h+U^\mu$, where

$$U^\mu(x) = \int \frac{d\mu(y)}{|x-y|^{m-2}}, \quad \mu \leqslant 0.$$

A modern proof (L. Schwartz) uses the theory of distributions and can be worked out by the reader after reading Chapter XIX.

Compact Convergence of Harmonic Functions. The maximum and minimum principle implies (in the same way as for holomorphic functions) the famous

THEOREM XV.5.12 (Weierstrass). 1° Let $(u_i)_1^\infty$ be a compactly convergent sequence of harmonic functions on a domain $\mathcal{O} \subset \boldsymbol{R}^m$. Then the limit $u := \lim_{i\to\infty} u_i$ is harmonic on $\mathcal{O}$.

2° Moreover, the sequence $(D^\alpha u_i)$ of derivatives is compactly convergent to $D^\alpha u$ (for any multi-index α).

3° Let M be a bounded domain and let $u_i \in \mathscr{H}(M)$, $i = 1, 2, \ldots$ be a sequence of harmonic functions continuous on $\overline{M}$ and such that the sequence (f_i) of boundary values

$$f_i := u_i|\partial M$$

is uniformly convergent (on ∂M) to f. Then (u_i) converges uniformly on $\overline{M}$ to $u \in \mathscr{H}(M)$ such that

$$u|\partial M = f.$$

PROOF. 1° Since (u_i) converges uniformly on any sphere ∂K such that $\overline{K} \subset \mathcal{O}$, u has the mean-value property on any such sphere and, hence, u is harmonic on K. Since K has been arbitrary, u is harmonic in $\mathcal{O}$. The formula

$$D^\alpha u_i(x) = \frac{1}{\omega_m} \int_{\boldsymbol{R}^m} u_i(y) D^\alpha \sigma(|y-x|)\,dy$$

implies 2°.

3° Let $\varepsilon > 0$ and $i, j > n(\varepsilon)$ such that $||f_i - f_j|| < \varepsilon$.

Therefore,

$$\sup|(u_i-u_j)(\overline{M})| < \varepsilon \quad \text{for } i,j > n(\varepsilon)$$

since (u_i-u_j) is harmonic on M and attains its maximum and minimum on ∂M. Therefore, $(u_i)_1^\infty$ converges uniformly on $\overline{M}$ to $u \in C(M)$. By 1° u is harmonic on M and, thus, $u \in \mathscr{H}(M)$. □

Remark. 1° asserts that $\mathscr{H}(\mathcal{O})$ is a Fréchet space seminormed as usual by $||u||_{V_n} = \sup|u|(V_n)$ for any sequence V_n of compact subsets of $\mathcal{O}$ such that $V_n \nearrow \mathcal{O}$.

One can prove a much stronger result.

THEOREM XV.5.13 (Montel–Stieltjes–Vitali). Let $\mathcal{O}$ be a domain in R^m; then the space $\mathscr{H}(\mathcal{O})$ of harmonic functions on $\mathcal{O}$ is a Montel–Fréchet space, i.e. any bounded closed subset is compact.

PROOF (similar to that of Theorem XV.4.5) follows from the

FUNDAMENTAL INEQUALITY. Let E be compact subset of a bounded domain $M \subset R^m$ and let $\sup|u(M)| \leqslant l$. Denote by d the distance of E from ∂M, then

$$(*) \qquad ||D^\alpha u||_E \leqslant l\left(\frac{m}{d}\right)^{|\alpha|} \cdot |\alpha|^{|\alpha|}.$$

PROOF. It is sufficient to prove $(*)$ for $|\alpha| = 1$. Since $\partial u/\partial x_k$ is harmonic, by the mean-value property of $\partial u/\partial x_k$ and the divergence theorem for $K(x, r) \subset M$, $x \in E$, we have

$$\frac{\partial u}{\partial x_k} = \frac{m}{\omega_m}\frac{1}{r^m}\int\limits_{|y|\leqslant r} \partial_k u = \frac{m}{\omega_m}\frac{1}{r^m}\int\limits_{|y|\leqslant r} u(e_k|n);$$

but $|(e_k|n)| \leqslant 1$ and, therefore,

$$\left|\frac{\partial u}{\partial x_k}\right| \leqslant \frac{m\omega_m r^{m-1}}{\omega_m r^m}\cdot l = \frac{m}{r}\cdot l.$$

Since we can approximate d by r, $(*)$ follows. □

6. GREEN'S FUNCTION. POISSON INTEGRAL FORMULA. HARNACK THEOREMS

We shall now prove the important

THEOREM XV.6.1. Let M be a finite domain in R^m, let $\varrho \in C^2(M)$, and let u be defined by $u(x) := \int\limits_M \psi(||x-\xi||)\varrho(\xi)\,d\xi$, where ψ is the fundamental solution of $\Delta v = 0$.

Then $u \in C^2(M)$ and

$$\Delta u = \varrho.$$

Proof. We can assume that $\operatorname{suppt} u \subset K = K(x_0, \delta)$, $\overline{K} \subset M$. Indeed, let $a \in C_0^\infty(K)$ and $a = 1$ on a neighbourhood of x_0; then $\varrho = a\varrho + (1-a)\varrho$,

$$u(x) = \int_M \psi(|x-\xi|)a\varrho(\xi)\,d\xi + \int_M \psi(|x-\xi|)(1-a)\varrho(\xi)\,d\xi$$

$$=: u_1(x) + u_2(x).$$

The integrand of u_2 is not singular in a neighbourhood $O(x_0)$ of x_0; therefore, we can differentiate under the integral sign. In u_1 we have $\operatorname{suppt}(a\varrho) \subset K$.

Since $\Delta_x \psi(|x-\xi|) = 0$, we have $\Delta u_2 = 0$ locally around x_0. If we can prove our theorem for $a\varrho$: $\Delta u_1 = a\varrho$, then we have

$$\Delta u(x) = \Delta u_1(x) + \Delta u_2(x) = a\varrho(x) = \varrho(x)$$

for $x \in O(x_0)$.

We can, therefore, assume that $\operatorname{suppt} \varrho \subset K$, $\overline{K} \subset M$ and we can now write

$$u(x) = \int_{R^m} \psi(|x-\xi|)\varrho(\xi)\,d\xi = \int_{R^m} \psi(|y|)\varrho(x+y)\,dy$$

$$= \int_0^R \int_{|s|=1} \psi(r)\cdot r^{m-1}\varrho(x+rs)\,ds\,dr,$$

for R so large that $\operatorname{suppt} \varrho \subset K(0, r)$, (we have introduced spherical coordinates).

But the last integrand is not singular in $r = 0$, and we can differentiate under the integral sign:

$$\Delta u = \int_0^R \int_{|s|=1} \psi(r) r^{m-1} \Delta_x \varrho(x+rs)\,ds\,dr$$

$$= \int_{|y| \leqslant R} \psi(|y|)\Delta_x \varrho(x+y)\,dy$$

$$= \int_K \psi(|\xi - x|)\Delta_\xi \varrho(\xi)\,d\xi \quad \text{(by the Gauss–Green formula)}$$

$$= \int_K \psi \Delta \varrho + \int_{\partial K} (\varrho \psi_n - \varrho_n \psi) = 0. \ \square$$

Remark. We could replace $\psi(|x-\xi)|$ by a kernel

$$(*) \qquad G(x, \xi) = \psi(|x-\xi|)+g(x, \xi),$$

where

$$g(\cdot, \xi) \in \mathscr{H}(M).$$

We have the following

COROLLARY XV.6.2. Let G be defined by $(*)$, and $\varrho \in C^2(M)$; then $G\varrho \in C^2(M)$ and $\Delta(G\varrho) = \varrho$, where

$$G\varrho(x) := \int_M G(x, \xi)\varrho(\xi)d\xi.$$

Remark. It is sufficient to assume $\varrho \in C^1(M)$: one differentiation can always be shifted onto the other factor, G or ψ.

Green's Function of a Domain. Let M be a finite domain in $\boldsymbol{R}^m$.

DEFINITION. The Green's function $G = G_M$ of $M \subset \boldsymbol{R}^m$ is defined by

$$(1) \qquad G(x, \xi) := \psi(|x-\xi|)+g(x, \xi),$$

where

$$g(\cdot, \xi) \in \mathscr{H}(M) \quad \text{for } \xi \in M$$

and

$$G(\cdot, \xi)|\partial M = 0, \quad \xi \in M.$$

Remark. Several authors take the Green's function with the opposite sign. Then $G(\cdot, \cdot) > 0$ (cf. Section XVI.11).

Remark. The Green's function does not exist for every domain M. To prove its existence one has to solve the following boundary-value problem

$$g(\cdot, \xi) \in \mathscr{H}(M), \quad g(\cdot, \xi)|\partial M = -\psi(\cdot, \xi)|\partial M$$

for $\xi \in M$.

LEMMA XV.6.3 (Symmetry of the Green's Function).

$$G(x, \xi) = G(\xi, x), \quad \xi, x \in M, \ x \neq \xi.$$

Let $\xi_1 \neq \xi_2$ and denote $u(\cdot) = G(\cdot, \xi_1)$, $v(\cdot) := G(\cdot, \xi_2)$. Let $K_j = K(\xi_j, r_j), j = 1, 2,$ where $r_1, r_2 > 0$ are so small that $K_1 \cap K_2 = \emptyset$ and $K_j \subset M, j = 1, 2$. Since u and v are harmonic on $M-(K_1 \cup K_2) =: \mathcal{O}$, the Green's formula for $\mathcal{O}$ yields

$$(*) \qquad 0 = \int_{\partial K_1} (uv_n - vu_n) + \int_{\partial K_2} (uv_n - vu_n).$$

Let $r_1, r_2 \to 0$; then $(*)$ tends to

$$0 = v(\xi_1) - u(\xi_2) = G(\xi_1, \xi_2) - G(\xi_2, \xi_1). \ \square$$

LEMMA XV.6.4. Let $w \in C(\overline{M})$ and let G be the Green's function of the domain M. Then the function

$$M \ni x \to u(x) := \int_M G(x, \xi) w(\xi) d\xi$$

vanishes on ∂M. More precisely: for any $x_0 \in \partial M$ $u(x) \to 0$ if $x \to x_0$.

PROOF. Since $G(\cdot, \xi)$ vanishes on ∂M (for $\xi \in M$) and on a small sphere with centre in ξ as negative in the same way as the fundamental solution ψ, we have (by the maximum principle) $G(\cdot, \xi) < 0$. Since (for $m \neq 2$) $g(\cdot, \xi) > 0$, we have the important inequalities

$$\psi(|x-\xi|) < G(x, \xi) < 0 \quad \text{for } m > 2,$$

$$\psi(|x-\xi|) - a < G(x, \xi) < 0 \quad \text{for } m = 2$$

(where a is a constant independent of x, ξ).

Since $\|w\| < l$, we can take $\delta > 0$ so small that if we denote $K_\delta \equiv K(x_0, \delta)$, $x_0 \in \partial M$, we have

$$\int_{M \cap K_\delta} |G(x, \xi) w(\xi)| d\xi < \int_{K_\delta} |\psi(|x-\xi|) - a| l < \varepsilon$$

for all $x \in K_\delta \cap M$. But

$$\int_{M - K_\delta} G(x-\xi) w(\xi) d\xi \to 0 \quad \text{for } x \to x_0,$$

since $G(\cdot, \cdot)$ is continuous on $K_{\delta/2} \times M - K_\delta$ and $G(x_0, \xi) = 0$. $\square$

We can now prove an important theorem, which elucidates the role of the Green's function in the Dirichlet problem:

THEOREM XV.6.5. Let $G_M = G$ be the Green's function of the bounded domain $M \subset R^m$. Assume that G is differentiable on $\overline{M}$. Then for each $f \in C(\partial M)$ the function

$$(3) \qquad u(\cdot) := \int_{\partial M} G_n(\cdot, \xi) f(\xi) d\xi$$

is the solution of the Dirichlet problem

$$\Delta u = 0 \text{ on } M, \quad u|\partial M = f.$$

PROOF. Let $\varphi \in C^4(\overline{M})$, and let v be defined by

$$v(\cdot) := \varphi(\cdot) - \int_M G(\cdot, \xi)\Delta\varphi(\xi)d\xi;$$

then Theorem XV.6.1 and Lemma XV.6.4 imply that $\Delta v = \Delta\varphi - \Delta\varphi = 0$, and $v(t) = \varphi(t)$ for $t \in \partial M$. Therefore, v is the solution of the Dirichlet problem. But from the Green's formula we have

$$v(\cdot) = \int_{\partial M} G_n(\cdot, t)\varphi(t)dt;$$

therefore,

$$u_j(\cdot) = \int_{\partial M} G_n(\cdot, t)\varphi_j(t)dt, \qquad \varphi_j \in C^4(\overline{M})$$

is the solution of $\Delta u_j = 0$, $u_j|\partial M = \varphi_j|\partial M$. But by virtue of the Weierstrass Theorem (XV.5.12) if $\varphi_j|\partial M \to f$ for $j \to \infty$ uniformly on ∂M, then $u_j \to u \in \mathscr{H}(M)$ and $u|\partial M = f$. □

Solution of the Dirichlet Problem for a Ball. The Poisson Integral. Harnack Theorems. In the case $M = K(x_0, R) \equiv K$ we can give a construction of the Green's function G_K and, hence, give the explicit formula for the solution of the Dirichlet problem for K. We take $x_0 = 0$. We define the function $G = G_K$ for K as follows:

$$G(x, \xi) := \psi(r) - \psi(s\bar{r}/R),$$

where $r := |x-\xi|$, $s := |\xi|$, $\bar{r} := |x-\bar{\xi}|$, $\bar{\xi} := \left(\frac{R}{|\xi|}\right)^2 \cdot \xi$. For $\xi = 0$ we put $G(x, 0) := \psi(r) - \psi(R)$. We immediately check that G has boundary values 0; since $\psi(s\bar{r}/R)$ is harmonic for $\xi \neq 0$, G is indeed the Green's function for K. But since $G_\xi(x, \xi) = G_n(x, \xi)$ on ∂K, we only have to compute G_n. But by virtue of the symmetry of the Green's function we can compute $G_x(x, \xi)$ and later permute x and ξ in order to obtain $G_\xi(x, \xi)$:

$$\omega_m G_\xi = \frac{\xi - x}{|\xi - x|^m} - \left(\frac{R}{|x|}\right)^{m-2}\left(\xi - \frac{R^2}{|x|^2}\cdot x\right)\cdot\left|\xi - \frac{R^2}{|x|^2}\cdot x\right|^{-m}.$$

But since $|\xi| = R$ on ∂K, we have

$$\left|\xi - \frac{R^2}{|x|^2}x\right| = \frac{R}{|x|}|\xi - x|;$$

therefore,

$$G_\xi(x, \xi) = \frac{R^2 - |x|^2}{\omega_m |x-\xi|^m} \cdot \frac{\xi}{R^2} =: P(x, \xi), \tag{4}$$

and we have proved the famous

THEOREM XV.6.6 (Poisson–Schwarz). Let $K = K(x_0, R)$ be a ball in R^m, $m \neq 2$. Then the solution of the Dirichlet problem $\Delta u = 0$, $u|\partial K = f$ is given by

$$u(\cdot) = \int_{\partial K} P(\cdot, z) f(z) d\sigma(z), \tag{5}$$

where $|z| = R$ and

$$P(x, z) = P_x(z) := R^{m-2} \frac{R^2 - |x-x_0|^2}{|x-z|^m} \tag{6}$$

and σ is an (invariant) measure on the sphere $S^{m-1}(x_0, R)$ such that $|\sigma| - 1$.

$P(\cdot, \cdot)$ is called the *Poisson kernel* and (5), the *Poisson integral formula.*

The Poisson kernel

$$P(x, \xi) = \frac{R^2 - |x|^2}{R\omega_m} \frac{1}{|\xi - x|^m}, \qquad x \in K, \ \xi \in \partial K,$$

satisfies for $|x| < R$ the obvious inequality

$$\begin{aligned} \frac{1}{R\omega_m}\left(\frac{1}{R+|x|}\right)^{m-2} \frac{R-|x|}{R+|x|} &\leqslant P(x, \xi) \\ &\leqslant \frac{1}{R\omega_m}\left(\frac{1}{R-|x|}\right)^{m-2} \frac{R+|x|}{R-|x|} \end{aligned} \tag{7}$$

since $|\xi| = R$ and $R - |x| \leqslant |\xi - x| \leqslant R + |x|$.

Remark. Both bounds in (7) are independent of ξ.

Consequences of the Poisson Formula.

COROLLARY XV.6.7 (Harnack Inequality).

Let $K = K(0, R) \subset R^m$. Then for any $0 \leqslant u \in \mathscr{H}(K)$ the following inequality is valid in K:

$$\left(\frac{R}{R+|x|}\right)^{m-2} \frac{R-|x|}{R+|x|} u(0) \leqslant u(x) \leqslant \left(\frac{R}{R-|x|}\right)^{m-2} \frac{R+|x|}{R-|x|} u(0). \tag{8}$$

PROOF. The Harnack inequality follows immediately from inequality (7), which is independent of ξ, for the Poisson kernel and from the mean-value property of u:

$$u(0) = \frac{1}{\omega_m R^{m-1}} \int_{|\xi|=R} u(\xi)\, d\xi . \ \square$$

The Liouville Theorem for harmonic functions follows from the Harnack inequality.

THEOREM XV.6.8 (Liouville). $(u \in \mathscr{H}(\boldsymbol{R}^m)$ and $|u(\boldsymbol{R}^m)| < M) \Rightarrow (u = \text{const})$.

PROOF. Consider $v = u+M$; then $0 < v \in \mathscr{H}(\boldsymbol{R}^m)$ and we can apply the Harnack inequality. But this inequality is valid for a fixed x for all R. If we let $R \to \infty$, then $v(0) \leqslant v(x) \leqslant v(0)$. $\square$

The Harnack inequality implies also

THEOREM XV.6.8 ("Harnack Principle"). Let $\mathcal{O}$ be a domain in $\boldsymbol{R}^m$ and let (u_n) be an isotonic sequence of harmonic functions in $\mathcal{O}$ which is convergent at a point $x_0 \in \mathcal{O}$. Then (u_n) is compactly convergent.

PROOF. For $n > k_0$ the function $v_n := u_n - u_{k_0}$ is positive and harmonic in $\mathcal{O}$. Let $\overline{K(x_0, 2R)} \subset \mathcal{O}$ and $|x-x_0| \leqslant R$, then by the Harnack inequality

$$0 \leqslant v_n(x) \leqslant 2^{m-2} \cdot 3 v_n(x_0).$$

Therefore, v_n and hence u_n are uniformly convergent on $\overline{K(x_0, R)}$. $\square$

7. SUBHARMONIC FUNCTIONS. PERRON'S SOLUTION OF THE DIRICHLET PROBLEM

In 1923 Oscar Perron discovered a beautiful method of solving the Dirichlet problem for a very extensive class of domains and applied his method also to the Dirichlet problem for Riemann surfaces. He worked with subharmonic functions and the Harnack principle is a pillar of his method.

By the Poisson solution of the Dirichlet problem we have the following equivalent definition of subharmonicity.

Let O be a domain in $\boldsymbol{R}^m$, K a ball such that $\overline{K} \subset O$ and $v \in C(O, \boldsymbol{R})$.

We define $\underline{K}v\colon O \to R$ by

$$\text{(1)} \qquad \underline{K}v(x) = \begin{cases} v(x) & \text{for } x \notin K, \\ H_v^K & \text{for } x \in K, \end{cases}$$

where H_v^K is the solution of the Dirichlet problem for K with boundary values v:

$$\text{(2)} \qquad H_v^K | \partial K = v | \partial K.$$

DEFINITION. A continuous function $v\colon O \to R$ on a domain $O \subset R^m$ is subharmonic (in O) if for every ball $\overline{K} \subset O$ we have

$$v \leqslant \underline{K}v.$$

If the opposite inequality $v \geqslant \underline{K}v$ is valid, then v is called *superharmonic* in O.

Remark. $v \to \underline{K}v$ is a linear monotonic operator:

$$(v \leqslant w) \Rightarrow (\underline{K}v \leqslant \underline{K}w).$$

This immediately follows from the maximum principle for harmonic functions.

The maximum principle is valid for subharmonic functions and the minimum principle is valid for superharmonic functions.

We shall list some properties of subharmonic functions in the following.

LEMMA XV.7.1. 1° Subharmonic (s.h.) functions (in O) form a convex cone: (v, w are s.h.) $\Rightarrow$ ($v+w$, λv are s.h. for $\lambda > 0$),

2° (v and w are s.h.) $\Rightarrow$ ($\max(u, v)$ is s.h.),

3° a locally s.h. function is s.h.,

4° (v is s.h. in O) $\Rightarrow$ ($\underline{K}v$ is s.h. for each ball $\overline{K} \subset O$).

PROOF. 1° is obvious.

2° The monotony of $\underline{K}$ implies

$$\underline{K}v - \underline{K}\max(v, w) \leqslant 0,$$
$$\underline{K}w - \underline{K}\max(v, w) \leqslant 0,$$

whereupon by the subharmonicity of v and w we have $\max(v, w) \leqslant \max(\underline{K}v, \underline{K}w) \leqslant \underline{K}\max(v, w)$.

3° Let $\overline{K} \subset O$ be a ball, then if v is a locally s.h. function, then $(v - \underline{K}v)|K$ is locally s.h. in K and, therefore, attains its maximum at the boundary. Hence, $v - \underline{K}v \leqslant 0$.

4° Let L be a ball and let v be s.h.; we have to prove that

$$(*)\qquad \underline{K}v - \underline{L}(\underline{K}v) \leqslant 0.$$

But it is sufficient to prove $(*)$ for $L \cap K$, since it is valid for $O-L$ and for $L-K$ we have $\underline{K}v = v \leqslant \underline{L}v = \underline{L}(\underline{K}v)$. But $u := \underline{K}v - \underline{L}(\underline{K}v)$ is harmonic on $K \cap L$ and on ∂L, $u = \underline{K}v - \underline{K}v = 0$ (cf. (1)). On ∂K we have $u = v - \underline{L}v \leqslant 0$; therefore, $u \leqslant 0$ on $\partial(K \cap L)$, and from the maximum principle it follows that $u \leqslant 0$ on $K \cap L$. But u is (*ex definitione*) the left-hand side of $(*)$. □

We can now prove the main result of this section.

THEOREM XV.7.2 (O. Perron). Let O be a bounded domain in $\boldsymbol{R}^m$ and let f be a continuous real and bounded function on ∂O: $f \in C(\partial O)$, $\sup|f(\partial O)| < \infty$. Denote $V(f) = \{v \text{ is s.h. on } O\colon \limsup_{x\to\xi} v(x) \leqslant f(\xi),\ \xi \in \partial O\}$. Then the upper envelope of $V(f)$

$$(3)\qquad u := \sup\{v\colon v \in V(f)\} \text{ is harmonic on } O.$$

Remark. The function u is the only candidate for the solution of the Dirichlet problem

$$(4)\qquad \Delta u = 0, \quad u|\partial O = f,$$

and, therefore, it is called the *generalized solution of the Dirichlet problem* (4). But since O is quite general we cannot expect that

$$(5)\qquad (O \ni x_n \xrightarrow[n\to\infty]{} \xi) \Rightarrow (u(x_n) \to f(\xi), \qquad \xi \in \partial O).$$

Points ξ of the boundary ∂O for which (5) is valid are called *regular*. There exists an extensive theory of regular points in the modern potential theory. We shall give below a simple reasonable criterion for regularity.

PROOF OF THE PERRON THEOREM. Let $M = \sup|f(\partial O)|$: then $u \leqslant M$; otherwise for some $x \in O$ there exists a $v \in V(f)$ such that $v(x) > M+\varepsilon$ with an $\varepsilon > 0$, which contradicts the maximum principle for s.h. functions. Let $\overline{K} \subset O$ be a ball and $x_0 \in K$. Then by virtue of (3) there exists a sequence $v_n \in V(f)$ such that $v_i(x_0) \to u(x_0)$. We can assume that $v_i \nearrow$ (otherwise replace v_i by $\max(v_1, \ldots, v_2)$) and that all v_i are harmonic on K, otherwise replace v_i by $\underline{K}v_i$. By virtue of the Harnack principle the sequence $(v_i|K)$, $i = 1, 2, \ldots$ converges compactly on K to a harmonic function v_∞. Take another $x \in K$. We have (by the same argument) a sequence (w_i) of harmonic functions on K such that $w_i(x) \to u(x)$, $w_i \in V(f)$ and $w_i \nearrow w_\infty$. Since we can assume that $w_i \geqslant v_i$, (otherwise replace w_i by $\max(v_i, w_i)$) we have $v_\infty \leqslant w_\infty \leqslant u|K$, $v_\infty(x_0) = w_\infty(x_0) = u(x_0)$, $w_\infty(x) = u(x)$. Therefore, the harmonic function $v_\infty - w_\infty$ attains in $x_0 \in K$ its maximum 0; hence, $v_\infty = w_\infty$ and $u = v_\infty$ is harmonic on K. Since K has been an arbitrary ball in O, the function u is harmonic in O. □

Barriers. We now provide a simple criterion for the solvability of the Dirichlet problem. To this end let us introduce the following notion.

DEFINITION. Let O be a bounded domain in R^m and $\zeta \in \partial O$. A *barrier* for ζ is a harmonic function b^ζ on O, such that

(6) $\quad b^\zeta(\xi) > 0$ for $\zeta \neq \xi \in \partial O$ and $b^\zeta(\zeta) = 0$.

The notion of a barrier was introduced by Lebesgue in 1912.

Example 1. Let $\zeta \in \partial O$ and suppose that there exists a ball $K = K(p, R)$ such that $K \cap \overline{O} = \zeta$; then $b^\zeta(x) := \psi(|x-p|) - \psi(R)$ (where ψ is the fundamental solution) is a barrier for ζ (cf. Fig. 5).

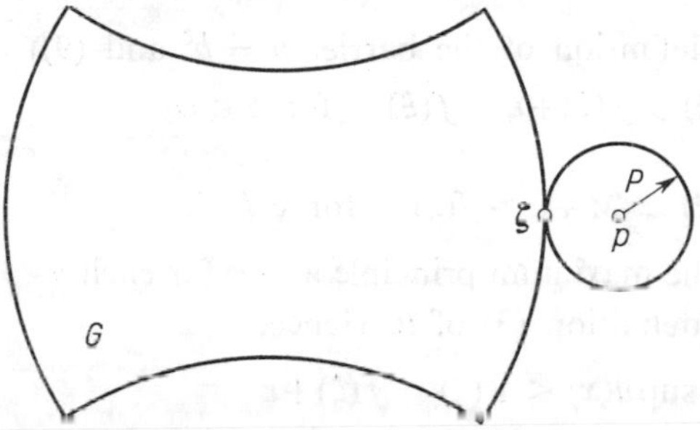

Fig. 5

Example 2. If G is a smooth submanifold of R^2 with boundary, then the criterion of Example 1 is always satisfied; make R less than the radius of the curvature of ∂G in ζ. The same is true for a submanifold in R^m, $m > z$. (It is sufficient to assume that G is of class C^2.)

The following theorem gives a criterion for the solvability of the Dirichlet problem for a bounded domain $O \subset R^m$.

THEOREM XV.7.3 (Perron–Lebesgue). 1° If a point $\zeta \in \partial O$ has a barrier b^ζ, then $V(f) \neq \emptyset$ and the harmonic function u defined by (3) can be continuously extended to ζ: $\lim_{x \to \zeta} u(x) = f(\zeta) =: u(\zeta)$.

2° Therefore, if every point of the boundary ∂O has a barrier, then the Dirichlet problem has the unique solution $u = H_f^O$ for any continuous boundary value $f \in C(\partial O)$.

PROOF. 1° Let $\infty > M = ||f|| = \sup|f(\partial O)|$. It suffices to prove the following inequalities:

$$(7) \qquad \limsup_{x\to\zeta} u(x) \leqslant f(\zeta)+\varepsilon,$$

$$(8) \qquad \liminf_{x\to\zeta} u(x) \geqslant f(\zeta)-\varepsilon$$

for any $\varepsilon > 0$. Let U be a neighbourhood of ζ such that

$$(9) \qquad |f(\xi)-f(\zeta)| < \varepsilon \quad \text{for } \xi \in U.$$

The barrier $b = b^\zeta$ has a minimum $b_0 > 0$ on $\overline{O}-U$ by virtue of the minimum principle. Consider the boundary values of the harmonic function w

$$(10) \qquad O \ni x \to w(x) := f(\zeta)+\varepsilon+\frac{b(x)}{b_0}(M-f(\zeta)).$$

Since (cf. the definition of the barrier $b = b^\zeta$ and (9))

$$w(\xi) \geqslant f(\zeta)+\varepsilon > f(\xi) \quad \text{for } \xi \in U$$

and

$$w(\xi) \geqslant M+\varepsilon > f(\xi) \quad \text{for } \xi \notin U,$$

we have from the maximum principle $w > v$ for each $v \in V(f)$; therefore, $w \geqslant u$, by the definition (3) of u. Hence,

$$\limsup_{x\to\zeta} u(x) \leqslant w(\zeta) = f(\zeta)+\varepsilon$$

and (7) is proved. In order to prove (8) we verify that

$$w_1 := f(\zeta)-\varepsilon-\frac{b}{b_0}(M+f(\zeta)) \in V(f).$$

In the same way we attain the following inequalities:

$$w_1(\xi) \leqslant f(\zeta)-\varepsilon < f(\xi) \quad \text{for } \xi \in U,$$
$$w_1(\xi) \leqslant -M-\varepsilon < f(\xi) \quad \text{for } \xi \notin U.$$

Therefore, $w_1 \in V(f)$ and $V(f) \neq \emptyset$; moreover, $w_1 \leqslant u$ and, hence,

$$\liminf_{x\to\zeta} u(x) \geqslant w_1(\zeta) = f(\zeta)-\varepsilon.$$

In this way (7) and (8) are proved and the proof is complete. □

Remark. We see that the set $V(f)$ is not empty if the boundary ∂O is not quite crazy: if ∂O has some ζ_0 with a barrier b^{ζ_0}.

Concluding Remarks. Harmonic Measures. The proof of the Perron theorem was not constructive but we can give it a constructive turn which probably has its origin in the famous "*alternating method*" of Hermann Amandus Schwarz.

COROLLARY XV.7.4 (Perron Schwarz). Let $v \in V(f)$ and let $(K_j)_1^\infty$ be a covering of O with balls $\overline{K_j} \subset O$; then the sequence

$$u_1 := \underline{K}_1 v, \quad u_2 := \underline{K}_2 u_1, \ldots, \quad u_{n+1} := \underline{K}_j u_n,$$

where the index j runs through N in the order $1, 2, 1, 2, 3, 1, 2, 3, 4, \ldots$, is convergent to the generalized solution u of the Dirichlet problem

$$\Delta u = 0, \quad u|\partial O = f.$$

PROOF. Since v is s.h. in O, all u_k are also s.h. and $v \leqslant u_1 \leqslant u_2 \leqslant \ldots$ Moreover, for each j there exists a subsequence (u_{j_n}) such that all u_{j_n} are harmonic in K_j. By virtue of the Harnack principle u_n are convergent to a harmonic function u (since by Harnack u is harmonic on each K_j). □

Harmonic Measures. A precompact set $O \subset R^m$ is *regular* if the Dirichlet problem has a (unique) solution H_f^O for any $f \in C(\partial O)$, in other words if any $f \in C(\partial O)$ can be extended in a unique way to such $u \in C(\overline{O})$ that its restriction H_f^O to O is harmonic.

DEFINITION. Let O be a regular subset of R^m; then for each $x \in O$ the map μ_x^O

$$C(\partial O) \ni f \to H_f^O(x) \in R$$

is a positive Radon measure on ∂O. The measure μ_x^O is called the *harmonic measure* corresponding to the set O and the point $x \in O$. The measure μ_x^O is defined by the identity

$$H_f^O(x) \equiv \int f d\mu_x^O, \quad f \in C(\partial O).$$

The reader may prove the following

Fact. Let U be a non-empty open subset of R^m. A lower semicontinuous function $u\colon U \to]-\infty, +\infty[$ is s.h. if and only if for any regular subset $O \subset U$

$$\int u\, d\mu_x^O \geqslant u(x).$$

Example. We know that any open ball $O = K(x_0, r)$ is a regular subset of $\boldsymbol{R}^m$, $m > 2$,

$$H_f^O(x) = \int f d\mu_x^O,$$

where

$$\mu_x^O = P_x \sigma, \quad P_x(z) := r^{m-2} \frac{r^2 - |x - x_0|^2}{|x - z|^m}$$

and σ is a canonical normed ($||\sigma|| = 1$) measure on the sphere $S^{m-1}(x_0, r) = \partial O$. $P(x, z) \equiv P_x(z)$ is of course the Poisson kernel:

$$H_f^O(x) = \int P(x, z) f(z) \, d\sigma(z).$$

Therefore $P_x \sigma$ is the harmonic measure corresponding to the ball $O = K(x_0, r)$ and the point x.

We plainly have $\mu_{x_0}^O = \sigma$.

Remark. The notion of subharmonic functions can be defined on a Riemann surface M, since subharmonicity on $\boldsymbol{C}^1$ is invariant with respect to conformal, i.e., biholomorphic, maps.

Exercise. Prove this.

The Perron method discovered independently by Remak, was extended and refined by M. Riesz, M. Brelot, O. Frostman, G. Anger and many others. The harmonic space theory of M. Brelot and H. Bauer, drew the ultimate consequences from the ideas of Perron and Nevanlinna.

The idea of harmonic measure can be traced to H. A. Schwarz, but was formulated by R. Nevanlinna. We shall return to potential theory in Ch. XVI and in Part III.

XVI. COMPLEX ANALYSIS IN ONE DIMENSION (RIEMANN SURFACES)

To Erich Kähler

> Zum ersten Male zeigten die mathematischen Formen ihren eigenen Willen; sie traten auf wie Gebilde der Natur, deren Formen nicht erfunden und erzwungen, sondern nur erforscht werden können. So entstand so etwas wie eine mathematische Ästhetik, eine Formenlehre des mathematischen Ausdrucks... Nils Henrik *Abel* heißt der Mathematiker, der zuerst von dieser Einsicht erfaßt wurde.
>
> *Erich Kähler*

Complex analysis in one dimension, i.e., the theory of Riemann surfaces, is a source of constant inspiration for the whole of mathematics.

Inaugurated by the great 19th-century mathematicians Cauchy, Gauss, Galois, Abel, Jacobi, and Weierstrass through Riemann it acquired a new dimension and a new quality, becoming the central part of mathematics. Orientation in this domain necessitated the creation of entirely new disciplines: topology (algebraic and general), theory of discontinuous transformation groups, Abelian differentials and integrals.

The theory of Riemann surfaces shows the unity of mathematics: it has woven into one inseparable whole differential and algebraic geometry, the theory of elliptic functions and arithmetics (the analytic and the algebraic theory of numbers), topology together with the theory of covering spaces, cohomology together with the potential theory (Green's functions, harmonic measure, harmonic forms) and functional analysis. The theory of Riemann surfaces, which arose out of specific natural problems, has now soared up into beautiful generality, returning again and again, like Antheus, to concrete problems of geometry and arithmetic. This wonderful creation is the work of the followers of Riemann and Weierstrass, H. A. Schwarz, L. Fuchs, Schottky, Klein, Koebe, Hilbert, and, above all, the greatest French scholar Poincaré. This theory is eternally young: it cannot be otherwise since it was developed by investigators of such rank as Weyl, Radó, R. Nevanlinna, Kähler, Hodge, Ahlfors, Bers, Behnke, and K. Stein, and is now being elaborated and enriched by such men as Serre, Kodaira, Shimura, Shafarevitch, P. A. Griffiths, Hirzebruch, Atiyah, Mumford, and others.

The present chapter, central in Part Two, aims at introducing the reader to this strange universe. The first part of present chapter (Sections 1–6) is of elementary nature: one investigates functions on the simplest Riemann surfaces, namely, domains in $\boldsymbol{C}$. The emphasis is on the examination of singularities and computations: computing integrals by the method of residua. The culminating point is the proof of the Koebe–Riemann Theorem on conformal mappings.

The fundamental notion of analytic continuation inevitably leads to the notion of homotopy of paths and to the fundamental group, covering surfaces (germs of meromorphic functions, the monodromy theorem) and culminates in what I have called the Poincaré theory (a worthy counterpart of the Galois theory). It forms the topological part of the theory of uniformization, one of the most magnificent edifices of mathematics. The analytical part of that theory requires entirely different measures. I have chosen a way which also derives from Riemann, namely, the method of the potential theory (the Green's function) following the Perron–Nevanlinna approach: the most important Riemann surfaces X, i.e., hyperbolic surfaces, have a conformal invariant—the Green's function g^X, characterizing a surface X to a high degree and giving, in the simply connected case, a biholomorphic mapping onto a unit disc $\boldsymbol{D}$. In the case of a domain in $\boldsymbol{C}$ we obtain the second proof of the Koebe–Riemann Theorem.

In the theory of non-compact Riemann surfaces a crucial role is played by the famous Behnke–Stein Theorem, which constitutes the fundamental generalization of the classical theorem of Runge. This theorem has been transferred by Malgrange in a beautiful manner onto arbitrary vector bundles and elliptic operators A: $E \to F$. If A is a Cauchy–Riemann operator d'', we obtain the Behnke–Stein Theorem; if A is Δ—the Laplace operator on a Riemann surface, we obtain the Pfluger Theorem (the proof of Malgrange's theorem will be given in Part III). However, in order not to disappoint the reader too much, in Section 21 I give a proof of Runge's Theorem (after Hörmander) and in Sections 14–15 applications of the Behnke–Stein Theorem to the Mittag-Leffler and the Weierstrass problems (known in the theory of several complex variables as the additive problem and the multiplicative problem of Cousin).

In Sections 16–19 we return to the problem of uniformization for particular Riemann surfaces, thus entering the enchanted world of elliptic functions and elliptic modular functions (cf. the motto); the watchwords are: the Weierstrass $\wp$-function, Eisenstein series, and Dedekind–Klein j-function. An original proof of the famous Picard Theorem by means of the j-function is given. I could not resist quoting the natural generalization of Picard's Theorem to hyperbolic manifolds which we owe to Kobayashi (Section 20).

Section 19 culminates in Abel's Theorem and Jacobi's inverse problem. (The original proof of Abel is reproduced (by Shafarevitch) in Section 20.)

Fortified by concrete experience, we come back in Section 20 to the general theory of uniformization. A classification of properly discontinuous transformation groups without fixed points of the three canonical simply connected Riemann surfaces ($\boldsymbol{P}(C)$, $\boldsymbol{C}$, $\boldsymbol{D} \cong \mathfrak{H}$—upper half-plane) shows from a higher point of view the place of rational functions, periodic and double-periodic (i.e., elliptic) and the theory of modular functions. We realize the necessity of the gigantic efforts of Poincaré and Klein towards the construction of the theory of automorphic functions. The role of the Gaussian curvature $(+1, 0, -1)$ in the theory of Riemann surfaces becomes clear, and so does the necessity of the Lobachevsky–Bolyai non-Euclidean geometry.

The theory of compact Riemann surfaces is governed by the Riemann–Roch Theorem (which is one of the most important facts in mathematics). We show the power of this theorem by means of examples of important applications (e.g., the existence of Abel differentials).

One cannot have a really deep understanding of mathematics without knowing its history. That is why Section 20 ends with a historical sketch, containing extensive quotations from two eminent contemporary mathematicians, I. Shafarevitch and L. Ahlfors.

Section 21 contains one more proof of the Koebe–Poincaré Theorem on the uniformization of simply connected surfaces, which is due to a young master of complex analysis, Otto Forster: as could be expected, since the Behnke–Stein Theorem is the main theorem on open Riemann surfaces, it must also lead to a proof of the theorem on uniformization. We owe both this proof and an approach to several essential facts in the theory of Riemann surfaces to O. Forster's *Riemannsche Flächen*, a book of classic beauty.

Readers well grounded in the literature of the subject will be able to see how much the present chapter owes to the monographs of such masters of complex analysis as A. Pfluger, L. Alhfors and Rolf Nevanlinna. The crucial sections, concerning modular functions and forms, owe a great deal to beautiful lectures of J. P. Serre.

In Section 21 I could not resist the temptation to supplement the historical sketch by mentioning H. A. Schwarz's wonderful and extremely important discovery of "triangle groups", comprising modular groups—a discovery which had a "catalytic effect" upon the fundamental works of Poincaré. I used the brilliant historical sketch contained in Chapter I of H. Lehner's classical work. The chapter ends with Theodor Schneider's famous theorem stating the relations between elliptic integrals and the transcendentality of numbers. I hope that this reference will stimulate the reader to study the little book *Transcendental Numbers* by one of the greatest mathematicians of our times, C. L. Siegel.

I dedicate this chapter, crucial in the book, to the man who initiated me into the the philosophical role of the modular function—to Erich Kähler.

1. ZEROS OF HOLOMORPHIC FUNCTIONS OF ONE VARIABLE

Sections 1–5 of this chapter are of a more elementary character than are later sections and the results presented here form the basis for the subsequent development, in fact for the whole theory in one variable. They need none of the heavy machinery developed earlier and they should also be accessible to readers who abhor differential forms on differentiable manifolds. However, we do freely use some of the basic (and elementary) facts established in the Chapter XV, e.g., the power series representation of holomorphic functions, and for the reader's convenience, in a few cases we state their one-dimensional versions as separate theorems.

Let Ω be an open, nonvoid subset of the complex plane $\boldsymbol{C}$. Usually we shall also assume that Ω is connected and we shall call such a set a *domain.*

For every function $f \in C^1(\Omega)$ and each $z_0 \in \Omega$ we have the formula

$$(1) \qquad f(z_0+\Delta z)-f(z_0) = \frac{\partial f}{\partial z}\Delta z+ \frac{\partial f}{\partial \bar{z}}\overline{\Delta z}+O(\Delta z).$$

Here, as in Chapter XV, $\frac{\partial f}{\partial z}$ and $\frac{\partial f}{\partial \bar{z}}$ stand for the formal complex derivatives which are defined by

$$\frac{\partial f}{\partial z} = \frac{1}{2}\left(\frac{\partial f}{\partial x} - i\frac{\partial f}{\partial y}\right), \qquad \frac{\partial f}{\partial \bar{z}} = \frac{1}{2}\left(\frac{\partial f}{\partial x} + i\frac{\partial f}{\partial y}\right).$$

DEFINITION. A function f defined in Ω is said to be *complex differentiable at a point* $z_0 \in \Omega$ if there exists a complex number l such that

$$(2) \qquad f(z_0+\Delta z)-f(z_0) = l\Delta z + O(\Delta z).$$

The complex mumber l is then unique and is seen to be the limit

$$l = \lim_{\Delta z \to 0} \frac{f(z_0+\Delta z)-f(z_0)}{\Delta z},$$

henceforth called the *complex derivative of f at z_0* and denoted by $f'(z_0)$. We shall say that a function f complex differentiable at each point of Ω is *continuously differentiable* if the function $\Omega \ni z \to f'(z) \in \mathbf{C}$ is continuous. The comparison of (1) and (2) reveals that a function f is continuously complex differentiable in Ω if and only if $f \in C^1(\Omega)$ and $\frac{\partial f}{\partial \bar{z}} = 0$, that is, if f is holomorphic in Ω. For a holomorphic function f we then have

$$f'(z_0) = \frac{\partial f}{\partial z}(z_0) = \lim_{\Delta z \to 0} \frac{f(z_0+\Delta z)-f(z_0)}{\Delta z}.$$

The space of all functions holomorphic in Ω is denoted by $A(\Omega)$. It follows from the results of Chapter XV that every continuously complex differentiable function has derivatives of all orders and, in fact, is even analytic. For an easy reference we sum this up in the form of

THEOREM XVI.1.1. Let f be a function defined in a domain Ω.

(i) f is holomorphic iff it is continuously complex differentiable.

(ii) If f is holomorphic then so is its complex derivative and, hence, f has complex derivatives of all orders.

(iii) Assume that f satisfies condition (i) and let $z_0 \in \Omega$. Then the Taylor series of f

$$\sum_{n=0}^{\infty} a_n(z-z_0)^n \quad \text{with} \quad a_n = \frac{f^{(n)}(z_0)}{n!}, \; n = 0, 1, \dots$$

converges to $f(z)$ for all z in the largest open disk contained in Ω.

It is worth noting that the class $A(\Omega)$ of holomorphic functions is the same as the class of complex differentiable functions. In other words, any complex differentiable function is in fact continuously complex differentiable and, consequently, holomorphic. This assumption of complex differentiability is traditionally taken as the starting point. We now state the Cauchy Theorem for the case of one variable

THEOREM XVI.1.2 (Cauchy). Let Ω be a contractible domain in $\boldsymbol{C}$. Then

$$\begin{aligned}&(f \in A(\Omega))\\ &\Rightarrow \Big(\int_\gamma f(\zeta)\,d\zeta = 0 \text{ for each 1-cycle } \gamma \text{ contained in } \Omega\Big).\end{aligned}$$

The Cauchy Theorem implies, in particular, the existence of holomorphic primitives of holomorphic functions (in contractible domains, of course). The proof of this relies on a result closely resembling the fundamental theorem of calculus.

PROPOSITION XVI.1.3. Let f be a continuous function in a domain $\Omega \subset \boldsymbol{C}$ such that

$$\int_\gamma f(\zeta)\,d\zeta = 0$$

for every 1-cycle in Ω. Then for any (fixed) $z_0 \in \Omega$ a function

$$F(z) = \int_{z_0}^{z} f(\zeta)\,d\zeta$$

is well defined (the integral depends only on the end-points of a path), holomorphic in Ω, and satisfies $F' = f$.

PROOF. That the integral defining F is independent of a path joining z_0 with z should be clear since if γ_1 and γ_2 are two such paths then $\gamma = \gamma_1 - \gamma_2$ is a cycle and thus $\int_\gamma f(\zeta)\,d\zeta = \int_{\gamma_1} f(\zeta)\,d\zeta - \int_{\gamma_2} f(\zeta)\,d\zeta = 0$. Now, given $z \in \Omega$, there is an $r > 0$ such that the disk $K(z, r)$ is contained in Ω and thus for all $z_1 \in K(z, r)$ we have

$$F(z_1) - F(z) = \int_{z_1}^{z} f(\zeta)\,d\zeta,$$

where a path joining z with z_1 may now be chosen to be an interval with end-points z and z_1. Then an easy estimate gives (putting $z_1 - z = \Delta z$)

$$\left| \frac{F(z+\Delta z)-F(z)}{\Delta z} - f(z) \right| = \left| \frac{1}{\Delta z} \int_z^{z_1} (f(\zeta)-f(z))d\zeta \right| \underset{\Delta z \to 0}{\to} 0. \ \square$$

COROLLARY XVI.1.4. Let $\Omega \subset \boldsymbol{C}$ be a contractible domain and $f \in A(\Omega)$. Then there is an $F \in A(\Omega)$ such that $F' = f$. (F is said to be a *primitive* of f.)

A number of useful consequences follow. First, since the set $\Omega = \boldsymbol{C} -]-\infty, 0]$ is contractible (it is starshaped with respect to 1; draw a picture!) primitives of the function $1/z$ exist in Ω. The reader may check that for any such primitive F, $\exp F(z) = cz$ for all $z \in \Omega$. Those F which satisfy $\exp F(z) = z$ are called *holomorphic branches of the logarithm*; it is clear that any two of them differ by an integral multiple of $2\pi i$.

More generally, we have the following

COROLLARY XVI.1.5. Let $\Omega \subset \boldsymbol{C}$ be a contractible domain (simply-connected would do also) and $f \in A(\Omega)$. If $f(z) \neq 0$ for all $z \in \Omega$ then

1° a holomorphic branch of $\log f(\cdot)$ exists in Ω, i.e., there is an $F \in A(\Omega)$, such that $\exp F(z) = f(z)$ for all $z \in \Omega$;

2° a holomorphic branch of an arbitrary (complex) power of f exists in Ω, i.e.,

$$f^a(z) := \exp(a \log f(z))$$

is holomorphic in Ω for any $a \in \boldsymbol{C}$.

PROOF. Since 2° is an obvious consequence of 1° it is sufficient to prove 1°. By Corollary XVI.1.4 the function $f'/f \in A(\Omega)$ has a primitive $G \in A(\Omega)$. By adding a constant we can get $\exp G(z_0) = f(z_0)$ at some point $z_0 \in \Omega$.

Since the derivative of $f(\cdot)\exp(-G(\cdot))$ vanishes in Ω, this function is constant, and by our choice of G this constant is 1. $\square$

Remark. The existence of a holomorphic branch of $\log f(\cdot)$ and also of a square root $f^{1/2}(\cdot)$ will be used in the proof of the Koebe Theorem (the fundamental theorem on conformal mappings) in Section 8.

A converse to the Cauchy Theorem can now be given.

THEOREM XVI.1.6 (Morera). Let $\Omega \subset C$ be a domain and let $f \in C(\Omega)$. If the integral $\int f dz$ is locally independent of a path, then f is holomorphic in Ω.

The assumption means that for any $z_0 \in \Omega$ there is a neighbourhood of z_0 such that for all paths γ contained in that neighbourhood and starting at z_0 the integral $\int_\gamma f dz$ depends only on the terminal point of γ. This is the same as saying that the integral $\int_\gamma f dz$ vanishes for all 1-cycles γ lying in that neighbourhood.

PROOF. For a point $z_0 \in \Omega$ choose an open disk $K(z_0, r) \subset \Omega$ such that the integral $\int_\gamma f dz$ vanishes for all 1-cycles contained in $K(z_0, r)$. Then, by Proposition XVI.1.3 there is an $F \in A\big(K(z_0, r)\big)$ such that $F' = f|_{K(z_0,r)}$. But then $f|_{K(z_0,r)}$, as the derivative of an holomorphic function, is itself holomorphic. □

Let us recall the definition: a function f is said to have a *zero of order k at a point* z_0 if

$$f^{(j)}(z_0) = 0, \qquad j = 0, 1, \ldots, k-1,$$
$$f^{(k)}(z_0) \neq 0.$$

This definition leads immediately to the following

COROLLARY XVI.1.7. $(f \in A(\Omega)$ has a zero of order k at a point $z_0)$ $\Leftrightarrow$ (there is an open disk $K(z_0, r) = K$ with the centre z_0 and a $g \in A(K)$ such that $f(z) = (z-z_0)^k g(z)$ for all $z \in K$ and $g(z_0) \neq 0\big)$.

PROOF. $\Leftarrow$: Obvious by calculation.

$\Rightarrow$: The first k terms drop out of the Taylor expansion and, hence,

$$f(z) = \sum_{j=0}^{\infty} (z-r_0)^{k+j} \frac{f^{(k+j)}}{(k+j)!}(z_0) = (z-z_0)^k g(z),$$

where

$$g(z) = \sum_{j=0}^{\infty} \frac{f^{(k+j)}}{(k+j)!}(z_0)(z-z_0)^j.$$

But $g(z_0) = f^{(k)}(z_0) \neq 0$. □

1. ZEROS OF HOLOMORPHIC FUNCTIONS

The zeros of a holomorphic function of one variable are isolated. It turns out that the zeros are never isolated in the case of n complex variables, $n > 1$; this fundamental difference is one reason why meromorphic functions of many complex variables must be defined in an entirely different manner. For one variable, on the other hand, we have:

THEOREM XVI.1.8. Let $\Omega \subset C$ be a domain and let a set $S \subset \Omega$ have at least one accumulation point in Ω. Then if a function $f \in A(\Omega)$ vanishes on S, it vanishes identically on Ω.

COROLLARY XVI.1.9. Under the same assumptions as in Theorem XVI.1.8 let $f, g \subset A(\Omega)$ and $f = g$ on S. Then $f = g$ identically on Ω.

Theorem XVI.1.8 is an immediate corollary to the following theorem:

THEOREM XVI.1.10 (On the Uniqueness of Power Series). Let

$$g(z) = \sum_{n=0}^{\infty} a_n(z-z_0)^n, \qquad h(z) = \sum_{n=0}^{\infty} b_n(z-z_0)^n$$

be series convergent in a disk $K(z_0, r)$ and let $(z_k)_{k=1}^{\infty}$ be a sequence of distinct points convergent to z_0. If $g(z_k) = h(z_k)$, $k = 1, 2, \dots$, then $g = h$ throughout $K(z_0, r)$.

PROOF. Let

$$f(z) := (g-h)(z) = \sum_{n=0}^{\infty} (a_n - b_n)(z-z_0)^n.$$

However, $f(z_k) = 0$. Suppose that we do not have $a_n = b_n$ for every n. Let s be the lowest index such that $a_s \neq b_s$. Thus

$$\begin{aligned} f(z) &= (z-z_0)^s\Big((a_s-b_s)+\sum_{j=1}^{\infty} (a_{s+j}-b_{s+j})(z-z_0)^j\Big) \\ &= (z-z_0)^s P(z), \end{aligned}$$

where $P(z) := \sum_{j=0}^{\infty} (a_{s+j}-b_{s+j})(z-z_0)^j$. Since $f(z_k) = 0$, $(z_k-z_0)^s \neq 0$ we have $P(z_k) = 0$, $k = 1, 2, \dots$ Accordingly

$$0 = \lim_{k\to\infty} P(z_k) = P(z_0) = (a_s - b_s), \quad \text{whereby } a_s = b_s.$$

Thus we have arrived at a contradiction and, hence, $a_n = b_n$, for all n, i.e. $g = h$. □

The preceding theorems reveal an important feature of holomorphic functions, namely, that they are uniquely determined by their values on any set with limit points in Ω, e.g. on an arbitrarily small arc contained in Ω.

It is important to observe that the connectedness of Ω is crucial for the conclusion of Theorem XVI.1.9 to hold. A trivial example can be obtained by taking Ω to be a sum of two disjoint open sets and a function equal to 0 on one of them and 1 on the other.

The following theorem is now a straightforward corollary.

THEOREM XVI.1.11 (On the Analytic Continuation of an Inverse Function). Let $\Omega \subset \boldsymbol{C}$ be a connected set, let $f \in A(\Omega)$, and let $g \in A(f(\Omega))$. Suppose that the set Ω contains an interval $]a, b[\subset \boldsymbol{R}$. Let $g \circ f |]a, b[= 1$. Then $g \circ f = 1$ on Ω.

We may now complete our discussion of the complex logarithm. The logarithm on $\boldsymbol{R}_+$ was defined as the function

$$]0, \infty[\ni x \to \log x := \int_1^x \frac{dt}{t}.$$

Therefore, the branch of the logarithm defined in $\Omega = \boldsymbol{C} -]-\infty, 0]$ by the formula

$$\operatorname{Log} z := \int_1^z \frac{d\zeta}{\zeta}, \qquad z \in \boldsymbol{C} -]-\infty, 0]$$

gives the values of the "real" logarithm for all real and positive z. This branch is customarily called the *principal branch of the logarithm* and the function

$$\operatorname{Log}_k z := \operatorname{Log} z + 2k\pi i, \qquad k = 0, \pm 1, \pm 2, \ldots,$$

is called the *k-th branch*.

From the theorem above we see that the logarithm is "the inverse" to the function $\exp(\cdot)$, that is,

$$\exp \operatorname{Log}(z) = z, \qquad z \in \boldsymbol{C} - \{]-\infty, 0]\}.$$

The occurrence of a countable number of branches is thus related to the periodicity of the function $\exp(\cdot)$. All the branches can be combined into one single function by extending the complex plane to a suitable complex manifold, the Riemann surface of the logarithm (cf. Section 7).

Now we give a brief proof of the fundamental theorem of algebra.

THEOREM XVI.1.12 (The Fundamental Theorem of Algebra). A polynomial $W_n(z) = a_0+a_1z+ \dots +a_nz^n$ of degree n ($a_n \neq 0$) with complex coefficients has n complex roots (not necessarily different).

PROOF. Suppose that a polynomial W_n does not have any roots. Then the function $f(z) := 1/W_n(z)$ is holomorphic throughout C. But

$$f(z) = \left|\frac{1}{W_n(z)}\right| = \frac{1}{|z^n||a_n+a_{n-1}/z+ \dots +a_0/z^n|} \underset{|z|\to\infty}{\to} 0,$$

and, hence, $|f|$ is bounded on C. By Liouville's Theorem it follows that $f = \text{const}$, which contradicts our hypothesis. Thus, there exists a point z_0 such that $W_n(z_0) = 0$. Dividing W_n by the monomial $z-z_0$, we obtain a polynomial W_{n-1}, to which we once again apply the procedure used previously, and so on. □

Remark. Other proofs of this theorem are given as Corollary XVI.5.5 and the exercise following Corollary XVI.6.8.

2. FUNCTIONS HOLOMORPHIC IN AN ANNULUS. THE LAURENT EXPANSION. SINGULARITIES

DEFINITION. An (*open*) *annulus* or *annular region* $R(z_0, r_1, r_2)$ centred at the point $z_0 \in C^1$, and of inner radius $r_1 > 0$ and outer radius $r_2 > r_1$, is the set

$$R(z_0, r_1, r_2) := \{z \in C^1 : r_1 < |z-z_0| < r_2\}.$$

THEOREM XVI.2.1 (Laurent). If a function f is holomorphic in an annulus $R(z_0, r_1, r_2)$, then it is the sum of the series (known as the Laurent series)

$$(1) \qquad f(z) = \sum_{n=-\infty}^{+\infty} a_n(z-z_0)^n, \qquad z \in R(z_0, r_1, r_2),$$

where

$$a_n = \frac{1}{2\pi i} \int\limits_{|\zeta-z_0|=r} \frac{f(\zeta)}{(\zeta-z_0)^{n+1}}\, d\zeta, \qquad r_1 < r < r_2$$

and the series (1) converges compactly and absolutely within $R(z_0, r_1, r_2)$.

PROOF. Let $r_1 < \varrho_1 < \varrho_2 < r_2$. Then $R(z_0, \varrho_1, \varrho_2) \subset R(z_0, r_1, r_2)$. We can use the Cauchy integral formula for the closed annulus $\overline{R(z_0, \varrho_1, \varrho_2)}$. We obtain

$$(2) \qquad f(z) = \frac{1}{2\pi i} \int\limits_{\partial R(z_0, \varrho_1, \varrho_2)} \frac{f(\zeta)}{\zeta - z} d\zeta, \qquad z \in R(z_0, \varrho_1, \varrho_2).$$

The boundary $\partial R(z_0, \varrho_1, \varrho_2)$ of the annulus $R(z_0, \varrho_1, \varrho_2)$ consists of two oppositely oriented circles $|\zeta - z_0| = \varrho_1$ and $|\zeta - z_0| = \varrho_2$ and thus we have

$$(3) \qquad f(z) = \frac{1}{2\pi i} \left\{ \int\limits_{|\zeta - z_0| = \varrho_2} \frac{f(\zeta)}{\zeta - z} d\zeta - \int\limits_{|\zeta - z_0| = \varrho_1} \frac{f(\zeta)}{\zeta - z} d\zeta \right\}.$$

We deal with the first integral first. Using an identity

$$\zeta - z = (\zeta - z_0)\left(1 - \frac{z - z_0}{\zeta - z_0}\right),$$

we write the function $\zeta \to \dfrac{1}{\zeta - z}$ (with $z \in R(z_0, \varrho_1, \varrho_2)$ regarded fixed) as the geometric series

$$(4) \qquad \frac{1}{\zeta - z} = (\zeta - z_0)^{-1} \sum_{n=0}^{\infty} \left(\frac{z - z_0}{\zeta - z_0}\right)^n = \sum_{n=0}^{\infty} \frac{(z - z_0)^n}{(\zeta - z_0)^{n+1}},$$

which converges provided $|\zeta - z_0| > |z - z_0|$.

Next we take a ϱ_2 such that $\varrho_1 < \varrho_2' < \varrho_2$ and consider the closed disk $\overline{K(z_0, \varrho_2')}$. For all $z \in \overline{K(z_0, \varrho_2')}$ and ζ from the circle $|\zeta - z_0| = \varrho_2$ the series on the right is dominated term by term by a convergent numerical series $\sum\limits_{n=0}^{\infty} \dfrac{(\varrho_2')^n}{(\varrho_2)^{n+1}}$, in particular since a function of ζ is uniformly convergent. By a well-known result on the integration of sequences of functions (cf. XIII.10.5) the integral

$$\frac{1}{2\pi i} \int\limits_{|\zeta - z_0| = \varrho_2} \frac{f(\zeta)}{\zeta - z} d\zeta = \frac{1}{2\pi i} \int\limits_{|\zeta - z_0| = \varrho_2} \sum_{n=0}^{\infty} f(\zeta) \frac{(z - z_0)^n}{(\zeta - z_0)^{n+1}} d\zeta$$

can be evaluated termwise to yield

$$(5)\qquad \sum_{n=0}^{\infty} a_n(z-z_0)^n \quad \text{with} \quad a_n = \frac{1}{2\pi i}\int_{|\zeta-z_0|=\varrho_2} \frac{f(\zeta)}{(\zeta-z_0)^{n+1}}d\zeta.$$

From the estimate given above it also follows that (5) is absolutely and uniformly convergent in the closed disk $|z-z_0| \leqslant \varrho_2'$.

To compute the second of the integrals in (3) we use an identity

$$\zeta - z = -(z-z_0)\left(1-\frac{\zeta-z_0}{z-z_0}\right)$$

and, expanding in a geometric series, we get (provided $|z-z_0| > |\zeta-z_0|$)

$$\frac{1}{\zeta-z} = -\frac{1}{(z-z_0)}\sum_{n=0}^{\infty}\frac{(\zeta-z_0)^n}{(z-z_0)^n} = -\sum_{n=1}^{\infty}\frac{(\zeta-z_0)^{n-1}}{(z-z_0)^n}.$$

Now, choosing a ϱ_1' such that $\varrho_1 < \varrho_1' < \varrho_2'$ and arguing as above, we obtain

$$(6)\qquad -\frac{1}{2\pi i}\int_{|\zeta-z_0|=\varrho_1}\frac{f(\zeta)}{\zeta-z}d\zeta = \sum_{n=1}^{\infty} a_n(z-z_0)^{-n}$$

with a_n given by

$$a_n = \frac{1}{2\pi i}\int_{|\zeta-z_0|=\varrho_1}(\zeta-z_0)^{n-1}f(\zeta)d\zeta, \qquad n = 1, 2, \ldots$$

and the series in (6) convergent absolutely and uniformly for $\varrho_1' \leqslant |z-z_0|$. Next, changing the signs of the summation indices in (6) and inserting (5) and (6) into (3), we get

$$f(z) = \sum_{n=-\infty}^{\infty} a_n(z-z_0)^n$$

uniformly and absolutely in the closed annulus $\varrho_1' \leqslant |z-z_0| \leqslant \varrho_2'$. Since each compact subset of $R(z_0, r_1, r_2)$ is contained in the closed annulus $\varrho_1' \leqslant |z-z_0| \leqslant \varrho_2'$ for suitable ϱ_1' and ϱ_2' we arrive at the desired conclusion on the convergence of the Laurent series. To see that (1) is valid, we observe that in the expressions for a_n obtained in the course of the proof the integrand is a holomorphic function inside the annulus $R(z_0, r_1, r_2)$ and, thus, the choice of the path of integration is immaterial. □

COROLLARY XVI.2.2. For the coefficients of the Laurent expansion we have

$$|a_n| \leqslant \frac{M(r)}{r^n}, \quad -\infty < n < \infty, \quad r_1 < r < r_2,$$

where

$$M(r) := \sup_{0\leqslant\varphi\leqslant 2\pi} |f(z_0 + re^{i\varphi})|.$$

PROOF. We know that

$$a_n = \frac{1}{2\pi i} \int_{|\zeta - z_0| = r} \frac{f(\zeta)}{(\zeta - z_0)^{n+1}}\, d\zeta.$$

Performing the parametrization $\zeta = z_0 + re^{i\varphi}$, we obtain

$$|a_n| \leqslant \frac{1}{2\pi} \int_0^{2\pi} \frac{|f(z_0 + re^{i\varphi})|}{r^n}\, d\varphi \leqslant \frac{M(r)}{r^n}. \ \square$$

COROLLARY XVI.2.3. If $M = \sup_{z\in R(z_0, r_1, r_2)} |f(z)|$, then

$$|a_n| \leqslant \begin{cases} \dfrac{M}{r_1^n} & \text{for } n < 0, \\ \dfrac{M}{r_2^n} & \text{for } n > 0, \\ M & \text{for } n = 0. \end{cases}$$

The theorem on the Laurent expansion of a holomorphic function in an annulus $R(z_0, r_1, r_2)$ is an extension of the Taylor expansion theorem for a holomorphic function. If

$$f(z) = \sum_{n=-\infty}^{\infty} a_n (z - z_0)^n,$$

then

$$f_1(z) := \sum_{n=-\infty}^{-1} a_n (z - z_0)^n$$

is called the *principal part* of the expansion of the function f. Clearly, $f_1(z) \equiv 0$ for a function holomorphic on $K(z_0, r_2)$. The principal part f_1 of the Laurent expansion of f can be regarded as a power series in the variable $(z - z_0)^{-1}$, and hence, turns out to be defined by this series outside of some disk $|z-z_0| < r$ or, it might be said, "in a disk with centre at ∞". We shall see later how to make the last phrase fully meaningful. Now we shall prove a certain characterization of the principal part of the Laurent expansion.

PROPOSITION XVI.2.4. If a function f is holomorphic in an annulus $R(z_0, r_1, r_2)$, then there exists a function f_2 holomorphic inside the disk $K(z_0, r_2)$ and a function f_1 holomorphic outside the disk $K(z_0, r_1)$ such that

$$f(z) = f_1(z)+f_2(z) \quad \text{for } z \in R(z_0, r_1, r_2).$$

Moreover, functions f_1 and f_2 are determined uniquely by demanding that f_1 tends to zero when $|z|$ tends to infinity.

PROOF. It suffices to take f_1 the principal part of the Laurent expansion of f and $f_2 = f-f_1$ is then given as

$$f_2(z) = \sum_{n=0}^{\infty} a_n(z-z_0)^n$$

with a_n given by (1). To prove the uniqueness observe first that the principal part f_1 vanishes as $|z| \to \infty$ since its development contains only negative powers of $z-z_0$ and assume that there is another pair g_1, g_2 of functions with these properties. Then from $f_1(z)+f_2(z) = g_1(z)+g_2(z)$ for all z in the annulus we find that the holomorphic functions f_1-g_1 and g_2-f_2 agree on the set $R(z_0, r_1, r_2)$ and, thus, there is an $h \in A(C)$ equal to g_2-f_2 on $K(z_0, r_2)$ and f_1-g_1 outside that disk. But that function would be bounded and would tend to 0 as $|z| \to \infty$, and, hence, by the Liouville Theorem it would be indentically 0. □

DEFINITION. We shall say that a point $z_0 \in C$ is an *isolated singular point* of a function f if for some open set $\Omega \ni z_0$ f is defined and holomorphic in $\Omega - \{z_0\}$. f may or may not be defined at z_0.

The Laurent expansion is an ideal tool for the classification of isolated singularities of holomorphic functions.

If z_0 is an isolated singular point for f, then for some $r > 0$ f is holomorphic in $R(z_0, 0, r)$ and thus, expanding f in the Laurent series

$$f(z) = \sum_{n=-\infty}^{\infty} a_n(z-z_0)^n, \quad z \in R(z_0, 0, r),$$

we arrive at the following three cases:

I. $a_n = 0$ for $n < 0$. Then, on setting $f(z_0) = a_0$, we have

$$f(z) = \sum_{n=0}^{\infty} a_n(z-z_0)^n \quad z \in K(z_0, r)$$

and thus obtain a function which is holomorphic on $K(z_0, r)$. The point z_0 is then called a *removable singularity* of f.

II. There exists a $k < 0$ such that $a_n = 0$ for $n < k$ and $a_k \neq 0$. Then

$$f(z) = \sum_{n=k}^{-1} a_n(z-z_0)^n + \sum_{n=0}^{\infty} a_n(z-z_0)^n, \quad z \in R(z_0, 0, r).$$

The point z_0 is then called a *pole of order* $-k$ of the function f.

III. For every $k < 0$ there exists a $k_1 < k$ such that $a_{k_1} \neq 0$. Then

$$f(z) = \sum_{n=-\infty}^{\infty} a_n(z-z_0)^n, \quad z \in R(z_0, 0, r).$$

The point z_0 is then called an *essential singularity*.

Examples. 1. Take the function

$$f(z) = \frac{\sin z}{z}, \quad z \neq 0.$$

This function has a Laurent expansion about the point $z_0 = 0$:

$$f(z) = \sum_{k=0}^{\infty} (-1)^k \frac{z^{2k}}{(2k+1)!}, \quad z \neq 0.$$

Putting $f(0) = 1$ yields

$$f(z) = \sum_{k=0}^{\infty} (-1)^k \frac{z^{2k}}{(2k+1)!}, \quad z \in \mathbf{C}.$$

The function f prescribed by the foregoing formula is an *entire* function, i.e., is holomorphic throughout the entire plane.

2. The Laurent expansion of the function $f(z) = (\cos z)/z$, $z \neq 0$, about the point $z_0 = 0$ is of the form

$$f(z) = \frac{1}{z} + \sum_{k=0}^{\infty} (-1)^{k+1} \frac{z^{2k+1}}{(2k+2)!}.$$

Thus f has a pole of order 1, i.e., a simple pole, at $z_0 = 0$.

3. The Laurent expansion of the function $f(z) = \exp(1/z)$, $z \neq 0$, about $z_0 = 0$ is of the form

$$f(z) = \sum_{k=0}^{\infty} \frac{1}{k!} \cdot \frac{1}{z^k}, \quad z \neq 0.$$

Hence, f has an essential singularity at $z_0 = 0$.

4. The functions

$$f(z) = \sin\frac{1}{z}, \quad g(z) = \cos\frac{1}{z}$$

have essential singularities at the point $z_0 = 0$, since their Laurent development around 0 are obtained by substituting $1/z$ for z in the Taylor series for $\sin z$ ($\cos z$ resp.).

5. The function $f(z) = \dfrac{1}{\sin(1/z)}$ and similarly the function $g(z) = \dfrac{1}{\cos(1/z)}$ have simple poles at the points $1/\pi n$ (resp. $1/(2n+1)\pi$), (cf. Proposition XVI.2.6 below) but a non isolated singularity at $z = 0$.

6. The function

$$f(z) = \frac{1}{\exp\left(\dfrac{1}{z^2}\right)+1}$$

have points $z_n = \pm(1\pm i)/\sqrt{2\pi(2k+1)}$ as simple poles, but again the point $z = 0$ is not an isolated singularity.

Now the question is whether the Laurent expansion is unique. The answer is provided by the following

THEOREM XVI.2.5. Let the function f be represented in the annulus $R(z_0, r_1, r_2)$ in terms of the series

$$f(z) = \sum_{k=-\infty}^{\infty} b_k(z-z_0)^k \tag{7}$$

which is compactly convergent in $R(z_0, r_1, r_2)$. The function f is then holomorphic on $R(z_0, r_1, r_2)$ and

$$b_k = \frac{1}{2\pi i} \int\limits_{|\zeta-z_0|=r} \frac{f(\zeta)}{(\zeta-z_0)^{k+1}}\, d\zeta, \quad r_1 < r < r_2.$$

In other words, (7) is the Laurent series of the function f.

PROOF. The function f is holomorphic in $R(z_0, r_1, r_2)$ as the compact limit of holomorphic functions. This function thus expands in a Laurent series with the coefficients:

$$a_n = \frac{1}{2\pi i} \int_{|\zeta - z_0| = r} \frac{f(\zeta)}{(\zeta - z_0)^{n+1}} \, d\zeta,$$

$$-\infty < n < \infty, \qquad r_1 < r < r_2.$$

Therefore,

$$a_n = \frac{1}{2\pi i} \int_{|\zeta - z_0| = r} \frac{\sum_{k=-\infty}^{\infty} b_k(\zeta - z_0)^k}{(\zeta - z_0)^{n+1}} \, d\zeta.$$

Integration term by term is permissible in view of the compact convergence of series (7). We obtain

$$(8) \qquad a_n = \frac{1}{2\pi i} \sum_{k=-\infty}^{\infty} \int_{|\zeta - z_0| = r} b_k(\zeta - z_0)^{k-n-1} d\zeta.$$

However,

$$(9) \qquad \int_{|\zeta - z_0| = r} (\zeta - z_0)^{k-n-1} d\zeta = i \int_0^{2\pi} r^{k-n} e^{i(k-n)\varphi} d\varphi$$

$$= \begin{cases} 0 & \text{if } k \neq n, \\ 2\pi i & \text{if } k = n. \end{cases}$$

Substitution of (9) into (8) yields $a_n = b_n$. □

PROPOSITION XVI.2.6. If a function f has a pole of order n at the point z_0, there exists an open neighbourhood U of z_0 such that the function f restricted to $U - \{z_0\}$ is of the form

$$f(z) = \frac{1}{(z - z_0)^n} g(z),$$

where $g \in A(U)$ and $g(z) \neq 0$ for $z \in U$.

Conversely, if there exists an open neighbourhood U of z_0 such that f can be written as

$$f(z) = \frac{1}{(z - z_0)^n} g(z), \qquad z \in (U - \{z_0\}),$$

where $g \in A(U)$ and $g(z_0) \neq 0$, then f has a pole of order n at z_0.

PROOF. Since z_0 is a pole of order n of the function f, the Laurent expansion leads to

$$f(z) = \frac{1}{(z-z_0)^n} g(z), \qquad z \in R(z_0, 0, r),$$

where

$$g(z) := a_{-n} + a_{-n+1}(z-z_0) + \ldots + a_{-1}(z-z_0)^{n-1} +$$
$$+ (z-z_0)^n \sum_{k=0}^{\infty} a_k (z-z_0)^k.$$

It is seen that $g \in A(K(z_0, r))$ and $g(z_0) = a_{-n} \neq 0$. By the continuity of g, it follows that there exists a neighbourhood U of the point z_0 such that $g(z) \neq 0$ for $z \in U$.

Conversely, if f can be reduced to

$$f(z) = \frac{1}{(z-z_0)^n} g(z) \quad \text{on } U - \{z_0\},$$

where $g \in A(U)$ and $g(z_0) \neq 0$, expansion of the function g in a Taylor series about z_0 gives us

$$f(z) = \sum_{k=0}^{\infty} a_k (z - z_0)^{k-n}.$$

The fact that the Laurent expansion is unique implies that f has a pole of order n at z_0 (because $a_0 = g(z_0) \neq 0$). □

COROLLARY XVI.2.7. If a point $z_0 \in C$ is a pole of order n of a function f, there exists an open neighbourhood U of z_0 such that the function g defined by the formulae

$$g(z) := (z-z_0)^n f(z), \qquad z \in (U - \{z_0\}),$$
$$g(z_0) := a_{-n},$$

is holomorphic on U and has no zeros on U.

DEFINITION. If z_0 is an isolated singularity of the function f, the coefficient a_{-1} of the Laurent expansion of this function about z_0 is called the *residue of the function f at z_0* and is denoted by the symbol $\mathrm{Res}_{z_0} f$.

From the formula (1) giving the coefficients of the Laurent expansion we have

$$\mathrm{Res}_{z_0} f := a_{-1} = \frac{1}{2\pi i} \int_{|\zeta - z_0| = r} f(\zeta)\, d\zeta,$$

where $r_1 > r > 0$ are such that $f \in A(R(z_0, 0, r_1))$. Knowledge of the residue at a given point thus makes it possible to evaluate an integral over a suitably chosen circle centred at z_0. A question thus arises: How to calculate the residue without expanding in a Laurent series? It turns out that the formulae required for this purpose can be given if a singularity is a pole.

PROPOSITION XVI.2.8. If z_0 is a pole of order n of the function f, then

$$\mathrm{Res}_{z_0} f = \frac{1}{(n-1)!} \lim_{z \to z_0} \frac{d^{n-1}}{dz^{n-1}} [(z - z_0)^n f(z)].$$

PROOF. By the previous proposition and the corollary, we know that the function

$$g(z) := (z - z_0)^n f(z), \quad z \neq z_0, \quad g(z_0) := a_{-n}$$

is holomorphic in a neighbourhood of the point z_0. This function is expressed by the formula

(10) $$g(z) = a_{-n} + a_{-n+1}(z - z_0) + \ldots + a_{-1}(z - z_0)^{n-1} + (z - z_0)^n \sum_{k=0}^{\infty} a_k (z - z_0)^k,$$

where a_k are the coefficients in the Laurent expansion of f. Formula (10) is the power-series expansion of the function g about z_0. From the uniqueness of the power-series expansion we see that

(11) $$a_{-1} = \frac{1}{(n-1)!} \cdot \frac{d^{n-1}}{dz^{n-1}} g(z_0).$$

The conclusion follows from formula (11) and the continuity of the derivatives of the function g. □

In the special case, when we have a simple pole at z_0, the formula takes on a very straightforward form,

$$\mathrm{Res}_{z_0} f = \lim_{z \to z_0} [(z - z_0) f(z)].$$

If f is a quotient of two holomorphic functions, $f = g/h$, $g(z_0) \neq 0$, and h has a simple zero at z_0, then

$$\mathrm{Res}_{z_0} f = \frac{g(z_0)}{h'(z_0)}.$$

In order to show this, note that f then has a simple pole at z_0 and

$$\mathrm{Res}_{z_0} f = \lim_{z \to z_0} [(z - z_0) f(z)] = \lim_{z \to z_0} \frac{g(z)}{\dfrac{h(z)}{z - z_0}} = \frac{g(z_0)}{h'(z_0)}.$$

LEMMA XVI.2.9. Suppose that f is a holomorphic function in the annulus $R(z_0, 0, r)$. If f is not bounded in any annulus $R(z_0, 0, r_1)$, $r_1 \leqslant r$, then z_0 is a pole or an essential singularity of f. On the other hand, if f is bounded in some annulus $R(z_0, 0, r_1)$, $r_1 \leqslant r$, then f is extendible to a holomorphic function on $K(z_0, r)$ (i.e., the point z_0 is a removable singularity).

PROOF. The first part of the lemma is obvious. Suppose, therefore, that

$$\sup_{z \in R(z_0, 0, r_1)} |f(z)| = M.$$

For any $r_2 < r_1$, by the Cauchy inequality (Corollary XVI.2.3)

$$|a_n| \leqslant \frac{M}{r_2^n}, \qquad -\infty < n < \infty.$$

For $n < 0$ we have $(1/r_2^n) \to 0$ as $r_2 \to 0$ and, hence, $a_n = 0$ for $n < 0$.

It is thus seen that the Laurent series of the function f reduces to Taylor's series. On putting $f(z_0) = a_0$, we obtain a function holomorphic on $K(z_0, r)$. □

Lemma XVI.2.9 leads immediately to the following characterization of the isolated singular points.

THEOREM XVI.2.10. Let f be a function holomorphic in an annulus $R(z_0, 0, r)$. Then

(i) z_0 is a removable singularity of f iff the limit $\lim_{z \to z_0} f(z)$ exists.

(ii) z_0 is a pole of f iff the following holds:

For every sequence $z_n \to z_0$ and every $M > 0$ there exists an $N > 0$ such that for every $n > N$ we have $|f(z_n)| > M$. (We then say that $\lim_{z \to z_0} f(z) = \infty$.)

PROOF. (i) It is clear from the form of the Laurent expansion at a removable singularity that $\lim_{z \to z_0} f(z)$ exists. The converse follows by the lemma above.

(ii) Suppose that f has a pole of order n at z_0. Then, by Proposition XVI.2.6 this function can be written as

$$f(z) = \frac{1}{(z-z_0)^n} g(z), \qquad z \in U - \{z_0\}, \tag{12}$$

where U is a neighbourhood of the point z_0, and $g \in A(U)$ and $g(z) \neq 0$ for $z \in U$. Formula (12) then implies $\lim_{z \to z_0} f(z) = \infty$.

Conversely, suppose that $\lim_{z \to z_0} f(z) = \infty$. It is seen then that there exists an $0 < r_1 < r$ such that the function f has no zeros in $R(z_0, 0, r_1)$. The function $g = 1/f$ thus is holomorphic in $R(z_0, 0, r_1)$ and has a limit equal to zero at the point $z_0 = 0$. In accordance with (i) above, on setting $g(z_0) = 0$ we get a function holomorphic on $K(z_0, r_1)$. The point z_0 is a zero of the finite order of the function g. Suppose that the order of this zero is k. The function g can then be written as

$$g(z) = (z-z_0)^k h(z), \qquad z \in V \subset K(z_0, r),$$

where V is a neighbourhood of the point z_0, $h \in A(V)$, $h(z) \neq 0$, $z \in V$. Therefore,

$$f(z) = \frac{1}{g(z)} = \frac{1}{(z-z_0)^k} \cdot \frac{1}{h(z)}.$$

The function $1/h$ is holomorphic on V and does not vanish at the point z_0; accordingly, by Proposition XVI.2.6, f has a pole of order k at z_0. □

Theorem XVI.2.10 leads directly to

THEOREM XVI.2.11. The point z_0 is an essential singularity for a function f holomorphic in the annulus $R(z_0, 0, r)$ iff no limit of that function (finite or equal to ∞) exists at that point.

Example. The function $f(z) = \exp(1/z)$ has an essential singularity at z_0. If we take the sequence $z_n = 1/n$, then $f(z_n) = e^n \to \infty$. On the other hand, if we take the sequence $z_k = -1/k$, then $f(z_k) = e^{-k} \to 0$. Thus we see that the function f does not have a limit at $z_0 = 0$.

We complete Theorem XVI.2.11 by the following

PROPOSITION XVI.2.12 (The Casorati–Weierstrass Theorem). Let f have an essential singularity at z_0 and let $w \in \boldsymbol{C}$. Then there exists a sequence (z_n) convergent to z_0, such that $f(z_n) \to w$.

In other words, the image under f of an arbitrarily small annulus around z_0 is dense in $\boldsymbol{C}$.

PROOF. If the assertion were false, there would be a neighbourhood U of z_0 and $\varepsilon > 0$, such that $|f(z)-w| \geqslant \varepsilon$ for all $z \in U$, $z \neq z_0$. Let $g(z) = \dfrac{1}{f(z)-w}$. Thus g is holomorphic on $U-\{z_0\}$ and bounded; then, by Lemma XVI.2.9, z_0 is a removable singularity of g. Let r be the order of zero of g at z_0 (set $r = 0$ if $g(z_0) \neq 0$). Hence

$$f(z) = w + \frac{1}{g(z)}$$

is either holomorphic (if $r = 0$) or has a pole of order r. This contradicts the assumption that f has an essential singularity. □

3. MEROMORPHIC FUNCTIONS

Now we describe an important class of functions of one complex variable which comprises all holomorphic functions and all rational functions, i.e., the quotients of two polynomials. The definition we give here cannot in that form be applied to functions of several complex variables. The difficulties with the notion of meromorphic function will be dealt with in one of the future chapters, where we intend to discuss the fundamental differences between theories of one and several complex variables.

DEFINITION. Let $\Omega \subset \boldsymbol{C}$ be an open set. A function f is said to be *meromorphic* in Ω if there exists a subset P of Ω such that P has no limit point in Ω, f is holomorphic in $\Omega - P$ and the points of P are poles of f.

Note that, in view of the assumption on P, its points are isolated and, hence, the definition is consistent. Moreover, from the second countability of the complex plane it follows that the set P is at most countable.

The set P is called the *set of poles* of the function f, and the set $\Omega - P$, its *set of regular points*.

A set of meromorphic functions in Ω is denoted by the symbol $\mathscr{M}(\Omega)$. Since in the last definition we can put $P = \emptyset$, we see that every function holomorphic on Ω is meromorphic on Ω. Thus we have $A(\Omega) \subset \mathscr{M}(\Omega)$.

PROPOSITION XVI.3.1. If $f \in \mathscr{M}(\Omega)$, K is a compact subset of the set Ω, and P is the set of poles of the function f, then $K \cap P$ is a finite set.

PROOF. Since P is discrete and closed in Ω, hence $K \cap P$ is compact and discrete and, therefore, finite.

It turns out that meromorphic functions form an algebra on Ω and form a division algebra on a connected set. This is the content of

THEOREM XVI.3.2. Let $\Omega \subset \boldsymbol{C}$ be open. Then:

(i) The set of meromorphic functions in Ω is an algebra over the field of complex numbers.

(ii) If Ω is a domain and if $f \in \mathscr{M}(\Omega)$, and does not vanish identically, then $1/f \in \mathscr{M}(\Omega)$. If f has a pole (zero) of order n at the point z_0, then $1/f$ has a zero (pole) of order n at the same point.

(iii) If Ω is a domain, $f \in \mathscr{M}(\Omega)$, $g \in \mathscr{M}(\Omega)$, and g does not vanish identically on Ω, then $f/g \in \mathscr{M}(\Omega)$.

PROOF. (i) Follows from the definition of a meromorphic function.

(ii) Let Z denote the set of zeros of the function f, and let P denote the set of its poles. It is easy to see that the set $\Omega - P$ is connected, P being discrete and Ω connected. Therefore, the set Z of zeros of f is also discrete and, thus, the function $h = 1/f$ is defined and holomorphic except at the discrete set $S = Z \cup P$. Consequently, each point of S is an isolated singularity of h. Now, with an aid of Proposition XVI.2.6 we can represent f in a neighbourhood of a $z_0 \in Z$ by the formula

$$f(z) = (z - z_0)^n g(z)$$

with g holomorphic and nonvanishing there. Thus we obtain a factorization

$$h(z) = (z - z_0)^{-n} \frac{1}{g(z)}$$

and it follows that h has a pole of order n at z_0 if f has a zero of order n there. Moreover, the poles of f are removable singularities of $1/f$ and become regular points (zeros of order n) after the value 0 is assigned to them.

(iii) Follows from (i) and (ii). □

In the special case when f and g are holomorphic functions on a connected set and $g \neq 0$, then f/g is a meromorphic function. The question is whether every meromorphic function can be written as a quotient of two holomorphic functions? An answer in the affirmative is given by the Poincaré Theorem XVI.14.6.

Examples of Meromorphic Functions 1. A rational function (i.e., the quotient of two polynomials, after cancellation)

$$R(z) = \frac{a_0+a_1z+\ldots+a_nz^n}{b_0+b_1z+\ldots+b_mz^m}, \qquad a_n \neq 0,\ b_m \neq 0.$$

This function has m poles, located at points which are zeros of the denominator.

2. $f_1(z) = \tan z$, $f_2(z) = \cot z$. The function $\tan z$ has first-order (simple) poles at the points $z_k = (2k+1)\frac{1}{2}\pi$, $k = 0, \pm 1, \pm 2, \ldots$ The function $\cot z$ has simple poles at the points $z_k = k$, $k = 0, \pm 1, \pm 2, \ldots$

3. The function $f(z) = 1/(e^{z^2}+1)$ has simple poles at the points

$$z_k = \pm(\tfrac{1}{2}\sqrt{2}\pm i\tfrac{1}{2}\sqrt{2})\sqrt{(2k+1)\pi}, \qquad k = 0, 1, 2, \ldots$$

(because $e^{z^2}+1$ has simple zeros).

4. The function $f(z) = (\cos z)/z$ has a simple pole at the point $z = 0$.

5. The function $f(z) = 1/(1-e^z)z$ has simple poles at the points $z_k = 2\pi ki$, $k = \pm 1, \pm 2, \ldots$, and a second-order pole at the point $z = 0$.

PROPOSITION XVI.3.3. If $f \in \mathscr{M}(\Omega)$, then $f' \in \mathscr{M}(\Omega)$. If the function f has a pole of order $k > 0$ at z_0, then f' has a pole of order $k+1$ at that point, whereas if f has a zero of order $n > 0$ at z_0, then f' has a zero of order $n-1$ at that point.

Remark. A zero of order 0 is a regular point at which $f \neq 0$.

PROOF. The derivative f' is defined and holomorphic on $\Omega - P$, where P is the set of poles of the function f. If f has a pole of order k at the point z_0, then by Proposition XVI.2.6 the function f can be represented in a certain disk $K(z_0,r)$ in the form

$$(1) \qquad f(z) = (z-z_0)^{-k}g(z)$$

with $k > 0$ and g holomorphic and nonvanishing there. Differentiation of (1) gives

$$(2) \qquad f'(z) = (z-z_0)^{-k-1}(-kg(z)+(z-z_0)g'(z)).$$

Let

$$h(z) := -kg(z)+(z-z_0)g'(z).$$

Thus formula (2) can be rearranged in the form

$$f'(z) = (z-z_0)^{-k-1}h(z), \qquad z \in K(z_0,r)-\{z_0\}.$$

Since $h(z)$ is obviously holomorphic in $K(z_0,r)$ and $h(z_0) \neq 0$, we see that f' has a pole of order $k+1$ at z_0.

Suppose now that f has a zero of order n at z_0. Then there is a factorization

$$(3) \qquad f(z) = (z-z_0)^n g(z)$$

with $n > 0$ and g holomorphic and nonvanishing in a punctured disk around z_0. Now arguing precisely as before, we obtain

$$f'(z) = (z-z_0)^{n-1}h(z)$$

with h holomorphic and $h(z_0) \neq 0$. That gives the desired conclusion.

COROLLARY XVI.3.4. If $\Omega \subset \boldsymbol{C}$ is a domain and if $f \in \mathcal{M}(\Omega)$ does not vanish identically on Ω, then $f'/f \in \mathcal{M}(\Omega)$, and the set of poles of the function f'/f is the union of the sets of the zeros and the poles of f. All the poles of the function f'/f are simple.

Moreover:

If z_0 is a pole of order k of the function f, then

$$\operatorname{Res}_{z_0}\frac{f'}{f} = -k.$$

If z_0 is a zero of order n of the function f, then

$$\operatorname{Res}_{z_0}\frac{f'}{f} = n.$$

PROOF. Theorem XVI.3.2 and Proposition XVI.3.3 imply that $f'/f \in \mathcal{M}(\Omega)$, that all the poles of this function are simple, and that the set of its poles is the union of the sets of the zeros and the poles of f. If z_0 is a pole of order k, then

$$f(z) = (z-z_0)^{-k}g(z), \quad z \in R(z_0, 0, r),$$

with g holomorphic and nonvanishing in $K(z_0, r)$. Then

$$f'(z) = (z-z_0)^{-k-1}(-kg(z)+(z-z_0)g'(z)),$$

$$\operatorname{Res}_{z_0}\frac{f'}{f} = \lim_{z\to z_0}\left[(z-z_0)\frac{f'(z)}{f(z)}\right] = -k.$$

If z_0 is a zero of order n of the function f, then arguing exactly as above, we obtain the desired formula. □

THEOREM XVI.3.5 (The Residue Theorem). Let $\Omega \subset C$ be an open set and let $P \subset \Omega$ have no limit point in Ω. Let $D \subset \Omega$ be a finite domain and let $\partial D \cap P = \emptyset$. Then the formula

$$\int_{\partial D} f(z)\,dz = 2\pi i \sum \operatorname{Res}_k f,$$

where the summation is extended over the finite set $P \cap D$, holds for every function $f \in A(\Omega - P)$.

PROOF. Let the points $z_k \in \Omega$, $k = 1, \ldots, m$, be all the points of $P \cap D$. Since $\partial D \cap P = \emptyset$, we have $z_k \in \operatorname{Int} D$. Proceeding as in the proof of the generalized Cauchy formula for one variable, we surround each point z_k with a disk $K(z_k, r_k)$ lying together with its closure in the interior of D. The choice of suitably small radii r_k enables us to obtain disjoint closures for these disks. When the Cauchy Theorem is applied to the set $V := D - \bigcup_{k=1}^{m} K(z_k, r_k)$ the result is

$$0 = \int_{\partial V} f(z)\,dz = \int_{\partial D} f(z)\,dz - \sum_{k=1}^{m} \int_{\partial K(z_k, r_k)} f(z)\,dz$$

(the minus sign follows from the orientation). Thus we have

$$\int_{\partial D} f(z)\,dz = \sum_{k=1}^{m} \int_{\partial K(z_k, r_k)} f(z)\,dz.$$

By the definition of residue, however,

$$\int_{\partial K(z_k, r_k)} f(z)\,dz = 2\pi i \operatorname{Res}_{z_k} f,$$

whence

$$\int_{\partial D} f(z)\,dz = 2\pi i \sum_{k=1}^{m} \operatorname{Res}_{z_k} f. \ \square$$

A particularly important application of the residue theorem is the case when f is a meromorphic function on Ω. For then we know a formula for calculating the residues. Thus we have the following

COROLLARY XVI.3.6. Let $\Omega \subset C$ be an open set, let f be a meromorphic function in Ω, and let $P \subset \Omega$ be the set of its poles. If $D \subset \Omega$ is a finite domain and if $\partial D \cap P = \emptyset$, then

$$\int_{\partial D} f(z)\,dz = 2\pi i \sum_{k=1}^{m} \operatorname{Res}_{z_k}, \quad z_k \in D \cap P.$$

The proof is immediate and may be omitted.

PROPOSITION XVI.3.7 (The "Argument Principle" — Formula for the Number of Zeros and Poles). Let f be a meromorphic function that is not identically zero in a domain Ω. Let $Z \subset \Omega$ be the set of its zeros, and let $P \subset \Omega$ be the set of its poles. Further, let D be a finite domain and let $\partial D \cap Z = \emptyset$ and $\partial D \cap P = \emptyset$.

Then

$$\frac{1}{2\pi i} \int_{\partial D} \frac{f'(z)}{f(z)}\,dz = N_z - N_p,$$

where N_z is the sum of the multiplicities of zeros of f in D, and N_p is the sum of the multiplicities of the poles of f in D.

PROOF. As a compact set, the finite domain contains a finite number of zeros and poles of the function f (cf. Proposition XVI.3.1). At the same time, we know from Corollary XVI.3.4 that the zeros and poles of f are the only singularities of the function f'/f and that the corresponding residues are equal to the multiplicities of the corresponding zeros or poles taken with suitable signs (plus for a zero, minus for a pole). Using the residue theorem, we thus have

$$\frac{1}{2\pi i} \int_{\partial D} \frac{f'(z)}{f(z)}\,dz = \sum_{k=1}^{n} \operatorname{Res}_{z_k} \frac{f'}{f} = N_z - N_p,$$

$z_k \in D \cap (P \cup Z)$. $\square$

4. APPLICATION OF THE CALCULUS OF RESIDUES TO THE EVALUATION OF INTEGRALS

The residue theorem provides one of the most effective methods for evaluating definite integrals without finding the primitive function. Let us consider the most typical applications of this method.

I. Integrals of the form

$$J = \int_0^{2\pi} R(\sin x, \cos x)\, dx,$$

where $R(y_1, y_2)$ is a rational function of two variables and is continuous on the circle $y_1^2 + y_2^2 = 1$. Setting $z = e^{ix}$, we have

$$dx = \frac{dz}{iz}, \qquad \sin x = \frac{z - 1/z}{2i}, \qquad \cos x = \frac{z + 1/z}{2}.$$

Integration over the circle $|z| = 1$ yields

$$J = \int_{|z|=1} \frac{1}{iz} R\left(\frac{z-1/z}{2i}, \frac{z+1/z}{2}\right) dz.$$

Hence

$$J = 2\pi i \sum \operatorname{Res}\left(\frac{1}{iz} R\left(\frac{z-1/z}{2i}, \frac{z+1/z}{2}\right)\right),$$

where we sum over all singularities inside the circle $|z| = 1$.

Example. Evaluate the integral

$$J = \int_0^{2\pi} \frac{dx}{1 - 2a\cos x + a^2}, \qquad 0 < a < 1.$$

The substitution $z = e^{ix}$ yields

$$J = \int_{|z|=1} \frac{dz}{i(1-az)(z-a)}.$$

The integrand has simple poles at the points $z_1 = a, z_2 = 1/a$. Only z_1 is of interest to us, since z_2 lies outside the unit circle. The residue at $z_1 = a$ is

$$\lim_{z \to a} \frac{z-a}{i(1-az)(z-a)} = \frac{1}{i(1-a^2)}.$$

Hence

$$\int_0^{2\pi} \frac{dx}{1-2a\cos x+a^2} = \frac{2\pi}{1-a^2}.$$

II. Integrals of the form

$$J = \int_{-\infty}^{\infty} R(x)\,dx,$$

where

(i) $R(z)$ is a rational function,
(ii) R has no poles on the real axis,
(iii) $\lim_{|z|\to\infty} zR(z) = 0$.

Observe that these assumptions guarantee the absolute convergence of the (improper Riemann) integral, i.e., the Lebesgue summability of $R(\cdot)$ on the real axis.

Choose the contour of integration C so that it consists of the interval $[-r, r]$ of the real axis and the semicircle K_r of radius r lying in the

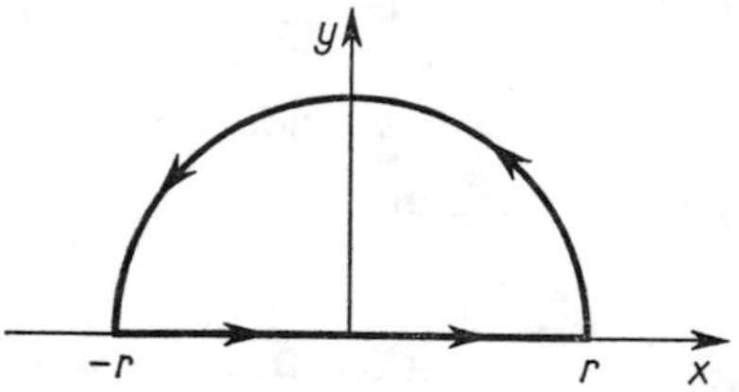

Fig. 6

upper half-plane (cf. Fig. 6) and enclosing all the poles of the function $R(\cdot)$ in the upper half-plane. Then

(1) $$\int_C R(z)\,dz = \int_{-r}^{r} R(x)\,dx + \int_{K_r} R(z)\,dz.$$

By the residue theorem

(2) $$\int_C R(z)\,dz = 2\pi i \sum \operatorname{Res}_z R,$$

where the summation extends over all the singularities of the function R which lies inside the contour C. On the other hand,

$$(2') \qquad \left|\int_{K_r} R(z)\,dz\right| = \left|\int_0^{\pi} ir\,R(re^{i\varphi})\,e^{i\varphi}d\varphi\right|$$
$$\leqslant \pi \sup_{0\leqslant\varphi\leqslant 2\pi} |r\cdot R(re^{i\varphi})| \underset{r\to\infty}{\to} 0 \qquad (\text{cf.(iii)}).$$

Thus, with r tending to ∞, by (1), (2), and (2′) we obtain

$$\int_{-\infty}^{\infty} R(x)\,dx = 2\pi i \sum \operatorname{Res}_z R,$$

where the summation extends over all the singular points of the function $R(\cdot)$ lying in the upper half-plane.

Example. Let us consider the integral

$$J = \int_{-\infty}^{\infty} \frac{dx}{(x^2+1)^3}.$$

Conditions (i)–(iii) are satisfied. The function $R(z) = 1/(z^2+1)^3$ has third-order poles at the points $z_1 = i, z_2 = -i$. Only z_1, lying in the upper half-plane, is of interest. The residue at this point is

$$\frac{1}{2!}\lim_{z\to i}\frac{d^2}{dz^2}\left(\frac{(z-i)^3}{(z-i)^3(z+i)^3}\right) = 6\lim_{z\to i}\frac{1}{(z+i)^5} = \frac{3}{16}i.$$

Hence $J = \frac{3}{8}\pi$.

Remark. If R is an even function, that is, $R(-x) = R(x)$ for $x \in \boldsymbol{R}$, then the identity

$$\int_0^{\infty} R(x)\,dx = \tfrac{1}{2}\int_{-\infty}^{\infty} R(x)\,dx$$

can be used to evaluate integrals of the type $\int_0^{\infty} R(x)\,dx$. Similar results can be obtained by integrating around a contour in the lower half-plane (cf. Fig. 7). Then

$$\int_{-\infty}^{\infty} R(x)\,dx = -2\pi i \sum \operatorname{Res}_z R$$

where the summation extends over all singularities of the function R which lie in the lower half-plane. Moreover, the rational function R can be replaced by any analytic function which has a finite number

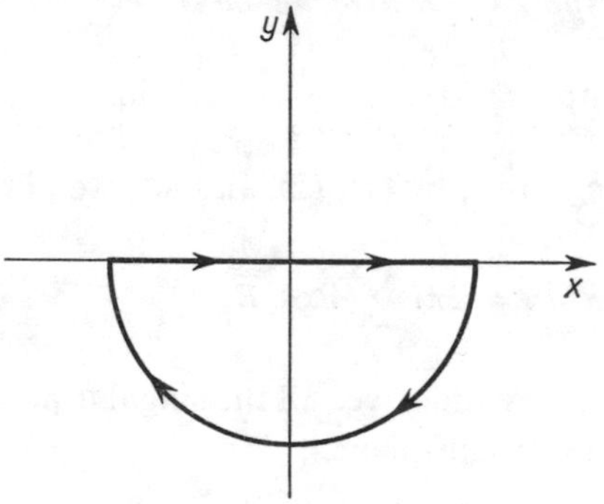

Fig. 7

of singularities in the upper (lower) half-plane and satisfies the condition $\lim_{|z|\to\infty} zR(z) = 0$.

III. Integrals of the form

$$\int_{-\infty}^{\infty} R(x)e^{iax}dx,$$

where

(i) $a > 0$,
(ii) $R(\cdot)$ is a rational function,
(iii) R has no poles on the real axis,
(iv) $\lim_{|z|\to\infty} R(z) = 0$.

The procedure is like that in II. Choosing a contour C in the upper half-plane as in Fig. 6, we have

$$\int_C R(z)e^{iaz}dz = \int_{-r}^{r} R(x)e^{iax}dx + \int_{K_r} R(z)e^{iaz}dz.$$

It turns out that

$$(*) \qquad \lim_{r\to\infty} \int_{K_r} R(z)e^{iaz}dz = 0.$$

Thus, on applying the residue theorem we have

$$\int_{-\infty}^{\infty} R(x)e^{iax}dx = 2\pi i \sum \operatorname{Res}\big(R(z)e^{iaz}\big),$$

where the summation encompasses all the singularities of $R(\cdot)$ which lie in the upper half-plane.

It remains to prove formula (*). It is a consequence of the following lemma:

Proposition XVI.4.1 (Jordan Lemma). Let f be a continuous function defined in the upper half-plane. If $a > 0$ and $\lim_{|z|\to\infty} f(z) = 0$, then

$$\lim_{r\to\infty} \int_{K_r} f(z)e^{iaz}dz = 0,$$

where $K_r := \{z \in C: |z| = r, \operatorname{Im} z \geqslant 0\}$.

Proof. To estimate the integral we shall use the following straightforward inequality. Let γ: $[a, b] \to C$ be a differentiable path; then

$$\Big|\int_{\gamma} f(z)dz\Big| \leqslant \int_a^b |f(\gamma(t))||\gamma'(t)|dt.$$

In our case γ, $[0, \pi] \ni \varphi \to re^{i\varphi} \in C$ and, denoting $M(r) = \sup_{\varphi\in[0,\pi]} |f(re^{i\varphi})|$, we obtain

$$\Big|\int_{K_r} f(z)e^{iaz}dz\Big| \leqslant M(r)r\int_0^{\pi} e^{-ar\sin\varphi}d\varphi. \tag{3}$$

Next we set

$$J(r) = r\int_0^{\pi} e^{-ar\sin\varphi}d\varphi = 2r\int_0^{\pi/2} e^{-ar\sin\varphi}d\varphi$$

and observe that, since by assumption $\lim_{r\to\infty} M(r) = 0$, it suffices to show that $J(r)$ is bounded by a constant independent of r. From the elementary inequality $\sin\varphi \geqslant 2\varphi/\pi$ valid for $0 \leqslant \varphi \leqslant \pi/2$ it follows

$$e^{-ar\sin\varphi} \leqslant e^{-2ar\varphi/\pi} \quad \text{for } 0 \leqslant \varphi \leqslant \pi/2.$$

Thus

$$J(r) \leqslant 2r\int_0^{\pi/2} e^{-2ar\varphi/\pi}d\varphi = \frac{\pi}{a}(1-e^{-ar}) \leqslant \frac{\pi}{a}.$$

Remark. If we assume $a < 0$ in hypothesis (i), the integration will have to be carried out around a contour in the lower half-plane, since then the (modification of) Jordan Lemma will also be valid. In that case

$$\int_{-\infty}^{\infty} R(x)e^{iax}dx = -2\pi i \sum \operatorname{Res}\left(R(z)e^{iaz}\right),$$

where the summation encompasses all the singularities of the function $R(\cdot)$ which lie in the lower half-plane.

In all honesty, it should be noted that the method described above leads to the evaluation of the "principal value" integral, i.e.,

$$\lim_{r\to\infty} \int_{-r}^{r} R(x)e^{iax}dx.$$

If we know beforehand that the integral $\int_{-\infty}^{\infty} R(x)e^{iax}dx$ exists as the Lebesgue integral (absolutely convergent improper Riemann integral), then that is not a real difficulty, since the principal value is the same as the true integral. However, the method is not sufficient to prove Lebesgue integrability under the stated assumptions and a different integration contour (a rectangular one) is needed for the purpose.

The method given in this section can be used to evaluate integrals of the type

$$\int_{-\infty}^{\infty} R(x)\cos ax\,dx, \qquad \int_{-\infty}^{\infty} R(x)\sin ax\,dx,$$

by expressing the functions $\cos ax$, $\sin ax$ in terms of an exponentia function and suitably choosing the contours of integration. If $R(x)$ is an even function, the formula

$$\int_{0}^{\infty} R(x)\cos ax\,dx = \tfrac{1}{2}\int_{-\infty}^{\infty} R(x)\cos ax\,dx$$

can be used to evaluate integrals of the type $\int_{0}^{\infty} R(x)\cos ax\,dx$.

Similarly, if $R(\cdot)$ is an odd function, we can evaluate integrals of the type $\int_{0}^{\infty} R(x)\sin ax\,dx$.

Example. Consider the function

$$J = \int_0^\infty \frac{x \sin ax}{x^2+b^2}\,dx, \quad a > 0,\ b > 0.$$

It is evident that

$$J = \frac{1}{2}\int_{-\infty}^{\infty} \frac{x \sin ax}{x^2+b^2}\,dx$$

$$= -\frac{1}{4}i\left(\int_{-\infty}^{\infty} \frac{x}{x^2+b^2}e^{iax}dx - \int_{-\infty}^{\infty} \frac{x}{x^2+b^2}e^{-iax}dx\right).$$

When we make the change of variables $x \to -x$ in the second integral, we obtain

$$J = -\frac{1}{2}i\int_{-\infty}^{\infty} \frac{x}{x^2+b^2}e^{iax}dx.$$

The function $\frac{z}{z^2+b^2}e^{iaz}$ has simple poles at the points $z_1 = ib$, $z_2 = -ib$. We are interested only in z_1, which lies in the upper half-plane. The residue at this point is

$$\lim_{z\to ib} \frac{z(z-ib)e^{iaz}}{(z-ib)(z+ib)} = \tfrac{1}{2}e^{-ab}.$$

Therefore, $J = \frac{1}{2}\pi e^{-ab}$.

IV. Integrals of the form

$$\int_{-\infty}^{\infty} x^{a-1}R(x)\,dx$$

where

(i) a is real and not an integer,

(ii) $R(\cdot)$ is a rational function without any poles on the positive part of the real axis (i.e. for $x > 0$),

(iii) $\lim\limits_{z\to 0}(z^a R(z)) = 0$,

(iv) $\lim\limits_{z\to\infty}(z^a R(z)) = 0$.

Take the contour Γ shown in Fig. 8. This contour is the boundary of a finite domain with the interior contained in a contractible open domain $\mathbf{C} - [0, \infty[$. Let us define a holomorphic branch of a power $z^a = e^{a \log z}$ by requiring that $0 < \operatorname{Im} \log z < 2\pi$. Integrating $z^{a-1}R(z)$

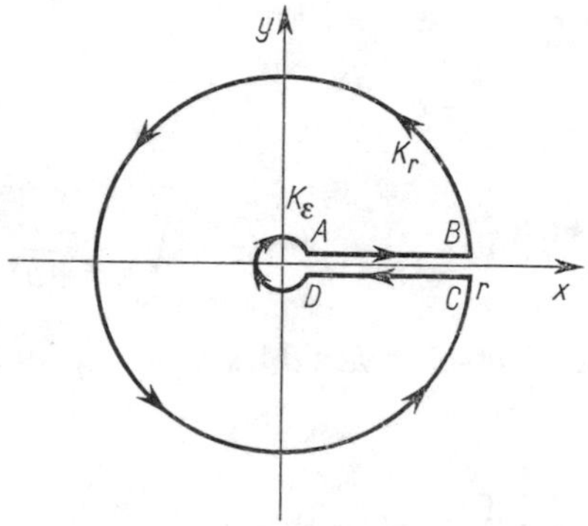

Fig. 8

along Γ we must be careful to choose an appropriate value of z^{a-1} on the segments $[A, B]$ and $[C, D]$ so to make the function continuous on Γ. It is easily seen that the right choice is

(5) $$z^a = e^{a \log |z|}$$

for $z \in [A, B]$ and

(5′) $$z^a = e^{a(\log |z| + 2\pi i)}$$

for $z \in [D, C]$.

Next, we have

(6) $$\int_{\Gamma} z^{a-1}R(z)\,dz = \int_{[A,B]} z^{a-1}R(z)\,dz + \int_{K_r} z^{a-1}R(z)\,dz + \int_{[C,D]} z^{a-1}R(z)\,dz + \int_{K_\varepsilon} z^{a-1}R(z)\,dz.$$

Estimates similar to those used in the proof of the Jordan Lemma show that, under the assumptions made about the function R, the integrals over K_r and K_ε tend to 0 as $r \to \infty$ (resp. $\varepsilon \to 0$). For the remaining integrals we have

$$\int_{[A,B]} z^{a-1}R(z)\,dz \to \int_0^\infty x^{a-1}R(x)\,dx,$$

$$\int_{[C,D]} z^{a-1}R(z)\,dz \to -e^{2\pi i(a-1)}\int_0^\infty x^{a-1}R(x)\,dx.$$

On the other hand, the integral on the left-hand side of (6) for appropriately large r and small ε is equal to the sum of all residues (there is only a finite number of them) of the function $z^{a-1}R(z)$. This point needs some justification since the residue theorem does not apply directly to the case at hand, but that can be safely left to the reader. Finally, we obtain

$$\int_0^\infty x^{a-1}R(x)\,dx = \frac{2\pi i \sum \operatorname{Res}(z^{a-1}R(z))}{1-e^{2\pi(a-1)i}},$$

where the summation runs over all singular points of the function $R(z)$.

Example. Consider the function

$$J = \int_0^\infty \frac{x^{a-1}}{1+x}\,dx, \qquad a > 0.$$

The function $z^{a-1}/(1+z)$ has a simple pole at $z = -1$; the residue at this point is

$$\lim_{z\to -1} \frac{z^{a-1}(z+1)}{(1+z)} = e^{\pi i(a-1)}.$$

We have

$$J = \frac{2\pi i e^{\pi i(a-1)}}{1-e^{2\pi i(a-1)}} = \frac{2\pi i}{e^{-\pi i(a-1)}-e^{\pi i(a-1)}} = \frac{\pi}{\sin(\pi - a\pi)}$$

$$= \frac{\pi}{\sin a\pi}.$$

V. Integrals of the form

$$\int_0^\infty \log x R(x)\,dx, \qquad \int_0^\infty R(x)\,dx,$$

where

(i) $R(z)$ is a rational function with no poles on the positive part of the real axis,

(ii) $\lim_{|z|\to\infty} zR(z) = 0$,

(iii) $R(x)$ is real-valued for all real x.

Take a contour as in Case IV and integrate the function $R(z)(\log z)^2$ along it (choose the branch of the logarithm as in Case IV). Since $\lim_{|z|\to 0} z(\log z)^2 = 0$, the integral around the small circle vanishes on passage to the limit. Since $\lim_{|z|\to\infty} zR(z) = 0$, we find that for $|z| > M$ the function $R(\cdot)$ is of the form $R(z) = \frac{1}{z^2} \cdot Q(z)$, where $\sup_{|z|>M} |Q(z)| < \infty$. Consequently,

$$\lim_{|z|\to\infty} z(\log z)^2 R(z) = \lim_{|z|\to\infty} \frac{1}{z}(\log z)^2 Q(z) = 0.$$

It thus follows that the integral along the large circle vanishes under the limiting process $r \to \infty$. On going to the limits $\varepsilon \to 0, r \to \infty$, we have

$$\int_0^\infty R(x)(\log x)^2 dx - \int_0^\infty R(x)(\log x + 2\pi i)^2 dx$$

$$= 2\pi i \sum \operatorname{Res}\big(R(z)(\log z)^2\big),$$

where the summation extends over all singularities of the function $R(\cdot)$.

Rearrangement of this formula yields

$$-4\pi i \int_0^\infty \log x R(x)\, dx + 4\pi^2 \int_0^\infty R(x)\, dx$$

$$= 2\pi i \sum \operatorname{Res}\big(R(z)(\log z)^2\big).$$

By hypothesis (iii), we can write

$$\int_0^\infty \log x \cdot R(x)\, dx = -\tfrac{1}{2}\operatorname{Re}\Big(\sum \operatorname{Res}\big(R(z)(\log z)^2\big)\Big),$$

$$\int_0^\infty R(x)\, dx = -\tfrac{1}{2}\operatorname{Im}\Big(\sum \operatorname{Res}\big(R(z)(\log z)^2\big)\Big),$$

where the summation encompasses all the singularities of $R(\cdot)$.

Remark. If R satisfies conditions (i) and (ii), then on integrating the function $\log z R(z)$ over the contour, we have

$$\int_0^\infty R(x)\,dx = -\sum \operatorname{Res}(R(z)\log z),$$

where the summation encompasses all the singularities of $R(\cdot)$.

Example. Consider the integral

$$J = \int_0^\infty \frac{\log x\,dx}{(x+a)^2+b^2}, \qquad a > 0,\ b > 0.$$

The function $\dfrac{\log z^2}{(z+a)^2+b^2}$ has simple poles at the points $z_1 = ib-a$, $z_2 = -ib-a$. The residue at z_1 is

$$\frac{1}{2ib}\left(\log\sqrt{a^2+b^2} + i\left(\pi-\arctan\frac{b}{a}\right)\right)^2.$$

The residue at z_2 is

$$-\frac{1}{2ib}\left(\log\sqrt{a^2+b^2} + i\left(\pi+\arctan\frac{b}{a}\right)\right)^2.$$

Hence

$$J = -\tfrac{1}{2}\operatorname{Re}\sum \operatorname{Res}(R(z)(\log z)^2) = \frac{\log\sqrt{a^2+b^2}}{b}\arctan\frac{b}{a}.$$

The types of integrals, discussed in Cases I–V, which can be evaluated by the residue method are, of course, only a few examples of applications of this very useful method. In all cases the successful application of the residue theorem demands judicious choice of the contour of integration.

5. APPLICATIONS OF THE ARGUMENT PRINCIPLE

An extremely important property of holomorphic functions is shown in the following theorem. We recall that a domain is, by definition, an open and connected subset of the complex plane.

THEOREM XVI.5.1 (The Open Mapping Theorem). If $\Omega \subset \boldsymbol{C}$ is a domain, $f \in A(\Omega)$ and $f \not\equiv$ const, then the set $f(\Omega)$ is also a domain.

Remark. The nontrivial part of the theorem is, of course, the assertion that $f(\Omega)$ is open, provided Ω is open and f holomorphic. Thus, the theorem may be stated as saying that a holomorphic map is open.

PROOF. That $f(\Omega)$ is connected follows from elementary topological considerations (cf. Theorem II.10.2). It thus suffices to show that for every point $w \in f(\Omega)$ there exists an $r > 0$ such that $K(w, r) \subset f(\Omega)$.

Let $w_0 \in f(\Omega)$. Then there is a $z_0 \in \Omega$ such that $f(z_0) = w_0$. The fact that the set Ω is open implies the existence of an $r_1 > 0$ such that $\overline{K(z_0, r_1)} \subset \Omega$.

The functions $g_w(z) := f(z) - w$ are holomorphic on Ω and do not vanish identically (because $f \not\equiv$ const). Thus, the formula for the number of zeros and poles is applicable to those functions. In particular

$$N(w_0) = \frac{1}{2\pi i} \int_{\partial K(z_0, r_2)} \frac{f'(z)dz}{f(z) - w_0},$$

is the number of zeros (counted with multiplicities) of g_{w_0} in $K(z_0, r_2)$, where $r_2 < r_1$ is so chosen that $f(z) \neq w_0$ on the boundary $\partial K(z_0, r_2)$ of the disk $K(z_0, r_2)$. This is possible since the set of zeros of the function g_{w_0} contained in $K(z_0, r_1)$ is finite. Now, since the boundary $\partial K(z_0, r_2)$ is compact, the set $f(\partial K(z_0, r_2))$ is also compact and does not contain w_0. It follows that there exists a disk $K(w_0, r)$ contained in the complement of $f(\partial K(z_0, r_2))$, in other words the function $z \to f(z) - w$ doesn ot vanish on $\partial K(z_0, r_2)$ for each $w \in K(w_0, r)$. Therefore, the integral

$$(1) \qquad N(w) := \int_{\partial K(z_0, r_2)} \frac{f'(z)}{f(z) - w} dz$$

is well defined for $w \in K(w_0, r)$ and as a function of w is continuous there (continuity of an integral with a parameter, Theorem XIII. 10.1). However, N is an integral-valued function $-N(w)$ equal to the number of zeros (each zero of order m counted m times) of the function $z \to f(z) - w$ inside $K(z_0, r_2)$, so $N =$ const on the whole $K(w_0, r)$. But since $N(w_0) \geqslant 1$ we see that the equation $f(z) = w$ has a solution $z \in K(z_0, r_2)$ for each $w \in K(w_0, r)$, i.e., $f(K(z_0, r_2)) \supset K(w_0, r)$. □

Indeed, we have obtained a more precise result:

THEOREM XVI.5.2. If $\Omega \subset C$ is a domain, $f \in A(\Omega)$, and f' has a zero of order $k \geqslant 0$ at a point $z_0 \in \Omega$, then there exist $r > 0$, $r_1 > 0$ such that for every point $w \in R(f(z_0), 0, r)$ there exist exactly $k+1$ distinct points $z_j \in K(z_0, r_1)$ such that $f(z_j) = w$.

Remark. We shall adopt the following convenient descriptive terminology. We shall say that a point z_0 is a root of the equation $f(z) = w$ with the multiplicity k if the function $z \to f(z) - w$ has a zero of order k at the point z_0. The multiplicity of a root is, therefore, given by the formula (1) above.

PROOF. We denote $w_0 = f(z_0)$. By the assumption the multiplicity of z_0 is equal to $k+1 = N(w_0)$. Let us choose a disk $K(z_0, r_2)$ such that f' does not vanish in $K(z_0, r_1)$ except at the point z_0. From the proof of the Theorem XVI.5.1 we know that there exists a disk $K(w_0, r)$ such that for each $w \in K(w_0, r)$ the equation $f(z) = w$ has $k+1$ roots belonging to $K(z_0, r_1)$, counted with multiplicities. But since $f'(z) \neq 0$ for each $z \in K(z_0, r_1)$ different from z_0, each root has a multiplicity of one. Therefore all roots are distinct and there are $k+1$ of them. □

This theorem shows that a holomorphic function can be locally invertible in the neighbourhood of a point z_0 only if $k+1 = 1$, that is, if $f'(z_0) \neq 0$. Hence, this is a converse of the theorem on the local invertibility of mappings in the case of functions of one complex variable.

A straightforward proof of the maximum principle for functions of a complex variable can be obtained from the open mapping theorem.

COROLLARY XVI.5.3. (The Maximum Principle). If $\Omega \subset C$ is an open connected set, if $f \in A(\Omega)$, and if there exists a $z_0 \in \Omega$ such that $|f(z)| \leqslant |f(z_0)|$ for $z \in \Omega$, then $f =$ const on Ω.

PROOF. Suppose that $f \neq$ const on Ω. The assertion is trivial if $f(z_0) = 0$, so we suppose that $w_0 = f(z_0) \neq 0$. By the open mapping theorem there exists an $r > 0$ such that $K(w_0, r) \subset f(\Omega)$. Then $w_1 = (1 + \frac{1}{2}r) w_0 \in K(w_0, r)$ and, thus, for a $z_1 \in \Omega$ such that $f(z_1) = w_1$ we have $|f(z_1)| > |f(z_0)|$, contradicting the maximality of $f(z_0)$. □

THEOREM XVI.5.4 (Rouché). Let $\Omega \subset C$ be an open connected set. Let $D \subset \Omega$ be a finite domain, and let f, g be holomorphic functions on Ω satisfying the following conditions:

(i) $f(z) \neq 0$ for $z \in \partial D$,

(ii) $|g(z)| < |f(z)|$ for $z \in \partial D$.

Then the function $f+g$ has in D as many zeros (each counted with its multiplicity) as the function f does.

PROOF. Let $0 \leqslant a \leqslant 1$. Consider the family of functions

$$h_a(z) := f(z)+ag(z), \quad z \in \Omega.$$

The functions of this family are holomorphic on Ω. By (ii),moreover, $h_a(z) \neq 0$ for $z \in \partial D$. When the formula for the number of zeros is applied to the function h_a, the result is

$$N(a) = \frac{1}{2\pi i} \int_{\partial D} \frac{f'(z)+ag'(z)}{f(z)+ag(z)} dz. \tag{2}$$

The function $a \to N(a)$ is continuous on the interval $[0, 1]$ as an integral with a parameter (Theorem XIII.10.1). On the other hand, this function assumes only integral values (the number of zeros of the function h_a in D) and, thus, is constant. Hence, $N(0) = N(1)$. But formula (2) implies that $N(0)$ is the number of zeros (each counted with its multiplicity) of the function f in D, and $N(1)$ is the number of zeros (each counted with its multiplicity) of the function $f+g$ in D. □

A simple proof of the fundamental theorem of algebra can be obtained from the Rouché Theorem.

COROLLARY XVI.5.5 (The Fundamental Theorem of Algebra). A polynomial of the n-th degree

$$W_n(z) = a_0 z^n + a_1 z^{n-1} + \ldots + a_n, \quad a_0 \neq 0,$$

with complex coefficients has n roots (counted with multiplicities) in C.

PROOF. Let

$$f(z) = a_0 z^n, \quad g(z) = \sum_{k=1}^{n} a_k z^{n-k}.$$

If we take a circle with centre at 0 and suitably large radius R, then

$$|a_0 z^n| > \left|\sum_{k=1}^{n} a_k z^{n-k}\right| \quad \text{for } |z| = R.$$

The function f has a zero of order n at the point $z = 0$ and no other zeros. Hence, by the Rouché Theorem the polynomial $W_n = f+g$ also has n zeros in the circle $K(0, R)$. □

As other important applications of the Rouché Theorem one obtains several results due to Adolf Hurwitz.

PROPOSITION XVI.5.6. If a sequence (f_n) of functions which are continuous on a closed set A and holomorphic on the interior $\mathring{A}$ of A is convergent uniformly to f on A and if f does not vanish at the boundary ∂A of A; then for $n \geqslant n_0$ all f_n have the same number of zeros on $\mathring{A}$ as f does.

PROOF. Let $m = \inf|f(\partial A)|$. Since $m > 0$, there exists such n_0, that for $n \geqslant n_0$

$$|f_n(z) - f(z)| < m \leqslant |f(z)|, \quad z \in \partial A.$$

But f is holomorphic on $\mathring{A}$ (Weierstrass!). Applying the Rouché Theorem to the pair f and $(f_n - f)$, we find that for $n \geqslant n_0$ the functions f and $f_n = (f_n - f) + f$ have the same number of zeros in $\mathring{A}$. □

This proposition immediately leads to

THEOREM XVI.5.7 (of Hurwitz). Let Ω be a domain in $\boldsymbol{C}$. If the sequence (f_n) of holomorphic functions on Ω is compactly convergent to f and if f does not vanish identically and has at least N different zeros, then there exists an n_0 such that for $n \geqslant n_0$ each f_n has at least N different zeros.

PROOF. Let $z_1, \ldots, z_N$ be different zeros of f. Since they are isolated, there exists an $r > 0$ such that the closed circles

$$\bar{K}_j = \{z\colon |z - z_j| \leqslant r\}, \quad j = 1, 2, \ldots, N$$

are disjoint and all the zeros of f are in $\bigcup \bar{K}_j$. By virtue of the preceding proposition for $n \geqslant n_0$, each f_n has at leas tone zero in every $\bar{K}_j$; thus, f_n has at least N zeros. □

The Hurwitz Theorem immediately yields a result which is important for a proof of the Riemann mapping theorem, the so-called

HURWITZ'S LEMMA XVI.5.8. If a sequence (f_n) of holomorphic injective functions in a domain $\Omega \subset \boldsymbol{C}$ is compactly convergent to f, then either $f =$ const or f is injective.

PROOF. If $f \neq$ const when $f(z_1) = f(z_2) = w$ for $z_1 \neq z_2$, then by applying the Hurwitz theorem to the sequence $(f_n - w)$ we find that for $n \geqslant n_0$ each function $f_n - w$ vanishes at least at two points of Ω, but this is in contraditcion with the injectivity of f_n. □

Remark. The reader who is eager to learn the Riemann–Koebe Theorem may pass the following 50 pages and go over to Section 8.

6. FUNCTIONS AND DIFFERENTIAL FORMS ON RIEMANN SURFACES

One-Point Compactification. If X is a locally compact space, X can be embedded in a compact space $\tilde{X}$ such that $\tilde{X}$ is formed from X by the adjunction of a single point, called the *point at infinity*. This is dealt with by the Alexandrov Theorem. Before actually stating the theorem, we recall that by a locally compact space we mean a Hausdorff space in which every point has a compact neighbourhood (it follows that it even has a basis of compact neighbourhoods). As an important class of examples we mention finite-dimensional differentiable manifolds.

THEOREM XVI.6.1 (Alexandrov). Let X be a locally compact space. Then there exists a compact space $\tilde{X}$ such that X is (homeomorphic to) a subspace of $\tilde{X}$ and the complement of X in $\tilde{X}$ reduces to a point. Moreover any two compact spaces $\tilde{X}_1$ and $\tilde{X}_2$ with these properties are homeomorphic. The space $\tilde{X}$ described above is called the *Alexandrov compactification* of X.

PROOF. Let ω be a point which does not belong to X. We define a topology in $\tilde{X} := X \cup \{\omega\}$ such that $\tilde{X}$ will be compact. Let $\mathcal{B}$ be the collection consisting of all open subsets of X, regarded as subsets of $\tilde{X}$ and all complements of compact subsets of X with respect to $\tilde{X}$. This collection satisfies the requirements to be a basis for a topology on $\tilde{X}$. In fact, if $\mathcal{U}, \mathcal{V} \in \mathcal{B}$ are both open subsets of X or both complements of compact subsets of X, then $\mathcal{U} \cap \mathcal{V}$ is also of the same type, open or the complement of a compact set. If $\mathcal{U}$ is open and $\mathcal{V} = \tilde{X} - K$ for a compact $K \subset X$, then $\mathcal{U} \cap \mathcal{V} = \mathcal{U} \cap (X - K)$ is an open subset of X. There is, therefore, a unique topology in $\tilde{X}$, for which $\mathcal{B}$ is a basis; it is not difficult to see that every member of this topology is either an open subset of X, the complement of a compact $K \subset X$, or the union of the two. Moreover, the topology induced on X is precisely the original one.

The separation (Hausdorff) property for points $x_1, x_2 \in \tilde{X}$, both of which belong to X, is clear by the assumption of the Hausdorff axiom for X and the definition of the topology in $\tilde{X}$. If one of the points, say $x_1 = \omega$, then there is a compact $K \ni x_2$ and an open set $\mathcal{O}_2 \subset X$ such that $x_2 \in \mathcal{O}_2 \subset K$ (local compactness!) and, thus, $\mathcal{O}_1 = \tilde{X} - K$ is an open neighbourhood of x_1 disjoint with $\mathcal{O}_2$.

It remains to prove that $\tilde{X}$ is compact. Let $\{\mathcal{O}_\alpha\}$ be an open covering of $\tilde{X}$ and let $\mathcal{O}_{\alpha_0} \ni \omega$. Then $\mathcal{O}_{\alpha_0} = \tilde{X} - K$ for some compact $K \subset X$ and, hence, there is a finite subcollection $\{\mathcal{O}_{\alpha_i}\}_1^n \subset \{\mathcal{O}_\alpha\}$ constituting a covering of K. Then $\{\mathcal{O}_{\alpha_0}\} \cup \{\mathcal{O}_{\alpha_i}\}_1^n$ is a finite covering of $\tilde{X}$.

Uniqueness (up to a homeomorphism) of the Alexandrov compactification is clear and is left for the reader. □

A typical application of the above-described construction is the compactification of the space $\boldsymbol{R}^n$. It turns out, somehow surprisingly, that this construction has a very strong geometric flavour and the result is: the Alexandrov compactification of $\boldsymbol{R}^n$ is (homeomorphic to) an n-dimensional sphere S^n. We sketch out the details in the case $n = 2$ and with the identification of $\boldsymbol{R}^2$ with the complex plane $\boldsymbol{C}$.

In this case it is customary to denote the Alexandrov compactification by $\bar{\boldsymbol{C}}$ and the point of the complement of $\boldsymbol{C}$ by ∞ so that $\bar{\boldsymbol{C}} = \boldsymbol{C} \cup \cup \{\infty\}$. A basis of neighbourhoods of ∞ may be chosen to consist of the sets $\{z \in \boldsymbol{C}\colon |z| > r\} \cup \{\infty\}$, with $r > 0$.

Let us consider the unit sphere S^2 in $\boldsymbol{R}^3$, i.e.,

$$(1) \qquad S^2 = \{x \in \boldsymbol{R}^3 \colon \|x\| = 1\},$$

$\|\cdot\|$ denoting the usual Euclidean norm and let us identify the complex plane $\boldsymbol{C}$ with the Ox_1x_2 plane in $\boldsymbol{R}^3$, i.e., with the plane defined by $x_3 = 0$. Let n and s denote the north and the south poles of S^2 respectively, i.e. $n = (0, 0, 1)$ and $s = (0, 0, -1)$. Let χ_+ be the stereographic projection of S^2 from the north pole onto the complex plane $x_3 = 0$, i.e., for $x \in S^2$, $x \neq n$ we form the straight line through x and n and let $\chi_+(x)$ be the point of intersection of the line with the plane $x_3 = 0$. It is a simple matter to compute the coordinates of $\chi_+(x)$; in complex terms we have

$$(2) \qquad S^2 - \{n\} \ni x \to \chi_+(x) := \frac{x_1}{1-x_3} + i\frac{x_2}{1-x_3}, \quad x = (x_1, x_2, x_3).$$

One also easily sees that χ_+ is a homeomorphism of $S^2 - \{n\}$ with $\boldsymbol{C}$ and also that χ_+ carries an open neighbourhood of n on the sphere with n deleted upon complements of compact sets in the plane. Therefore, S^2 is homeomorphic with the Alexandrov compactification $\bar{\boldsymbol{C}}$ of the plane $\boldsymbol{C}$, the north pole corresponding to the point ∞.

In the same way one may also consider the stereographic projection from the south pole s and this again will be a homeomorphism of $S^2 - \{s\}$ with the plane $\boldsymbol{C}$. However, for reasons which will become clear shortly, it is preferable to assign to each point $x \in S^2 - \{s\}$ the complex conjugate of its image under the stereographic projection form s. We then obtain the map

$$(3) \qquad S^2 - \{s\} \ni x \to \chi_-(x) := \frac{x_1 - ix_2}{1 + x_3}, \qquad x = (x_1, x_2, x_3).$$

For $x \in S^2 - \{s, n\}$ the images $\chi_+(x)$ and $\chi_-(x)$ are inverse to each other, namely.

$$(4) \qquad \chi_+(x)\chi_-(x) = 1.$$

Exercise. Show that the straight lines in $\boldsymbol{C}$ are images under χ_+ of circles passing through the north pole.

The sphere S^2 endowed with the above-defined maps χ_+ and χ_- will be called the *Riemann sphere*. We shall often blur the distinction between the Riemann sphere S^2 and the extended (compactified) complex plane $\overline{\boldsymbol{C}}$, provided that the identification is given via the map χ_+ and the north pole corresponds to the point ∞. As the Riemann sphere will be indispensable model in the sequel, the reader is advised to make every attempt to visualize the geometry of the Riemann sphere.

The Riemann sphere is just a very special, but nevertheless important, case of the notion of a Riemann surface which, in turn, is a special case of a general notion of a complex manifold. Formally we introduce

DEFINITION. A one-dimensional complex connected manifold is called a *Riemann surface*.

For the definition of a complex manifold of an arbitrary dimension (and some others to follow) the reader should consult Chapter XIV. We merely recall here that, to define a structure of a 1-dimensional complex manifold on a Hausdorff space M, it suffices to single out an open covering $\{U_i\}$ of M and a family of homeomorphisms $\{\chi_i\}$, $\chi_i\colon U_i \to \boldsymbol{C}$, mapping U_i onto an open subset of the complex plane $\boldsymbol{C}$ such that the following compatibility condition holds: for each pair of these homeomorphisms χ_i, χ_j, if $U_i \cap U_j \neq \emptyset$, then the map $\chi_i \circ \chi_j^{-1}\colon \chi_j(U_i \cap U_j) \to \chi_i(U_i \cap U_j)$ is holomorphic (in fact, it is then a biholomorphism of these sets, i.e., a holomorphic homeomorphism).

The homeomorphisms χ_i or, better *still*, pairs (U_i, χ_i) are called *local maps* (*coordinate systems* or *charts*) of the manifold. It is always possible to enlarge the collection of local maps given initially by adjoining all the local maps compatible, in the sense given above, with all maps of the original collection. Such an enlarged family is called the *maximal atlas of the Riemann surface*.

A final remark concerns the notion of a holomorphic mapping between Riemann surfaces. A continuous mapping $f: X \to Y$ between Riemann surfaces is said to be *holomorphic* if for each point $x \in X$ and any local maps χ and $\varkappa$ around x and $f(x)$, resp. the function $\varkappa \circ f \circ \chi^{-1}$ (defined and valued in the complex plane), is holomorphic. Two Riemann surfaces are said to be *conformally equivalent* (*biholomorphic*) if there is a homeomorphism of one onto another which is also holomorphic. Using the knowledge about the local behaviour of holomorphic functions one sees easily that this condition implies that the inverse is also holomorphic (cf. also Theorem XVI.6.5).

Examples. 1. The complex plane $\boldsymbol{C}$, a domain $\Omega \subset \boldsymbol{C}$, or slightly more generally an open connected submanifold of a Riemann surface are (trivial) examples of Riemann surfaces.

2. Define an equivalence relation $\sim$ in the complex plane by $z \sim z'$ if and only if $z - z'$ is an integer. The quotient space $M = \boldsymbol{C}/\sim$ is a Hausdorff space with respect to the (unique) strongest topology in M making the projection $p: \boldsymbol{C} \ni z \to [z] \in M$ continuous. Moreover, each point $z \in \boldsymbol{C}$ is contained in a disk K such that no two distinct points of K are equivalent and, therefore, $p|K$ is a homeomorphism onto $p(K) \subset M$. A local coordinate in M around $q = p(z)$ is defined by taking $U = p(K)$ and $\chi = (p|K)^{-1}$. These maps are easily seen to fulfil the compatibility condition.

3. A complex torus $T(\omega_1, \omega_2)$ can be defined in much the same way as in 2. Let $\omega_1, \omega_2 \in \boldsymbol{C}$ be linearly independent over the field $\boldsymbol{R}$, i.e., they form a basis of $\boldsymbol{C}$ over the field $\boldsymbol{R}$. A discrete additive group of complex numbers generated by ω_1 and ω_2 will be denoted by $L = L(\omega_1, \omega_2)$ and called a *lattice*. The quotient space $T(\omega_1, \omega_2) = \boldsymbol{C}/L$ of equivalence classes of the relation $\sim_L$ defined by $z \sim_L z'$ iff $z - z' \in L$ with the quotient topology and the complex structure defined analogously to the definition in 2 is a Riemann surface.

4. The maps χ_+ and χ_- define a structure of a Riemann surface on the sphere S^2; by reference to that structure it is called the Riemann sphere.

Since a Riemann surface is, by definition, locally biholomorphic to a domain in $\boldsymbol{C}$, many theorems of Sections 1–6 have their counterparts for arbitrary Riemann surfaces. In the remainder of the present section we shall collect some of these theorems and show how the Riemannian point of view simplifies and unifies them.

THEOREM XVI.6.2 (Identity Theorem). Let X, Y be Riemann surfaces and let f_1, f_2: $X \to Y$ be holomorphic maps equal on a subset $A \subset X$ (i.e. $f_1|A = f_2|A$) with an accumulation point in X. Then $f_1 = f_2$.

The proof is obtained by a reduction to the corresponding Corollary XVI.1.9 in the plane: Let

$$G = \{x \in X\colon \bigvee_{O_x \ni x} f_1|O_x = f_2|O_x,\ O_x\text{—neighbourhood of } x\}. \tag{5}$$

Plainly, G is open and not empty. The proof will be completed if we verify that G is also closed, since then $G = X$ by the connectedness of X. Let b be a point of the closure $\overline{G}$; by the continuity of f_1, f_2 we have $f_1(b) = f_2(b)$. Therefore there exist (local) charts φ: $U \to V$ of X and ψ: $U' \to V'$ of Y such that $b \in U$ and $f_k(U) \subset U'$. We can assume that U is connected. Since f_k are holomorphic, the mappings

$$g_k := \psi \circ f_k \circ \varphi^{-1} : V \to V' \subset \boldsymbol{C}, \quad k = 1, 2 \tag{6}$$

are holomorphic. Since $U \cap G \neq \emptyset$, g_1 and g_2 are identical on a domain $\subset \boldsymbol{C}$ (by Corollary XVI.1.9). Therefore, $f_1|U = f_2|U$ and, hence, $b \in G$, and G is closed. □

Remark. Theorem XVI.6.2 is valid if we assume only that X is a connected complex manifold.

Meromorphic Functions on a Riemann Surface. We can now give a definition of meromorphic functions which is a natural extension of the notion introduced in Section 3.

DEFINITION. Let M be a Riemann surface and $\overline{\boldsymbol{C}}$ the Riemann sphere. A holomorphic mapping f: $M \to \overline{\boldsymbol{C}}$ such that $f \not\equiv \infty$, is called a *meromorphic function on M*. The space of meromorphic functions on M will be denoted by $\mathcal{M}(M)$. Points of the set $f^{-1}(\infty)$ are called *poles* of f.

COROLLARY XVI.6.3. $(f \in \mathcal{M}(M)) \Rightarrow (f^{-1}(\infty)$ is discrete), i.e., the poles of a meromorphic function are isolated.

PROOF (indirect). Otherwise $f \equiv \text{const} \equiv \infty$ by the uniqueness theorem.

Let p be a pole of a meromorphic function f and let $\varkappa$ be a map defined in such small a neighbourhood of p that it does not contain any other pole of f. Then, using the map χ_+ on $\bar{C}$, we find the corresponding local expression for f, i.e., the function $\chi_+ \circ f \circ \varkappa^{-1}$, is meromorphic in a disk with centre at $\varkappa(p)$ and has the only one pole at that point. It is easy to see that by changing the map $\varkappa$ we may move the location within the complex plane of the pole of a local expression for f, but we cannot change the order of this pole. Hence, the order of a pole of a meromorphic function is uniquely defined and can be found by using an arbitrary chosen map around that pole.

Exercise. Verify that the above definition of a meromorphic function in the case of an open subset of the complex plane is equivalent to the "elementary" one given in Section 4.

The corresponding theorem in the plane leads to

THEOREM XVI.6.4 (Riemann's Extension Theorem). Let $\mathcal{O}$ be an open subset of a Riemann surface and let $a \in \mathcal{O}$. If $f \in A(\mathcal{O} - \{a\})$ and f is bounded in a neighbourhood of a, then there exists a unique holomorphic extension $\tilde{f} \in A(\mathcal{O})$ of f, i.e., $\tilde{f}|(\mathcal{O} - \{a\}) = f$.

Remark 1. The point a is then called a *removable singularity of f*.

Remark 2. The usual four algebraic operations can be performed on meromorphic functions on a Riemann surface, once again yielding meromorphic functions. The apparently troublesome cases of $\infty - \infty$, $0 \cdot \infty$, $0/0$, and ∞/∞ can be handled by using local expressions and/or the Riemann extension theorem. The result is that the set $\mathcal{M}(M)$ of meromorphic functions on M with so defined addition and multiplication is a field, containing as a subfield the field of constants isomorphic with C.

Theorem XVI.5.1 immediately implies

THEOREM XVI.6.5 (The Open Mapping Theorem). Let X, Y be Riemann surfaces and let $f: X \to Y$ be a non-constant holomorphic mapping. Then f is open, i.e., $f(\mathcal{O})$ is open for each open $\mathcal{O}$.

THEOREM XVI.6.6. Let X, Y be Riemann surfaces, X being compact, and let $f: X \to Y$ be a non-constant holomorphic mapping.

Then Y is compact and f is surjective.

PROOF. By the elementary topology argument $f(X)$ is compact, hence closed, and by Theorem XVI.6.5 it is also open. Since $f(X)$ is nonvoid, it must be equal to Y. □

An important consequence is the following

PROPOSITION XVI.6.7. Let X, Y be Riemann surfaces and suppose X is compact. Then any injective holomorphic mapping $f: X \to Y$ is in fact invertible and its inverse is also holomorphic. In short, f is an holomorphic isomorphism (biholomorphism) of X with Y.

PROOF. By the preceding theorem the mapping f is surjective and, thus, is invertible. The inverse is continuous by the general topology. It follows that f^{-1} is holomorphic if it is observed that, for any coordinate systems $\varkappa_1$ around $p \in X$ and $\varkappa_2$ around $f(p) \in Y$ the function $\varkappa_2 \circ f \circ \varkappa_1{}^{-1}$ (defined in some domain in C) is injective and, thus its derivative does not vanish (Theorem XVI.5.2) and if the inverse mapping theorem for holomorphic functions is then applied (Chapter XV). □

COROLLARY XVI.6.8. $(X\text{—compact}, f \in A(X)) \Rightarrow (f = \text{const})$.

The proof (indirect) follows from Theorem XVI.6.6 since $\boldsymbol{C}$ is not compact. Very short proofs of some classical results, the Liouville Theorem and the fundamental theorem of algebra, can be obtained from the above theorems. We ask the reader to do this as an exereise.

Exercise. (i) By extending a polynomial to a holomorphic mapping of the Riemann sphere into itself, prove that it has at least one root.

(ii) Give a proof of the Liouville Theorem.

Hint: use the Riemann extension theorem and then Corollary XVI.6.8.

We now give an elegant characterization of rational functions.

PROPOSITION XVI.6.9. Every meromorphic function f on $\bar{\boldsymbol{C}}$ is a rational function, i.e., a quotient of two polynomials.

PROOF. f has only a finite number of poles since a discrete compact set is finite (cf. Corollary XVI.6.3). We can assume that f has no pole at ∞ (otherwise, we would consider the function $1/f$). Let $a_1, a_2, \ldots, a_n \in C$ be the poles of f and let

$$(7) \qquad h_j(z) = \sum_{l=-k_j}^{-1} c_{jl}(z-a_j)^l, \quad j = 1, \ldots, n,$$

be the principal part of f at a_j. Then the function

$$g := f-(h_1+ \ldots +h_n) \text{ is holomorphic on } \bar{C}$$

and, by the Liouville Theorem, is a constant. Therefore, f is rational as a sum of rational functions. □

In particular, we have the following characterization of polynomials:

LEMMA XVI.6.10. 1° A polynomial of the n-th degree

$$f(z) = z^n+a_1 z^{n-1}+ \ldots +a_n, \quad n \geqslant 1$$

can be considered as a meromorphic function on $\bar{C}$ which has a unique pole at ∞ of order n.

2° Conversely, this property is characteristic of polynomials of order n.

PROOF. 1° is obvious. 2° Let $f \in \mathscr{M}(\bar{C})$; there then exists an $r > 0$ such that for $|z| > r$

$$f(z) = b_n z^n+ \ldots +b_1 z+ \sum_{k=0}^{-\infty} b_k z^k.$$

Hence, $h(z) := f-(b_n z^n+ \ldots +b_1 z)$ is holomorphic and bounded on $\bar{C}$ and, therefore, by the Liouville Theorem $h = \text{const}$. Hence

$$f = \sum_{k=1}^{n} b_k z^k+\text{const}. \; □$$

COROLLARY XVI.6.11. Every rational function f on $\bar{C}$ can be decomposed (in a unique way) into partial fractions

$$f(z) = \sum_{j=1}^{p} h_j(z)+\sum_{k=0}^{n} b_k z^k,$$

where h_j is the principal part of f at a pole a_j given by $h_j(z) = \sum_{l=1}^{n_j} c_{jl}(z-a_j)^{-l}$ and $b_0+b_1 z+ \ldots +b_k z^k$ is the principal part of p at ∞.

Another important class of functions to be dealt with are harmonic functions. Although it is possible to define globally a Laplace operator (said now Laplace–Beltrami) Δ on any Riemann surface and then define harmonic functions by the condition $\Delta u = 0$, it will better suit our purposes to give another definition in terms of local coordinates.

Let $p \in M$ and let χ be a coordinate system defined in a neighbourhood U of p. If a function $u \in C^2(U)$ satisfies the condition $\Delta(U \circ \chi^{-1}) = 0$, where Δ is the usual Laplace operator in $\chi(U) \subset \boldsymbol{C}$, then an easy computation shows that $\Delta(U \circ \psi^{-1}) = 0$ for any other coordinate system ψ around p. The following definition, therefore, has an intrinsic character.

Definition. A function $u \in C^2(\mathcal{O})$, where $\mathcal{O} \subset M$ is open, is said to be *harmonic* if for each point $p \in \mathcal{O}$ and any coordinate system χ around p (hence, all in view of the preceding remark) $\Delta(u \circ \chi^{-1}) = 0$. The set of harmonic, complex-valued functions on $\mathcal{O}$ is denoted $\mathscr{H}(\mathcal{O}, \boldsymbol{C})$. Harmonic functions will be discussed in more detail in Section 9.

Exercise. Let $\omega_1, \omega_2 \in \boldsymbol{C}$ be linearly independent over $\boldsymbol{R}$ and put $L := Z\omega_1 + Z\omega_2$. A meromorphic function $f: \boldsymbol{C} \to \overline{\boldsymbol{C}}$ is said to be *L-periodic* if $f(z) = f(z+\omega)$ for every $z \in \boldsymbol{C}$ and $\omega \in L$. Prove the following

1° Each L-periodic holomorphic function is constant.

2° Every non-constant L-periodic meromorphic function $f: \boldsymbol{C} \to \overline{\boldsymbol{C}}$ assumes every value $c \in \overline{\boldsymbol{C}}$.

Hint: Consider f as a meromorphic function $[f]$ on the torus $T := \boldsymbol{C}/L$. Note that T is compact and apply Theorem XVI.6.6.

Harmonic and Holomorphic Forms on a Riemann Surface. In Chapter XV we considered complex-valued differential forms $\omega \in \Lambda^{p,q}(M)$ of type (p, q) on a complex manifold M. In the case of a Riemann surface M we have forms of order only up to 2. The most important, besides harmonic forms, are holomorphic forms, also called Abelian differentials of the first kind:

Definition. Let $\mathcal{O}$ be an open subset of a Riemann surface M. A differential 1-form ω defined on $\mathcal{O}$ is said to be *holomorphic*, also said to be an *Abelian differential of the first kind*, if it is of type $(1, 0)$ and if $d''\omega = 0$.

In terms of local coordinates, holomorphic differential 1-forms are precisely those which have the form $f dz$ with holomorphic f. The space of holomorphic differential 1-forms on $\mathcal{O}$ is denoted $A^1(\mathcal{O})$, the conjugate space $\overline{A}^1(\mathcal{O})$ is said to be the *space of antiholomorphic* 1-*forms*.

We recall the reader that there is a globally defined conjugation of forms on a complex manifold. In terms of local coordinates it is given by $\overline{(dz)} = d\bar{z}$, $\overline{(d\bar{z})} = dz$ and the usual conjugation of functions.

We have the following simple

THEOREM XVI.6.12. Let $\mathcal{O}$ be an open subset of a Riemann surface M; then

1° Every holomorphic 1-form ω is closed.

2° Every closed form $\omega \in \Lambda^{1,0}(\mathcal{O})$ is holomorphic,

3° $(u \in \mathscr{H}(\mathcal{O}, C)) \Rightarrow (d'u \in A^1(\mathcal{O}))$.

PROOF. 1°, 2°. Let $\omega \in \Lambda^{1,0}(\mathcal{O}, C)$; then, since in the local coordinates we have $\omega = f dz$,

$$d\omega = df \wedge dz = -\frac{\partial f}{\partial \bar{z}} dz \wedge d\bar{z}.$$

For forms of type $(1,0)$ we therefore have $(d\omega = 0) \Leftrightarrow (\partial f/\partial \bar{z} = 0)$.

3° $0 = \Delta u\, dx \wedge dy = 2id'd''u = 2id''(d'u)$; therefore, $d'u$ is holomorphic. □

Harmonic Forms. It is proper to consider complex-valued harmonic forms on a Riemann surface M. For 1-forms we define the $*$-operator of Hodge in the following way:

DEFINITION. Let $\omega \in \Lambda^1(M, C)$; then, decomposing $\omega = \omega_1 + \omega_2$, where $\omega_1 \in \Lambda^{1,0}(M)$, $\omega_2 \in \Lambda^{0,1}(M)$ we put

$$(8) \qquad *\omega := i(\omega_1 - \overline{\omega}_2).$$

A form $\omega \in \Lambda^1(M, C)$ is called *harmonic* if

$$(9) \qquad d\omega = d*\omega = 0.$$

The space of complex harmonic 1-forms is denoted by $\mathscr{H}^1(M, C)$. Suppose that $\omega_1 \in \Lambda^{1,0}(M)$, $\omega_2 \in \Lambda^{0,1}(M)$. Then we check that the following holds:

$$(10) \qquad **\omega = -\omega, \quad \overline{*\omega} = *\overline{\omega} \quad \text{for } \omega \in \Lambda^1(M, C),$$

$$(11) \qquad d*(\omega_1 + \omega_2) = id'\overline{\omega}_1 - id''\overline{\omega}_2,$$

(12) $\quad *d'f = id''\bar{f}, \quad *d''f = -id'\bar{f} \quad$ for $f \in C^\infty(M)$,

(13) $\quad d*df = 2id'd''\bar{f} = \Delta\bar{f}.$

THEOREM XVI.6.13. For $\omega \in \Lambda^1(M, C)$ the following conditions are equivalent:

(a) $\omega \in \mathscr{H}^1(M, C)$ (i.e., $d\omega = d*\omega = 0$).

(b) $d'\omega = 0 = d''\omega$,

(c) $\omega = \omega_1 + \omega_2$, where $\omega_1 \in A^1(M)$, $\omega_2 \in \overline{A^1}(M)$.

(d) For every $x \in M$ there exists an open neighbourhood $\mathcal{O} \ni x$, and an $f \in \mathscr{H}(\mathcal{O}, C)$ such that $\omega = df$.

PROOF. The equivalence of (a)–(c) follows from (9)–(13).

(d) $\Rightarrow$ (a). Let $\omega = df$ and $f \in \mathscr{H}(\mathcal{O}, C)$; then

$$d\omega = ddf = 0 \quad \text{and} \quad d*\omega = d*df = \Delta f = 0.$$

(a) $\Rightarrow$ (d). Let $\omega \in \mathscr{H}^1(M, C)$; then $d\omega = 0$ and by the Poincaré Lemma we have locally $\omega = df$. But $0 = d*\omega = d*df = 2id'd''f = \Delta f$. Hence, f is harmonic. □

DEFINITION. The complex dimension of the space $A^1(M)$ for a compact Riemann surface M is called the *genus* of M and is denoted by g. Usually, the genus is defined by topological means and later shown to be equal to the dimension of the space of Abelian differentials of the first kind. We shall discuss this point later on.

COROLLARY XVI.6.14. 1° We have

(14) $\quad \mathscr{H}^1(M, C) = A^1(M) \oplus \overline{A^1}(M).$

In particular, for a compact Riemann surface M,

(15) $\quad \dim \mathscr{H}^1(M, C) = 2 \dim A^1(M) = 2g.$

2° Every real harmonic 1-form $h \in \mathscr{H}^1(M, C)$ is the real part of onc and only one Abelian differential $\omega \in A^1(M)$.

PROOF. (14) has already been proved (Theorem XVI.6.13(c)). Plainly, (14) $\Rightarrow$ (15). We only have to verify 2°.

Existence: By (14) we have $h = \omega_1 + \bar{\omega}_2$, where $\omega_k \in A^1(M)$. But $h = \bar{h} = \bar{\omega}_1 + \bar{\bar{\omega}}_2 = \bar{\omega}_1 + \omega_2$; therefore, $\omega_1 = \omega_2$ and $h = 2\mathrm{Re}(\omega_1)$.

Uniqueness: Let $\omega \in A^1(M)$ and $\mathrm{Re}(\omega) = 0$. Since $d\omega = 0$, we have locally $\omega = df$, where f is holomorphic. Hence, $\mathrm{Re}(f) =$ const and, therefore, $f =$ const and $\omega = 0$. □

We specialize the definition of a scalar product of 1-form to the case at hand.

DEFINITION. Let M be a compact Riemann surface

(16) $\quad (\omega_1|\omega_2) := \int\limits_M \omega_1 \wedge *\omega_2, \quad$ where $\omega_k \in \Lambda^1(M, C)$.

Plainly, (16) defines a scalar product on $\Lambda^1(M, C)$: if $\omega = f dz + g d\bar{z}$, then $*\omega = i(\bar{f} d\bar{z} - \bar{g} dz)$ and

(17) $\quad \omega \wedge *\omega = i(|f|^2 + |g|^2) dz \wedge d\bar{z} = 2(|f|^2 + |g|^2) dx \wedge dy.$

Therefore, $((\omega|\omega) = 0) \Rightarrow (\omega = 0)$.

For Riemann surfaces we have the following version of the Hodge Theorem:

THEOREM XVI.6.15 (Hodge). Let M be a compact Riemann surface; then

(18) $\quad \Lambda^1(M, C) = *dC^\infty(M) \oplus dC^\infty(M) \oplus \mathscr{H}^1(M, C),$

where $\oplus$ denotes the orthogonal sum,

(19) $\quad \mathscr{H}^1(M, C) \cong H^1(M, C),$

where

$$H^1(M, C) := \frac{\ker(\Lambda^1(M, C) \xrightarrow{d''} \Lambda^2(M, C))}{\operatorname{Im}(C^\infty(M, C) \xrightarrow{d''} \Lambda^1(M, C))},$$

(20) $\quad \Lambda^{0,1}(M) = d''C^\infty(M, C) \oplus \bar{A}^1(M).$

PROOF. We only have to check (20) since (18) and (19) are special cases of a general theorem of Hodge and Kodaira (Chapter XIV).

The spaces $d'C^\infty(M, C)$, $d''C^\infty(M, C)$, $A^1(M)$, and $\bar{A}^1(M)$ are pairwise orthogonal. Since $d = d' + d''$, we obtain (20) from (18) and the decomposition

$$\Lambda^1(M, C) = \Lambda^{1,0}(M) \oplus \Lambda^{0,1}(M). \ \square$$

Singularities of Differential Forms. Let $\mathcal{O}$ be an open subset of a Riemann surface M and let $a \in \mathcal{O}$. Let ω be a holomorphic form on $\mathcal{O} - \{a\}$. If (z, V) is a local coordinate at a with $V \subset \mathcal{O}$ and $z(a) = 0$, then on $V - \{a\}$ we have

(21) $\quad \omega = f dz, \quad$ where $f \in A(V - \{a\})$.

We classify the types of singularity of the form ω at the point a according to the type of singularity of its local representative f. However, we should check beforehand that the type of singularity of f does not depend on the local coordinate chosen, but this can be safely left to the (patient) reader. Thus we say that ω has a removable singularity, a pole of order k, or an essential singularity at a if the same is true of f. Moreover, if f is expanded in its Laurent series

$$(22) \qquad f(z) = \sum_{-\infty}^{\infty} c_n z^n,$$

it is tempting to define c_{-1} as the residuum of ω at a. The independence of coordinates now relies on an integration argument.

LEMMA XVI.6.16 (and definition). Let V be a relatively compact open subset of M with a differentiable boundary, containing a and no other singularities of ω, and let c_{-1} be defined by means of the expansion (22) which is valid in some coordinate system (V_0, z) with $V_0 \subset V$ and $z(a) = 0$. Then

$$(23) \qquad \frac{1}{2\pi i} \int_{\partial V} \omega = c_{-1}.$$

Therefore, we define the *residuum of* ω *at the point* a to be the common value of the both sides in the equality (23).

PROOF. We may assume that V_0 is so chosen that $z(V_0) = K(0, r)$ and $\overline{V}_0 \subset V$. Then we put $U := V - \overline{V}_0$. Since ω is closed on U (because it is holomorphic there), by the Stokes Theorem we have

$$0 = \int_U d\omega = \int_{\partial U} \omega = \int_{\partial V} \omega - \int_{\partial V_0} \omega.$$

Therefore,

$$\int_{\partial V} \omega = \int_{\partial V_0} \omega = 2\pi i c_{-1}$$

since we can integrate the series (22) termwise and

$$\int_{|z|=r} z^k dz = i r^{k+1} \int_0^{2\pi} e^{(k+1)it}\, dt = \begin{cases} 0 & \text{for } k \neq -1, \\ 2\pi i & \text{for } k = -1. \end{cases} \qquad \square$$

Meromorphic Differential Forms. Abelian Differentials. We can now define the following fundamental notion:

DEFINITION. Let $\mathscr{X}$ be an open subset of a Riemann surface M. A *meromorphic form* ω on $\mathscr{X}$ is such a holomorphic form on an open subset $\mathscr{U} \subset \mathscr{X}$ that

1° the set $\mathscr{X} - \mathscr{U}$ has no limit points in $\mathcal{O}$

2° ω has a pole at every $a \in \mathscr{X} - \mathscr{U}$.

The set of meromorphic differential forms on $\mathscr{X}$ will be denoted by $\mathscr{M}^1(\mathscr{X})$. Meromorphic forms are also called *Abelian differentials*:

1° ω is of the *first kind* if ω is holomorphic,

2° ω is of the *second kind* if $\mathrm{Res}_a \omega = 0$ at every pole a,

3° otherwise ω is of the *third kind*

In particular, if ω has only two poles $a_1 \neq a_2$ such that

$$\mathrm{Res}_{a_1}(\omega) = -\mathrm{Res}_{a_2}(\omega) = 1,$$

then ω is called an *elementary* (or *normal*) *differential of the third kind.*

Remark. Abelian differentials on compact Riemann surfaces were introduced and investigated by Riemann. He was probably inspired by physical considerations (this is a suggestion of Felix Klein): an elementary differential of the third kind corresponds to two sources at a_1, a_2 with equal but opposite strengths. Such differentials play an important role in the theory of uniformization (cf. Sections 10 and 20).

THEOREM XVI.6.17 (The Residue Theorem). Let M be a compact Riemann surface and let $a_1, \ldots, a_n$ be distinct points of M. Then, for any holomorphic form ω in $M' := M - \{a_1, \ldots, a_n\}$ we have

$$\sum_{j=1}^{n} \mathrm{Res}_{a_j}(\omega) = 0. \tag{26}$$

Therefore, the sum of the residues of a meromorphic differential ω on a compact Riemann surface is zero.

PROOF. Let V_j be small circular disjoint neighbourhoods of a_j, $j = 1, \ldots, n$. Let $U := M - \bigcup_j \overline{V}_j$; then, since $d\omega = 0$ on U, we have

$$\sum_{j=1}^{n} \mathrm{Res}_{a_j}(\omega) = \sum_{j=1}^{n} \frac{1}{2\pi i} \int_{\partial V_j} \omega$$

$$= -\frac{1}{2\pi i} \int_{\partial U} \omega = -\frac{1}{2\pi i} \int_{U} d\omega = 0. \ \square$$

COROLLARY XVI.6.18. A non-constant meromorphic function f on a compact Riemann surface assumes every value the same number of times. In particular f has the same number of zeros as poles.

PROOF. The differential form $\omega := df/f$ is holomorphic outside the zeros and poles of f. If a is a zero (resp. pole) of order r of f, then $\mathrm{Res}_a(\omega) = r$ (resp. $-r$). If $c \in C^*$, then we consider the form $\omega := df/(f-c)$. The residue theorem gives the conclusion.

By the same method we obtain

PROPOSITION XVI.6.19 (The "Argument Principle"). Let G be a precompact region in a Riemann surface and let ω be a meromorphic differential in $\bar{G}$ which has a finite number of zeros and poles inside of G and is regular on ∂G. Then we have

$$\int_{\partial G} \frac{df}{f} = 2\pi i(N-P),$$

where N (resp. P) is the number of zeros (resp. poles) of f in G.

7. ANALYTIC CONTINUATION. COVERINGS. FUNDAMENTAL GROUP. THE THEORY OF POINCARÉ

The uniqueness principle was for Weierstrass a starting point for a magnificent theory which is characterized by such notions as "function element" (or "germ"), "analytic continuation", "monodromy theorem". The Weierstrass theory led to the important notion of covering surface (H. A. Schwarz, Poincaré, Klein, Koebe), which was codified by H. Weyl in his monumental *Idee der Riemannschen Fläche*. Covering surfaces of a given X were classified by H. Poincaré by means of a group $\pi_1(X)$ which he discovered and called the *fundamental* (*Poincaré*'s by others), *group* of X. All these notions had grown from complex analysis and became indispensable in modern topology, analysis, and differential geometry. All of them bear the stamp of their origin: the theory of Riemann surfaces.

Presheaves. In complex analysis one has to do with functions in varying domains of definition. The notion of (pre)sheaf (French: *faisceau*, German: *Garbe*) developed by J. Leray and H. Cartan, is a natural tool for handling such a situation.

DEFINITION. Let X be a topological space and $\mathscr{T}$ its topology. A *presheaf of vector spaces (Abelian groups, rings, sets,...)* over X is a pair $(\mathscr{F}, \varrho)$ consisting of

1° a family $\mathscr{F} = \{\mathscr{F}(\mathscr{O})\}$, $\mathscr{O} \in \mathscr{T}$ of vector spaces,

2° a family $\varrho = \{\varrho^{\mathscr{O}}_{\mathscr{V}}\colon \mathscr{O}, \mathscr{V} \in \mathscr{T}$ and $\mathscr{V} \subset \mathscr{O}\}$ of linear maps (homomorphisms), called *restriction maps*,

$$\varrho^{\mathscr{O}}_{\mathscr{V}}\colon \mathscr{F}(\mathscr{O}) \to \mathscr{F}(\mathscr{V}),$$

such that

$$\varrho^{\mathscr{O}}_{\mathscr{O}} = \mathrm{id}_{\mathscr{F}(\mathscr{O})} \quad \text{for all } \mathscr{O} \in \mathscr{T},$$
$$\varrho^{\mathscr{V}}_{\mathscr{W}} \circ \varrho^{\mathscr{O}}_{\mathscr{V}} = \varrho^{\mathscr{O}}_{\mathscr{W}} \quad \text{for } \mathscr{W} \subset \mathscr{V} \subset \mathscr{O}.$$

One often writes $\mathscr{F}$ for $(\mathscr{F}, \varrho)$ and

$$f|V \quad \text{for } \varrho^{\mathscr{O}}_{\mathscr{V}}(f),\ f \in \mathscr{F}(\mathscr{O}).$$

Remark. An algebraic structure on $\mathscr{F}(\mathscr{O})$ is entirely irrelevant in most of what follows; we have given the definition for the sake of the reader's education and for the future use. For that matter, we could have restricted ourselves to considering simply presheaves of sets, where $\mathscr{F}(\mathscr{O})$ is assumed merely to be a set and the restriction mappings are simply maps of sets. This may seen trivial but it will sometimes allow greater flexibility, since there are cases where $\mathscr{F}(\mathscr{O})$ is not defined for some $\mathscr{O}$, and then we have the possibility of taking the empty set as $\mathscr{F}(\mathscr{O})$ for these $\mathscr{O}$.

Standard Examples. 1. Let X be a topological space. For each open $\mathscr{O} \subset X$ let $C(\mathscr{O})$ be the vector space of continuous functions $f\colon \mathscr{O} \to \mathbf{C}$ and let $\varrho^{\mathscr{O}}_{\mathscr{V}}\colon C(\mathscr{O}) \to C(\mathscr{V})$ be the usual restriction map. Then $\{C, \varrho\}$ is a presheaf of vector spaces (or even rings).

2. Suppose that X is a complex manifold, that $A(\mathscr{O})$ is the vector space (a ring) of holomorphic functions on $\mathscr{O} \subset X$, $f\colon \mathscr{O} \to \mathbf{C}$, and that $\varrho^{\mathscr{O}}_{\mathscr{V}}$ is again the restriction mapping. The presheaf $\{A, \varrho\}$ will be denoted by $\mathscr{A}$ or by $\mathscr{A}_X$ and called the *presheaf of holomorphic functions on X*.

Germs (of Functions). Let $\mathscr{F}$ be a presheaf over X and let a be a point of X.

On the disjoint sum

$$\coprod_{\mathscr{T} \ni \mathscr{O} \ni a} \mathscr{F}(\mathscr{O})$$

we introduce the following equivalence relation $\underset{a}{\sim}$: $f \underset{a}{\sim} g$ for $f \in \mathscr{F}(\mathcal{O})$, $g \in \mathscr{F}(\mathscr{V})$ if there exists an open set $\mathscr{W}$ such that $a \in \mathscr{W} \subset \mathcal{O} \cap \mathscr{V}$, and $f|\mathscr{W} = g|\mathscr{W}$. The set $\mathscr{F}_a$ of all equivalence classes is called a *stalk of* $\mathscr{F}$ *over* a:

$$\mathscr{F}_a := \Big(\coprod_{\mathcal{O} \ni a} \mathscr{F}(O)\Big) \Big/ \underset{a}{\sim} .$$

For an open subset $\mathcal{O} \subset X$ let us define a map $\varrho_a: \mathscr{F}(\mathcal{O}) \to \mathscr{F}_a$, sending each function $f \in \mathscr{F}(\mathcal{O})$ upon its equivalence class modulo $\underset{a}{\sim}$, i.e., $\varrho_a(f) := [f]_a$; then $\varrho_a(f)$ is called the *germ of f at a.*

Remark. If our presheaf is $\mathscr{A} = \mathscr{A}_X$, where X is a domain in $\mathbf{C}$, then each germ at a is uniquely determined by a power series $\sum_{n=0}^{\infty} c_n(z-a)^n$ with a positive radius of convergence, namely, the Taylor series at a of its representatives. These germs were called by Weierstrass the *function elements.*

Covering Defined by a Presheaf. Let X be a topological space and $\mathscr{F}$ a presheaf over X. We define the *total space* also called the *sheaf of germs of* $\mathscr{F}$, denoted $|\mathscr{F}|$, as the disjoint union of all the stalks

$$|\mathscr{F}| := \coprod_{x \in X} \mathscr{F}_X .$$

The projection $p: |\mathscr{F}| \to X$, defined by $p\big(\varrho_x(f)\big) = x$ or, more formally, the triple $(|\mathscr{F}|, p, X)$, is called the *covering defined by the presheaf* $\mathscr{F}$.

With a view to introducing a complex structure on $|\mathscr{A}_M|$ for a Riemann surface M, we now turn to the problem of topologizing the total space $|\mathscr{F}|$ of a given presheaf $\mathscr{F}$ over X.

For each open $\mathcal{O} \subset X$ and a given $f \in \mathscr{F}(\mathcal{O})$ we define

$$[\mathcal{O}, f] := \{\varrho_x(f) \subset |F|: x \in \mathcal{O}\}.$$

We have a simple but important

THEOREM XVI.7.1 (H. Weyl). Let $\mathscr{F} = \mathscr{F}_X$ be a presheaf over a topological space X. Then

1° $\mathscr{B} = \{[\mathcal{O}, f]: \mathcal{O}$ is open in $X, f \in \mathscr{F}(\mathcal{O})\}$ is a basis for a topology on $|\mathscr{F}|$.

2° The projection $p: |\mathscr{F}| \to X$ is a local homeomorphism with respect to the topology defined in 1°, i.e. each point in $|\mathscr{F}|$ has an open neighbourhood $\tilde{U}$ such that $p(\tilde{U})$ is open and $p|\tilde{U}$ is a homeomorphism.

PROOF. The reader can verify 1°.

2° Let $\varphi \in |\mathscr{F}|$ then $\varphi = \varrho_x(f)$ for some $x \in X$ and $f \in \mathscr{F}(\mathcal{O})$ with $x \in \mathcal{O}$. Therefore $\tilde{\mathcal{O}} = [\mathcal{O}, f]$ is, by virtue of 1°, an open neighbourhood of φ and $p(\tilde{\mathcal{O}}) = \mathcal{O}$ is open. Now observe that if $[\mathcal{O}_1, f_1]$ is another basic open set and $\varrho_y(f) \in [\mathcal{O}, f] \cap [\mathcal{O}_1, f_1]$, then f coincides with f_1 on a certain neighbourhood of y and so the topology induced on $[\mathcal{O}, f]$ consists of all the sets $[\mathscr{U}, f|\mathscr{U}]$ with open $\mathscr{U} \subset \mathcal{O}$. This makes it clear that $p|\tilde{\mathcal{O}}$ is a homeomorphism. □

The following question arises: when $|\mathscr{F}_X|$ is a Hausdorff space and particularly is this true in the most interesting for us case of the presheaf $\mathscr{A}_M$ of holomorphic functions on M. A positive answer to the second question was already given in Weyl's *Idee der Riemannschen Fläche* (1913). It is interesting to note that, the book presented a definition of a Hausdorff space for the first time and gave the proof of Theorem XVI.7.1 along the same lines. Moreover, Weyl proved that $|\mathscr{A}_X|$ (for a domain X in $\boldsymbol{C}$) is a Hausdorff space.

Let us introduce the following useful

DEFINITION. A presheaf $\mathscr{F}_X$ *satisfies the identity principle* if, for any open connected $\mathcal{O} \subset X$ and each $f, g \in \mathscr{F}(\mathcal{O})$, the equality $\varrho_a(f) = \varrho_a(g)$ satisfied at a point $a \in \mathcal{O}$ implies $f = g$ (throughout $\mathcal{O}$).

Example 1. Let X be a complex manifold and let $\mathscr{A}$ be the presheaf of holomorphic functions; then $\mathscr{A}$ satisfies the principle of identity.

Example 2. Let X be an open subset of $\boldsymbol{R}^n$. Let $\mathscr{H}(\mathcal{O})$ be the vector space of harmonic functions in an open $\mathcal{O} \subset X$; then $\mathscr{H} = \{\mathscr{H}(\mathcal{O})\}$ is a presheaf which satisfies the principle of identity.

Example 3. Let X be a Riemann surface and $\mathscr{M}(\mathcal{O})$ the space of meromorphic functions in $\mathcal{O}$; then the sheaf $\mathscr{M}_X = \{\mathscr{M}(\mathcal{O})\}$, of meromorphic functions in X (denoted also by $\mathscr{M}_X$) satisfies the principle of identity.

As a complement to Theorem XVI.7.1, we have a theorem proved by Weyl in case $\mathscr{F} = \mathscr{A}_X$:

THEOREM XVI.7.2. Let X be a locally connected Hausdorff space and let $\mathscr{F}$ be a presheaf on X which satisfies the principle of identity. Then $|\mathscr{F}|$ with the topology defined in Theorem XVI.7.1 is a Hausdorff space.

PROOF. Let $\varphi_1, \varphi_2 \in |\mathscr{F}|$; we have to construct disjoint neighbourhoods U, V of φ_1, φ_2. Let us consider the case of $p(\varphi_1) = p(\varphi_2) =: x$. If the germ φ_k is represented by $f_k \in \mathscr{F}(\mathcal{O}_k)$, $\mathcal{O}_k \ni x$, $k = 1, 2$, then for a connected neighbourhood $U \subset \mathcal{O}_1 \cap \mathcal{O}_2$ of x $[\mathcal{O}_k, f_k | \mathcal{O}_k]$ is an open neighbourhood of φ_k. Suppose that $\psi \in [\mathcal{O}_1, f_1 | \mathcal{O}_1] \cap [\mathcal{O}_2, f_2 | \mathcal{O}_2]$ and let $y := p(\psi)$, then $\psi = \varrho_y(f_1) = \varrho_y(f_2)$. By the identity principle, $f_1 | U = f_2 | U$ and, hence, $\varphi_1 = \varphi_2$, which is a contradiction.

The case $p(\varphi_1) =: x \neq y := p(\varphi_2)$ is easier and is left to the reader. □

The following simple fact allows us to introduce in a natural way a differentiable (complex) structure on objects with similar properties to $|\mathscr{F}|$.

PROPOSITION XVI.7.3. Let X be a complex (resp. differentiable) manifold of the dimension n, let Y be a Hausdorff space, and let $p: Y \to X$ be a local homeomorphism.

Then there exists on Y a unique structure of a complex (resp. differentiable) manifold of dimension n such that $p: Y \to X$ is holomorphic (resp. differentiable). Hence, p is a local biholomorphism (resp. diffeomorphism).

PROOF. *Existence*: For a given $y \in Y$ let $U \subset Y$ be such an open neighbourhood of y that $p|U$ is a homeomorphism and $p(U)$ is contained in the domain of a coordinate map $(V, \varkappa)$ on X around $p(Y)$. Then $\chi = \varkappa \circ (p|U)$ is a homeomorphism of U onto an open subset of an n-dimensional space. The family of χ so defined for all points of Y is a family of local maps satisfying the compatibility condition and, thus, defines a unique structure of a complex (differentiable) manifold on Y. Evidently, with respect to this structure p becomes a holomorphic, and even a locally biholomorphic (resp. differentiable) mapping.

Uniqueness: If $\mathscr{B}$ is another complex atlas on Y such that $p: (Y, \mathscr{B}) \to X$ is holomorphic and, hence, locally biholomorphic, then the identity

$$\mathrm{id}_Y: (Y, \mathscr{A}) \to (Y, \mathscr{B})$$

is locally biholomorphic, and hence, biholomorphic. Therefore $\mathscr{A}$ and $\mathscr{B}$ define the same complex structure. □

As a straightforward corollary we obtain

PROPOSITION XVI.7.4. Let X be a Riemann surface. Let $\tilde{X}$ denote any connected component of $|\mathscr{A}_X|$; then $\tilde{X}$ is a Riemann surface and $\pi := p|\tilde{X}$ is a holomorphic (and locally biholomorphic) map $\pi: \tilde{X} \to X$.

The construction of $p: |\mathscr{F}| \to X$ preceding Theorem XVI.7.1 leads to an important topological notion.

DEFINITION. Let X, Y be topological spaces; a map $p: Y \to X$ is called a *covering* if it is

1° continuous, 2° open, and 3° discrete in the sense that every fibre $p^{-1}(x)$ is a discrete subset of Y.

Two coverings $p_k: Y_k \to X$, $k = 1, 2$, are called *isomorphic* if there exists a homeomorphism $h: Y_1 \to Y_2$ such that $p_2 \circ h = p_1$. A point $y \in Y$ is a *branch point* if in no neighbourhood of y the map p is injective. A map $p: Y \to X$ is called *unbranched* (or *unramified*) if it has no branch points.

Remark. In algebraic topology the name covering is used in a much more restrictive sense, which here goes under the name of a complete covering (to be defined later).

THEOREM XVI.7.5. Let Y, X be topological spaces. (A map $p: Y \to X$ is an unbranched covering) $\Leftrightarrow$ ($p: Y \to X$ is locally topological).

PROOF. $\Leftarrow$: If $p: Y \to X$ is a local homeomorphism, then it is open and continuous. But p is also discrete, since if $y \in p^{-1}(x)$ and U is an open neighbourhood of y which is topologically mapped onto an open subset of X, then $U \cap p^{-1}(x) = \{y\}$.

$\Rightarrow$: Let $p: Y \to X$ be an unbranched covering and let $y \in Y$ be arbitrary. Since y is not a branch point, there exists an open neighbourhood U of y such that $p|U$ is an injection. Since p is continuous and open, p maps U topologically onto $V := p(U)$. □

Example 1. Let $L \subset \boldsymbol{C}$ be a lattice and let $p: \boldsymbol{C} \to \boldsymbol{C}/L$ be the canonical projection; then p is an unbranched covering (this map is of paramount importance in the theory of elliptic functions).

Example 2. $\exp: \boldsymbol{C} \to \boldsymbol{C}^*$ is locally homeomorphic since exp is injective on each subset U which does not contain points which are equivalent modulo $2\pi i$.

Example 3. Let $k \in \boldsymbol{N}$, $k \geqslant 2$; denote $p_k = p_k: \boldsymbol{C} \to \boldsymbol{C}$, $p_k(z) := z^k$. Then p_k is a covering and $0 \in \boldsymbol{C}$ is a unique branch point of p_k. Therefore, $p_k|\boldsymbol{C}^*: \boldsymbol{C}^* \to \boldsymbol{C}$ is unbranched.

PROPOSITION XVI.7.6. Let X, Y be Riemann surfaces. If $p: Y \to X$ is a non-constant holomorphic mapping, then p is a covering.

PROOF. Plainly, p is continuous and open by Theorem XVI.6.5. Let us assume that for some $x, p^{-1}(x)$ is not discrete; then p is constant by the Identity Theorem XVI.6.2 which is a contradiction. □

Remark. The older literature often speaks of "multi-valued" functions or mappings. This is done mostly in the following context: given a covering $p: Y \to X$; then a map $f: Y \to E$ is regarded as a multi-valued map of X into E. We encountered this situation in Section 1 when we defined log: $\boldsymbol{C}^* \to \boldsymbol{C}$ by the following diagram

$$\begin{array}{ccc} \boldsymbol{C} & \xrightarrow{\text{id}} & \boldsymbol{C} \\ {\scriptstyle \exp}\downarrow & \swarrow {\scriptstyle \log} & \\ \boldsymbol{C}^* & & \end{array}$$

Since Riemann wanted to avoid "multi-valued" maps, he created his surfaces as covering surfaces. This notion was made precise several decades later, namely, about 1880, by H. A. Schwarz, to whom we owe the idea of a "universal covering surface" (cf. below). This idea was further developed by H. Poincaré and F. Klein and codified by Hermann Weyl in his classic *Idee der Riemannschen Fläche* in 1913.

DEFINITION. Let X, Y, Z be topological spaces, let $p: Y \to X$ be a covering and let $f: Z \to X$ be a continuous map. Then a continuous map $g: Z \to Y$ such that $f = p \circ g$, i.e., that the diagram

$$\begin{array}{ccc} & & Y \\ & {\scriptstyle g}\nearrow & \downarrow {\scriptstyle p} \\ Z & \xrightarrow[f]{} & X \end{array}$$

is commutative, is said to be a *lift of f with respect to p*. Generally, a lift need not exist at all and if it does exist, need not be unique but the following simple fact does hold true:

THEOREM XVI.7.7 (Uniqueness of the Lift). Let X, Y be Hausdorff and let $p: Y \to X$ be an unbranched covering. Suppose that Z is a connected space, that $f: Z \to X$ is a continuous map, and that g_1, g_2: $Z \to Y$ are such lifts of f that $g_1(z_0) = g_2(z_0)$ at a point $z_0 \in Z$. Then $g_1 = g_2$.

Hence, either $g_1 = g_2$ or $g_1(z) \neq g_2(z)$ for all $z \in Z$.

PROOF. Let $A = \{z \in Z: g_1(z) = g_2(z)\}$. Then $z_0 \in A$ and A is a closed subset of Z as the inverse image of the diagonal of $Y \times Y$ by the map (g_1, g_2): $Z \to Y \times Y$. It is sufficient to verify that A is also open, since then $Z = A$ by the connectedness of Z. Let $z \in A$ and $g_1(z) = g_2(z) =: y$. Since p is locally topological, there exists a neighbourhood $\mathcal{O}$ of y which is mapped topologically onto a neighbourhood $\mathcal{U}$ of $p(y) = f(z)$. Since g_1, g_2 are continuous, there exists a neighbourhood $\mathcal{V}$ of z such that $g_1(\mathcal{V}), g_2(\mathcal{V}) \subset \mathcal{O}$. Let h: $\mathcal{U} \to \mathcal{O}$ denote the map $p(\mathcal{O})^{-1}$. By $p \circ g_k = f$ we have $g_k(\mathcal{V}) = h \circ (f|\mathcal{V})$, $k = 1, 2$. Therefore, $g_1|\mathcal{V} = g_2|\mathcal{V}$ and, hence, $\mathcal{V} \subset A$, whereby A is open. □

PROPOSITION XVI.7.8. Let X, Y, Z be Riemann surfaces, let p: $Y \to X$ be a holomorphic unbranched covering, and let f: $Z \to X$ be a holomorphic map. Then every lift g: $Z \to Y$ of f is holomorphic.

The proof is left to the reader. □

COROLLARY XVI.7.9. Let X, Y, Z be Riemann surfaces and let p: $Y \to X$ and π: $Z \to X$ be holomorphic unbranched coverings. Then every continuous mapping f: $Y \to Z$ such that $p = \pi \circ f$ is holomorphic.

The proof is obvious since π is locally biholomorphic. □

We now arrive at a result which occupies a central position in the theory of covering spaces. Since it generalizes the classical monodromy theorem of Weierstrass, it is called the

Monodromy Principle. (Topologists call it the "theorem on lifting of homotopic arcs".) We start with a definition: an *arc* (or a *path*) in a topological space X is a continuous mapping γ: $[0, 1] \to X$; the points $\gamma(0)$ and $\gamma(1)$ are called *end-points* of the arc γ and γ is said to *join* $\gamma(0)$ with $\gamma(1)$. Usually, we shall denote the unit interval $[0, 1]$ by I.

Suppose that p: $Y \to X$ is a covering. The problem which we wish to consider is the following. Let $a, b \in X$ be given and let $\tilde{a} \in Y$ cover a (i.e., $p(\tilde{a}) = a$). Then, if we choose an arc γ with end-points a and b and its lift $\tilde{\gamma}$ starting at $\tilde{a}$, to what extent do the end-point $\tilde{\gamma}(1)$ depend on the chosen γ? This purely topological problem arose from a very concrete question of complex analysis: Weierstrass asked when analytic continuations of a function element (a power series) along curves with the same end-points lead to the same function element. To deal with that question we need some machinery, namely, the notion of the *homotopy of arcs*. And since in all topological consideration of this section the assumption that topological spaces are Hausdorff spaces is essential, we introduce the convention that

all topological spaces are assumed Hausdorff spaces

unless the contrary is explicitly stipulated.

DEFINITION. Let X be a topological space. Two arcs $\gamma_0: I \to X$ and $\gamma_1: I \to X$ such that $\gamma_0(0) = \gamma_1(0)$ and $\gamma_0(1) = \gamma_1(1)$ are said to be *homotopic* if there exists a continuous function $H: I \times I \to X$ such that

1° $H(t, 0) = \gamma_0(t)$ for all $t \in I$,

2° $H(t, 1) = \gamma_1(t)$ for all $t \in I$,

3° $H(0, s) = \gamma_0(0)$, $H(1, s) = \gamma_0(1)$ for all $s \in I$.

Any function H with properties as above is called a *homotopy from* γ_0 *to* γ_1. The function H above defines a family of arcs in X indexed by points of the interval I by means of the formula $\gamma_s: I \ni t \to \gamma_s(t) := H(t, s) \in X$, and the conditions on H intuitively mean that the family $\{\gamma_s\}_{s \in I}$ continuously joins γ_0 with γ_1, while keeping the common end-points of γ_0 and γ_1 fixed.

Exercise. Check that the relation of being homotopic is an equivalence relation in the set of arcs in X with given end-points.

In rough terms the answer to the problem raised above is as follows: the end-point of a lift $\tilde{\gamma}$ depends only on the homotopy class of a given arc. As already pointed out, however, a lift of a given arc need not exist at all. (Think of lifting arcs in the covering $p:]0, 10[\to S^1$ where $p(t) = e^{2\pi i t}$.) We must, therefore, first make certain of the existence of lifts for any arc $\tilde{\gamma}$ from every point over the initial point of γ. In the following theorem we shall simply assume the existence of sufficiently many lifts among the hypotheses and later we shall discuss a sufficient condition for that (completeness).

THEOREM XVI.7.10 (Monodromy Principle. Covering Homotopy Theorem). Suppose that $p: Y \to X$ is an unbranched covering and let $a \in X$. Assume that γ_0, γ_1 are homotopic arcs in X with the initial point a and that there is a point $\tilde{a} \in Y$ over a (i.e., $p(\tilde{a}) = a$) and a homotopy $H: I \times I \to X$ from γ_0 to γ_1, such that each arc $\gamma_s: I \ni t \to \gamma\ (t) := H(t, s)$ can be lifted to the arc $\tilde{\gamma}_s$ from $\tilde{a}$. Then $\tilde{\gamma}_0$ and $\tilde{\gamma}_1$ are homotopic and, in particular, $\tilde{\gamma}_0(1) = \tilde{\gamma}_1(1)$.

PROOF. Let $H\colon I\times I\to X$ be a homotopy from γ_0 to γ_1 such that $\gamma_s(t) := H(t,s)$ has a lift $\tilde{\gamma}_s$ with the initial point $\tilde{\gamma}_s(0) = \tilde{a}$. We define $\tilde{H}(t,s) = \tilde{\gamma}_s(t)$ and we claim that $\tilde{H}$ is a required homotopy from $\tilde{\gamma}_0$ to $\tilde{\gamma}_1$. In fact properties 1° and 2° of the definition are immediate and 3° will follow once the continuity of $\tilde{H}$ is established. Indeed, we set $b = \gamma_0(1) = \gamma_1(1)$ and note that since $H(\{1\}\times I) = \{b\}$ and $p\circ\tilde{H} = H$, we have $\tilde{H}(\{1\}\times I)\subset p^{-1}(\{b\})$. If $\tilde{H}$ is continuous, then by the discreteness of $p^{-1}(\{b\})$ and the connectedness of $\{1\}\times I$ the set $\tilde{H}(\{1\}\times I)$ consists of a single element.

The continuity of $\tilde{H}$ is proved in a natural way: If $R\subset I\times I$ is a small rectangle such that a local inverse of p, say q, can be defined on $H(R)$, then by an appropriate choice of q we obtain $\tilde{H}|R = q\circ H|R$. In view of the compactness of the square $I\times I$ and of the continuity of H a finite covering of $I\times I$ by such squares R_i exists and to get the continuity of $\tilde{H}$ we only have to match the images of $\tilde{H}|R_i$ on overlapping sets. Let us make this idea rigorous: We first prove that there exists an $\varepsilon_0 > 0$ such that $\tilde{H}$ is continuous on $[0,\varepsilon_0[\times[0,1]$. Indeed, let U be a neighbourhood of $\tilde{a}$ and V a neighbourhood of a such that $p|U\to V$ is topological. Let $q := (p|U)^{-1}$. Since H is continuous and $H(\{0\}\times[0,1]) = \{a\}$, there exists such an $\varepsilon_0 > 0$ that $H([0,\varepsilon_0]\times[0,1])\subset V$. By the uniqueness of the lifts of γ_s we have

$$\tilde{\gamma}_s|[0,\varepsilon_0] = q\circ\gamma_s|[0,\varepsilon_0] \qquad \text{for all } s\in[0,1],$$

i.e., $\tilde{H} = q\circ H$ on $[0,\varepsilon_0]\times[0,1]$.

Now, let $\tau\in I$ be the greatest number with the property that $\tilde{H}$ is continuous on $[0,\tau]\times I$; we have shown above that $\tau\geqslant\varepsilon_0>0$ and we intend to show now that $\tau = 1$. Assume, contrary, that $\tau<1$ and let $x = H(\tau,0)$, $\tilde{x} = \tilde{H}(\tau,0) = \tilde{\gamma}_0(\tau)$ and let $\varepsilon_0>0$, $\delta_0>0$ be such that p has a local inverse defined on a neighbourhood V of the set $H([\tau-\varepsilon_0,\tau+\varepsilon_0]\times[0,\delta_0])$. Choose a branch of that inverse, say q_0, such that $q_0(x) = \tilde{x}$; this branch is then uniquely determined. Then, by the uniqueness of the lifts, for all $s\in[0,\delta_0]$ and $t\in[\tau-\varepsilon_0,\tau+\varepsilon_0]$ we have $\tilde{H}(t,s) = \tilde{\gamma}_s(t) = q_0\circ H(t,s)$; in particular, $\tilde{H}(t,s)$ is continuous on $[\tau-\varepsilon_0,\tau+\varepsilon_0]\times[0,\delta_0]$. Now we continue this for $x_1 = H(\tau,\delta_0)$ and $\tilde{x}_1 = \tilde{H}(\tau,\delta_0)$, etc., arriving finally (by the compactness of the interval I) at a sequence $0 = \delta_{-1} < \delta_0 < \delta_1 < \ldots < \delta_k = 1$ and a sequence ε_i, $0\leqslant i\leqslant k$ of positive numbers with the property that

$\tilde{H}$ is continuous on each of the rectangles $[\tau-\varepsilon_i, \tau+\varepsilon_i]\times[\delta_{i-1}, \delta_i]$. Since the values of $\tilde{H}$ coincide on the horizontal edges ($s = \delta_i$) of these rectangles, we see that $\tilde{H}$ is continuous on the rectangle $[0, \tau+\varepsilon]\times I$, with $\varepsilon = \min(\varepsilon_i)$, contrary to the assumed property of τ. Therefore, $\tau = 1$ and $\tilde{H}$ is continuous on $I\times I$. □

Analytic Continuation. Let us now apply this topological machinery to the classical problem of Weierstrass, namely, to the problem of an analytic continuation of germs of analytic functions (which he called *Funktionselement*). We first give the original, Weierstrass definition of an analytic continuation and reformulate it in the topological language later on (naturally, Weierstrass used the definition for domains in the plane rather than Riemann surfaces). We remind the reader that $\varrho_a(f)$ is the germ of an f at a point a.

DEFINITION (version 1). Let X be a Riemann surface, let γ: $[0, 1] \to X$ be a path, and let $a := \gamma(0)$, $b := \gamma(1)$. Let ψ be a germ of analytic functions at the point b (i.e., $\psi \in \mathscr{A}_b$) and φ a germ of analytic functions at a. We shall say that ψ is an *analytic continuation of φ along the path γ* if the following conditions are satisfied. There is a partition

$$0 = t_0 < t_1 < \dots < t_{n-1} < t_n = 1$$

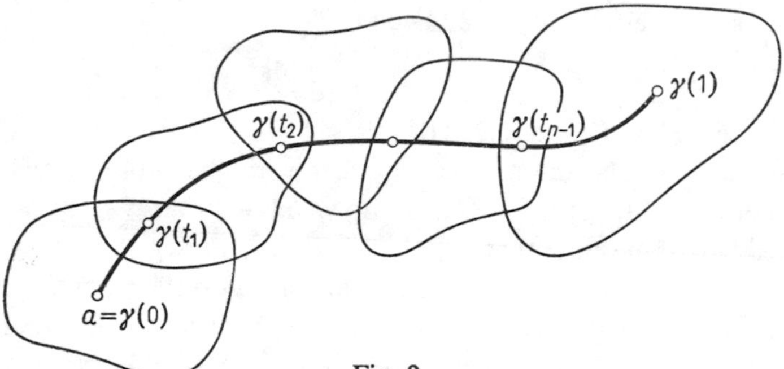

Fig. 9

of $[0, 1]$, a domains $\mathcal{O}_j \subset X$ such that $\gamma([t_{j-1}, t_j]) \subset \mathcal{O}_j$ and functions $f_j \in A(\mathcal{O}_j)$, $j = 1, \dots, n$, such that:

(I) $\varphi = \varrho_a(f_1)$ is the germ of f_1 at a, and $\psi = \varrho_b(f_n)$,

(II) $f_j|\mathscr{V}_j = f_{j+1}|\mathscr{V}_j$ for $j = 1, \dots, n-1$, where $\mathscr{V}_j$ is the connected component of $\mathcal{O}_j \cap \mathcal{O}_{j+1}$ which contains $\gamma(t_j)$.

Obviously, the preceding definition can be reformulated as

DEFINITION (version 2). There exists a family of germs of analytic functions $\varphi_t \in \mathscr{A}_{\gamma(t)}$, $t \in [0, 1]$ such that $\varphi_0 = \varphi, \varphi_1 = \psi$ and for each $s \in [0, 1]$ there exists a neighbourhood $V_s \subset [0, 1]$ of s and an open subset $\mathcal{O} \subset X$ such that $\gamma(V_s) \subset \mathcal{O}$ and an $f \in A(\mathcal{O})$ such that

$$\varrho_{\gamma(t)}(f) = \varphi_t \quad \text{for all } t \in V_s.$$

A third way of defining an analytic continuation is contained in the following

LEMMA XVI.7.11. Let X be a Riemann surface and let $p\colon |\mathscr{A}_X| \to X$ be the covering defined by the presheaf of analytic functions on X. Further, let $\gamma\colon I \to X$, be a path in X with end-points $a = \gamma(0)$, $b = \gamma(1)$. Then

$$\begin{aligned}&(\psi \in \mathscr{A}_b \text{ is an analytic continuation of } \varphi \in \mathscr{A}_a \text{ along } \gamma)\\ &\Leftrightarrow (\text{there exists a lift } \tilde{\gamma}\colon [0, 1] \to |\mathscr{A}_X| \text{ of } \gamma \text{ such}\\ &\text{that } \tilde{\gamma}(0) = \varphi \text{ and } \tilde{\gamma}(1) = \psi).\end{aligned}$$

PROOF. $\Leftarrow$: We set $\varphi_t := \tilde{\gamma}(t)$; then $\varphi_t \in \mathscr{A}_{\gamma(t)}$ and $\varphi_0 = \varphi, \varphi_1 = \psi$. Let $s \in [0, 1]$ and choose an open neighbourhood $[\mathcal{O}, f] \subset |\mathscr{A}_X|$ of $\tilde{\gamma}(s)$. Then there exists a neighbourhood $V_s \subset [0, 1]$ of s such that $\tilde{\gamma}(V_s) \subset [\mathcal{O}, f]$. Hence, $\tilde{\gamma}(V_s) \subset \mathcal{O}$ and $\varphi_t = \tilde{\gamma}(t) = \varrho_{\gamma(t)}(f)$.

$\Rightarrow$: Let $\varphi_t \in \mathscr{A}_{\gamma(t)}$, $t \subset [0, 1]$ be a family of germs as in the second version of the definition of an analytic continuation. It then follows from the construction of the topology of $|\mathscr{A}_X|$ that $\tilde{\gamma}\colon [0, 1] \to |\mathscr{A}_X|$ defined by

$$[0, 1] \ni t \to \varphi_t =: \tilde{\gamma}(t) \in |\mathscr{A}_X|$$

is a lift of γ and we have $\tilde{\gamma}(0) = \varphi_0 = \varphi$ and $\tilde{\gamma}(1) = \varphi_1 = \psi$. □

The preceding lemma, together with the fact that for a Riemann surface X the mapping $p\colon |\mathscr{A}_X| \to X$ is an unbranched covering (Theorems XVI.7.5 and XVI.7.1) makes it clear how to apply the general topological considerations to the problem of an analytic continuation. As a result we obtain Weyl's version of the Weierstrass monodromy principle.

THEOREM XVI.7.12 (Monodromy Theorem). Let X be a Riemann surface and let $\gamma_0, \gamma_1\colon [0, 1] \to X$ be two homotopic paths from a to b. Let $H(\cdot, s) = \gamma_s(\cdot)$ be a homotopy connecting γ_0 with γ_1 and $\varphi \in \mathscr{A}_a$ a germ which can be analytically continued along every path γ_s. Then both continuations of φ, along γ_0 and γ_1, produce the same germ $\psi \in \mathscr{A}_b$.

The following special case is very useful, and often called the monodromy theorem:

COROLLARY XVI.7.13 (Weierstrass' Monodromy Theorem). Let X be a Riemann surface and let $\varphi \in A_a$ be a germ (function element) which can be analytically continued along any path starting from a. Then there exists a unique $f \in A(X)$ such that $\varrho_a(f) = \varphi$.

Remark 1. This version of the monodromy theorem is of paramount importance: it gives sufficient conditions for a germ to grow into a full holomorphic function on the whole Riemann surface.

Remark 2. The idea of analytic continuation is vividely described by H. Weyl as follows: starting from a power series centred at $a \in \mathbf{C}$, "the plan is to conquer a largest possible domain of a complex plane. But the single-valuedness of the function is usually lost in the process of analytic continuation." One should remember that the concept of a Riemann surface was not known to Weierstrass or to anybody at that time. The unification of the precise notion of Weierstrass' analytic function (constructed by all possible analytic continuations of a function element) with the rich and pregnant but still not precise idea of a Riemann surface was the work of such giants as H. A. Schwarz, Klein, Poincaré, Schottky, and Koebe, and accomplished in H. Weyl's monumental *Idee der Riemannschen Fläche*. The unusually rich ideas of the masterpiece of young Hermann Weyl were difficult to absorb. They were made available to a wider mathematical public by the beautiful monograph *Uniformization* of the great Rolf Nevanlinna (1953) (cf. "A Short Historical Sketch" in Section 20).

Deck transformations. Given a covering $p: Y \to X$, the natural objects to consider are those homeomorphisms of Y which permute among themselves the points of $p^{-1}(\{x\})$ for each $x \in X$. They are known as deck transformations.

DEFINITION. Let X, Y be topological spaces and $p: Y \to X$ a covering. A homeomorphism

$$h: Y \to Y \quad \text{such that } p \circ h = p$$

is called a *deck transformation* or else *cover transformation*. Deck transformations are simply automorphisms of (Y, p, X). The set of all deck transformations forms a group denoted by Deck(Y/X) or, more precisely, Deck($Y \xrightarrow{p} X$). (This group is also called the *group of automorphisms of the covering* (Y, p, X).)

In most applications one deals with the so-called normal coverings, but let us first define another broader class of coverings.

DEFINITION. Let X, Y be topological spaces. A covering $p: Y \to Y$ is called *complete* (or *unbranched and unlimited*) if every $x \in X$ has an open neighbourhood V such that $p^{-1}(V) = \bigcup_{j \in J} U_j$, where $U_j, j \in J$ are disjoint open subsets of Y and all mappings $p|U_j \to V$ are topological. The reader should note that it is implicitly assumed that $p^{-1}(V)$ is not empty and, hence, that p is surjective.

Remark 1. Topological textbooks consider primarily complete coverings.

Remark 2. This terminology stems, of course, from the theory of analytic continuation: sometimes one encounters the following situation: one tries to continue analytically a germ φ along a path γ: $[0, 1] \to X$ but the continuation is possible only along the part $\gamma|[0, t_0[$ of γ but not up to t_0. Weyl says: "... before reaching the end of γ one bumps into the border of the covering surface over the point $\gamma(t_0)$."

Example. The map $p: R \to S^1$ defined as $p(t) = e^{2\pi i t}$ is a complete covering of S^1. The same is true of the map p_n: $S^1 \to S^1$ given by $p_n(z) = z^n$ (we regard elements of S^1 as complex numbers with modulus 1). However, the map p: $]-10, 10[\ni t \to e^{2\pi i t} \in S^1$ is an unbranched but not unlimited (i.e. uncomplete) covering.

DEFINITION. Let X, Y be connected Hausdorff spaces. A complete covering $p: Y \to X$ is called *normal* (or a *Galois covering* or else *regular*) if for each pair $y_0, y_1 \in Y$ such that $p(y_0) = p(y_1)$ there exists a deck transformation $h: Y \to Y$ such that $h(y_0) = y_1$.

We remind the reader that a group of homeomorphisms of a space X is said to be *transitive* if for any two points of X there is an element of the group sending one of these points upon another. Hence, the condition in the definition of a Galois covering can be phrased as "the group Deck(Y/X) is transitive on each fibre $p^{-1}(x)$". But there is more than the transitivity on fibres, since each deck transformation is a lift of the projection p, by the theorem on the uniqueness of lifts (Theorem XVI.7.7) there is only one deck transformation sending a given $y_0 \in Y$ upon a given y_1 or, equivalently, no deck transformation other than identity has a fixed point. This property together with transitivity is usually called *simple transitivity*. In descriptive terms the points of a given fibre are indistinguishable.

Example. For each $k \in \boldsymbol{N}^*$ the map p: $\boldsymbol{C}^* \to \boldsymbol{C}^*, p(z) := z^k$, is a Galois covering. Indeed: let $z_1, z_2 \in \boldsymbol{C}^*$ be such that $z_1^k = z_2^k$; then $z_2 = az_2$ where a is a k-th root of unity and $\boldsymbol{C}^* \ni z \to az \in \boldsymbol{C}^*$ is a deck transformation.

Exercise. Determine the group of deck transformations of the covering p: $\boldsymbol{R}^1 \ni t \to e^{2\pi it} \in S^1$. (Answer: $\mathrm{Deck}(\boldsymbol{R}^1/S^1) \cong \boldsymbol{Z}$.)

Remark. If p: $Y \to X$ is a Galois covering, then Weyl calls $\mathrm{Deck}(Y/X)$ "the *Galois group* of Y". This terminology is well founded: Deck $(Y \xrightarrow{p} X)$, for normal p, is in fact an analogue of the Galois group of a normal algebraic field over a base field. The normal coverings are in turn analogues of Galois field extensions. Cf. Chapter XX.1.

The term "normal" will be substantiated below in connection with the fundamental group $\pi_1(X)$ of X.

Among all connected coverings p: $Y \to X$ of a given space there is one naturally distinguished by the property that, somewhat imprecisely, all closed arcs in Y are homotopic to constant arcs. This property of a space is (together with certain additional assumptions, cf. the definition below) usually called the *simple connectivity* and a covering p: $Y \to X$ with simply connected Y is called the *universal covering*. The concept of a universal covering surface was discovered by Hermann Amandus Schwarz about 1880 and he communicated it to Klein in 1882. Klein wrote about it to Poincaré soon after and the notion of universal covering surface became the most important tool in the marvellous papers on the uniformization of Riemann surfaces by Poincaré, Klein, and Koebe.

We shall give several characterizations of this fundamental notion, but in order to have a deeper insight into the jungle of coverings of a manifold X it is necessary to study

The Fundamental Group $\pi_1(X)$. We have seen that in the theory of analytic continuation of a germ $\varrho_a(f)$ from $a \in X$ an important role is played by homotopy classes of closed paths starting from a. Poincaré made in 1883 the great discovery that the set of these homotopy classes forms a group with respect to an appropriately defined multiplication.

The idea of path multiplication is consecutively to follow one path and then the next one, but at "twice the speed". Of course we get a (continuous) path only if the terminal point of the first path to follow is the initial point of the next. In formal terms let α, β be such arcs (paths) in a space X that $\alpha(1) = \beta(0)$. The product of α with β (in that order) is a path denoted $\alpha\beta$ and defined by

$$\alpha\beta(t) = \begin{cases} \alpha(2t), & 0 \leqslant t \leqslant \frac{1}{2}, \\ \beta(2t-1), & \frac{1}{2} \leqslant t \leqslant 1. \end{cases}$$

The multiplication of paths so defined has none of the usual properties demanded of algebraic operations, e.g., it is not defined for all pairs of paths, it obviously lacks associativity, etc. But an idea of Poincaré, which greatly improves the situation is to consider homotopy classes of paths instead of the paths themselves. To denote that paths α and β in X are homotopic we use the notation $\alpha \sim \beta$ and we remind the reader that $\sim$ is an equivalence relation on the set of paths in X with common end-points. Now the following is true.

Lemma XVI.7.15. Let X be a topological space and α, β, γ be paths in X. If the product of paths is defined as above, then the following properties hold:

1° If $\alpha \sim \alpha_1$ and $\beta \sim \beta_1$ and if $\alpha(1) = \beta(0)$, then $\alpha\beta \sim \alpha_1\beta_1$.

2° The multiplication of paths is homotopy associative, i.e., if α, β, γ are such that the products $\alpha\beta$ and $\beta\gamma$ are defined, then both $(\alpha\beta)\gamma$ and $\alpha(\beta\gamma)$ are defined and $(\alpha\beta)\gamma \sim \alpha(\beta\gamma)$.

3° For a given point $x \in X$ let ε_x denote the constant path at x, i.e., $\varepsilon_x(t) \equiv x$. Then for all paths α starting at x (resp. terminating at x) $\varepsilon_x\alpha \sim \alpha$ (resp. $\alpha\varepsilon_x \sim \alpha$).

The proof of all this is straightforward but perhaps slightly tedious. We illustrate the method by proving 3° and leave the rest for the (interested) reader who, in the event of doubts, should consult any standard textbook on algebraic topology, e.g., W. Massey, *Algebraic Topology: An Introduction*, Harcourt, Brace & World, or a lovely textbook by Singer and Thorpe.

According to the definition the path $\varepsilon_x \alpha$ is parametrized as

$$\varepsilon_x \alpha(t) = \begin{cases} x, & 0 \leqslant t \leqslant \frac{1}{2}, \\ \alpha(2t-1), & \frac{1}{2} \leqslant t \leqslant 1. \end{cases}$$

A homotopy from $\varepsilon_x \alpha$ to α can be taken as the following

$$H(t, s) = \begin{cases} x, & 0 \leqslant t \leqslant \frac{1}{2}s, \\ \alpha\left(\dfrac{2t-s}{2-s}\right), & \frac{1}{2}s \leqslant t \leqslant 1. \end{cases}$$

Interpreting the formula by means of a picture of a suitably divided square $I \times I$ if necessary, the reader can convince himself that H has all the required properties. □

In particular, for any two closed paths α, β based at the same point (i.e., $\alpha(0) = \alpha(1) = \beta(0) = \beta(1)$) the products $\alpha\beta$ and $\beta\alpha$ are defined. Therefore, we restrict the multiplication of paths to the set of all closed paths based at a given point $x \in X$ and form the quotient set modulo the relation of homotopy $\sim$. Then, in view of point 1° of Lemma XVI.7.5 above the multiplication induces the multiplication of homotopy classes. We obtain

THEOREM XVI.7.16 (Poincaré). *Let X be a topological space and let $x \in X$. The set of homotopy classes of closed paths based at x (that is to say, paths $\gamma\colon I \to X$ such that $\gamma(0) = \gamma(1) = x$) is a group with the multiplication (induced from the multiplication of paths and) given by*

$$(*) \qquad [\gamma_1][\gamma_2] := [\gamma_1 \gamma_2].$$

The neutral element of the group is the class $[\varepsilon_x]$ of the constant path at x; it is usually denoted by 1.

For a class $[\gamma]$ the inverse $[\gamma]^{-1}$ is the class of a path γ^{-1} defined by $\gamma^{-1}(t) := \gamma(1-t)$, $t \in I$.

This group is denoted by $\pi_1(X, x)$ and is called the fundamental group *based at x or the* first homotopy group *of the pair (X, x).*

PROOF. The independence of the product in $(*)$ of the chosen representatives follows from 1° of Lemma XVI.7.15. Similarly, the associativity follows from 2° and the assertion that $[\varepsilon_x]$ is the neutral element is contained in 3° of the lemma. It remains to prove that

(i) if $\alpha \sim \beta$ then $\alpha^{-1} \sim \beta^{-1}$,

(ii) the paths $\gamma\gamma^{-1}$ and $\gamma^{-1}\gamma$ are both homotopic to the constant path ε_x, for any closed path γ based at x.

The easy verification of (i) is left for the reader; we merely sketch out the argument for (ii). Define H: $I\times I \to X$ by the formula

$$H(t, s) = \begin{cases} \gamma(2t), & 0 \leqslant t \leqslant \frac{1}{2}s, \\ \gamma(s), & \frac{1}{2}s \leqslant t \leqslant 1-\frac{1}{2}s, \\ \gamma^{-1}(2t-1), & 1-\frac{1}{2}s \leqslant t \leqslant 1, \end{cases}$$

it is readily verified that H is indeed a homotopy from ε_x to $\gamma\gamma^{-1}$. A homotopy from ε_x to $\gamma^{-1}\gamma$ is constructed in a similar manner. □

Remark. The fundamental group is, in general, noncommutative. Actual computation of the fundamental group for a given space is a highly difficult and intricate task. We shall determine fundamental groups of several simple spaces as we procede further.

DEFINITIONS. 1° A topological space X is said to be *arcwise connected* if for each pair x_1, x_2 of points from X there is an arc joining them. A subset $S \subset X$ is arcwise connected if the space S (induced topology) is such.

2° A topological space X is said to be *locally arcwise connected* if each point of X has a basis of neighbourhoods consisting of arcwise connected sets.

We state the following simple

PROPOSITION XVI.7.17. 1° An arcwise connected space is connected (but not conversely).

2° A locally arcwise connected and connected space is arcwise connected.

The proof is simple and can be left to the reader. As a consequence, we point out that since manifolds are obviously locally arcwise connected, all connected manifolds are arcwise connected.

The fundamental groups based at different points are not the same. However, they are all isomorphic for arcwise connected spaces, although not in an intrinsic way.

PROPOSITION XVI.7.18. If X is arcwise connected, then for any two points $a, b \in X$ the fundamental groups $\pi_1(X, a), \pi_1(X, b)$ are isomorphic. Therefore, we write simply $\pi_1(X)$ for $\pi_1(X, a)$.

PROOF. Let γ be a path connecting a and b ($\gamma(0) = a, \gamma(1) = b$). Then the mapping

$$f_\gamma\colon \pi_1(X, a) \to \pi_1(X, b),$$

defined by

$$f_\gamma([\alpha]) := [\gamma^{-1}\alpha\gamma]$$

is an isomorphism. Indeed, f_γ is well defined in view of Lemma XVI.7.15 and is a homomorphism since

$$\begin{aligned} f_\gamma([\alpha\beta]) &= [\gamma^{-1}\alpha\beta\gamma] = [\gamma^{-1}\alpha\gamma\gamma^{-1}\beta\gamma] \\ &= [\gamma^{-1}\alpha\gamma][\gamma^{-1}\beta\gamma] = f_\gamma([\alpha])f_\gamma([\beta]) \end{aligned}$$

again by virtue of the same lemma. Since the map $f_{\gamma-1}\colon \pi_1(X, b) \to \pi_1(X, a)$, $f_{\gamma-1}([\alpha]) = [\gamma\alpha\gamma^{-1}]$ is the inverse of f_γ, we see that f_γ (and also $f_{\gamma-1}$) is an isomorphism. □

The result of the preceding proposition can be supplemented by noting that the isomorphism f_γ depends only on the homotopy class of γ and that, in fact, f_γ and f_{γ_1} differ by a conjugation (an inner automorphism) of the group $\pi_1(X, b)$. On the other hand, for spaces which are not arcwise connected the fundamental groups attached to different points may be totally different.

DEFINITION. A topological arcwise connected space X is said to be *simply connected* if every closed path is homotopic to a constant path. Therefore,

$$(X \text{ is simply connected}) \Leftrightarrow (\pi_1(X) = \{1\}),$$

i.e., the fundamental group at any point has only one element. The pair (X, a), where $a \in X$, is called a *pointed space* and the distinguished point a, the *base point*. The correspondence (functor in the language of category theory) which assigns the fundamental group to a pointed space is an important tool in the study of various topological problems. The natural behaviour of this correspondence under the action of continuous mappings makes it even more important. With some deliberate inaccuracy we shall denote this correspondence by

$$\pi_1\colon \{\text{pointed topological spaces}\} \to \{\text{groups}\}.$$

LEMMA XVI.7.19. Let $f\colon Y \to X$ be a continuous map. Then

1° (the paths $\gamma_1, \gamma_2\colon I \to Y$ are homotopic) $\Rightarrow$ ($f \circ \gamma_1, f \circ \gamma_2$ are homotopic).

2° The map

$$f_*: \pi_1(Y, y) \to \pi_1(X, f(y))$$

induced by virtue of 1° by the formula

$$f_*([\gamma]) = [f \circ \gamma]$$

is a homomorphism, i.e.,

$$f_*([\gamma_1][\gamma_2]) = f_*([\gamma_1])f_*([\gamma_2]).$$

3° For any continuous map $g: X \to Z$

$$(g \circ f)_* = g_* \circ f_* \quad (\text{"covariance of } \pi_1\text{"}).$$

4° If $\mathrm{id}: X \to X$ is the identity, then id_* is the identity map for every group $\pi_1(X, x)$.

The proof is a straightforward verification and is left to the reader. □

We shall see later that for a complete covering $p: Y \to X$ the induced map p_* is injective.

Remark. The preceding lemma allows us to construct a subgroup $\Gamma \subset \pi_1(X)$ for every covering $(\tilde{X}, p, X)$ of X as follows: $\Gamma := p_*\pi_1(\tilde{X}, x)$. The "map" $G := p_* \circ \pi_1$ will be called the *Galois map* since it is an analogue of the Galois map from the Galois theory of field extensions (cf. Theorem XVI.7.20). A natural question arises as to whether for each subgroup $\Gamma \subset \pi_1(X, x)$ it is possible to construct a covering $(\tilde{X}_\Gamma, p, X)$ of X such that $p_*\pi_1(\tilde{X}_\Gamma, \tilde{x}) = \Gamma$, where $p(\tilde{x}) = x$. A positive answer (and the construction) is given in the proof of Proposition XVI.7.20. We call the "map" $P: \Gamma \to X_\Gamma$ the *Poincaré map*. Such a map "projects" the equivalence classes of corresponding objects (conjugate classes of subgroups resp. isomorphy classes of coverings) upon each other and gives the desired isomorphism of corresponding objects.

We recall some standard algebraic terminology. Two subgroups G_1, G_2 of a group G are conjugate if there exists such a $g \in G$ that $G_1 = g^{-1}G_2 g$.

The *normalizer* $N(S)$ of a subset $S \subset G$ is defined as follows: $N(S) = \{g \in G: gSg^{-1} = S\}$ and, thus, $N(\{e\}) = G$. The set $Z(S) = \{x \in G: xsx^{-1} = s \text{ for every } s \in S\}$ is called the *centralizer* of S. Both $N(S)$ and $Z(S)$ are subgroups of G which, in general, are distinct.

$Z(G)$ is called the *centre of the group G*.

If $N(S) = S$, then S is a subgroup of G and is called the *normal subgroup*. Therefore, a subgroup S is normal if $S = gSg^{-1}$ for all $g \in G$, i.e., $Sg = gS$ for all $g \in G$. We now come to the formulation and proof of a fundamental duality, discovered by Poincaré, between subgroups of the fundamental group $\pi_1(X)$ and the coverings $p\colon \tilde{X} \to X$ of a given space. Since we are not striving for the greatest possible generality (and with a view to later applications to the analysis on Riemann surfaces), we shall be concerned mostly with the case of (topological) manifolds, that is, Hausdorff spaces in which every point has an open neighbourhood homeomorphic with an open set in a Euclidean space $\boldsymbol{R}^n$. As remarked before, topological manifolds are locally arcwise connected, and connected manifolds are arcwise connected. Moreover, if X is a manifold and $p\colon Y \to X$ is an unbranched covering, then Y is also a manifold and a structure of holomorphic (differentiable) manifold on X is naturally carried over (lifted) to Y (cf. Proposition XVI.7.3). To avoid repetitions of cumbersome adjectives, all coverings will be assumed connected, that is, the phrase: "$p\colon Y \to X$ is a covering" contains the statement that both X and Y are connected unless the contrary is explicitly stated.

The proof and the consequences of the following theorem will occupy the rest of this section.

THEOREM XVI.7.20 (The Theory of Poincaré). Suppose that X is a connected manifold and $a \in X$.

1° For every subgroup $\Gamma \subset \pi_1(X, a)$ there is a complete covering $p_\Gamma = p\colon \tilde{X}_\Gamma \to X$ such that Γ is isomorphic with the fundamental group $\pi_1(\tilde{X}, \tilde{a})$ for $\tilde{a} \in p^{-1}(a)$.

2° The correspondence in 1° induces bijective correspondence between the classes of conjugate subgroups $\Gamma \subset \pi_1(X, a)$ and the isomorphy classes of complete covering $(\tilde{X}, p, X)$ of X.

3° We say that a covering $p_2\colon \tilde{X}_2 \to X$ is *stronger* than $p_1\colon \tilde{X}_1 \to X$ if there exists a mapping $p_{21}\colon \tilde{X}_2 \to \tilde{X}_1$ such that $p_2 = p_1 \circ p_{21}$. This relation defines a partial order in the set of (isomorphy classes of) coverings of X, and is denoted $\tilde{X}_1 \prec \tilde{X}_2$.

The natural partial ordering (inclusion) between (conjugacy classes of) subgroups of $\pi_1(X, a)$ is reversed by this correspondence, i.e., if Γ_k corresponds to $\tilde{X}_k$, $k = 1, 2$, then

$$(\Gamma_1 \subset \Gamma_2) \Leftrightarrow (\tilde{X}_1 \succ \tilde{X}_2).$$

4° The two extreme cases:

$$\text{I}: \Gamma = \{1\}, \qquad \text{II}: \Gamma = \pi_1(X, a)$$

correspond to the so-called *universal covering* (case I) $p: \hat{X} \to X$ and to the X itself (case II).

5° A normal (Galois) covering corresponds to any normal subgroup Γ of $\pi_1(X, a)$.

6° For the covering $p: \tilde{X}_\Gamma \to X$ associated with Γ the group Deck $(\tilde{X}_\Gamma/X)$ is isomorphic to the quotient group $N(\Gamma)/\Gamma$; here, Γ is a group conjugated to $p_*\pi_1(\tilde{X}_\Gamma, a)$ and $N(\Gamma)$ is the normalizer of Γ in $\pi_1(X, a)$.

7° $\text{Deck}(\hat{X}/X) \cong \pi_1(X)$, where $\hat{X} \to X$ is the universal covering.

Remark 1. 4° defines the universal covering $(\hat{X}, p, X)$ of X as a maximal one. This maximal covering does indeed possess the following universal property:

8° For every complete covering $q: Z \to X$ and for every two points $\hat{x}_0 \in \hat{X}$, $z_0 \in Z$ with $p(\hat{x}_0) = q(z_0)$ there exists one and only one continuous map $f: \hat{X} \to Z$ such that $p = qf$ and $f(\hat{x}_0) = z_0$.

Remark 2. We have tried to formulate the theorem in such a way that the amazing analogy between the Poincaré theory of covering manifolds and the Galois theory of extension of algebraic fields should catch the eye: We have the following correspondences:

Poincaré theory	Galois theory (field extensions)
manifold X	field k
covering of X	extension of k
Galois (normal) covering	Galois extension
$\text{Deck}(\hat{X}/X)$	Galois group $\text{Gal}(K/k)$
($p: \hat{X}_\Gamma \to X$ is normal)	(K is a normal extension of k)
$\Leftrightarrow$(Γ is normal subgroup)	$\Leftrightarrow$ Gal (K/k) is a normal subgroup of G
fundamental group $\pi_1(X)$	Galois group G
Poincaré map	Galois map

We have named the content of Theorem XVI.7.20 the Poincaré theory as a topological counterpart to the Galois theory. Now we proceed to the proof. But since 1°–8° are of general topological interest, we shall subdivide the proof into several lemmata and theorems which are interesting in themselves.

We have proved above two important theorems about liftings for an unbranched covering $p\colon Y \to X$:

1° The Unique Lifting Theorem (XVI.7.7),

2° The Monodromy Principle (XVI.7.10) about the lifting of homotopic paths: $H(\cdot, s) := \gamma_s\colon I \to X, s \in I$, where it was assumed that each path γ_s had a lift $\tilde{\gamma}_s$ from the same point over $\gamma_s(0)$. The conclusion was that the homotopy H had a continuous lift

$$\tilde{H}\colon I \to Y.$$

The proof of the monodromy principle is also valid in a more general situation: as in 1°, we consider a family of maps $\gamma_s\colon Z \to X, s \in I$ of an arbitrary compact and connected space Z for which the lifts $\tilde{\gamma}_s$ exist

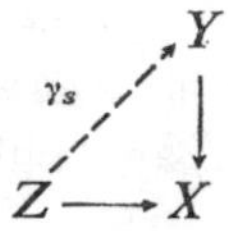

and we obtain the existence of the continuous lift $\tilde{H}$ for the "homotopy" $H(\cdot, s) = \gamma_s(\cdot)$

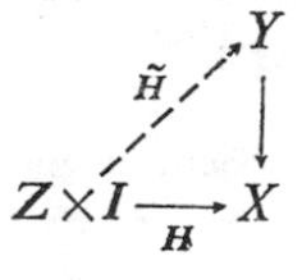

This generalized form of the Monodromy Principle is called the *Covering Homotopy Theorem.*

In both cases 1° and 2°, it was assumed that the respective lifts $\tilde{\gamma}$ and $\tilde{\gamma}_s$ do exist. The natural question is when a covering $p\colon Y \to X$ has the "path-lifting property", i.e., when for every path $\gamma\colon I \to X$ and every point $\tilde{y}_0 \in p^{-1}(\gamma(0))$ does there exist a lift $\tilde{\gamma}\colon I \to Y$ such that $\tilde{\gamma}(0) = \tilde{y}_0$.

The following theorem gives the answer:

THEOREM XVI.7.21. Every complete (i.e., unbranched and unlimited) covering $p\colon Y \to X$ has the path-lifting property.

PROOF. Let γ: $[0, 1] \to X$ be a path and $y_0 \in Y$ such that $p(y_0) = \gamma(0)$. By the compactness of $[0, 1]$ there exists a partition of $[0, 1]$

$$0 = t_0 < t_1 < \dots < t_n = 1$$

and open subsets $\mathcal{O}_k \subset X, k = 1, \dots, n$ such that

(a) $\gamma([t_{k-1}, t_k]) \subset \mathcal{O}_k$,

(b) $p^{-1}(\mathcal{O}_k) = \bigcup_{j \in J_k} V_{kj}$,

where $V_{kj} \subset Y$ are open and disjoint for $j \in J_k$ and $p|V_{kj} \to \mathcal{O}_k$ are homeomorphisms. We prove the existence of a lift $\tilde{\gamma}|[0, t_k] \to X, \tilde{\gamma}(0) = y_0$ by induction with respect to $k = 0, 1, \dots, n$: For $k = 0$ this is obvious. Let $\tilde{\gamma}|[0, t_{k-1}] \to Y$ be constructed and let $y_{k-1} := \tilde{\gamma}(t_{k-1})$. Since $p(y_{k-1}) = \gamma(t_{k-1}) \in \mathcal{O}_k$, there exists a $j \in J_k$ such that $y_{k-1} \in V_{kj}$. We set

$$\tilde{\gamma}|[t_{k-1}, t_k] := (p|V_{kj})^{-1} \circ \gamma|[t_{k-1}, t_k].$$

In this way we have constructed a continuous extension of the lift $\tilde{\gamma}$ onto $[0, t_k]$. □

Remark. The same proof works in the general case: we can replace the interval I by a compact connected space Z.

COROLLARY XVI.7.22. Let X, Y be Hausdorff arcwise connected spaces p: $Y \to X$ a complete covering and $x_1, x_2 \in X$. Then

1° The fibres $p^{-1}(x_1)$ and $p^{-1}(x_2)$ have the same cardinality.

2° If $p^{-1}(x)$ is a single point for some $x \in X$, then p is a homeomorphism.

DEFINITION. The cardinality of a fibre $p^{-1}(x)$ is called the *multiplicity of the covering* or the *number of sheets of the covering p*.

Proof. 1° Let γ: $[0, 1] \to X$ be a path connecting x_1 with x_2, let $y \in p^{-1}(x_1)$, and let $\tilde{\gamma}$ be the unique lift of γ such that $\tilde{\gamma}(0) = p^{-1}(x_1)$. Let us denote

$$h(y) := \tilde{\gamma}(1) \in p^{-1}(x_2).$$

From the unique lifting theorem it follows that h: $p^{-1}(x_2) \to p^{-1}(x_2)$ is bijective. □

The following lemma is, in a sense, a converse to Theorem XVI.7.21:

LEMMA XVI.7.23. Let X be a manifold, let Y be a Hausdorff space, and let $p\colon\ Y \to X$ be an unbranched covering with the path-lifting property. Then p is complete.

PROOF. Let $x_0 \in X$ and let $\{y_j\colon \in J\}$ be the fibre $p^{-1}(x_0)$. Let $\mathcal{O}$ be an open neighbourhood of x_0 homeomorphic to a ball (of the model of X) and let $\varkappa\colon\ \mathcal{O} \to X$ be the canonical injection. We can prove (cf. the exercise below) that for each $j \in J$ there exists a lift $\tilde{\varkappa}_j\colon\ \mathcal{O} \to Y$ of $\varkappa$ such that $\tilde{\varkappa}_j(x_0) = y_j$. Let $V_j := \tilde{\varkappa}_j(\mathcal{O})$. We also check that $p^{-1}(\mathcal{O}) = \bigcup_{j\in J} V_j$, where the V_j are disjoint open subsets and each $p|V_j \to \mathcal{O}$ is a homeomorphism.

Exercise. Complete the preceding proof: Let $x \in \mathcal{O}$ and let $\gamma\colon\ [0, 1] \to \mathcal{O}$ be a path from x_0 to x. Take the lift $\tilde{\gamma}_j$ of γ such that $\tilde{\gamma}_j(0) = y_j$ and set

$$(*) \qquad \tilde{\varkappa}_j(x) := \tilde{\gamma}_j(1).$$

Verify that $(*)$ is independent of the choice of the path γ from x_0 to x. Now, prove the remaining assertions given in the proof of the lemma.

The path-lifting property leads to the following fundamental result:

PROPOSITION XVI.7.24. Suppose that $p\colon\ \tilde{X} \to X$ is a complete covering and let $x \in X$ and $\tilde{x} \in p^{-1}(\{x\})$. Then the homomorphism $p_*\colon\ \pi_1(\tilde{X}, \tilde{x}) \to \pi_1(X, x)$ induced by the projection p is injective.

PROOF. This is an immediate consequence of the path-lifting property of complete coverings and the monodromy principle.

The proposition above allows us to identify the fundamental group of the covering space $\tilde{X}$ with a subgroup (the image under p_*) of the fundamental group of X. The theorem to be proved now shows that every subgroup of $\pi_1(X, x)$ can be obtained that way:

THEOREM XVI.7.25 (Construction of $\tilde{X}_\Gamma$ and the Poincaré map). Suppose that X is a connected manifold and let $x_0 \in X$. Then for every subgroup $\Gamma \subset \pi_1(X, x_0)$ there exist a complete covering manifold $p\colon\ \tilde{X}_\Gamma \to X$ and a point $\tilde{x}_0 \in p^{-1}(\{x_0\})$ such that

$$p_*\colon\ \pi_1(\tilde{X}_\Gamma, \tilde{x}_0) \to \Gamma$$

is an isomorphism of groups.

PROOF. Let A be the set of all paths in X which start at x_0. For $\alpha, \beta \in A$ write $\alpha \underset{\Gamma}{\sim} \beta$ if $\alpha(1) = \beta(1)$ and $[\alpha\beta^{-1}] \in \Gamma$. Plainly, $\underset{\Gamma}{\sim}$ is an equivalence relation because Γ is a group. Let

$$\tilde{X}_\Gamma := A/\underset{\Gamma}{\sim}.$$

If $\alpha \in A$, then $[\alpha]_\Gamma$ denotes the class of α with respect to the relation $\underset{\Gamma}{\sim}$ and we define

$$p = p_\Gamma\colon \tilde{X}_\Gamma \to X \quad \text{by } p([\alpha]_\Gamma) := \alpha(1).$$

Since X is arcwise connected, $p(\tilde{X}_\Gamma) = X$. We now construct a basis for a topology on $\tilde{X}_\Gamma$. (It is the weakest topology for which p is a local homeomorphism.) Let V be a neighbourhood of $\alpha(1) \in X$; then $V_\alpha = \{[\beta]_\Gamma \in \tilde{X}_\Gamma\colon$ there exists a path γ in V with $\alpha\gamma \sim \beta\}$ ($\sim$ means homotopy of paths, as always).

$$\mathscr{B} := \{V_\alpha\colon \alpha \in A \text{ and } V \text{ ranges over a basis of neighbourhoods of } \alpha(1)\}.$$

We check that $\mathscr{B}$ is a basis of neighbourhoods of $[\alpha]_\Gamma$. Indeed, let $[\gamma]_\Gamma \in V_\alpha \cap U_\beta$; then $\gamma(1) \in V \cap U$; now, choosing an open, simply connected neighbourhood $W \subset V \cap U$ of $\gamma(1)$ we check that

$$[\gamma]_\Gamma \in W_\gamma \subset V_\alpha \cap U_\beta.$$

Hence, $\mathscr{B}$ is a basis of neighbourhoods of $[\alpha]_\Gamma$ for a topology on $\tilde{X}_\Gamma$. When $\tilde{X}_\Gamma$ is endowed with this topology the mapping $p\colon \tilde{X}_\Gamma \to X$ is locally topological since $p|V_\alpha \to V$ is a homeomorphism.

$(\tilde{X}_\Gamma, \mathscr{B})$ is a Hausdorff space, Indeed, it is sufficient to verify that for two paths $\alpha, \beta \in A$, such that $\alpha(1) = \beta(1)$ but $[\alpha]_\Gamma \neq [\beta]_\Gamma$, the points $[\alpha]_\Gamma$ and $[\beta]_\Gamma$ have open disjoint neighbourhoods. We choose an open, simply connected neighbourhood V of the point $\alpha(1) = \beta(1)$ and let V_α and V_β be the neighbourhoods of $[\alpha]_\Gamma$ and $[\beta]_\Gamma$, respectively, constructed as before. We claim that $V_\alpha \cap V_\beta = \emptyset$, for otherwise for $[\gamma]_\Gamma \in V_\alpha \cap V_\beta$ there would exist a path δ in V, joining $\alpha(1)$ with $\gamma(1)$ and such that

$$\alpha\delta \sim \gamma \sim \beta\delta.$$

Hence,

$$[\alpha\delta]_\Gamma = [\beta\delta]_\Gamma, \quad \text{i.e.} \quad \Gamma \ni [\alpha\delta(\beta\delta)^{-1}] = [\alpha\beta^{-1}]$$

and, therefore, $[\alpha]_\Gamma = [\beta]_\Gamma$, which is a contradiction.

We now prove that the space $\tilde{X}_\Gamma$ is arcwise connected and $p\colon \tilde{X}_\Gamma \to X$ is a complete covering. For any arc γ in X, starting at x_0, we denote $\gamma_s(t)$

$:= \gamma(st)$, $s \in R$. If α is any closed path based at x_0, then the map $\tilde{\gamma}$: $[0, 1] \to \tilde{X}_\Gamma$ defined by $\tilde{\gamma}(t) := [\alpha\gamma_t]_\Gamma$ is continuous and is a lift of γ such that $\tilde{\gamma}(0) = [\alpha]_\Gamma$. Therefore, p: $\tilde{X}_\Gamma \to X$ has the path-lifting property. Thus, by Lemma XVI.7.21, p: $\tilde{X}_\Gamma \to X$ is a complete covering. The arcwise connectivity of $\tilde{X}_\Gamma$ also follows, since, if α is taken to be (homotopic to) the constant path ε_{x_0}, the $\tilde{\gamma}$ constructed above joins $[\alpha]_\Gamma$ with $[\gamma]_\Gamma$.

Let σ be a (closed) path in $\tilde{X}_\Gamma$ based at $\tilde{x} := [\varepsilon_{x_0}]_\Gamma$. Then, by the definition of $(p_\Gamma)_* = p_*$ we have $p_*\pi_1(\tilde{X}_\Gamma, \tilde{x}) = \Gamma$ (we recall that $p_*[\sigma] := [p \circ \sigma]$). □

Corollary XVI.7.26. Every connected manifold has a simply connected (universal) covering manifold.

This is the covering constructed in the Theorem XVI.7.25 corresponding to the subgroup $\Gamma = \{1\}$ of $\pi_1(X, x_0)$. We shall later show that a simply connected covering is unique up to an isomorphism and that it has the property of universality, hence, is rightly called the universal covering.

We now determine to what extent the group $p_*\pi_1(\tilde{X}, \tilde{x}) = \Gamma(\tilde{x})$, the image of the fundamental group of a complete covering p: $\tilde{X} \to X$, depends on the choice of the point $\tilde{x}$ from the fibre $p^{-1}(x)$, with $x = p(\tilde{x})$. The simple answer is given by

Proposition XVI.7.27. Let p: $Y \to X$ be a complete covering manifold, $y \in Y$, $x = p(y)$. Then $\Gamma(y') := p_*\pi_1(Y, y')$ ranges over the set of all conjugates of $\Gamma(y)$ in the group $\pi_1(X, x)$ as y' ranges over the fibre $p^{-1}(x)$.

Proof. Given $y' \in p^{-1}(x)$, we choose a path $\tilde{\sigma}$ from y to y'. Its projection $\sigma := p \circ \tilde{\sigma}$ is a closed arc based at x. If γ is a closed arc from x, then ($\sigma\gamma\sigma^{-1}$ is lifted to a closed arc from y) $\Leftrightarrow$ (γ is lifted to a closed arc from y').

Therefore,

$$(*) \qquad \Gamma(y') = [\sigma]^{-1}\Gamma(y)[\sigma].$$

Conversely, every conjugate group of $\Gamma(y)$ can be obtained in this manner: Take the lift $\tilde{\sigma}$ of σ such that $\tilde{\sigma}(0) = y$. □

Remark. If by σ_* we denote the conjugation (the inner automorphism) of $\pi_1(X, x)$ defined by σ in terms of $(*)$ and if by $\tilde{\sigma}_*$ we denote the map

$\tilde{\sigma}_*: \pi_1(Y, y) \ni [\gamma] \to [\tilde{\sigma}^{-1}\gamma\tilde{\sigma}] \in \pi_1(Y, y')$ (cf. Proposition XVI.7.18), then we have the following commutative diagram:

$$\begin{array}{ccc} \pi_1(Y, y) & \xrightarrow{\tilde{\sigma}_*} & \pi_1(Y, y') \\ \downarrow{\scriptstyle p_*} & & \downarrow{\scriptstyle p_*} \\ \pi_1(X, x) & \xrightarrow[\sigma_* = (p\tilde{\sigma})]{} & \pi_1(X, x) \end{array}$$

We recall that, given two coverings $p_i: \tilde{X}_i \to X$, $i = 1, 2$, a continuous map $h: \tilde{X}_1 \to \tilde{X}_2$ is called a *homomorphism of coverings* if $p_2 \circ h = p_1$. It is an isomorphism (equivalence) if, moreover, it is a homeomorphism of $\tilde{X}_1$ with $\tilde{X}_2$. Note that then h^{-1} is also an isomorphism. We shall now investigate the question of the existence of homomorphisms of coverings. The key result is the following

THEOREM XVI.7.28. Suppose that $p_i: \tilde{X}_i \to X$ for $i = 1, 2$ are complete coverings of a connected manifold X and $\tilde{x}_i \in \tilde{X}_i$, $i = 1, 2$ and have the same projections, i.e., $p_1(\tilde{x}_1) = p_2(\tilde{x}_2)$.

Then a homomorphism $h: \tilde{X}_1 \to \tilde{X}_2$ such that $h(\tilde{x}_1) = \tilde{x}_2$ exists iff

$$(*) \qquad p_{1*}\pi_1(\tilde{X}_1, \tilde{x}_1) \subset p_{2*}\pi_1(\tilde{X}_2, \tilde{x}_2).$$

PROOF. The necessity is clear if we consider the equality $p_{2*} \circ h_* = p_{1*}$. We prove the existence of such h. Given a point $\tilde{y} \in \tilde{X}_1$, take a path $\tilde{\gamma}_1$ from $\tilde{x}_1$ to $\tilde{y}$, project it by means of p_1 onto the path $\gamma = p_1 \circ \tilde{\gamma}_1$ in X and then lift the path γ to the (unique) path $\tilde{\gamma}_2$ starting at $\tilde{x}_2$. We would like to define the image of the point y under h to be the terminal point of $\tilde{\gamma}_2$, but to do so we must check the consistency of the proposed definition, i.e., that the point obtained is independent of the choice of $\tilde{\gamma}_1$. Let $\tilde{\beta}_1$ be another path in $\tilde{X}$ from $\tilde{x}_1$ to $\tilde{y}$; then $\tilde{\gamma}_1\tilde{\beta}_1^{-1}$ is a closed path based at $\tilde{x}_1$ and from the assumption $(*)$ it follows that the lift $\tilde{\gamma}_2\tilde{\beta}_2^{-1}$ of $\gamma\beta^{-1}$ (where $\beta = p_1 \circ \tilde{\beta}_1$ and $\tilde{\beta}_2$ is the lift of β starting at $\tilde{x}_2$) also is a closed path. Thus $\tilde{\beta}_2(1) = \tilde{\gamma}_2(1)$, and that allows us to define

$$h(\tilde{y}) = \tilde{\gamma}_2(1) \quad \text{for } \tilde{y} \in \tilde{X}_1$$

where $\tilde{\gamma}_2$ is the path constructed above.

To prove the continuity of h we follow a standard argument using the local arcwise connectivity of X and X_i, $i = 1, 2$. Let U be an open neighbourhood of $h(\tilde{y}) \in \tilde{X}_2$; we can assume that U is arcwise connected and $p_2|U$ is a homeomorphism since such neighbourhoods form a basis at each point of $\tilde{X}_2$. We now choose an open neighbourhood W of $\tilde{y}$ in $\tilde{X}_1$ with the same properties as U and finally choose a neighbourhood $V \subset W$ (open, arcwise connected) such that $p_1(V) \subset p_2(U)$. Then it is easily seen that $h(V) \subset U$. □

As a corollary we obtain the previously mentioned universality property of a simply connected covering.

COROLLARY XVI.7.29. A simply connected covering $p: \hat{X} \to X$ of a connected manifold X (which exists by Theorem XVI.7.25) is a covering of any covering $p_1: \tilde{X} \to X$. More precisely, any homomorphism $h: \hat{X} \to \tilde{X}$ is a covering of $\tilde{X}$.

We leave the verification of this to the reader. A natural question about the condition for equivalence of coverings is also obtained with the aid of Theorem XVI.7.20. It is readily seen that equivalent coverings lead to conjugate subgroups $\Gamma_i \subset \pi_1(X, x)$, $i = 1, 2$. But the converse is also true:

THEOREM XVI.7.30. Let X be a connected manifold. Then (complete coverings $p_i: Y_i \to X$, $i = 1, 2$ are equivalent) $\Leftrightarrow$ ($p_{i*}\pi_1(Y_i, y_i)$ are conjugate subgroups of $\pi_1(X, x)$) where $y_i \in p_i^{-1}(x)$.

PROOF. $\Rightarrow$: Let $h: Y_1 \to Y_2$ and $p_2 \circ h = p_1$ let $y' := h(y_1)$; then $h_*: \pi_1(Y_1, y_1) \to \pi_1(Y_2, y_2')$ is an isomorphism and $p_{1*}\pi_1(Y_1, y_1) = p_{2*}h_*\pi_1(Y_1, y_1) = p_{2*}\pi_1(Y_2, y')$ is conjugate to $p_{2*}\pi_1(Y_2, y_2)$ by Proposition XVI.7.27.

$\Leftarrow$: immediately follows from Proposition XVI.7.27 and Theorem XVI.7.20. □

The last point of the Poincaré theory is the relation between the group $\mathrm{Deck}(\tilde{X}_\Gamma \xrightarrow{p} X)$ of deck transformations and the normalizer $N(\Gamma)$ of Γ in $\pi_1(X, x)$. Let us recall that a deck transformation of a covering $p: Y \to X$ is a homeomorphism $h: Y \to Y$ such that $p \circ h = p$. Hence a deck transformation sends fibres of p into fibres. We now prove an important property of these transformations.

THEOREM XVI.7.31. Let $p: Y \to X$ be a complete covering. Then:

1° A deck transformation other than identity has no fixed points. Hence the group $\mathrm{Deck}(Y/X)$ acts freely, i.e., without fixed points on Y.

2° Moreover, every point $y \in Y$ has a neighbourhood V such that

$$\{h \in \mathrm{Deck}(Y/X)\colon\ h(V) \text{ meets } V \text{ is finite}\}.$$

Remark 1. If a transformation group $\Gamma \subset \mathrm{Aut}(Y)$ of a topological space Y has property 2°, then Γ is said to *act properly discontinuously on* Y. Hence the content of the theorem is that the group $\mathrm{Deck}(Y/X)$ acts freely and properly discontinuously on the covering manifold Y of X.

Remark 2. In Section 17 (Theorems XVI.17.8–XVI.17.12) we shall prove important properties of properly discontinuous transformation groups.

PROOF OF THEOREM XVI.7.31. The assertion in 1° has been already seen to follow immediately from the theorem on the uniqueness of lifts.

2° If $h \neq \mathrm{id}_Y$, then $V \cap h(V) = \emptyset$. Indeed, if $y' \in V \cap h(V)$, then $y' = h(y)$ with $y, y' \in V$; hence $p(y') = p(h(y)) = p(y)$, which is possible only if $y' = y$; but then y' is a fixed point of h, and so (by 1°) $h = \mathrm{id}_Y$ (a contradiction). □

THEOREM XVI.7.32. Let X be a connected manifold and let $p\colon Y \to X$ be a complete covering. If $x \in X$, $y \in p^{-1}(x)$, let $\Gamma := p_*\pi_1(Y, y)$.

1° Deck $(Y \xrightarrow{p} X) = N(\Gamma)/\Gamma$ $(= \Gamma\backslash N(\Gamma))$, where $N(\Gamma)$ is the normalizer of Γ in $\pi_1(X, x)$.

2° If Γ is the normal subgroup, i.e., if $N(\Gamma) = \pi_1(X, x)$, then Deck $(Y/X) \cong \pi_1(X, x)/\Gamma$.

In this case $p\colon Y \to X$ is a Galois covering and $\mathrm{Deck}(Y/X)$ simply acts transitively on every $p^{-1}(x)$.

3° If $p\colon \hat{X} \to X$ is the universal covering of X, then

$$\mathrm{Deck}(\hat{X}/X) = \pi_1(X, x) \cong \pi_1(X).$$

The universal covering is a Galois covering.

Remark. 3° points out that we can consider the fundamental group $\pi_1(X)$ of the connected manifold as the group of deck transformations of the universal covering $p\colon \hat{X} \to X$. The Poincaré group is, therefore, a close analogue of the Galois group of a field extension.

PROOF. 1° Recall that

$$N(\Gamma) = \{g \in \pi_1(X, x)\colon g\Gamma = \Gamma g\}.$$

For each closed path γ based at x such that $[\gamma] \in N(\Gamma)$ we define a mapping $h_\gamma\colon Y \to Y$ as follows: join $y \in p^{-1}(x)$ to y' by a path $\tilde{\sigma}$, if $\sigma = p \circ \tilde{\sigma}$

set $h(y') :=$ end-point of the lift $(\gamma\sigma)^\sim$ of $\gamma\sigma$. This definition is independent of the choice of σ. Replace $\tilde{\sigma}$ by $\tilde{\sigma}_1$; then $[\sigma\sigma_1^{-1}] \in \Gamma$ and, hence, $[\gamma\sigma\sigma_1^{-1}\gamma^{-1}] \in \Gamma$. Thus, $(\gamma\sigma_1)^\sim$ has the same end-point as $(\gamma\sigma)^\sim$. Moreover, since $p \circ h_\gamma(y') = \gamma\sigma(1) = \sigma(1) = p\tilde{\sigma}(1) = p(y')$, the mapping h_γ is a (continuous) deck transformation. We check that

$$h_\gamma \circ h_{\gamma_1} = h_{\gamma\gamma_1}$$

and

$$(h_\gamma = \mathrm{id}_Y) \Leftrightarrow ([\gamma] \in \Gamma).$$

Hence, we have an injective homomorphism μ:

$$N(\Gamma)/\Gamma \ni \gamma \xrightarrow{\mu} h_\gamma \in \mathrm{Deck}(Y/X).$$

We ought to prove that μ is surjective. Let $h \in \mathrm{Deck}(Y/X)$ and let $\tilde{\gamma}$ be a path connecting y with $h(y) =: y'$ such that $p \circ \tilde{\gamma} =: \gamma$; then $h_\gamma(y) = h(y)$. Hence, $h_\gamma \circ h^{-1}$ has y as the fixed point and so by Theorem XVI.7.31. 1°, $h_\gamma \circ h^{-1} = \mathrm{id}_Y$, i.e., $h_\gamma = h$. Since 3° follows from 2°, we are left with the proof of 2°.

2° If $p_*\pi_1(Y, y) = \Gamma$ is a normal subgroup of $\pi_1(X, x)$ for some (hence, every) $y \in Y$ with $x = p(y)$, then $N(\Gamma) = \pi_1(X, x)$. Therefore, a deck transformation h_γ corresponds to every closed path γ from x. Moreover, for each pair $y, y_1 \in p^{-1}(x)$ there exists a deck transformation $h = h_\gamma$ such that $h(y) = h(y_1)$.

As was remarked earlier, such a deck transformation, being a lift of p, is unique by the unique lifting property.

Hence, p: $Y \to X$ is a Galois covering. □

We have incidentally also proved the following

COROLLARY XVI.7.33. Let X be a connected manifold and let p: $Y \to X$ be a complete covering. Then the following conditions are equivalent:

1° p is a Galois covering.

2° The group $\Gamma(y) := p_*\pi_1(Y, y)$ is a normal subgroup of $\pi_1(X, x)$, where $x = p(y)$.

3° A closed path in X has every lift closed or has no closed lift.

We have seen that if p: $Y \to X$ is a normal connected covering of a connected manifold, then the group $\mathrm{Deck}(Y \xrightarrow{p} X)$ of deck transformations acts freely and properly discontinuously on Y.

Let us now reverse this situation, but in order to avoid unnecessary repetition let us adopt the following

Convention. If a manifold Y is topological (resp. differentiable, resp. real analytic, resp. complex), then an "automorphism" means a homeomorphism (resp. diffeomorphism, resp. bi-analytic map, resp. biholomorphism) of Y. By $\mathrm{Aut}(Y)$ we shall denote the corresponding group of automorphisms of the manifold Y.

DEFINITION. Let Γ be a subgroup of $\mathrm{Aut}(Y)$ where Y is a connected manifold Y. Then the *quotient space* or the *orbit space* $\Gamma\backslash Y$ is the set of all Γ-orbits

$$\Gamma \cdot y = \{\gamma(y) \colon \gamma \in \Gamma\},$$

and the *natural projection* $p\colon Y \to \Gamma\backslash Y$ is given by $p(y) := \Gamma \cdot y$. We impose on $\Gamma\backslash Y$ the quotient topology, i.e., the strongest (finest) topology for which p is continuous. Let us prove the following important

THEOREM XVI.7.34. 1° Let Γ be a subgroup of $\mathrm{Aut}(Y)$ acting freely and properly discontinuously on a connected manifold Y.

Then the natural projection $p\colon Y \to \Gamma\backslash Y$ is a Galois covering whose group of deck transformations is Γ, i.e.,

$$\mathrm{Deck}(Y \xrightarrow{p} \Gamma\backslash Y) = \Gamma.$$

2° Conversely, if Y, X are connected manifolds and $p\colon Y \to X$ is a complete covering then the group $\mathrm{Deck}(Y \xrightarrow{p} X)$ of deck transformations acts freely and properly discontinuously on Y.

If, moreover, $p\colon Y \to X$ is a Galois covering, then the quotient space $\mathrm{Deck}(Y \xrightarrow{p} X)\backslash Y$ is naturally isomorphic to X.

Remark. It is sufficient to assume in 2° that the deck group acts transitively on some fibre $p^{-1}(x)$.

PROOF. 1° Since Γ acts freely and properly discontinuously on Y, each y has an open (arcwise) connected neighbourhood V such that

$$\{\gamma \in \Gamma \colon \gamma(\overline{V}) \cap \overline{V} \neq \emptyset\} = \{1\}.$$

But for every $\gamma \in \Gamma$, the covering $p\colon \gamma(V) = p(V)$ and $\gamma(V)$ are components of $p^{-1}(p(V))$. Hence, $p\colon Y \to \Gamma\backslash Y$ is a complete covering. Plainly, Γ is a subgroup of Deck $(Y \xrightarrow{p} \Gamma\backslash Y)$ which is transitive on each fibre $p^{-1}((y)) = \Gamma \cdot y$. But the group of deck transformations acts freely on each fibre and, hence, Γ is the whole group of deck transformations and the covering $p\colon Y \to \Gamma\backslash Y$ is a Galois covering since Γ acts transitively on each fibre.

2° It is sufficient to prove the last statement.

Denote $\mathrm{Deck}(Y/X)$ by Γ. We now define the map $\pi\colon \Gamma\backslash Y \to X$ as follows. If $[y]$ is a Γ-orbit, then $\pi([y]) := p(y)$. This makes sense since $d(\gamma \cdot y) = p(y)$ for all $\gamma \in \Gamma$, because γ is a deck transformation. Since Y is connected, $\Gamma\backslash Y$ is connected. The mapping π is surjective since $p\colon Y \to X$ is surjective. But π is a complete covering (check this). Since p is a Galois covering, Γ acts transitively on $p^{-1}(x)$ for some point $x \in X$. Therefore, $\pi^{-1}(x)$ consists of one point and $\pi\colon \Gamma\backslash Y \to X$ is an isomorphism (being one-sheeted) by Corollary XVI.7.20.2°. Therefore, $\Gamma\backslash Y$ and X are isomorphic. □

We have thus completed the proof of all the facts comprising what we have named "the Poincaré theory", and have (we hope) provided the reader with some deeper insight in its content. Some applications to concrete problems in complex analysis are given in the next sections.

Problems of Separability. Before we leave the general theory of coverings, let us briefly discuss two facts. The first one is almost obvious:

PROPOSITION XVI.7.35. Let X, Y be topological spaces and let $f\colon Y \to X$ be a continuous, open surjection. If Y is second countable, then the same is true of X.

PROOF. Let $\mathscr{B}_Y$ be a countable basis of the topology of Y; then $\mathscr{B}_X = \{f(\mathscr{O})\colon \mathscr{O} \in \mathscr{B}_Y\}$ is a countable open covering of X. We check that $\mathscr{B}_X$ is a basis of the topology of X. Let $\mathscr{U}$ be an open subset of X and $x \in \mathscr{U}$; we have to show that there exists a $\mathscr{V} \in \mathscr{B}_X$ such that $x \in \mathscr{V} \subset \mathscr{U}$. Since f is a surjection, there is an $y \in Y$ with $f(y) = x$ and $f^{-1}(\mathscr{U})$ is an open neighbourhood of y. Therefore, there exists an $\mathscr{O} \in \mathscr{B}_Y$ such that $y \in \mathscr{O} \subset f^{-1}(\mathscr{U})$. For

$$\mathscr{V} := f(\mathscr{O}) \qquad \text{we have } x \in \mathscr{V} \subset \mathscr{U}. \ \square$$

The following famous result, due in a special case to V. Volterra and H. Poincaré, is much deeper.

THEOREM XVI.7.36 (Poincaré–Volterra). Let X be a connected, second countable manifold. Let $f\colon Y \to X$ be a continuous, discrete map. Then Y also satisfies the second countability axiom.

Hints for the proof: Let $\mathscr{B}_X$ be a countable basis of the topology of X. Denote by $\mathscr{B}_Y$ all open sets $V \subset Y$ such that

1° V is second countable,

2° V is a connected component of a set $f^{-1}(\mathcal{O})$, $\mathcal{O} \in \mathscr{B}_X$.

(a) Prove first that $\mathscr{B}_Y$ is a basis of the topology of Y.

(b) Prove next that for each $V_0 \in \mathscr{B}_Y$ there are at most countably many $V \in \mathscr{B}_Y$ such that $V_0 \cap V \neq \emptyset$.

(c) Now prove that $\mathscr{B}_Y$ is countable. Take a $U \in \mathscr{B}_Y$ and define for each $n \in N$ the set $\mathscr{B}_n \subset \mathscr{B}_Y$ as follows: $\mathscr{B}_n = \{V \in \mathscr{B}_Y$: there exist $V_0, V_1, \ldots, V_n \in \mathscr{B}_Y$ such that $V_0 = U, V_n = V$ and $V_{k-1} \cap V_k \neq \emptyset$ for $k = 1, 2, \ldots, n\}$. Since Y is connected, we have $\bigcup_{n \in N} \mathscr{B}_n = \mathscr{B}_Y$.

By induction prove the countability of each $\mathscr{B}_n$.

Remark. Theorem XVI.7.28 and Corollary XVI.7.29 may be used to prove that every Riemann surface is second countable (Theorem of Radó). But we shall prove this famous theorem in another manner (cf. Theorem XVI.9.10).

Let us give now

Weierstrass-Weyl's Construction of the Riemann Surface of a Function-Germ. Let X and Y be Riemann surfaces, let $\mathscr{A}_X$ and $\mathscr{A}_Y$ be the corresponding sheaves of holomorphic functions, and let $p: Y \to X$ be a holomorphic unbranched covering. Since p is locally biholomorphic, then for each $y \in Y$ there exists an isomorphism

$$p^*: \mathscr{A}_{X,x} \to \mathscr{A}_{Y,p(x)}.$$

Next, by

$$p_*: \mathscr{A}_{Y,p(x)} \to \mathscr{A}_{X,x}$$

we denote the inverse map ($p_* := (p^*)^{-1}$). We can now give the general notion of analytic continuation:

DEFINITION. Let X be a Riemann surface, $a \in X$ and $\varphi \in \mathscr{A}_a$ a holomorphic germ. A quadruplet (Y, p, f, b) is an *analytic continuation* of φ if:

1° Y is a Riemann surface and $p: Y \to X$ is a holomorphic unbranched covering,

2° $f \in A(Y)$,

3° $b \in Y$ is such that $p(b) = a$ and $p_*(\varrho_b(f)) = \varphi$.

An analytic continuation (Y, p, f, b) of φ is *maximal* if it has the following universal property:

If (Z, q, g, c) is another analytic continuation of φ, then there exists a holomorphic (homo)morphism $F\colon Z \to Y$ such that $F(c) = b$ and $F^*(f) := f \circ F = g$.

Remark 1. We could replace holomorphic germs by ***meromorphic*** germs: $\varphi \in \mathscr{M}_a$.

Remark 2. A maximal analytic continuation is unique (up to an isomorphism). Indeed, if both (Y, p, f, b) and (Z, q, g, c) are maximal continuations of φ, then there exists a holomorphic homomorphism $G\colon Y \to Z$ such that $G(b) = c$ and $G^*(g) = f$. But $F \circ G\colon Y \to Y$ is a holomorphic morphism of $p\colon Y \to X$, with $p(b) = b$. By the uniqueness of a lift (Theorem XVI.7.7) $F \circ G = \mathrm{id}_Y$; similarly, $G \circ F = \mathrm{id}_Z$. Therefore, $G\colon Y \to Z$ is biholomorphic.

The following lemma shows that the preceding notion of analytic continuation is a generalization of the Weierstrass analytic continuation along a path:

LEMMA XVI.7.37. Let X be a Riemann surface, $a \in X$, $\varphi \in \mathscr{A}_a$ and let (Y, p, f, b) be an analytic continuation of φ. If $v\colon [0, 1] \to Y$ is a path such that $v(0) = b$, $v(1) =: y$, then the germ

$$\psi := p_*\big(\varrho_y(f)\big) \in \mathscr{A}_{p(y)}$$

is an analytic continuation of φ along the path $u := p \circ v$.

PROOF. Let $t \in [0, 1]$, put

$$\varphi_t := p_*\big(\varrho_{v(t)}(f)\big) \in \mathscr{A}_{p(v(t))} = \mathscr{A}_{u(t)}.$$

We have: $\varphi_0 = \varphi$ and $\varphi_1 = p_*(f_y) = \psi$.

Let $t_0 \in [0, 1]$; since $p\colon Y \to X$ is unbranched, there exist open neighbourhoods $V \subset Y$ and $U \subset X$ of $v(t_0)$ and of $p\big(v(t_0)\big) = u(t_0)$ such that $p|V \to U$ is biholomorphic. Let $q := (p|V)^{-1}$ and $g := q^*(f|V) \in A(U)$. Then

$$p_*\big(\varrho_\eta(f)\big) = \varrho_{p(\eta)}(g) \quad \text{for all } \eta \in V.$$

There exists a neighbourhood $T \subset [0, 1]$ of t_0 such that $v(T) \subset V$ and, hence, $u(T) \subset U$. But for all $t \in T$ we have

$$\varrho_{u(t)}(g) = p_*\big(\varrho_{v(t)}(f)\big) = \varphi_t$$

and, hence, ψ is an analytic continuation of φ along u. □

We can now prove a theorem which is due to Weyl and into which the Weierstrass and Riemann ideas of analytic continuation and holomorphic covering coalesce.

THEOREM XVI.7.38 (Weyl). Let X be a Riemann surface, $a \in X$, $\varphi \in \mathscr{A}_a$. Then there exists a maximal analytic continuation (Y, p, f, b) of φ.

Remark. The same is valid *mutatis mutandis* for a meromorphic germ $\varphi \in M_a$.

PROOF. The construction is new immediate. Let Y be a connected component of $|\mathscr{A}|$ which contains φ. We also denote by p the restriction of p: $|\mathscr{A}| \to X$ to Y. As we know, Y is in a natural way a Riemann surface and p: $Y \to X$ is a holomorphic regular covering. We now have to construct a holomorphic function f: $Y \to \mathbf{C}$. We know, that every $y \in Y$ is (by definition) a germ at $p(y)$. We put

$$f(y) := y(p(y)).$$

Plainly, f is holomorphic and $p_*(\varrho_y(f)) = y$ for all $y \in Y$. If we set $b := \varphi$ we see that (Y, p, f, b) is an analytic continuation of φ. We only have to verify, that it is a maximal one. Let (Z, q, g, c) be another analytic continuation of φ. We define F: $Z \to Y$ in the following way. Let $z \in Z$, $q(z) =: x$ We know by the preceding lemma that the germ $q_*(\varrho_z(g)) \in \mathscr{A}_x$ can be obtained by analytic continuation of φ along a path from a to x. But Lemma XVI.7.11 asserts that Y consists of all germs which can be obtained by analytic continuation of φ along a path. Therefore, there exists exactly one $y \in Y$ such that $q_*(\varrho_z(g)) = y$. We define $F(z) := y$. From this construction follows, that F: $Z \to Y$ is a holomorphic morphism: $q = p \circ F$, with $F(c) = b$ and $F^*(f) = g$. □

8. THE KOEBE–RIEMANN THEOREM. NON-EUCLIDEAN GEOMETRY. MÖBIUS TRANSFORMATIONS

In this section we want to give a proof of the famous "Fundamental Conformal Mapping Theorem". This theorem is often called the "Riemann Mapping Theorem" since it was formulated by Riemann. The first successful proof is due to Paul Koebe (1900), although a similar theorem had been proved earlier by W. F. Osgood. (For the history of this theorem cf. the second part of Section 20.) A generalization of the Koebe–Riemann Theorem to any simply connected Riemann surface was proved independently by Poincaré and Koebe in 1907. This "Uniformization Theorem" is proved in Section 13 and the general "Uniformization Principle" in Section 20. The theorem reads as follows: A simply connected domain $\Omega \subset \boldsymbol{C}$ which is not the whole $\boldsymbol{C}$ can be mapped biholomorphically onto the unit disk $\boldsymbol{D}$. We shall see that the

A Table of Some Useful Conformal Mappings

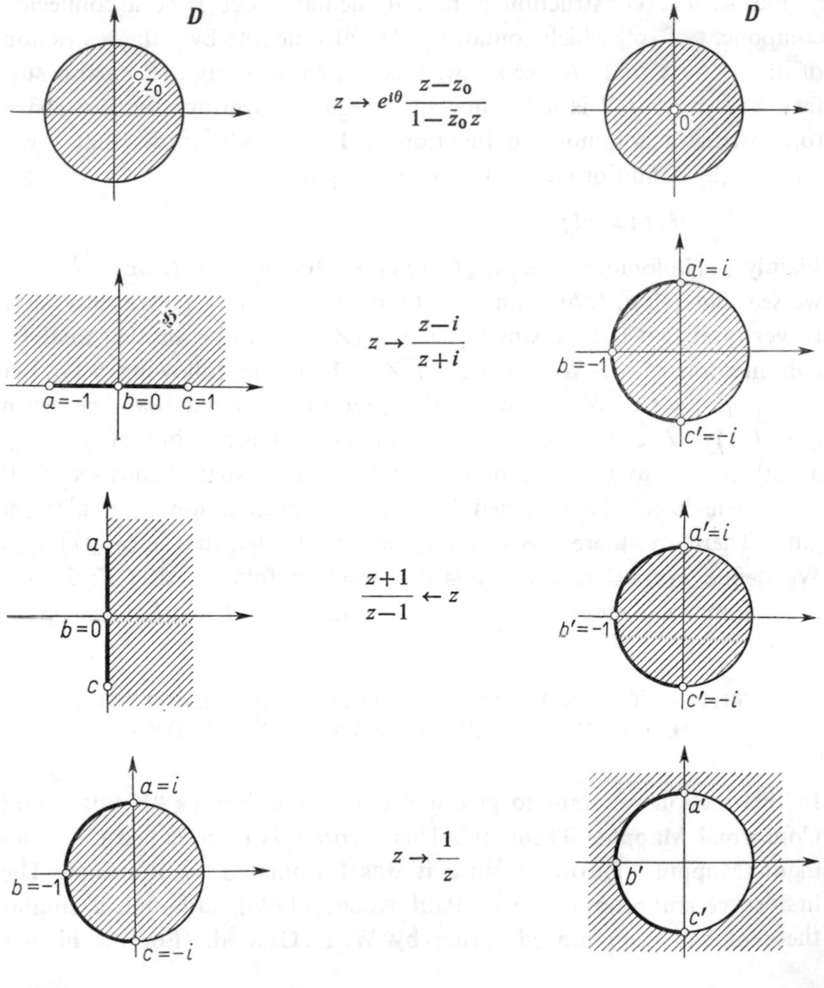

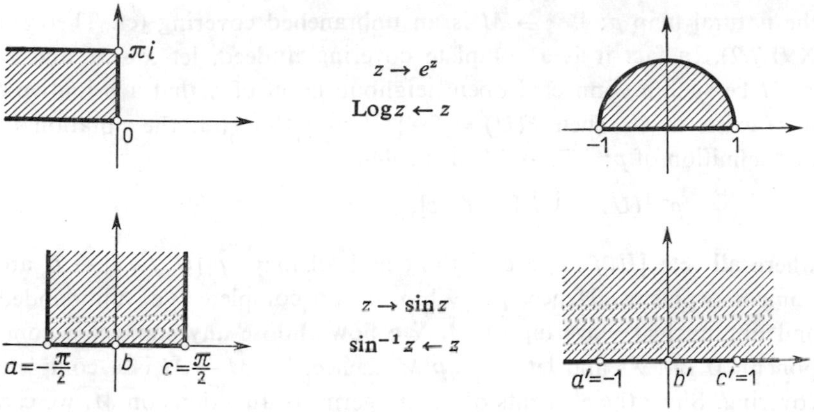

simple connectivity of Ω is essential for the theorem to hold, but is not sufficient: the biholomorphicity of domains is a much stronger condition than their topological equivalence: $\boldsymbol{C}$ and $\boldsymbol{D}$ are homeomorphic but not biholomorphic (by the Liouville Theorem every holomorphic map $f\colon \boldsymbol{C} \to \boldsymbol{D}$ is a constant). The boundary $\partial\Omega$ of Ω plays a paramount role: $\boldsymbol{C}$, considered as a subset of the Riemann sphere $\boldsymbol{P}(\boldsymbol{C}) \equiv \bar{\boldsymbol{C}}$, has only one point ∞ as its boundary whereas the boundary of $\partial\boldsymbol{D}$ is a circle. The condition $\pi_1(\Omega) = \{1\}$ under the form of the contractibility of Ω was also essential for the Poincaré Lemma: the existence of a primitive of a closed differential form. We now establish a corresponding fact for a simply connected Riemann surface:

THEOREM XVI.8.1. Let M be a Riemann surface and $\omega \in \Lambda^1(M)$ a closed differential form, i.e. $d\omega = 0$. Then there exists a connected complete covering $\pi\colon \tilde{M} \to M$ and a primitive $g \in C^\infty(\tilde{M})$ of $\pi^*\omega$, i.e.

$$dg = \pi^*\omega.$$

PROOF. For an open $\mathcal{O} \subset M$ we denote by $F(\mathcal{O})$ the set of primitives of ω defined in $\mathcal{O}$, i.e., $F(\mathcal{O}) = \{f \in C^\infty(\mathcal{O})\colon df = \omega|\mathcal{O}\}$. Since each point of M has an open neighbourhood biholomorphic to the unit disk the Poincaré Lemma guarantees that for all "nice" (e.g., contractible) $\mathcal{O}$ the set $F(\mathcal{O})$ is not empty. The presheaf (of sets) $\mathscr{F} = \{F(\mathcal{O})\colon \mathcal{O}$ open in $M\}$ satisfies the principle of uniqueness, since for each domain $U \subset M$ and each $f_1, f_2 \in F(U)$ $df_1 = df_2$ implies $f_1 - f_2$ equal to a constant. The space $|\mathscr{F}|$ of germs of the presheaf $\mathscr{F}$, therefore, is Hausdorff and the natural map $p\colon |\mathscr{F}| \to M$ is an unbranched covering (cf. Theorem XVI.7.2). In fact it is a complete covering. Indeed, let $x \in M$ and $U \subset M$ be such a connected open neighbourhood of x that a primitive f of ω exists in U. Then $F(U) = f + \{C\}$ and thus (cf. the notation in the definition of $p\colon |\mathscr{F}| \to M$ on p. 446)

$$p^{-1}(U) = \bigcup_{c \in C} [U, f+c],$$

where all sets $[U, f+c]$ are disjoint and all maps $p|[U, f+c] \to U$ are homeomorphisms. Hence, $p\colon |\mathscr{F}| \to M$ is a complete (i.e., unbounded and unbranched) covering of M. We now choose any connected component $\tilde{M}$ of $|\mathscr{F}|$ and let $\pi := p|\tilde{M}$; hence, $\pi\colon \tilde{M} \to M$ is a complete covering. Since the elements of $\tilde{M}$ are germs of functions on M, we can define $g\colon \tilde{M} \to C$ by $g(\varphi) := \varphi(p(\varphi))$. From the construction of g it follows that $dg = \pi^*\omega$. □

Remark. We add some explanation for the reader who may feel uneasy about our construction of germs and our use of the Weyl Theorem in a situation when some of the $F(\mathcal{O})$ are empty. A little more careful examination of the construction of germs and of the proof of the Weyl Theorem shows that all that is needed is the following:

The family of sets U, for which $F(U)$ is not empty, and the functions from $F(U)$ have to satisfy the following:

(i) For every such U and U', every $p \in U \cap U'$ and every $f \in F(U)$ there is an $f' \in F(U')$ such that f and f' agree in a neighbourhood of the point p.

(ii) For $f \in F(U)$ and $f' \in F(U')$ if $U \cap U' \neq \emptyset$, then the set $\{p \in U \cap U'\colon f(p) = f'(p)\}$ is closed with respect to $U \cap U'$.

The reader undoubtedly noticed that (ii) is a form of the identity principle and (i) assures the existence of sufficiently many restrictions of a given $f \in F(U)$. These conditions are seen to be satisfied in the case under consideration. As an immediate consequence we obtain

THEOREM XVI.8.2. 1° Let $p\colon \hat{X} \to X$ be the universal covering surface of a Riemann surface X and let $\omega \in Z^1(X)$ (a closed form on X). Then there exists on $\hat{X}$ a primitive $f \in C^\infty(\hat{X})$ of $p^*\omega$.

2° On a simply connected Riemann surface X every closed 1-form is exact. In other words, the first cohomology space of X vanishes,

(1) $$H^1(X, C) = \{0\} = H^1(X),$$

or equivalently

(2) $\quad Z^1(X) = B^1(X) = d(C^\infty(X)).$

PROOF. 1° Let $\pi\colon \tilde{X} \to X$ be a complete connected covering and let $g \in C^\infty(\tilde{X})$ be such that $dg = \pi^*\omega$ as constructed in Theorem XVI.8.1. Since $p\colon \hat{X} \to X$ is the universal covering of X, there exists a homomorphism $\varrho\colon \tilde{X} \to \hat{X}$ such that $\pi \circ \varrho = p$. Then for $f := \varrho^* g (\equiv g \circ \varrho)$ we have $df = d(\varrho^* g) = \varrho^* dg = \varrho^*(\pi^*\omega) = p^*\omega$.

2° Since $p = \mathrm{id}_X\colon X \to X$ is the universal covering of X, $df = p^*\omega = \omega$. □

In Nevanlinna's proof of the uniformization theorem (Section 13) it is very important to know about the

Existence of log *and of a Harmonic Conjugate Function.* As a corollary of the point 2° of the preceding theorem we obtain

PROPOSITION XVI.8.3. Let X be a simply connected Riemann surface. Then:

1° We have $A^1(X) = dA(X) = d'A(X)$.

2° For every holomorphic nonvanishing function f on X there exist branches of $\log f$ and of $\sqrt{f}$, i.e., there exist $g, g_1 \in A(X)$ such that $e^g = f$ and $g_1^2 = f$.

We know that the real part of a holomorphic function is harmonic, but for simply connected X the converse is also true:

3° Every harmonic function $h \in \mathscr{H}_R(X)$ is the real part of a holomorphic function $f \in A(X)$, i.e.,

$$\mathscr{H}_R(X) = \mathrm{Re}(A(X)).$$

4° Every harmonic function $h \in \mathscr{H}_R(X)$ has a harmonic conjugate function: i.e., a function $h_1 \in \mathscr{H}_R(X)$ such that, $h + ih_1 \in A(X)$.

PROOF. 1° We know (Theorem XVI.6.12) that every holomorphic form is closed. Since $\pi_1(X) = \{1\}$, there exists an $f \in C^\infty(X)$ such that $\omega = df = d'f + d''f$, but $\omega \in A^{1,0}(X)$ and, hence, $d''f = 0$, i.e., $f \in A(X)$ and $\omega = d'f$.

2° Since df/f is a holomorphic form, we have by 1° a $g \in A(X)$ such that $\quad f^{-1}df = dg.$

We can assume that, for some x_0,

(3) $\quad \exp g(x_0) = f(x_0).$

But

$$d(fe^{-g}) = (df)e^{-g} - fdge^{-g} = (df)(e^{-g} - ff^{-1}e^{-g}) = 0.$$

Hence, $fe^{-g} = \text{const} = 1$ by (3). Thus

$$f = e^g.$$

Putting $g_1 := e^{g/2}$, we obtain $g_1^2 = e^g = f$.

3° Since dh is a harmonic form $(0 = d(dh) = d*(dh))$, we know by Corollary XVI.6.14 that there exists an $\omega \in A^{1,0}(X)$ such that $dh = \operatorname{Re}(\omega)$. But by 1° $\omega = dg$ for some $g \in A(X)$. Therefore,

$$dh = \operatorname{Re}(dg); \quad \text{hence } h = \operatorname{Re}(g) + \text{const} = \operatorname{Re}(f)$$

for some $f \in A(X)$.

4° Put $h_1 := \operatorname{Im}(f)$, with f as in 3°. □

Remark. 2° is a natural generalization of a corresponding fact for $X \subset \boldsymbol{C}$ (Section 1).

COROLLARY XVI.8.4. Let X be a simply connected Riemann surface and $h \in \mathscr{H}_R(X)$. Then there exists a holomorphic function f such that

$$\log|f| = h.$$

PROOF. Take such an $f \in A(X)$ that $f = h + ih_1$, $h_1 \in \mathscr{H}_R(X)$. Then $\log|f| = h$. □

Remark. This corollary is used in a proof of the uniformization theorem for hyperbolic Riemann surfaces. A Riemann surface X is (by definition) hyperbolic if its universal covering surface has a Green's function (cf. Sections 13 and 20 and the exercises at the end of the present section).

The Group of Möbius Transformations Möb = Aut$(\boldsymbol{P}(\boldsymbol{C}))$. In the sequel we use "biholomorphic" and "conformal" as synonyms since for domains in $\boldsymbol{C}$ both terms are equivalent (cf. below). The group of all biholomorphic maps of a complex manifold M will be denoted by Aut(M) since biholomorphic maps are automorphisms of the complex structure of M.

We now collect some notions concerning the action of groups which were scattered throughout the preceding sections. Let us assume we have a group G acting on a space E; G may be thought of as a group of automorphisms of some structure on E, e.g., of a vector space structure, topology, differentiable manifold structure, etc. Given a point $e \in E$, the set of all G-translates of e will be called an *orbit* of e under G. In other words, an orbit in E is a set $G \cdot e = \{g \cdot e \colon g \in G\}$. If there is a point $e \in E$ such that its orbit equals the whole space E (i.e., $G \cdot e = E$) (the same is then true for any other point), then G is said to be *transitive* (or *act transitively on E*). For a given $e \in E$ the set of all $g \in G$ which fix the point e, i.e., such that $g \cdot e = e$, is seen to be a subgroup of G called the *isotropy* (*stability*) *subgroup* of the point e. We denote the isotropy subgroup of e by G_e.

Exercise. Show that if G is transitive then any two isotropy subgroups are conjugate.

One says that *G acts freely* if no $g \in G$ other than identity has a fixed point. This is the case precisely when $G_e = \{\mathrm{id}\}$ for all $e \in G$. The following simple fact is often useful.

LEMMA XVI.8.5. *Let G be a subgroup of $\mathrm{Aut}(E)$ such that*
(a) G acts transitively;
(b) there exists an $e \in E$ such that

$$(\mathrm{Aut}(E))_e \subset G.$$

Then $G = \mathrm{Aut}(E)$, i.e., G is the whole group $\mathrm{Aut}(E)$.

PROOF. Let $h \in \mathrm{Aut}(E)$; then by (a) there exists a $g \in G$ such that $g \cdot e = h \cdot e$. Hence, $(g^{-1}h) \cdot e = e$ and, by (b), $g^{-1}h \in G$. Therefore $h = g(g^{-1}h) \in G$. □

One of the main domains of mathematical research is the investigation of automorphisms of interesting objects. It is an immense and probably an infinite task to determine $\mathrm{Aut}(E)$ for interesting E.

The Koebe–Riemann Theorem allows us to determine the group $\mathrm{Aut}(X)$ for every simply connected Riemann surface in the following way:

Each simply connected Riemann surface is isomorphic to a canonical subdomain of $\boldsymbol{P}(\boldsymbol{C})$:

1° the Riemann sphere $\boldsymbol{P}(\boldsymbol{C}) = \boldsymbol{C}$,

2° the complex plane $\boldsymbol{C}$,

3° the unit disk $\boldsymbol{D} = \{z \in \boldsymbol{C} \colon |z| < 1\}$ which, in turn, is equivalent to $\mathfrak{H} := \{z \in \boldsymbol{C} \colon \operatorname{Im} z > 0\}$.

Therefore, if we can determine the three groups

$$\mathrm{Aut}(\boldsymbol{C}), \quad \mathrm{Aut}(\overline{\boldsymbol{C}}), \quad \mathrm{Aut}(\boldsymbol{D}) \cong \mathrm{Aut}(\mathfrak{H}),$$

then we "in principle" know the group $\mathrm{Aut}(X)$ for any simply connected Riemann surface X.

The following theorem describes the groups $\mathrm{Aut}(\boldsymbol{C})$ and $\mathrm{Aut}(\overline{\boldsymbol{C}})$:

THEOREM XVI.8.6. (α) $\mathrm{Aut}(\boldsymbol{C}) = \mathrm{Aff}(\boldsymbol{C}) = \{$affine transformations $z \to az+b,\ a, b \in \boldsymbol{C},\ a \neq 0\}$;

(β) $$\mathrm{Aut}(\overline{\boldsymbol{C}}) = \left\{\overline{\boldsymbol{C}} \ni z \to \frac{az+b}{cz+d} \in \overline{\boldsymbol{C}}: a, b, c, d \in \boldsymbol{C},\ ad-cb \neq 0\right\}.$$

DEFINITION. The group on the right-hand side in (β) is denoted by Möb and is called the *group of Möbius transformations* or the *group of homographies*.

Therefore, (β) can be rewritten as:

$$\mathrm{Aut}\big(\boldsymbol{P}(\boldsymbol{C})\big) = \text{Möb}.$$

PROOF. (α) Let $f \in \mathrm{Aut}(\boldsymbol{C})$; then the point ∞ is an isolated singularity of f and by the Casorati–Weierstrass Theorem it follows that f cannot have an essential singularity at ∞. Indeed, since $f\colon \boldsymbol{C} \to \boldsymbol{C}$ is injective, we have

$$f(\{z\colon |z| > 1\}) \subset \big(\boldsymbol{C} - f(\boldsymbol{D})\big).$$

The set $f(\{z\colon |z| > 1\})$, therefore, cannot be dense in $\boldsymbol{C}$. Thus, f has a pole at ∞. But by virtue of the bijectivity of f this pole is of order 1; hence

$$f(z) = az+b, \quad \text{where } a \neq 0,$$

since $f \neq$ const.

(β) Let $T\colon \overline{\boldsymbol{C}} \to \overline{\boldsymbol{C}}$ be defined as

(*) $$T(z) = \frac{az+b}{cz+d}, \quad ad-bc \neq 0$$

for all $z \in \overline{\boldsymbol{C}}$ different from ∞ and $-d/c$. If we put $T(\infty) = a/c$ and $T(-d/c) = \infty$, then the T so extended is seen to be holomorphic on $\overline{\boldsymbol{C}}$. Let $\mathrm{GL}_2(\boldsymbol{C}) := \left\{\begin{bmatrix} a & b \\ c & d \end{bmatrix}\colon a, b, c, d \in \boldsymbol{C} \text{ and } ad-cb \neq 0\right\}$ be the group of invertible complex two-by-two matrices. To every matrix $g = \begin{bmatrix} a & b \\ c & d \end{bmatrix}$

$\in GL_2(\boldsymbol{C})$ there corresponds a transformation $T(g)\colon \overline{\boldsymbol{C}} \to \overline{\boldsymbol{C}}$ given by the formula (*).

A simple computation shows that the superposition of two such transformations given by the matrices $g_1, g_2 \in GL_2(\boldsymbol{C})$ corresponds to the product of these matrices or, in other words, that

$$GL_2(\boldsymbol{C}) \ni g \to T(g) \in \text{Möb}$$

is a homomorphism of groups. (Determine its kernel!). Since for

$$GL_2(\boldsymbol{C}) \ni g = \begin{bmatrix} a & b \\ c & d \end{bmatrix}$$

we have

$$g^{-1} = (\det g)^{-1} \begin{bmatrix} d & -b \\ -c & a \end{bmatrix},$$

we see that for a given $w \in \overline{\boldsymbol{C}}$ the equation

$$w = \frac{az+b}{cz+d}$$

has the solution

$$z = \frac{dw-b}{-cw+d}$$

and, thus, Möb is transitive on $\overline{\boldsymbol{C}}$. It therefore suffices to show that it contains a stability group of some point (cf. Lemma XVI.8.5). But obviously $\operatorname{Aut}(\overline{\boldsymbol{C}})_\infty = \operatorname{Aut}(\boldsymbol{C})$ and by (α) we have $\operatorname{Aut}(\boldsymbol{C}) \subset$ Möb. □

The Lemmata of Schwarz and Pick. In order to determine the group $\operatorname{Aut}(\boldsymbol{D})$ we need a very important result, due in essence to H. A. Schwarz. The classical formulation and proof was given much later by Constantin Carathéodory.

THEOREM XVI.8.7 (The Schwarz Lemma). Let $f\colon \boldsymbol{D} \to \boldsymbol{D}$ be such a holomorphic map that $f(0) = 0$. Then

1°

(4) $\quad |f(z)| \leqslant |z| \quad \text{and} \quad |f'(0)| \leqslant 1$

2° there exists $z_0 \in \boldsymbol{D} - \{0\}$ such that

$$\big(|f(z)| = |z_o|\big) \Leftrightarrow \big(f(z_o) = e^{i\theta} z, \text{ where } \theta \in \boldsymbol{R}\big).$$

PROOF. 1° We apply the maximum principle to the function

$$f_1(z) := \begin{cases} f(z)/z & \text{for } z \neq 0, \\ f'(0) & \text{for } z = 0. \end{cases}$$

On the circle $\{z: |z| = r\}$, $|f_1(z)| \leqslant r^{-1}$ and, hence, $|f_1(z)| \leqslant r^{-1}$ for $|z| \leqslant r$. If we let $r \to 1$, then we find that $|f_1(z) \leqslant 1$ for all $z \in \boldsymbol{D}$. Thus $|f(z)| \leqslant |z|$.

2° $\Rightarrow$: By assumption $|f_1|$ attains its maximum at $z_0 \in \boldsymbol{D}$; therefore, $f_1 = \text{const}$. Hence, $f(z) = \text{const}\, z$, but this constant has a modulus of 1. □

Exercise. Prove the *Schwarz lemma for a polydisk*. Let f be a holomorphic function on a polydisk $\boldsymbol{D}_n(0, r) \subset \boldsymbol{C}^n$ such that

(5) $\qquad |f| \leqslant M \quad$ and f has a zero of order k at 0.

Then

(6) $$|f(z)| \leqslant M\left(\frac{|z|}{r}\right)^k \qquad \text{for all } z \in \boldsymbol{D}(0, r).$$

Hint: Consider the map $t \to g_z(t) := \left(\frac{t}{|z|}\right)^{-k} f\left(\frac{tz}{|z|}\right)$, where $t \in \boldsymbol{C}$, $|t| < r$ and $z \in \boldsymbol{D}_n(0, r)$, $z \neq 0$ is considered fixed. Use the maximum principle to show that the holomorphic function $g_z(t)$ satisfies

$$|g_z(t)| \leqslant M\left(\frac{r}{|z|}\right)^{-k} \qquad \text{for all } t \in \boldsymbol{D}_1(0, r).$$

In particular this estimate applies to $g_z(|z|) = f(z)$.

COROLLARY XVI.8.8. Let $f: \boldsymbol{D} \to \boldsymbol{D}$ be holomorphic and assume that for $z_0 \in \boldsymbol{D}$ $f(z_0) = 0$; then

$$|f(z)| \leqslant \left|\frac{z - z_0}{1 - \bar{z}_0 z}\right|.$$

If the equality holds for some point different from z_0, then

$$f(z) = e^{i\theta} \frac{z - z_0}{1 - \bar{z}_0 z}.$$

PROOF. The proof is reduced to the case considered earlier. Let $z \to T(z)$ be a Möbius transformation which maps $\boldsymbol{D}$ onto $\boldsymbol{D}$ with $T(z_0) = 0$. It is easily checked that T defined by

$$T(z) = \frac{z - z_0}{1 - \bar{z}_0 z}$$

satisfies these conditions. Then $g(z) := f(T^{-1}(z))$ satisfies the assumption of the Schwarz Lemma. Hence, if we put $\zeta = T(z)$, we get

$$|g(\zeta)| = |f(T^{-1}(\zeta))\leqslant |\zeta|$$

or

$$|f(z)| \leqslant |T(z)| = \left|\frac{z-z_0}{1-\bar{z}_0 z}\right|. \ \square$$

COROLLARY XVI.8.9.

$$\operatorname{Aut}(\boldsymbol{D}) = \left\{\boldsymbol{D} \ni z \to e^{i\theta}\frac{z-z_0}{z-\bar{z}_0 z} \in \boldsymbol{D}\colon \theta \in \boldsymbol{R},\ |z_0| < 1\right\}.$$

Exercises. 1. Show that each of the transformations from Aut($\boldsymbol{D}$) transforms the circumference of the disk $\boldsymbol{D}$ into itself and the exterior of the disk into itself.

2. Show that the Möbius transformation given as

$$T(z) = \frac{z-i}{z+i}$$

is an equivalence of $\mathfrak{H} = \{z \in \boldsymbol{C}\colon \operatorname{Im} z > 0\}$ with the unit disk.

As was shown by Pick, one can give a geometrical formulation of the Schwarz Lemma.

THEOREM XVI.8.10 (Pick). Given a holomorphic mapping $f\colon \boldsymbol{D} \to \boldsymbol{D}$. Then

$$(7)\qquad \frac{|f'(z)|}{1-|f(z)|^2} \leqslant \frac{1}{1-|z|^2} \qquad \text{for } z \in \boldsymbol{D}.$$

We can also rewrite (7) as

$$(8)\qquad \frac{|df(z)|}{1-|f(z)|^2} \leqslant \frac{dz}{1-|z|^2}, \qquad z \in \boldsymbol{D}.$$

The equality in (7) at a single point implies that $f \in \operatorname{Aut}(\boldsymbol{D})$.

PROOF. Consider the automorphisms g and h of $\boldsymbol{D}$ defined by

$$g(\zeta) := \frac{\zeta+z}{1+\bar{z}\zeta}, \qquad h(\zeta) := \frac{\zeta-f(z)}{1-\overline{f(z)}\,\zeta}, \qquad \zeta \in \boldsymbol{D},$$

where z is an arbitrarily chosen point of D. Then $F := h \circ f \circ g$ maps D into D and $F(0) = 0$. But

$$F'(0) = \frac{1-|z|^2}{1-|f(z)|^2} f'(z).$$

Hence by Schwarz Lemma

$$\frac{1-|z|^2}{1-|f(z)|^2} |f'(z)| \leqslant 1. \; \square$$

We shall now consider the Schwarz–Pick Lemma from the differential geometric point of view.

DEFINITION. The Hermitian metric

$$(9) \qquad ds_D^2 \equiv h_D(z) := \frac{dz \otimes d\bar{z}}{1-|z|^2}, \qquad z \in D,$$

is called the *Poincaré* (*Hermitian*) *metric* (or the *Poincaré–Bergman metric*) on the unit disk D.

The Poincaré metric on D (which is not the metric in the meaning of a distance) defines a metric (i.e., distance) on D in the following way. For a piecewise differentiable curve $\gamma: I \to D$ we define the length of γ by the formula

$$(10) \qquad |\gamma| = \int_0^1 \sqrt{\frac{|f'(\gamma(t))|}{1-|f(\gamma(t))|^2}}\, dt.$$

This is a familiar procedure used in (Riemannian) differential geometry: we integrate the length of the tangent vector along a curve. Then for any two points $z_1, z_2 \in D$ we define the distance $d_D(z_1, z_2)$ by means of the formula

$$d_D(z_1, z_2) = \inf\{|\gamma| : \text{piecewise smooth } \gamma \text{ joining } z_1 \text{ with } z_2\}.$$

The reader may check that d_D satisfies the requirements of the metric; d_D is called the *Poincaré distance* or the *hyperbolic metric* of the unit disk D.

Theorem XVI.8.10 may now be stated as follows:

THEOREM XVI.8.11. Let D be the unit disk in C with the Poincaré Hermitian metric h_D and the hyperbolic metric d_D. Then every holomorphic mapping $f: D \to D$ is distance-decreasing, i.e.,

$$d_D(f(z_1), f(z_2)) \leqslant d_D(z_1, z_2).$$

Moreover, Aut($\boldsymbol{D}$) is the group of isometries of the Riemannian manifold $(\boldsymbol{D}, d_{\boldsymbol{D}})$.

DEFINITION. Let X be a Riemannian manifold provided with the Hermitian metric $2h = 2h_X dz \otimes d\bar{z}$. Then the Gaussian curvature of the metric $2h$ is defined by

$$(11) \qquad K_h := -\frac{1}{h}\frac{\partial^2 \log h}{\partial z \partial \bar{z}}.$$

Example. The Gaussian curvature of $(\boldsymbol{D}, h_{\boldsymbol{D}})$, where $h_{\boldsymbol{D}}$ is the Poincaré metric given by (9), is equal to -1.

The disk as a model for a non-Euclidean (hyperbolic) geometry. An important numerical quantity to be connected with the geometric features of Möbius transformations is the cross ratio.

DEFINITION. The *cross ratio* (z_1, z_2, z_3, z_4) of four points $z_k \in \boldsymbol{P}(\boldsymbol{C})$, $k = 1, \ldots, 4$, three of which are different, is defined by

$$(z_1, z_2, z_3, z_4) := \frac{z_1 - z_3}{z_1 - z_4} \Big/ \frac{z_2 - z_3}{z_2 - z_4}.$$

In order to investigate the cross ratio let us first prove

LEMMA XVI.8.12. A Möbius transformation has at most two fixed points. If it has three distinct fixed points, it is an identity.

PROOF. The fixed points of

$$T(z) = \frac{az + b}{cz + d}$$

are roots of the equation $cz^2 + (d - a)z - b = 0$.

Since there are only two roots of this equation, unless all the coefficients vanish, the assertion follows.

THEOREM XVI.8.13. 1° For any two triplets of distinct points z_1, z_2, z_3 and z_1', z_2', z_3' of the Riemann sphere $\boldsymbol{P}(\boldsymbol{C}) = \overline{\boldsymbol{C}}$, there exists exactly one $g \in \mathrm{Aut}(\boldsymbol{P}(\boldsymbol{C}))$ such that $g \cdot z_k = z_k'$, $k = 1, 2, 3$.

2° Let z_2, z_3, z_4 be three distinct points of $\boldsymbol{P}(\boldsymbol{C})$ and let $z_1 \subset \boldsymbol{P}(\boldsymbol{C})$ be any point. Then for every Möbius transformation g we have

$$(z_1, z_2, z_3, z_4) = (gz_1, gz_2, gz_3, gz_4),$$

i.e., the cross ratio is invariant under Möbius transformations.

3° The cross ratio of four distinct points $z_1, z_2, z_3, z_4 \in \boldsymbol{P}(\boldsymbol{C})$ is real iff the points lie on a circle or on a straight line.

Remark. The distinction made in 3° between lines and circles is unimportant when they are viewed as subsets of the Riemann sphere (the projective plane $\boldsymbol{P}(\boldsymbol{C})$). The point is that they are all images under the stereographic projection of circles on the Riemann sphere, straight lines being the images of circles passing through the north pole. Henceforth we shall use the term "circle" to denote both circles and straight lines in the complex plane.

PROOF. 1° Let us define two homographies g_1, g_2 by

$$(*) \qquad g_1 z := \frac{z-z_2}{z-z_3} \cdot \frac{z_1-z_3}{z_1-z_2} = (z, z_1, z_2, z_3),$$

$$(**) \qquad g_2 z := \frac{z-z_2'}{z-z_3'} \cdot \frac{z_1'-z_3'}{z_1'-z_2'} = (z, z_1', z_2', z_3').$$

Then $(g_1 z_1, g_1 z_2, g_1 z_3) = (1, 0, \infty)$; similarly, g_2 sends (z_1', z_2', z_3') onto $(1, 0, \infty)$. Hence, $g_2^{-1} \circ g_1 =: g$ is the desired homography. The map g is the unique map with this property. For if $k \in$ Möb and $k \cdot z_j = z_j'$, $j = 1, 2, 3$, then $k^{-1} \circ g$ has three distinct fixed points: $0, 1, \infty$. By virtue of Lemma XVI.8.12 $k^{-1} \cdot g = \mathrm{id}$; therefore $k = g$.

2° Let g be a given Möbius map and let $k \in$ Möb be defined by

$$kz = \frac{z-z_3}{z-z_4} \cdot \frac{z_2-z_4}{z_2-z_3} = (z, z_2, z_3, z_4).$$

From 1° it follows that k is the only map sending z_2 to 1, z_3 to 0 and z_4 to ∞. Therefore, $h = kg^{-1}$ satisfies

$$h(gz_2) = 1, \qquad h(gz_3) = 0, \qquad h(gz_4) = \infty$$

and by the uniqueness h is given as $hz = (z, gz_2, gz_3, gz_4)$. Then for $z = gz_1$ we have

$$\begin{aligned} h(gz_1) &= (gz_1, gz_2, gz_3, gz_4) = kg^{-1}gz_1 = kz_1 \\ &= (z_1, z_2, z_3, z_4). \end{aligned}$$

3° Take $k \in$ Möb as above: $kz = (z, z_2, z_3, z_4)$. It will suffice to show that $k^{-1}(\overline{\boldsymbol{R}})$ is a circle in $\boldsymbol{P}(\boldsymbol{C})$, where $\overline{\boldsymbol{R}} = \boldsymbol{R} \cup \{\infty\}$. To do so, let us consider a Möbius transformation in its general form

$$gz = \frac{az+b}{cz+d}, \qquad ad-bc \neq 0.$$

If $z \in \overline{\boldsymbol{R}}$ and $w = g^{-1}z$, then $gw = \overline{gw}$ which is the same as

$$\frac{aw+b}{cw+d} = \frac{\bar{a}\bar{w}+\bar{b}}{\bar{c}\bar{w}+\bar{a}}.$$

This gives the equation

$$(a\bar{c}-c\bar{a})|w|^2+(a\bar{d}-c\bar{b})w+(b\bar{c}-d\bar{a})\bar{w}+b\bar{d}-db = 0.$$

If $a\bar{c}$ is real, then the equation defines a line in $\boldsymbol{C}$ (check this). If $a\bar{c}$ is not real, then we can rewrite this in the form

$$|w|^2+\bar{\alpha}w+\alpha\bar{w}-\beta = 0$$

with β real and α complex; this is the equation of a circle. □

As a corollary, we obtain

COROLLARY XVI.8.14. 1° A Möbius transformation carries circles in $\boldsymbol{P}(\boldsymbol{C})$ onto circles. Moreover, if Γ and Γ' are orthogonal circles then their images are also orthogonal.

2° For any two circles Γ_1, Γ_2 in $\boldsymbol{P}(\boldsymbol{C})$ there exists a Möbius transformation g such that $g\Gamma_1 = \Gamma_2$ (g is not unique).

PROOF. 1° is an easy consequence of 2° and 3° above, except for the orthogonality statement. That follows in turn from the fact that holomorphic maps preserve angles (are conformal).

3° follows by taking $g_1, g_2 \in$ Möb, which sends Γ_1 (resp. Γ_2) onto $\overline{\boldsymbol{R}}$, and then taking $g = g_2 g_1^{-1}$. Observe the nonuniqueness of both g_1 and g_2. How can it be remedied?

A remarkable connection of the cross ratio with hyperbolic geometry is shown in the following theorem.

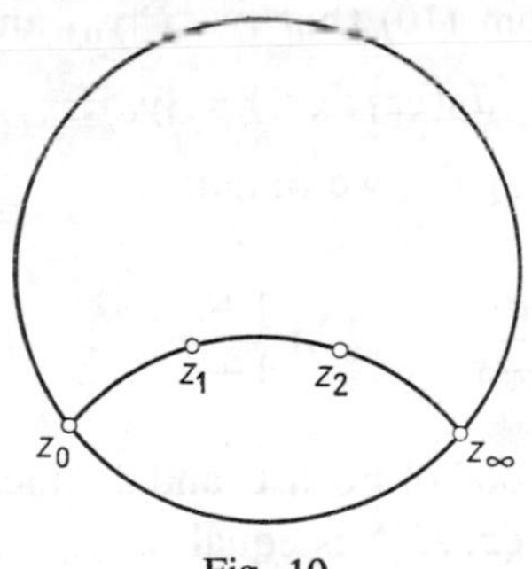

Fig. 10

THEOREM XVI.8.15. Let z_1, z_2 be two distinct points of $\boldsymbol{D}$. Let T denote the unit circle, the circumference of $\boldsymbol{D}$, i.e., $T = \{z: |z| = 1\}$.

Further, let z_0 and z_∞ be the points of intersection of T with the (unique) circle orthogonal to T and passing through z_1 and z_2. Notation is chosen so that z_1 lies in between z_0 and z_2 whereas z_2 lies in between z_1 and z_∞ (cf. Fig. 10). Then

(1) $d_D(z_1, z_2) = \frac{1}{2}\ln(z_1, z_2, z_\infty, z_0)$,

(2) $d_D(z_1, z_2)$ is equal to the length of the segment $\langle z_1, z_2\rangle$ cut by z_1, z_2 off the circle through z_0, z_1, z_2, z_∞.

Remark. In the differential geometry a curve is called a *geodesic curve* if it has the property that for any two of its points (not too far apart) the distance between these points is equal to the length of the segment of that curve between these points. The assertion of (2) thus shows that the circles orthogonal to the unit circle T are geodesic curves in the unit disk $\boldsymbol{D}$ endowed with the Poincaré metric h_D. Speaking somewhat loosely, they are the shortest lines between two points.

PROOF. Let $g \in$ Möb be such that it sends z_∞ to 1, z_0 to -1, and z_1 to 0. Then g is an automorphism of $\boldsymbol{D}$ and, consequently, it preserves the metric d_D. To prove 1°, it, therefore, suffices to show that

$$(*) \qquad d_D(0, gz_2) = \tfrac{1}{2}\ln(0, gz_2, 1, -1).$$

But $gz_2 = r \in \boldsymbol{R}$ and $0 < r < 1$, so the right-hand side of $(*)$ is

$$\tfrac{1}{2}\ln(0, r, 1, -1) = \tfrac{1}{2}\ln\frac{1+r}{1-r}.$$

If γ is any path from 0 to r and if γ_0 is the path defined by $\gamma_0(t) = tr$, $t \in I$, then it is clear from (10) that $|\gamma| \geqslant |\gamma_0|$ and, consequently,

$$d_D(z_1, z_2) = d_D(gz_1, gz_2) = |\gamma_0|.$$

Using the formula (10) again, we obtain

$$|\gamma_0| = \int_0^r \frac{ds}{1-s^2} = \tfrac{1}{2}\ln\frac{1+r}{1-r}$$

which concludes the proof of both 1° and 2° since, obviously, the length of the circular segment $\langle z_1, z_2\rangle$ is equal to $|\gamma_0|$. □

It is of some interest to give explicit conditions for the coefficients of those Möbius transformations which give automorphisms of $\boldsymbol{D}$, i.e. which leave $\boldsymbol{D}$ invariant (and, consequently, also leave the unit circle and the exterior of $\boldsymbol{D}$ invariant). Using the formula for automorphisms of $\boldsymbol{D}$ given in Corollary XVI.8.9, we obtain easily

PROPOSITION XVI.8.16. A Möbius transformation defined by

$$gz = \frac{az+b}{cz+d}, \qquad \text{where } ad-bc = 1$$

is an automorphism of $\boldsymbol{D}$ if and only if $d = \bar{a}$, $c = \bar{b}$.

Remark. The condition $ad-bc = 1$ is nothing but a normalization condition, imposed in order to get rid of the (trivial) nonuniqueness due to the fact that matrices differing by a numerical multiple give rise to the same Möbius transformation.

The upper half-plane $\mathfrak{H}$ *and the group* $\mathrm{SL}_2(\boldsymbol{R})$. We have seen that Möb = Aut$(\bar{\boldsymbol{C}})$ and we have constructed a surjective homomorphism of the group $\mathrm{GL}_2(\boldsymbol{C})$ onto Möb by means of

$$\mathrm{GL}_2(\boldsymbol{C}) \ni \begin{bmatrix} a & b \\ c & d \end{bmatrix} = g \to T(g) \in \text{Möb},$$

where

$$T(g)(z) = \frac{az+b}{cz+d}.$$

To facilitate notation we shall continue to use the same symbol g to denote both a matrix and the Möbius transformation defined by that matrix. Since the kernel of $g \to T(g)$ is precisely the set $\{\lambda I: \lambda \in \boldsymbol{C}_*, I \text{ is the unit matrix}\}$, we can restrict our attention to the subgroup $\mathrm{SL}_2(\boldsymbol{C}) = \{g \in \mathrm{GL}_2(\boldsymbol{C}): \det g = 1\}$.

We have already remarked that the upper half-plane $\mathfrak{H} := \{z \in \boldsymbol{C}: \operatorname{Im} z > 0\}$ is conformally equivalent to the unit disk $\boldsymbol{D}$. In many instances, however, it is easier to deal with $\mathfrak{H}$ than with the disk $\boldsymbol{D}$. Thus we shall now give some results concerning the upper half-plane which are analogous to the results established for $\boldsymbol{D}$.

THEOREM XVI.8.17. Let $\boldsymbol{D} = \{z \in \boldsymbol{C}: |z| < 1\}$ be the unit disk and $\mathfrak{H} = \{z \in \boldsymbol{C}: \operatorname{Im} z > 0\}$ the upper half-plane.

1° The map

$$(12) \qquad \mathfrak{H} \ni z \to \varphi(z) = \frac{z-i}{z+i} \in \boldsymbol{D}$$

with the inverse given by

$$\boldsymbol{D} \ni z \to \varphi^{-1}(z) = \frac{1}{i}\frac{z+1}{z-1} \in \mathfrak{H}$$

is an isomorphism of $\mathfrak{H}$ with $\boldsymbol{D}$.

2° Denote $\mathrm{SL}_2(\boldsymbol{R}) = \left\{\begin{bmatrix} a & b \\ c & d \end{bmatrix} : a, b, c, d \in \boldsymbol{R} \text{ and let } \begin{bmatrix} a & b \\ c & d \end{bmatrix} = 1\right\}$.
Then every automorphism of $\mathfrak{H}$ is of the form

$$\mathfrak{H} \ni z \to gz = \frac{az+b}{cz+d} \in \mathfrak{H}$$

with $g = \begin{bmatrix} a & b \\ c & d \end{bmatrix} \in \mathrm{SL}_2(\boldsymbol{R})$. Moreover, this establishes an isomorphism between the group $\mathrm{Aut}(\mathfrak{H})$ and the group $\mathrm{SL}_2(\boldsymbol{R})/\{I, -I\}$, where I is the unit matrix in $\mathrm{SL}_2(\boldsymbol{R})$.

3° The Hermitian metric (of Poincaré and Klein) on $\mathfrak{H}$ which is left-invariant under $\mathrm{Aut}(\mathfrak{H})$ is given by

$$ds_{\mathfrak{H}}^2 = \tfrac{1}{2}\frac{dz \otimes d\bar{z}}{y^2}, \qquad \text{where } z = x+iy \in \mathfrak{H}.$$

The Gaussian curvature of that metric is $K_{\mathfrak{H}} = -4$.

4° The geodesics of the Poincaré–Klein metric on $\mathfrak{H}$ are precisely circles (including the straight lines) orthogonal to the axis $\mathrm{Im}\, z = 0$.

PROOF. 1° Since φ defined by (12) takes the points $0, 1, \infty$ upon $-1, -i, 1$, it thus maps the real line upon the unit circle. In those circumstances the upper half-plane has to be mapped either onto the unit disk $\boldsymbol{D}$ or onto the exterior of $\boldsymbol{D}$. Checking that $\varphi(i) = 0$, we see that $\varphi(\mathfrak{H}) = \boldsymbol{D}$.

2° One way to prove this part is to use Proposition XVI.8.16 and then part 1°. We shall proceed in a different way. First note that if a, b, c, d are real, then the formula

$$\mathrm{Im}\left(\frac{az+b}{cz+d}\right) = \frac{ad-bc}{|cz+d|^2}\,\mathrm{Im}\, z$$

shows that $\mathfrak{H}$ is mapped upon itself by the map $gz = \dfrac{az+b}{cz+d}$ provided that $ad-bc > 0$. Since matrices differing by a constant give rise to the same Möbius transformation, we can always assume that $ad-bc = 1$, i.e., $\det g = 1$.

Now the group $\mathrm{SL}_2(\boldsymbol{R})$ acts transitively in $\mathfrak{H}$ by means of Möbius transformations, since for a suitably chosen $g = \begin{bmatrix} a & b \\ c & d \end{bmatrix} \in \mathrm{SL}_2(\boldsymbol{R})$ the point

$$gi = \frac{ai+b}{ci+d}$$

will be equal to any chosen point from $\mathfrak{H}$. To get the result, we need to find the isotopy group of some point $z_0 \in \mathfrak{H}$. We choose $z_0 = i$. If $f \in \mathrm{Aut}(\mathfrak{H})_i$ then $f_1 = \varphi \circ f \circ \varphi^{-1} \in \mathrm{Aut}(\boldsymbol{D})_0$ and so it is of the form

$$f_1(z) = e^{2i\theta} z, \quad \theta \in \boldsymbol{R};$$

in particular f_1 is a Möbius transformation. Then $f = \varphi^{-1} \circ f_1 \circ \varphi$ is also a Möbius transformation and is computed to be

$$fz = \frac{z\cos\theta - \sin\theta}{z\sin\theta + \cos\theta}.$$

The remaining part of 2° is clear.

Parts 3° and 4° follow immediately from the corresponding results valid for $\boldsymbol{D}$. We omit the details. □

Remark 1. Part 1° is a particular instance of the Koebe–Riemann Theorem (XVI.8.19) to be proved later.

Remark 2. Part 2° of the preceding theorem shows in particular that every automorphism of $\mathfrak{H}$ can be extended to an automorphism of $\boldsymbol{P}(\boldsymbol{C})$, a fact which is not obvious at first sight. We have made use of the Schwarz Lemma to prove the corresponding result for $\boldsymbol{D}$ (and also implicitly in the proof above). The group $\mathrm{SL}_2(\boldsymbol{R})$ is one of the most important groups in mathematics, perhaps even more important for the development of mathematics than the group of Euclidean notions of $\boldsymbol{R}^2$. We give here an important parametrization of $\mathrm{SL}_2(\boldsymbol{R})$.

PROPOSITION XVI.8.18. Let A, K, N denote the following subgroups of the group $\mathrm{SL}_2(\boldsymbol{R})$:

$$A = \left\{\begin{bmatrix} e^t & 0 \\ 0 & e^{-t} \end{bmatrix} : t \in \boldsymbol{R}\right\}, \quad K = \left\{\begin{bmatrix} \cos\varphi & -\sin\varphi \\ \sin\varphi & \cos\varphi \end{bmatrix} : \varphi \in \boldsymbol{R}\right\},$$

$$N = \left\{\begin{bmatrix} 1 & x \\ 0 & 1 \end{bmatrix} : x \in \boldsymbol{R}\right\}.$$

Then the map

$$A \times N \times K \ni (a, n, k) \to ank \in \mathrm{SL}_2(\boldsymbol{R})$$

is a diffeomorphism onto $\mathrm{SL}_2(\boldsymbol{R})$.

In loose terms, every element of $SL_2(\boldsymbol{R})$ can be uniquely decomposed (in a "smooth way") into a product of elements from A, N, K. Moreover, K is precisely the isotopy subgroup of the point $i \in \mathfrak{H}$.

We shall not give the proof; the interested reader may take it as a challenging exercise. □

Exercise. Compute the Hurwitz measure on $SL_2^{\pm}(\boldsymbol{R})$ by using the Iwasawa decomposition.

We now turn to

THEOREM XVI.8.19 (The Koebe–Riemann Mapping Theorem). Let Ω be a simply connected domain in $\boldsymbol{P}(\boldsymbol{C})$ with the boundary containing at least two distinct points a, b. Then there exists a biholomorphic map $f\colon \Omega \to \boldsymbol{D}$ of Ω onto the unit disk.

From this theorem and Theorem XVI.8.6 (β) follows

THEOREM XVI.8.20. Any simply connected domain in $\boldsymbol{P}(\boldsymbol{C})$ is biholomorphically isomorphic to one of three canonical domains:

1° $\boldsymbol{D}$ ($\cong \mathfrak{H}$), a disk (upper half-plane),

2° $\boldsymbol{C}$, a plane ($\cong$ punctured sphere),

3° $\boldsymbol{P}(\boldsymbol{C})$ ($= \overline{\boldsymbol{C}}$), a sphere.

The proof of the Koebe–Riemann Theorem which we give now is, in principle, due to Koebe, but its final shape comes from Carathéodory.

The proof is carried out in two steps:

I. Construction of a holomorphic injection $h_1\colon \Omega \to \boldsymbol{D}$.

II. Construction of a holomorphic bijection $f_0\colon \Omega \to \boldsymbol{D}$ as a function with the following extremal property. Choose $z_0 \in \Omega$ and define $F := \{f \in A(\Omega)\colon 1° f$ injective, $2° |f(z)| \leqslant 1$ in Ω, $3° |f'(z_0)| \geqslant |h_1'(z_0)|\}$, where h_1 is a function constructed in step I. For $\mu\colon F \to \boldsymbol{R}$ defined as $\mu(f) = |f'(z_0)|$ let f_0 be a function from F at which μ assumes its maximum value. (Since F is shown to be a compact subset of $A(\Omega)$ and μ is continuous there exist such functions by virtue of the theorem of Weierstrass XII.8.7.).

Ad I. By assumption, Ω has at least two boundary points, z_1 and z_2. Then, by Corollary XVI.1.5 there is a function $h \in A(\Omega)$ such that

$$h^2(z) = \frac{z - z_1}{z - z_2}, \qquad z \in \Omega,$$

i.e., h is a holomorphic branch of the square root of the function $z \to \frac{z-z_1}{z-z_2}$. Now, h is injective (check this) and if $w \in h(\Omega)$, then $-w \notin h(\Omega)$. Indeed, let $h(\zeta) = w$, $h(\eta) = -w$, where $\zeta, \eta \in \Omega$; then by the definition of h

$$w^2 = \frac{\zeta-z_1}{\zeta-z_2} = \frac{\eta-z_1}{\eta-z_2}, \qquad \text{hence } \zeta = \eta.$$

From the open mapping theorem (XVI.5.1) it follows that $h(\Omega)$ is open: if $w_0 \in h(\Omega)$, there exists a $\delta > 0$ such that $K(w_0, \delta) \subset h(\Omega)$; thus, by the remark above the set $K(-w_0, \delta) = \{w \in \boldsymbol{C}\colon |w+w_0| < \delta\}$ is in $\complement(h(\Omega))$. Thus, $|h(z)+w_0| \geqslant \delta$ for $z \in \Omega$. Therefore

$$h_1(z) := \frac{\delta}{h(z)+w_0}$$

is holomorphic. But h_1 is also injective and, plainly, $|h_1(z)| \leqslant 1$. Having constructed a holomorphic injection $h_1\colon \Omega \to \boldsymbol{D}$, we define the family $F \subset A(\Omega)$ as in II, above, using the function h_1. Since h_1 obviously belongs to F this family is not empty. To prove that it is compact, it suffices by the Montel–Stieltjes–Vitali Theorem (XV.4.5) to check that F is closed. Let $f = \lim f_n$ where $f_n \in F$ and f_n converges compactly. Obviously, f satisfies condition 2° of the definition of the family F as well as 3° since the derivatives f_n' form a compactly convergent sequence with the limit f'. By the Hurwitz Lemma (XVI.5.8) f is injective since it is not constant and, thus, finally we get $f \in F$.

Let $f_0 \in F$ be a function with the maximum value of $\mu(f) = |f'(z_0)|$ among all the functions in F. The proof will be completed if we verify that $f_0(\Omega) = \boldsymbol{D}$. Assume that $w_0 \in \boldsymbol{D}$ and $w_0 \notin f_0(\Omega)$. This assumption allows us to construct a $g \in F$ such that $|g'(z_0)| > |f_0'(z_0)|$, which will contradict the definition of f_0. We prove first that $f_0(z_0) = 0$. If this were not the case, let

$$g_1(z) := \frac{f_0(z)-f_0(z_0)}{1-\overline{f_0(z_0)}f_0(z)};$$

then $g_1 \in F$ and

$$g_1'(z_0) = \frac{(1-|f_0(z_0)|^2)f_0'(z_0)}{(1-|f_0(z_0)|^2)^2} = \frac{f_0'(z_0)}{1-|f_0(z_0)|^2}.$$

This implies that $|g_1'(z_0)| > |f_0'(z_0)|$, provided that $f_0(z_0) \neq 0$, which contradicts the definition of f_0; thus $f_0(z_0) = 0$. Next we put

$$h(z) = \left(\frac{f_0(z) - w_0}{1 - \overline{w}_0 f_0(z)}\right)^{1/2}, \quad z \in \Omega$$

which is well defined, since $f_0(z) \neq w_0$ for all $z \in \Omega$, and is injective and satisfies $|h(z)| \leqslant 1$. To get a function from F we form a superposition of h with an automorphism of $\boldsymbol{D}$, which sends $h(z_0)$ to 0. Specifically, if we put

$$g(z) = \frac{h(z) - h(z_0)}{1 - \overline{h(z_0)} h(z)}$$

then we verify that conditions 1° and 2° are satisfied. As for 3°, after some calculations we obtain

$$g'(z_0) = \frac{h'(z_0)}{1 - |h(z_0)|^2} = \frac{1 + |w_0|^2}{2\sqrt{w_0}} f_0'(z_0),$$

which yields

$$|g'(z_0)| > |f_0'(z_0)| > |h_1'(z_0)|;$$

we thus have 3° and a contradiction with the maximality of f_0. □

Remark. Riemann's original idea of the proof was "potential-theoretic": Let $g = g_{z_0}^{\Omega}$ be the Green's function of the simply connected region $\Omega \subset \boldsymbol{C}$ with pole at z_0. One can determine a holomorphic function $f = f_{z_0}$ such that

$$(*) \qquad \log|f_{z_0}(z)| = -g_{z_0}^{\Omega}(z).$$

Then $f_{z_0}\colon \Omega \to \boldsymbol{D}$ and f_{z_0} is biholomorphic, $f_{z_0}(z_0) = 0$. The reader now has all the tools to make Riemann's idea precise:

Exercise. Let Ω have a regular boundary $\partial\Omega$. Apply the existence theorem for the Green's function from Chapter XV and the theorem on the existence of conjugate harmonic functions. Prove the injectivity of f_{z_0} in a way similar to that followed in the pıoof of the Koebe–Riemann Theorem.

Remark. In Section 13 the Green's function method is used in the proof of the "general uniformization theorem".

The table on pages 480–481 and Theorem XVI.21.27 may convince the reader that conformal mappings exist, in spite of the difficult proof of the Riemann mapping theorem.

A Method of Effective Analytic Continuation. The Schwarz Symmetry Principle. Now we shall give a theorem concerning a method of analytic continuation. This theorem is the one-dimensional case of the edge-of-the-wedge theorem. In particular, the familiar symmetry principle of Schwarz follows from it (see also Chapter XVII, Section 27).

THEOREM XVI.8.21 (Painlevé). Let $\Omega_1 \subset \boldsymbol{C}$ be an open set lying in the upper half-plane $\mathfrak{H}$, and having an interval $]a, b[$ of the real axis as part of its boundary. Let Ω_2 be an open set lying in the lower half-plane ($\operatorname{Im} z < 0$) and having the same interval $]a, b[$ as part of its boundary. Let

(i) $F_1 \in A(\Omega_1)$, $F_2 \in A(\Omega_2)$,

and

(ii) suppose that the limits

$$F_1(x) = \lim_{y \to 0+} F_1(x+iy), \qquad F_2(x) = \lim_{y \to 0+} F_2(x-iy)$$

exist and the convergence is compact on $]a, b[$. Moreover, suppose that $F_1(x) = F_2(x)$ for all $x \in]a, b[$.

Then there exists a holomorphic function F on $\Omega_1 \cup \Omega_2 \cup]a, b[$ such that

$$F(z) = \begin{cases} F_1(z), & z \in \Omega_1, \\ F_2(z), & z \in \Omega_2, \\ F_1(x) = F_2(x), & x \in]a, b[. \end{cases}$$

PROOF. Let $a < a_1 < b_1 < b$. Construct a rectangle $a_1 b_1 dc$ lying in $\bar{\Omega}_1$ (see Fig. 11). Let z lie in this rectangle, but not on the real axis.

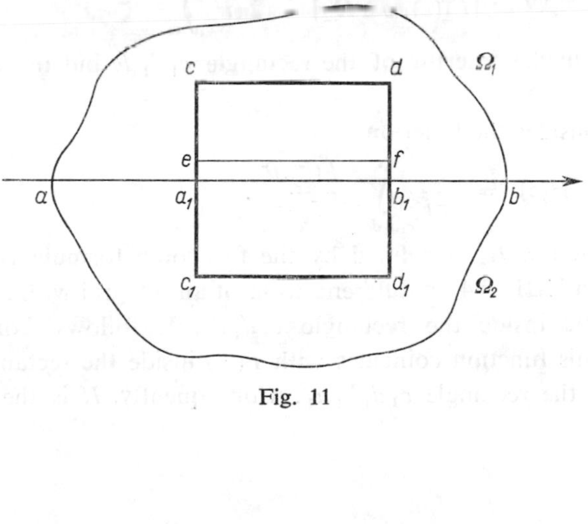

Fig. 11

It is then possible to construct a rectangle *efdc* containing z as an internal point. By Cauchy's Theorem we have

$$(13)\qquad F_1(z) = \frac{1}{2\pi i}\int_{efac e} \frac{F_1(\zeta)\,d\zeta}{\zeta - z}$$

for this rectangle. On passing to the limit with the segment *ef*, by hypothesis (ii) we have

$$(14)\qquad F_1(z) = \frac{1}{2\pi i}\int_{a_1b_1dca_1} \frac{F_1(\zeta)\,d\zeta}{\zeta - z}$$

for all z in the interior of the rectangle $a_1 b_1 dc$. Clearly, for z lying outside rectangle $a_1 b_1 dc$, formula (13), and hence formula (14) as well, yields

$$(15)\qquad \frac{1}{2\pi i}\int_{a_1b_1dca_1} \frac{F_1(\zeta)\,d\zeta}{\zeta - z} = 0.$$

On constructing another rectangle $c_1 d_1 b_1 a_1$ in similar fashion, we have

$$(16)\qquad F_2(z) = \frac{1}{2\pi i}\int_{c_1d_1b_1a_1c_1} \frac{F_2(\zeta)\,d\zeta}{\zeta - z}$$

for z inside the rectangle $c_1 d_1 b_1 a_1$ and

$$(17)\qquad \frac{1}{2\pi i}\int_{c_1d_1b_1a_1c_1} \frac{F_2(\zeta)\,d\zeta}{\zeta - z} = 0$$

for z outside it.

When (14) and (16) are added together and then (15) and (17) are taken into account, the result is

$$(18)\qquad F(z) = \begin{Bmatrix} F_1(z), & z \in \Omega_1 \\ F_2(z), & z \in \Omega_2 \end{Bmatrix} = \frac{1}{2\pi i}\int_{c_1d_1cc_1} \frac{F(\zeta)\,d\zeta}{\zeta - z}$$

for z lying in the interior of the rectangle $c_1 d_1 dc$ but not on the real axis.

Now consider the function

$$(19)\qquad H(z) := \frac{1}{2\pi i}\int_{c_1d_1dcc_1} \frac{F(\zeta)\,d\zeta}{\zeta - z}.$$

The function $H(\cdot)$ defined by the foregoing formula is, by virtue of Theorem XIII.10.4 on differentiation of an integral with a parameter holomorphic inside the rectangle $c_1 d_1 dc$. It follows from formula (18) that this function coincides with $F(\cdot)$ inside the rectangle $a_1 b_1 dc$ and inside the rectangle $c_1 d_1 b_1 a_1$. Consequently, H is the only analytic continuation of the function F to the rectangle $c_1 d_1 dc$. But by hypothesis (ii), the function F can be continued to a function continuous on $c_1 d_1 dc$ by setting $F(x) = \lim_{y\to 0+} F(x \pm iy)$.

Thus, $F = H$ in the interior of the rectangle $c_1 d_1 dc$ and F is holomorphic in the interior of $c_1 d_1 dc$ and, hence, in particular on the interval $]a_1, b_1[$ of the real axis. Such procedure can be carried out for every interval $]a_1, b_1[\subset]a, b[$, whereby F is holomorphic on $\Omega_1 \cup \Omega_2 \cup \cup]a, b[$. □

COROLLARY XVI.8.22 (The Symmetry Principle of Schwarz). If Ω_1 is an open set lying in the upper half-plane and has an interval $]a, b[$ of the real axis as part of its boundary and if F_1 belongs to $A(\Omega_1)$, is continuous over $\Omega_1 \cup]a, b[$, and assumes real values on $]a, b[$, then the function

$$F(z) := \begin{cases} F_1(z): & z \in \Omega_1 \cup]a, b[, \\ \bar{F}_1(\bar{z}): & z \in \tilde{\Omega}_1 \end{cases}$$

(where $\tilde{\Omega}_1 := \{z \in \boldsymbol{C}: \bar{z} \in \Omega\}$) is the analytic continuation of the function F_1 to the set $\Omega = \Omega_1 \cup \Omega_2 \cup]a, b[$.

PROOF. On setting $F_2(z) = \bar{F}_1(\bar{z})$ for $z \in \Omega_2 = \tilde{\Omega}_1$, we see that

$$\frac{\partial F_2}{\partial \bar{z}}(z) = \frac{\partial \bar{F}_1}{\partial z}(\bar{z}) = \overline{\frac{\partial F_1}{\partial \bar{z}}(\bar{z})} = 0.$$

Hence the functions F_1 and F_2 are holomorphic. If use is made of the fact that F_1 is real-valued on $]a, b[$, it becomes readily evident that

$$\lim_{y\to 0+} F_1(x + iy) = \lim_{y\to 0+} F_2(x - iy)$$

compactly with respect to $x \in]a, b[$. Application of the preceding theorem yields the conclusion. □

9. THE PERRON METHOD FOR RIEMANN SURFACES. THE RADÓ THEOREM

Sections 9–16 are devoted to the study of holomorphic and harmonic functions on Riemann surfaces. In the present section we shall show that the Perron method works for a large class of Riemann surfaces, the so-called hyperbolic surfaces. It is a remarkable and beautiful feature of the Perron method that it uses only local constructions and global countability never enters the picture (Ahlfors). Moreover, the second countability of Riemann surfaces is proved by this method. For hyperbolic surfaces one can introduce in a natural way the notions of the Green's function, of harmonic measure, and so on. Our aim is the fantastic Uniformization Theorem of Koebe and Poincaré (1907), which can be considered as a far-reaching generalization of Koebe's conformal mapping theorem. We prove here "only" the most important part of this theorem, namely, the case of a simply connected surface, which for hyperbolic surfaces reads: The simply connected hyperbolic Riemann surface M is a biholomorphic image of the unit disk $\boldsymbol{D}_1$.

In order to make the reading easier we recall here some basic facts from the theory of harmonic and subharmonic functions in $O \subset \boldsymbol{C}$ and give the corresponding theorems for Riemann surfaces. A parametric disk (or coordinate patch) with centre $p \in M$ is a coordinate system $(V, \varkappa)$ of M such that $\varkappa$ is holomorphic in a neighbourhood of $\overline{V}$, and $\varkappa(\overline{V}) = \overline{\boldsymbol{D}}_1 = \{z \in \boldsymbol{C}\colon |z| \leqslant 1\}$.

Since, for any two maps $\varkappa_j$, $\varkappa_k$ of M, $\varkappa_k \circ \varkappa_j^{-1}$ satisfies the Cauchy–Riemann equations, the (locally defined) Laplace operator Δ_z transforms according to the rule: if $z' := \varkappa_k \circ \varkappa_j^{-1}(z)$, then

$$\Delta_z = \left|\frac{dz'}{dz}\right|^2 \Delta_{z'}.$$

Thus, the Laplace equation $\Delta u = 0$ has an invariant global meaning and we can speak of harmonic functions on a Riemann surface M: u is *harmonic* on M if $u \in C^2(M)$ and satisfies $\Delta u = 0$. The space of harmonic functions on M is denoted by $\mathscr{H}(M)$. As before harmonic functions on M are characterized by Gauss's

MEAN VALUE THEOREM XVI.9.1. $(u \in \mathscr{H}(M)) \Leftrightarrow$ (for any coordinate disk $(V, \varkappa)$ the following holds:

$$(1) \qquad u(p) = \frac{1}{2\pi}\int_0^{2\pi} u(\varkappa^{-1}(re^{i\varphi}))\,d\varphi, \qquad 0 < r \leqslant 1.$$

For the same reasons the notion of subharmonicity can be introduced for functions on a Riemann surface.

DEFINITION. Let M be a Riemann surface. Then a function u is said to be *subharmonic* on M if

1° $u: M \to [-\infty, \infty[$, $u \not\equiv -\infty$;

2° u is upper semicontinuous in M, i.e.,

$$\limsup_{p' \to p} u(p') \leqslant u(p), \quad p \in M;$$

3° for each $x \in M$ there is a coordinate patch $(V, \varkappa)$ with centre x such that for each $0 < r \leqslant 1$ we have the mean-value inequality

$$(2) \qquad u(x) \leqslant \frac{1}{2\pi} \int_0^{2\pi} u\big(\varkappa^{-1}(re^{i\varphi})\big)\, d\varphi.$$

If $-v$ is subharmonic on M, then v is said to be *superharmonic*.

We shall denote sub(super) harmonic functions on M by $\underline{\mathscr{H}}(M)$ $(\overline{\mathscr{H}}(M))$. Plainly,

$$(u \text{ is harmonic on } M) \Leftrightarrow (u \in \overline{\mathscr{H}}(M) \cap \underline{\mathscr{H}}(M)).$$

Examples. 1. A harmonic function is both subharmonic and superharmonic.

2. If h is holomorphic on M, then $\log|h|$ is subharmonic. Therefore,

3. $z \to \log \dfrac{1}{|z-\zeta|}$ is superharmonic on $\boldsymbol{C} - \{\zeta\}$.

4. Let φ be a convex increasing function on $\boldsymbol{R}$ and set $\varphi(-\infty) := \lim\limits_{x \to -\infty} \varphi(x)$; then $\varphi \circ u \in \underline{\mathscr{H}}(M)$ if $u \in \underline{\mathscr{H}}(M)$.

In order to have an elegant formulation of several theorems let us introduce the following

DEFINITION. Let M be a metrizable space which is locally compact but not compact. We say that a sequence $(a_n)_1^\infty$ of points of M *converges* to the ideal boundary of M: $a_n \to \partial_{\text{id}} M$ if every compact subset of M contains only (at most) finite number of points a_n.

We now give two versions of the maximum principle.

THEOREM XVI.9.2 (Maximum Principle). Let M be a Riemann surface and $u \in \underline{\mathscr{H}}(M)$.

(a) If u attains its supremum (least upper bound), i.e., there is a $x_0 \in M$ such that $u(x_0) = \sup u(M)$, then $u \equiv$ const. Hence

$$(\text{const} \not\equiv u \in \underline{\mathscr{H}}(M)) \Rightarrow (u < \sup u(M)).$$

(b) Let $W = \lim\limits_{x \to \partial_{\text{id}} M} \sup u(x)$. Then

$$u \leqslant W \text{ or more precisely, either } u < W \text{ or } u \equiv W.$$

PROOF. (a) Let $L = \sup u(M)$ and $A = \{x \in M\colon\ u(x) = L\}$. Then $\emptyset \neq A$ and A is closed since u is upper semicontinuous. Let V be a coordinate disk with centre at $x \in A$. Then

$$L = u(x) \geqslant \frac{1}{2\pi}\int_0^{2\pi} u(\varkappa^{-1}(re^{i\varphi}))d\varphi, \quad 0 < r \leqslant 1;$$

but 3° of the definition implies that $u \equiv L$ on V. Therefore, $V \subset A$ and A is open. Since M is connected, $A = M$.

(b) Obviously $L \geqslant W$; choose a sequence (a_n), such that $\lim u(a_n) = L$. By (a) no compact subset of M can contain infinitely many points a_n and, hence, $a_n \to \partial_{\text{id}} M$. Then $L = \lim\limits_{n\to\infty} u(a_n) \leqslant \lim\limits_{x\to\partial_{\text{id}} M} \sup u(x) = W$ and both inequalities together give the desired result.

Poisson Integral. As we know (cf. Theorem XV.6.6) in the complex plane the Poisson integral formula has the following simple form:

THEOREM XVI.9.3 (Schwarz). The Poisson kernel for the disk $\boldsymbol{D}_R = \{z \in \boldsymbol{C}\colon |z| < R\}$ is of the form

$$P(z, \zeta) = \mathrm{Re}\,\frac{\zeta+z}{\zeta-z} = \frac{R^2-r}{R^2+r^2-2Rr\cos(\theta-\varphi)}, \tag{3}$$

where $\zeta = Re^{i\theta}$, $z = re^{i\varphi}$.

For any $f \in C(\partial\boldsymbol{D}_R)$ the Poisson integral

$$u(z) = \frac{1}{2\pi}\int_0^{2\pi} f(Re^{i\theta})\,\frac{R^2-r^2}{R^2+r^2-2Rr\cos(\theta-\varphi)}\,d\theta \tag{4}$$

is the solution of the Dirichlet problem

$$\Delta u = 0, \quad u|\partial\boldsymbol{D}_R = f.$$

We denote by P_z the integral operator with Poisson kernel $P_z(\zeta) \equiv P(z, \zeta)$ defined by (4). The Harnack inequalities and the Harnack principle are valid for any Riemann surface M.

THEOREM XVI.9.4 (Harnack). 1° Let $S = (u_n)$ be a monotone sequence of harmonic functions on M which is bounded at a point z_0; then (u_n) is compactly convergent to a harmonic function ("Harnack Principle" or "The Second Harnack Theorem").

2° Every compactly convergent sequence of harmonic functions converges to a harmonic function ("The First Harnack Theorem", also called the "Weierstrass Theorem").

We shall also need the following extension of the Harnack principle to cover the case of (upper) directed families of harmonic functions.

3° Let $\mathscr{H}$ be a family of harmonic functions on M such that for every $u_1, u_2 \in \mathscr{H}$ there exists a $u \in \mathscr{H}$ with $u \geqslant \sup(u_1, u_2)$. Then the function $H = \sup_{h \in \mathscr{H}} h$ is either harmonic or $\equiv \infty$.

The proofs of 1° and 2° were given as XV.6.8 and XV.5.12: as we know, 1° follows from the Harnack inequalities for nonnegative $u \in \mathscr{H}(D_R)$

$$\frac{R-|z|}{R+|z|}\,u(0) \leqslant u(z) \leqslant \frac{R+|z|}{R-|z|}\,u(0). \tag{5}$$

For completeness we sketch the proof of 3°. We choose $z_0 \in M$ and consider a sequence $(u_n) \subset \mathscr{H}$ such that $\lim u_n(z_0) = H(z_0)$. Since $\mathscr{H}$ is assumed to be upper directed, we can always take (u_n) to be a nondecreasing sequence and we thus see that the function $H_0 := \lim u_n$ is either harmonic or identically $+\infty$ by the First Harnack Theorem. Now we take another point $z_1 \in M$ and a sequence $(w_n) \subset \mathscr{H}$ converging at z_1 to $H(z_1)$. We can arrange things so that (w_n) is nondecreasing and $w_n \geqslant u_n$ (by taking the sup of the two if necessary) for all n. Then $H_1 := \lim w_n \geqslant H_0$ in general, whereas at z_0 we have $H_1(z_0) = H_0(z_0)$. Therefore, if H_0 is identically $+\infty$, the same is true of H_1 and since $H_1(z_1) = H(z_1)$ and z_1 was an arbitrary point from M, the function H is identically $+\infty$ as well. On the other hand, if H_0 and H_1 are finite, then the harmonic function $H_0 - H_1$ has a maximum equal to 0 at the point z_0 and thus is constantly equal to 0. In particular, $H_0(z_1) = H_1(z_1) = H(z_1)$ and in view of the arbitrariness of z_1 it follows that $H_0 = H$ throughout M.

DEFINITION. We say that a set $A \subset M$ is of *measure zero* if, for each coordinate disk V, $\lambda^2(A \cap V) = 0$ (the Lebesgue measure of $A \cap V$ vanishes). We say that a (numerical) function $f\colon M \to \bar{R}$ is *locally integrable* if each restriction $f|V$ to a coordinate disk is (Lebesgue) integrable.

Exercise. Prove that each super(sub)harmonic function is locally summable and hence is finite almost everywhere.

COROLLARY XVI.9.5.
1° $(u_1, u_2 \in \underline{\mathscr{H}}(M)) \Rightarrow (u_1 \cup u_2 := \max(u_1, u_2) \in \underline{\mathscr{H}}(M))$.

2° Let $S = (u_n) \searrow$ be a sequence of subharmonic functions which is bounded from below at a point $p_0 \in M$; then $\lim_{n\to\infty} u_n$ is subharmonic.

COROLLARY XVI.9.6. Let $S \subset \underline{\mathscr{H}}(M)$ be a lower directed set (i.e., for $s_1, s_2 \in S$ there exists an $s \in S$ such that $s \leqslant s_1$, $s \leqslant s_2$); then its lower envelope inf S is either identical with $-\infty$ or subharmonic.

We now prove a simple

LEMMA XVI.9.7. Let $u \in \underline{\mathscr{H}}(M)$, $u \not\equiv -\infty$ and let K be a parametric disk in M. Then there exists in K a harmonic function h satisfying $u|_K \leqslant h$ and $u|_{\partial K} = h|_{\partial K}$.

For obvious reasons such h is called the *best harmonic majorant* of u.

PROOF. Let (f_n) be a sequence of continuous functions on ∂K such that $f_n \searrow u$ on ∂K. Let $h_n \in \mathscr{H}(K)$ and $h_n|_{\partial K} = f_n$. Then (h_n) is a monotone sequence of harmonic functions and by virtue of the maximum principle $h_n \geqslant u$. By Harnack principle $h = \lim h_n$ is harmonic and plainly has the desired properties. □

We shall define a linear map $\underline{K}$ on the space $\underline{\mathscr{H}}(M)$. Let u be subharmonic on M

$$\underline{K}u := \begin{cases} u & \text{on } M-K, \\ h & \text{on } K, \end{cases}$$

where h is the best harmonic majorant of u on K (cf. Lemma XVI.9.6). Then $\underline{K}u \geqslant u$ and the following simple lemma shows that $\underline{K}u$ is subharmonic.

LEMMA XVI.9.8. Let v be subharmonic on M, $G \subset M$ a domain and s be such a function that $s|G \in \underline{\mathscr{H}}(G)$, $s \geqslant v$ on G and $s = v$ on $M-G$, then s is subharmonic on M.

PROOF. We have only to check the mean-value inequality and it is sufficient to examine only the points of ∂G. Let $x \in \partial G$ and let (K, φ) be a parametric disk with centre at x. Then for every $0 < r \leqslant 1$

$$s(x) = v(x) \leqslant \frac{1}{2\pi}\int_0^{2\pi} v(\varphi^{-1}(re^{i\theta}))d\theta \leqslant \frac{1}{2\pi}\int_0^{2\pi} s(\varphi^{-1}(re^{i\theta}))d\theta . \square$$

By examination of the proof of Theorem XV.7.2 we arrive at the following general

THEOREM XVI.9.8 (Perron). Let M be a Riemann surface and let v be a superharmonic function on M.

If $\mathfrak{P} \equiv \mathfrak{P}(v)$ is such a family of subharmonic functions on M that

1° $((u_1, u_2) \in \mathfrak{P}) \to (\sup(u_1, u_2) \in \mathfrak{P})$,

2° $(u \in \mathfrak{P}) \Rightarrow (\underline{K}u \in \mathfrak{P}$ for any coordinate disk $K)$,

3° $u \leqslant v$ for every $u \in \mathfrak{P}$,

then $H := \sup_{u \in \mathfrak{P}} u$ is harmonic on M.

Remark. A family which satisfies 1°–3° is usually called a *Perron family*. We shall modify these conditions slightly, however, for greater flexibility.

DEFINITION. A non-void set $\mathfrak{P} \subset \underline{\mathscr{H}}(M)$ is a Perron family if

1°′ $(u_1, u_2 \subset \mathfrak{P}) \Rightarrow (\bigvee_{u \in \mathfrak{P}} u_1, u_2 \leqslant u)$, i.e. $\mathfrak{P}$ is upper directed;

2°′ $\bigwedge_{p \in M} \bigvee_{K_p \ni p} \bigwedge_{u \in \mathfrak{P}} \bigvee_{\tilde{u} \in \mathfrak{P}} \tilde{u}|K_p \geqslant P(u|\partial K_p)$, where K_p is a parametric disk centred at p;

3°′ there exists a superharmonic majorant v of $\mathfrak{P}$, i.e., $u \leqslant v$ for every $u \in \mathfrak{P}$.

PROOF OF THE PERRON THEOREM. From 3° it follows that $H \not\equiv +\infty$. Let $\mathfrak{P}_{K_p} = \{u \in \mathscr{H}(K_p): u = P(s|\partial K_p), s \in \mathfrak{P}\}$ then

1°′ $\Rightarrow \mathfrak{P}_{K_p}$ is upper directed;

2°′ $\Rightarrow \sup \mathfrak{P}_{K_p} = H_{|}K_p$,

and hence, either $H|K_p \equiv +\infty$ or $H|K_p$ is harmonic. Observe that this argument shows that the subset of M where H is infinite is open. Since it is obviously closed, it must be either empty or the whole M. But the latter is impossible since $H \leqslant v$ and v is not identically infinite. $\square$

As a corollary of the Perron Theorem we obtain the following theorem which is very important for modern potential theory:

THEOREM XVI.9.9 (The Principle of the Best Subharmonic Minorant). A superharmonic function which possesses a subharmonic minorant has the best subharmonic minorant; this best subharmonic minorant is harmonic.

As an important application of the Perron method we shall prove the

Second Countability of Riemann Surfaces. Radó's Theorem. Our next task is to construct a function $f \in C^1(M)$ such that df vanishes only at isolated points. This in turn allows us to metrize the surface M. And by virtue of a theorem of P. Alexandrov stating that "a connected locally compact metric space is separable", we obtain an important result of Tibor Radó (Theorem XVI.9.13): Any Riemann surface satisfies the second axiom of countability.

Let G be a relatively compact domain of M and assume that $\partial G \neq \emptyset$. Let f be an arbitrary numerical function on ∂G; we define a function H_f *the (lower) Peron solution of the Dirichlet problem* with the boundary values f by the formula

$$H_f := \sup_{v \in \mathfrak{P}} v,$$

where the family $\mathfrak{P} = \mathfrak{P}(f)$ is defined as

$$\mathfrak{P} = \{u \in \underline{\mathscr{H}}(G) \colon \limsup_{x \to \xi} u(x) \leqslant f(\xi) \text{ for all } \xi \in \partial G\}.$$

We remark that unless it is identically $+\infty$ or identically $-\infty$, H_f is harmonic.

To consider the boundary behaviour of H_f we introduce the slightly generalized notion of a barrier.

DEFINITION. A function $b = b^a$ defined in G will be called a *barrier* at $a \in \partial G$ if

1° $b \in \overline{\mathscr{H}}(G)$,

2° $\lim\limits_{G \ni x \to a} b(x) = 0$,

3° $\liminf\limits_{x \to \xi} b(x) > 0$ for all $\xi \neq a$, $\xi \in \partial G$.

We leave it to the reader to verify that just as in XV.7.3 the existence of a barrier at a assures that a is a regular point: $\lim\limits_{x \to a} H_f(x) = f(a)$ for every continuous $f \in C(\partial G)$ and conversely. Therefore, for any relatively compact domain in M with only regular boundary points the Dirichlet problem with continuous boundary values f has a (unique) solution.

We shall now give quite a general condition for the existence of a barrier.

THEOREM XVI.9.10. The point $a \in \partial G$ has a barrier if the component F of $M-G$ which contains a does not reduce to $\{a\}$.

PROOF. Since the problem is a local one, it is sufficient to consider the case of plane regions. An auxiliary linear transformation permits us to choose $a = 0$ a nd$z = 1 \subset F$. Let V be a disk with centre at $a = 0$. Denote by G_j the connected components of $V-F$. On each G_j we have a single-valued branch of $\log z$. If $\partial G_j \cap \partial V \neq \emptyset$, we take a $z_j \in \partial G_j \cap \cap \partial V$ and denote by w_j that branch of $\log z$ on G_j for which

$$-\pi \leqslant \operatorname{Im} w_j(z_j) < \pi .$$

Let us define s^V:

$$s^V(z) := \begin{cases} 0 & \text{for such } G_j \text{ that } \partial G_j \cap \partial V = \emptyset, \\ 0 < \arg \dfrac{w_j(z)-3\pi i}{w_j(z)+3\pi i} < 1 & \text{on al other } G_j, \\ 1 & \text{on } G-V. \end{cases}$$

We check that s^V is superharmonic, $0 \leqslant s^V \leqslant 1$, and $\lim s^V(b) = 0$ for $G \ni b \to a$.

Let $V_n \downarrow \{a\}$ be a sequence of disks with centres at a; then

$$h^a := \sum_{n=1}^{\infty} 2^{-n} s^{V_n}$$

is a barrier at a. □

A much more primitive criterion is given in the

Exercise. Let G be a relatively compact domain in $\mathbf{C}$. Prove that $a \in \partial G$ has a barrier b^a if there exists a closed interval $[a, \zeta]$ such that

$$]a, \zeta] \subset \mathbf{C}-\overline{G}.$$

Hint. Since a holomorphic branch of $((z-a)/(z-\zeta))^{1/2}$ can be defined outside the segment $[a, \zeta]$, verify (after a proper determination of the angle θ) that the function

$$G \ni z \to \operatorname{Im}\left(e^{-i\theta}\sqrt{\frac{z-a}{z-\zeta}}\right)$$

is a barrier for a.

COROLLARY XVI.9.11. Let $M_1 := M - \overline{V}_0 \cup \overline{V}_1$ be a complement of two disjoint closed parametric disks of a Riemann surface M. Then there exists a non-constant harmonic function h on M_1 such that:

$$h|\partial V_0 \equiv 0, \quad h|\partial V_1 \equiv 1.$$

PROOF. The existence of such a harmonic h follows from Theorem XVI.9.8: we take the Perron family $\mathfrak{P}(1)$ of bounded by 1 subharmonic functions u on M_1 which vanish on ∂V_0 and are 1 on ∂V_1. Then $h := \sup_{u \in \mathfrak{P}(1)} u$ is a non-constant harmonic function on M_1.

(The boundary $\partial V_0 \cup \partial V_1$ is regular and, hence, $h|\partial V_0 = 0, h|\partial V_1 = 1$.) □

We now prove an important

PROPOSITION XVI.9.12. The existence of a non-constant harmonic function on a Riemann surface M implies that M is metrizable and, therefore, M has a countable basis (by the cited theorem of Alexandrov; see Theorem XVI.9.14 below).

PROOF. Let h be a non-constant harmonic function on M; then $f := |\operatorname{grad} h|$ vanishes only at isolated points (otherwise, h would have to be constant). We define a distance d on M in the following way: denote by $W(p, q)$ the set of piecewise differentiable arcs γ from p to q and put $\gamma\colon [0, 1] \to M$, $\gamma(0) = p, \gamma(1) = q$. Let $l(\gamma) := \int_0^1 f(\gamma(t))\,dt$, and put

$$(*) \qquad d(p, q) := \inf_{\gamma \in W(p,q)} l(\gamma).$$

Noting that every connected manifold is arcwise connected since it is connected and locally arcwise connected, we see that d is well defined on $M \times M$.

We check that $d(\cdot, \cdot)$ is a distance on M:

1° $d(p, q) = d(q, p)$ is obvious,

2° $d(p, r) \leqslant d(p, q) + d(q, r)$.

Let $w(p, q)$ be a path from p to q and $w(q, r)$ a path from q to r; since these paths give a path $w(p, r)$ from p to r, from $(*)$ we have $l(w(p, q)) + l(w(q, r)) \geqslant d(p, r)$ and 2° is satisfied.

3° If $p = q$, then $d(p, q) = 0$: take a constant path $\gamma([0, 1]) = p$; then $l(\gamma) = 0$.

If $p \neq q$, then $d(p, q) > 0$. Indeed, let (V_p, z) be such a coordinate patch with centre at p that $q \notin V_p$; then every path $\gamma \in W(p, q)$ contains a segment γ_1 such that $z(\gamma_1) \in K(0, 1) \subset C$. Let $\gamma([0, \varepsilon]) \subset V_p$, $\varepsilon > 0$ and let

$$\varepsilon_0 = \sup \{\varepsilon \colon \gamma([0, \varepsilon]) \subset V_p\}.$$

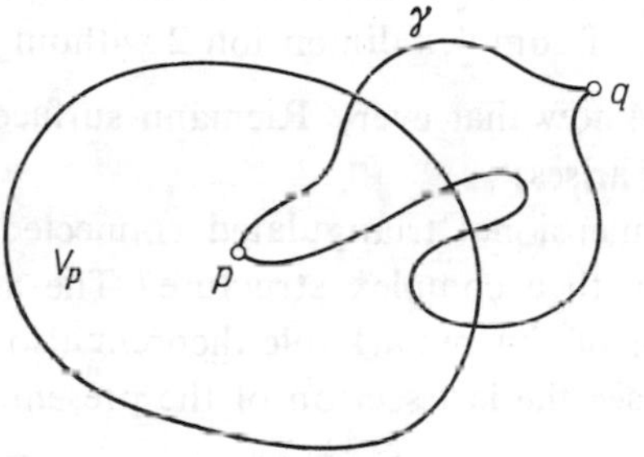

Fig. 12

Since $\overline{V}_p$ is compact, $\gamma(\varepsilon_0) \in \overline{V}_p - V_p = \partial V_p$ and we have $|z(\gamma(\varepsilon_0))| = 1$. But f vanishes on a discrete set; therefore, there exists an annulus $A = \{z \in C \colon r_1 \leqslant |z| \leqslant r_2 \leqslant 1\}$ and $f(z^{-1}(A)) \geqslant c > 0$. Consequently, $l(\gamma) > c(r_2 - r_1)$ for every $\gamma \in W(p, q)$ and, hence, $d(p, q) > 0$.

We now have to check that $\mathscr{U}(p, \varrho) = \{x \in M \colon d(p, x) < \varrho\}$ form a basis of the topology of the Riemann surface M. We have proved above that $\mathscr{U}(p, \varepsilon_0) \subset V_p$. Since f is bounded on V_p: $0 \leqslant f(x) < m$, $x \in V_p$ and since for each $\varrho > 0$ there exists a neigbourhood $\mathcal{O}(p)$ of p such that $|z(q)| < \varrho/m$ for $q \in \mathcal{O}(p)$, we have $d(p, q) < m \cdot \varrho/m = \varrho$ for $q \in \mathcal{O}(p)$. Therefore, $\mathcal{O}(p) \subset \mathscr{U}(p, \varrho) \subset V_p$.

Since we have proved that M is metrizable by the theorem of Alexandrov, it follows that M has a countable basis. □

The preceding two facts easily imply the famous

THEOREM XVI.9.13 (Radó). Every Riemann surface M has a countable basis.

PROOF. We remove two coordinate disks V_0, V_1. By Corollary XVI.9.11 and Proposition XVI.9.12 $M_1 = M - (V_0 \cup V_1)$ has a countable basis and, therefore, the same is true for $M = M_1 \cup V_0 \cup V_1$. □

Remark 1. Tibor Radó proved his theorem by an entirely different method, namely, by the uniformization theorem. But the uniformization theorem is much more profound and its proof poses quite different difficulties (see below). The question arises as to whether connected complex manifolds of complex dimension greater than 1 are second countable. The method of proof plainly does not extend to higher dimensions since a non-constant harmonic function may vanish on a non-discrete subset. In fact, E. Calabi and M. Rosenlicht constructed (in 1953) connected complex manifolds of complex dimension 2 without a countable basis.

Remark 2. We know that every Riemann surface is orientable. The following question arises:

Can every 2-dimensional triangulated connected topological manifold be provided with a complex structure? The answer is "yes". We owe the first proof of this remarkable theorem also to T. Radó (1924); for another proof see the last section of the present chapter.

Remark 3. A modern proof of Alexandrov's Theorem is given in the last section of Chapter XVII, but since it seems instructive to have an elementary proof of this important fact, we provide one below. We recall that a compact metric space is separable.

THEOREM XVI.9.14 (of Alexandrov). Let (X, d) be a locally compact connected metric space. Then X can be covered by a sequence of compact subsets.

PROOF. Let $U(p, \varepsilon) = \{x \in X\colon d(p, x) \leqslant \varepsilon\}$. Plainly, $U(p, \varepsilon)$ is compact in X for small ε. Denote $r(p) := \sup \varepsilon$ where $U(p, \varepsilon)$ is compact in X. If $r(p) = \infty$, the proof is complete, since $\{U(p, n)\}_n$, $n = 1, 2, \ldots$ is a covering of X by compact subsets. If, for a certain p, $r(p) < \infty$, then the same is valid for each point of X. Indeed, we have for all $p, p' \in M$

$$(*) \qquad r(p') \leqslant r(p) + d(p, p').$$

Otherwise, there would exist p, p', $p \neq p'$ and $\varepsilon > 0$ such that

$$r(p') = r(p) + d(p, p') + \varepsilon,$$

and, since $d(p', x) \leqslant d(p', p) + d(p, x) < r(p') - \varepsilon$, we would have $U(p, r) \subset U(p', r(p') - \varepsilon)$, contradicting the extremal property of $r(p)$.

The sets

$$V(p, \lambda) := \{x \in X\colon d(x, p) \leqslant \lambda r(p),\ 0 < \lambda < 1\}$$

are compact and form a basis of compact neighbourhoods at p.

Take a fixed $p_0 \in X$ and set

$$K_0 := V(p_0, \tfrac{1}{2}), \quad K_{n+1} := \bigcup_{x \in K_n} V(x, \tfrac{1}{2}), \quad n = 0, 1, 2, \ldots$$

Plainly, $X = \bigcup_{n=0}^{\infty} K_n$ since (by the connectedness of X) we can connect each point p of X with p_0 by a finite chain $V(p_0, \frac{1}{2}), \ldots, V(p_k, \frac{1}{2})$. (The non-void set A of such p is closed and open, whence $A = X$.) We have merely to check that K_n are compact. Plainly, K_0 is compact; we assume that K_n is compact and prove that K_{n+1} is compact, too. Let $(x_j)^\infty$ be a sequence in K_{n+1}, then (by virtue of the definition of K_{n+1}) for each x_j there is such an $x'_j \in K_n$ that $x_j \in V(x'_j, \frac{1}{4})$. But since K_n is compact, there exists a convergent subsequence (x'_{j_p}): $x'_{j_p} \to x' \in K_n$. But

$$\begin{aligned} d(x_j, x') &\leqslant d(x_j, x'_j) + d(x'_j, x') \leqslant \tfrac{1}{4} r(x'_j) + d(x'_j, x') \\ &\leqslant \tfrac{1}{4} r(x') + \tfrac{5}{4} d(x'_j, x'). \end{aligned}$$

Therefore, for sufficiently large j_p all x_{j_p} are contained in the compact set $V(x', \frac{1}{2})$ and thus have an accumulation point x in $V(x', \frac{1}{2})$. But $x \in K_{n+1}$ and, hence, K_{n+1} is compact. □

DEFINITION. A domain G in a Riemann surface M is *normal* if

1° $M - G$ contains no compact components,

2° $\overline{G}$ is compact in M,

3° ∂G is a sum of finite disjoint sectionally analytic Jordan curves.

Alexandrov's Theorem implies the important

PROPOSITION XVI.9.15 (On Normal Exhaustion). Every open Riemann surface can be exhausted by a sequence $(G_n)_1^\infty$ of normal domains such that $\overline{G}_n \subset G_{n+1}$.

The proof is left to the reader.

Remark. Many proofs in complex analysis use normal exhaustion, e.g., all classical proofs of the uniformization theorem. Those procedures are jocularly called "oil speck methods". The reader may check that the proof of the uniformization principle given below avoids oil specks.

10. RESOLUTIVE FUNCTIONS. HARMONIC MEASURES. BRELOT'S THEOREM

In the preceding chapter, for a regular domain $B \subset \boldsymbol{R}^n$ and every point $x \in B$ we defined the harmonic measure of x with respect to B, denoted by μ_x^B. This measure was concentrated on the boundary ∂B of B and defined by H_f^B. The solution of the Dirichlet problem with the boundary values f, i.e.,

$$H_f^B \in \mathscr{H}(B), \quad H_f^B | \partial B = f, \quad f \in C(\partial B),$$

is given by the formula

$$H_f^B(x) =: \int_{\partial B} f(p)\, d\mu_x^B(p) = \mu_x^B(f).$$

The Perron method works, as we have already remarked, for a relatively compact domain G on a Riemann surface M with a boundary ∂G consisting only of regular points, e.g., points possessing a barrier.

Therefore, given a regular domain $G \subset M$ and $x \in G$, we define the harmonic measure of x with respect to G, denoted μ_x^G, as the mapping $C(\partial G) \ni f \to H_f(x) \in \boldsymbol{R}$. Since every μ_x^G is a Radon measure (i.e., a linear positive functional on $C(\partial G)$) we can apply the Daniell–Stone procedure to extend the functional μ_x^G to integrable functions $\mathscr{L}^1 = \mathscr{L}^1(\mu_x^G)$.

It is interesting to note that the space $\mathscr{L}^1$ is independent of $x \in G$.

Oskar Perron developed his method in order to solve the Dirichlet problem. We have seen that this method works if there exists a non-constant positive superharmonic function. This leads to the following classification of non-compact (in classical terminology, open) Riemann surfaces.

DEFINITION. An open Riemann surface M is:

(P) *parabolic* if every negative subharmonic function is constant,

(H) *hyperbolic* if it is not parabolic.

Thus an open Riemann surface M is hyperbolic if

(H′) there exist on M non-constant positive superharmonic functions or, what is the same, there exist bounded from above non-constant subharmonic functions.

A domain G of a Riemann surface is *hyperbolic* if G, considered as a Riemann surface, is hyperbolic.

A point b belonging to the boundary of a relatively compact domain $G \subset M$ is regular if for each $f \in C(\partial G)$ the lower Perron solution H_f of the Dirichlet problem with the boundary values f satisfies

$$\lim_{G \ni a \to b} H_f(a) = f(b).$$

If each point of ∂G is regular we call G a *regular domain.*

Example 1. Since the Dirichlet problem has a solution for any regular relatively compact domain, we see that a regular relatively compact domain, whose boundary consists of at least two points is hyperbolic. One point will not suffice, as the next example shows.

Example 2. Every positive superharmonic function on C is a constant: the complex plane is parabolic.

PROOF. Let u be a negative subharmonic function on C and let $z_1 \neq z_2$; take $\varepsilon > 0$ such that $u(z_1) < -\varepsilon$, then there exists such δ with $0 < \delta < |z_1 - z_2|$ that $u(z) < u(z_1) + \varepsilon$ for $|z - z_1| < \delta$. But the harmonic function h_R

$$h_R(z) := (u(z_1)+\varepsilon)\frac{\log R - \log|z-z_1|}{\log R - \log\delta}$$

majorizes u in the annulus $\delta \leqslant |z-z_1| \leqslant R$ for any R. Since h_R converges to the constant $u(z_1)+\varepsilon$ as $R \to \infty$, we have $u(z) \leqslant u(z_1)+\varepsilon$ for each $\varepsilon > 0$ and, hence, $u(z_2) \leqslant u(z_1)$. In the same way $u(z_1) \leqslant u(z_2)$ and, therefore, $u = \text{const}$.

Later in Section 11 we shall need the following simple characterization of hyperbolicity.

Remark. Let $p \in M$ and let S_p^M be the set of positive superharmonic functions v on M such that, for every coordinate patch V_p with centre at p, we have $v(z) \geqslant \log \dfrac{1}{|z|}$. Then

$$(S_p^M \neq \emptyset) \Leftrightarrow (M \text{ is hyperbolic}).$$

In fact, $\Rightarrow$ is obvious by virtue of the definition of hyperbolicity.

$\Leftarrow$: Let M be hyperbolic and V_p a parametric disk with centre at p. Then Lemma XVI.10.11 asserts the existence of a positive superharmonic s such that $s(z) - \log\dfrac{1}{|z|}$ is harmonic on V_p; it can be made positive by adding a constant, if necessary; hence, $s \in S_p^M$.

Resolutive Functions. Theorem of Brelot. Let $B \subset M$ be a relatively compact domain in M. Let f be an arbitrary (real) numerical function on ∂B. Denote by $\overline{\mathscr{H}}_f^B \equiv \overline{\mathscr{H}}_f$ the set of superharmonic functions v on B such that

1° v is bounded from below,

2° $\liminf\limits_{x \to y \in \partial B} v(x) \geqslant f(y)$ for all $y \in \partial B$, and by $\underline{\mathscr{H}}_f \equiv \underline{\mathscr{H}}_f^B$ denote the set of subharmonic function u on B such that

1°′ u is bounded from above,

2°′ $\limsup\limits_{x \to y \in \partial B} u(x) \leqslant f(y)$ for all $y \in \partial B$.

Plainly, $\overline{\mathscr{H}}_f^B = -\underline{\mathscr{H}}_{(-f)}^B$ and $u \leqslant v$ for $u \in \underline{\mathscr{H}}_f$, $v \in \overline{\mathscr{H}}_f$, since $u-v$ is subharmonic and

$$\limsup_{x \to y \in \partial B} (u-v)(x) \leqslant 0.$$

Since $\underline{\mathscr{H}}_f$ and $-\overline{\mathscr{H}}_f$ are directed families,

$$\overline{H}_f^B := \inf_{v \in \overline{\mathscr{H}}_f} v \quad \text{and} \quad \underline{H}_f^B := \sup_{u \in \underline{\mathscr{H}}_f} u$$

are either harmonic or infinite. The following relations are immediate.

LEMMA XVI.10.1. Let B be a relatively compact domain of M; then for any real numerical functions f, g on B we have

(1) For $c \in \boldsymbol{R}$, $\underline{H}_c = \overline{H}_c = c$,

(2) $-\overline{H}_f^B = \underline{H}_{-f}^B$, $\underline{H}_f^B \leqslant \overline{H}_f^B$,

(3) $(f \leqslant g) \Rightarrow (\overline{H}_f^B \leqslant \overline{H}_g^B)$,

(4) $\alpha \in \boldsymbol{R}_+ \Rightarrow \overline{H}_{\alpha f}^B = \alpha \overline{H}_f^B$,

(5) $\overline{H}_{f+g}^B \leqslant \overline{H}_f^B + \overline{H}_g^B$,

where we adopt the convention: $\infty - \infty = -\infty + \infty = \infty$.

DEFINITION. A numerical function f on ∂B is called *resolutive* (for B) if $\overline{H}_f^B$, $\underline{H}_f^B$ are harmonic and equal: $\overline{H}_f^B = \underline{H}_f^B$. We then denote their common value by H_f^B. The harmonic function H_f^B is called the *generalized solution of the Dirichlet problem* (*in the sense of Perron–Wiener–Brelot*).

An open set B of M is called *resolutive* if every continuous function with a compact support on ∂B is resolutive. We denote by $\mathscr{L}^1(\partial B) \equiv \mathscr{L}^1$ the space of resolutive functions (for B). Plainly, any regular set B is resolutive and $C_0(\partial B) \in \mathscr{L}^1(\partial B)$.

Our goal is to prove the "Resolutivity Theorem of Brelot", namely, that each resolutive function is integrable with respect to all harmonic measures μ_x^B, $x \in B$. To this end we prove the following

LEMMA XVI.10.2. Let $(f_n)_1^\infty$ be a monotone sequence of numerical functions on ∂B such that all $\overline{H}_{f_n}$ are harmonic (in B in the sequel we omit the index B). Then

$$(6) \qquad \overline{H}_{\sup(f_n)} = \sup(\overline{H}_{f_n}).$$

PROOF. Let $f := \sup(f_n)$. Then $f_n \leqslant f$ and, hence, (by (3)) $\overline{H}_{f_n} \leqslant \overline{H}_f$. We now prove the opposite inequality. Let $y \in B$ and $\varepsilon > 0$; then there exist $u_n \in \overline{\mathscr{H}}_{f_n}$ such that

$$0 \leqslant u_n(y) - \overline{H}_{f_n}(y) \leqslant \varepsilon 2^{-n}.$$

Let

$$w := \sup(\overline{H}_{f_n}) + \sum_{n=1}^{\infty} (u_n - \overline{H}_{f_n});$$

then w is superharmonic in B and

$$w \geqslant \overline{H}_{f_n} + u_n - \overline{H}_{f_n} = u_n, \qquad n = 1, 2, \ldots$$

Therefore, $w \in \overline{\mathscr{H}}_{f_n}$ for all $n = 1, 2, \ldots$ Moreover, $w \in \overline{\mathscr{H}}_f$, hence $w \geqslant \overline{H}_f$. In fact, for every $y_0 \in \partial B$ from $w \geqslant u_n \in \overline{\mathscr{H}}_{f_n}$ it easily follows that $\liminf_{y \to y_0} w(y) \geqslant f_n(y_0)$ and, hence, $\liminf_{y \to y_0} w(y) \geqslant f(y_0)$. Further we have

$$\overline{H}_f(y) \leqslant w(y) \leqslant \sup \overline{H}_{f_n}(y) + \varepsilon.$$

Since y was arbitrary, $\overline{H}_f(y) \leqslant \sup \overline{H}_f(y)$ for all $y \in B$. □

THEOREM XVI.10.3 (Brelot). Let $B \subset M$ be a relatively compact domain. Then: (a) For every numerical function on ∂B we have

$$(7) \qquad \overline{H}_f^B(x) = \int^* f d\mu_x^B \equiv (\mu_x^B)^*(f) \quad \text{for all } x \in B.$$

(b) (A numerical function f on ∂B is resolutive) $\leftrightarrow$ (f is μ_x^B integrable for all $x \in B$) and for each resolutive f we have

$$(7') \qquad H_f^B(x) = \int f d\mu_x^B, \quad x \in B.$$

PROOF. (a) For $f \in C_0(\partial B)$ (7) is satisfied. For a lower semicontinuous f bounded from below we can prove (7) in the following way: since M is second countable (the Radó Theorem), there exists a monotone sequence (φ_n) in $C(\partial B)$ such that $f = \sup \varphi_n$. By virtue of (6) we have

$$\overline{H}_f(x) = \sup H_{\varphi_n}(x) = \sup \int \varphi_n d\mu_x^B$$
$$= \int f d\mu_x^B = \int^* f d\mu_x^B \quad \text{for all } x \in B.$$

For an arbitrary f we denote by F the set of all lower semicontinuous majorants of f bounded from below. Then

$$(8)\qquad \int^* f d\mu_x^B = \inf_{\psi \in F} \int \psi d\mu_x^B = \inf_{\psi \in F} \overline{H}_\psi(x) \geqslant \overline{H}_f(x)$$

for all $x \in B$. For $u \in \overline{\mathscr{H}}_f$, we define $\psi(y) := \liminf_{x \to y \in \partial B} u(x)$. Then $\psi \in F$ and $u \in \overline{H}_\psi$ and we have for $x \in B$

$$\int^* f d\mu_x^B \leqslant \int \psi d\mu_x^B = \overline{H}_\psi(x) \leqslant u(x).$$

Taking the lower envelope of all such u, we obtain

$$(9)\qquad \int^* f d\mu_x^B \leqslant \overline{H}_f(x).$$

From (8) and (9) follows (7).

(b) Applying (7) to f and $-f$, we have

$$(10)\qquad \overline{H}_f(x) = \int^* f d\mu_x^B, \qquad \underline{H}_f(x) = \int_* f d\mu_x^B$$

for all $x \in B$. Plainly, (10) gives (b). □

COROLLARY XVI.10.4. For every $u \in \mathscr{H}^*(M)$

$$(11)\qquad \int u d\mu_x^B \leqslant u(x), \qquad x \in B.$$

DEFINITION. A subset $E \subset \partial B$ is *harmonic integrable* if its indicator function is resolutive, i.e., $\mathbf{1}_E \in L^1(\partial B)$; the value $\mu_x^B(E)$ is called the *harmonic measure* of the set E at a point x and is sometimes denoted by $H(x, E, B)$.

Exercise. Let B be a regular domain; then each point $q \in \partial B$ is of harmonic measure 0 for all x.

We can now prove the following generalization of the maximum principle

THEOREM XVI.10.5 (Generalized Maximum Principle). Suppose $B \subset M$ is a relatively compact domain and let u be subharmonic and bounded from above on B with a bound N:

$$(12)\qquad u \leqslant N.$$

Assume that there is a subset $E \subset \partial B$ such that $\mu_x^B(E) = 0$ for all $x \in B$ and

$$(13) \qquad \limsup_{x \to q} u(x) \leqslant 0 \quad \text{for all } q \in \partial B - E.$$

Then

$$(14) \qquad u \leqslant 0 \quad \text{on } B.$$

PROOF. For $x \in B$ and $\varepsilon > 0$ we take such a $v \in \overline{H}_{1_E}$ that $v(x) < \varepsilon$. By virtue of (12) and (13) we have

$$\limsup_{y \to q} \big(u(y) - Nv(y)\big) \leqslant 0 \quad \text{for all } q \in \partial B.$$

Therefore, by the maximum principle for subharmonic functions we have $u - Nv \leqslant 0$ on B. Hence, $u(x) \leqslant N\varepsilon$ for all $\varepsilon > 0$ and, therefore, $u(x) \leqslant 0$. □

COROLLARY XVI.10.6. Let h be harmonic and bounded on B. If

$$\lim_{x \to q} h(x) = 0 \quad \text{for all } q \in \partial B - E,$$

where E has the harmonic measure 0 at all points of B, then $h \equiv 0$.

PROOF. By virtue of Theorem XVI.10.5 we have h and $-h \leqslant 0$.

An argument similar to that which led to Lemma XVI.10.2 gives the following useful

LEMMA XVI.10.7. Let B be a (hyperbolic) domain and f such a function on ∂B that $\overline{\mathscr{H}}_f^B, \underline{\mathscr{H}}_f^B \neq \emptyset$. Then there exists an $0 \leqslant h \in \mathscr{H}(B)$ such that for every $\varepsilon > 0$

$$\overline{H}_f^B + \varepsilon h \in \overline{\mathscr{H}}_f^B \quad \text{and} \quad \underline{H}_f^B - \varepsilon h \in \underline{\mathscr{H}}_f^B.$$

LEMMA XVI.10.8. Let $G_1 \subset G$ be two open sets in M, and $f \in \mathscr{L}^1(\partial G)$. We define a numerical function f_1 on ∂G_1 by

$$f_1(x) = \begin{cases} H_f^G(x) & \text{for } x \in G \cap \partial G_1, \\ f(x) & \text{for } x \in \partial G \cap \partial G_1. \end{cases}$$

Then f_1 is a resolutive for G_1, i.e., $f_1 \in \mathscr{L}^1(\partial G_1)$ and

$$(15) \qquad H_{f_1}^{G_1} = H_f^G \quad \text{on } G_1.$$

PROOF. It is sufficient to consider connected G and G_1. For a (harmonic) function h from Lemma XVI.10.7 we have on G_1

$$H_f^G - \varepsilon h \leqslant \underline{H}_{f_1}^{G_1} \leqslant \overline{H}_{f_1}^{G_1} \leqslant H_f^G + \varepsilon h.$$

Since $\varepsilon > 0$ was arbitrary, we have

$$\underline{H}_{f_1}^{G_1} = \overline{H}_{f_1}^{G_1} = H_f^G | G_1. \ \square$$

LEMMA XVI.10.9. Let G be such a domain in a hyperbolic M that $M - G$ is compact. Then

$$(16) \qquad H_1^G < 1.$$

PROOF. Let s be a positive non-constant superharmonic function on M. By virtue of the lower semicontinuity of s on $M - G$ there is an $x_0 \in M - G$ such that

$$(*) \qquad 0 \leqslant a := s(x_0) = \inf s(M - G) > \inf s(M)$$

(the inequality on the right-hand side is valid since $s \neq \text{const}$). Therefore, $\liminf_{x \to y \in \partial G} s(x) \geqslant a$ and $\liminf_{x \to y \in \partial G} \left(\frac{1}{a} s\right)(x) \geqslant 1$ and thus

$$\frac{1}{a} s \in \overline{\mathscr{H}}_1^G,$$

what implies

$$H_1^G = \inf(\overline{\mathscr{H}}_1^G) \leqslant \frac{s}{a}.$$

It follows

$$\Rightarrow \inf H_1^G \leqslant \frac{1}{a} \inf s < 1 \qquad (\text{by } (*)).$$

But the harmonic function H_1^G is $0 \leqslant H_1^G \leqslant 1$; therefore, $H_1^G < 1$ (otherwise, $H_1^G(x_0) = 1 = \sup H_1^G$ and by the maximum principle $H_1^G \equiv \text{const} = 1$). $\square$

An interesting characterization of hyperbolic surfaces follows from Lemma XVI.10.9.

THEOREM XVI.10.10 (Nevanlinna). (A Riemann surface M is hyperbolic) $\Leftrightarrow$ (For each domain $G \subset M$ such that $M - G$ is compact there exists a $H_1^G < 1$).

PROOF. Since $\Rightarrow$ is proved we have only to establish

$\Leftarrow$ (indirect): Assume that M is not hyperbolic and G is a domain in M with compact complement. Define a function

$$u := \begin{cases} H_1^G & \text{in } G, \\ 1 & \text{on } M-G; \end{cases}$$

then u is superharmonic and positive and, therefore,

$$u \equiv \text{const}, \quad \text{and whence } H_1^G = 1. \; \Box$$

Remark. If $H_1^G < 1$ for any domain G such that $M-G$ is compact, then following R. Nevanlinna one says that the ideal boundary p_∞ of M has a positive harmonic measure or that M s "*positivberandet*" ("positively bordered"). Theorem XVI.10.10 can, therefore, be formulated more concisely: M is hyperbolic iff its ideal boundary has a positive harmonic measure (is positively bordered).

The following lemma is fundamental for the construction of a Green's function of a hyperbolic surface:

LEMMA XVI.10.11. Let M be hyperbolic, let K be a disk in M, and let s be a superharmonic function in a neighbourhood of $\overline{K}$. Then there exists such a positive superharmonic function h on M that

$$(h-s)|K \in \mathscr{H}(K).$$

PROOF. Let K_1, K_2 be such disks in M that (cf. Fig. 13)

$$\overline{K} \subset K_1 \subset \overline{K_1} \subset K_2,$$

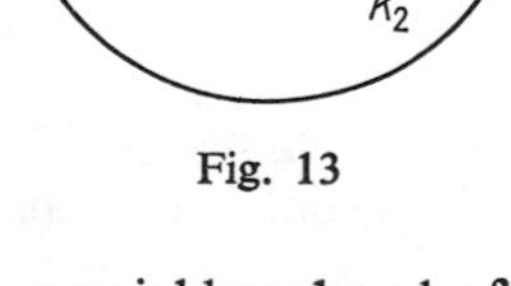

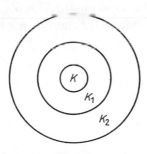

Fig. 13

and that s is defined on a neighbourhood of $\overline{K_2}$. We can assume (by the addition of a constant) that s is positive on K_2. If we define

$$f = \begin{cases} s & \text{on } \partial K, \\ 0 & \text{on } K_2 - \overline{K}, \end{cases}$$

$$s_1 = \begin{cases} s & \text{on } \overline{K}, \\ H_f^{K_2-\overline{K}} & \text{on } K_2 - \overline{K}, \end{cases}$$

then $0 < s_1$ is harmonic in $K_2 - \overline{K}$ and $s_1 | \partial K_2 = 0$. Let $u := H_1^{M-\overline{K}_1}$ and $v := H_u^{K_2}$. By Lemma XVI.10.9, $u < 1$ and $v < u$ in $K_2 - \overline{K}_1$. But $a(u-v) \geqslant s_1$ for sufficiently large a. If

$$h := \begin{cases} au & \text{in } M - K_2, \\ s_1 + av & \text{in } K_2, \end{cases}$$

then $h \in \overline{\mathscr{H}}(M)$, hence $h \in \overline{\mathscr{H}}(K_2)$ and, finally, $(h - s_1) | K_2 = av \in \mathscr{H}(K_2)$. □

Concluding Remark. All considerations of Section 10 are also valid for non-relatively compact $B \subset M$; one has only to modify the definition of $\overline{\mathscr{H}}_f^B$, adding to 1° and 2° a third condition:

3° $\liminf_{x \to \partial_{\text{id}} B} v(x) \geqslant 0$.

11. THE GREEN'S FUNCTION OF A RIEMANN SURFACE

In this section we shall construct a Green's function g^M for a Riemann surface M. Since the Green's function g^M is by definition a non-constant positive superharmonic function, only an (open) hyperbolic surface M can possess a Green's function. Our aim will be to prove that every hyperbolic surface possesses a unique Green's function.

We shall define the Green's function by a minimum property.

DEFINITION. Let M be a hyperbolic Riemann surface, $p \in M$. Denote by S_p^M the set of positive superharmonic functions v on M such that for every disk K_p with centre p

(1) $$v(z) \geqslant \log \frac{1}{|z|}.$$

(Recall that in a remark at the beginning of Section 10 it was observed that $S_p^M \neq \emptyset$ iff M is hyperbolic (cf. also Lemma XVI.10.11).)

The function

(2) $$g_p^M := \inf S_p^M$$

is called the *Green's function* of M with pole p. We shall also write g_p or g for g_p^M.

PROPOSITION XVI.11.1. The Green's function g_p^M is characterized by the following two properties:

I.

$$(3) \qquad 0 < g_p^M, \qquad g_p^M \in \mathscr{H}(M-\{p\})$$

with the logarithmic singularity with coefficient 1 at the point p, i.e., the function $g_p^M(z)+\log|z|$ is harmonic in a parametric disk centre at p.

II. If G is a domain, $p \notin \overline{G}$, then

$$(4) \qquad g_p = H_{g_p}^G \quad \text{in } G.$$

PROOF. I. Since $-S_p^M$ is a Perron family of negative functions away from p, the Green's function g_p^M is harmonic in $M-\{p\}$ and $0 \leqslant g_p^M$.

Let us choose a parametric disk K around p; we define

$$v(z) = \begin{cases} H_{g_p}^K(z) + \log \dfrac{1}{|z|} & \text{in } K, \\ g_p(z) & \text{in } M-K \end{cases}$$

and observe that $0 \leqslant v \leqslant g_p$ and that v is superharmonic. This implies that $v \in S_p^M$ and, hence, by (2) $v = g_p$. Therefore, $g_p(z)+\log|z| = H_{g_p}^K(z)$ is harmonic in K.

II. Consider first $G := M-\overline{K}_p$ then the function

$$s = \begin{cases} g_p & \text{on } \overline{K}_p, \\ H_{g_p}^G & \text{on } G \end{cases}$$

s superharmonic and satisfies (1). Since $s \leqslant g_p$, we have $s \equiv g$ (by (2)). For an arbitrary domain G_1 with $p \notin \overline{G}_1$, take such a disk K_p that $G_1 \subset M-\overline{K}_p$, and by Lemma XVI.10.8 we arrive at

$$H_{gp}^{G_1} = H_{gp}^G = g_p \quad \text{on } G_1.$$

Finally, since $g_p^M(p) = +\infty$ and $0 \leqslant g_p^M \neq \text{const}$, the minimum principle implies $g_p^M > 0$, so we have proved that g_p^M has properties I and II.

But I and II determine the Green's function of M; if γ satisfy I and II, then $(g_p-\gamma) \in \mathscr{H}(M)$, and by II $(g_p-\gamma)$ attains its infimum, whence $g_p-\gamma \equiv \text{const} = a$. But by Lemma XVI.10.9 if $a \neq 0$, then $a = H_a^{M-\overline{K}_p} < a$ (a contradiction). □

Other important properties of the Green's functions are collected in

THEOREM XVI.11.2. Let M be an arbitrary Riemann surface.

(a) If $G \subset G'$ are hyperbolic domains in M then

(6) $$g_p^G \leqslant g_p^{G'} \quad \text{for each } p \in G.$$

(b) If G_n are hyperbolic and $G_n \nearrow M$ is a normal exhaustion of M, then either

(7) $$g_p^{G_n} \nearrow g_p^M \quad \text{for hyperbolic } M,$$

or

(8) $$g_p^{G_n} \nearrow +\infty \quad \text{for parabolic } M.$$

(c) Let G be a hyperbolic domain in M with regular boundary G; then

(9) $$\lim_{x \to \xi} g_p^G(x) = 0 \quad \text{for any point } \xi \text{ of } \partial G.$$

Moreover, if ∂G is arbitrary, then (9) is satisfied for regular $\xi \in \partial G$.

(d) Let M be hyperbolic. Then for $p \neq q$

(10) $$g_p^M(q) = g_q^M(p) \quad \text{(symmetry of the Green's function).}$$

(e) Let $f\colon \tilde{M} \to M$ be a holomorphic injection of a Riemann surface $\tilde{M}$ in a hyperbolic surface M; then $\tilde{M}$ is hyperbolic and thus possesses the Green's function.

(f) Moreover, if $f\colon \tilde{M} \to M$ is a biholomorphism, then

(11) $$g_{\tilde{p}}^{\tilde{M}} = g_{f(\tilde{p})}^M \circ f \quad \text{for each } \tilde{p} \in \tilde{M}.$$

Hence "the Green's function is a conformal invariant".

Remark. For a regular domain G in $\boldsymbol{R}^m$ we have defined (in Chapter XV) the Green's function g_p^G by the boundary condition (9). Since in the present section we are dealing with another definition of g_p^G, we have to check (10). Conformal invariance ((11)) and monotonicity of the Green's function ((b)) will be used in the construction of Green's functions and in the proof of the uniformization theorem.

PROOF. (a) follows immediately from the definition since $S_p^{G'}|G \subset S_p^G$.

(b) follows from (a) and the (non)existence of g_p^M for (non)hyperbolic M.

(c) follows from II. Let $h = H_{g_p}^G$; then $h \in \mathscr{H}(G)$ and $h \leqslant g_p$. Indeed, by II, $h = g_p$ in a $G - K_p$. Take a disk V_p such that $g_p \geqslant \sup h(\partial G)$.

(Since ∂G is compact, $h \geqslant 0$ on ∂G, and $g_p(x) \uparrow +\infty$ for $x \to p$, such a choice of V_p is possible.) Then $g_p - h \geqslant 0$ on $G - \overline{V_p}$ by virtue of the maximum principle applied to the function $g_p - h$ (harmonic on $G - \overline{V_p}$). Since $h \leqslant g_p$ on $G - \overline{V_p}$ for arbitrary small disk V_p (with centre p), $h \leqslant g_p$ on G.

The function $\gamma_p := g_p - h \in S_p^G$ and is harmonic in $G - \{p\}$ and $\gamma_p \geqslant g_p$; therefore, γ_p is identical with g_p, but γ_p vanishes at the boundary ∂G. Hence, (c) is proved. The same proof applies to the general case (arbitrary boundary ∂G).

(d) If G is a regular domain with analytic boundary ∂G; then the symmetry of g^G follows from the Green's formula in the same way as in XV.6.

Let $G_n \nearrow M$ be a normal exhaustion of M; then by virtue of (7) and the symmetry of $g_p^{G_n}$ we have

$$g_p^M(q) = \lim_{n\to\infty} g_p^{G_n}(q) = \lim_{n\to\infty} g_q^{G_n}(p) = g_q^M(p).$$

(e) Define the function $\gamma(\cdot\,,\,\cdot)$ by

(12) $$\gamma(\tilde{q}, \tilde{p}) := g\big(f(\tilde{q}), f(\tilde{p})\big).$$

Plainly, γ is positive, superharmonic and has at $f(\tilde{p})$ the singularity of a Green's function. Therefore, by virtue of Proposition XVI.11.1 on $\tilde{M}$ there exists a Green's function $g^{\tilde{M}}(\cdot\,,\,\cdot)$ which satisfies

(13) $$g^{\tilde{M}}(\tilde{q}, \tilde{p}) \leqslant g\big(f(\tilde{q}), f(\tilde{p})\big) \quad \text{for } \tilde{p}, \tilde{q} \in \tilde{M}.$$

(f) Since the roles of $\tilde{M}$ and M are now quite symmetrical, by (e) we have

(14) $$g^M(q, p) \leqslant g^{\tilde{M}}\big(f^{-1}(q), f^{-1}(p)\big).$$

From (13) and (14) follows (11). □

We now give important

Examples of Green's Functions. The most important example (by virtue of Koebe's conformal mapping theorem) is

Example 1. $M = \boldsymbol{D} = \{z \in \boldsymbol{C}\colon |z| < 1\}$, the unit disk in $\boldsymbol{C}$

(15) $$g_0^D(z) = \log \frac{1}{|z|}.$$

Example 2. $M = \boldsymbol{D}^* = \{z \in \boldsymbol{C}\colon 0 < |z| < 1\} = \boldsymbol{D} - \{0\}$ (punctured unit disk in $\boldsymbol{C}$). Then

$$g_0^{\boldsymbol{D}^*}(z) = \log\frac{1}{|z|} = g_0^{\boldsymbol{D}}(z).$$

Since $\lim_{z\to 0} g_0^{\boldsymbol{D}^*}(z) \neq 0$, 0 is not a regular point of the boundary of $\boldsymbol{D}^*$.

Remark. This example, given in 1911 by S. Zaremba, started a new epoch in potential theory: the theory of regular points.

Example 3. Denote by W^- the half-plane

$$W^- = \{z \in \boldsymbol{C}\colon \operatorname{Re} z < 0\}.$$

Since the function $f = f_a$ given by

$$(16)\qquad f_a(z) = \frac{z-a}{z+\bar{a}}$$

maps W^- biholomorphically onto $\boldsymbol{D}$, and since $f_a(a) = 0$, we have

$$(17)\qquad g_a^{W^-}(z) = -\log\left|\frac{z-a}{z+\bar{a}}\right|, \qquad z \in W^-.$$

Example 4. Every Riemann surface $\tilde{M}$ which possesses a holomorphic non-constant bounded function f is hyperbolic. The proof follows from Theorem XVI.11.2 (e) for all such $\zeta \in \tilde{M}$ that $f'(\zeta) \neq 0$, since

$$\gamma(z, \zeta) := \log\frac{2m}{|f(z)-f(\zeta)|} + \text{harmonic function}$$

is a superharmonic function bounded from below. If f' has at ζ a zero of order $n-1$ then replace γ by $n^{-1}\gamma$.

Example 5. Every bounded domain G of $\boldsymbol{C}$ is hyperbolic. In Example 4 take, $f(z) \equiv z$. In this way we have proved (once more) that every relatively compact domain has a Green's function.

Example 6. Let M be a simply connected Riemann surface which is biholomorphically mapped by f onto $\boldsymbol{D}$; then

$$(18)\qquad g_p^M(q) = \log\frac{1}{|f(q)-f(p)|}.$$

This last example suggests a proof of the uniformization theorem for hyperbolic simply connected surfaces presented in the next section.

COROLLARY XVI.11.3. 1° Let G be a bounded simply connected domain in $\boldsymbol{C}$; then (9) is valid.

2° Let G, G_1 be domains in $\boldsymbol{C}$ such that $G \subsetneqq G_1$ and G is bounded and simply connected; then $g_p^G \leqslant g_p^{G_1}$ and for some $a \in G$, $g_p^G(a) < g_p^{G_1}(a)$ (i.e., the Green's function is a strictly increasing map).

PROOF. 1° By a preliminary mapping we can reduce to the case when $G \subset D$ and a point $\partial G \in \xi = 0$. Since G is simply connected, we can choose a single-valued branch of logarithm denoted $\log: G \to \boldsymbol{C}$,

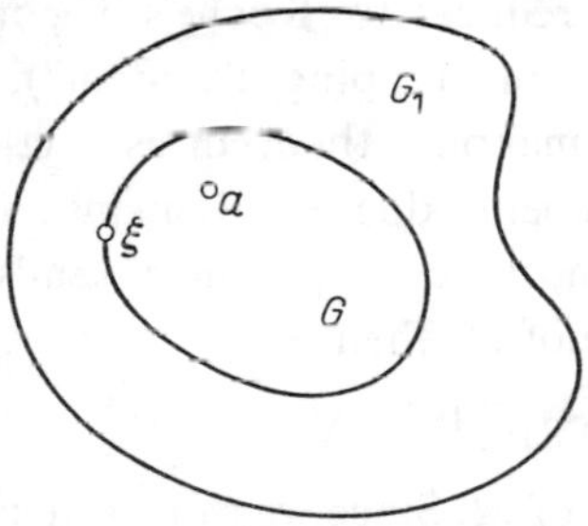

Fig. 14

$w = \log z$, $z \in G$. Since $\log G \subset W^- = \{w \in \boldsymbol{C}: \operatorname{Re} w < 0\}$, from Theorem XVI.11.2 (a) and (f) and (17) (Example 3) we have

$$0 < g_{z_0}^G(z) \leqslant g_{w_0}^{W^-}(w) = -\log\left|\frac{w-w_0}{w+\overline{w}_0}\right|.$$

For $z \to 0$ $\operatorname{Re}(\log z) \to -\infty$; therefore, $g_{w_0}^{W^-}(\log z) \to 0$ and, hence, $g_{z_0}^G(z) \to 0$.

2° Let $\partial G \ni \xi \in G_1$. Then $g^{G_1}(\xi) = b > 0$. But by 1° there exists an $a \in G$ such that $g^G(a) < \frac{1}{3}g^{G_1}(\xi)$. Also by the continuity of g^{G_1} there exists an $a \in G$ such that

$$|g^{G_1}(\xi)-g^{G_1}(a)| < \tfrac{1}{3}g^{G_1}(\xi)$$

and by 1°

$$g^G(a) < \tfrac{1}{3}g^{G_1}(\xi).$$

Therefore,

$$\begin{aligned} g^{G_1}(a)-g^G(a) &> |g^{G_1}(a)|-|g^G(a)| \\ &> \tfrac{2}{3}g^{G_1}(\xi)-\tfrac{1}{3}g^{G_1}(\xi) = \tfrac{1}{3}g^{G_1}(\xi) > 0. \ \square \end{aligned}$$

12. THE UNIFORMIZATION THEOREM

In this section we shall prove the most important case of the famous Koebe and Poincaré uniformization theorem:

THEOREM XVI.12.1. Every simply connected hyperbolic Riemann surface M is biholomorphically isomorphic (it is often said: conformally equivalent) to the disk $\boldsymbol{D}$. A parabolic M is equivalent to $\boldsymbol{C}$ and a simply connected compact M is equivalent to S^1.

This theorem is "perhaps the single most important theorem in the whole theory of analytic functions of one variable" (Ahlfors). For plane regions this theorem reduces to Koebe's Theorem (sometimes called erroneously the "Riemann Mapping Theorem"). In this way we obtain another proof of the mapping theorem as a corollary.

The heuristic argument, due to Riemann, is as follows. Let w_p: $M \to \boldsymbol{D}$ be a biholomorphic map which sends the (arbitrary) point $p \in M$ to 0, the centre of $\boldsymbol{D}$; then

$$(*) \qquad g_p^M := -\log|w_p|$$

is the Green's function of M. Since we can construct the Green's function for a hyperbolic surface M, we shall try to construct such a harmonic function h_p conjugate to g_p on $M-\{p\}$ that $w_p = \exp(-g_p - ih_p)$ is injective and holomorphic.

LEMMA XVI.12.2. Let $g = g^M$ be the Green's function of a simply connected hyperbolic surface M with singularity at (arbitrary) $q \in M$ then there is a holomorphic on M function $f(\cdot, q)$ which vanishes at q and satisfies

$$(1) \qquad \log|f(p, q)| = -g(p, q) \quad \text{for } p \neq q$$

and

$$(2) \qquad |f(p, q)| < 1.$$

PROOF. For a fixed q

$$(3) \qquad \varphi := dg + i*dg$$

is a differential form which is analytic in $M-\{q\}$ and at q has a logarithmic singularity with residue $2\pi i$: $\operatorname{Res}(\varphi) = 2\pi i$. Since M is simply connected, by virtue of the residue theorem we have for every 1-cycle $c \notin q$, the integral $\langle c, \varphi\rangle = 0 \pmod{2\pi i}$. Let $x_0 \neq q$ be a fixed point and let γ

$= \gamma(x_0, p)$ be a path connecting x_0 with $p \neq q$ such that $q \notin \gamma$; then the integral

$$(4) \qquad \int_{\gamma(x_0, p)} \varphi =: h(p)$$

depends on a choice of a path γ up to an additive constant, an integer multiple of $2\pi i$. Therefore, the formula

$$(5) \qquad p \to f(p, q) := \exp(-g(x_0, p) - h(p))$$

defines an honest (one-valued) function in $M - \{q\}$ and a simple argument shows that it is holomorphic. From (3)–(5) we obtain

$$\mathrm{Re}\, h = g(p, q) - g(x_0, p), \qquad \log|f(p, q)| = -g(p, q). \ \square$$

PROOF OF THEOREM XVI.12.1. We are going to show that the function $f(\cdot, \cdot)$ constructed in the preceding lemma has the following properties:

1° $f(\cdot, q)$ is injective for each $q \in M$,

2° $f(M, q) = \boldsymbol{D}$.

It then follows that for each q, $f_q := f(\cdot, q)$ maps M biholomorphically onto the unit disk and solves the uniformization problem in the hyperbolic case.

PROOF. 1° We first check the identity

$$(6) \qquad f(p, q') = e^{i\alpha} \frac{f(p, q) - f(q', q)}{1 - \overline{f(q', q)} f(p, q)}.$$

Let us denote the right-hand side by $F(p, q, q')$. Then $F(\cdot, q, q')$ is holomorphic, $|F(\cdot, q, q')| < 1$, and has a unique zero at q'. But Example 4 (§11) asserts that $-\log|F(\cdot, q, q')| \in S_{q'}^M$ and by the minimum property of the Green's function

$$(7) \qquad -\log|F(\cdot, q, q')| \geqslant g_{q'}^M = -\log|f(\cdot, q')|.$$

For $p = q$ we have $|F(q; q, q')| = |f(q', q)|$, since $f(q, q) = 0$, and (7) implies $|f(q', q)| \leqslant |f(q, q')|$. A permutation of q and q' gives

$$|f(q, q')| \leqslant |f(q', q)| \leqslant |f(q, q')|;$$

hence, $|f(q', q)| = |f(q, q')|$ and in fact the equality holds in (7) (by the maximum principle for subharmonic functions) $|F(\cdot, q, q')|$

$= |f(\cdot, q')|$ and (6) is proved. Since $f(\cdot, q')$ has zero only at $p = q'$, by (6) we have

$$f(p, q) = f(q', q) \quad \text{only for } p = q'.$$

In this way 1° is proved.

2° By virtue of the conformal invariance of the Green's function (Theorem XVI.11.2 (*f*)) we have

$$g_0^W = g_p^M = \log \frac{1}{|f(p, \cdot)|} = \log \frac{1}{|w|} = g_0^D$$

where $W := f(M, q) \subset \boldsymbol{D}$.

Since the Green's functions of W and $\boldsymbol{D}$ are identical, $W = \boldsymbol{D}$ by virtue of Corollary XVI.11.3.2°. □

We now give an outline of the proof of the uniformization theorem in the parabolic and the elliptic cases.

THEOREM XVI.12. 3 (a) Let M be a simply connected parabolic surface. Then M is conformally equivalent with the complex plane.

(b) Any simply connected compact Riemann surface is conformally equivalent to the sphere $\boldsymbol{P}(\boldsymbol{C})$.

As we know, parabolic and elliptic Riemann surfaces do not possess Green's functions. But there exists a substitute for the Green's function, so-called normal potentials of the third kind.

DEFINITION. A *normal potential of the third kind* on a Riemann surface M is a harmonic function $u(\cdot, a, a')$ in $M - \{a, a'\}$ which, for certain parametric disks K_a and $K_{a'}$,

1° is bounded on $M - (K_a \cup K_{a'})$,

2° is singular at a as $-\log|z|$ and at a' as $+\log|z'|$.

Hence, $u - \log \left| \dfrac{z-a}{z-a'} \right|$ is harmonic on M.

Therefore, the differential form

$$(8) \qquad \psi := du + \sqrt{-1} * du$$

is holomorphic in $M - \{a, a'\}$ and has the residue $-2\pi i$ (resp. $+2\pi i$) in a (resp. a'):

$$\operatorname{Res}_a \psi = -2\pi i, \quad \operatorname{Res}_{a'} \psi = 2\pi i.$$

In the same way as in the hyperbolic case we construct a primitive h of the form ψ, i.e., $dh = \psi$. This function h is (by virtue of the residue theorem and the simple connectedness of M) uniquely defined modulo $k2\pi i, k \in \boldsymbol{Z}$. Therefore,

$$(9) \qquad w(\cdot, a, a') := e^{-h}$$

is (for fixed a, a') a well-defined holomorphic function on $M - \{a'\}$, w has a simple pole at a' and a simple zero at a. In $M - (K_a \cup K_{a'})$, w satisfies the inequality

$$\varepsilon < |w| < \varepsilon^{-1} \quad \text{for some } \varepsilon > 0.$$

We now check that the mapping

$$(10) \qquad f = f_{a, a'} := w(\cdot, a, a') \to \boldsymbol{P}(\boldsymbol{C})$$

is

1° injective,
2° in the parabolic case $f(M) = \boldsymbol{C}\ (= \boldsymbol{P}(\boldsymbol{C}) - \{\infty\})$,
3° in the elliptic case $f(M) = \boldsymbol{P}(\boldsymbol{C})$.

Ad 1°. The injectivity of f' is demonstrated much more easily than in the hyperbolic case.

In small neighbourhoods of a and a', f is obviously injective since a and a' are of order one.

Let $c \neq a, a'$ and consider the function $v(p)$

$$(11) \qquad v(p) := \frac{w(p, a, a') - w(c, a, a')}{w(p, c, a')}.$$

Since w is holomorphic on M and bounded outside small neighbourhoods U_a, $U_{a'}$, U_c of a, a', c (resp.), v is bounded on M. Therefore, $\log|v|$ is a bounded from above subharmonic function on a non-hyperbolic surface M and, hence, $\log|v| \equiv \text{const}$; therefore, v is also constant and $v \neq 0$ since $w(\cdot, a, a')$ is non-constant. Therefore, $(w(p, a, a') = w(c, a, a')) \Rightarrow (p = c)$ and the injectivity of f is proved.

2° and 3° follow immediately from the following simple

LEMMA XVI.12.4. The only simply connected non-hyperbolic domains of the Riemann sphere $\boldsymbol{P}(\boldsymbol{C})$ are $\boldsymbol{P}(\boldsymbol{C})$ or the punctured $\boldsymbol{P}(\boldsymbol{C})$, i.e., $\boldsymbol{C}$.

PROOF (indirect). Assume that a simply connected non-hyperbolic domain G_w of the Riemann w-sphere contains $\{+\infty\}$ and *another* boundary point a ($\neq \infty$). The differential form $(w-a)^{-1}dw$ is holomorphic in G_w and has the primitive $w \to \log(w-a) =: z$ which maps G_w onto a domain G_z. Denote $K(z_0, r) = \{z \in \boldsymbol{C}: |z-z_0| < r\} \subset G_z$; then $K(z_0+2\pi i, r) \cap G_z = \emptyset$ since the inverse function $w-a = e^z$ maps both disks into G_w. Therefore, the non-constant function u

$$u(z) := -\log|z-(z_0+2\pi i)|$$

is bounded from above and subharmonic on G_z; hence, is G_z hyperbolic and, thus, G_w is also hyperbolic (contradiction). □

In this way the proof of the uniformization theorem for simply connected surfaces is completed.

Remark. Our proof has a gap, namely, we have assumed the existence of a normal potential of the third kind on a (non-hyperbolic) Riemann surface. The proof of this deep fact requires other instruments, e.g., the Dirichlet principle and will be provided in Part III. An idea of such a proof is given in Section 17.

COROLLARY XVI.12.6 (the Koebe–Riemann Mapping Theorem). Each simply connected domain of the Riemann sphere $\boldsymbol{P}(\boldsymbol{C})$ is conformally equivalent to one of the standard domains: $\boldsymbol{D}$, $\boldsymbol{C}$ or $\boldsymbol{P}(\boldsymbol{C})$.

13. RUNGE'S THEOREM. THEOREM OF BEHNKE AND STEIN. THEOREM OF MALGRANGE

In these section we shall continue the analysis on open Riemann surfaces. The starting point of a beautiful theory was a famous classical theorem of Carl Runge from 1885 about compact approximation of a holomorphic function in a canonical domain $F \subset \boldsymbol{C}$ by polynomials. This theorem easily implies the classical theorem of Mittag-Leffler which gives a construction of a meromorphic function with prescribed principal ($\equiv$ singular) parts, a far-reaching generalization of decomposition in partial fractions and also the famous theorem of Weierstrass on the existence of a meromorphic function with prescribed zeros and poles.

The fundamental role of theorems of the Runge type was realized step by step. In the theory of open Riemann surfaces Heinrich Behnke and Karl Stein succeeded in 1943 in proving their famous theorem of the Runge type. Corresponding theorems of the Mittag-Leffler and the Weierstrass type were deduced.

In several complex variables (complex manifolds) the Mittag-Leffler and Weierstrass problems are known as the first and the second Cousin problem, respectively. Attempts to solve them gave rise new notions and tools from topology and analysis (sheaf theory, cohomology with values in a sheaf) and gave a tremendous stimulus to the development of the theory of vector bundles and the theory of Stein manifolds (and Stein spaces).

Behnke and Stein were able to solve the famous Carathéodory conjecture:

On every open Riemann surface there exists a non-constant holomorphic function.

The original proof of Behnke and Stein is natural but rather involved: it works with normal exhaustions and modifies Runge's method of shifting singularities.

In his famous (doctor's) thesis *Existence et approximation de solutions des équations aux derivées partielles et des équations de convolutions*, Ann. Inst. Fourier VI (1955–56), 271–354, in 1954–55 Bernhard Malgrange succeeded in giving a general theorem of the Runge type for approximation by solutions of elliptic differential equations. The theorem of Behnke and Stein and corresponding theorem of Pfluger for harmonic functions are its corollaries. In the present section we shall formulate the theorem of Malgrange. Its proof is postponed to Part III, where the theory of elliptic operators in vector bundles will be developed. Here it will be shown how the Behnke–Stein and Pfluger theorems are deduced.

Important consequences: solutions of Cousin problems for open surfaces given are also here. In order not to frustrate the reader a proof of the Runge Theorem is given in Section 16.

Elliptic Equations. In this section M denotes a smooth manifold of dimension m which is assumed to be connected and second countable (hence, paracompact). We assume also that M is oriented. In Chapter XIII we considered tensor bundles $\mathscr{T} \to M$ over M. The notion of vector bundle (v.b.) $E \to M$ over M of rank p is a natural generalization of tangent and tensor bundles over M: it is a triplet (E, π, M), where E and M are differentiable manifolds and $\pi\colon E \to M$ is a smooth surjective map such that each fibre $E_x := \pi^{-1}(x)$ has a structure of a vector space of dimension p and E is locally diffeomorphic to the product of an open subset of M and $\boldsymbol{R}^p$. This means that for each $x \in M$ there exists an open neighbourhood $\mathcal{O} \ni x$ and a smooth injective mapping

$$(1) \qquad \varphi\colon \pi^{-1}(\mathcal{O}) \xrightarrow{\text{onto}} \mathcal{O} \times \boldsymbol{R}^p$$

such that φ^{-1} is smooth, i.e., φ is a diffeomorphism. Moreover, we require that for every $x \in \mathcal{O}$ the mapping φ maps E_x onto $\{x\} \times \boldsymbol{R}^p$ and $\varphi|_{E_x}$ composed with p_2 (the canonical projection onto the second factor)

$$(2) \qquad E_x \xrightarrow{\varphi} \{x\} \times \boldsymbol{R}^p \xrightarrow{p_2} \boldsymbol{R}^p$$

is linear (hence, a linear isomorphism). The space $\boldsymbol{R}^p$ is called a *typical fibre* and, M is the base of the vector bundle.

For an open subset $\mathcal{O} \subset M$ the (open) subset $\pi^{-1}(\mathcal{O})$ will be denoted by $E_{\mathcal{O}}$. Note that $(E_{\mathcal{O}}, \pi|_{E_{\mathcal{O}}}, \mathcal{O})$ is also a vector bundle.

Remark 1. If M is a complex manifold, then by repleacing $\boldsymbol{R}^p$ by $\boldsymbol{C}^q$ and smooth by holomorphic in the above one arrives at the notion of holomorphic vector bundles. In the same way one defines real analytic vector bundles.

Remark 2. In complex analysis holomorphic vector bundles with typical fibre $\boldsymbol{C}$, called *holomorphic line bundles* (or *complex line bundles*), are of paramount importance (cf. below).

Let $\{\mathcal{O}_j\}_{j \in J}$ be an open covering of M which locally "trivializes the bundle $E \to M$", i.e., for each j there is a differentiable map $\varphi_j\colon \pi^{-1}(\mathcal{O}_j) \xrightarrow{\text{onto}} \mathcal{O}_j \times \boldsymbol{R}^p$ with properties listed above. When that happens we shall also say that the bundle E is trivial over each of the sets $\mathcal{O}_j$. Then

$$(3) \qquad f_{jk} \equiv \varphi_j \circ \varphi_k^{-1}\colon \mathcal{O}_j \cap \mathcal{O}_k \to \mathrm{GL}_p(\boldsymbol{R}) \subset \boldsymbol{R}^{p^2}$$

is a differentiable map on $\mathcal{O}_j \cap \mathcal{O}_k$ with values in the (Lie) group $\mathrm{GL}_p(\boldsymbol{R})$ of invertible $p \times p$ matrices with real coeficients.

Here, differentiability means precisely the differentiability of matrix coefficients. We have

$$(4) \qquad f_{ij} \circ f_{ji} = \text{identity in } \mathcal{O}_i \cap \mathcal{O}_j \text{ for all } i, j \in J,$$

$$(5) \qquad f_{ij} \circ f_{jk} \circ f_{ki} = \text{identity on } \mathcal{O}_i \cap \mathcal{O}_j \cap \mathcal{O}_k, \text{for all } i, j, k \in J.$$

Such a family $\{f_{jk}\}$ of matrix-valued functions defined in $\mathcal{O}_j \cap \mathcal{O}_k$ with differentiable coeficients is called a *system of transition maps (functions)* subordinate to the covering $(\mathcal{O}_j)_{j\in J}$.

DEFINITION. A pair consisting of a covering $\{\mathcal{O}_j\}_{j\in J}$ and a system of transition maps subordinate to it is called a $\{\mathcal{O}_j\}_{j\in J}$*-cocycle*.

It is not difficult to show that the bundle $E \to M$ can be recovered from the system of transition maps (cocycle) (3).

Exercise. Prove that. The proof follows the same line as in the definition of the tangent bundle $T(M)$. We take the set B of all triples $(j, x, w) \in J \times M \times R^n$ and declare that two elements (j, x, w) and (j', x', w') are equivalent if $x = x'$ and $w' = f_{j'j}(x)w$. This is an equivalence relation $\sim$ by virtue of (4) and (5). Denote $E := B/\sim$ and let the projection π: $E \to M$ be induced by the map $B \ni (j, x, w) \to x \in M$. Now prove that (E, π, M) is a differentiable vector bundle.

DEFINITION. Let $\mathcal{O} \subset M$ be an open subset of M. A C^r*-section* of E over $\mathcal{O}$ is a C^r-map

$$(6) \qquad \mathcal{O} \ni x \to u(x) \in E$$

such that $\pi(u(x)) = x$. In terms of a local trivialization $\{\mathcal{O}_j\}_{j\in J}$ as above this means precisely that

$$(8) \qquad \varphi_j \circ u =: u_j \text{ is a } C^r \text{ map of } \mathcal{O} \cap \mathcal{O}_j \text{ into } R^p$$

such that

$$(9) \qquad u_j = f_{jk} \circ u_k \quad \text{in} \quad \mathcal{O} \cap \mathcal{O}_j \cap \mathcal{O}_k.$$

Conversely, every system of C^r maps u_j: $\mathcal{O} \cap \mathcal{O}_j \to R^p$ which satisfy (9) defines (via (8)) precisely one C^r section of E over $\mathcal{O}$. The space of sections over is denoted by $C^r(\mathcal{O}, E)$ or $C_{\mathcal{O}}(E)$.

Although sections defined locally (over an appropriately chosen neighbourhood of a given point) always exist (they can be constructed with the aid of the maps in (1)), the existence of global sections (i.e. defined over the whole base) of a given vector bundle is a highly intricate problem. We shall see later that the problem of constructing a nontrivial (i.e., non-zero) section of an arbitrary holomorphic line bundle over an (open) Riemann surface is equivalent to solving of the second Cousin problem (e.g., in the case of the open complex plane to the Weierstrass problem of constructing analytic function with prescribed zeros, cf. Section 14 below).

After this preparation we shall define

Differential Operators in V. B. Let $E \to M$, $F \to M$ be v.b. over M. A linear differential operator of the order r from E to F with smooth (analytic) coefficients is a linear map $A\colon C^\infty(E) \to C^\infty(F)$ with the property that for each covering $\{\mathcal{O}_j\}_{j\in J}$ of M such that E and F are trivial over each $\mathcal{O}_j$ and each section $u \in C^\infty(E)$ the section Au is given locally (in the notation of (8) above) by

$$(Au)_j = \sum_{|\alpha| \leqslant r} a_\alpha(\cdot)D^\alpha u_j,$$

where $a_\alpha \in C^\infty(\mathcal{O}, L(R^p, R^q))$ (or a_α is holomorphic), $D^\alpha = \dfrac{\partial^{|\alpha|}}{\partial x^{\alpha_1} \dots \partial x^{\alpha_n}}$, and $\alpha = (\alpha_1, \dots, \alpha_n) \in N^n$, $|\alpha| = \alpha_1 + \dots + \alpha_n$.

A is said to be *elliptic* if for every $x \in \mathcal{O}$ and any non-zero $\xi = (\xi_1, \dots \dots, \xi_m)$ the map

$$\sigma(A)(x, \xi) := \sum_{|\alpha| = r} a_\alpha(x)\xi^\alpha \in L(R^p, R^q)$$

is invertible, i.e., $\det A(x, \xi) \neq 0$.

Let us note that a differential operator A from E to F uniquely determines a differential operator from $E_\mathcal{O}$ to F for every open $\mathcal{O} \subset M$.

We can now formulate the approximation theorem of Malgrange.

THEOREM XVI.13.1. Let M be a real oriented (paracompact) manifold and let E and F be real analytic vector bundles over M. Let A be an elliptic operator of order r from E to F with analytic coefficients. Let G be such open subset of M that $M - G$ has no compact connected components.

Then every section $f \in C^\infty(G, E)$ (over G!) satisfying $Au = 0$ is a limit, in the sense of the compact convergence of sections and their derivatives of all orders, of a sequence of (globally defined) sections $u_n \in C^\infty(M, E)$ satisfying $Au_n = 0$ on M.

Remark. Malgrange proved that the topological condition for G that $M - G$ has no compact connected components is also necessary for the approximation theorem to hold (compare the proof of Runge's Theorem in the last section).

We now give several very important applications (corollaries) of Theorem XVI.13.1.

COROLLARY XVI.13.2. (Behnke–Stein Theorem). Let M be an open Riemann surface and G an open subset of M. In order that each function f holomorphic on G can be compactly approximated on G by functions holomorphic on M it is sufficient (and necessary) that $G = \hat{G}$, where $\hat{G}$ is the union of G and all compact connected components of $M-G$.

PROOF. Take $E = M \times \mathbf{C}$, $F = \Lambda^{(0,1)}$, $C^{\infty}(\Lambda^{(0,1)}) = \Lambda^{(0,1)}(M)$ is the space of differential forms of type (0, 1). Take $A = d''$ $(= \bar{\partial})$—the Cauchy–Riemann operator $d''f = \frac{\partial f}{\partial \bar{z}} d\bar{z}$, then $\operatorname{Ker} d'' = A(M)$ is the space of holomorphic functions on M. Since d'' is elliptic, the proof has been completed. □

If M is an open Riemann surface, $E = M \times \mathbf{C}$, $F = M \times \mathbf{C}$, $A = \Delta$ (Laplace operator), we obtain

COROLLARY XVI.13.3 (Pfluger). Let M be an open Riemann surface and let G be an open subset of M such that $G = \hat{G}$. Then every $h \in \mathscr{H}(G)$ can be compactly approximated by harmonic functions on M.

As a corollary of the Behnke–Stein Theorem we obtain the famous

THEOREM XVI.13.4 (Runge). Let G be an open subset of $\mathbf{C}$ and let K be a compact subset of G. The following two conditions on G and K are equivalent:

1° Every function which is holomorphic in a neighbourhood of K can be approximated uniformly on K by functions in $A(G)$.

2° The set $G-K = G \cap \complement K$ has no relatively compact component in G.

COROLLARY XVI.13.5 (Behnke–Stein). The conjecture of Carathéodory is true: a non-constant holomorphic function exists on every open Riemann surface M.

Hint for the proof. Take G to be a parametric disk and let u be a local coordinate on G.

A very important class of non-compact complex manifolds was investigated by Karl Stein and Henri Cartan called them Stein manifolds in his honour.

DEFINITION. A complex manifold M is called a *Stein manifold* if

(a) $A(M)$ separates points of M, i.e., for $x, y \in M$, $x \neq y$ there exists an $f \in A(M)$ such that $f(x) \neq f(y)$,

(b) for each $x \in M$ there exist $h_1, \ldots, h_n \in A(M)$, where $n = \dim M$, which define a system of coordinates for a suitable neighbourhood of x,

(c) for every compact subset $K \subset M$ the envelope of K,

$$K' = \{x \in M : \bigwedge_{f \in A(M)} |f(x)| \leqslant \sup |f(K)|\},$$

is compact.

COROLLARY XVI.13.6 (Behnke–Stein). Every open Riemann surface is a Stein manifold.

PROOF (sketch). (a) Let K_x be a small disk around x, let K_y be a small disk around y, and let $K_x \cap K_y = \emptyset$. In Corollary XVI.13.2 take $G = K_x \cup K_y$ and

$$f = \begin{cases} 0 & \text{on } K_x, \\ 1 & \text{on } K_y. \end{cases}$$

(b) Take f precisely as in the proof of Corollary XVI.13.5.

(c) From the Behnke–Stein Theorem (XVI.13.2) it follows that K' is contained in the envelope of each open relatively compact neighbourhood of K; therefore, K' is compact (and $= \hat{K}$).

14. COUSIN PROBLEMS FOR OPEN RIEMANN SURFACES. THEOREMS OF MITTAG-LEFFLER AND WEIERSTRASS

The plan of the present section is the following:

1° we formulate Cousin problems in complete generality;

2° we prove that, in the case of an open subset M of $\mathbf{C}$, they are equivalent to the following classical problems:

(I) Mittag-Leffler problem (the first Cousin problem),

(II) Weierstrass problem (the second Cousin problem);

3° we solve (I) and (II) by the Runge Theorem:

4° we finally remark that the method developed in 3° applies to general open Riemann surfaces.

Remark. Some very important examples are given in Section 15.

DEFINITION (The First Cousin Problem). Let M be a complex manifold and let $\{O_j\}_1^\infty$ be an open covering of M. Let

(1) $g_{jk} \in A(\mathcal{O}_j \cap \mathcal{O}_k)$

be an additive cocycle of holomorphic functions, i.e.,

(2) $g_{jk}+g_{kj} = 0, \quad g_{ij}+g_{jk}+g_{ki} = 0$
in $\mathcal{O}_i \cap \mathcal{O}_j \cap \mathcal{O}_k, \ i, j, k \in J = \{1, 2, \ldots\}$.

It is required to find such functions $g_j \in A(\mathcal{O}_j)$ that

(3) $g_{jk} = g_k - g_j$ in $\mathcal{O}_j \cap \mathcal{O}_k$ for all j, k.

This problem is called the *first* (or *additive*) *Cousin problem* and (1) and (2) are called the *first Cousin data*. Let us recall that the space of non-vanishing holomorphic functions in Ω is denoted by $A^*(\Omega)$. We now give

DEFINITION (The Second Cousin Problem). Let $\{\mathcal{O}_j\}_1^\infty$ be an open covering of a complex manifold M. Let

(4) $f_{jk} \in A^*(\mathcal{O}_j \cap \mathcal{O}_k)$

be a (multiplicative) cocycle of non-vanishing holomorphic functions, i.e.,

(5) $f_{jk}f_{kj} = 1, \quad f_{ij}f_{jk}f_{ki} = 1$ in $\mathcal{O}_i \cap \mathcal{O}_j \cap \mathcal{O}_k$ for all i, j, k.

We are required to find $f_j \in A^*(\mathcal{O}_j)$ such that

(6) $f_{jk} = f_k f_j^{-1}$ in $\mathcal{O}_j \cap \mathcal{O}_k$ for all j, k.

This problem is called the *second* (or *multiplicative*) *Cousin problem*, and (4) and (5) are called the *second Cousin data*.

Remark. Comparing the discussion in Section 13 on sections of holomorphic vector bundles, we see that a solution of the multiplicative Cousin problem may be viewed as giving a local expression for a section of the line bundle defined by the cocycle $(\{\mathcal{O}_j\}_1^\infty, \{f_{ik}\}_{j,k=1}^\infty)$. Hence to solve the second Cousin problem for arbitrary data means precisely to prove the existence of (nontrivial) holomorphic sections for holomorphic line bundles. An additive Cousin problem for an open Riemann surface was solved by Behnke and Stein. A solution of this problem will be sketched out at the end of the present section. The problem is not always solvable for $\dim_C M > 1$. A sufficient condition of solvability of the second Cousin problem for a Stein manifold M is that the second cohomology group $H^2(M, \mathbf{Z}) = 0$ vanishes. This will be proved in Part III. This condition is always fulfilled by any open Riemann surface.

Remark. We have chosen a formulation of Cousin problems which avoids the notion of a meromorphic function since in the case of ($\dim_C M > 1$) several complex variables the definitio nof meromorphic functions presents some difficulties. But in order to have a nice transition to the classical (Mittag-Leffler and Weierstrass) situation let us give another formulation of the Cousin problems for Riemann surfaces, using meromorfic functions to do so.

In the sequel we shall write $\mathcal{O}_{jk}$ for $\mathcal{O}_j \cap \mathcal{O}_k$, and $\mathcal{O}_{jkl}$ for $\mathcal{O}_j \cap \mathcal{O}_k \cap \mathcal{O}_l$.

DEFINITION. An *additive Cousin data* on a Riemann surface M is a family $I = \{\mathcal{O}_j, f_j\}_{j \in J}$, where $\{\mathcal{O}_j\}_{j \in J}$ is an open covering of M and $f_j \in \mathcal{M}(\mathcal{O}_j)$ (f_j are meromorphic on M), such that

(7) $\qquad f_i|\mathcal{O}_{ij} - f_j|\mathcal{O}_{ij} \in A(\mathcal{O}_{ij}) \quad$ for all $i, j \in J$.

A solution is then a meromorphic function $f \in \mathcal{M}(M)$ such that

(8) $\qquad f|\mathcal{O}_j - f_j \in A(\mathcal{O}_j), \quad j \in J.$

The corresponding formulation of the multiplicative Cousin problem is left to the reader.

PROPOSITION XVI.14.1. A solution of the first formulation ((1)–(3)) of the additive Cousin problem gives a solution of the additive Cousin problem in the second formulation.

PROOF. Set $g_{jk} := f_j - f_k$, where $\{\mathcal{O}_j, f_j\}_{j \in J}$ satisfy (7). Then (g_{jk}) satisfy (1) and (2). Let (g_j) be the solution of the first Cousin problem; $f_j - f_k = g_{jk} = g_k - g_j$ in $\mathcal{O}_{jk}$ for all j, k. Hence, $f_j + g_j = f_k + g_k$ in $\mathcal{O}_{jk}$; threrefore, there exists a meromorphic function $f \in \mathcal{M}(M)$ such that $f = f_j + g_j$ in $\mathcal{O}_j$ for every j. Since $f - f_j = g_j \in A(\mathcal{O}_j)$, the proof is complete. □

We shall now prove the famous

THEOREM XVI.14.2 (Mittag-Leffler). Let $(a_j)_{j \in \mathbf{N}}$ be a discrete sequence of different points in the open set $M \subset \mathbf{C}$ and let f_j be given the "principal parts"

(9) $$f_j(z) := \sum_{\nu=1}^{n_j} b_{j\nu}(z - a_j)^{-\nu} \quad \text{for } j \in \mathbf{N}.$$

(We may consider that as first Cousin data. Let $\mathcal{O}_j := M - \{a_k : k \neq j\}$; then $\{\mathcal{O}_j\}_{j\in N}$ is an open covering of M, $f_j \in \mathcal{M}(\mathcal{O}_j)$ and $f_j|\mathcal{O}_{jk} - f_k|\mathcal{O}_{jk} \in A(\mathcal{O}_{jk})$ for all $j, k \in N$.)

There then exists an $f \in \mathcal{M}(M)$ such that f is holomorphic outside from $\{a_j, j \in N\}$ and f_j is the principal part of f, i.e.,

$$(10) \qquad f|\mathcal{O}_j - f_j \in A(\mathcal{O}_j).$$

Hence, f is a solution of the additive Cousin problem $= \{\mathcal{O}_j, f_j\}_{j\in N}$.

PROOF. Let $F_j \nearrow M$ be a normal exhaustion of M. We may assume that $a_k \notin F_j$ for $k \geqslant j$ since (a_k) has no accumulation point in M. By virtue of the Runge Theorem there exists a $u_j \in A(M)$ such that

$$|f_j(z) - u_j(z)| < \frac{1}{2^j} \quad \text{for } z \in \overline{F}_j.$$

Hence, the series

$$\sum_{j=k}^{\infty} (f_j - u_j)$$

coverges compactly on F_k to a function which is holomorphic in F_k. Therefore,

$$f := \sum_{j=k}^{\infty} (f_j - u_j) \in \mathcal{M}(M)$$

and this is a solution of the first Cousin problem. □

The Weierstrass Factorization Theorem solves the second Cousin problem for an open subset $M \subset C$. We have to construct a meromorphic function on M with given zeros and poles or, to put it briefly, to construct a meromorphic function with a given divisor.

DEFINITION. Let $(a_j)_{j\in N}$ be a discrete sequence of different points in an open subset $M \subset C$ (or more generally M may be a Riemann surface) and let $n_j \in Z$ be arbitrary integers. The family $\vartheta = \{a_j, n_j\}_{j\in N}$ is called a *divisor* and is written as the formal sum $\sum n_j a_j$. We shall say that a meromorphic f has a *divisor* $\sum n_j a_j$ if $f(z)(z - a_j)^{-n_j} \neq 0$ and is holomorphic in a neighbourhood of a_j for every j.

THEOREM XVI.14.3 (Weierstrass). For each divisor $\vartheta = \{a_j, n_j\}_{j\in N}$ in an open set $M \subset C$ there exists a meromorphic $f \in \mathcal{M}(M)$ with divisor ϑ. The totality of such functions is $\{fg : g \in A^*(M)\}$.

It is well known that if $a_1, \ldots, a_l$ is a finite set in C, then $\prod_{j=1}^{l} (z-a_j)^{n_j}$ solves the problem. The Weierstrass' construction is a natural generalization of this procedure. It works with "infinite products" and, therefore, we first introduce this notion.

DEFINITION. For a sequence of complex numbers (u_n) we say that an infinite product

$$\prod_{n=1}^{\infty} u_n \tag{11}$$

is *convergent* if there exists an $m \in N$ such that, for any $n > m$, $u_n \neq 0$ and the limit

$$U_m := \lim_{n\to\infty} u_{m+1} \cdot u_{m+2} \cdot \ldots \cdot u_n$$

exists and is not zero. The number $U := u_1 \cdot \ldots \cdot u_m \cdot U_m$ is called the *value of the infinite product* (11).

COROLLARY XVI.14.4. $(\prod u_n = 0) \Leftrightarrow$ (for at least one $j \in N$, $u_j = 0$).

The following proposition is obtained by taking the principal branch of the logarithm and applying the Cauchy test.

PROPOSITION XVI.14.5. Suppose that $|z_n| < 1$ for all $n \in N$, then

$$\Big(\prod_{n=1}^{\infty} (1+z_n) \text{ converges}\Big) \Leftrightarrow \Big(\sum_{n=1}^{\infty} \log(1+z_n) \text{ converges}\Big). \ \Box \tag{12}$$

PROOF OF THEOREM XVI.14.3. Let $F_j \nearrow M$ be a normal exhaustion of M. Since every F_j contains only a finite number of points a_j, it is possible to find a rational f_j with desired zeros and poles in F_j. We shall inductively define rational functions f_j with the prescribed zeros and poles in F_j and analytic in M functions g_j such that

$$\left|\frac{f_{j+1}}{f_j} e^{g_j} - 1\right| < \varepsilon_j \text{ on } F_j, \text{ where } \sum \varepsilon_j < \infty. \tag{13}$$

If $f_1, \ldots, f_j$ and $g_1, \ldots, g_{j-1}$ are already constructed in that way, we let f be a rational function with the desired poles and zeros in F_{j+1}. Then

$$\frac{f(z)}{f_j(z)} = a \prod (z-z_k)^{m_k} \quad \text{(finite product!)}, \tag{14}$$

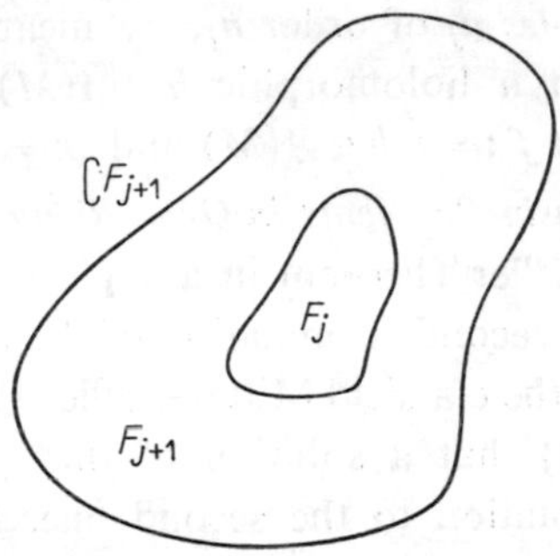

Fig. 15

where $z_k \in \complement K_j$ for every k and $m_k \in \mathbf{Z}$. Since no component of $M - K_j$ is relatively compact in M, a point $w_k \in \complement K_{j+1}$ in the same component of $\complement K_j$ as z_k exists for every k. Therefore,

$$(15) \qquad f_{j+1}(z) := f(z) \prod (z - w_k)^{-m_k}$$

has the prescribed zeros and poles in K_{j+1}. Since z_k and w_k are in the same component of $M - K_j$, a uniquely defined holomorphic branch of $\log((z - z_k)(z - w_k)^{-1})$ can be chosen in a neighbourhood of K_j. Therefore, by (14) and (15)

$$\log(f_{j+1}/f_j)(z) = \log a + \sum m_k \log \frac{z - z_k}{z - w_k}$$

is holomorphic in a neighbourhood of K_j. Hence, by the Runge Theorem, there exists a $g_j \in A(M)$ such that

$$|\log(f_{j+1}/f_j) + g_j| < \log(1 + \varepsilon_j) \quad \text{in } K_j,$$

and (13) is satisfied as desired. From (13) we can see that

$$f_1 \prod_{j=1}^{\infty} \frac{f_{j+1}}{f_j} e^{g_j} := \lim f_{n+1} \prod_{l=1}^{n} e^{g_l}$$

is a meromorphic function with a given divisor: indeed, the product $\prod_j^{\infty} (\ldots)$ converges to a non-vanishing holomorphic function in F_j. □

As a corollary we obtain a proof of

THEOREM XVI.14.6 (Poincaré). Let M be an open subset of $\mathbf{C}$. Then each meromorphic $m \in \mathcal{M}(M)$ is a quotient f/h, where $f, h \in A(M)$.

PROOF. If m has *poles* a_j of order n_j, by means of the Weierstrass Theorem we construct a holomorphic $h \in A(M)$ with *zeros* in a_j of orders n_j, $j \in \boldsymbol{N}$. Then $f := mh \in A(M)$ and $m = f/h$. □

Solution of the Cousin Problems in Open Riemann Surfaces. We now come to the Mittag-Leffler Theorem in a slightly more general setting. We have seen that the second formulation of the additive Cousin problem closely resembled the classical Mittag-Leffler Theorem and we noted (Proposition XVI.14.1) that a solution of that problem in its initial formulation gave a solution to the second one. Here, following Hörmander, we shall prove the existence of a solution for the first formulation of the (additive) Cousin problem. We demonstrate first a simple but important fact about the inhomogeneous Cauchy–Riemann equation.

THEOREM XVI.14.7. Let M be an open subset of $\boldsymbol{C}$; then for every $f \in C^\infty(M)$ the equation $\dfrac{\partial u}{\partial \bar{z}} = f$ has a solution $u \in C^\infty(M)$.

PROOF. Let $F_n \nearrow M$ be a normal exhaustion of M and let $\alpha_j \in C_0^\infty(M)$ be equal to 1 in a neighbourhood of F_j. Set $\varphi_1 = \alpha_1$, $\varphi_j = \alpha_j - \alpha_{j-1}$ for $j > 1$. We plainly have $\varphi_j = 0$ in a neighbourhood of $\bar{F}_{j-1}$ and

$$(16) \qquad \sum_{j=1}^{\infty} \varphi_j \equiv 1 .$$

This is easy to check by using the generalized Cauchy formula that the function u_j defined by

$$u_j(\zeta) := -\frac{1}{2\pi i} \int (\varphi_j f)\,(\zeta - z) z^{-1} dz \wedge d\bar{z}$$

satisfies $\partial u_j / d\bar{z} = \varphi_j f$ and $u_j \in C^\infty(\boldsymbol{R}^2)$. Since φ_j vanishes in a neighbourhood $\mathcal{O}$ of $\bar{F}_{j-1}$, this means that $u_j \in A(\mathcal{O}_{j-1})$. The Runge Theorem asserts the existence of $f_j \in A(M)$ such that $|u_j - f_j| < 2^{-j}$ in $\bar{F}_{j-1}$. Hence, the sum

$$(17) \qquad u := \sum_{j=1}^{\infty} (u_j - f_j)$$

is compactly convergent on M. But, for $j > k$, $u_j - f_j$ is holomorphic in a neighbourhood of $\bar{F}_k$; therefore,

$$\sum_{j=k+1}^{\infty} (u_j - f_j)$$

converges uniformly on $\overline{F}_k$ to a function which is holomorphic on F_k. Thus $u \in C^\infty(M)$, and since we can differentiate (17) term by term, we obtain (by (16))

$$\frac{\partial u}{\partial \bar{z}} = \sum_{j=1}^{\infty} \left(\frac{\partial u_j}{\partial \bar{z}} - \frac{\partial f_j}{\partial \bar{z}} \right) = \sum_{j=1}^{\infty} \frac{\partial u_j}{\partial \bar{z}} = \sum_{j=1}^{\infty} \varphi_j f = f. \ \square$$

THEOREM XVI.14.8 (strengthened form of Theorem XVI. 14.2). Let M be an open subset of C, let $\{\mathcal{O}_j\}_{j \in N}$ be an open covering of M, and let $g_{jk} \in A(\mathcal{O}_{jk})$ satisfy, for all $i, j, k \in N$,

(18) $$g_{jk} + g_{kj} = 0, \quad g_{ij} + g_{jk} + g_{ki} = 0 \quad \text{in } \mathcal{O}_{ijk}.$$

Then there exist $g_j \in A(\mathcal{O}_j)$ such that

(19) $$g_{jk} = g_k - g_j \quad \text{in } \mathcal{O}_{jk} \text{ for all } j, k \in N.$$

Hence, the additive Cousin problem has a solution.

PROOF. Let $\{\beta_j\}$ be a smooth partition of unity subordinate to the covering $\{\mathcal{O}_j\}_{j \in N}$. Put

(20) $$h_k := \sum \beta_\nu g_{\nu k}.$$

Thus, (by virtue of $g_{k\nu} + g_{\nu j} + g_{jk} = 0$) we have

$$h_k - h_j = \sum \beta_\nu (g_{\nu k} - g_{\nu j}) = \sum \beta_\nu g_{jk} = g_{jk}.$$

Since $\partial g_{jk}/\partial \bar{z} = 0$, we obtain

$$\frac{\partial h_k}{\partial \bar{z}} = \frac{\partial h_j}{\partial \bar{z}} \quad \text{in } \mathcal{O}_{jk}.$$

Therefore, there exists a $\varphi \in C^\infty(M)$ such that

$$\varphi | \mathcal{O}_k = \frac{\partial h_k}{\partial \bar{z}} \quad \text{for all } k \in N.$$

Now taking $u \in C^\infty(M)$ to be a solution of the equation

$$\frac{\partial u}{\partial \bar{z}} = -\varphi$$

and putting $g_k := h_k + u$, we obtain $g_k - g_j = h_k - h_j = g_{jk}$ in $\mathcal{O}_{jk}$, but since

$$\frac{\partial g_k}{\partial \bar{z}} = \frac{\partial h_k}{\partial \bar{z}} + \frac{\partial u}{\partial \bar{z}} = \varphi + \frac{\partial u}{\partial \bar{z}} = \varphi - \varphi = 0,$$

we have

$$g_k \in A(\mathcal{O}_k). \quad \square$$

Remark. We see that for the solution of the first Cousin problems two facts are of decisive importance:

A) a theorem of the Runge type,

B) solvability of the equation

$$d''u = f \quad \text{for } f \in \Lambda^{(0,1)}(\mathcal{O}),$$

such that $d''f = 0$, for arbitrary relatively compact open sets $\mathcal{O} \subset M$.

A) is implied by theorem of Malgrange.

B) is true for any open Riemann surface and even for any Stein manifold. We postpone the proofs of these facts to Part III, where the necessary machinery of elliptic operators, vector bundles, and Stein manifolds will be prepared. Here we only announce the result, due also to Behnke and Stein.

THEOREM XVI.14.9 (Behnke–Stein). For any open Riemann surfaces both Cousin problems have solutions.

15. EXAMPLES OF PARTIAL FRACTIONS AND FACTORIZATIONS. FUNCTIONS $\cos \pi z$, $\pi^2/\sin^2 \pi z$, $\Gamma(z)$. MELLIN AND HANKEL FORMULAE. CANONICAL PRODUCTS

In this section we shall give some famous "examples" of Mittag-Leffler and Weierstrass theorems. Our starting point is the following development in partial fractions

$$(1) \qquad \frac{\pi^2}{\sin^2 \pi z} = \sum_{n \in \boldsymbol{Z}} \frac{1}{(z-n)^2}.$$

To establish its validity, note that the left-hand side of (1) has double poles at $z \in \boldsymbol{Z}$. The principal part at the origin is z^{-2} and since $\sin^2 \pi z = \sin^2 \pi (z-n)$, the principal part at $z = n$ is $(z-n)^{-2}$; the series at the right-hand side of (1) is convergent for $z \notin \boldsymbol{Z}$: we see that by making a comparison with $\sum_{1}^{\infty} \frac{1}{n^2}$. Since

$$(*) \qquad \sum_{n \in \boldsymbol{Z}} \frac{1}{(z-n)^2}$$

is uniformly convergent on any compact subset K of $\boldsymbol{C}$, after omission of the finite number of terms which are infinite on K we can write

$$(2) \qquad \frac{\pi^2}{\sin^2 \pi z} = \sum_{n \in \boldsymbol{Z}} \frac{1}{(z-n)^2} + h(z)$$

where $h \in A(\boldsymbol{C})$. We now show that $h \equiv 0$. Since both sides of (1) are periodic with the period 1, we see that $h(\cdot)$ has the same period. If $z = x+iy$, then a simple computation yields

$$|\sin \pi z|^2 = \cos h^2 y - \cos^2 x,$$

and, thus, $(\pi/\sin \pi z)^2 \to 0$ uniformly for $|y| \to \infty$. But the series $(*)$ has the same property; in fact, the series converges uniformly for $|y| \geqslant 1$ and the limit can be obtained term by term.

Hence, $h(z) \to 0$ uniformly for $|y| \to \infty$, which implies that $|h(\cdot)|$ is bounded in the period strip $0 \leqslant x \leqslant 1$, and by the periodicity h is bounded on the whole $\boldsymbol{C}$ and, thus, by the Liouville Theorem h is constant. Since the only constant tending to 0 must be 0 itself, we finally get $h = 0$. □

By integrating both sides of (1) and adding to each term $1/n$ to obtain convergence we obtain

$$(3) \qquad \pi \cot \pi z = \frac{1}{z} + \sum_{n \neq 0} \left(\frac{1}{z-n} + \frac{1}{n} \right) + a$$

We shall show that the integration constant $a = 0$. If we add together the terms corresponding to n and $-n$, we can rewrite (3) as

$$(4) \qquad \pi \cot \pi z = \lim_{k \to \infty} \sum_{n=-k}^{k} \frac{1}{z-n} + a = \frac{1}{z} + \sum_{n=1}^{\infty} \frac{2z}{z^2 - n^2} + a.$$

Since both extreme sides of (4) are odd functions of z, the constant a must vanish. By the use of the identity

$$\frac{\pi}{2} \cot \frac{\pi z}{2} - \frac{\pi}{2} \cot \frac{\pi(z-1)}{2} = \frac{\pi}{\sin \pi z},$$

we obtain from (3) an important formula

$$(5)\qquad \frac{\pi}{\sin \pi z} = \lim_{k\to\infty} \sum_{n=-k}^{k} (-1)^n \frac{1}{z-n}.$$

In a similar manner we obtain

$$(6)\qquad \frac{\pi}{\cos \pi z} = \pi + \sum_{n=-\infty}^{\infty} (-1)^n \left(\frac{1}{z - \frac{2n-1}{2}} + \frac{1}{\frac{2n-1}{2}} \right).$$

Canonical Products. An important special case of Theorem XVI.14.3 concerns entire functions, that is, functions analytic in the entire complex plane $\boldsymbol{C}$. Thus, an entire f has no poles but only zeros.

THEOREM XVI.15.1 (Weierstrass). Let $a_n \in \boldsymbol{C}$, $n = 1, 2, \ldots$ be the zeros of a function $f \in A(\boldsymbol{C})$. Assuming $a_n \to \infty$ in the case of infinitely many zeros, there exist certain integers $m, m_n, n = 1, 2, \ldots$ and a function $g \in A(\boldsymbol{C})$ such that

$$(7)\qquad f(z) = z^m e^{g(z)} \prod_{n=1}^{\infty} \left(1 - \frac{z}{a_n}\right) \times$$
$$\times \exp\left[\frac{z}{a_n} + \frac{1}{2}\left(\frac{z}{a_n}\right)^2 + \ldots + \frac{1}{m_n}\left(\frac{z}{a_n}\right)^{m_n}\right]$$

where all the factors corresponding to $a_n = 0$ are omitted in the product. The proof can be obtained by combining Theorem XVI.14.3 and the following

LEMMA XVI.15.2

$$(8)\qquad \big(h \in A^*(\boldsymbol{C})\big) \Leftrightarrow \Big(\bigvee_{g\in A(\boldsymbol{C})} h(z) = \exp\big(g(z)\big)\Big).$$

PROOF. $\Leftarrow$ is obvious.

$\Rightarrow$: Since $h'/h \in A(\boldsymbol{C})$, therefore, h'/h is the derivative of an entire function g_1, which implies, with the aid of simple computation, that

$$\big((f(z)e^{-g_1(z)})' \equiv 0\big)$$

and, hence, $fe^{-g_1} = \text{const}$ and, finally, $f = \text{const}\, e^{g_1} = e^g$. □

The representation (7) is much more useful if it is possible to take all m_n equal to each other. The reader can verify that the product

$$(9) \qquad \prod_{n=1}^{\infty}\left(1-\frac{z}{a_n}\right)\exp\left[\frac{z}{a_n}+\frac{1}{2}\left(\frac{z}{a_n}\right)^2+\ldots+\frac{1}{h}\left(\frac{z}{a_n}\right)^h\right]$$

converges if

$$(10) \qquad \sum_{n=1}^{\infty}\frac{1}{|a_n|^{h+1}} < \infty .$$

If h is the smallest integer for which (10) is convergent, then (9) is called the *canonical product* associated with the sequence (a_n) and h is called the *genus* of the canonical product. We give now some important

Examples of Product Developments.

$$(11) \qquad \sin \pi z = \pi z \prod_{n\neq 0}\left(1-\frac{z}{n}\right)e^{z/n} .$$

In order to prove (11) we note that the zeros of $\sin \pi z$ are all integers $n \in \boldsymbol{Z}$. Since $\sum 1/n$ is divergent and $\sum n^{-2}$ converges the canonical product associated with the integers (different from 0) is

$$\prod_{n\neq 0}\left(1-\frac{z}{n}\right)e^{z/n} .$$

It follows that

$$(12) \qquad \sin \pi z = z e^{g(z)} \prod_{n\neq 0}\left(1-\frac{z}{n}\right)e^{z/n} .$$

To determine g we compute the logarithmic derivatives of both sides of (12), leaving the easy justification to the reader, and obtain

$$\pi \cot \pi z = \frac{1}{z}+g'(z)+\sum_{n\neq 0}\left(\frac{1}{z-n}+\frac{1}{n}\right).$$

Comparison with (3) gives $g'(z) \equiv 0$ and, hence, $g = \text{const}$. But $\lim\limits_{z\to 0}\dfrac{\sin \pi z}{z} = \pi$, whence $e^{g(z)} = \pi$ and (11) is proved.

The Gamma Function is defined by

$$(13) \qquad \Gamma(z) := \frac{e^{-\gamma z}}{z}\prod_{n=1}^{\infty}\left(1+\frac{z}{n}\right)^{-1} e^{z/n} ,$$

where γ is Euler's constant (cf. Chapter VI). This function is meromorphic without zeros and with simple poles at $z = 0, -1, -2, \ldots$ One can check the following basic formulae more or less easily:

$$\Gamma(z+1) = z\Gamma(z), \tag{14}$$

$$\Gamma(z)\Gamma(1-z) = \frac{\pi}{\sin \pi z}, \tag{15}$$

$$\Gamma(\tfrac{1}{2}) = \sqrt{\pi}\,. \tag{16}$$

Hence, the function Γ defined by (13) satisfies the same functional equation which was satisfied by a real-valued function defined in Chapter VI by the Euler integral

$$\Gamma(s) := \int_0^\infty t^{s-1}e^{-t}dt, \qquad s > 0. \tag{17}$$

Moreover, the expression (13) agrees with (17) for real values of z (although the proof of that is not completely trivial) and, hence, the function Γ defined by (13) may be considered as a continuation of the Euler integral (16) into the whole complex plane. The definition (13) is due to Gauss. A question arises: is Euler's definition (17) valid for complex s? The answer is partially affirmative: the integral in (17) converges for all $z \in \boldsymbol{C}$ satisfying $\operatorname{Re} z > 0$ and, hence, (by the principle of analytic continuation) we have the Mellin representation

$$\Gamma(z) = \int_0^\infty t^{z-1}e^{-z}dt \quad \text{for } \operatorname{Re} z > 0. \tag{18}$$

Since the integral (18) converges only for $\operatorname{Re} z > 0$, it is of some interest to obtain a related integral valid for all complex values of z. This can be done by the loop integral

$$G(z) = \frac{1}{2\pi i} \int_{-\infty}^{(0+)} w^{-z}e^{w}dw, \tag{19}$$

representing a limiting case of a contour integral (on the Riemann sphere), with the closed contour passing through ∞ and encircling the point 0 in the positive direction. To be precise, we choose the principal branch of the function $w \to w^{-z}$ in the plane with the negative real half-axis omitted, i.e., the branch given by

$$w^{-z} = e^{-z(\log|w| + i\arg w)} \quad \text{with } |\arg w| < \pi,$$

and extend it by continuity to the "boundary of the cut plane" by inclusion of the "edges" $\arg w = \pm\pi$. The integral in (19) is defined as the

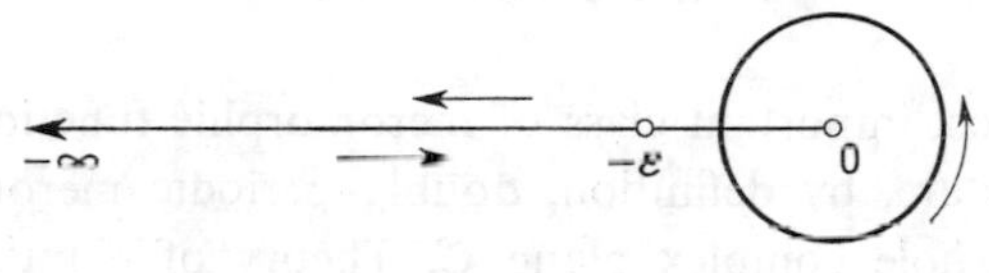

Fig. 16

limit of the contour integrals along the curve consisting of three following pieces (cf. Fig. 16):

1° a (real) half-line from $-\infty$ to $-\varepsilon$ on the lower edge of the cut, i.e., with $\arg w = -\pi$,

2° a positively directed circle of radius ε about 0,

3° a half-line on the upper edge of the cut ($\arg w = +\pi$) from $-\varepsilon$ to $-\infty$.

Calculation which is not very difficult gives

$$G(z) = \frac{\sin \pi z}{z} \int_{\varepsilon}^{\infty} x^{-z} e^{-x} dx +$$

$$+ \frac{\varepsilon^{1-z}}{2\pi} \int_{-\pi}^{\pi} \exp(\varepsilon e^{i\varphi}) \exp\big((1-z) i\varphi\big) d\varphi.$$

The last integral converges to 0 as $\varepsilon \to 0$ for $\mathrm{Re}\, z < 1$. The first integral converges, with the same proviso, to

$$\int_0^{\infty} e^{-x} x^{-z} dx.$$

In this way we obtain by (18) and (15)

$$G(z) = \frac{\sin \pi z}{\pi} \Gamma(1-z) = \frac{1}{\Gamma(z)}$$

and finally the famous Hankel formula

$$\frac{1}{\Gamma(z)} = \frac{1}{2\pi i} \int_{-\infty}^{(0+)} w^{-z} e^{w} \, dw. \tag{20}$$

16. ELLIPTIC FUNCTIONS. EISENSTEIN SERIES. THE FUNCTION $\wp$

Perhaps the most important class of meromorphic functions are elliptic functions: they are, by definition, doubly-periodic meromorphic functions on the whole complex plane $\boldsymbol{C}$. Theory of elliptic functions is a vast domain of intensive research because of its importance in algebra, number theory, algebraic geometry and classical analysis. The natural place of the study of elliptic functions is the theory of compact Riemann surfaces, which will be further developed in Part III. In the present section we want to give some short glimpses at this beautiful subject.

We recall that a mapping $f\colon X \to Y$ defined on an Abelian group X has a period ω $(\neq 0)$ if $f(x+\omega) = f(x)$ for all $x \in X$. Plainly if ω is a period, then $n\omega$ is a period for any $n \in \boldsymbol{Z}$ as well.

DEFINITION. A function $f\colon \boldsymbol{C} \to \boldsymbol{C}$ is called *doubly periodic* if it has two periods ω_1, ω_2 whose ratio ω_2/ω_1 is not real. Since $-\omega$ is also a period if ω is a period for f, then one can always arrange matters so that $\operatorname{Im}(\omega_1/\omega_2) > 0$. A pair (ω_1, ω_2) of periods of a function $f\colon \boldsymbol{C} \to \boldsymbol{C}$ is called a *fundamental pair* if every period of f is of the form $m\omega_1 + n\omega_2$, with $m, n \in \boldsymbol{Z}$. If the set $Z\omega_1 + Z\omega_2$ is denoted by L or by $[\omega_1, \omega_2]$ or by $L(\omega_1, \omega_2)$, it is called the *lattice* (of periods) generated by ω_1 and ω_2.

Two pairs (ω_1, ω_2), (ω_1', ω_2') of complex numbers, each with non-real ω_1/ω_2, ω_1'/ω_2', are called *equivalent* if they generate the same lattice of periods.

The reader may easily prove the following

PROPOSITION XVI.16.1. (Two pairs (ω_1, ω_2), (ω_1', ω_2') are equivalent) $\Leftrightarrow$ (there exists a 2×2 matrix $\begin{bmatrix} a & b \\ c & d \end{bmatrix}$, such that $a, b, c, d \in \boldsymbol{Z}$, and $\det\begin{bmatrix} a & b \\ c & d \end{bmatrix} = \pm 1$ and $\omega_1' = a\omega_1 + b\omega_2$, $\omega_2' = c\omega_1 + d\omega_2$). More briefly:

Two pairs (ω_1, ω_2), (ω_1', ω_2') such that $\operatorname{Im}(\omega_1/\omega_2)$, $\operatorname{Im}(\omega_1', \omega_2') > 0$ define the same lattice iff they are congruent modulo $\mathrm{SL}_2(\boldsymbol{Z})$. If (ω_1, ω_2)

is a fundamental pair and x any number, the parallelogram with vertices x, $x+\omega_1$, $x+\omega_1+\omega_2$, $x+\omega_2$ is called a *period parallelogram*. Note that all period parallelograms have the same area, although in general they are not of the same shape.

DEFINITION. A function f: $\boldsymbol{C} \to \boldsymbol{C}$ is *elliptic* if

1° f is meromorphic $-f \in \mathcal{M}(\boldsymbol{C})$,

2° f is doubly periodic.

Remark 1. Since a meromorphic function has only finite numbers of poles and zeros in any bounded subset of $\boldsymbol{C}$, a period parallelogram can always be translated to a (congruent) parallelogram with no poles and no zeros on its boundary.

Remark 2. Holomorphic doubly-periodic functions are constant: they are entire and bounded in the compact period parallelogram: hence, by periodicity they are bounded in the whole plane and, thus, by the Liouville Theorem, are constant.

Remark 3. Let L be a lattice (of periods). L is a subgroup of the Abelian group $\boldsymbol{C}$ $(=R^2)$ and the quotient group $\boldsymbol{C}/L$ is the familiar two-dimensional torus isomorphic (as a group) to the product of two circles. With a natural complex structure obtained by requiring the canonical projection π: $\boldsymbol{C} \to \boldsymbol{C}/L$ to be holomorphic, $\boldsymbol{C}/L$ is a compact Riemann surface. Therefore, the theory of elliptic functions is a theory of meromorphic functions on the torus $\boldsymbol{C}/L$. Compact Riemann surfaces will be considered in greater detail in Section 20 and in Part III.

In the theory of theta functions and, more generally, in the theory of automorphic functions of several variables we shall need a more general notion of a lattice.

DEFINITION. A lattice L in a real vector space V of finite dimension is a *subgroup* L of V which satisfies one of the following equivalent conditions:

1° L is discrete and V/L is compact.

2° L is discrete and generates the $\boldsymbol{R}$-vector space V.

3° There exists an $\boldsymbol{R}$-basis $(e_1, \ldots, e_n)$ of V which is a $\boldsymbol{Z}$-basis of L, i.e.,

$$L = \boldsymbol{Z}e_1 \oplus \ldots \oplus \boldsymbol{Z}e_n.$$

Exercise 1. Prove the equivalence of 1°–3°.

Exercise 2. Prove the following. Let L and M be two lattices in $\boldsymbol{C}$. Then the (complex) tori $\boldsymbol{C}/L$ and $\boldsymbol{C}/M$ are conformally isomorphic iff there exists an $\alpha \in \boldsymbol{C}^*$ such that $L = \alpha M$.

We shall now collect some elementary properties of elliptic functions discovered by Liouville:

THEOREM XVI.16.2 (Liouville). 1° If an elliptic function f has no poles in some closed period parallelogram then f is constant.

2° An elliptic function f without zeros is constant.

3° Let Ω be a period parallelogram of an elliptic function f such that there are no zeros and poles of f on $\partial\Omega$; then

$$\int_{\partial\Omega} f dz = 0. \tag{2}$$

4° In each period parallelogram the number of zeros of an elliptic function is equal to the number of its poles (each counted with multiplicity).

5° The sum of residues of an elliptic function f in any period parallelogram is zero.

6° Let f have order m_j at a_j (recall that if a is neither a zero nor a pole of f then the order of f at a is zero) then

$$\sum m_j a_j \equiv 0 \pmod{L}.$$

PROOF. 1° By the periodicity f is holomorphic in $\boldsymbol{C}$ hence by virtue of the Remark 2 $f =$ const.

2° Apply 1° to $1/f$ (2° follows also from 1° and 4°).

3° The integrals along parallel sides of $\partial\Omega$ cancel because of periodicity.

4° follows immediately from the formula

$$N_z - N_p = \frac{1}{2\pi} \int_{\partial\Omega} \frac{f'(z)}{f(z)} dz \tag{3}$$

giving the difference of the number of zeros and poles of f in Ω. But f'/f is elliptic with the same periods as f and, hence, by virtue of 3° the right-hand side of (3) vanishes.

5° follows from the residue theorem and (3).

6° We have the relation

$$\int_{\partial\Omega} z\frac{f'(z)}{f(z)}\,dz = 2\pi i \sum m_j a_j,$$

which follows from

$$\mathrm{Res}_{a_j} zf'(z)/f(z) = m_j a_j.$$

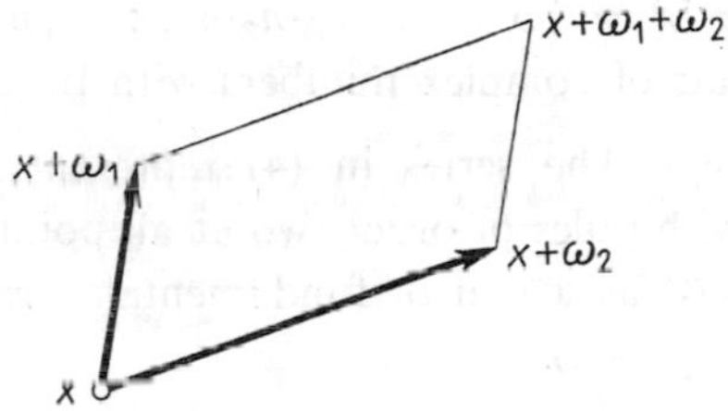

Fig. 17

Let us compute the integral over $\partial\Omega$ by adding contributions from two opposite sides (cf. Fig. 17). For one pair of such integrals this is equal to

$$\int_{x}^{x+\omega_1} z\frac{f'(z)}{f(z)}\,dz - \int_{x+\omega_2}^{x+\omega_1+\omega_2} z\frac{f'(z)}{f(z)}\,dz$$

$$= -\omega_2 \int_{x}^{x+\omega_1} \frac{f'(t)}{f(t)}\,dt = 2\pi i k\omega_2$$

for an integer k. The computation for the opposite side is performed in the same way. □

DEFINITION. The number of poles (or zeros) of an elliptic function f in any period parallelogram is called the *order* of f.

COROLLARY XVI.16.4. Every non-constant elliptic function has order $\geqslant 2$.

Weierstrass Approach. The $\wp$ *Function.* Since the "simplest" elliptic function has order 2 then if we want to construct an eliptic function with a prescribed lattice of periods L we need two simple poles or a second-order pole in each period parallelogram. The first possibility leads to the σ-functions of Jacobi, the second is the approach of Weierstrass and leads to his famous $\wp$ (read "p") function. It is introduced and by the following

DEFINITION. The *Weierstrass $\wp$ function* is the sum of the series

$$\wp(z) := \frac{1}{z^2} + \sum_{0 \neq \omega \in L} \left(\frac{1}{(z-\omega)^2} - \frac{1}{\omega^2} \right), \tag{4}$$

where ω ranges over $L^* = \{\omega = n_1\omega_1 + n_2\omega_2, : n_1, n_2 \in \boldsymbol{Z},\ n_1 \neq 0 \neq n_2\}$ and (ω_1, ω_2) is a pair of complex numbers with $\operatorname{Im}(\omega_1/\omega_2) \neq 0$.

THEOREM XVI.16.5. The series in (4) represents a function holomorphic in $\boldsymbol{C}-L$, with poles of order two at all points of L and doubly periodic with (ω_1, ω_2) as a pair of fundamental periods. Moreover, $\wp$ is even, i.e., $\wp(z) = \wp(-z)$.

PROOF. We have to verify that the series in (4) converges. If $|\omega| > 2|z|$, then

$$\left| \frac{1}{(z-\omega)^2} - \frac{1}{\omega^2} \right| = \left| \frac{z(2\omega - z)}{\omega^2(z-\omega)^2} \right| \leqslant \frac{10|z|}{|\omega|^3}.$$

Therefore, the series (4) converges compactly if

$$\sum_{\omega \neq 0} \frac{1}{|\omega|^3} < \infty. \tag{5}$$

But (5) is satisfied. Indeed, since $\operatorname{Im}(\omega_2/\omega_1) \neq 0$, there exists $a > 0$ such that

$$|n_1\omega_1 + n_2\omega_2| \geqslant a(|n_1| + |n_2|) \quad \text{for all real } n_1, n_2.$$

If we consider only $n_1, n_2 \in \boldsymbol{Z}$, then there are $4n$ pairs (n_1, n_2) with $|n_1| + |n_2| = n$. Therefore,

$$\sum_{\omega \neq 0} |\omega|^{-3} \leqslant 4a^{-3} \sum_{n=1}^{\infty} n^{-2} < \infty.$$

In order to prove the periodicity of $\wp$ we first show that $\wp'$ is periodic: by termwise differentiation we obtain

$$\wp'(z) = -\frac{2}{z^3} - \sum_{\omega \neq 0} \frac{2}{(z-\omega)^3} = -2 \sum_{\omega \in L} \frac{1}{(z-\omega)^3}, \tag{6}$$

but the last sum is plainly doubly periodic. Therefore, $\wp(z+\omega_1)-\wp(z)$ and $\wp(z+\omega_2)-\wp(z)$ are constants. Since evidently $\wp(-z) = \wp(z)$, upon putting $z = -\omega_1/2$ and $z = -\omega_2/2$ we see that these constants are 0. The rest of the proposition is clear. □

Eisenstein Series. The Eisenstein series play the paramount role in the arithmetic and the theory of elliptic functions. Gotthold Eisenstein (1823–1852) a one of greatest number-theoreticians of all times (so said Gauss), starved to death at the age of 29.

DEFINITION. Let a lattice L be given. For $n \geqslant 3$ the series

$$G_n := \sum_{\omega \neq 0} \frac{1}{\omega^n} \tag{6'}$$

converges and is called the *Eisenstein series* of weight n.

Remark. We shall discuss the Eisenstein series in more detail in Section 17 and also several times in Part III in the theory of group representations and automorphic forms. We merely note here that the Eisenstein series depends on the lattice L.

THEOREM XVI.16.6. We have the Laurent expansion of $\wp$:

$$\wp(z) = \frac{1}{z^2} + \sum_{n=1}^{\infty} (2n+1) G_{2n+2} z^{2n} \tag{7}$$

for $0 < |z| < r := \min\{|\omega| : \omega \in L^*\}$;

$$G_{2n+1} = 0 \quad \text{for } n = 1, 2, \ldots \tag{8}$$

PROOF. If $0 < |z| < r$, then $|z/\omega| < 1$ and we have

$$\frac{1}{(z-\omega)^2} = \frac{1}{\omega^2\left(1-\dfrac{z}{\omega}\right)^2} = \frac{1}{\omega^2}\left(1+\sum_{n=1}^{\infty}(n+1)\left(\frac{z}{\omega}\right)^n\right),$$

whence

$$\frac{1}{(z-\omega)^2} - \frac{1}{\omega^2} = \sum_{n=1}^{\infty} \frac{n+1}{\omega^{n+2}} z^n. \tag{9}$$

From (9) and (4), summing over all $\omega \neq 0$ we obtain

$$\wp(z) = \frac{1}{z^2} + \sum_{n=1}^{\infty} (n+1) \sum_{\omega \neq 0} \frac{1}{\omega^{n+2}} z^n$$

$$= \frac{1}{z^2} + \sum_{n=1}^{\infty} (n+1) G_{n+2} z^n.$$

Since $\wp$ is an even function, the coefficients of z^{2k+1} must vanish and we obtain (7) and (8). □

The Differential Equation of the $\wp$ Function. We shall now derive the famous differential equation satisfied by $\wp$. This equation points out the connection of elliptic functions and elliptic integrals (cf. also Section 20).

THEOREM XVI.16.7 (Weierstrass). The function $\wp$ satisfies the differential equation

$$\left(\frac{d}{dz}\wp\right)^2 = 4\wp^3(z) - 60G_4\wp(z) - 140G_6 \tag{10}$$

or putting

$$g_2 := 60G_4, \quad g_3 := 140\, G_6$$

$$\left(\frac{d}{dz}\wp\right)^2 = 4\wp^3(z) - g_2\wp - g_3. \tag{10'}$$

PROOF. We simply check (10) by writing down the members of both sides of (10): by (7) we have (omitting terms of higher order which are relevant to the argument)

$$\wp(z) = \frac{1}{z^2} + 3G_2 z^2 + 5G_6 z^4 + \dots,$$

$$\wp'(z) = -\frac{2}{z^3} + 6G_4 z + 20G_6 z^3 + \dots,$$

$$\wp'(z)^2 = \frac{4}{z^6} - \frac{24G_4}{z^2} - 80G_6 + \dots,$$

$$60G_4\wp(z) = \frac{60}{z^2} G_4 + \dots,$$

$$4\wp^3(z) = \frac{4}{z^6} + \frac{36G_4}{z^2} + 60G_6 + \dots$$

Summing both sides, we obtain

$$\wp'(z)^2 - 4\wp(z)^3 + 60G_4\wp(z) = -140G_6 + \dots$$

Since the left-hand side is a doubly-periodic function and the right-hand side has no poles it must be a constant by Theorem XVI.5.4. 1° and this constant is $140G_6$. □

We can solve equation (10′) explicitly: we denote $w = \wp(z)$; then

$$(11)\qquad z - z_0 = \int_{\wp(z_0)}^{\wp(z)} \frac{dw}{\sqrt{4w^3 - g_2 w - g_3}},$$

where the path of integration is the $\wp$-image of a path from z to z_0 that avoids the poles and zeros of $\wp'$, and where the sign of the square root is chosen so as to equal $\wp'$.

We now discuss certain simple notions from algebraic geometry in order to interpret the differential equation (10). Let $P = P(x, y)$ be an irreducible polynomial in two variables, then the equation $P(x, y) = 0$ defines a subset of $\boldsymbol{C}\times\boldsymbol{C}$ called an *algebraic relation*. Functions contained in this relations are called *algebraic functions* defined by $P(x, y) = 0$; thus f is algebraic if $P(z, f(z)) \equiv 0$ for $z \in \mathcal{O} \subset \boldsymbol{C}$.

A simple, perhaps even naive, example may help. The equation

$$(12)\qquad x^2 + y^2 - 1 = 0$$

defines an algebraic function

$$(13)\qquad y = \sqrt{1 - x^2}.$$

Geometers speak of *algebraic curves* rather than of algebraic functions: equation (12) defines the unit circle in $\boldsymbol{R}^2$. The relation given by (12) defines a two-valued function (13). The pair of functions

$$(14)\qquad x = \sin t, \quad y = \cos t = (\sin)'(t)$$

uniformizes the circle (12). After this elementary digression let us return to elliptic functions. Consider an algebraic equation

$$(15)\qquad y^2 - (4x^3 - g_2 x - g_3) = 0,$$

Theorem XVI.16.6 asserts that the algebraic curve of the third order defined by (15) is uniformized by the pair $(\wp, \wp')$ or, in other words, the algebraic curve defined by (15) has the parametrization

$$x = \wp(z), \quad y = \wp'(z).$$

The connection with the problem of uniformization is obvious. We shall return to it in Sections 19 and 20.

The Weierstrass equation (10) can be written as

(16) $$\wp'(z)^2 = 4(\wp(z)-e_1)(\wp(z)-e_2)(\wp(z)-e_3)$$

where e_1, e_2, e_3 are the roots of the polynomial

(*) $$4x^3-g_2x-g_3.$$

The discriminant of this cubic polynomial is proportional to

(17) $$\Delta = g_2^3-27g_3^2.$$

We recall the notion of the discriminant of a polynomial.

DEFINITION. Let

(**) $$P(z) = z^n+a_1z^{n-1}+ \dots +a_n$$

be a polynomial (with complex coefficients) and by $x_1, \dots, x_n$ let us denote the roots of P, appearing according to their multiplicity. The product

$$D(x_1, \dots, x_n) = \prod_{i<j} (x_i-x_j)^2$$

is called the *discriminant* of the polynomial P. Since D is symmetric, then according to a well-known result from algebra it is a polynomial of the so-called *elementary symmetric polynomials*:

$$\sigma_1 = x_1+ \dots +x_n,$$

$$\sigma_2 = x_1x_2+x_1x_3+ \dots +x_{n-1}x_n,$$

$$\dots\dots\dots\dots\dots\dots\dots\dots$$

$$\sigma_n = x_1 \dots x_n.$$

Note also the relations

$$a_1 = -\sigma_1, \quad a_2 = \sigma_2, \quad \dots, \quad a_n = (-1)^n\sigma_n.$$

The vanishing of the discriminant is a necessary and sufficient condition for the existence of multiple roots of the polynomial (**).

PROPOSITION XVI.16.8. The roots e_1, e_2, e_3 of the polynomial (*) are distinct, and hence, $\Delta \neq 0$. We have $e_k = \wp\left(\frac{\omega_k}{2}\right)$, $k = 1, 2, 3$ where $\omega_3 := \omega_1+\omega_2$. Explicitly

(18) $$4\wp^3(z)-g_2\wp(z)-g_3 = 4(\wp(z)-e_1)(\wp(z)-e_2)(\wp(z)-e_3)$$

PROOF. We determine the zeros of $\wp'$ (cf. (16)). Since $\wp$ is even, $\wp'$ is odd. We check that the half-periods of an odd elliptic function are either poles or zeros: By periodicity we have, where ω stands for ω_i, $i = 1, 2, 3$,

$$\wp'(-\tfrac{1}{2}\omega) = \wp'(\omega - \tfrac{1}{2}\omega) = \wp'(\tfrac{1}{2}\omega) = -\wp'(-\tfrac{1}{2}\omega).$$

Hence, $\wp'(\frac{1}{2}\omega) = 0$ if $\wp'(\frac{1}{2}\omega)$ is finite. Since $\wp'$ has no poles at $\frac{1}{2}\omega_k$, these points must be zeros of $\wp'$. But, as we know, $\wp'$ is of order 3 and, thus, $\frac{1}{2}\omega_k$ must be simple zeros of $\wp'$. Hence, $\wp'$ have no other zeros in the period parallelogram with vertices $0, \omega_1, \omega_2, \omega_3$. Equation (10′) shows that these points are zeros of the cubic polynomial (*) and we have proved (18). We have to check that the roots e_1, e_2, e_3 are distinct. Since $\wp$ assumes each of its values with multiplicity 2 and has only one $\operatorname{mod} L$ pole of order 2, we see that $e_k \neq e_j$ for $k \neq j$. □

The Addition Theorem for the Function $\wp$. Since $\wp$ bears some similarity to the function sin and since the addition formula is valid, for sin,

$$\sin(u_1 + u_2) = \sin u_1 \sin' u_2 + \sin' u_1 \sin u_2,$$

we may expect a similar, perhaps more complicated, formula to be valid. Indeed, we have the famous

THEOREM XVI.16.9 (Addition Theorem for $\wp$). For $u_1 \not\equiv u_2 (\operatorname{mod} L)$ we have

$$\wp(u_1 + u_2) = -\wp(u_1) - \wp(u_2) + \frac{1}{4}\left(\frac{\wp'(u_1) - \wp'(u_2)}{\wp(u_1) - \wp(u_2)}\right)^2; \tag{19}$$

in particular,

$$\wp(2u) = -2\wp(u) + \frac{1}{4}\left(\frac{\wp''(u)}{\wp'(u)}\right)^2. \tag{20}$$

PROOF. Let us take $L \not\ni u_1, u_2 \in C$, let us assume that $u_1 \not\equiv u_2$ $(\operatorname{mod} L)$, and let $y = ax + b$ be the line in C through $(\wp(u_1), \wp'(u_1))$ and $(\wp(u_2), \wp'(u_2))$. In other words, let $a, b \in C$ be such that

$$\wp'(u_1) = a\wp(u_1) + b, \quad \wp'(u_2) = a\wp(u_2) + b. \tag{21}$$

Then, since $\wp' - (a\wp + b)$ has a pole of order 3 at 0, by the Liouville Theorem (XVI.16.2.4°), it has three zeros, two of which are u_1 and u_2.

If u_2 has a multiplicity of 2, then by the same theorem (6°) we have

(22) $\quad 2u_1 + u_2 \equiv 0 \pmod{L}.$

If we fix u_1, equation (22) can be satisfied by only one (mod L) value of u_2. Let us assume that this is not the case; then both u_1, u_2 are of multiplicity 1 and the third zero is

(∗) $\quad u_3 \equiv -(u_1+u_2) \pmod{L}$

(by Theorem XVI.16.2. 6°). Therefore, we have

(21′) $\quad \wp'(u_3) = a\wp(u_3)+b.$

From (21), (21′) and (18) we know that the equation

(23) $\quad 4x^3 - g_2x - g_3 - (ax+b)^2 = 0$

has three roots, $\wp(u_1)$, $\wp(u_2)$, $\wp(u_3)$ and we can factorize the left-hand side as

(24) $\quad 4(x-\wp(u_1))(x-\wp(u_2))(x-\wp(u_3)).$

Comparing the coefficients of x^2, we obtain

(25) $\quad a^2 = 4(\wp(u_1)+\wp(u_2)+\wp(u_3)).$

But (21) yields

$$a = \frac{\wp'(u_1)-\wp'(u_2)}{\wp(u_1)-\wp(u_2)}.$$

Since $\wp(u_2) = \wp(-(u_1+u_2)) = \wp(u_1+u_2)$, we get the addition formula (19) from (25). We have obtained (19) with fixed u_1 for all but a finite number of $u_2 \not\equiv u_1 \pmod{L}$ and, hence, for all $u_2 \not\equiv u_1 \pmod{L}$ by analytic continuation. Equation (20) is obtained from (19) by passing to the limit $u_2 \to u_1$. □

Exercise 1. Prove the formula

(26) $$\begin{vmatrix} \wp(z) & \wp'(z) & 1 \\ \wp(u) & \wp'(u) & 1 \\ \wp(u+z) & -\wp'(u+z) & 1 \end{vmatrix} = 0.$$

The Weierstrass Sigma Function is defined by

(27) $$\sigma(z, L) \equiv \sigma(z) := z \prod_{0 \neq \omega \in L} \left(1-\frac{z}{\omega}\right) \exp\left[\frac{z}{\omega}+\frac{1}{2}\left(\frac{z}{\omega}\right)^2\right],$$

and, hence, σ has zeros of order 1 at all lattice points. Plainly,

$$(28) \qquad \sigma(\lambda z, \lambda L) = \lambda\sigma(z, L), \qquad \lambda \in \boldsymbol{C}.$$

The Weierstrass zeta function is obtained from (27) by logarithmic derivation:

$$(29) \qquad Z(z, L) \equiv Z(z) := \frac{\sigma'(z)}{\sigma(z)} = \frac{1}{z} + \sum_{0 \neq \omega \in L} \left(\frac{1}{z-\omega} + \frac{1}{\omega} + \frac{z}{\omega^2} \right).$$

We leave it to the reader to verify the validity of the operation performed. The right hand side of (29) converges absolutely and uniformly for z in every compact set not containing any lattice points. Differentiation of (29) gives

$$(30) \qquad Z'(z) = -\frac{1}{z^2} - \sum_{0 \neq \omega \in L} \left(\frac{1}{(z-\omega)^2} - \frac{1}{\omega^2} \right) = -\wp(z).$$

Plainly, σ and Z are odd functions.

Remark. The Weierstrass Z has nothing to do with the famous Riemann zeta function defined by the series

$$\zeta(z) = \sum \frac{1}{n^z}.$$

Since $\wp$ is periodic, differentiation of

$$Z(z+\omega) - Z(z)$$

gives 0 for any $\omega \in L$, thus, there exists a constant $\eta(\omega)$ such that

$$Z(z+\omega) = Z(z) + \eta(\omega).$$

Plainly, η is $\boldsymbol{Z}$-linear in ω.

PROPOSITION XVI.16.10. The function σ is a theta function, that is to say, it satisfies an identity

$$(31) \qquad \sigma(z+\omega) = \sigma(z) \exp(\eta(\omega)z + c(\omega)), \qquad \omega \in L.$$

PROOF. We have

$$\frac{d}{dz} \log \frac{\sigma(z+\omega)}{\sigma(z)} = \eta(\omega)$$

and, therefore,

$$\log \frac{\sigma(z+\omega)}{\sigma(z)} = \eta(\omega)z + c(\omega),$$

whereupon (31) follows by exponentiation.

DEFINITION. A theta function on $\boldsymbol{C}$ with respect to a lattice L is an entire function $\theta \in A(\boldsymbol{C})$ satisfying the (automorphy) condition

$$(32) \qquad \theta(z+\omega) = \theta(z)\exp[2\pi i(l(z,\omega)+c(\omega))], \qquad z \in \boldsymbol{C},\ \omega \in L,$$

where $(z,\omega) \to l(z,\omega)$ is $\boldsymbol{C}$-linear in z and $\boldsymbol{R}$-linear in ω.

Remark. The general theta function and automorphic forms will be considered in connection with group representations in Part III.

17. MODULAR FUNCTIONS AND FORMS. THE MODULAR FIGURE, DISCONTINUOUS GROUPS OF AUTOMORPHISMS

In this section we shall continue to use the notation introduced in Section 8 of this chapter. In particular, we use $\mathfrak{H}$ to denote the upper half-plane $\{z \in \boldsymbol{C}\colon \operatorname{Im} z > 0\}$ and we recall that $\mathfrak{H}$ is conformally equivalent with the unit disk $\boldsymbol{D}$. In Theorem XVI.8.18 we proved that the group $SL_2(\boldsymbol{R}) = \left\{\begin{bmatrix} a & b \\ c & d\end{bmatrix}\colon a,b,c,d \in \boldsymbol{R} \text{ and } \det\begin{bmatrix} a & b \\ c & d\end{bmatrix} = 1\right\}$ acting in $\mathfrak{H}$ by means of the Möbius transformations

$$(1) \qquad gz := \frac{az+b}{cz+d}, \qquad g = \begin{bmatrix} a & b \\ c & d\end{bmatrix} \in SL_2(\boldsymbol{R})$$

gives the group $\operatorname{Aut}(\mathfrak{H})$. More specifically, if we let $I = \begin{bmatrix} 1 & 0 \\ 0 & 1\end{bmatrix}$ be the identity matrix in $SL_2(\boldsymbol{R})$, then the group

$$(2) \qquad PSL_2(\boldsymbol{R}) := SL_2(\boldsymbol{R})/\{\pm I\}$$

is isomorphic with $\operatorname{Aut}(\mathfrak{H})$, the isomorphism given by the formula (1). (Observe that matrices from $SL_2(\boldsymbol{R})$ which give the same Möbius transformation differ at most by a sign.) The reader can check easily that

$$(1') \qquad \operatorname{Im}(gz) = \frac{\operatorname{Im}(z)}{(cz+d)^2}, \qquad g = \begin{bmatrix} a & b \\ c & d\end{bmatrix} \in SL_2(\boldsymbol{R}).$$

We now introduce the following

DEFINITION. The subgroups $\Gamma_+ := SL_2(\boldsymbol{Z})$ and $\Gamma := SL_2(\boldsymbol{Z})/\{\pm I\}$ are called *modular groups*. Here, $SL_2(\boldsymbol{Z})$ denotes the group of matrices from $SL_2(\boldsymbol{R})$ with integral entries. They are discrete subgroups of $SL_2(\boldsymbol{R})$ and $PSL_2(\boldsymbol{R})$, resp. Γ, is also called a *homogeneous modular group*.

It is often important to parametrize the space $\Gamma\backslash\mathfrak{H}$ of Γ-orbits. This leads to the important general notion of the fundamental domain which we discuss in a fairly general setting.

Let G be a topological (usually locally compact) group and let X be a locally compact space. (We adopt the convention that topological spaces are Hausdorff.) We say that *G acts continuously on X* or that G is a *topological transformation group of the space X* if there is a mapping

$$G \times X \ni (g, x) \to gx \in X$$

which is continuous (jointly) and satisfies the following conditions: 1° $g_1(g_2x) = (g_1g_2)x$ for all $g_1, g_2 \in G$ and all $x \in X$ and 2° $ex = x$ for all $x \in X$, where e denotes the identity of G. In this situation, to every element $g \in G$ there corresponds a homeomorphism of X, namely, τ_g: $X \ni x \to gx \in X$ and the mapping $G \ni g \to \tau_g \in \mathrm{Aut}(X)$ is a homomorphism. One should also note that the isotropy group G_x of any element $x \in X$ is a closed subgroup of G. Two points $x_1, x_2 \in X$ are called *equivalent under G* if there exists a $g \in G$ such that $gx_1 = x_2$, i.e., if x_1, x_2 belong to the same orbit in X. We denote by $G\backslash X$ the set of all orbits in X and by π: $X \to G\backslash X$ the canonical projection defined by $\pi(x) = Gx$. A subgroup Γ of a topological group is called *discrete* if it is discrete in the induced topology. It can be shown that for a discrete subgroup Γ we have $\bar{\Gamma} = \Gamma$ and, thus, Γ has no accumulation points in G.

Exercise. Prove the last statement.

DEFINITION. Suppose Γ is a discrete group acting on a locally compact space X. A connected subset $F \subset X$ is called a *fundamental domain* of the orbit space $\Gamma\backslash X$ (or of the group Γ) if the following conditions are satisfied:

1° For every point $x \in X$ there is at least one point in F equivalent to x.

2° The interior of F does not contain any pairs of distinct equivalent points.

It follows from the definition that each orbit of the group meets F in at least one point and if an orbit contains more than one point from F, they have to belong to the boundary of F.

Remark. A slightly different definition of a fundamental domain is often adopted in the literature, namely, the interior of the set which we called a *fundamental domain.* This is clearly a matter of convenience. Also the precise choice of F and the inclusion or omission of some parts of the boundary is largely a matter of convenience.

Example 1. Let $X = \boldsymbol{C}$ and let $\Gamma = [\omega_1, \omega_2]$ be the lattice generated by two R-independent numbers ω_1, ω_2. Then the fundamental parallelogram $F = \{t\omega_1 + s\omega_2 \colon 0 \leqslant t, s \leqslant 1\}$ is a fundamental domain of Γ.

Example 2. $X = \boldsymbol{C}$, $\Gamma = Z\omega$, where $0 \neq \omega \in \boldsymbol{C}$. Then the infinite strip $F = \{t\omega \colon 0 \leqslant t < 1\}$ is a Γ-fundamental domain.

It is an important, but unfortunately in most cases a very difficult, task to determine a fundamental domain. Therefore, one often works with the space X/Γ when no fundamental domain is known. The following theorem is of paramount importance for the theory of elliptic modular functions:

THEOREM XVI.17.1. Let Γ be the modular (homogeneous) group. Denote

$$(3) \qquad T = \begin{bmatrix} 1 & 1 \\ 0 & 1 \end{bmatrix}, \quad S = \begin{bmatrix} 0 & -1 \\ 1 & 0 \end{bmatrix},$$

$$(4) \qquad F = \{z \in \mathfrak{H} \colon \operatorname{Re} z \in [-\tfrac{1}{2}, +\tfrac{1}{2}], |z| \geqslant 1\}$$

(cf. Fig. 18). Then

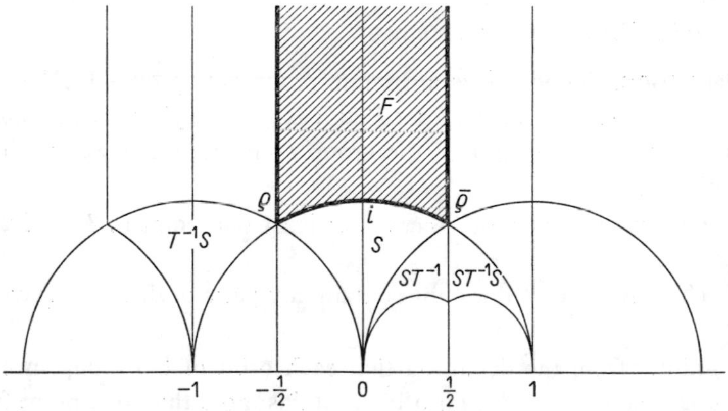

Fig. 18

1° F is a fundamental domain for the modular group.

2° The images of S, T in Γ generate Γ.

3° If $\Gamma_z = \{\gamma \in \Gamma\colon \gamma z = z\}$ is the stabilizer of z, then $\Gamma_z = \{1\}$ for $z \in F$, except in the following three cases:

1. for $z = i$, Γ_z is a two-element group generated by (the image of) S,

2. for $z = \varrho = e^{2\pi i/3}$, Γ_z is a group of order 3 (three elements) generated by ST,

3. for $z = -\bar{\varrho} = e^{\pi i/3}$, Γ_z is a group of order 3 generated by TS.

Before we begin the proof let us remark that

$$(5) \qquad Sz = -\frac{1}{z}, \qquad Tz = z+1, \qquad S^2 = 1, \qquad (ST)^3 = 1,$$

$$(6) \qquad ST = \begin{bmatrix} 0 & -1 \\ 1 & 1 \end{bmatrix}, \qquad ST\varrho = \varrho, \qquad Si = i, \qquad TS(-\bar{\varrho}) = -\bar{\varrho}.$$

Figure 18 represents the transforms of F by the elements

$$\{1, T, TS, ST^{-1}S, S, ST, STS, T^{-1}S, T^{-1}\}$$

of the modular group Γ.

PROOF. 1° Let Γ' denote the subgroup of Γ generated by S and T and let $z \in \mathfrak{H}$. It is sufficient to show that there exists a $\gamma' \in \Gamma'$ such that $\gamma' z \in F$. If $\gamma = \begin{bmatrix} a & b \\ c & d \end{bmatrix} \in \Gamma'$, then we have (1'); $\operatorname{Im} \gamma z = \operatorname{Im}(z)|cz+ +d|^{-2}$. Since c and d are integers, the number of pairs (c, d) such that $|cz+d|$ is less than a given number is finite. Hence, there exists a $\gamma \in \Gamma'$ such that $\operatorname{Im}(\gamma z)$ is maximum. Now, take an integer n such that $-\frac{1}{2} \leqslant \operatorname{Re}(T^n\gamma z) \leqslant \frac{1}{2}$. We check that $z' := T^n\gamma z \in F$, i.e., that $|z'| \geqslant 1$. But if $|z'| < 1$, then $\operatorname{Im}(Sz') = \operatorname{Im}(-1/z') > \operatorname{Im}(z')$, which is impossible since $S \in \Gamma'$. Therefore, $\gamma' := T^n\gamma$ is the desired element of Γ'.

2° We have to show that if $z, z' \in F$ are on the same orbit of Γ', then only two situations may occur: either they lie on the vertical sides and are translates of each other by ± 1, or they lie on the base arc and are transforms of each other by S $(z' = -z^{-1})$.

If $\gamma z = z'$, where $\gamma = \begin{bmatrix} a & b \\ c & d \end{bmatrix}$, then, since, we can replace the pair (γ, z) by $(\gamma^{-1}, \gamma z)$, we can assume that $\operatorname{Im} \gamma z \geqslant \operatorname{Im} z$, i.e., by (1) we can

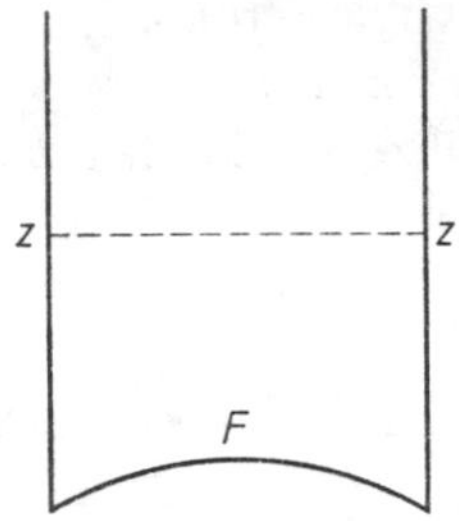

Fig. 19

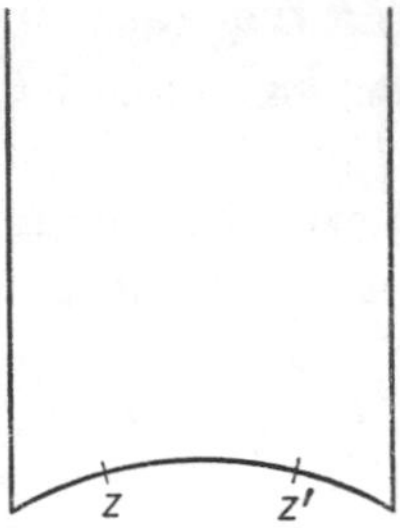

Fig. 20

assume that $|cz+d| \leqslant 1$ and, hence, $|c| < 2$. Therefore, we have only three possibilities: $c = 0$, $c = -1$, $c = 1$.

I. $(c = 0) \Rightarrow (d = \pm 1$ and $\gamma z = z \pm b)$. But $\mathrm{Re}(z)$, $\mathrm{Re}(\gamma z) \in [-\frac{1}{2}, \frac{1}{2}]$; hence, either $b = 0$ and $\gamma = 1$ or $b = \pm 1$ and $\mathrm{Re}(z) = -1/z$, $\mathrm{Re}(\gamma z) = 1/z$ (or vice versa). This is situation I.

II. $(c = 1$, i.e., $|z+d| \leqslant 1) \Rightarrow (d = 0$, except $z = \varrho$, (resp., $z = -\bar{\varrho}$)); in the last cases we have $d = 0, 1$ (resp. $d = 0, -1$). The case $d = 0$ gives $|z| \leqslant 1$ and, hence, $|z| = 1$ since $|z|_1 \geqslant 1$. But $(ad-bc = 1) \Rightarrow (b = -1)$, whereby $\gamma z = a - 1/z$, and $a = 1$ (if $|\mathrm{Re}\, z| < \frac{1}{2}$) and, hence, $z' = \gamma z = -1/z$ and the proof is complete. If $\mathrm{Re}\, z = \pm\frac{1}{2}$, i.e., if $z = \varrho$ or $z = -\bar{\varrho}$, then $a = 0, -1$ or $a = 0, 1$. The case $z = \varrho$, $d = 1$ gives (since $c = 1$) $a - b = 1$ and $\gamma\varrho = a - 1/(1+\varrho) = \alpha + \varrho$ and $a = 0, 1$, i.e., $\gamma\varrho = \varrho$ or $\gamma\varrho = 1 + \varrho$.

III. The case $c = -1$ is reduced to the case $c = 1$, by replacing γ by $-\gamma$. From (5) and (6) we have 2°.

3° We have to prove that $\Gamma' = \Gamma$. Let $\gamma \in \Gamma$, take a point $z_0 \in \mathrm{int} F$ (for instance $z_0 = 2i$), and let $z := \gamma z_0$. We have proved in 1° that there exists a $\gamma' \in \Gamma$ such that $\gamma' z \in F$. Since the points z_0 and $g'z = (g'g)z_0$ lie on the same Γ-orbit and one of them is interior to F, then by 1° and 2° they coindice: $z_0 = g'gz_0$ and $g'g = 1$, whereby $g = g'^{-1} \in \Gamma'$. But g was an arbitrary element of Γ, and therefore $\Gamma' = \Gamma$. □

COROLLARY XVI.17.2. The canonical map $\pi\colon F \to \Gamma\backslash\mathfrak{H}$ is surjective and its restriction to the interior of F is injective.

Remark. It is sometimes more convenient to take as a fundamental region the set $F_1 := \{z \in \mathfrak{H}\colon \frac{1}{2} \leqslant \mathrm{Re}\, z < \frac{1}{2},\ |z| > 1\} \cup P$, where $P = \{e^{t\pi i/3}\colon \frac{3}{2} \leqslant t \leqslant 2\}$ (Fig. 21).

Modular Forms. We shall now introduce the fundamental notions of modular functions and modular forms. Let us assume that a function f is holomorphic in a half-plane $\mathfrak{H}_\alpha = \{z \in \mathbf{C}: \operatorname{Im} z > \alpha\}$ for some $\alpha > 0$,

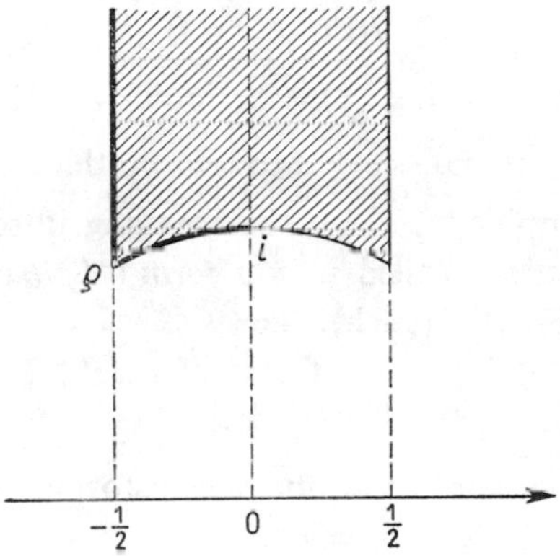

Fig. 21

and is periodic with period 1, i.e., $f(z+1) = f(z)$ for all z in that half-plane. The function $\tau \to e^{2\pi i\tau}$ maps $\mathfrak{H}_\alpha$ onto the punctured disk $\{q \in \mathbf{C}: 0 < |q| < e^{-2\pi\alpha}\}$ so that all points differing by an integer are mapped upon one point of that disk. Through this mapping such periodic f determines a holomorphic function in the punctured disk by the formula

$$(7) \qquad \hat{f}(q) = f(\tau), \quad \text{where } q = e^{2\pi i\tau}, \ \tau \in \mathfrak{H}_\alpha.$$

Under these circumstances we shall say that f is *holomorphic* (*meromorphic*) *at* ∞ if the function $\hat{f}$ has a removable singularity (resp., a pole) at the point 0 and the value of $\hat{f}$ at the point 0 (or the order of $\hat{f}$ at 0 if it is a zero or a pole of $\hat{f}$) will be said to be the *value of f at* ∞ (resp., the order of f at ∞). We should warn the reader that this usage of the notion of holomorphicity at ∞ is different from that introduced earlier, when we dealt with functions defined on the Riemann sphere. In the present case f is not defined in any neighbourhood of the infinity of the Riemann sphere (i.e., the north pole). The notion introduced above is an anticipation of later treatment with the use of a suitably compactified Riemann surface (cf. Theorem XVI.17.12.7°).

DEFINITION. Let k be an integer. A function $f\colon \mathfrak{H} \to \boldsymbol{C}$ is called a *modular form of weight* $2k$ if it satisfies the following conditions:

1° $f \in A(\mathfrak{H})$ is holomorphic on $\mathfrak{H}$,

2° for all $\gamma = \begin{bmatrix} a & b \\ c & d \end{bmatrix} \in \Gamma = \mathrm{SL}_2(\boldsymbol{Z})$

(8) $\qquad f(\gamma z) = (cz+d)^{2k} f(z), \quad z \in \mathfrak{H}.$

Since any function satisfying (8) is periodic with the period 1 (use (8)) for $\gamma = T = \begin{bmatrix} 1 & 1 \\ 0 & 1 \end{bmatrix}$, it makes sense to require that

3° f be holomorphic at ∞ (in the meaning introduced above). If f is 0 at infinity, then f is called a *cusp form* ("*Spitzenform*" in German, "*forme parabolique*" in French).

The space of modular (cusp) forms of weight $2k$ will be denoted by $M(k, \Gamma)$ (resp., $M^0(k, \Gamma)$).

Remarks. 1. Consider the Möbius transformation of $\mathfrak{H}$ defined by a matrix $g \in \mathrm{SL}_2(\boldsymbol{R})$, i.e.,

$$z \to gz = \frac{az+b}{cz+d}, \qquad g = \begin{bmatrix} a & b \\ c & d \end{bmatrix}.$$

Then

$$\frac{d(gz)}{dz} = \frac{1}{(cz+d)^{-2}}$$

and (8) can be rewritten as

$$\frac{f(\gamma z)}{f(z)} = \frac{d(gz)^{-k}}{dz}.$$

2. Let f be a modular form of weight $2k$ and let $\hat{f}$ be the function corresponding to it via (7). Then, by virtue of the condition 3°, in the definition the function $\hat{f}$ is represented by a power series and we have

(9) $\qquad f(\tau) = \hat{f}(e^{2\pi i\tau}) = \sum_{n=0}^{\infty} a_n e^{2\pi i n\tau} \quad$ for all $\tau \in \mathfrak{H}$.

Conversely, if the power series in (9) converges for all $\tau \in \mathfrak{H}$ and if its sum f satisfies the condition

(10) $\qquad f(-\tau^{-1}) = \tau^{2k} f(\tau) \quad$ for all $\tau \in \mathfrak{H}$,

then it is easily seen by using Theorem XVI.17.1.2° that f is a modular form of weight $2k$.

Moreover, the condition for f to be a cusp form is equivalent to the equality $a_0 = 0$, where a_0 is the 0-th coefficient in the expansion (9).

3. Equation (8) with $2k$ replaced by $2k+1$ forces f to be identically 0; in fact, for $\gamma = -I$ we obtain $f(z) = (-1)^{2k+1}f(z)$.

DEFINITION. A function $g \in \mathcal{M}(\mathfrak{H})$ is called a *modular function* if it is Γ-invariant; i.e., for all $\gamma \in \Gamma$

$$f(\gamma\tau) = f(\tau)$$

and is meromorphic at ∞.

If a *meromorphic* function $f \in \mathcal{M}(\mathfrak{H})$ satisfies (8) and is meromorphic at ∞, it is called a *modular function of weight* $2k$; the space of these functions is denoted by $\mathcal{M}(k, \Gamma)$. Hence, modular functions are elements of $\mathcal{M}(0, \Gamma)$.

The preceding definitions immediately imply

LEMMA XVI.17.3. 1° If f and g are modular forms of weight $2k_1$ and $2k_2$, then fg has the weight $2(k_1+k_2)$.

2° If f, g are modular forms of the same weight, then f/g is a modular function.

The easy proof is left to the reader.

Let $L = L(\omega_1, \omega_2)$ be a lattice in C where (ω_1, ω_2) is a fundamental pair chosen so that $\omega_1/\omega_2 \in \mathfrak{H}$. Let $G_k(\omega_1, \omega_2)$ be the Eisenstein series of order k associated with the lattice L, $k > 2$, i.e.,

$$G_k(\omega_1, \omega_2) = \sum_{(m,n) \in Z^2 - \{(0,0)\}}' (m\omega_1 + n\omega_2)^{-k}$$

(cf. Section 16 of this chapter). G_k is a homogeneous function of degree $-k$, i.e.,

$$G_k(\lambda\omega_1, \lambda\omega_2) = \lambda^{-k} G_k(\omega_1, \omega_2), \quad C \ni \lambda \neq 0,$$

and, consequently, the functions $g_2 = 60G_4$, $g_3 = 140G_6$ are homogeneous of degree -4 and -6, respectively. We recall that the discriminant Δ of the Weierstrass equation (Theorem XVI.16.6) is given as up to a constant

$$\Delta = g_2^3 - 27g_3^2.$$

All of these functions are in a natural way interpreted as modular functions. In fact, when the lattice $L(\omega_1, \omega_2)$ is allowed to vary, the Eisenstein series G_k becomes a function of its period pairs (ω_1, ω_2). Since two pairs (ω_1, ω_2) and (ω_1', ω_2') generate the same lattice iff there exists a matrix $\gamma \in \mathrm{SL}_2(\boldsymbol{Z})$ such that

$$\begin{aligned} \omega_1' &= a\omega_1 + b\omega_2, \\ \omega_2' &= c\omega_1 + d\omega_2, \end{aligned} \qquad \gamma = \begin{bmatrix} a & b \\ c & d \end{bmatrix};$$

this function will depend only on classes of equivalent pairs. If we wish to take the homogeneity of the function, into account, it is natural to set $\tau = \omega_1/\omega_2$ and to transform the function to the form

$$G_k(\omega_1, \omega_2) = \omega_2^{-k} G_k(\tau, 1).$$

Now, using the independence of G_k on the chosen basis, we get

$$G_k(\omega_1', \omega_2') = (\omega_2)^{-k}(c\tau + d)^{-k} G_k(\gamma\tau, 1) = \omega_2^{-k} G_k(\tau, 1).$$

We thus see that the function $\tau \to G_k(\tau, 1)$, which we shall also denote by $G_k(\tau)$, satisfies the condition (8) in the definition of a modular form. By verifying that it has the right regularity properties we shall prove it is in fact a modular function. We also define the so-called Dedekind–Klein modular function

$$J := \frac{g_2^3}{\Delta}, \qquad j := 12^3 J = 12^3 \frac{g_2^3}{\Delta};$$

j is sometimes called a *modular invariant* or an *elliptic modular function.*

THEOREM XVI.17.4. We have the following facts:

1° The Eisenstein series $G_k(\tau)$ is a modular form of weight k (recall that for any odd integer k G_k vanishes identically).

2° The discriminant Δ is a cusp form of weight 12 ($\Delta(\infty) = 0$).

3° j is a modular function which is holomorphic in $\mathfrak{H}$ and has a simple pole at infinity.

PROOF. We only determine the regularity properties of those functions.

1° We have ($*$ means that we omit the term with $m = n = 0$)

$$(11) \qquad G_{2k}(\tau) = \sum_{m,n}^{*} \frac{1}{(m\tau + n)^{2k}}, \qquad k > 1.$$

Since F is a Γ-fundamental domain, the sets γF, $\gamma \in \Gamma$ cover $\mathfrak{H}$. Let us assume first that $\tau \in F$; then, here $\varrho = e^{2\pi i/3}$,

$$(12) \qquad \begin{aligned} |m\tau + n|^2 &= m|\tau|^2 + 2mn \operatorname{Re}\tau + n^2 \\ &\geqslant m^2 - mn + n^2 = |m\varrho - n|^2. \end{aligned}$$

But the series $\sum^*|m\varrho-n|^{-2k}$ converges (cf. the proof of Corollary XVI.16.4); therefore, the series G_{2k} converges normally in F and, hence, G_{2k} is holomorphic in F. Therefore, $G_{2k}\circ\gamma^{-1}$ is holomorphic in γF for any $\gamma\in\Gamma$ and, hence, $G_{2k}\in\mathscr{A}(\mathfrak{H})$. We merely have to check that G_{2k} is holomorphic at infinity. Let $\operatorname{Im}(\tau)\to\infty$; we can suppose that τ remains in F and, by virtue of the uniform convergence in F, we can pass to the limit term by term.

For $m\neq 0$, $\lim(m\tau+n)^{-2k}\to 0$. The other terms give n^{-2k} and, therefore,

$$\text{(13)}\qquad \lim_{\operatorname{Im}\tau\to\infty} G_{2k}(\tau)=\sum_{n\in Z}^{*}\frac{1}{n^{2k}}=2\sum_{n=1}^{\infty}\frac{1}{n^{2k}}=2\zeta(2k).$$

Here, $\zeta(z)=\sum_{n=1}^{\infty}\frac{1}{n^z}$ is the ζ-function of Riemann.

2° Let us define the Bernoulli numbers B_k by the expansion

$$z\cot z=1-\sum_{k=1}^{\infty}B_k\frac{2^{2k}z^{2k}}{(2k)!}.$$

Since, on the other hand (cf. Section 15)

$$z\cot z=1-2\sum_{n=1}^{\infty}\sum_{k=1}^{\infty}\frac{z^{2k}}{n^{2k}\pi^{2k}},$$

we obtain

$$\text{(14)}\qquad \zeta(2k)=\frac{2^{2k-1}}{(2k)!}B_k\pi^{2k}.$$

Using the known values of $B_1=\frac{1}{6}$, $B_2=\frac{1}{30}$ and $B_3=\frac{1}{42}$, by virtue of (13), (14) from $g_2=60G_4$ and $g_3=140G_6$ we obtain

$$\zeta(4)=\frac{\pi^3}{2\cdot 3^2\cdot 5},\qquad \zeta(6)=\frac{\pi^6}{3^3\cdot 5\cdot 7}.$$

Therefore, by (13) we have

$$\text{(15)}\qquad g_2(\infty)=\tfrac{4}{3}\pi^4,\quad g_3(\infty)=\tfrac{8}{27}\pi^6,$$
$$\Delta(\infty)=g_2^3(\infty)-27g_3^2(\infty)=0.$$

That ∞ is, in fact, a simple zero of Δ will follow from the Fourier expansion given as Theorem XVI.17.6.

3° This follows from 1° and 2° since Δ and g_2^3 are both of weight 12. We know (Theorem XVI.16.7) that $\Delta \neq 0$ on $\mathfrak{H}$ and has a simple zero at ∞, while $g_2(\infty) \neq 0$. □

The Fourier Expansions of g_2, g_3, Δ *and* j. We shall now investigate the Fourier expansion (9) of the Eisenstein series in greater detail. We tart with the following technical lemma.

LEMMA XVI.17.5. If we set $q = q_\tau = e^{2\pi i\tau}$, then

$$\sum_{m\in Z}(\tau+m)^{-4} = \tfrac{8}{3}\pi^4\sum_{k=1}^{\infty}k^3q^k, \tag{16}$$

$$\sum_{m\in Z}(\tau+m)^{-6} = -\tfrac{8}{15}\pi^6\sum_{k=1}^{\infty}k^5q^k. \tag{17}$$

The proof follows from the expansion of $\cot\pi\tau$:

$$\begin{aligned}\pi\cot\pi\tau &= \frac{1}{\tau}+\sum_{0\neq m\in Z}\left(\frac{1}{\tau+m}-\frac{1}{m}\right)\\ &= -\pi i\left(1+2\sum_{k=1}^{\infty}e^{2\pi ik\tau}\right),\end{aligned}$$

where the latter series is obtained from

$$\pi\cot\pi\tau = i\pi\frac{e^{2\pi i\tau}+1}{e^{2\pi i\tau}-1}$$

by expansion in a geometric series.

By repeated differentiation we arrive at

$$\frac{1}{\tau^2}-\sum_{0\neq m\in Z}(\tau+m)^{-2} = -(2\pi i)^2\sum_{k=1}^{\infty}kq^k,$$

$$-3!\sum_{m\in Z}(\tau+m)^{-4} = -(2\pi i)^4\sum_{k=1}^{\infty}k^3q^k,$$

$$-5!\sum_{m\in Z}(\tau+m)^{-6} = -(2\pi i)^6\sum_{k=1}^{\infty}k^5q^k.$$

This gives the lemma. □

A remarkable fact about the Eisenstein series is that their Fourier expansions have integer coefficients. This brings in a close connection with the number theory, where the Eisenstein series are a very important tool. To provide the reader with a glimpse of this area we give the following theorem with a sketch of the proof:

THEOREM XVI.17.6. If $\tau \in \mathfrak{H}$, we have the following Fourier expansions (we set $q = e^{2\pi i\tau}$):

$$g_2(\tau) = \tfrac{4}{3}\pi^4\Big(1+240\sum_{k=1}^{\infty}\sigma_3(k)q^k\Big), \tag{18}$$

$$g_3(\tau) = \frac{8\pi^6}{27}\Big(1-504\sum_{k=1}^{\infty}\sigma_5(k)q^k\Big), \tag{19}$$

where $\sigma_a(k) = \sum_{d|k} d^a$ and $d|k$ means that the sum is extended over all integers d such that d divides k;

$$\Delta(\tau) = (2\pi)^{12}\sum_{n=1}^{\infty}\tau(n)\,q^n, \tag{20}$$

where the coefficients $\tau(n)$ are integers, with $\tau(1) = 1$, $\tau(2) = -24$ ($\tau(\cdot)$ is called the *Ramanujan tau function*). Moreover,

$$j(\tau) = \frac{1}{q}+744+\sum_{n=1}^{\infty}c(n)q^n, \tag{21}$$

where $c(n)$ are integers. We have

$$c(1) = 2^2\cdot 3^3\cdot 1823,\qquad c(2) = 2^{11}\cdot 5\cdot 2099.$$

Remark. Equation (21) is one reason why we work with $j(\tau) = 12^3J(\tau)$: we obtain a Fourier expansion with integer coefficients and, moreover, j has $\operatorname{Res}_\infty j = 1$.

PROOF (of (18) and (19)). We evaluate the sum (11) for $G_{2k}(\tau)$ by summing separately over n with $m = 0$ and then again over n with $m \neq 0$. We then obtain

$$G_4(\tau) = \sum_{n\neq 0} n^{-4} + \sum_{m\geq 1}\sum_{n\in Z}(m\tau+n)^{-4} + $$
$$+\sum_{m\leq -1}\sum_{n\in Z}(m\tau+n)^{-4}$$

and, similarly,

$$G_6(\tau) = \sum_{n\neq 0} n^{-6} + \sum_{m\geqslant 1}\sum_{n\in Z} (m\tau+n)^{-4} +$$

$$+ \sum_{m\leqslant -1}\sum_{n\in Z} (m\tau+n)^{-6}.$$

The first sum gives $2\zeta(4)$ (resp., $2\zeta(6)$) and the two remaining sums are equal since the exponents are even. When the summation is performed by using (16) (resp., (17)) with τ replaced by $m\tau$, $m \geqslant 1$, we get

$$G_4(\tau) = 2\zeta(4)+2\frac{(2\pi i)^4}{3!}\sum_{m=1}^{\infty}\sum_{k=1}^{\infty} k^3 q^{mk},$$

$$G_6(\tau) = 2\zeta(6)+2\frac{(2\pi i)^6}{5!}\sum_{m=1}^{\infty}\sum_{k=1}^{\infty} k^5 q^{mk}.$$

Using the explicit expressions for $\zeta(4)$ and $\zeta(6)$ and making the "change of variables" in the double sums, we obtain (18) and (19), respectively. The detailed proof can be found in any textbook of arithmetic or modular functions (cf. T. M. Apostol, *Modular Functions and Dirichlet Series in Number Theory*, Springer-Verlag, 1976, pp. 20–21, or the monograph of S. Lang, *Elliptic Functions*, pp. 44–45).

Discontinuous Groups of Automorphisms. We shall now look at modular functions from Riemann's point of view. We notice that $\Gamma\backslash\mathfrak{H}$ has a natural structure of a Riemann surface such that the canonical projection π: $\mathfrak{H} \to \Gamma\backslash\mathfrak{H}$ is holomorphic and, indeed, $\Gamma\backslash\mathfrak{H}$ is conformally equivalent to the entire complex plane $\boldsymbol{C}$. The one-point compactification of $\Gamma\backslash\mathfrak{H}$, therefore, gives us a compact Riemann surface conformally equivalent to $\boldsymbol{P}(\boldsymbol{C})$. Moreover, modular functions become simply meromorphic functions on this compactification $\overline{\Gamma\backslash\mathfrak{H}}$ and, by the familiar result concerning meromorphic functions on $\boldsymbol{P}(\boldsymbol{C})$, they are rational functions of a uniformizing parameter λ, which gives the above-mentioned isomorphism $\lambda: \overline{\Gamma\backslash\mathfrak{H}} \to \boldsymbol{P}(\boldsymbol{C})$. It turns out that with the proper normalization of λ (obtained by requiring that $\lambda(\infty) = \infty$, $\mathrm{Res}_\infty\lambda = 1$ and $\lambda(\pi(\varrho)) = 0$) the only such uniformizing parameter is the Dedekind–Klein modular function j. The reader should note that this normalization uniquely determines an isomorphism of $\overline{\Gamma\backslash\mathfrak{H}}$ with $\boldsymbol{P}(\boldsymbol{C})$.

This points to the fundamental role of the Dedekind–Klein modular invariant. It is reflected in the theorem given below, whose proof will be the aim of an exposition to follow.

THEOREM XVI.17.7. Modular functions (on $\mathfrak{H}$) form a field. Every modular function is a rational function of j; in brief,

$$\mathcal{M}(0, \Gamma) = C(j).$$

A non-constant modular function f takes on every value in $\overline{F}_\Gamma$ with the same multiplicity.

Before we shall carry on this program let us introduce an important notion which was discovered by Poincaré and Klein in connection with their famous work on automorphic functions.

DEFINITION. We say that the action of a discrete group Γ on a locally compact space X is *properly discontinuous* if for every pair of compact subsets C, C' of X the set

$$\{\gamma \in \Gamma\colon \gamma \cdot C \cap C' \neq \emptyset\}$$

is finite. The property is crucial for the space $\Gamma \backslash X$ to satisfy the usual separation assumptions. In the case of the upper half-plane $\mathfrak{H}$ this is assured by the following result, together with the fact that $\mathfrak{H} \simeq SL_2(R)/SO_2$, where SO_2 is the compact group of matrices of the form $\begin{bmatrix} \cos\varphi & -\sin\varphi \\ \sin\varphi & \cos\varphi \end{bmatrix}$ with real φ (cf. Theorem XVI.8.17).

LEMMA XVI.17.8. Let G be a locally compact group and let X be a compact subgroup of G. Then the action of any discrete subgroup $\Gamma \subset G$ on the space $X := G/K$ is properly discontinuous.

PROOF. It is well known (or easy to prove otherwise) that, given two compact subsets $A, B \subset G$, the set $AB = \{ab\colon a \in A, b \in B\}$ is also compact. If C is a compact subset of G/K, then its inverse image $C^\# := \pi^{-1}C$ (where $\pi\colon C \to G/K$) is also compact (verify this!). Therefore, $C^\#(C^\#)^{-1}$ is compact as well (here $(C^\#)^{-1}$ is the set of inverses of elements from $C^\#$) and, hence, $\Gamma \cap C^\#(C^\#)^{-1}$ is finite. But every $\gamma \in \Gamma$ satisfying $(\gamma \cdot C) \cap C \neq \emptyset$ is contained in $\Gamma \cap C^\#(C^\#)^{-1}$. □

LEMMA XVI.17.9. Let X be any locally compact space and let Γ be a properly discontinuous group of automorphisms of X. Then $\Gamma \backslash X$ is Hausdorff.

PROOF. Let x and y be non-Γ-equivalent points in X and let N_x and N_y be compact neighbourhoods of x, y. Then there exist only a finite number of $\gamma \in \Gamma$, say, $\gamma_1, \ldots, \gamma_r$, such that $\gamma x \in N_y$.

Since $y \notin \Gamma_x$, we have $\gamma_i x \neq y$, for $i = 1, \ldots, r$. Therefore, we can choose neighbourhoods $\mathcal{O}_i$ of $\gamma_i x$ and neighbourhoods U_i of y such that $\mathcal{O}_i \cap U_i = \emptyset$. Let

$$\mathcal{O} := N_x \cap \gamma_1^{-1}(\mathcal{O}_1) \cap \ldots \cap \gamma_r^{-1}(\mathcal{O}_r),$$

$$U := N_y \cap U_1 \cap \ldots \cap U_r;$$

then $\gamma(\mathcal{O}) \cap U = \emptyset$ and, hence, $\Gamma \backslash X$ is Hausdorff.

COROLLARY XVI.17.10. Let G be a locally compact group and let K and Γ, respectively, be a compact and a discrete subgroup of G. Then the space $\Gamma \backslash G/K$ of double $\Gamma - K$ cosets equipped with the strongest topology (quotient topology) which makes the natural projection π: $G \to \Gamma \backslash G/K$ continuous is a locally compact Hausdorff space.

Remark. A double $\Gamma - K$ coset is a set of the form $\Gamma \times K$, $x \in G$. The space $\Gamma \backslash G/K$ may be also viewed as the space of orbits of Γ acting via the left translations τ_γ: $G/K \ni gK \to \gamma gK \in G/K$ in G/K.

COROLLARY XVI.17.11. Let X be a complex manifold and let Γ be a properly discontinuous group of automorphisms of X without fixed points. Then $\Gamma \backslash X$ can be given (uniquely) a complex structure, so that the natural projection π: $X \to \Gamma \backslash X$ is a holomorphic mapping which is locally biholomorphic.

PROOF. Let $x_0 \in X$ and $y_0 := \pi(x_0)$ and let N_0 be such an open neighbourhood of x_0 that $\gamma(x_0) \notin N_0$ for all $\gamma \in \Gamma - \{e\}$. Plainly, $W_0 := \pi(N_0) = \pi(\Gamma(N_0))$ is a neighbourhood of y_0. Moreover, if $W_1 = \pi(N_1)$ is such a neighbourhood of $y_1 \neq y_0$ that $W_0 \cap W_1 \neq \emptyset$, then there exists a $\gamma \in \Gamma$ such that $\gamma(N_0) \cap N_1 \neq \emptyset$ and the overlap transformation is of the form $\gamma | N_0 \cap \gamma^{-1}(N_1)$: $N_0 \cap \gamma^{-1}(N_1) \to \gamma(N_0) \cap N_1$ and is biholomorphic. □

We shall now use the above results to carry on the program outlined. The first step will be to introduce the structure of a Riemann surface on the quotient space $\Gamma \backslash \mathfrak{H}$ of Γ-orbits in $\mathfrak{H}$.

THEOREM XVI.17.12. Let $\mathfrak{H}$ be the upper half-plane and Γ the (homogeneous) modular group. Then the following holds.

1° The transformation group Γ acts properly discontinuously on $\mathfrak{H}$.

2° The quotient space $X := \Gamma\backslash\mathfrak{H}$ (with the quotient topology) is locally compact.

3° The canonical projection $p = \pi|F$ of the fundamental domain $F_\Gamma = F$ onto $\Gamma\backslash\mathfrak{H}$ is proper ($p^{-1}(K)$ is compact for any compact K).

4° Let $\sim$ be the equivalence relation induced on F by the equivalence under Γ. Then the canonical projection $F_\Gamma \to \Gamma\backslash\mathfrak{H}$ induces a homeomorphism of $F/\sim$ onto $\Gamma\backslash\mathfrak{H}$.

5° The space $F/\sim$ and, consequently, $\Gamma\backslash\mathfrak{H}$ is homeomorphic to the entire complex plane $\boldsymbol{C}$.

6° The one point compactification $\overline{\Gamma\backslash\mathfrak{H}}$ of $\Gamma\backslash\mathfrak{H}$ is homeomorphic with the Riemann sphere $\boldsymbol{P}(\boldsymbol{C})$.

7° There exists a structure of a Riemann surface on $\Gamma\backslash\mathfrak{H}$ such that the natural projection $\pi\colon \mathfrak{H} \to \Gamma\backslash\mathfrak{H}$ is holomorphic; then $\Gamma\backslash\mathfrak{H}$ is conformally equivalent with $\boldsymbol{C}$ (and, consequently, $\overline{\Gamma\backslash\mathfrak{H}}$ is conformally equivalent with $\boldsymbol{P}(\boldsymbol{C})$). Moreover, $\pi\colon \mathfrak{H} \to \Gamma\backslash\mathfrak{H}$ is a covering with branching points at $\pi(\varrho)$ and $\pi(i)$.

PROOF. 1° and 2° are contained in Theorems XVI.17.8 and XVI.17.9.

3° Let K be a compact subset of $\mathfrak{H}$; we have to demonstrate that $B := (\Gamma K) \cap F$ is compact. But, since ΓK is closed, we only have to verify that B is bounded. For this purpose it suffices to prove that

$$(22) \qquad \bigvee_{n_0} \bigwedge_{z\in K} \bigwedge_{\gamma\in\Gamma} \operatorname{Im}(\gamma z) < n_0 .$$

From the formula

$$(*) \qquad I(\gamma z) = \frac{\operatorname{Im}(z)}{|cz+d|^2}, \quad \text{where } \gamma = \begin{bmatrix} a & b \\ c & d \end{bmatrix}$$

(22) follows immediately since $\operatorname{Inf}\{|cz+d|\} > 0$ when z runs through K and c and d run through all pairs of relatively prime integers.

4° The map $F/\sim \to \Gamma\backslash\mathfrak{H}$ is bijective (by Theorem XVI.17.1), continuous, and proper (by 3°). Since $F/\sim$ and $\Gamma\backslash\mathfrak{H}$ are locally compact, this map is a homeomorphism.

5° That $\Gamma\backslash\mathfrak{H}$ is homeomorphic with the complex plane $\boldsymbol{C}$ or, equivalently, with the punctured sphere $\boldsymbol{P}(\boldsymbol{C}) - \{\infty\}$ may be verified geometrically by determining points equivalent under $\sim$ on the boundary of the fundamental domain F. First the pairs of opposite points on the half-lines $\operatorname{Re} z = \pm\frac{1}{2}$ are equivalent; thus, by identifying points of these pairs, we get an infinite cylinder. Then the points lying on the part of the circle $|z| = 1$ belonging to F are equivalent iff they are symmetric with respect to the line $\operatorname{Re} z = 0$. When these are identified, the resulting surface is seen to be homeomorphic with the punctured sphere. (For an analytic proof see Exercise 1 at the end of this section.)

6° Follows immediately from 5°.

7° From the definition of the topology on $\Gamma\backslash\mathfrak{H}$ it follows that at all points from $\mathfrak{H}$, except those equivalent to ϱ or i, π restricted to a suitable neighbourhood U is a homeomorphism onto its open image. Therefore, at all points of $\Gamma\backslash\mathfrak{H}$ different from $\pi(\varrho)$ and $\pi(i)$ we can take $(\pi|U)^{-1}$ as a local parameter. At points $\pi(\varrho)$ and $\pi(i)$ we have to proceed differently. First, note that if the order of the isotropy subgroup $G_x (x = \varrho$ or $i)$ is denoted by e_x, then in every sufficiently small disk around x there are precisely e_x equivalent points. In fact, with respect to the local parameter $\tau = \dfrac{z-i}{z+i}$ $\left(\text{resp. } \tau = \dfrac{z-\varrho}{z+\varrho}, \right)$ the transformation $z \to Sz$ (resp. $z \to STz$) is a rotation of period 2 (resp. 3) around $z = i$ (resp. $z = \varrho$). Therefore, one-half of the parametric disk (resp. one-third) is mapped bijectively by π onto an open neighbourhood of the point $\pi(i)$ (resp. $\pi(\varrho)$). Thus we may take τ^2 (resp. τ^3) as a local parameter about $\pi(i)$ (resp. $\pi(\varrho)$). It is obvious that with the complex structure so obtained, $\Gamma\backslash\mathfrak{H}$ is biholomorphic with the plane $\boldsymbol{C}$ (cf. the proof of 5°) and $\pi\colon \mathfrak{H} \to \Gamma\backslash\mathfrak{H}$ is a covering with branching points of index 2 and 3, respectively, at the points $\pi(i)$ and $\pi(\varrho)$.

Now, let $\bar{X} = X \cup \{\infty\}$ be the one-point compactification of X: we want to extend the complex structure of X to $\bar{X}$. Let $\mathfrak{H}_1$ be the half-plane $\operatorname{Im}(\tau) > 1$. But it follows from this that the equivalence relation $\sim$ induced on $\mathfrak{H}_1$ by Γ is given by the translations T^n, $n \in \boldsymbol{Z}$:

$$T^n\tau = \tau + n, \quad T = \begin{bmatrix} 1 & 1 \\ 0 & 1 \end{bmatrix}.$$

We thus have $\pi(\mathfrak{H}_1) \cong \mathfrak{H}_1/\sim$. But the map $\tau \to q = e^{2\pi i\tau}$ is a holomorphic isomorphism of $\mathfrak{H}_1/\sim$ onto the punctured unit disk $\boldsymbol{D}^*$. Putting $q(\infty) = 0$, we extend q to $\pi(\mathfrak{H}_1) \cup \{\infty\}$ and take this function as the local parameter near ∞ on $\bar{X}$.

Since the space $\overline{\Gamma\backslash\mathfrak{H}}$ with that holomorphic structure is obviously isomorphic to $\boldsymbol{P}(\boldsymbol{C})$, the proof is complete. □

COROLLARY XVI.17.13. Suppose $\pi\colon \mathfrak{H} \to \Gamma\backslash\mathfrak{H}$ is a canonical projection. Then:

(1) the mapping $\mathcal{M}(\overline{\Gamma\backslash\mathfrak{H}}) \ni f \to f \circ \pi \in \mathcal{M}(0, \Gamma)$ is an isomorphism;

(2) if $\lambda\colon \overline{\Gamma\backslash\mathfrak{H}} \to \boldsymbol{P}(\boldsymbol{C})$ is the (unique) isomorphism such that $\lambda(\infty) = \infty$, $\lambda(\pi(\varrho)) = 0$, $\operatorname{res}_\infty \lambda = 1$, then each $f \in \mathcal{M}(0, \Gamma)$ is a rational function of λ;

(3) every modular function assumes all values with the same multiplicity, provided the multiplicity is divided by 2 at i and by 3 at ϱ.

PROOF. (1) is obvious by virtue of part 7° of Theorem XVI.17.12.

(2) follows from (1) by virtue of the normalization of λ and by the fact that all meromorphic functions on $\boldsymbol{P}(\boldsymbol{C})$ are rational.

(3) Because of the compactness of $\overline{\Gamma\backslash\mathfrak{H}}$ and the residue theorem, a modular function f has the same number of zeros as poles. But $f-c$ has the same number of poles for any $c \in \boldsymbol{C}$, and, hence, the multiplicity of c is the same for all c. □

Remark. The corollary above contains nearly the entire content of Theorem XVI.17.7 except for the identification of λ with j. This will be supplied in the next section.

Exercise 1. Prove 5°, i.e., that $\Gamma\backslash\mathfrak{H} \simeq \boldsymbol{C}$, in the following analytic way. The "half fundamental domain" of Γ

$$F_\Gamma = \{\tau \in \mathfrak{H}\colon 0 < \operatorname{Re}(\tau) < \tfrac{1}{2}, |\tau| > 1\}$$

is isomorphic to a half-plane; then apply the Schwarz symmetry principle.

Exercise 2. The proof of Theorem XVI.17.7, i.e.,

$$\mathcal{M}(0, \Gamma) = \boldsymbol{C}(j),$$

can be accomplished in the following elementary way: Let $f \in \mathcal{M}(0, \Gamma)$ and suppose that f has zeros at $z_1, \ldots, z_n$ and poles at $p_1, \ldots, p_n$ (with the usual convention concerning multiplicities). Put

$$h(\tau) := \prod_{k=1}^{n} \frac{j(\tau)-j(z_k)}{j(\tau)-j(p_k)},$$

where the factor 1 is inserted whenever z_k or p_k is ∞. Since f and h have the same zeros and poles in the closure of $F = F_\Gamma$, f/h is holomorphic and has no zeros and no poles and, hence, is a constant c. Now, $f = ch$ implies that f is a rational function of j. □

18. THE MULTIPLICITY FORMULA FOR ZEROS OF A MODULAR FORM. DIMENSION OF VECTOR SPACES $M^0(k,\Gamma)$ OF CUSP FORMS

We have denoted by $M(k,\Gamma)$ and $M^0(k,\Gamma)$ the vector spaces of Γ-modular forms of weight $2k$ and that of cusp forms of weight $2k$, respectively. We recall that for a non-zero meromorphic function $f\in\mathscr{M}(\mathfrak{H})$ and $x\in\mathfrak{H}$ an integer $k\in\boldsymbol{Z}$ such that the function $\tau\to(\tau-x)^{-k}f(\tau)$ is holomorphic and non-zero at x is called the *order* of f at x and is denoted by $\operatorname{ord}_x(f)$. Our goal is to determine the dimensions of $M^0(k,\Gamma)$ and $M(k,\Gamma)$. The key result is

THEOREM XVI.18.1 (Multiplicity Formula). Let $f\not\equiv 0$ be any modular function of weight $2k$; then

$$\text{(1)}\qquad \tfrac{1}{2}\operatorname{ord}_i(f)+\tfrac{1}{3}\operatorname{ord}_\varrho(f)+\operatorname{ord}_\infty f+\sum_{i,\varrho\neq\tau\in F}\operatorname{ord}_\tau(f)=\frac{k}{6},$$

where $\varrho=e^{2\pi i/3}$.

Remark. From the definition of a modular function f it follows that $\operatorname{ord}_\tau(f)=\operatorname{ord}_{\gamma\tau}(f)$, $\gamma\in\Gamma$; hence, $\operatorname{ord}_\tau(f)$ depends only on the image

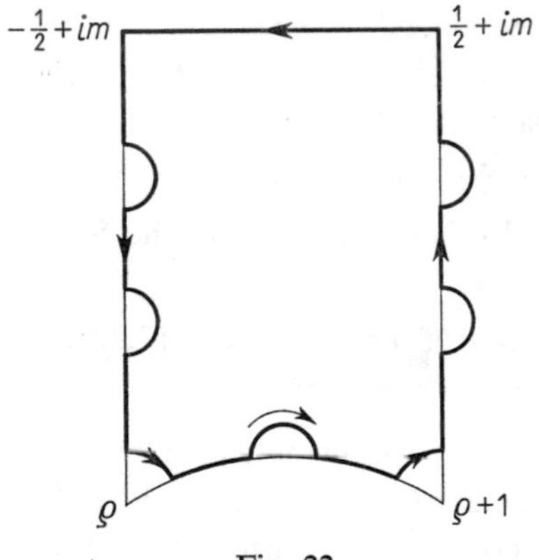

Fig. 22

of τ in $\Gamma\backslash\mathfrak{H}$. If by e_τ we denote the order of Γ, the stabilizer of τ, we have $e_\tau=2,3$ if τ is congruent modulo Γ to i, ϱ. Therefore, we could recast (1) in a condensed form

$$\text{(2)}\qquad \sum_{x\in\Gamma\backslash\mathfrak{H}}\frac{1}{e_x}\operatorname{ord}_x(f)=\frac{k}{6}.$$

Exercise. Prove the multiplicity formula in the following, elementary way:

Integrate $\frac{1}{2\pi i}\frac{df}{f}$ over the boundary of F as indicated in Fig. 22.

PROOF of (1). If $h, g \in \mathcal{M}(k, \Gamma)$ then h/g is meromorphic on $\overline{X} = \overline{\Gamma\backslash\mathfrak{H}}$ and (by the residue theorem) has as many zeros as poles on $\overline{X}$, counted, as always, according to multiplicities.

We now take $f \in \mathcal{M}(k, \Gamma), f \neq 0$ and consider the modular function $h = \frac{f^6}{\Delta^k}$ as a meromorphic function over $\overline{X}$. Since Δ has no zeros in $\mathfrak{H}$ and a simple pole at ∞, we have

$$\operatorname{ord}_\infty(h) = 6\operatorname{ord}_\infty f - k,$$

$$\operatorname{ord}_\tau(h) = 6\operatorname{ord}_\tau(f), \qquad \tau \in F,\ \tau \neq \varrho, i,$$

$$\operatorname{ord}_\varrho(h) = 2\operatorname{ord}_\varrho(f), \qquad \operatorname{ord}_i(h) = 3\operatorname{ord}_i(f).$$

For the last two equalities let us recall how the local maps on $\Gamma\backslash\mathfrak{H}$ around the points ϱ and i were defined. Now, since

$$\operatorname{ord}_\infty(h) + \operatorname{ord}_\varrho(h) + \operatorname{ord}_i(h) + \sum_{\tau \in F} \operatorname{ord}_\tau(h) = 0,$$

the desired formula follows. □

By definition, $M^0(k, \Gamma) = \ker \varepsilon_k$, where

$$\varepsilon_k\colon M(k, \Gamma) \to \mathbf{C}, \qquad \varepsilon_k(f) := f(\infty).$$

From the above we can draw some information concerning the multiplicity of zeros and poles of our old friends, functions g_2, g_3, and Δ.

COROLLARY XVI.18.2. (1) The function g_2 (resp. g_3) has zeros only at points equivalent to ϱ (resp. i) and all these zeros are simple (of order one).

(2) The discriminant Δ has the only zero at the point ∞ and this zero is simple.

Remark. Thus far our knowledge of the behaviour of Δ has relied on Theorem XVI.17.6, which was not proved in all its details. It is, therefore, of some interest to give an independent derivation of this. The one to follow depends only on the multiplicity formula.

PROOF. Since the Eisenstein series G_k is a modular form of weight k, by applying formula (1) to g_2 we get

$$\tfrac{2}{6} = \operatorname{ord}_\infty(g_2)+\tfrac{1}{2}\operatorname{ord}_i(g_2)+\tfrac{1}{3}\operatorname{ord}_\varrho(g_2)+\sum \operatorname{ord}_\tau(g_2).$$

Since $\operatorname{ord}_p(g_2) \geqslant 0$ and are integers, the only possibilities are $\operatorname{ord}_\infty(g_2) = \operatorname{ord}_i(g_2) = \operatorname{ord}_\tau(g_2) = 0$, $\tau \neq \varrho$ and $\operatorname{ord}_\varrho(g_2) = 1$. A similar argument gives the result for g_3, a modular form of weight 6. For Δ we know that $\Delta \not\equiv 0$ and by Theorem XVI.17. 3 we know that $\Delta(\infty) = 0$ and that Δ is a modular form of weight 6. The right-hand side of (1), therefore, is 1 and so we are left with $\operatorname{ord}_\infty(\Delta) = 1$, all others being 0. □

We can now complete the gap left in the proof of Theorem XVI.17.7.

COROLLARY XVI.18.3. The Dedekind–Klein modular function j is the only modular function such that (1) the only pole of j is at infinity and $\operatorname{res}_\infty j = 1$ and (2) the only zeros (of order 3) are at the points equivalent to ϱ.

PROOF. That j satisfies (1) was already proved and (2) follows from the defining formula for j and the corollary above. The uniqueness follows from Corollary XVI.17.13, since if λ is another such function, then $g_2^3/\Delta = a\lambda+b$. Using $\lambda(\pi(\varrho)) = 0$ and $g_2(\varrho) = 0$, we get $b = 0$. Then, comparing coefficients in the Fourier expansion, we obtain $a = 2^{-6}3^{-3}$, i.e., $\lambda = 2^6 3^3 g_2^3/\Delta = j$. □

PROPOSITION XVI.18.4. 1° $M(k,\Gamma) = M^0(k,\Gamma)\oplus \mathbf{C}G_{2k}$, for $k \geqslant 0$; hence,

(3) $\quad \dim M^0(k,\Gamma) = \dim M(k,\Gamma)-1.$

2° Multiplication by Δ defines an isomorphism of $M(k-6,\Gamma)$ onto $M^0(k,\Gamma)$.

3°

(4) $\quad M(k,\Gamma) = 0 \quad$ for $k < 0$ and $k = 1$.

(5) $\quad \dim M(k,\Gamma) = 1$ for $k = 0, 2, 3, 4, 5$ and G_{2k} is a basis of $M(k,\Gamma)$.

PROOF. Since $M^0(k,\Gamma) = \ker \varepsilon_k$, we have $\dim M(k,\Gamma)/M^0(k,\Gamma) \leqslant 1$. But for $k \geqslant 2$ $G_{2k} \in M(k,\Gamma)$ and $G_{2k}(\infty) \neq 0$, whence we have 1°.

2° If $f \in M^0(k,\Gamma)$ and if we set $h = f/\Delta$, then h is of weight $k-6$.

But we have the formula

$$(*) \qquad \operatorname{ord}_p(h) \equiv \operatorname{ord}_p\left(\frac{f}{\Delta}\right)$$

$$= \operatorname{ord}_p(f) - \operatorname{ord}_p(\Delta) = \begin{cases} \operatorname{ord}_p(f) & \text{if} \quad p \neq \infty, \\ \operatorname{ord}_p(f) - 1 & \text{if} \quad p = \infty, \end{cases}$$

since Δ has a simple 0 at ∞ and Δ does not vanish on $\mathfrak{H}$. Therefore, $\operatorname{ord}_p(h) \geqslant 0$ for all p and, hance, $h \in M(k-6, \Gamma)$.

3° Let $0 \not\equiv f \in M(k, \Gamma)$; since all the terms on the left-hand side of (1) are $\geqslant 0$, we have $k \geqslant 0$ and also $k \neq 1$ since $1/6$ cannot be written in the form $n + n'/2 + n''/3$ with $n, n', n'' \geqslant 0$ and, hence, we have (4). If $k \leqslant 5$, so that $k-6 < 0$, we have $M^0(k, \Gamma) = 0$ by 1° and 2°. Therefore $\dim_C M(k, \Gamma) \leqslant 1$. Since G_{2l} are non-zero elements of $M(l, \Gamma)$ for $l = 0, 2, 3, 4, 5$, we have $\dim M(l, \Gamma) = 1$ for these l's and (5) is also verified. □

From the preceding result we see that the dimensions of spaces of modular forms are finite. In fact, we have more precise information available.

THEOREM XVI.18.5. If $k \geqslant 0$, then

$$(6) \qquad \dim M(k, \Gamma) = \begin{cases} \left[\dfrac{k}{6}\right] & \text{if } k \equiv 1 \pmod 6, \\ \left[\dfrac{k}{6}\right] + 1 & \text{if } k \not\equiv 1 \pmod 6 \end{cases}$$

(where $[x]$ denotes the largest integer n such that $n \leqslant x$),

$$(7) \qquad \dim M^0(k, \Gamma) = \dim M(k, \Gamma) - 1.$$

Remark. Similar formulas are valid for $\dim \mathscr{M}(k, \Gamma)$ (cf. Section 20).

PROOF. If $k \geqslant 6$, by Theorem XVI.18.2.2° we have

$$(8) \qquad \dim M^0(k, \Gamma) = \dim M(k-6, \Gamma).$$

Formulas (8) and (3) imply

$$(9) \qquad \dim M(k, \Gamma) = 1 + \dim M(k-6, \Gamma) \quad \text{for } k \geqslant 6.$$

Equation (6) follows from (9), (4) and (5).

PROPOSITION XVI.18.6. Let $f \in M(k, \Gamma)$, then f is an (isobaric) polynomial in G_4 and G_6, i.e.,

$$(10) \qquad f = \sum_{k=4a+6b} c_{a,b} G_4^a G_6^b, \qquad \text{where } c_{a,b} \in \boldsymbol{C}.$$

COROLLARY XVI.18.7. Let $M := \bigoplus_{k=0}^{\infty} M(k, \Gamma)$ be the graded algebra which is the direct sum of $M(k, \Gamma)$. Then one can identify M with the polynomial algebra $\boldsymbol{C}(G_4, G_6)$.

19. MAPPING PROPERTIES OF j. PICARD THEOREM. ELLIPTIC CURVES. JACOBI'S INVERSION PROBLEM. ABEL'S THEOREM

Let us return to Theorem XVI.17.12. We demonstrated there the existence of holomorphic isomorphisms $\Gamma\backslash\mathfrak{H} \to \boldsymbol{C}$ and $\overline{\Gamma\backslash\mathfrak{H}} \to \boldsymbol{P}(\boldsymbol{C})$ and denoted a normalized isomorphic map by j. We identified that j with the Dedekind–Klein modular function defined (analytically) by

$$j = \frac{12^3 g_2^3}{\Delta}, \qquad \text{where } \Delta = g_2^3 - 27 g_3^2$$

(we recall that $g_2 := 60G_4, g_3 := 140G_6$). We list the important properties of the function j in

THEOREM XVI.19.1. 1° The function j is modular.

2° $j \in A(\mathfrak{H})$ and j has a simple pole at ∞, $\operatorname{Res}_\infty j = 1$,

$$j(i) = 12^3 \text{ (twice)}, \qquad j(\varrho) = 0 \text{ (triple zero)}.$$

3° By passing to the quotient the function j defines a (holomorphic) bijection of $\Gamma\backslash\mathfrak{H}$ onto $\boldsymbol{C}$.

4° The mapping j: $\mathfrak{H} \to \boldsymbol{C}$ is a branched covering with countably many sheets and the branch points at 0 and 1.

The proof of 1°–3° has been given before (Theorems XVI.17.3, XVI.17.6, XVI.8.2 and XVI.18.3). But it is instructive to prove 3° in another way: we have to check that the modular function $f_c := 12^3 g_2^3 - c\Delta$ has a unique zero modulo Γ. To check this we apply the weight formula (1) of Section 18. Since f_c is of weight 12, the only decompositions of $k/12 = 1$ into the sum $n + n'/2 + n''/3$ with $n, n', n'' \geqslant 0$ correspond to $(n, n', n'') = (1, 0, 0)$, or $(0, 2, 0)$ or $(0, 0, 3)$. This demonstrates that f_c is zero (i.e., that j attains c) at one and only one point of $\Gamma\backslash\mathfrak{H}$.

4° Since $j'(\tau) \neq 0$ if $\tau \neq \varrho$ or $\tau \neq i$ and since $j'(\varrho) = j'(i) = 0$, 4° follows from 3°. □

(Small) Picard Theorem. The function j was discovered in 1877 by Richard Dedekind and one year later, by Felix Klein. Two years later (1880) Charles Émile Picard had the great inspiration for the famous "small" Picard Theorem, which asserts that an entire function omits at most one finite value. Many other proofs of this celebrated theorem were later found "which are more elementary in that they need less preparation, but none is as penetrating as the original proof" (Ahlfors).

Remark. One beautiful proof of the Picard Theorem follows from the famous Ahlfors Lemma.

We shall now reproduce the original proof of

THEOREM XVI.19.2 ("Small Picard"). Every non-constant entire function attains every complex value with at most one exception. In "other" words: each holomorphic map $f\colon \mathbf{C} \to \mathbf{C} - \{a, b\}$ (with distinct a, b) is a constant.

PROOF. By 3°, 4° of Theorem XVI.19.1 we know that the inverse function j^{-1} maps the Riemann sphere back onto the closure of the fundamental region $F = F_\Gamma$ and (since $j'(\tau) \neq 0$ except $\tau = i, \varrho$) each single-valued branch of j^{-1} is locally holomorphic except at $0 = j(\varrho)$, $12^3 = j(i)$ and $\infty = j(\infty)$.

Assume that $f \in A(\mathbf{C})$ omits $a, b, b \neq a$ and let

$$(*) \qquad g := \frac{f-a}{b-a}, \qquad \text{whence} \qquad f = a+(b-a)g.$$

Then g is entire and omits the values 0 and 1. For each single-valued branch of j^{-1} the composite function

$$(**) \qquad h := \left(\frac{j}{12^3}\right)^{-1} \circ g$$

is single-valued and is locally holomorphic at finite points since g omits 0 and 1. Hence, h can be analytically continued onto the whole of $\mathbf{C}$. By the monodromy theorem this extension is a holomorphic and single-valued entire function. Hence $h \in A(\mathbf{C})$ and therefore $u := \exp(ih) \in A(\mathbf{C})$. But $\operatorname{Im} h(z) > 0$ since, by $(**)$, $h(z) \in \mathfrak{H}$ and, therefore,

$$|u(z)| = e^{-\operatorname{Im} p(z)} < 1 \qquad \text{for all } z \in \mathbf{C}.$$

Thus, u is a bounded entire function which (by the Liouville Theorem) must be constant. Hence, $h \equiv$ const and g is also constant because $g = 12^3 j \circ h$. Finally (cf.(*)), since $f = a+(b-a)g$, it is also constant. □

Remark. Another proof of the Picard Theorem is given in Section 20.5.

The modular function j allows us to solve

The Inversion Problem for Eisenstein Series.

THEOREM XVI.19.3. Given two complex numbers c_2, c_3 such that

(*) $\qquad c_2^3 - 27c_3^2 \neq 0$

then there exists a lattice $L = [\omega_1, \omega_2]$ such that $g_2(L) = c_2, g_3(L) = c_3$.

PROOF. By Theorem XVI.19.1.3°, there exists a $\tau \in \mathfrak{H}$ such that $j(\tau) = 12^3 c_2^3/(c_2^3 - 27c_3^2)$. Let M denote the lattice $[\tau, 1]$. If $c_2 = 0$, then $c_3 \neq 0$ by (*) and $j(\tau) = 0$ and $\tau = \varrho\ (= e^{2\pi i/3})$. Let $z \in \boldsymbol{C}^*$ be such that $z^{-6}g_3(M) = c_3 \neq 0$. Putting $L := zM$, we have

$$g_2(L) = z^{-4}g_2(M) = z^{-4}g_2(\varrho) = 0 = c_2,$$

and $g_3(L) = z^{-6}g_3(M) = c_3$ and the proof is complete.

If $c_2 \neq 0$, take such a $w \in \boldsymbol{C}^*$ that $w^{-4}g_2(M) = c_2$ (since $g_2(M) \neq 0$). If $L := wM$, then $g_2(L) = c_2$. Therefore, since j is homogeneous of degree 0, we have

$$12^3 \frac{c_2^3}{c_2^3 - 27c_3^2} = j(\tau) = j(wM) = j(L) = 12^3 \frac{g_2^3(L)}{g_2^3(L) - 27g_3^2(L)}$$

$$= 12^3 \frac{c_2^3}{c_2^3 - 27g_3^2(L)}.$$

Hence, $g_3^2(L) = c_3^2$ and $g_3(L) = \pm c_3$.

Replace w by iw if necessary: this does not change $g_3^2(L)$ but it does change $g_3(L)$ by -1 and the theorem is proved. □

Finally, we have the following fundamental result:

THEOREM XVI.19.4.1° An arbitrary "elliptic curve" A given by the equation

(*) $\qquad y^2 = 4x^3 - c_2 x - c_3$

with a non-vanishing discriminant ($c_2^3 - 27c_3^2 \neq 0$) can always be parametrized by Weierstrass functions: i.e., there exists such a lattice (of periods) L that $c_2 = g_2(L)$, $c_3 = g_3(L)$ and the map $C/L - \{0\} \ni z \to (\wp_L(z), \wp'_L(z)) \in A \subset C^2$ is an isomorphism.

2° C/L is isomorphic with a projective curve $A = A_C$ given by (*).

3° Denote by j_A the value $j(L)$ for any lattice L such that the elliptic curve A is isomorphic to C/L (j_A is called the *j-invariant* of A). Then

$$(\text{elliptic curves } A, B \text{ are isomorphic}) \Leftrightarrow (j_A = j_B) \Leftrightarrow (L \cong M),$$

where L, M are lattices defining A, B.

PROOF. 1° is already checked.

2° is a projective version of 1°: let us define π: $C \to P(C^2)$ by

$$\pi(z) = \begin{cases} (0, 0, 1) & \text{if } z \in L, \\ (1, \wp(z), \wp'(z)) & \text{if } z \notin L. \end{cases}$$

Then π factors through C/L and gives the desired isomorphism $[\pi]$: $C/L \to A_C$ (= a projective cubic given by (*)). □

3° We can view j as a function of lattices. But since j is homogeneous of degree 0, we see that for the lattice $L = [\omega_1, \omega_2]$ we can write $j(L) = j(\tau)$ if ω_1, ω_2 are chosen so that $\omega_1/\omega_2 =: \tau \in \mathfrak{H}$. If $L = aM$, where $a \in C^*$, then $j(L) = j(M)$. But the fact (Theorem XVI.19.1.3°) that j: $\Gamma\backslash\mathfrak{H} \to C$ is a bijection may be stated as:

$$(\text{two lattices } L, M \text{ in } C \text{ are equivalent}) \Leftrightarrow (j(L) = j(M)).$$

But

$$(C/L \cong C/M) \Leftrightarrow (L = aM, \text{ for some } a \in C^*). \ \square$$

Remark. In this way we have proved another important property of the modular function j:

> j parametrizes isomorphism classes of elliptic curves (complex tori). This is the reason why the Dedekind–Klein function is often called the *elliptic modular function.*

We shall return once more to elliptic curves from another point of view: the elliptic integrals. We shall see that the inverse isomorphism E: $A_C \to C/L$ is given by the elliptic integral

$$E(P) = \int_{P_0}^{P} \omega \quad \text{where } \omega = \frac{dx}{\sqrt{4x^3 - c_2 x - c_3}}.$$

As a converse of Theorem XVI.19.4 we have the following

THEOREM XVI.19.5. The elliptic integral $E = \int \omega$ defines a holomorphic isomorphism

$$E\colon A_C \to C/L,$$

where L is a lattice of periods of ω.

PROOF. If γ is a 1-cycle on $X = A_C$, i.e., $\partial\gamma = 0$, then the period of ω on γ is defined by $\pi(\gamma) := \langle \gamma, \omega \rangle$. If γ is a boundary, i.e., $\gamma = \partial\beta$, where β is a 2-chain, then from the Stokes Theorem we have

$$\pi(\gamma) = \int_\gamma \omega = \int_{\partial\beta} \omega = \int_\beta d\omega = 0.$$

But $\pi(\gamma_1 + \gamma_2) = \pi(\gamma_1) + \pi(\gamma_2)$. Hence, π gives a homomorphism $\pi\colon H_1(X) \to C$ from the first homology group $H_1(X)$ to C. Let $\hat{X}$ be the simply-connected covering surface (i.e., the universal covering surface, cf. Section 7) of X, then E globally defines a map

$$\hat{E}\colon \hat{X} \to C \quad \text{or} \quad E\colon X \to C/L,$$

where

$$L = \{\pi(\gamma)\colon \gamma \in H_1(X)\} \text{ is the group of periods of } \omega.$$

We first prove the following

LEMMA XVI.19.6. L is a lattice in C.

PROOF. We know that $H_1(X) \cong Z \oplus Z$. Suppose that L is not a lattice; then L is contained in a real 1-dimensional subspace of C and for some $a \in C^*$ we would have $\operatorname{Re}(a\pi(\gamma)) = 0$ for all γ. Therefore, $\operatorname{Re}(aE)$ would have no periods; hence, $\operatorname{Re}(aE)$ is a harmonic function on the compact manifold and, thus, $\operatorname{Re}(aE) = \text{const}$ and aE is also constant $\Rightarrow (0 = d(aE) = \omega)$. Therefore, $\omega = 0$, which is absurd. □

Therefore, C/L is a holomorphic torus and the proof of Theorem XVI.19.5 is continued as follows: Since $dE = \omega$ is nowhere zero, E is a (local) chart of X: it gives a local homeomorphism from X to C/L. Since X is compact, $E\colon X \to C/L$ is a finite-sheeted covering of C/L. The theory of covering spaces (cf. Section 7) asserts that all such coverings correspond (bijectively) to subgroups of $\pi_1(C/L)$, the fundamental group of C/L. Therefore, they can be considered as maps, $C/M \to C/L$, where $M \subset L$ is a sublattice of L. Thus, E factors

$$X \xrightarrow{\text{topol}} C/M \to C/L.$$

Therefore, its lifting $\hat{E}: \hat{X} \to \boldsymbol{C}$ satisfies $\hat{E}(x) - \hat{E}(y) \in M$ for all $x, y \in \hat{X}$ over the same point of $\hat{X}$. Hence, all periods of ω lie in M. Therefore, $L \subset M$ and, hence, $L = M$. □

COROLLARY XVI.19.7. An elliptic curve can be provided with the structure of an Abelian group in which the addition $+ : X \times Y \to X$ and the inverse $- : X \to X$ are regular.

PROOF. $\boldsymbol{C}$ is a group under $+$ and L is a subgroup and, therefore, $\boldsymbol{C}/L$ is a commutative group. □

Remark. Theorem XVI.19.5 implies the celebrated "Addition Theorem" for the elliptic integral due to Abel (cf. Section 20).

Jacobi's Inversion Problem. Abel's Theorem. We have constructed an isomorphism of a *compact* Riemann surface X of genus 1 (elliptic curve) onto the complex torus, $\boldsymbol{C}^1/\mathrm{Per}(\varphi)$, where $\mathrm{Per}(\varphi)$ is the lattice of periods of the holomorphic 1-form φ (an Abelian differential of the first kind). Two natural questions arise:

1° Is it possible to construct a lattice $\mathrm{Per}(\varphi_1, \ldots, \varphi_g)$ of periods for a basis $\varphi_1, \ldots, \varphi_g$ of the space of Abelian differential of the first kind on a compact Riemann surface of genus $g > 1$?

2° Is the compact Riemann surface X of genus g isomorphic to $\boldsymbol{C}^g/\mathrm{Per}(\omega^1, \ldots, \omega^g)$?

These two questions constitute the famous Jacobi inversion problem. This problem was solved by Riemann and Weierstrass. We give here a modern version of the solution.

The answer to the first question is given by

THEOREM XVI.19.8. Let X be a closed (i.e. compact) Riemann surface of genus $g \geqslant 1$ and let $\varphi_1, \ldots, \varphi_g$ be a basis of $A^1(X)$—the space of Abelian differentials of the first kind on X; then there exist $2g$ cycles (closed curves) $\alpha_1, \ldots, \alpha_{2g}$ such that the vectors

$$(1) \qquad \omega_j := \Big(\int_{\alpha_j} \varphi_1, \ldots, \int_{\alpha_j} \varphi_g\Big) \equiv \int_{\alpha_j} \Phi \in \boldsymbol{C}^g, \quad j = 1, \ldots, 2g$$

are $\boldsymbol{R}$-linear-independent and the period-group

$$(2) \qquad \mathrm{Per}(\varphi_1, \ldots, \varphi_g) := \Big\{\omega \in \boldsymbol{C}^g : \omega = \int_{\alpha} \Phi\Big\} = \boldsymbol{Z}\omega_1 \oplus \ldots \oplus \boldsymbol{Z}\omega_{2g},$$

where α ranges over the homology group $H^1(X)$, is a lattice in $\boldsymbol{C}^g$.

As in the elliptic case, we now define a g-dimensional complex torus.

DEFINITION. Let X be a compact Riemann surface of genus $g \geqslant 1$ and let $\varphi_1, \ldots, \varphi_g$ be a basis of $A^1(X)$. Then the torus

$$(3) \qquad \operatorname{Jac}(X) := \boldsymbol{C}^g/\operatorname{Per}(\varphi_1, \ldots, \varphi_g)$$

is called the *Jacobi variety* (or *jacobian*) of X. An answer to the second question and a solution of the Jacobi inversion problem is given by the so-called Jacobi Theorem.

THEOREM XVI.19.9. Let $X = X^g$, $\varphi_1, \ldots, \varphi_g$ be as in the preceding theorem. For fixed $a_1, \ldots, a_g$ X denote

$$(4) \qquad E(x_1, \ldots, x_g) := \Big(\sum_{j=1}^{g} \int_{a_j}^{x_j} \varphi_k\Big)_{1 \leqslant k \leqslant g} \operatorname{mod} \operatorname{Per}(\varphi_1, \ldots, \varphi_g).$$

Then $E\colon X^g \to \operatorname{Jac}(X)$ is surjective. The following natural question arises: what is the kernel of E? The answer is given by the famous theorem of Abel. In order to investigate this question let us discuss the connection between

Divisors and Homology. A 1-chain in X by definition is a formal combination with integral coefficients of smooth curves in X, i.e.,

$$c := \sum_{j=1}^{k} n_j c_j, \qquad n_j \in \boldsymbol{Z}, \quad \text{where } c_j\colon [0, 1] \to X.$$

The set $C_1(X)$ of 1-chains forms an Abelian group. The boundary operator ∂ can be considered as the map of the group $C_1(X)$ into the group $C_0(X)$ of 0-chains which will be called also *divisors*:

$$(5) \qquad \partial\colon C_1(X) \to C_0(X) = \operatorname{Div}(X),$$

where ∂c is the "divisor" defined in the following way: if $c_j\colon [0, 1] \to X$, then $\partial c_j(1) = 1$, $\partial c_j(0) = -1$ and $\partial c_j(t) = 0$ for $t \neq 0, 1$, $\partial c := \sum n_j \partial c_j$ if $c = \sum n_j c_j$.

Thus we introduce the following

DEFINITION. Let Y be an arbitrary Riemann surface (compact or not). A *divisor* is a map: $Y \to \boldsymbol{Z}$ such that for each compact $K \subset Y$ there exist only finite points $y \in K$ with $(y) \neq 0$.

The Abelian group of divisors on Y (pointwise addition) is denoted by $\mathrm{Div}(Y)$. Divisors were originally denoted by small Gothic letters $\mathfrak{d}_1, \mathfrak{d}_2$; later on we shall find it convenient to depart from this tradition.

If Y is *compact* (i.e. closed), then, for any $\mathfrak{d} \in \mathrm{Div}(Y)$, $\mathfrak{d}(y) \neq 0$ only for a finite number of $y \in Y$ and, therefore, the mapping

$$\deg\colon \mathrm{Div}(Y) \to Z, \quad \deg(\mathfrak{d}) := \sum_{y \in Y} \mathfrak{d}(y)$$

is well defined and is called a *degree*. Its value at $\mathfrak{d}$: $\deg(\mathfrak{d})$ is called the *degree* of $\mathfrak{d}$.

Every meromorphic function f on Y (or a meromorphic differential 1-form ω on Y) determines a divisor on Y, namely, the function which assigns the order of zero (resp. pole) to points at which f (resp., ω) has a zero (resp., pole) and zero to all other points. This divisor will be denoted by (f) or (ω). The divisors of meromorphic functions on Y are called *principal divisors*. They satisfy $\deg(f) = 0$.

Plainly, ∂c is a divisor on X with vanishing degree:

$$\text{(6)} \qquad \deg(\partial c) = 0 \quad \text{for every } c \in C_1(X).$$

Conversely, on a compact Riemann surface X for every divisor $\mathfrak{d}$ with $\deg(\mathfrak{d}) = 0$ there exists a 1-chain c such that $\partial c = \mathfrak{d}$.

$$Z_1(X) := \ker\left(C_1(X) \xrightarrow{\partial} \mathrm{Div}(X)\right)$$

is the Abelian group of 1-cycles. Closed curves are the most important example of 1-cycles. Two 1-cycles $c, c' \in Z_1(X)$ are homologous if

$$\int_c \omega = \int_{c'} \omega \quad \text{for every closed 1-form } \omega,$$

i.e., all such ω have the same periods on c and c'. Homology classes form an Abelian group

$H_1(X)$ is the first homology group of X.

As we know, one of the most important problems of complex analysis, the *Weierstrass problem*, also called the *second Cousin problem* is to "solve a divisor $\mathfrak{d}$", i.e., to find for a divisor $\mathfrak{d}$ a meromorphic function f with given $\mathfrak{d}$, $(f) = \mathfrak{d}$. On an open, i.e., non-compact Riemann surface, the Florack–Behnke Stein Theorem asserts that this problem has a solution for every divisor. On a closed, i.e., compact, Riemann surface the situation is quite different: not every divisor is solvable. Plainly, $\deg \mathfrak{d} = 0$ is a *necessary* condition.

THEOREM XVI.19.10 ("Abel's Theorem"). Let X be a compact Riemann surface and let $\mathfrak{d} \in \mathrm{Div}(X)$ such that $\deg \mathfrak{d} = 0$. Then $\mathfrak{d}$ is divisor of a meromorphic function iff there exists such a chain $c \in C_1(X)$ that $\partial c = \mathfrak{d}$ and for all holomorphic 1-forms on X

$$\int_c \varphi = 0. \tag{7}$$

Remark. Since (Condition (7) is satisfied) $\Leftrightarrow$ $\left(\int_c \varphi_j = 0\right.$ for a basis $\varphi_1, \ldots, \varphi_g$ of $\left.A^1(X)\right)$ we can reformulate (7) in the following way.

If $\gamma \in C_1(X)$ such that $\partial \gamma = \mathfrak{d}$, then there exists a cycle $\alpha \in Z_1(X)$ (namely, $\alpha = \gamma - c$) such that

$$\int_\gamma \varphi_j = \int_\alpha \varphi_j, \quad j = 1, \ldots, g, \text{ for a basis } \varphi_1, \ldots, \varphi_g \tag{7'}$$

of $A^1(X)$ (Abelian differentials of the first kind).[1]

Let us return to Jacobi's inversion problem. Denote

$$\mathrm{Div}_0(X) = \{\mathfrak{d} \in \mathrm{Div}(X) \colon \deg \mathfrak{d} = 0\},$$

i.e., the subgroup of divisors of degree 0, and let $\mathrm{Div}_P(X)$ be the subgroup of principal divisors, i.e., divisors of meromorphic functions. Since X is compact, $\mathrm{Div}_P(X) \subset \mathrm{Div}_0(X)$ (the "Residue Theorem").

DEFINITION. The Picard group of the compact Riemann surface X is the quotient group

$$\mathrm{Pic}(X) := \mathrm{Div}_0(X)/\mathrm{Div}_P(X).$$

We now define a map $F\colon \mathrm{Div}_0(X) \to \mathbf{C}^g$ (g is the genus of X) in the following way: let $\mathfrak{d} \in \mathrm{Div}_0(X)$, $c \in C_1(X)$, such that $\partial c = \mathfrak{d}$,

$$F(\mathfrak{d}) := \left(\int_c \varphi_1, \ldots, \int_c \varphi_j\right) \in \mathbf{C}^g \tag{8}$$

then $F(\mathfrak{d})$ is defined by the divisor $\mathfrak{d}$ modulo $\mathrm{Per}(\varphi_1, \ldots, \varphi_g)$. Abel's Theorem asserts that

$$\ker F = \mathrm{Div}_P(X) \quad \text{(Abel's Theorem)}. \tag{9}$$

Therefore F can be lifted to the map

$$[F]\colon \mathrm{Pic}(X) \to \mathrm{Jac}(X). \tag{10}$$

Jacobi's inversion problem is the following: is F surjective? As we know, Jacobi's Theorem XVI.19.9 answers this question positively. We can now put together these fundamental facts in

[1] Cf. also pages 621–623.

THEOREM XVI.19.11 ("Abel–Jacobi Theorem"). For every compact Riemann surface the mapping $[F]$: $\mathrm{Pic}(X) \to \mathrm{Jac}(X)$ is an isomorphism.

Concluding Remarks. We have seen that the Jacobi inversion problem leads in a natural way to the theory of complex functions of g complex variables (where g = genus of X): to meromorphic functions on the complex torus $\mathrm{Jac}(X)$ of g—complex dimensions. Such functions considered as functions on $\boldsymbol{C}^g$ have $2g$ periods and are called *Abelian functions.* The inversion problem for Abelian integrals of the first kind on the surface of genus $g = 1$ leads to elliptic functions: they uniformize elliptic curves. The situation for $g > 1$ is not only more complicated but quite different.

1° Abelian integrals of kind I have no single-valued inverses, since they have $2g$ (> 2) periods and, as Jacobi early remarked, a function of one complex variable may have at most 2 periods.

2° Complex tori $\boldsymbol{C}^g/L$ are not, in general, algebraic (projective) varieties. In order that $\boldsymbol{C}^g/L$ be an Abelian variety it is necessary (and sufficient) that the lattice $L = \mathrm{Per}(\varphi_1, \ldots, \varphi_g)$ of periods satisfy the so-called Frobenius–Riemann relations. Denote by $\Omega := (\omega_{jk})$, $j = 1, \ldots, g$; $k = 1, \ldots, 2g$ the matrix of periods of the lattice L.

THEOREM XVI.19.12 (Frobenius). A necessary and sufficient condition that the torus $\boldsymbol{C}^g/\mathrm{Per}(\varphi_1, \ldots, \varphi_g)$ be an Abelian variety is: There exists such skew-symmetric $2g \times 2g$ matrix P that

(A) $\quad \Omega P^t\Omega = 0,$

(B) $\quad i\,\Omega P^t\overline{\Omega}$ is positive definite.

(A) and (B) are the famous Frobenius–Riemann conditions. But this leads deep into the beautiful theory of theta functions and Abelian varieties which will be considered in Part III.

20. UNIFORMIZATION PRINCIPLE. AUTOMORPHIC FORMS. RIEMANN–ROCH THEOREM AND ITS CONSEQUENCES. HISTORICAL SKETCH

This final section of our central chapter has two objects:

1° to give some complements of the theories developed in the preceding sections: to show that these theories are developing vigorously and to announce several theorems which will be considered more extensively and in more detail in Part III;

2° and to give a short historical sketch of the developement of Abelian integrals (and elliptic functions) and of the "concept of Riemann surfaces".

We have thought it much better, in 2°, to let great masters speak for themselves.

The uniformization theorem of Koebe and Poincaré together with Poincaré's theory of covering surfaces gives the famous uniformization principle, which is one of the greatest achievements of mathematics:

THEOREM XVI.20.1 (Koebe–Poincaré General Uniformization Principle). Let M be an arbitrary Riemann surface. Then there exists a

(a) universal covering surface $p\colon \hat{M} \to M$ of M,

(b) a properly discontinuous subgroup $\Gamma \subset \operatorname{Aut}(\hat{M})$ acting without fixed points such that M is isomorphic to $\Gamma\backslash\hat{M}$. Therefore, the theory of meromorphic functions on M becomes the theory of Γ-automorphic functions on $\hat{M}$.

(c) There are (essentially) only three types of universal covering surfaces

$$\hat{M} \cong \begin{cases} 1^\circ \text{ upper half-plane } \mathfrak{H} \cong \boldsymbol{D} \text{ (hyperbolic case)},\\ 2^\circ\ \boldsymbol{C} \text{ (parabolic case)},\\ 3^\circ\ \boldsymbol{P}(\boldsymbol{C}) \text{ (elliptic case)}.\end{cases}$$

Remark. The three types of Riemann surfaces can be characterized by the Gaussian K_X curvature:

Since

$$K_{\mathfrak{H}} = K_D = -4, \qquad K_C = 0, \qquad K_{S^1} = 1,$$

the corresponding quotient spaces $\Gamma\backslash\hat{M}$ can be provided with a metric with a constant Gaussian curvature:

hyperbolic surface has curvature -4,

parabolic surface has curvature 0,

elliptic surface has curvature 1.

Therefore, a classification of Riemann surfaces reduces to a classification of properly discontinuous groups of automorphisms of $\mathfrak{H}$, $\boldsymbol{C}$ and $\boldsymbol{P}(\boldsymbol{C})$ acting without fixed points. The following proposition gives the complete information in the non-hyperbolic case:

PROPOSITION XVI.20.2. 1° Every automorphism of $\boldsymbol{P}(\boldsymbol{C})$ has fixed points and, therefore, if $\hat{M} = \boldsymbol{P}(\boldsymbol{C})$ then $M \cong \hat{M}$.

2° In the case of a complex plane $\boldsymbol{C}$ we have three groups Γ:

(i) $\Gamma = \{\mathrm{id}\}$,

(ii) $\Gamma = \{z \to z+k\omega;\ k \in \boldsymbol{Z},$ where $\omega \in \boldsymbol{C}^*\}$,

(iii) $\Gamma = L = \{z \to z+k\omega_1+l\omega_2 \colon k, l \in \boldsymbol{Z}, \omega_1, \omega_2 \in \boldsymbol{C}^*, \mathrm{Im}(\omega_1/\omega_2) \neq 0\}$.

3° In the case $\hat{M} = \mathfrak{H}$, $\mathrm{Aut}(\mathfrak{H}) = \mathrm{SL}_2(\boldsymbol{R})$, and Γ are discrete subgroup of $\mathrm{SL}_2(R)$ acting without fixed points.

Accordingly, we have three large domains of function theory:

1° Rational functions, i.e. quotients of polynomials on $\boldsymbol{C}$ (meromorphic functions on $\boldsymbol{P}(\boldsymbol{C})$).

2° (ii) Theory of periodic functions ($\sin z$, $\cos z$, ...).

2° (iii) Theory of elliptic functions.

3° The vast theories of modular functions and modular forms and, more generally, the theory of automorphic functions.

The following theorem shows that the class of hyperbolic Riemann surfaces is by far the richest .

THEOREM XVI.20.3. All Riemann surfaces are hyperbolic with four exceptions:

1° $\boldsymbol{P}(\boldsymbol{C})$—elliptic

2° $\boldsymbol{C}$, $\boldsymbol{C}^*$, tori $\boldsymbol{C}/L$.

PROOF. Let $\hat{M}$ be the simply connected covering surface of a non-hyperbolic M, then

either $\hat{M} = \boldsymbol{P}(\boldsymbol{C})$, and hence $M \cong \boldsymbol{P}(\boldsymbol{C})$, by Proposition XVI.20.2.1°,

or $\hat{M} = \boldsymbol{C}$ and we have to do with the three cases enumerated in 2° of Proposition XVI.20.2:

(i) M biholomorphic with $\boldsymbol{C}$,

(ii) M biholomorphic with $\boldsymbol{C}^*$ since the covering is isomorphic to $p\colon \boldsymbol{C} \to \boldsymbol{C}^*$, $z \to \exp(2\pi iz/\omega)$,

(iii) M biholomorphic with $\boldsymbol{C}/L$, where $L = [\omega_1, \omega_2]$, and the covering is isomorphic to $p\colon \boldsymbol{C} \to \boldsymbol{C}/L$. □

COROLLARY XVI.20.4. The following Riemann surfaces are hyperbolic:

1. $\mathfrak{H} \cong \boldsymbol{D}$,

2. $\boldsymbol{C}-\{a, b\}$,
3. compact surfaces of genus $g > 1$.

A simple consequence of the hyperbolicity of $\boldsymbol{C}-\{a, b\}$ is the following generalization of the Picard Theorem:

THEOREM XVI.20.5 (Picard, Kobayashi). 1° Let X be a hyperbolic Riemann surface, then every holomorphic mapping $f\colon \boldsymbol{C} \to X$ is constant. Hence

2° Every holomorphic function $\hat{f}\colon \boldsymbol{C} \to \boldsymbol{C}-\{a, b\}$ is constant (Picard's Theorem).

PROOF. Since $p\colon \boldsymbol{D} \to X$ is the universal covering of X, we can lift $f\colon \boldsymbol{C} \to X$ to the holomorphic map $\hat{f}\colon \boldsymbol{C} \to \boldsymbol{D}$, which is constant by the Liouville Theorem. Therefore, f is also constant. □

Remark. The reader may compare the present proof of Picard's Theorem with the original one by means of an elliptic modular function reproduced in Section 17: the modular function there gives the explicit construction of the lift $\tilde{f}\colon \boldsymbol{C} \to \boldsymbol{D}$ of f. Extending the beautiful idea of Carathéodory, Kobayashi introduced an important class of complex manifolds which he called *hyperbolic manifolds*. They are a natural generalization of hyperbolic Riemann surfaces. Let us be more explicit.

Hyperbolic Manifolds of Kobayashi. Let $(\boldsymbol{D}, d_{\boldsymbol{D}})$ be the unit disk provided with the Poincaré metric $d_{\boldsymbol{D}}$. For a complex manifold M let us define an intrinsic pseudo-distance d_M in the following way: Given two points $p, q \in M$, we choose finite sequences of points

$$p = p_0, p_1, \ldots, p_{k-1}, p_k = q \quad \text{in } M$$

and points

$$a_1, \ldots, a_k, b_1, \ldots, b_k \quad \text{in } \boldsymbol{D}$$

and holomorphic mappings

$$f_1, \ldots, f_k\colon \boldsymbol{D} \to M$$

such that

$$f_i(a_i) = f_{i-1}(b_{i-1}) = p_{i-1} \quad \text{for } i = 1, \ldots, k \text{ and } f_k(b_k) = p_k.$$

We set

$$d_M(p, q) := \inf\Big(\sum_{i=1}^{k} d_{\boldsymbol{D}}(a_i, b_i)\Big),$$

where the infimum is taken with respect to all possible choices of p_i, a_i, b_i, f_i as above.

Exercise 1. Prove that d_M is a pseudo-distance.

Exercise 2. Check that $d_{C^k} \equiv 0$.

Exercise 3. The Kobayashi pseudo-distance d_D for the unit disk is identical with the Poincaré distance.

DEFINITION. If d_M is a distance, then the manifold M is called a *hyperbolic manifold* or a *Kobayashi manifold.*

PROPOSITION XVI.20.6 (Kobayashi). Let M, N be hyperbolic manifolds and let $f\colon M \to N$ be a holomorphic map, then

$$(*) \qquad d_N(f(p), f(q)) \leqslant d_M(p, q) \quad \text{for } p, q \in M.$$

The proof follows immediately from the definition of d_M and d_N.

PROPOSITION XVI.20.7 (Kobayashi). Let M be a complex manifold and $\pi\colon \tilde{M} \to M$ a covering manifold of M. Then

$$(M \text{ is hyperbolic}) \Leftrightarrow (\tilde{M} \text{ is hyperbolic}).$$

The proof is left to the reader as an exercise.

COROLLARY XVI.20.8. A Riemann surface X is hyperbolic iff it is a Kobayashi hyperbolic manifold.

PROOF. The proof follows from Exercise 3 and Proposition XVI.20.7.

We now immediately obtain the following generalization of Picard's Theorem.

THEOREM XVI.20.9 (Picard–Kobayashi). Let M be a hyperbolic manifold. Then every holomorphic map $f\colon C^k \to M$ is constant.

PROOF. $d_M(f(p), f(q)) \leqslant d_{C^k}(p, q) = 0$, whereby $f(p) = f(q)$. □

From Proposition XVI.20.6 it follows that any automorphism f of a hyperbolic manifold M is an isometry:

$$d_M(f(p), f(q)) \leqslant d_M(p, q) \leqslant d_M(f(p), f(q)).$$

Hence $\mathrm{Aut}(M)$ is a subgroup of $I(M)$, the group of isometries of (M, d_M), and so is a Lie group by the following general theorems:

THEOREM XVI.20.10 (Bochner–Montgomery). Let G be a locally compact group of differentiable transformations of a manifold M. Then G is a Lie (transformation) group.

THEOREM XVI.20.11 (van Danzig and van der Waerden). Let $I(M)$ be the group of isometries of a locally connected metric space M. Then

1° $I(M)$ is locally compact with respect to the compact-open topology.

2° For each $x \in M$ the isotropy subgroup $I(M)_x$ is compact.

3° If M is compact, then $I(M)$ is also compact.

Let us return to hyperbolic manifolds, especially to hyperbolic Riemann surfaces. We may expect that compact hyperbolic manifolds have a discrete, and hence (by Theorem XVI.20.10.3°) a finite group of automorphisms. Indeed, there is a quite general theorem of Kobayashi (1971):

THEOREM XVI.20.12 (Kobayashi). Let M be a compact hyperbolic manifold, then the group $\mathrm{Aut}(M)$ is finite.

This is a generalization of the following famous and very precise result of Hurwitz (1893) (a proof of which appears as XVI.21.22):

THEOREM XVI.20.13 (Hurwitz). Let M be a compact Riemann surface of genus $g > 1$. Then the order of $\mathrm{Aut}(M)$ is at most $84(g-1)$.

Remark. This result of Hurwitz cannot be improved, since the upper bound $84(g-1)$ is attained in some important cases. (That $\mathrm{Aut}(M)$ is finite for $g > 1$ was first proved by Schwarz.)

On the other hand, we know that for the following classes of Riemann surfaces $\mathrm{Aut}(M)$ is obviously a Lie group:

1° $\boldsymbol{P}(\boldsymbol{C})$, a Riemann sphere (a surface of genus zero),
2° $\boldsymbol{C}$,
3° $\boldsymbol{C}^*$ (i.e., $\boldsymbol{P}(\boldsymbol{C})$ with two points removed),
4° T (tori, i.e., surfaces of genus 1),
5° $\boldsymbol{D}$,
6° $\boldsymbol{D}^*$ (once-punctured disk),
7° annuluses.

We know that tori and annuluses contain infinite class of distinct conformal types (cf. p. 556, Exercise 2); 5°–7° are hyperbolic surfaces. The following natural question, put first by Felix Klein, arises: are these seven types the only surfaces with continuous automorphism groups? The answer is yes.

THEOREM XVI.20.11 (Klein). The group of automorphisms of a Riemann surface is always (properly) discontinuous, except the seven cases enumerated above.

Remark. The preceding theorem is due in essence to Klein, but the first precise proof was given by Poincaré in 1883 in the case of compact surfaces. But, as Weyl remarks, "Poincaré's proof, word for word, remains valid for open surfaces". We now leave this fascinating subject and advise the reader to consult the great master himself:

Shishichi Kobayashi has written the following two beautiful monographs: *Transformation Groups in Differential Geometry*, Springer, 1972, and *Hyperbolic Manifolds and Holomorphic Mappings*, Dekker, 1970.

Discrete Subgroups Γ *of* $SL_2(\boldsymbol{R})$. *Cusps. Compactification of* $\Gamma\backslash\mathfrak{H}$. *Fuchsian Groups.* In Sections 17 and 18 we considered fundamental domain, F_Γ, modular forms, and modular functions of weight $2k$ when the discrete subgroup $\Gamma \subset SL_2(\boldsymbol{R})$ was the modular group $SL_2(\boldsymbol{Z})$. We have seen the paramount role of the fundamental domain $F_\Gamma \cong \Gamma\backslash\mathfrak{H}$ and a compactification $\overline{\Gamma\backslash\mathfrak{H}}$, constructed by attaching the point $\{\infty\}$ to F. In this way we have obtained a compact Riemann surface equivalent to the sphere $\boldsymbol{P}(\boldsymbol{C})$. The point $\{\infty\}$ (denoted also by $\{i\infty\}$ is called the *cusp* of Γ (or of F_Γ). A modular form vanishing at the cusp was called a *cusp form*.

The situation is much more complicated when we have to do with a proper subgroup of $SL_2(\boldsymbol{Z})$: then we arrive at Riemann surfaces of genus $g > 0$. We can here only give a definition and formulate important theorems. The interested reader may consult the following excellent monograph of a great specialist in this field: Goro Shimura, *Introduction to the Arithmetic Theory of Automorphic Functions*, Iwanami Shoten and Princeton U.P., 1971.

DEFINITION. Let $g \in SL_2(\boldsymbol{R})$, $g \neq \pm I$. Then we say that:
g is *parabolic* if g has only one fixed point (in $\boldsymbol{R} \cup \{\infty\}$),
g is *elliptic* if g has one fixed point $z \in \mathfrak{H}$ and another $\bar{z}$,
g is *hyperbolic* if g has two fixed points in $\boldsymbol{R} \cup \{\infty\}$.

Equivalently one can describe these types of transformations by condition on matrices.

DEFINITION Let $g \in \mathrm{SL}_2(\boldsymbol{R})$ and $g \neq \pm I$. Then
g is *parabolic* iff $\operatorname{trace}(g) = \pm 2$,
g is *elliptic* iff $|\operatorname{trace}(g)| < 2$,
g is *hyperbolic* iff $|\operatorname{trace}(g)| > 2$.

Exercise. Prove that these two conditions are equivalent.

Let us now fix a discrete subgroup $\Gamma \subset \mathrm{SL}_2(\boldsymbol{R})$.

DEFINITION. A point $s \in \boldsymbol{R} \cup \{\infty\}$ is called a *cusp* of Γ if there exists a parabolic element $g \in \Gamma$ such that $g \cdot s = s$. A point $z \in \mathfrak{H}$ is an *elliptic point* of Γ if there exists an elliptic element of Γ such that $g \cdot z = z$.

Plainly, if s is a cusp (resp. elliptic point), then γs is also a cusp (elliptic point).

The reader will undoubtedly notice that the cusps for the modular group Γ are precisely the rational points on $\boldsymbol{R}$ and the point $\{\infty\}$.

Remark. This definition elucidates why cusps are called "*points paraboliques*" by French mathematicians.

Throughout this section $\mathfrak{H}^*$ will denote the union of $\mathfrak{H}$ and the cusps of Γ.

We know that $\Gamma \backslash \mathfrak{H}$ is a Hausdorff space. Since Γ acts on $\mathfrak{H}^*$, the Γ-orbit space $\Gamma \backslash \mathfrak{H}^*$ is meaningful: $\Gamma \backslash \mathfrak{H}^*$ is the union of $I \backslash \mathfrak{H}$ and the equivalence classes of cusps. The following natural question arises: what is the topology of $\Gamma \backslash \mathfrak{H}^*$? We must first define a topology of $\mathfrak{H}^*$: As the basis of neighbourhoods of every $z \in \mathfrak{H}$ we take the usual one. Let s be a cusp $\neq \infty$; then we take the following sets: $\{s\} \cup \{$the interior of a circle in $\mathfrak{H}$ tangent to $\boldsymbol{R}$ at $s\}$. If $s = \infty$ is a cusp (as in the modular case), we take the sets:

$$\{\infty\} \cup \{z \in \mathfrak{H} : \operatorname{Im}(z) > c \text{ for all } c > 0\}.$$

THEOREM XVI.20.12. 1° The quotient space $\Gamma \backslash \mathfrak{H}^*$ is Hausdorff,

2° $I \backslash \mathfrak{H}^*$ is locally compact,

3° if $\Gamma \backslash \mathfrak{H}^*$ is compact, then the number of Γ-inequivalent cusps (resp. elliptic points) is finite,

4° if $\Gamma \backslash \mathfrak{H}$ is compact then Γ has no parabolic elements.

DEFINITION. If $\Gamma\backslash\mathfrak{H}^*$ is compact, then Γ is called a *Fuchsian group of the first kind* (in German also *Hauptkreisgruppe erster Art*).

Lazarus Fuchs (1833–1902) was professor of mathematics in Heidelberg and later in Berlin. His work was fundamental for the theory of linear differential equations with singular coefficients (elements of this important theory will be considered in Part III). His paper of 1880 had great influence on the 26-years-old Poincaré. This was the starting point of Poincaré's fundamental investigations, which gave birth to a class of automorphic functions which he called *Fuchsian functions*.

We shall now introduce a complex structure on $\Gamma\backslash\mathfrak{H}^*$. By Lemma XVI.17.8 we know that for each $\tau \in \mathfrak{H}$ there exists an open neighbourhood $\mathcal{O}$ of z such that the stabilizer of τ is

$$\Gamma_\tau = \{\gamma \in \Gamma \colon \gamma(\mathcal{O}) \cap \mathcal{O} \neq \emptyset\}.$$

The same property holds for cusps of Γ as well (cf. Shimura, *loc. cit*). Therefore, we have an injection

$$\Gamma_\tau\backslash\mathcal{O} \to \Gamma\backslash\mathfrak{H}^*$$

and $\Gamma_\tau\backslash\mathcal{O}$ is an open neighbourhood of $\pi(\tau)$, where $\pi\colon \mathfrak{H}^* \to \Gamma\backslash\mathfrak{H}^*$ is the canonical projection. If τ is neither a cusp nor an elliptic element, then we take $(\Gamma\backslash\mathcal{O}, \pi^{-1})$ as a (local) chart of the complex structure of $\Gamma\backslash\mathfrak{H}^*$, since then Γ_τ contains I and, possibly, $-I$.

If τ is an elliptic point, we denote by $'\Gamma_\tau$ the group $(\Gamma_\tau \cdot \{\pm I\})/\{\pm I\}$. Let φ be a biholomorphism: $\mathfrak{H} \to \boldsymbol{D}$ such that $\varphi(\tau) = 0$. If $'\Gamma_\tau$ is of order n, then

$$\varphi'\Gamma_\tau\varphi^{-1} = \{v \to \zeta^k v\colon k = 0, \ldots, n-1,\ \zeta := e^{2\pi i/n}\};$$

thus define $p\colon \Gamma_\tau\backslash\mathcal{O} \to \boldsymbol{C}$ via $p\big(\pi(z)\big) := \varphi(z)^n$. We see that p is a homeomorphism onto an open subset of $\boldsymbol{C}$. We take $(\Gamma_\tau\backslash\mathcal{O}, p)$ as further charts of the complex structure.

Finally, if s is a cusp, and we have $g \in \mathrm{SL}_2(\boldsymbol{R})$ such that $g(s) = \infty$, then

$$g\Gamma_s g^{-1} \cdot \{\pm I\} = \left\{\pm\begin{bmatrix}1 & b\\ 0 & 1\end{bmatrix}^m : m \in \boldsymbol{Z}, b > 0\right\},$$

then $p|(\Gamma_s\backslash\mathcal{O})\colon \Gamma_s\backslash\mathcal{O} \to \Gamma_s\backslash\mathcal{O} \subset \boldsymbol{C}$ is a homeomorphism and we include $(\Gamma_s \backslash \mathcal{O}, p)$ into the complex structure of $\Gamma \backslash \mathfrak{H}^*$. We check that in this way $\Gamma \backslash \mathfrak{H}^*$ is a Riemann surface.

An important class of Fuchsian groups are

Congruence Subgroups of $\mathrm{SL}_2(\boldsymbol{Z})$. For every positive integer N we set

$$\text{(1)} \qquad \Gamma(N) = \{\gamma \in \mathrm{SL}_2(\boldsymbol{Z})\colon \gamma \equiv I \bmod (N)\}$$
$$= \left\{\begin{bmatrix} a & b \\ c & d \end{bmatrix} \in \mathrm{SL}_2(\boldsymbol{Z})\colon a \equiv d \equiv 1, b \equiv c \equiv 0 \bmod (N\boldsymbol{Z})\right\}.$$

DEFINITION. $\Gamma(N)$ is called a *principal congruence subgroup of level* N (in German *der Stufe* N). Then $\Gamma(N)$ is a normal subgroup of $\mathrm{SL}_2(\boldsymbol{Z})$. (Plainly, $\Gamma(1) = \mathrm{SL}_2(\boldsymbol{Z})$.)

An f is called a *modular function f of weight k and level N* if

1° f is meromorphic on $\mathfrak{H}$;

$$\text{(2)} \qquad 2° \ f(\gamma\tau) = (c\tau + d)^k f(\tau) \ \text{ for } \tau \in \mathfrak{H}, \gamma = \begin{bmatrix} a & b \\ c & d \end{bmatrix} \in \Gamma(N),$$

3° f is meromorphic at every cusp of $\Gamma(N)$.

Functions of weight $k = 0$ are called *simply modular functions of level N on $\mathfrak{H}$*.

Similarly to the case of the modular group Γ (cf. Section 17) the condition 3° must be interpreted with the use of a local parameter at the cusp points (we take $k = 0$ for simplicity, so

$$f \text{ is } \Gamma(N)\text{-invariant}\colon f(\gamma\tau) = f(\tau), \gamma \in \Gamma(N)).$$

Let $q := e^{2\pi i \tau}$ and $\mathfrak{H}_B = \{\tau \in \mathfrak{H}\colon \mathrm{Im}(\tau) > B\}$. Then the map $\tau \to q^{1/N}$ defines a holomorphic surjection $h_B\colon \mathfrak{H}_B \to \boldsymbol{D}_r^*$, a punctured disk, and h_B is defined on $\mathfrak{H}$ modulo the translation by N. Since $\begin{bmatrix} 1 & N \\ 0 & 1 \end{bmatrix} \in \Gamma(N)$ and acts as translation by N on $\mathfrak{H}$, it follows that f induces a meromorphic function $\hat{f}$ on $\boldsymbol{D}_r^*$. If there exists a positive power q^m such that $|\hat{f}(q)q^m|$ is bounded near 0, then $\hat{f}$ is meromorphic on $\boldsymbol{D}_r^*$ and $\hat{f}$ has a power series expansion in $q^{1/N}$ (a Fourier series in τ/N) with a finite number of negative terms. Now 3° reads as follows:

3′° if for every $\gamma \in \mathrm{SL}_2(\boldsymbol{Z})$ the function $(f \circ \gamma)\hat{}$ has a power series expansion in $q^{1/N}$ with at most a finite number of negative terms, then f is called a *modular function of level N on $\mathfrak{H}$*.

4° If, moreover, $(f\circ\gamma)\hat{}$ has the power series expansion at each cusp

$$(f\circ\gamma)\hat{}(q)=\sum_{n=0}^{\infty}c_nq^{1/n}\quad\text{and}\quad c_0=0,$$

then f is called a *cusp function of level N*.

Cusp forms of level N and weight k are defined similarly. We have seen, in the case of the modular group $\Gamma(1)$, the fundamental role of the elliptic modular function j,

$[j(\tau):=12^3\cdot g_2(\tau)^3/\Delta(\tau)$, where $\Delta(\tau)=g_2(\tau)^3-27g_3(\tau)^2$ and g_k are the Eisenstein series]:

every meromorphic function of the Riemann surface $\Gamma(1)\backslash\mathfrak{H}^*$ is a rational function of j. In other words,

(3) $\quad\mathcal{M}(1,\Gamma(1))=C(j)$, a field over C generated by j.

The question arises whether the field of modular functions of level $N>0$ is generated in a similar way.

The answer is positive but much more difficult; it was given by Heinrich Weber (1842–1913) in the monumental third volume of his *Lehrbuch der Algebra* (1908). Let us introduce the

DEFINITION. Let $\wp(u)=\wp(u;L)=\wp(u;\omega_1,\omega_2)$ be the Weierstrass function and $L=[\omega_1,\omega_2]$ be the period lattice $L=Z\omega_1\oplus Z\omega_2$. The first *Weber function* is

(4) $$f_a(\omega_1,\omega_2):=\frac{g_2(\omega_1,\omega_2)g_3(\omega_1,\omega_2)}{\Delta(\omega_1,\omega_2)}\wp(a_1\omega_1+a_2\omega_2;\omega_1,\omega_2)$$

where $a=(a_1,a_2)\in Q^2$.

We can now state Weber's result:

THEOREM XVI.20.13 (H. Weber). For every positive integer N the field $\mathcal{M}(0,\Gamma(N))$ of modular functions of level N is equal to $C(j,f_a)$, $a\in N^{-1}Z^2$, $a\notin Z^2$, where f_a are defined by (4).

Remark. Plainly, there is only a finite number of such $a\in Q^2$. If $N=1$, we obtain the former result, $\mathcal{M}(0,\Gamma(1))=C(j)$. We see that the Weber function plays as fundamental role as the elliptic invariant j.

Automorphic Forms. Poincaré Series. We know (Theorem XVI.20.1) that the theory of meromorphic functions on a (hyperbolic) Riemann surface M is equivalent to the theory Γ automorphic functions on $\mathfrak{H}$ (or $\boldsymbol{D}$) where Γ is a properly discontinuous subgroup of Aut($\mathfrak{H}$) such that $M = \Gamma\backslash\mathfrak{H}$. We know also that modular functions can be represented as quotients of modular forms. It is much easier to construct modular forms than modular functions. We shall now define general Γ-automorphic forms and give the classical construction of automorphic forms first proposed by Poincaré in 1880.

DEFINITION. Let D be a bounded domain in $\boldsymbol{C}^n$ and let Γ be a properly discontinuous subgroup of Aut(D) acting without fixed points on D and such that

$$\Gamma\backslash D \text{ is compact.}$$

A holomorphic function on D is a Γ-automorphic form of weight k if

$$(5) \qquad f(\gamma x) = J_\gamma(x)^k f(x) \quad \text{for all } x \in D, \gamma \in \Gamma,$$

where $J_\gamma(x) := \dfrac{\partial}{\partial x}(\gamma x)$ is the Jacobi determinant of the map γ at the point x.

Example. Let $D = \boldsymbol{D}$ (or $\mathfrak{H}$), and let Γ be a Fuchsian group of first kind. Denote

$$(6) \qquad j_\gamma(z) = (cz+d), \quad \text{where } \gamma = \begin{bmatrix} a & b \\ c & d \end{bmatrix} \in \Gamma;$$

then

$$(7) \qquad j_{\sigma\gamma}(z) = j_\sigma(\gamma(z))\, j_\tau(z),$$

$$J_\gamma(z) = \frac{\partial}{\partial z}\gamma(z) = \det(\gamma)\cdot j_\gamma(z)^{-2} = j_\gamma(z)^{-2},$$

the last equality following from $\Gamma \subset \mathrm{SL}_2(\boldsymbol{C})$. Thus, the $\Gamma(N)$-automorphic form of weight k is a modular form of weight $2k$ and level N. For the following definition and proposition we shall confine ourselves to this case.

DEFINITION. Let h be a bounded holomorphic function on the unit disk $\boldsymbol{D}$, then

$$(8) \qquad \sum_{\gamma \in \Gamma} \gamma^*(h) J_\gamma^{-k},$$

where $\gamma^* h := h \circ \gamma$ is called a *Poincaré series*.

PROPOSITION XVI.20.14 (Poincaré). A Poincaré series converges for $k \geqslant 2$ absolutely and compactly and defines a Γ-automorphic form of weight k.

PROOF. By the fundamental inequality

$$(9) \qquad |h(0)| \leqslant \frac{1}{\pi r^2} \int\limits_{|z| \leqslant r} |h(z)|^2 dx\, dy, \qquad z = x+iy,$$

for any $h \in \mathscr{A}(D)$ we see that, since $|h|$ is bounded, it is sufficient to prove the compact convergence of the series

$$(10) \qquad \sum_{\gamma \in \Gamma} |J_\gamma|^2 = \sum_{\gamma \in \Gamma} \left| \det \frac{d}{dz} (\gamma(z)) \right|^2.$$

Let $\mathcal{O} = K(z_0, \delta)$ with such δ that $\gamma(\mathcal{O}) \cap \mathcal{O} = \emptyset$ for $\gamma \neq \mathrm{id}$. Then if $|A|$ denotes the λ^2-measure of $A \subset \mathbf{C}$, we have

$$(11) \qquad \sum_{\gamma \in \Gamma} |J_\gamma(z_0)|^2 \leqslant \frac{1}{\pi r^2} \sum_{\gamma \in \Gamma} \int\limits_{\mathcal{O}} |J_\gamma(z)|^2 dx\, dy$$

$$= \frac{1}{\pi r^2} \sum_{\gamma \in \Gamma} |\gamma(\mathcal{O})| \leqslant \frac{|D|}{r^2} = \frac{1}{r^2}.$$

Let A be a fixed compact subset of D and $\varepsilon > 0$, $d(A, \complement D) =: 2r$. Let A_r be a compact neighbourhood of A such that $d(A_r, \complement D) < r$. Take $0 < l < 1$ such that $A_r \subset D_l$ and $|D - D_l| < \varepsilon$. Let m be the number of $\gamma \in \Gamma$ in the set $\{\gamma \in \Gamma\colon \gamma A_r \cap A_r \neq \emptyset\}$. For almost all $\gamma \in \Gamma$ we have $\gamma A_r \subset D - D_l$. Denote by $\sum'$ the sum extended over such γ. As in (9) we obtain

$$\sum_{\gamma}{}' |J_\gamma(z)|^2 \leqslant \frac{1}{\pi r^2} \sum_{\gamma}{}' |\gamma \mathcal{O}| < \frac{m}{\pi r^2} |D - D_l| < \varepsilon \frac{m}{\pi r^2}.$$

Hence the series (8) converges compactly for $k \geqslant 2$. □

That the sum of the Poincaré series satisfies (5) can be verified directly and is left to the reader. □

We have, as in the modular case, the important problems of determining the dimension of vector spaces of modular forms, modular cusp forms of level N and, more generally, of spaces of Γ-automorphic forms. The Riemann–Roch Theorem (R–R) is necessary for precise results.

Riemann–Roch (R–R) *Theorem and Its Consequences.* The R–R is the most important theorem in the theory of compact Riemann surfaces: it is a global theorem which connects the topological invariant g, the genus of the surface X, and the dimensions of very important vector spaces of differential forms on X. We know the importance of the existence of Abelian differentials with prescribed poles and meromorphic functions and forms which have poles (zeros) of order not greater (lower) described by a given divisor A on X. Such functions and forms are called *multiples* of A. The R–R allows us to prove the existence of such functions and forms.

We shall now give a formulation of the R–R and several important applications of it. Three proofs (two of them analytic and one algebraic) of this fundamental theorem will be given in Part III.

Assume M is a compact Riemann surface. For any divisor $A \in \mathrm{Div}(M)$ and $x \in M$ we set $\mathrm{ord}_x(A) := \nu_x(A) := A(x)$ and we write $A = \sum \nu_x(A)x$, $\deg(A) = \sum \nu_x(A)$. We say

$$A \geqslant 0 \text{ if } A(x) = \nu_x(A) \geqslant 0$$

for all $x \in M$. Then $B \geqslant A$ if $B - A \geqslant 0$.

Principal divisors are divisors of meromorphic functions; the linear space of principal divisors is denoted by $\mathrm{Div}_P(M)$. We write (f) or $\mathrm{div}(f)$ for the divisor of $f \in \mathcal{M}(M)$. We can also define a divisor of any meromorphic differential form $\omega \in \mathcal{M}^1(M)$ on M in the following way: Let us identify $\boldsymbol{C}$ with the subfield of $\mathcal{M}(M)$ consisting of all constant functions. Denote by $\mathcal{M}(M)^*$ the group of all invertible (i.e., non-zero) elements of $\mathcal{M}(M)$. We know that every $\omega \in \mathcal{M}^1(M)$ can be written as

$$\omega = h\,df, \quad \text{where } h \in \mathcal{M}(M) \text{ and } f \in \mathcal{M}(M) - \boldsymbol{C},$$

i.e.,

$$\mathcal{M}^1(M) = \mathcal{M}(M) \cdot df \quad \text{for some } f \in \mathcal{M}(M) - \boldsymbol{C}.$$

For each $x \in M$ we take such $t \in \mathcal{M}(M)$ that $\nu_x(t) = 1$ and put $\nu_x(\omega) := \nu_x(\omega/dt)$; this is independent of the choice of t.

DEFINITION. For $\omega \in \mathcal{M}^1(M)$ we put

$$(\omega) \equiv \mathrm{div}(\omega) := \sum \nu_x(\omega)x.$$

Then

$$(f\omega) = (f) + (\omega) \quad \text{for any } f \in \mathcal{M}(M) \text{ and } \omega \in \mathcal{M}^1(M).$$

Let us introduce the following vector spaces

$$M_{-D} = \{f \in \mathcal{M}(X)\colon (f) \geqslant -D\}, \quad \text{where } D \in \mathrm{Div}(X)$$

(the condition means that $\mathrm{ord}_x(f) \geqslant -D(x)$, $x \in X$), and

$$M_D^1 = \{\omega \in \mathcal{M}^1(X)\colon (\omega) \geqslant -D\}.$$

THEOREM XVI.20.15 (Riemann–Roch). Let X be a compact Riemann surface of genus g, $D \in \mathrm{Div}(X)$. Then $\dim M_{-D}$, $\dim M^1_{-D} < \infty$ and

$$\text{(R–R)}\quad \dim M_{-D} - \dim M_D^1 = 1 - g - \deg D,$$

or, what is the same,

$$\text{(R–R)}\quad \dim M_D - \dim M^1_{-D} = 1 - g + \deg D.$$

Remark. $\dim M^1_{-D} =: i(D)$ is called the *speciality index* of the divisor D.

Therefore, we can write the R–R theorem in the following form

$$\text{(R–R)}'\quad \dim M_D = 1 - g + \deg D + i(D).$$

COROLLARY XVI.20.16.

$$(\deg D < 0) \Rightarrow (M_D = \emptyset) \Rightarrow \big(i(D) = g - 1 - \deg D\big).$$

PROOF. Let $f \in M_D, f \neq 0$, then $(f) \geqslant -D$; hence, $\deg(f) \geqslant -\deg D > 0$, but we know that $\deg(f) = 0$. □

Another simple consequence of (R–R) is

THEOREM XVI.20.17. Let X be of genus g and let $x_0 \in X$. Then there exists a non-constant meromorphic function $f \in \mathcal{M}(X)$ which has at x_0 a pole of order $\leqslant g+1$ and which is holomorphic on $X - \{x_0\}$.

PROOF. Let $D \in \mathrm{Div}(X)$ be such that $D(x_0) = g+1$, $D(x) = 0$ for $x \neq x_0$. Then by (R-R)′ we have

$$\dim M_D \geqslant 1 - g + \deg D = 1 - g + (g+1) = 2,$$

which yields the assertion. □

The function f in the preceding theorem gives a covering $f\colon X \to \boldsymbol{P}(\boldsymbol{C})$ of the Riemann sphere which has at most $g+1$ sheets, since f assumes the value ∞ with multiplicity $\leqslant g+1$. In this way we have proved

PROPOSITION XVI.20.18. For every Riemann surface X of genus g there exists a holomorphic covering

$$f: X \to P(C)$$

with $g+1$ sheets at most.

As a corollary we obtain the elliptic case of the uniformization theorem.

THEOREM XVI.20.19 (Conformal Mapping, The Elliptic Case).

1° Every Riemann surface of genus 0 is biholomorphically isomorphic to the Riemann sphere.

2° $(X \text{ compact}, \pi_1(X) = \{1\}) \Rightarrow (X \cong P(C))$.

PROOF. 1° By the preceding proposition we know that there exists a one-sheeted, and hence biholomorphic, map $f: X \to P(C)$.

2° follows from 1° since in this case the genus of X is 0. □

In order to obtain further interesting applications of (R–R) let us prove the following simple

LEMMA XVI.20.20. Let ω be a non-vanishing meromorphic 1-form on a compact Riemann surface X. Denote by K its divisor; then we have an isomorphism

$$M_{D+K} \xrightarrow{\sim} M_D^1, \quad f \to f\omega;$$

hence $\dim M_{D+K} = \dim M_D^1$, for any $D \in \mathrm{Div}(X)$, and thus

$$(*) \qquad \dim M_K = \dim M_0^1 = \dim A^1 = g.$$

PROOF. Let h be a non-constant meromorphic function on X and $\omega := dh$; then $f \to f\omega$ gives the required isomorphism. From the preceding lemma follows an important

THEOREM XVI.20.21. Let X be of genus g. Then, for each $\omega \in \mathscr{M}^1(X)$, $\omega \neq 0$

$$\deg(\omega) = 2(g-1).$$

PROOF. Let $K := (\omega)$. Set in (R–R) $D = K$; then from $(*)$ we obtain

$$g - \dim M_{-K}^1 = 1 - g + \deg K,$$

but $\dim M_{-K}^1 = \dim M_{-K+K} = \dim M_0 = 1$ since $M_0 = A(X) \cong C$. Therefore, $g-1 = 1-g+\deg K$. Hence

$$\deg K = 2(g-1). \ \square$$

Remark. The last theorem allows us to "replace" the genus g in the preceding formulations of (R–R).

Let ω be a non-vanishing meromorphic form on a compact Riemann surface X; then for any $D \in \mathrm{Div}(X)$ we have

(R–R)″ $\dim M_D - \dim M^1_{-D} = \deg D - \frac{1}{2}\deg(\omega)$.

COROLLARY XVI.20.22. Let X be of genus g; then

$$\big(M^1_{-D}(X) = \emptyset\big) \Leftarrow \big(\deg D > 2(g-1)\big).$$

PROOF. Let $0 \neq \omega \in \mathscr{M}^1(X)$ and $K := (\omega)$; then, since $\deg(K-D) = 2(g-1) - \deg D < 0$, we have

$$M^1_D(X) \cong M_{K-D}(X) = \emptyset \quad \text{by Corollary XVI.20.16. } \square$$

COROLLARY XVI.20.23. For every lattice $L \subset \boldsymbol{C}$ the torus $\boldsymbol{C}/L$ has the genus $g = 1$.

PROOF. The differential form dz on $\boldsymbol{C}$ induces a 1-form on $\boldsymbol{C}/L$ which has neither zeros nor poles. Hence, $0 = \deg(\omega) = 2(g-1)$. Therefore, $g - 1$. $\square$

The Riemann–Roch Theorem allows us to prove the

Existence of Abelian Differentials of the Second and Third Kind. We recall that meromorphic differential forms ω are called *Abelian differentials.* If ω has a single pole a of order two and $\mathrm{Res}_a(\omega) = 0$, then ω is called an *elementary differential of the second kind.* And if $\eta \in \mathscr{M}^1(X)$ has only two poles of order one at points u_+, u_-, then ω is an *elementary* (or *normal*) *differential of the third kind.*

LEMMA XVI.20.24. Let C be a divisor such that $\deg C = n$, $C \geqslant 0$. Then we have

$$\dim M^1_C = g + n - 1 > 0.$$

Therefore, there exist differential forms ω such that $(\omega) \geqslant -C$, i.e., which have poles of order not greater than $C(x)$, $x \in X$.

PROOF. If we set $K = (\omega)$ and $D = K + C$, we have

$$\deg(K-D) = \deg(-C) = -n < 0; \quad \text{hence } M_{-C} = \emptyset.$$

Therefore, from (R–R) we obtain

$$-\dim M^1_C = 1 - g + \deg(-C) = -(g+n) + 1. \ \square$$

From this lemma we immediately get

THEOREM XVI.20.25. Let X be a compact Riemann surface. Then
1° There exists an elementary differential of the second kind,
2° There exists an elementary differential of the third kind.

PROOF. 1° Set $C = 2a$.
2° Set $C = a_+ + a_-$. □

Lemma XVI.20.24 also implies the existence of Abelian differentials with given polar parts. The following two theorems are valid.

THEOREM XVI.20.26. Let X be a compact Riemann surface. Then for every $x_0 \in X$ and a given principal (i.e., polar) part of order n there exists an Abelian differential of the second kind with a given polar part.

THEOREM XVI.20.27. Let X be a compact Riemann surface, let $a_1, \ldots, a_k \in X$ and let $\omega_j = h_j(\zeta)d\zeta$ be a principal part at $a_j, j = 1, \ldots, k$ such that

$$\sum_{j=1}^{k} \operatorname{Res}_{a_j}(\omega_j) = 0.$$

Then there exists an Abelian differential ω with these principal parts. The form ω is unique modulo Abelian differentials of the first kind.

PROOF OF THEOREM XVI.20.26. If we set $C_\nu = \nu \cdot x_0$, $\nu = 2, \ldots, n$, then by Lemma XVI.20.24

$$\dim M_{C_\nu} = (g-1)+\nu, \quad \nu = 2, \ldots, n.$$

This means that there exist Abelian differentials of the second kind which at the given point x_0 have poles of order $2, 3, \ldots, n$. Indeed, we check by induction that $g+k-1$ differentials which at x_0 have poles of order $\leqslant k-1$ without residua are linearly dependent. But differentials of the second kind with poles of different orders are plainly linearly independent. Therefore, by forming linear combinations one can construct an Abelian differential ω of the second kind with given polar part of order $n \geqslant 2$ at x_0 with $\operatorname{Res}_{x_0}(\omega) = 0$. □

PROOF OF THEOREM XVI.20.27. By addition of Abelian differentials of second and third kind one can obtain an Abelian differential ω with given polar parts such that ω has a vanishing sum of residua. Plainly such an ω is unique modulo holomorphic differential. □

Remark. Riemann–Roch also allows us to compute dimensions of spaces of automorphic forms. Of course this problem is much more difficult (cf. books of C. L. Siegel and G. Shimura).

The Hurwitz–Riemann Formula (The Ramification Index). In the theory of algebraic functions the following notion is of great importance:

DEFINITION. Let X, Y be compact Riemann surfaces and $f\colon X \to Y$ a non-constant holomorphic map. By $m(f, x)$ we denote the multiplicity of the value of $f(x)$ at $x \in X$. Then $b(f, x) := m(f, x) - 1$ is called the *index of ramification of f at x*. Hence,

$$(b(f, x) = 0) \Leftrightarrow (f \text{ is non-ramified (unbranched) at } x).$$

$$b(f) := \sum_{x \in X} b(f, x)$$

is the *total index of ramification of f*.

Remark. Since X is compact, there is only a finite number of $x \in X$ with $b(f, x) \neq 0$ and, hence, $b(f)$ is well defined.

The following important formula, already introduced by Riemann but proved by Hurwitz, gives a relation between the general ramification index of X, Y and the total ramification index of $f\colon X \to Y$.

THEOREM XVI.20.28 (The Hurwitz–Riemann Formula). Let $f\colon X \to Y$ be an n-sheeted holomorphic covering, where X, Y are compact Riemann surfaces of genera g_X, g_Y. Then the following formula is valid:

$$g_X = \frac{b(f)}{2} + n(g_Y - 1) + 1,$$

where $b(f)$ is the total ramification index of f.

PROOF. If $0 \neq \omega \in \mathcal{M}^1(Y)$, then

$$\deg(\omega) = 2(g_Y - 1), \quad \deg(f^*\omega) = 2(g_X - 1).$$

Let $x \in X$ and $y = f(x)$. Hence, we have local charts (O, z), (U, w) at x, y with $z(x) = 0$ and $w(y) = 0$. Therefore, we can describe f as $w = z^k$, where $k = m(f, x)$. Let $\omega = \varphi(w)\,dw$; then in O we have

$$f^*(\omega) = \varphi(z^k)\,d(z^k) = kz^{k-1}\varphi(z^k)\,dz.$$

Therefore,

$$(*) \qquad \operatorname{ord}_x(f^*\omega) = b(f, x) + m(f, z)\operatorname{ord}_y(\omega).$$

Since f is n-sheeted, we have

$$\sum\{m(f, x):\ x \in f^{-1}(y)\} = n \qquad \text{for every } y \in Y;$$

therefore, by virtue of (∗), we obtain

$$\sum_{x \in f^{-1}(y)} \operatorname{ord}_x(f^*\omega) = \sum_{x \in f^{-1}(y)} b(f, x) + n \operatorname{ord}_y(\omega);$$

hence

$$\begin{aligned} 2(g_X - 1) &= \deg(f^*\omega) := \sum_{x \in X} \operatorname{ord}_x(f^*\omega) \\ &= \sum_{y \in Y} \sum_{x \in f^{-1}(y)} \operatorname{ord}_x(f^*\omega) \\ &= \sum_{x \in X} b(f, x) + n \sum_{y \in Y} \operatorname{ord}_y(\omega) = b(f) + n \deg(\omega) \\ &= b(f) + n2(g_Y - 1). \ \square \end{aligned}$$

Coverings of the Riemann Sphere (*Hyperelliptic Case*). Let us consider the most important case of f: $X \to Y \equiv \boldsymbol{P}(\boldsymbol{C})$. Since $g_{\boldsymbol{P}} = 0$, the Hurwitz–Riemann formula becomes

$$g_X = \frac{b(f)}{2} - n + 1 \qquad \text{(the Riemann formula).}$$

DEFINITION. A 2-sheeted covering space π: $X \to \boldsymbol{P}(C)$ with genus $g_X > 1$ is called *hyperelliptic.*

In this case the Riemann–Hurwitz formula reads

$$\text{(R)} \qquad g = \frac{b}{2} - 1.$$

We now give a very important

EXAMPLE. Let f: $X \to \boldsymbol{P}(C)$ be the Riemann surface of

$$\sqrt{(z-a_1)\ \dots\ (z-a_n)} =: \sqrt{P(z)},$$

where $a_1, \dots, a_k$ are distinct zeros of P.

Since b has to be even, by (R) we have thus proved that f: $X \to \boldsymbol{P}(C)$ is ramified over ∞ if and only if k is even. The genus of X is $g = [(k-1)/2]$, where $[x]$ denotes the "entire of x". We can construct a basis $\varphi_1, \dots, \varphi_g$ of $A^1(X)$ on X:

$$\varphi_j := \frac{z^{j-1}dz}{\sqrt{P(z)}}, \qquad 1 \leqslant j \leqslant g = \left[\frac{k-1}{2}\right].$$

A Brief Historical Sketch

Bei oberflächlichem Blick auf die Mathematik mag man den Eindruck haben sie sei das Ergebnis der getrennten persönlichen Bemühungen von vielen Gelehrten, die über Länder und Zeiten verstreut waren. Jedoch die innere Logik der mathematischen Entwicklung erinnert einen viel mehr an das Werk eines einzigen Intellektes, der seinen Gedanken systematisch und beständig entwickelt und dabei die Verschiedenheit menschlicher Individualitäten nur als Mittel benutzt. Gleichwie bei der Aufführung einer Symphonie durch ein Orchester das Thema von einem Instrument zum anderen hinüber geht, und wenn einer der Mitwirkenden mit einem Teil zu Ende ist, so folgt ein anderer mit fehlerfreier Genauigkeit.

I. R. Shafarevich (Göttingen 1973)

We have already interpolated brief historical comments several times but it seems that some historical consideration is indispensable for a more understanding the "necessity" of developing the ideas and notions which have been considered in the present chapter. There are some excellent books on this subject: written with great charm, there is the classic of great Felix Klein, who was one of the proponents of the developement of Riemann's ideas:

1. F. Klein, *Vorlesungen über Entwicklung der Mathematik in XIX Jahrhundert*, Springer, 1923.

Klein was no historian and his work is very biased, but perhaps this is one of its major charms.

2. An excellent booklet by the great Jean Dieudonné, *Cours de géometrie algébrique I* (*Aperçu historique sur le développement de la géométrie algébrique*), Paris, 1974.

After long hesitation I decided to borrow six pages from a great master of our science, from Igor Shafarevich's *Basic Algebraic Geometry*. His beautifully written history of Abel's Theorem is a masterpiece and I thought that it would be a great pity to "improve" on it.

"1. *Elliptic Integrals.* They were an object of study as early as the XVII century, as an example of integrals that cannot be expressed in terms of elementary functions and lead to new transcendental functions.

"At the very end of the XVII century Jacob, and later Johann, Bernoulli came up against a new interesting property of these integrals. In their investigations they considered integrals expressing the arc length of certain curves. They found certain transformations of one curve into another that preserve the arc length of the curve, although the corresponding arcs cannot be superposed one to another. It is clear that analytically this leads to the transformation of one integral into another. In some cases there arise tranformations of an integral into itself. In the first half of the XVIII century many examples of such transformations were found by Fagnano.

"In general form the problem was raised and solved by Euler. He communicated his first results in this direction in a letter to Goldbach in 1752. His investigations on elliptic integrals were published from 1756 to 1781.

"Euler considers an arbitrary polynomial $f(x)$ of degree 4 and asks for the relations between x and y if

$$\frac{dx}{\sqrt{f(x)}} = \frac{dy}{\sqrt{f(y)}}. \tag{1}$$

He regards this as a differential equation connecting x and y. The required relation is the general integral of this equation. He finds this relation: it turns out to be algebraic of degree 2 both in x and in y. Its coefficients depend on the coefficients of the polynomial $f(x)$ and on one independent parameter c.

"Euler formulates this result in another form: the sum of the integrals $\int_0^\alpha \frac{dx}{\sqrt{f(x)}}$ and $\int_0^\beta \frac{dx}{\sqrt{f(x)}}$ is equal to a single integral:

$$\int_0^\alpha \frac{dx}{\sqrt{f(x)}} + \int_0^\beta \frac{dx}{\sqrt{f(x)}} = \int_0^\gamma \frac{dx}{\sqrt{f(x)}}, \tag{2}$$

and γ can be expressed rationally in terms of α and β. Euler also brings forward arguments why such a relation cannot hold if the degree of the polynomial $f(x)$ is greater than 4.

"For arbitrary elliptic integrals of the form $\int \frac{r(x)dx}{\sqrt{f(x)}}$ Euler proves a relation that generalizes (2):

$$\int_0^\alpha \frac{r(x)dx}{\sqrt{f(x)}} + \int_0^\beta \frac{r(x)dx}{\sqrt{f(x)}} - \int_0^\gamma \frac{r(x)dx}{\sqrt{f(x)}} = \int_0^\delta V(y)dy, \tag{3}$$

where γ is the same rational function of α and β as in (2), and where δ and V are also rational functions.

"The reason for the existence of an integral of the equation (1) and of all its special cases discovered by Fagnano and Bernoulli is the presence of a group law on an elliptic curve with the equation $s^2 = f(t)$ and the invariance of the everywhere regular differential form $s^{-1}dt$ under translations by elements of the group. The relations found by Euler that connect x and y in (1) can be written in the form

$$(x, \sqrt{f(x)}) \oplus (c, \sqrt{f(c)}) = (y, \sqrt{f(y)}),$$

where $\oplus$ denotes addition of points on the elliptic curve. Thus, these results contain at once the group law on an elliptic curve and the existence of an invariant differential form on this curve.

"The relation (2) is also an immediate consequence of the invariance of the form $\varphi = \dfrac{dx}{\sqrt{f(x)}}$. In it

$$(\gamma, \sqrt{f(\gamma)}) = (\alpha, \sqrt{f(\alpha)}) \oplus (\beta, \sqrt{f(\beta)})$$

and

$$\int_0^\alpha \varphi + \int_0^\beta \varphi = \int_0^\alpha \varphi + \int_\alpha^\gamma t_g^* \varphi = \int_0^\alpha \varphi + \int_\alpha^\gamma \varphi = \int_0^\gamma \varphi,$$

where t_g is the translation by $g = (\alpha, \sqrt{f(\alpha)})$. Observe that we write here the equation between integrals formally, without indicating the paths of integration. Essentially this is an equation "to within a constant of integration", that is, an equation between the corresponding differential forms. This is how Euler understood them.

"Finally, the meaning of the relation (3) will become clear later, in connection with Abel's theorem.

"2. *Elliptic Functions*. After Euler the theory of elliptic integrals was developed mainly by Legendre. His investigations, beginning in 1786, are collected in the three-volume "Traité des fonctions elliptiques et des intégrales Eulériennes".* In his preface to the first supplement published in 1828 Legendre writes: 'So far the geometers have hardly taken part in investigations of this kind. But no sooner had this book seen the light of day, no sooner had it become known to scholars abroad,

* Legendre called elliptic functions what we now call elliptic integrals. The contemporary terminology became accepted after Jacobi.

than I learned with astonishment as well as joy that two young geometers, Herr Jacobi in Königsberg and Herr Abel in Christiania, have achieved in their works substantial progress in the highest branches of this theory'.

"Abel's papers on the theory of elliptic functions appeared in 1827–1829. He starts out from the elliptic integral

$$\theta = \int_0^\lambda \frac{dx}{\sqrt{(1-c^2x^2)(1-e^2x^2)}},$$

where c and e are complex numbers; he regards it as a function of the upper limit and introduces the inverse function $\lambda(\theta)$ and the function $\Delta(\theta) = \sqrt{(1-c^2\lambda^2)(1-e^2\lambda^2)}$. From the properties of elliptic integrals known at that time [essentially, from Euler's relations (2) in 1] he deduces that the functions $\lambda(\theta \pm \theta')$ and $\Delta(\theta \pm \theta')$ can be simply expressed in the form of rational functions of $\lambda(\theta)$, $\lambda(\theta')$, $\Delta(\theta)$, and $\Delta(\theta')$. Abel shows that both these functions have in the complex domain two periods 2ω and $2\tilde{\omega}$:*

$$\omega = 2\int_0^{1/c} \frac{dx}{\sqrt{(1-c^2x^2)(1-e^2x^2)}},$$

$$\tilde{\omega} = 2\int_0^{1/e} \frac{dx}{\sqrt{(1-c^2x^2)(1-e^2x^2)}}.$$

He finds representations of the functions introduced by him in the form of infinite products extended over their zeros.

"As an immediate generalization of the problem with which Euler had been occupied, Abel raises the question: 'To list all the cases in which the differential equation

$$(1) \qquad \frac{dy}{\sqrt{(1-c_1^2y^2)(1-e_1^2y^2)}} = \pm a \frac{dx}{\sqrt{(1-c^2x^2)(1-e^2x^2)}}$$

can be satisfied by taking for y an algebraic function of x, rational or irrational'.

---

* As E. I. Slavutin has remarked, already Euler drew attention to the fact that the function $\int_0^y \frac{dx}{\sqrt{1-x^4}}$ has in the real domain a "modulus of multi-valuedness" similar to the inverse trigonometric functions.

"This problem became known as the transformation problem for elliptic functions. Abel showed that if the relation (1) can be satisfied by means of an algebraic function y, then it can also be done by means of a rational function. He showed that if $c_1 = c, e_1 = e$, then a must either be rational or a number of the form $\mu' + \sqrt{-\mu}$, where μ' and μ are rational numbers and $\mu > 0$. In the general case he showed that the periods ω_1 and $\tilde{\omega}_1$ of the integral of the left-hand side of (1), multiplied by a common factor, must be expressible in the form of an integral linear combination of the periods ω and $\tilde{\omega}$ of the integral of the right-hand side.

"Somewhat later than Abel, but independently, Jacobi also investigated the function inverse to the elliptic integral, proved that it has two independent periods, and obtained a number of results on the transformation problem. Transforming into series the expressions for elliptic functions that Abel had found in the form of products, Jacobi arrived at the concept of θ-functions* and found numerous applications for them, not only in the theory of elliptic functions but also in number theory and in mechanics.

"Finally, after Gauss's posthumous works were published, especially his diaries, it became clear that long before Abel and Jacobi he had mastered some of these ideas to a certain extent.

"The first part of Abel's results requires hardly any comment. The mapping $x = \lambda(\theta)$, $y = \Lambda(\theta)$ determines a uniformization of the elliptic curve $y^2 = (1-c^2x^2)(1-e^2x^2)$ by elliptic functions. Under the corresponding mapping $f\colon \mathbf{C}^1 \to X$ the regular differential form $\varphi = \dfrac{dx}{y}$ goes over into a regular differential form on $\mathbf{C}^1$ that is invariant under translations by the vectors of the lattice $2\omega\mathbf{Z}+2\tilde{\omega}\mathbf{Z}$. This form differs by a constant factor from $d\theta$, and we may assume that $d\theta = f^*\dfrac{dx}{y}$, that is, $\theta = \displaystyle\int \frac{dx}{y}$.

"3. *Abelian Integrals.* The transition to arbitrary algebraic curves proceeded entirely within the framework of analysis: Abel showed that

---

* θ-functions occurred first in 1826 in a book by Fourier on the theory of heat.

the basic properties of elliptic integrals can be generalized to integrals of arbitrary algebraic functions. These integrals later became known as Abelian integrals.

"In 1826 Abel wrote a paper which was the beginning of the general theory of algebraic curves. He considers in it an algebraic function y determined by two equations

$$(1) \qquad \chi(x, y) = 0,$$

and

$$(2) \qquad \theta(x, y) = 0,$$

where $\theta(x, y)$ is a polynomial that depends, apart from x and y, linearly on some parameters $a, a', \ldots$, the number of which is denoted by α. When these parameters are changed, some simultaneous solutions of (1) and (2) may not change. Let $(x_1, y_1), \ldots, (x_\mu, y_\mu)$ be variable solutions, and $f(x, y)$ an arbitrary rational function. Abel shows that

$$(3) \qquad \int_0^{x_1} f(x, y)\,dx + \ldots + \int_0^{x_\mu} f(x, y)\,dx = \int V(g)\,dg,$$

where $V(t)$ and $g(x, y)$ are rational functions depending also on the parameters $a, a', \ldots$ Abel interpreted this result by saying that the left-hand side of (3) is an elementary function.

"Using the freedom in choosing the parameters $a, a', \ldots$ Abel shows that the sum of any number of integrals $\int^{x_i} f(x, y)\,dx$ can be expressed in terms of $\mu - \alpha$ such integrals and a term of the same type as that on the right-hand side of (3). He establishes that the number $\mu - \alpha$ depends only on (1). For example, for $y^2 + p(x)$, where the polynomial p is of degree $2m$, we have $\mu - \alpha = m - 1$.

"Next Abel investigates for what functions f the right-hand side of (1) does not depend on the parameters $a, a', \ldots$ He expresses f in the form $\dfrac{f_1(x, y)}{f_2(x, y)\chi'_y}$ and he shows that $f_2 = 1$, and f_1 satisfies a number of restrictions as a consequence of which the number γ of linearly independent ones among the required functions f is finite. Abel shows that $\gamma \geqslant \mu - \alpha$ and that $\gamma = \mu - \alpha$, for example, if (using a much later terminology) the curve $\chi(x, y) = 0$ has no singular points.

"The discussion of the solutions $(x_1, y_1), \ldots, (x_\mu, y_\mu)$ of the system consisting of (1) and (2) leads us at once to the contemporary concept of equivalence of divisors. Namely: let X the curve with the equation (1) and D_λ the divisor cut out on it by the form θ_λ (in homogeneous coordinates), where λ is the system of parameters $a, a', \ldots$ By hypothesis, $D_\lambda = \overline{D}_\lambda + D_0$, where D_0 does not depend on λ. Therefore all the $\overline{D}_\lambda = (x_1, y_1) + \ldots + (x_\mu, y_\mu)$ are equivalent to each other. The problem with which Abel was concerned reduces to the investigation of the sum $\int_{\alpha_1}^{\beta_1} \varphi + \ldots + \int_{\alpha_\mu}^{\beta_\mu} \varphi$, where φ is a differential form on X, α_i and β_i are points on X, $(\alpha_1) + \ldots + (\alpha_\mu) \sim (\beta_1) + \ldots + (\beta_\mu)$. We give a sketch of a proof of Abel's theorem that is close in spirit to the original proof. We may assume that

$$(\alpha_1) + \ldots + (\alpha_\mu) - (\beta_1) - \ldots - (\beta_\mu) = (g), \quad g \in C(X),$$
$$(\alpha_1) + \ldots + (\alpha_\mu) = (g)_0, \quad (\beta_1) + \ldots + (\beta_\mu) = (g)_\infty.$$

We consider a morphism $g: X \to \boldsymbol{P}^1$ and the corresponding extension $C(X)/C(g)$. For simplicity we assume that this is a Galois extension (the general case easily reduces to this), and we denote its Galois group by G. The automorphisms $\sigma \in G$ act on the curve X and the field $C(X)$ and carry the points $\alpha_1, \ldots, \alpha_\mu$ into each other, because $\{\alpha_1, \ldots, \alpha_\mu\} = g^{-1}(0)$. Therefore $\{\alpha_1, \ldots, \alpha_\mu\} = \{\sigma\alpha: \sigma \in G\}$, where α is one of the points α_i. Similarly $\{\beta_1, \ldots, \beta_\mu\} = \{\sigma\beta: \sigma \in G\}$. Representing φ in the form $u\,dg$, we see that

$$(4) \qquad \sum_{i=1}^{\mu} \int_{\alpha_i}^{\beta_i} \varphi = \sum_{\sigma \in G} \int_{\sigma\alpha}^{\sigma\beta} u\,dg = \int_{\alpha}^{\beta} \Big(\sum_{\sigma \in G} \sigma u\Big)\,dg.$$

The function $v = \sum_{\sigma \in G} u$ is contained in $C(g)$, and Abel's theorem follows from this.

"We see that every sum of integrals $\sum_i \int_0^{x_i} f(x, y)\,dx$ can be expressed as a sum of l integrals $\sum_{j=1}^{l} \int_0^{x'_j} f(x, y)\,dx + \int V(g)\,dg$ if the equivalence

$$(5) \qquad \sum_i ((\alpha_i) - o) \sim \sum_{j=1}^{l} ((\alpha'_j) - o)$$

holds, where $\alpha_i = (x_i, y_i)$, $\alpha'_j = (x'_j, y'_j)$, and o is the point with $x = 0$.

From the Riemann–Roch theorem it follows at once that the equivalence (5) holds (for arbitrary α_i and certain α'_j corresponding to them), with $l = g$. Thus, the constant $\mu - \alpha$ introduced by Abel is the same as the genus.

"If $\varphi \in \Omega^1[X]$, then also $v\,dg \in \Omega^1[P^1]$, where $v = \sum_{\sigma \in G} \sigma u$ in (4). Since $\Omega^1[P^1] = 0$, in this case the term on the right-hand side of (3) disappears. Hence it follows that $\gamma \geqslant g$. In some cases arising naturally the two numbers coincide.

"We see that this paper of Abel's contains the concept of the genus of an algebraic curve and the equivalence of divisors and gives a criterion for equivalence in terms of integrals. In the last relation it leads to the theory of Jacobian varieties of algebraic curves."

We now arrive at perhaps the biggest revolution in mathematical thought: Bernhard Riemann's inaugural dissertation on December 26, 1851, in Göttingen.

A hundred years later a "Conference on Riemann Surfaces" was held in Princeton in commemoration of this event. On that occasion a great master of complex analysis, Lars V. Ahlfors, held an impressive lecture: *Development of the Theory of Conformal Mapping and Riemann Surfaces Through a Century*. We reproduce here a part of it:

"*Geometric Function Theory*. Riemann's paper marks the birth of geometric function theory. At the time of its appearance Cauchy had already laid the foundation of analytic function theory in the modern sense, but its use was not widespread. It is clear that complex integration had introduced a certain amount of geometric content in analysis, but it would be wrong to say that Riemann's ideas were in any way anticipated. Riemann was the first to recognize the fundamental connection between conformal mapping and complex function theory: to Gauss, conformal mapping had definitely been a problem in differential geometry.

"The most astonishing feature in Riemann's paper is the breath-taking generality with which he attacks the problem of conformal mapping. He has no thought of illustrating his methods by simple examples which to lesser mathematicians would have seemed such an excellent preparation and undoubtedly would have helped his paper to much earlier recognition. On the contrary, Riemann's writings are full of almost cryptic messages to the future. For instance, Riemann's mapping theorem is ultimately formulated in terms which would defy any attempt of proof, even with modern methods.

"*Riemann Surfaces.* Among the creative ideas in Riemann's thesis none is so simple and at the same time so profound as his introduction of multiply covered regions, or Riemann surfaces. The reader is led to believe that this is a commonplace convention, but there is no record of anyone having used a similar device before. As used by Riemann it is a skillful fusion of two distinct and equally important ideas: 1) A purely topological notion of covering surface, necessary to clarify the concept of mapping in the case of multiple correspondence; 2) An abstract conception of the space of the variable, with a local structure defined by a uniformizing parameter. The latter aspect comes to the foreground in the treatment of branch points.

"From a modern point of view the introduction of Riemann surfaces foreshadows the use of arbitrary topological spaces, spaces with a structure, and covering spaces.

"*Existence and Uniqueness Theorems.* There is a characteristic feature of Riemann's thesis which should not be underestimated. The whole paper is built around existence and uniqueness theorems. To us, this seems the most natural thing in the world, for this is what we expect from a paper which introduces a new theory. But we must realize that Riemann is, to say the least, one of the earliest and strongest proponents of this point of view. Again and again, explicitly and between the lines, he emphasizes that a function can be defined by its singularities. This approach calls for existence and uniqueness theorems, in contrast to the classical conception of a function as a closed analytic expression. There is no doubt that Riemann's point of view has had a decisive influence on modern mathematics.

"*Potential Theory.* Next to the geometric interpretation, the leading mathematical idea in Riemann's paper is the importance attached to Laplace's equation. He virtually puts equality signs between two-dimensional potential theory and complex function theory. Riemann's aim was to make complex function theory a powerful tool in real analysis, especially in the theory of partial differential equations and thereby in mathematical physics. It must be remembered that Riemann was in no sense confined to a mathematical hothouse atmosphere; his broad mind was prone to accept all the inspiration he could gather from his unorthodox, but suggestive, conception of contemporary physics.

"*Dirichlet's Principle.* Riemann's proof of the fundamental existence theorems was based on an uncritical use of the Dirichlet principle. It is perhaps wrong to call Riemann uncritical, for he made definite attempts to exclude a degenerating extremal function. In any case, if he missed the correct proof, he made up for it by giving a very general formulation. In modern language, his approach, which is clearer in the paper on Abelian integrals, is the following:

"Given a closed differential α with given periods, singularities and boundary values, he assumes the existence of a closed differential β such that $\alpha+\beta^*$ (β^* denotes the conjugate differential) has a finite Dirichlet norm. Then he determines an exact differential ω_1, with zero boundary values, whose norm distance form $\alpha+\beta^*$ is a minimum. But this is equivalent to an orthogonal decomposition

$$\alpha+\beta^* = \omega_1+\omega_2^* \qquad (\omega_1 \text{ exact}, \omega_2 \text{ closed})$$

from which it follows that

$$\alpha-\omega_1 = \omega_2^*-\beta^*$$

is simultaneously closed and co-closed, that is to say harmonic. Hence the existence theorem: there exists a harmonic differential with given boundary values, periods, and singularities.

"The easiest way to make the reasoning exact is to complete the differentials to a Hilbert space. Closed and exact differentials can be defined by orthogonality, and in the final step one needs a lemma of Hermann Weyl for which there is now a very short proof. Riemann, who did not have these tools, was nevertheless able to choose this beautiful approach which unifies the problems of boundary values, periods, and singularities.

"*Schwarz–Neumann.* Riemann's failure to provide a rigorous proof for the Dirichlet principle was beneficial in causing a flurry of attempts to prove the main existence theorems by other methods. The first to be successful was H. A. Schwarz who devised the alternating method. Minor improvements were contributed by C. Neumann who is most notable as a popularizer of Riemann's ideas. The alternating method proved to be sufficient to dispose of the existence problems on closed surfaces.

"The method of Schwarz is a linear method, and in principle it amounts to solving a linear integral equation by iteration. The difficulties in adapting the method to the existence theorems are of a practical nature, but they are considerable. The advantages are that the method is successful, and constructive, but in simplicity and elegance it does not compare, even remotely, with Riemann's method.

"*Poincaré–Klein.* In the next generation the leaders were Poincaré and F. Klein. Their important innovation is the introduction of the problem of uniformization of algebraic and general analytic curves. The study of Riemann surfaces in this light leads to automorphic functions and the use of non-euclidean geometry. The method is extremely beautiful; as H. Weyl puts it, the nature of Riemann surfaces is reflected in the non-euclidean crystal. It has also the advantage of leading to very explicit representations by way of Poincaré's theta-series and their generalizations.

"The disadvantage is that the existence proofs are quite difficult. For compact surfaces Poincaré produces a correct proof based on a continuity method which is simple in principle, but technically very involved. It is interesting to note that Poincaré concentrates his efforts on proving the general uniformization theorem. In spite of its generality this theorem would not even include the Riemann mapping theorem. Poincaré has been extremely influential in developing methods which ultimately led to proofs of the mapping and uniformization theorems, but he did not himself produce a complete proof until 1908, having been preceded by Osgood who proved the Riemann mapping theorem in 1900, and Koebe who proved the uniformization theorem in 1907.

"Osgood's proof is very remarkable, because it is so clear and concise and does not leave any room for doubt. It is based on an idea of Poincaré, but Osgood deserves full credit for making the idea work. The proof uses the modular function, and is thus not elementary.

"*Koebe.* The crowning glory was achieved by Koebe when he proved the general uniformization theorem. This is the theorem which asserts that every simply connected Riemann surface is conformally equivalent with the sphere, the disk, or the plane. It immediately takes care of the uniformization of the most general surface, for it is sufficient to map the universal covering surface. As a tool he uses his famous 'Verzerrungssatz'.

"The stage was now set for deeper investigations of the problem of conformal mapping. The standard theorems which concern the canonical mappings of multiply connected regions are from this time. Koebe was an undisputed leader, and Leipzig a center for conformal mapping.

"Looking back one cannot help being impressed by Koebe's life work. His methods were completely different from those of his predecessors, and when the initial difficulties were conquered he did not hesitate to attack new problems of ever increasing complexity....

"*Idee der Riemannschen Fläche.* In the classical literature no clear definition of a Riemann surface is ever given. Primarily, the classical authors thought in terms of multiply covered regions with branch points, but applications to surfaces in space are not uncommon. It is of course true that F. Klein had a general conception of a Riemann surface which is quite close to modern ideas, but his conception is still partially based on geometric evidence.

"H. Weyl's book "*Die Idee der Riemannschen Fläche*", first published in 1913, was the real eye-opener. Pursuing the ideas of Klein it brings, for the first time, a rigorous and general definition of a Riemann surface, and it marks the death of the glue-and-scissors period. The pioneer qualities of this book should not be forgotten. It is a forerunner which has served as a model for the axiomatization of many mathematical topics. For his definition of the abstract Riemann surface Weyl uses the power series approach. The equivalent definition of Radó is perhaps a little smoother, and Radó added the important recognition that the separability is a consequence of the conformal structure.

"Weyl was able to base the existence theorems on the Dirichlet principle which had been salvaged by Hilbert. The book is a reminder of how Riemann's original idea still provides the easiest access to the existence theorems. It has exerted a strong implicit influence by its change of emphasis which has led to a strengthening of the ties between the theory of Riemann surfaces and differential geometry...

"*Topological Aspects.* The abstract approach to Riemann surfaces, with all its advantages, tends to neglect the covering surface aspect.

With the advances made in topology the notions of fundamental group and universal covering space had become thoroughly familiar, and accordingly the case of smooth covering surfaces was well covered in Weyl's book. Stoilow filled the gap with a study of covering surfaces with branch-points. Whyburn completed the work of Stoilow and based it more firmly on pure topology.

"*Higher Dimensions.* Finally, the important question of generalizing Riemann's work to several dimensions has made enormous strides in the last decades. The greenest laurels belong to Hodge for his pioneer research on harmonic integrals on Riemannian manifolds. Through his initiative, and the parallel work of de Rham in topology, it was discovered that the problems of Riemann have a significant counterpart on more-dimensional closed manifolds with a Riemannian metric. The existence theorems presented initial difficulties, but it was finally found that Hilbert's integral equation method as well as Riemann's own method of orthogonal projection can both be made successful (Weyl, de Rham, Kodaira). For the purpose of pursuing the function-theoretic analogy Kählerian manifolds have the most desirable properties, and for such manifolds the problem of singularities has been successfully attacked (A. Weil, Kodaira)."

With this strong accent let us finish this section. A theory of Kähler manifolds will be developed in Part III.

21. APPENDICES. EXERCISES (PROOFS OF THEOREMS OF RUNGE, FLORACK, KOEBE, AND HURWITZ. TRIANGLE GROUPS. ELLIPTIC INTEGRALS AND TRANSCENDENTAL NUMBERS)

Before we give the promised proof of Runge's Theorem, let us provide a simple but important criterion for the density of a subspace in a locally convex space.

PROPOSITION XVI.21.1 (Banach). Let E be a locally convex vector space and let $A \subset B \subset E$ be subspaces. If $\mu|B = 0$ for each linear continuous functional $\mu\colon E \to \boldsymbol{C}$ such that $\mu|A = 0$, then A is dense in B.

PROOF (indirect) follows from the Hahn–Banach Theorem: Assume that A is not dense in B; then there exists such a $b_0 \in B$ that $b_0 \notin \overline{A}$. Let

$E_0 := \overline{A} \oplus C b_0$. We now define a linear functional

$$\mu_0 \colon E_0 \to \boldsymbol{C}, \quad \mu_0(a+\lambda b_0) := \lambda \quad \text{for } a \in \overline{A},\ \lambda \in \boldsymbol{C}.$$

We check that μ_0 is continuous; hence by the Hahn–Banach Theorem it can be extended to a continuous linear functional $\mu\colon E \to \boldsymbol{C}$. But plainly

$$\mu | A = 0 \text{ and } \mu | B \neq 0, \text{ a contradiction. } \square$$

THEOREM XVI.21.2 (Runge). Let Ω be an open subset of $\boldsymbol{C}$ and K a compact subset of Ω. Then the following conditions are equivalent:

(a) The open set $\Omega - K \equiv \Omega \cap \complement K$ has no component relatively compact in Ω.

(b) Every holomorphic function in a neighbourhood of K can be approximated uniformly on K by elements of $A(\Omega)$.

(c) For every $z \in \Omega - K$ there is an $f \in A(\Omega)$ such that

(1) $\quad |f(z)| > \sup |f(K)|.$

PROOF. (a) ⇒ (b). By Theorem XVI.21.1 it is sufficient to prove that every (Radon) measure μ on K which vanishes on $A(\Omega)$ vanishes also at each f which is holomorphic in a neighbourhood of K. Let

$$h(\zeta) := \int (z-\zeta)^{-1} d\mu(z), \quad \zeta \in \complement K.$$

Then $h \in A(\complement K)$ and for $\zeta \in \complement \Omega$ we have

$$h^{(k)}(\zeta) = k! \int (z-\zeta)^{-k-1} d\mu(z) = P \quad \text{for every } k,$$

since the function $z \to (z-\zeta)^{-k-1}$ is holomorphic in Ω if $\zeta \in \complement \Omega$. Therefore, $h = 0$ in every component of $\complement K$ which intersects $\complement \Omega$. But

$$\int z^n d\mu(z) = 0 \quad \text{for every } n$$

and since $(z-\zeta)^{-1}$ can be expanded in a power series in z which converges uniformly on K if $|\zeta| > \sup |K|$, we also have $h = 0$ in the unbounded component of $\complement K$. But now (a) ⇒ ($h = 0$ in $\complement K$).

Let U be a neighbourhood of K where f is holomorphic; choose $\varphi \in C_0^\infty(U)$ with $\varphi = 1$ on K. Then

$$f(z) = \varphi(z) f(z) = \frac{1}{2\pi i} \iint \frac{f(\zeta)}{\zeta - z} \frac{\partial \varphi}{\partial \bar{\zeta}} d\zeta \wedge d\bar{\zeta}, \quad z \in K.$$

Since $\partial\varphi/\partial\bar{\zeta} = 0$ on U, by the Fubini Theorem we have

$$\int f(z)d\mu(z) = -\frac{1}{2\pi i}\iint f(\zeta)\frac{\partial\varphi}{\partial\bar{\zeta}}h(\zeta)d\zeta \wedge d\bar{\zeta} = 0,$$

and (a) $\Rightarrow$ (b) is proved.

PROOF (INDIRECT) OF (c) $\Rightarrow$ (a) and (b) $\Rightarrow$ (a). Assume that $\Omega - K$ has a component $\mathcal{O}$ such that $\bar{\mathcal{O}}$ is compact and $\subset \Omega$. Then $\partial\mathcal{O} \subset K$ and by virtue of the maximum principle we have

$$\text{(2)} \qquad \sup|f(\bar{\mathcal{O}})| \leqslant \sup|f(K)|, \quad f \in A(\Omega),$$

which contradicts (c).

Now assume (a) is satisfied. For every $f \in A(U)$ (U is a neighbourhood of K) we can choose $f_n \in A(\Omega)$ such that $f_n \to f$ uniformly on K.

By (2) applied to $(f_n - f_m)$ we see that (f_n) converges uniformly on $\bar{\mathcal{O}}$ to a limit f_0. But $f_0 = f$ on $\partial\mathcal{O}$ and $f_0 \in A(\mathcal{O})$ and $f_0 \in C(\bar{\mathcal{O}})$. If we take $f(z) = (z-\zeta)^{-1}$ for $\zeta \in \mathcal{O}$, then $(z-\zeta)f_0(z) = 1$ on $\partial\mathcal{O}$, hence (by the maximum principle) $(z-\zeta)f_0(z) \equiv 1$ in $\mathcal{O}$, which gives a contradiction when $z = \zeta$. ⊔

The proof of (a) $\Rightarrow$ (c) is left to the reader as an exercise.

In this way the equivalence of (a), (b), (c) is established.

The first and the second Cousin problems were solved for non-compact Riemann surfaces by Herta Florack in 1948 by means of methods developped by Behnke and Stein.

Before we give a proof of the Mittag-Leffler Theorem on open Riemann surfaces, let us prove

THEOREM XVI.21.3. Let X be an open Riemann surface; then for every $\omega \in \Lambda^{0,1}(X)$ there exists an $f \in C^\infty(X)$ such that $d''f = \omega$. Hence

$$\text{(3)} \qquad \Lambda^{0,1}(X) = d''C^\infty(X).$$

PROOF. For every relatively compact open subset $Y \subset X$ there exists a $g \in C^\infty(Y)$ such that $d''g = \omega|Y$. Let us exhaust X by Runge domains $Y_j \nearrow X$

$$Y_0 \Subset Y_1 \Subset \dots \qquad (\text{“canonical exhaustion of } X\text{”}).$$

By induction (with respect to n) we construct an $f_n \in C^\infty(Y_n)$ such that

(a) $d''f_n = \omega|Y_n$,

(b) $\sup|(f_{n+1} - f_n)(Y_{n-1})| =: \|f_{n+1} - f_n\|_{Y_{n-1}} \leqslant \dfrac{1}{2^n}$.

Let f_0 be any solution of $d''f_0 = \omega|Y_0$. If $f_0, \ldots, f_n$ are already constructed, then there exists a $g_{n+1} \in C^\infty(Y_{n+1})$ such that $d''g_{n+1} = \omega|Y_{n+1}$.

But $d''g_{n+1} = d''f_n$ on Y_n and hence $(g_{n+1}-f_n) \in A(Y_n)$. By virtue of the Behnke–Stein approximation theorem (XVI.13.2) there exists an $h \in A(Y_{n+1})$ such that

$$\|(g_{n+1}-f_n)-h\|_{Y_{n-1}} \leqslant 2^{-n}.$$

Set $f_{n+1} := g_{n+1}-h$. Then

$$d''f_{n+1} = d''g_{n+1} = \omega|Y_{n+1}$$

and

$$\|f_{n+1}-f_n\|_{Y_{n-1}} \leqslant 2^{-n}.$$

Therefore, (f_n) converges to an $f \in C^\infty(Y)$ with $d''f = \omega$. □

Remark. We shall prove in Part III that for the sheaf $\mathscr{A} = \big(A(\mathcal{O})\big)$ of holomorphic functions on X one can form the cohomology space

$$H^1(X, A) \cong \Lambda^{0,1}(X)/d''A(X);$$

hence we obtain the important

COROLLARY XVI.21.4. For any non-compact Riemann surface X the cohomology space is

(4) $\qquad H^1(X, A) = 0.$

We have seen that from Theorem XVI.21.3 follows

The Florack–Behnke–Stein Theorem:

On an open Riemann surface every Mittag-Leffler (i.e., first Cousin) datum is solvable.

The Weierstrass problem for an open Riemann surface will be solved by an elegant method of Otto Forster in two steps:

(i) every divisor D has a weak solution,

(ii) for a non-compact surface X every $D \in \mathrm{Div}(X)$ is a divisor of a meromorphic function $f \in \mathscr{M}(X)$.

DEFINITION. Let X be a Riemann surface and $D \in \mathrm{Div}(X)$. Let

$$X_D = \{x \in X \colon D(x) \geqslant 0\}.$$

A *weak solution* of D is an $f \in C^\infty(X_D)$ such that for each $a \in X$ there exists a chart (U, z) with $z(a) = 0$ and $\psi \in C^\infty(U)$ with $\psi(a) \neq 0$ such that

(5) $\quad f = \psi z^k$ in $U \cap X_D$, where $k = D(a)$.

Remark. Plainly

$$(f \text{ is a weak solution of } D) \Leftrightarrow (f \in A(X_D)).$$

If f and g are weak solutions of D, then there exists a $\varphi \in C^\infty(X)$ without zeros such that $f = \varphi g$.

Exercise. Prove the following

LEMMA XVI.21.5. Let X be a Riemann surface, c: $[0, 1] \to X$ a path and let U be a relatively compact neighbourhood of $c([0, 1])$. Then there exists a weak solution f of the divisor ∂c such that

$$(6) \qquad f|(X-U) = 1 \quad \text{and} \quad \int_c \omega = \frac{1}{2\pi i}\int_X \frac{df}{f} \wedge \omega$$

for every closed form $\omega \in A^{(1)}(X, C)$.

PROOF (sketch). Consider first the case where (U, z) is a coordinate neighbourhood, such that $z(U) = \boldsymbol{D}$. We can thus identify U with the unit disk $\boldsymbol{D}$. Let $a := c(0)$, $b := c(1)$. If $c([0, 1]) \subset \{z \in \boldsymbol{C}\colon |z| < r\}$, then $\log \dfrac{z-b}{z-a}$ has a branch in the annulus $\{z\colon r < |z| < 1\}$. Take $\psi \in C^\infty(U)$ such that $\psi|\{|z| < r\} = 1$ and $\psi|\{|z| \geqslant r'\} = 0$, $r < r' < 1$

$$f_0(z) := \begin{cases} \exp \psi \log \dfrac{z-b}{z-a} & \text{for } r < |z| < 1, \\[2ex] \dfrac{z-b}{z-a} & \text{for } |z| \leqslant r. \end{cases}$$

Extend f_0 by 1 onto $X - U$ to an $f \in C^\infty(X - \{a\})$. Check that f satisfies (5) and is a weak solution of ∂c.

Reduce the general case to the local case by a subdivision $0 = t_0 < t_1 < \ldots < t_n = 1$ of $[0, 1]$. $\square$

PROOF OF (i). Let $(K_j)_1^\infty$ be a sequence of compact subsets of X such that

(a) $K_j = \hat{K}_j$, $j = 1, 2, \ldots$, ($\hat{K}$ is the envelope of holomorphy of K),

(b) $K_j \subset \mathring{K}_{j+1}$ ($\equiv \operatorname{int} K_{j+1}$) for all $j \geqslant 1$,

(c) $\bigcup K_j = X$.

Let $a_0 \in X - K_j$ and let $B_0 \in \operatorname{Div}(X)$ with $B_0(a_0) = 1$ and $B_0(x) = 0$ for $x \neq a_0$. Since $K_j = \hat{K}_j$, a_0 is in a non-relatively compact connected component U of $X-K_j$. Then there exists a point $a_1 \in U-K_{j+1}$ and a path c_0 in U from a_1 to a_0. By Lemma XVI.21.5 there exists a weak solution φ_0 of the divisor ∂c_0 such that $\varphi_0|K_{j+1} = 1$. Repeating this construction, we obtain a sequence $a_l \in X-K_{j+l}$, $l \in N$, and paths c_l in $X-K_{j+l}$ with endpoints a_{l+1}, a_l and weak solutions φ_l of the divisors ∂c_l such that $\varphi_l|K_{j+l} = 1$. Plainly, $\partial c_l = B_l - B_{l+1}$, where $B_l \in \mathrm{Div}(X)$ such that $B_l(a_l) = 1$ and $B_l(x) = 0$ for $x \neq a_l$. Therefore, $\varphi_0 \dots \varphi_n$ is a weak solution of the divisor $B_0 - B_{n+1}$. The infinite product

$$\varphi := \prod_{l=0}^{\infty} \varphi_l$$

converges since on each compact subset of X only a finite number of factors $\neq 1$. Hence φ is a weak solution of B_0.

Let D be an arbitrary divisor on X. For $n \in N$ we set

$$D_n(x) := \begin{cases} D(x) & \text{for } x \in K_{n+1}-K_n, \\ 0 & \text{for } x \notin K_{n+1}-K_n, \end{cases}$$

(we set $K_0 := \emptyset$). Then

$$D = \sum_{n=0}^{\infty} D_n.$$

Since $D_n(x) \neq 0$ only for a finite number of x, there exists a weak solution ψ_n of D_n such that $\psi_n|K_n = 1$.
Therefore,

$$\psi := \prod_{n=0}^{\infty} \psi_n$$

is a weak solution of the divisor D. □

We can now give a proof of a solution of the Weierstrass (second Cousin) problem for an open Riemann surface.

THEOREM XVI.21.6 (Florack). For a non-compact Riemann surface X every $D \in \mathrm{Div}(X)$ is the divisor of a meromorphic function $\mathscr{M}^*(X)$

$$D = (f).$$

PROOF. We know that this problem is locally solvable: there exists an open covering $(\mathcal{O}_j)_{j\in J}$ of X and meromorphic functions $f_j \in \mathscr{M}^*(\mathcal{O}_j)$ such that $D|\mathcal{O}_j = (f_j)$. We can assume that all $\mathcal{O}_j$ are simply connected.

Since, over $\mathcal{O}_i \cap \mathcal{O}_j, f_i$ and f_j have the same poles and zeros, then $f_i/f_j \in A^*(\mathcal{O}_i \cap \mathcal{O}_j)$ for all $i, j \in J$. Let ψ be a weak solution of D. On $\mathcal{O}_j$ we have $\psi = \psi_j f_j$ where $\psi_j \in C^\infty(\mathcal{O}_j)$ and ψ_j has no zeros. Since $\mathcal{O}_j$ is simply connected, there exists a $\varphi_j \in C^\infty(\mathcal{O}_j)$ such that $\psi_j = \exp(\varphi_j)$, i.e., $\psi = \exp(\varphi_j) f_j$ on $\mathcal{O}_j$. We therefore have

$$(6) \qquad \exp(\varphi_j - \varphi_i) = \frac{f_i}{f_j} \in A^*(\mathcal{O}_i \cap \mathcal{O}_j).$$

Hence $\varphi_{ij} := \varphi_j - \varphi_i \in A(\mathcal{O}_i \cap \mathcal{O}_j)$. Plainly, $\varphi_{ij} + \varphi_{jk} = \varphi_{ik}$ on $\mathcal{O}_i \cap \mathcal{O}_j \cap \mathcal{O}_k$; hence (φ_{ij}) is a holomorphic 1-cocycle. But by virtue of (4) there exists a holomorphic function $g_j \in A(\mathcal{O}_j)$ such that

$$\varphi_{ij} = \varphi_j - \varphi_i = g_j - g_i \quad \text{on } \mathcal{O}_i \cap \mathcal{O}_j \quad \text{for } i, j \in J.$$

From (6) it follows that

$$\exp(\varphi_j - \varphi_i) = f_i/f_j,$$

i.e.,

$$\exp(g_j) f_j = \exp(g_i) f_i \quad \text{on } \mathcal{O}_i \cap \mathcal{O}_j.$$

Therefore, there exists a meromorphic function $f \in \mathcal{M}^*(X)$ such that $f = \exp(g_j) f_j$ on $\mathcal{O}_j$ for all $j \in J$. Since f and f_j have the same divisor on $\mathcal{O}_j$, we have $(f) = D$. □

COROLLARY XVI.21.7. On every open Riemann surface X there exists an Abelian differential of the first kind $\omega \in A^1(X)$ without zeros.

PROOF. Let const $\neq g \in \mathcal{M}(X)$ and let $f \in \mathcal{M}^*(K)$ with $(f) = -(dg)$. Then $\omega := f dg$ is a holomorphic 1-form without zeros.

Exercise. Prove the following

PROPOSITION XVI.21.8. Let X be a non-compact Riemann surface and let $(a_n), n \in N$, be a discrete sequence of different points. Then for given $c_n \in C, n \in N$, there exists an $f \in A(X)$ such that $f(a_n) = c_n$ for all $n \in N$.

The Koebe–Riemann Mapping Theorem. We have proved the mapping theorem in the parabolic case by means of the elementary differential of the second kind. But the existence of such a differential was proved only for compact Riemann surfaces. For a non-compact surface one can reduce the proof of the compact case by means of the ingenious construction (due to Schottky) of a double of a surface with a boundary. We shall provide this missing link in Part III. Here we shall outline another proof of the mapping theorem for open simply connected surfaces by the oil speck method following closely the beautiful exposition of O. Forster.

We know that on a simply connected surface every Abelian differential of the first kind ω has a holomorphic primitive $f \in A(X)$, i.e.,

$$A^1(X) = d''A(X). \tag{7}$$

Therefore, if (following Forster) we introduce for any Riemann surface X a holomorphic de Rham cohomology space

$$\mathrm{Rh}_A^1(X) := A^1(X)/dA(X), \tag{8}$$

we have the following relation between the homology and cohomology of a Riemann surface.

LEMMA XVI.21.9. Let X be a Riemann surface; then

$$(\pi_1(X) = 0) \Rightarrow (\mathrm{Rh}_A^1(X) = 0). \square$$

Remark. The Koebe–Riemann Theorem gives the opposite implication. Indeed, we shall prove this theorem under an apparently weaker assumption: $\mathrm{Rh}_A^1(X) = 0$. But since the three standard Riemann surfaces $\boldsymbol{P}(\boldsymbol{C})$, $\boldsymbol{C}$, $\boldsymbol{D}$ are simply connected and $\pi_1(\cdot)$ is an invariant of homeomorphisms, we obtain the implication

$$(\mathrm{Rh}_A^1(X) = 0) \Rightarrow (\pi_1(X) = 0): \tag{9}$$

The reader will have observed that the proof of the existence of branches of log and the square root assumed only that $\mathrm{Rh}_A^1(X) = 0$:

LEMMA XVI.21.10. Let $\mathrm{Rh}_A^1(X) = 0$. Then, for every holomorphic function f: $X \to \boldsymbol{C}^*$, there exist $g, h \in A(X)$ with

$$e^g = f, \quad h^2 = f. \tag{10}$$

The proof of the mapping theorem in the hyperbolic case also used only the vanishing of the space $\mathrm{Rh}_A^1(X)$.

THEOREM XVI.21.11. Let X be a non-compact Riemann surface and $Y \Subset X$ a domain such that $\mathrm{Rh}_A^1(X) = 0$ and ∂Y is regular (with respect to the Dirichlet problem). Then there exists a biholomorphic map of Y onto the unit disk $\boldsymbol{D}$.

As a special case of the de Rham Theorem we have

LEMMA XVI.21.12 (de Rham). Let X be a Riemann surface; then (A closed form $\omega \in A^{(1)}(X)$ has a primitive $f \in A(X)$) $\Leftrightarrow$ (all periods of ω vanish, i.e., $\int_\alpha \omega \equiv 0$, $\alpha \in Z_1(X)$).

The Behnke–Stein Theorem and the preceding results imply

LEMMA XVI.21.13. Let X be a non-compact Riemann surface such that $\mathrm{Rh}^1_A(X) = 0$ and let $Y \subset X$ be a Runge domain. Then $\mathrm{Rh}^1_A(Y) = 0$.

PROOF. Let $\omega \in A^1(Y)$; then we choose (cf. Corollary XVI.21.7) an $\omega \in A^1(Y)$ without zeros. Then $\omega = f\omega_0$ with $f \in A(Y)$. By virtue of the Behnke–Stein Theorem there exists a sequence $f_n \in A(X)$, $n \in N$, which converges compactly on Y to f. For any closed curve α in Y we have

$$(11) \qquad \int_\alpha f_n\omega_0 \underset{n\to\infty}{\to} \int_\alpha f\omega_0 = \int_\alpha \omega .$$

But since $\mathrm{Rh}^1_A(X) = 0$, we have (by the Stokes Theorem)

$$0 = \int_\alpha f_n\omega_0 \quad \text{and hence} \quad \int_\alpha \omega = 0 \quad \text{by (11)}.$$

Therefore, by Lemma XVI.21.12, there exists a $g \in A(Y)$ such that $\omega = d''g$. □

Denote by D_R the disk $\{z \in C\colon |z| < R\}$, $R \in]0, \infty]$.

Exercise. The reader may prove the following

LEMMA XVI.21.14. Let $R \in]0, \infty]$ and let $Y \neq D_R$ be a subdomain of D_R such that $\mathrm{Rh}^1_A(Y) = 0$. Then there exist an $r < R$ and a holomorphic mapping

$$f\colon Y \to D_r \quad \text{such that } f(0) = 0, f'(0) = 1 .$$

PROOF. The proof was already given in Section 8 under the assumption that $\pi_1(Y) = 0$, but we used only $\mathrm{Rh}^1_A(Y) = 0$. The inequality $r < R$ is obtained from the simple fact that

$$(12) \qquad (f\colon D_{r_1} \to D_{r_2} \text{ is holomorphic}) \Rightarrow \left(|f'(0)| \leqslant \frac{r_2}{r_1}\right),$$

which is a simple consequence of the Cauchy formula.

In the course of proof of Theorem XVI.8.20 we proved the following

LEMMA XVI.21.15. The family F of all injective holomorphic functions $f\colon \boldsymbol{D} \to \boldsymbol{C}$ such that

$$f(0) = 0, \quad f'(0) = 1,$$

is compact in $A(\boldsymbol{D})$.

We can now prove the following version of the

KOEBE–RIEMANN MAPPING THEOREM XVI.21.16. Let X be a Riemann surface such that $\mathrm{Rh}_A^1(X) = 0$. Then

1° if X is compact, then $X \cong \boldsymbol{P}(\boldsymbol{C})$;

2° if X is non-compact, then X can be biholomorphically mapped onto $\boldsymbol{D}_R$ with $R \in]0, \infty]$, i.e.,

(a) either $X \cong$ disk $\boldsymbol{D}_r$, $r < \infty$ or

(b) $X \cong \boldsymbol{C}$ $(\equiv \boldsymbol{D}_\infty)$.

PROOF. 1° Since X is compact, every holomorphic function on X is a constant and hence

$$dA(X) = 0,$$

whereby

$$\big(\mathrm{Rh}_A^1(X) = 0\big) \Rightarrow \big(A^1(X) = 0\big).$$

Hence X has genus $g = 0$. By virtue of Theorem XVI.20.19, however, $X \cong \boldsymbol{P}(\boldsymbol{C})$.

2° We have an exhaustion of X by Runge domains $Y_0 \Subset Y_1 \Subset \dots$ such that ∂Y_j, $j = 0, 1, \dots$ are regular. By virtue of Lemmas XVI.21.13 and XVI.21.14 every Y_j can be mapped biholomorphically onto $\boldsymbol{D}$. Let $a \in Y_0$ and let (U, z) be a coordinate neighbourhood of a. Then there exist real numbers $r_n > 0$ and biholomorphic maps

$$(13) \qquad f_n\colon Y_n \to \boldsymbol{D}_{r_n} \quad \text{such that } f_n(a) = 0, \ \frac{df_n}{dz}(a) = 1.$$

We have

$$(14) \qquad r_n \leqslant r_{n+1} \quad \text{for all } n = 0, 1, \dots$$

Indeed, for the holomorphic map

$$h := f_{n+1} \circ f_n\colon \boldsymbol{D}_{r_n} \to \boldsymbol{D}_{r_{n+1}}$$

we have $h(0) = 0$ and $h'(0) = 1$. But by (12) we have $1 = h'(0) \leqslant r_{n+1}/r_n$. Let $\lim r_n =: R \in]0, \infty]$. We have to prove that

(15) $\quad X \cong D_R$.

To this end we prove that there is a subsequence $(f_{n_k})_{k \in N}$ of $(f_n)_{n \in N}$ such that for every m the sequence $(f_{n_k}|Y_m)_{k \geqslant m}$ converges compactly on Y_m. The mapping

$$D \ni z \to f_0^{-1}(r_0 z) \in Y_0$$

is biholomorphic. If we set

$$g_n(z) := \frac{1}{z_0} f_n(f_0^{-1}(r_0 z)), \qquad n \geqslant 0,$$

then the map $g_n\colon D \to C$ is holomorphic, injective with $g_n(0) = 0$ and $g_n'(0) = 1$. Then by Lemma XVI.21.15 there exists a subsequence $(f_{n_{0k}})_{k \in N}$ of (f_n) which converges compactly on Y_0. The same reasoning allows us to extract from this subsequence another subsequence $(f_{n_{1k}})$, which converges compactly on Y_1. Repeating this procedure, we obtain for every m a subsequence $(f_{n_{mk}})$ of the preceding subsequence which converges compactly on Y_m. We now form the "diagonal sequence" $f_{n_k} := f_{n_{kk}}$, $k = 0, 1, \ldots$ The sequence $(f_{n_k})_{k \in N}$ has the required property. Let $f := \lim f_{n_k}$, $k \to \infty$. Then $f \in A(X)$ is on each Y_m the limit of $(f_{n_k}|Y_m)_{k \geqslant m}$. Hence $f\colon X \to C$ is injective, $f(a) = 0$ and $f'(a) = 1$. We have to prove that f maps X biholomorphically onto D_R. Plainly, $f(X) \subset D_R$. We have to check that $f\colon X \to D_R$ is surjective. Otherwise, by virtue of Lemma XVI.21.14, there is an $r < R$ and a holomorphic mapping

$$g\colon f(X) \to D_r \text{ such that } g(0) = 0 \text{ and } g'(0) = 1.$$

Choose an n such that $r_n > r$; then for the map

$$h := g \circ f \circ f_n^{-1}\colon D_{r_n} \to D_r$$

we have $h(0) = 0$ and $h'(0) = 1$, but this contradicts (12), hence $f\colon X \to D_R$ is surjective and the proof is complete. □

COROLLARY XVI.21.17. For any Riemann surface we have

$$(\pi_1(X) = 0) \Leftrightarrow (\mathrm{Rh}_A^1(X) = 0).$$

Remark. Forster's proof of the conformal mapping theorem shows once more that the Behnke–Stein approximation theorem is the central and most important theorem in the theory of Riemann surfaces.

The Measure of $\Gamma\backslash\mathfrak{H}$. We recall that the differential form

$$d\mu := \frac{1}{y^2} dx \wedge dy = \frac{i}{2y^2} dz \wedge d\bar{z}$$

defines an $SL_2(\boldsymbol{R})$-invariant Radon measure μ on the upper half-plane $\mathfrak{H}$.

If $\Gamma \subset SL_2(\boldsymbol{R})$ is a Fuchsian group of the first kind, i.e., if $\Gamma\backslash\mathfrak{H}^*$ is compact (where $\mathfrak{H}^* = \mathfrak{H} \cup \{\text{cusps of } \Gamma\}$), then the measure μ induces a Radon measure on the space $\Gamma\backslash\mathfrak{H}^*$. More precisely: we have the following construction of the measure on $\Gamma\backslash\mathfrak{H}$. Let Γ be a discrete subgroup of $SL_2(\boldsymbol{R})$. Denote

$$m(A) := \int_A y^2 dx \wedge dy \quad \text{for a subset } A \subset \mathfrak{H}.$$

Denote by $p\colon\ \mathfrak{H}^* \to \Gamma\backslash\mathfrak{H}^*$ the projection map. Let $\Gamma_x = \{\gamma \in \Gamma\colon \gamma(x) = x\}$ and let $\mathcal{O}$ be an open neighbourhood of x such that $\gamma(\mathcal{O}) = 0$ for all $\gamma \in \Gamma_x$ and $\Gamma_x = \{\gamma \in \Gamma\colon \gamma(\mathcal{O}) \cap \mathcal{O}\} \neq \emptyset$. Then $\Gamma_x\backslash\mathcal{O}$ can be identified with an open neighbourhood of $p(x)$ in $\Gamma\backslash\mathfrak{H}^*$. Let $(\mathcal{O}_j)$ be a covering of $\Gamma\backslash\mathfrak{H}^*$ with sets of the form $\Gamma_x\backslash\mathcal{O}$ and let (φ_j) be a partition of unity subordinate to this covering.

Exercise. Prove that for an $f \in C(\Gamma\backslash\mathfrak{H}^*)$

$$(16) \qquad \int_{\Gamma\backslash W^*} f d\mu := \sum_{j=1}^{\infty} \int_{\mathcal{O}_j} f\varphi_j dm$$

does not depend on the choice of $(\mathcal{O}_j)$ and (φ_j).

Here is an important result of Carl Ludwig Siegel:

THEOREM XVI.21.18 (Siegel).

$$(\mu(\Gamma\backslash\mathfrak{H}) < \infty) \Leftrightarrow (\Gamma\backslash\mathfrak{H}^* \text{ is compact}).$$

PROOF. To prove $\Leftarrow$ is not very difficult and is left to the reader as an exercise.

The proof of $\Rightarrow$ is difficult (cf. C. L. Siegel, *Ges. Abh. III*, p. 118, and C. L. Siegel, *Some remarks on discontinuous groups*, Ann. of Math. 46 (1945), pp. 708–718).

If the Riemann surface $\Gamma\backslash\mathfrak{H}^*$ is compact and of genus g, it can be represented as a "normal form", i.e., as a $4g$-sided polygon $a_1 b_1 a_1^{-1} b_1^{-1} \dots a_g b_g a_g^{-1} b_g^{-1}$ whose vertices are all identified and whose boundary consists of $2g$ curves a_i, b_i traced once in each direction in the above order and such that the elliptic points and cusps are in the interior of the polygon. This polygon can be mapped onto a polygon π in $\mathfrak{H}$. Actually, one can construct a Γ-fundamental domain F_Γ which is a polygon whose sides are hyperbolic straight lines (i.e., geodesics). A careful construction allows us to obtain for the measure of F_Γ (or $\Gamma\backslash\mathfrak{H}$) the following beautiful formula

THEOREM XVI.21.19. Let g be the genus of the Riemann surface $\Gamma\backslash\mathfrak{H}^*$, let m be the number of equivalent cusps of Γ, and let $e_1, \ldots, e_r$ be the orders of the non-equivalent elliptic points of Γ. Then

$$(17) \qquad \frac{1}{2\pi}|F_\Gamma| = \frac{1}{2\pi}\int\limits_{\Gamma\backslash\mathfrak{H}} y^{-2}dx\,dy$$

$$= 2(g-1)+m+\sum_{j=1}^{r}\left(1-\frac{1}{e_j}\right).$$

COROLLARY XVI.21.20. Assumptions as in Theorem XVI.20.19. Then

$$(18) \qquad 2(g-1)+m+\sum_{j=1}^{r}(1-e_j^{-1}) > 0.$$

If $g = 1$, then $m+r \geqslant 1$.

If $g = 0$, then $m+\sum_{j=1}^{r}(1-e_j^{-1}) > 2$. Hence $m+r \geqslant 3$.

Exercise. Prove that the case

$$(19) \qquad g = 0, \quad m = 0, \quad (e_1, e_2, e_3) = (2, 3, 7)$$

gives Γ with the smallest $|F_\Gamma|$. Hence we obtain the following important formula, also due to Siegel:

THEOREM XVI.21.21 (Siegel). The (hyperbolic) measure of the fundamental region of the Fuchsian group Γ of the first kind satisfies the inequality

$$(20) \qquad |F_\Gamma| \equiv \int\limits_{\Gamma\backslash\mathfrak{H}} y^{-2}dx\,dy \geqslant \frac{\pi}{21}.$$

The equality is attained in the case (19), the so called "triangle group of Schwarz". From (20) it is not difficult to obtain the famous Hurwitz Theorem announced in Section 19:

THEOREM XVI.21.22 (Hurwitz). The group Aut(X) of automorphisms of any compact Riemann surface X of genus g has an order of at most

(21) $\quad 84(g-1).$

PROOF (sketch). As we know from the Poincaré Theory, $X \cong \Gamma \backslash \mathfrak{H}$, where $\Gamma \cong \text{Deck}(\mathfrak{H} \to X) \cong \pi_1(X)$. $\text{Aut}(X) \cong N(\Gamma)/\Gamma$, where $N(\Gamma)$ is the normalizer of Γ in $SL_2(\boldsymbol{R})$. Since X is hyperbolic (for its genus $g > 1$ cf. Section 19), Γ has no elliptic points and no (parabolic) cusps. Therefore, formula (17) gives

(22) $\quad |F_\Gamma| = 4\pi(g-1).$

But the index of Γ in $N(\Gamma)$

$$[N(\Gamma)\colon \Gamma] = \frac{|F_{N(\Gamma)}|}{|F_\Gamma|} \leqslant \frac{4\pi(g-1)}{\pi/21} = 84(g-1)$$

(by virtue of (20) and (22)) and the proof is complete.

Remark. The upper bound in (21) is attained if $N(\Gamma)$ is the triangle group $(2, 3, 7)$; thus $|F_{N(\Gamma)}| = \dfrac{\pi}{21}$.

We have already twice mentioned the

Triangle Groups of Schwarz. One of the most important sources of discontinuous groups and automorphic functions is the theory of linear differential equations with singular coefficients, so-called equations of the Fuchs type (this theory will be considered in some detail in Part III). The one that has been best investigated is the *hypergeometric differential equation*

$$\frac{d^2\eta}{dw^2} + p_1(w)\frac{d\eta}{dw} + p_2(w)\eta = 0, \tag{23}$$

where

$$p_1 = \frac{(\alpha+\beta+1)w-\delta}{w(w-1)}, \qquad p_2 = \frac{\alpha\beta}{w(w-1)}.$$

This equation embraces many very important equations, e.g., the Bessel equation. (Our exposition follows closely the beautiful Chapter I of a monumental monograph by J. Lehner, *Discontinuous Groups and Automorphic Functions*, Providence, 1964.)

The following important problem was considered by H. A. Schwarz in 1872 (and before him by Kummer in 1836 and by Riemann in 1857; Riemann's work was discovered much later in the files at Göttingen and published in 1902): find all the cases of equation (23) which can be integrated by algebraic functions in w. Let η_1, η_2 be a fundamental system (basis) of solutions of (23) which are regular at $w = w_0$, a point distinct from the singular points $0, 1, \infty$. Let $z(w) := \eta_1(w)/\eta_2(w)$. By analytic continuation of z around a closed curve, the solutions η_j remain solutions of (23) but can not return to their initial values. However, since any solution of (23) is a linear combination of η_1, η_2, we can see that the final value of z is

$$z_g = \frac{az+b}{cz+d}, \quad [g] \in \pi_1(X).$$

We thus have a homomorphism $T\colon \pi_1(X) \ni [g] \xrightarrow{T} GL_2(C)$ of the fundamental group of $X = C - \{0, 1, \infty\}$ into $GL_2(C)$. The group $\Gamma := T(\pi_1(X))$ is called the *monodromy group* of equation (23).

Writing $z = \zeta(w)$, $w = f(z)$, we obtain

$$\text{(24)} \qquad w = f(z) = f(\gamma z) \quad \text{for each } \gamma \in \Gamma.$$

Hence f is a Γ-automorphic function. Schwarz immediately obtained a differential equation of the third order for z:

$$\text{(25)} \qquad \frac{z'''}{z'} - \frac{3}{2}\left(\frac{z''}{z'}\right)^2 = \frac{1-\lambda^2}{2w^2} + \frac{1-\mu^2}{2(1-w)^2} + \frac{\lambda^2+\mu^2-\nu^2-1}{2w(w-1)},$$

where $\lambda = 1-\delta$, $\mu = \delta - \alpha - \beta$, $\nu = \beta - \alpha$. The left-hand side of (25) is called the *Schwarz derivation* and is invariant with respect to Möbius transformations. Schwarz proved that for real λ, μ, ν the function $z = \zeta(w)$ maps $\mathfrak{H}$ biholomorphically onto a triangle with angles $\lambda\pi$, $\mu\pi$, $\nu\pi$ bounded with geodesic arcs (circular or line segments). Since η has to be algebraic in w, z has to be algebraic in w. Then $f = \eta^{-1}$ will be single-valued in the neighbourhood of all points except branch points. We can prove that λ, μ, ν must be either 0 or the reciprocal of a positive integer. We now continue z analytically by the Schwarz reflection principle: we reflect on the side c, which is equivalent to extending z to the lower half-plane across the segment $]1, \infty[$. Next we reflect upon the side b, or on the side a. The new triangle is left unshaded. We continue the process indefinitely, always reflecting on the free side of the figure in the z-plane. Since reflection is a conformal mapping, all triangles thus obtained have the same angles $\lambda\pi$, $\mu\pi$, $\nu\pi$. Because of the condition on λ, μ, ν, there will be no overlapping of triangles. The plane or a portion of the plane is covered without gaps and without overlapping by a network (tesselation of $\mathfrak{H}$) of triangles. We obtain a discontinuous group of reflections: the triangle groups.

There are three different cases:

(i) $\lambda+\mu+\nu = 1$, we have a straight-sided, i.e., an Euclidean triangle.

Exercise. Prove that there are four possibilities including a degenerate one: $(\lambda, \mu, \nu) = (1/2, 1/2, 0)$, a half-stripe. We obtain simply and doubly periodic functions. The groups contain only elliptic and parabolic transformations.

(ii) $\lambda+\mu+\nu > 1$, an elliptic (or spherical) triangle. The triangles can be mapped stereographically on the sphere $\boldsymbol{P}(\boldsymbol{C})$, yielding a covering with a finite number of triangles. The groups are finite: they are groups of symmetrics of regular solids.

Examples. 1° $(\frac{1}{3}, \frac{1}{3}, \frac{1}{2})$ is called the *tetrahedral group.*

2° $(\frac{1}{2}, \frac{1}{3}, \frac{1}{4})$, the *octahedral group.*

3° $(\frac{1}{2}, \frac{1}{3}, \frac{1}{5})$, the *icosahedral group.*

(iii) $\lambda+\mu+\nu < 1$, a hyperbolic (i.e., non-Euclidean) triangle.

Because the angle sum is less than π, if we take the triangle ABC with A at the centre of the disk $\boldsymbol{D}$, two straight sides AB and AC and a circular arc BC.

Exercise. Prove that a reflection upon any side of the triangle leaves $\partial\boldsymbol{D}$ fixed and that the reflection process leads to an infinite network of triangles all lying in $\boldsymbol{D}$. They fill $\boldsymbol{D}$. In this way we obtain an infinite discrete subgroup of $\mathrm{SL}_2(\boldsymbol{C})$ denoted by

$$(\lambda, \mu, \nu) \quad \text{or by} \quad T\left(\frac{1}{\lambda}, \frac{1}{\mu}, \frac{1}{\nu}\right).$$

Exercise. Show that for the modular group $\Gamma(1)$

$$\Gamma(1) = T(\tfrac{1}{3}, \tfrac{1}{2}, 0).$$

We have reproduced both solutions of $\mathfrak{H}$ and $\boldsymbol{D}$ (corresponding to this group).

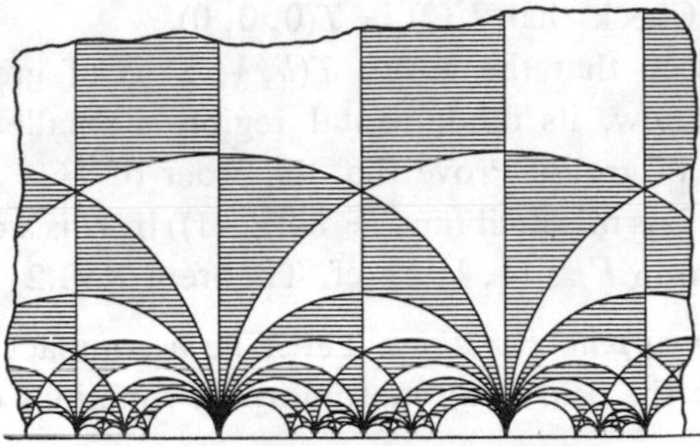

Fig. 23. The modular function.

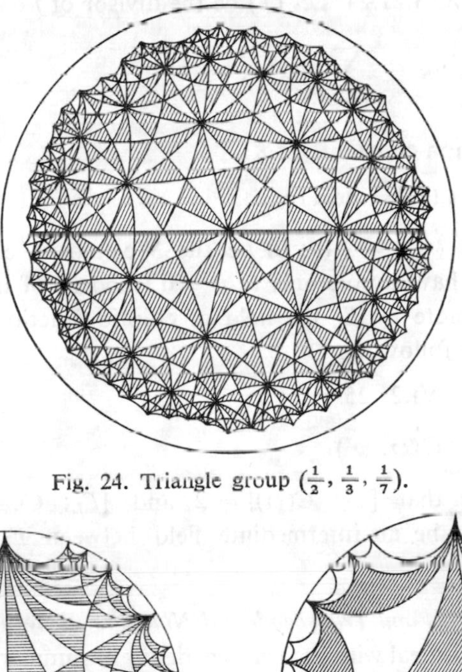

Fig. 24. Triangle group $(\frac{1}{2}, \frac{1}{3}, \frac{1}{7})$.

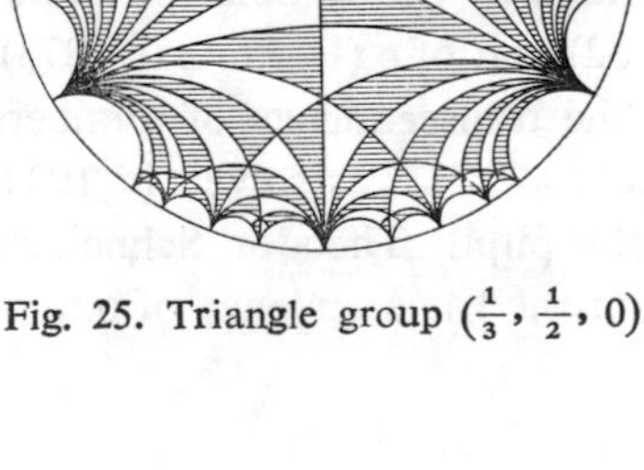

Fig. 25. Triangle group $(\frac{1}{3}, \frac{1}{2}, 0)$.

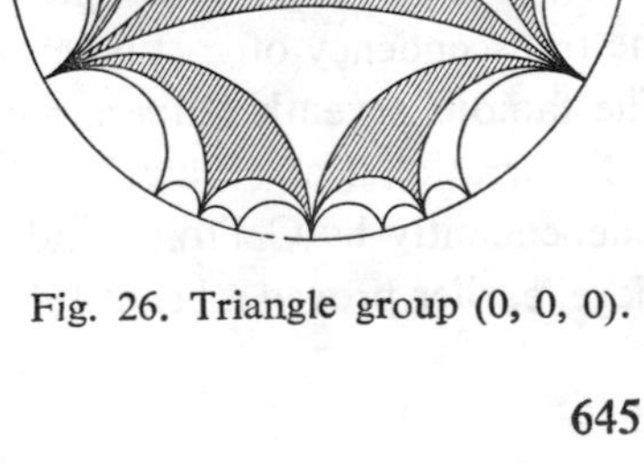

Fig. 26. Triangle group $(0, 0, 0)$.

Exercises. 1. Check that $\Gamma(2) = T(0, 0, 0)$.

1. We have seen that the group $T(\frac{1}{2}, \frac{1}{3}, \frac{1}{7})$ is of great importance because, as we know, its fundamental region is smaller than that of any other Fuchsian group. Prove that the order of $H = \mathrm{Aut}(X)$, where X is of genus $g \geqslant 2$, is maximal (and $= 84(g-1)$) iff H is a quotient group of a Fuchsian group $\Gamma \cong (\frac{1}{2}, \frac{1}{3}, \frac{1}{7})$ (cf. Theorem XVI.21.22).

Fields of Automorphic Functions. Let X be a compact Riemann surface and let K be the field of all meromorphic functions on X.

Exercise. Prove the following

PROPOSITION XVI.21.24. Let (f) be the divisor of $f \in K$. Then if

$$(f)_\infty := -\sum_{n(x)<0} n(x)x,$$

then for every non constant $f \in K$

$$[K: C(f)] = \deg(f)_\infty.$$

Exercise. Let $\wp(u) = \wp(u; L)$ be the Weierstrass elliptic function. Then $\wp$ and $\wp'$ have a pole only at $u = 0 \pmod L$ of degree 2 and 3, respectively. Denote by E_L the field of elliptic functions with periods in L. Prove the following

PROPOSITION XVI.21.25.

$$E_L = C(\wp, \wp').$$

PROOF. Note that $[E_L: C(\wp)] = 2$ and $[E_L: C(\wp')] = 3$; hence $C(\wp, \wp')$ cannot be an intermediate field between E_L and $C(\wp)$ or $C(\wp')$. □

Elliptic Integrals and Transcendental Numbers. Several famous mathematical problems deal with the transcendency of numbers (we recall that a real number is transcendent if it is not a root of any polynomial with rational coefficients). Examples are provided by the famous proofs of the transcendency of π (Lindemann in 1882) and of e (Hermite in 1873). The famous seventh Hilbert problem of the transcendency of numbers a^b for irrational algebraic b and algebraic $a \neq 0, 1$ was solved in 1934, independently by Gel'fond and by Siegel's pupil, Theodor Schneider. Siegel earlier proved a beautiful theorem on the transcendency of one of periods of the function $\wp$ for algebraic values of the invariants g_2 and g_3 in the differential equation for:

$$(\wp')^2 = 4\wp^3 - g_2\wp - g_3.$$

Siegel's result was extended by Schneider in 1937 to the famous

THEOREM XVI.21.26 (Schneider). Let $p_k := p(\xi_k, \eta_k)$, $k = 1, 2$, be two points on the Riemann surface of the curve $\varphi(\xi, \eta) = 0$ of genus 1 whose coordinates ξ_k, η_k, $k = 1, 2$ are algebraic numbers. Let $w(p)$ be an indefinite elliptic integral of the second kind which is regular at p_1 and p_2 and does not reduce to a rational function of ξ and η. Then the value of the definite elliptic integral

$$(26) \qquad w(p_2) - w(p_1) := \int_{p_1}^{p_2} dw$$

is transcendental except when p_1 and p_2 coincide and the path of integration is homologous to 0.

Examples for Schneider's Theorem are the following exercises:

Exercise 1. Let (ξ_1, η_1) and (ξ, η) be two real points on the ellipse

$$\frac{\xi^2}{a^2} + \frac{\eta^2}{b^2} = 1, \qquad 0 < b < a.$$

Check that the arc length

$$s(\xi, \xi_1) = \int_{\xi_1}^{\xi} \sqrt{\frac{a^2 - \varepsilon\xi^2}{a^2 - \xi^2}}\, d\xi, \qquad \text{where } \varepsilon^2 = 1 - \frac{b^2}{a^2},$$

is then an elliptic integral of the second kind and not a rational function of ξ and η.

Therefore, s is transcendental if a, b, ξ_1, ξ_2 are algebraic numbers (except the trivial case $s = 0$). The corresponding result for the circle, $a = b$, gives the transcendency of π.

Exercice 2. The arc length of the lemniscate

$$(\xi^2 + \eta^2)^2 = 2a^2(\xi^2 - \eta^2) \qquad (a > 0)$$

is given by

$$s(\xi, \xi_1) = a\sqrt{2} \int_{t_1}^{t} \frac{dt}{\sqrt{1 - t^4}}, \qquad t := \frac{\xi^2 - \eta^2}{\xi^2 + \eta^2}.$$

As a function of t, this is an elliptic integral of the first kind. It follows that for algebraic a the arc length of the lemniscate between points with algebraic coordinates ξ, η is transcendental (except the trivial case $s = 0$). We hope that the reader will consult a charming little book by the great master—C. L. Siegel, *Transcendental Numbers*, Princeton, 1949.

Schwarz–Christoffel formulas give conformal map of $\mathfrak{H}$ onto a polygon Π_n with n sides and angles $\alpha_1\pi, \ldots, \alpha_n\pi$, where $\alpha_1 + \ldots + \alpha_n = n-2$. (If $\alpha_m < 1$ for $m = 1, 2, \ldots, n$, then Π_n is convex.)

Prove the following

THEOREM XVI.21.27 (Schwarz–Christoffel). Let $a_1 < a_2 < \ldots < a_n$ be real numbers and $\alpha_k \in [-2, 2]$, $k = 1, \ldots, n$, where $\sum_{k=1}^{n} \alpha_k = n-2$. Then

$$\text{(S–C)} \qquad z \to w(z) := c\int_{z_0}^{z} (t-a_1)^{\alpha_1-1} \ldots (t-a_n)^{\alpha_n-1}\,dt + c_1$$

gives a conformal map of $\mathfrak{H}$ on an n-sided polygon with angle $\alpha_k\pi$ at the k-th vertex.

PROOF (idea). The argument of derivative of the function w given by (S–C):

$$\arg\frac{dw}{dz} = \arg C + \sum_{k=1}^{n} (\alpha_k - 1)\arg(z-a_k)$$

is constant on each interval $]a_k, a_{k+1}[$, $k = 1, \ldots, n-1$ and $w' \neq 0$ there. Therefore w maps $]a_k, a_{k+1}[$ onto interval $A_k A_{k+1}$. The same is true for the interval $]a_n, a_1[$. Therefore w maps $\mathfrak{H}$ onto a polygon $A_1 \ldots A_n$, where several vertices may be at infinity (for $\alpha_k \leqslant 0$). □

Remark. Prove that the theorem is valid also in case $\sum \alpha_k \neq n-2$. In that case there appears as A_{n+1} corresponding to $z = \infty$. A_{n+1} is finite if $\sum \alpha_k < n-2$ and $A_{n+1} = \{\infty\}$ if $\sum \alpha_k > n-2$.

(S–C) *and Elliptic Integrals*. A very important example is the mapping $z \to w(z)$ given by

$$w(z) = \int_{0}^{z} \frac{dt}{\sqrt{(1-t^2)(t-k^2t^2)}}, \qquad 0 < k < 1$$

(the elliptic Legendre integral of first special type) of $\mathfrak{H}$ onto the rectangle $-K, K, K+iK', -K+iK'$ with

$$K := \int_0^1 \frac{dt}{\sqrt{(1-t^2)(1-k^2t^2)}}, \quad K' := \int_0^{1/k} \frac{dt}{\sqrt{(1-t^2)(1-k^2t^2)}}.$$

The inverse function w^{-1} is the famous Jacobi elliptic function $\operatorname{sn}(w, k)$.

Exercise. Prove that $\operatorname{sn}(\cdot\,, k)$ has the periods

$$\omega_1 = 4K, \quad \omega_2 = 2iK'.$$

XVII. NORMAL AND PARACOMPACT SPACES PARTITION OF UNITY

In this chapter we shall concern ourselves with important types of Hausdorff spaces, the paracompact spaces introduced in 1944 by Dieudonné. It turns out that the spaces most often met with in analysis belong to this very class. In particular, the following are paracompact:

(i) metric spaces (somewhat more generally: metrizable), hence Banach spaces, Hilbert spaces, L^p spaces familiar from integral theory, where $p \geqslant 1$.

(ii) discrete spaces,

(iii) compact spaces,

(iv) locally compact spaces countable at infinity, hence in particular:

(v) locally compact spaces having a countable basis.

Dieudonné observed that *partitions of unity* play an extremely important role in mathematical analysis. Characterization of Hausdorff spaces which have a partition of unity led him to the concept of paracompactness.

In analysis on differentiable manifolds the partition of unity is today such an important tool that manifolds are usually assumed to be paracompact (cf. Chapter XIV).

All topological spaces considered in this chapter are Hausdorff spaces.

1. LOCALLY COMPACT SPACES COUNTABLE AT INFINITY

DEFINITION. A locally compact space X is said to be *countable at infinity* if it is the (set-theoretical) union of a countable number of compact sets, i.e., if

$$X = \bigcup_{i=1}^{\infty} K_i, \qquad K_i \text{ — compact.}$$

Examples. 1. A discrete space is locally compact. It is countable at infinity if and only if it consists of a countable number of points.

2. The space $\boldsymbol{R}^1$ is locally compact and countable at infinity since it is the sum of compact intervals $K_n := [-n, n]$ (n is a natural number).

3. The space $\boldsymbol{R}^n$ is locally compact and countable at infinity, being the sum of compact balls with radii which are natural numbers.

4. A closed subspace of a locally compact space countable at infinity is also locally compact and countable at infinity.

THEOREM XVII.1.1. X is a locally compact space countable at infinity iff there exists a sequence $\{\mathcal{O}_n\}_{n\in N}$ of open sets, relatively compact in X and such that

(i) $\mathcal{O}_n \subset \mathcal{O}_{n+1}, \quad n \in N,$

(ii) $X = \bigcup_{n=1}^{\infty} \mathcal{O}_n$.

PROOF. The condition given above is clearly sufficient since the sets $\overline{\mathcal{O}}_n$ are compact and their union is the whole space. Let us prove that this condition is necessary. Suppose that $X = \bigcup_{n=1}^{\infty} K_n$. Let $\mathcal{O}_1 := \mathcal{U}_1(K_1)$ be an open, relatively compact neighbourhood of the set K_1. We define $\mathcal{O}_n$ recurrently for $n > 1$ as an open, relatively compact neighbourhood of the set $\overline{\mathcal{O}}_{n-1} \cup K_n$. It is easily seen that conditions (i) and (ii) are satisfied. □

COROLLARY XVII.1.2. With the same assumptions as in Theorem XVII.1.1 for every compact set $K \subset X$ there exists a number $m \in N$ such that $K \subset \mathcal{O}_m$.

PROOF. A finite open covering corresponding to indices i_k, $k = 1, 2, \ldots, p$, can be selected from the open covering $\{\mathcal{O}_n\}_{n\in N}$ of the set K. Let $m := \max_{k=1,\ldots,p} i_k$. It is easily seen that $K \subset \mathcal{O}_m$, since $\mathcal{O}_{i_k} \subset \mathcal{O}_m$ is also true for all sets $\mathcal{O}_{i_k}$ (because $i_k \leqslant m$). □

THEOREM XVII.1.3. The Cartesian product of a finite number of locally compact spaces countable at infinity is itself a locally compact space countable at infinity with respect to the Tychonoff topology.

PROOF. The Tychonoff Theorem (XIII.19.2) implies that $\bigtimes_{i=1}^{n} X_i$ is a locally compact space if all X_i were locally compact.

Let $\bigcup_{j=1}^{\infty} K_j^i$ be a covering of the space X_i with compact sets K_j^i with the property that $K_j^i \subset K_{j+1}^i$, for all i and j. The Tychonoff theorem implies that the sets $K_j := \bigtimes_{i=1}^{n} K_j^i$ are compact in the space $X := \bigtimes_{i=1}^{n} X_i$, and

$\bigcup_{j=1}^{\infty} K_j = X$, and thus X is a locally compact space countable at infinity. □

For instance, it follows from the foregoing theorem that $\boldsymbol{R}^n$ is a locally compact space countable at infinity (another proof of this fact was given in Example 3 on p. 653).

THEOREM XVII.1.4. Let X be a locally compact space, and let $\tilde{X} := X \cup \{p_\infty\}$ be the compact space obtained by adjunction of the point "at infinity" (cf. Alexandrov Theorem XVI.6.1). Then X is countable at infinity if and only if $\tilde{X}$ has a countable basis of neighbourhoods at p_∞.

We recall that a family $\{\mathscr{U}_i(x)\}_{i\in I}$ of neighbourhoods of a point x is called a *basis of neighbourhoods* of x if for every neighbourhood $\mathscr{U}(x)$ of x there exists an $i \in I$ such that $\mathscr{U}_i(x) \subset \mathscr{U}(x)$.

PROOF. $\Leftarrow$: If $\{\mathscr{U}_n(p_\infty)\}$ is a basis of open neighbourhoods of the point p_∞, then the sets $K_n := \tilde{X} - \mathscr{U}_n(p_\infty)$ are compact in X by the definition of the topology in $\tilde{X}$. To show that $X = \bigcup_{n=1}^{\infty} K_n$ let $x \in X$ and let $\mathscr{U}(x)$ be an open neighbourhood of x with compact closure. Then $\tilde{X} - \overline{\mathscr{U}(x)}$ is an open neighbourhood of p_∞ so that it contains $\mathscr{U}_n(p_\infty)$ for some $n \subset N$. It follows that $x \in \tilde{\mathscr{U}}(x) \subset X - \mathscr{U}_n(p_\infty) = K_n$.

$\Rightarrow$: Suppose that the sets $\mathcal{O}_n$ possess the properties stated in Theorem XVII.1.1. Then by Corollary XVII.1.2 the sets $\tilde{X} - \overline{\mathcal{O}}_n$ constitute a basis of neighbourhoods of the point p_∞. □

2. NORMAL SPACES. URYSOHN'S LEMMA

Before we define paracompact spaces, we shall introduce a broader class of Hausdorff spaces which was studied by Urysohn, viz., normal spaces.

DEFINITION I. A Hausdorff space $(X, \mathscr{T})$ is called *normal* if for every two closed disjoint sets A_1 and A_2 there exist disjoint open sets $\mathcal{O}_1$ and $\mathcal{O}_2$ such that $A_1 \subset \mathcal{O}_1$, $A_2 \subset \mathcal{O}_2$.

We now give several general examples of normal spaces:

LEMMA XVII.2.1. A metric space is normal.

PROOF. Let A_1, A_2 be closed disjoint sets. For $x \in X$ we denote

$$d(x, A_2) := \inf_{y \in A_2} d(x, y)$$

(the distance of the point x from the set A_2). Plainly, $d(x, A_2) > 0$ for $x \in A_1$ since if $d(x, A_2) = 0$, then $x \in \overline{A_2} = A_2$, but $A_1 \cap A_2 = \emptyset$. Hence for $x \in A_1$ the ball $K(x, \frac{1}{2}d(x, A_2))$ is an open set, disjoint from A_2. Take

$$\mathcal{O}_1 := \bigcup_{x \in A_1} K(x, \tfrac{1}{2}d(x, A_2)).$$

Clearly, $A_1 \subset \mathcal{O}_1$. Moreover, $\mathcal{O}_1$ is an open set as a sum of open balls. We shall show that $\overline{\mathcal{O}_1} \cap A_2 = \emptyset$. Suppose that $y_0 \in \overline{\mathcal{O}_1} \cap A_2$ then $\varrho := d(y_0, A_1) > 0$, by the same argument as above. Since $y_0 \in \overline{\mathcal{O}_1}$ there exists a point $z \in \mathcal{O}_1$ such that $d(y_0, z) \leqslant \frac{1}{4}\varrho$.

The point z, however, must lie in one of the balls constituting the set $\mathcal{O}_1$, say, $z \in K(x_0, \frac{1}{2}d(x_0, A_2))$. But then

$$\begin{aligned} d(x_0, y_0) &\leqslant d(x_0, z) + d(z, y_0) \leqslant \tfrac{1}{2}d(x_0, A_2) + \tfrac{1}{4}d(y_0, A_1) \\ &= \tfrac{1}{2}\inf_{y \in A_2} d(x_0, y) + \tfrac{1}{4}\inf_{x \in A_1} d(y_0, x) \\ &\leqslant \tfrac{1}{2}d(x_0, y_0) + \tfrac{1}{4}d(y_0, x_0) = \tfrac{3}{4}d(x_0, y_0). \end{aligned}$$

The resulting contradiction proves that $\overline{\mathcal{O}_1} \cap A_2 = \emptyset$, which means that $A_2 \subset X - \overline{\mathcal{O}_1}$. Hence the set $\mathcal{O}_2 := X - \overline{\mathcal{O}_1}$ is open, and contains A_2, and we have $\mathcal{O}_1 \cap \mathcal{O}_2 = \emptyset$. □

LEMMA XVII.2.2. Compact spaces are normal.

PROOF. Let A_1 and A_2 be closed and disjoint in a compact space $(X, \mathcal{T})$. As closed subsets of a compact space, they are compact sets.

Let $x \notin A_2$. For every $y \in A_2$ there are disjoint open sets $\mathcal{O}_{xy}$ and $\mathcal{O}_y$ such that $x \in \mathcal{O}_{xy}$, $y \in \mathcal{O}_y$ (a compact space is a Hausdorff space by definition). It is seen that $A_2 \subset \bigcup_{y \in A_2} \mathcal{O}_y$. Since the set A_2 is compact, a finite covering

$$A_2 \subset \bigcup_{i=1}^{n} \mathcal{O}_{y_i} =: \mathcal{O}_2^x$$

can be selected from this covering. The set $\mathcal{O}_2^x$ is open. Moreover,

$$\overline{\mathcal{O}_2^x} = \overline{\bigcup_{i=1}^{n} \mathcal{O}_{y_i}} = \bigcup_{i=1}^{n} \overline{\mathcal{O}_{y_i}}.$$

Now, x does not belong to any of the sets $\overline{\mathcal{O}}_{y_i}$ (since it has a neighbourhood $\mathcal{O}_{xy_i}$ disjoint from $\mathcal{O}_{y_i}$), and hence $x \notin \overline{\mathcal{O}_2^x}$. The set $\mathcal{O}'_x := X - \overline{\mathcal{O}_2^x}$ therefore is an open neighbourhood of the point x and is disjoint from $\mathcal{O}_2^x$.

We have thus shown that if $x \notin A_2$, there exist disjoint open neighbourhoods: $\mathcal{O}'_x$ of the point x and $\mathcal{O}_2^x$ of the set A_2. The family $\{\mathcal{O}'_x\}_{x\in A_1}$ is an open covering of A_1 and since the set A_1 is compact, the finite covering $\{\mathcal{O}'_{x_j}\}_{j=1}^m$ can be chosen. The set $\mathcal{O}_1 := \bigcup_{j=1}^{m} \mathcal{O}'_{x_j}$ is open. Moreover, $\overline{\mathcal{O}}_1 = \overline{\bigcup_{j=1}^{m} \mathcal{O}'_{x_j}} = \bigcup_{j=1}^{m} \overline{\mathcal{O}}'_{x_j}$.

The set A_2 does not intersect any of the sets $\overline{\mathcal{O}'}_{x_j}$ (since it has a neighbourhood $\mathcal{O}_2^{x_j}$ disjoint from $\mathcal{O}'_{x_j}$); consequently, $A_2 \cap \overline{\mathcal{O}}_1 = \emptyset$. Hence the set $\mathcal{O}_2 := X - \overline{\mathcal{O}}_1$ is an open neighbourhood of the set A_2 and is disjoint from $\mathcal{O}_1$. □

The definition of a normal space can be recast in a slightly different (equivalent) form:

Definition II. A Hausdorff space is said to be *normal* if the following condition is satisfied:

If A is any closed set and $\mathcal{U}$ is an open neighbourhood of A (i.e., $A \subset \mathcal{U}$), there exists an open neighbourhood $\mathcal{V}$ of A such that $\overline{\mathcal{V}} \subset \mathcal{U}$ (i.e. $A \subset \mathcal{V} \subset \overline{\mathcal{V}} \subset \mathcal{U}$).

Let us prove that the two conditions are equivalent:

Lemma XVII.2.3. Definitions I and II are equivalent.

Proof. (i) If X is normal in the sense of Definition I, we take $A_1 := A$, $A_2 := X - \mathcal{U}$. Having obtained sets $\mathcal{O}_1$ and $\mathcal{O}_2$ mentioned in Definition I, we set $\mathcal{V} = \mathcal{O}_1$.

Since $\mathcal{O}_1 \cap \mathcal{O}_2 = \emptyset$, we have $\mathcal{V} = \mathcal{O}_1 \subset X - \mathcal{O}_2$, where $X - \mathcal{O}_2$ is a closed set. Hence $\overline{\mathcal{V}} = \overline{\mathcal{O}}_1 \subset \overline{X - \mathcal{O}_2} = X - \mathcal{O}_2$. However, $X - \mathcal{U} = A_2 \subset \mathcal{O}_2$, and thus $X - \mathcal{O}_2 \subset \mathcal{U}$, or $\overline{\mathcal{V}} = \overline{\mathcal{O}}_1 \subset \mathcal{U}$.

(ii) If the space X is normal in the sense of Definition II then for closed and disjoint A_1 and A_2 we set $A := A_1$, $\mathcal{U} := X - A_2$. Having obtained a set $\mathcal{V}$ mentioned in Definition II, we take $\mathcal{O}_1 = \mathcal{V}$, $\mathcal{O}_2 = X - \overline{\mathcal{V}}$.

The set $\mathcal{O}_1$ is an open neighbourhood of the set A_1. The set $\mathcal{O}_2$ is also open, and is disjoint from $\mathcal{O}_1$. We shall show that it contains the set A_2: $\overline{\mathscr{V}} \subset \mathscr{U} = X - A_2$, and hence $A_2 \subset X - \overline{\mathscr{V}} = \mathcal{O}_2$. □

Normal spaces are characterized by the following

THEOREM XVII.2.4 (Urysohn's Lemma). If A_0 and A_1 are closed disjoint sets in the normal space $(X, \mathscr{T})$, then there exists a continuous function f, defined in X, with the following properties:

(i) $0 \leqslant f(x) \leqslant 1$,

(ii) if $x \in A_0$, then $f(x) = 0$,

(iii) if $x \in A_1$, then $f(x) = 1$.

PROOF. Let $\mathcal{O}_1 = X - A_1$. The normality of the space X implies the existence of an open set $\mathcal{O}_{1/2}$ such that $A_0 \subset \mathcal{O}_{1/2} \subset \overline{\mathcal{O}_{1/2}} \subset \mathcal{O}_1$. Similarly, there exist open sets $\mathcal{O}_{1/4}$, $\mathcal{O}_{3/4}$ such that

$$A_0 \subset \mathcal{O}_{1/4} \subset \overline{\mathcal{O}_{1/4}} \subset \mathcal{O}_{1/2}, \qquad \overline{\mathcal{O}_{1/2}} \subset \mathcal{O}_{3/4} \subset \overline{\mathcal{O}_{3/4}} \subset \mathcal{O}_1 .$$

Continuing this reasoning, for every irreducible fraction of the form $r = m/2^n$, $0 < m < 2^n$, $n = 1, 2, \ldots$, we define an open set $\mathcal{O}_r$ such that the following conditions are satisfied:

$$A_0 \subset \mathcal{O}_r, \qquad \overline{\mathcal{O}}_r \subset \mathcal{O}_1, \qquad \overline{\mathcal{O}}_r \subset \mathcal{O}_s \quad \text{for } r < s.$$

We put

$$f(x) := \begin{cases} 1 & \text{if } x \notin \bigcup_r \mathcal{O}_r, \\ \inf\{r : x \in \mathcal{O}_r\} & \text{otherwise.} \end{cases}$$

It follows immediately from the difinition that $f(x) = 1$ for $x \in A_1$ and $f(x) = 0$ for $x \in A_0$ and that $f(X) \subset [0, 1]$. It thus remains to prove that f is continuous. Note that if $0 < b \leqslant 1$ and $f(x) < b$, then $x \in \mathcal{O}_r$ for some $r < b$. Conversely: if $x \in \mathcal{O}_r$ for $r < b$, then $f(x) < b$. Hence, $\{x : f(x) < b\} = \bigcup_{r<b} \mathcal{O}_r$ is an open set, being the union of open sets.

Now let $0 \leqslant a < 1$. If $a < f(x)$, there exist two irreducible fractions r_1 and r_2 which have a power of two for the denominator and satisfy the condition: $a < r_1 < r_2 < f(x)$. It follows from the definition of f that $x \notin \mathcal{O}_{r_2}$ and hence also $x \notin \overline{\mathcal{O}}_{r_1} \subset \mathcal{O}_{r_2}$, that is, $x \in X - \overline{\mathcal{O}}_{r_1}$.

Conversely: if there is a number $r > a$ such that $x \in X - \overline{\mathcal{O}}_r$, then $x \notin \mathcal{O}_s$ for $s \leqslant r$, that is, $f(x) \geqslant r > a$.

Thus, finally, $\{x: f(x) > a\} = \bigcup_{a<r} (X-\bar{\mathcal{O}}_r)$ is an open set since it is the union of open sets. Hence the inverse images of the intervals $[0, b[$ and $]a, 1]$ are open sets. Since such intervals and their intersections constitute a basis of the topology of the space $[0, 1]$, the inverse image of an open set is open, so f is a continuous function. □

COROLLARY XVII.2.5. Let X be a locally compact space. Let $K \subset X$ be a compact set and let $\mathscr{U} \subset X$ be an open neighbourhood of this set. Then there exists a continuous function f: $X \to [0, 1]$ such that $f(x) = 1$ for $x \in K$, and $f(x) = 0$ for $x \notin \mathscr{U}$.

PROOF. Suppose that $\tilde{X} = X \cup p_\infty$ is the Alexandrov compactification of X. As it is compact, $\tilde{X}$ is normal. Applying the Urysohn's Lemma to the closed sets K and $\tilde{X}-\mathscr{U}$, we obtain the result. □

Remark 1. Clearly, if the assertion of the Urysohn's Lemma holds, the space is normal and hence the Urysohn's Lemma completely characterizes normal spaces.

Remark 2. Normal spaces possess a host of continuous functions in the sense that continuous functions separate every pair of disjoint closed subsets.

By Lemma XVII.2.3, the Urysohn's Lemma can be reformulated in the following equivalent form:

THEOREM XVII.2.6. If A is a closed set in a normal space $(X, \mathscr{T})$ and $\mathscr{U}$ is an open neighbourhood of it, then there exists a continuous function f: $X \to [0,1]$ such that $f(x) = 1$ on some neighbourhood $\mathscr{V}$ of A, whereas $f(x) = 0$ everywhere beyond $\mathscr{U}$.

3. EXTENDIBILITY OF CONTINUOUS FUNCTIONS ON NORMAL SPACES

Now we give another extremely important theorem characterizing normal spaces. It is also due to Urysohn (1925); the special case of this theorem, for metric spaces, was proved much earlier by Tietze (1915). This theorem concerns the extendibility of continuous functions in normal spaces.

THEOREM XVII.3.1 (Tietze–Urysohn). Suppose that X is a normal space and that M is a closed subset in X.

(i) Every continuous function $f\colon M \to \mathbf{R}^1$ can be extended to a continuous function $F\colon X \to \mathbf{R}^1$.

(ii) Moreover, if $f\colon M \to [-c, c]$, then there exists a continuous extension F such that $F\colon X \to [-c, c]$.

COROLLARY XVII.3.2. If the assertion of the Tietze–Urysohn Theorem holds for a Hausdorff space X, then X is a normal space.

PROOF of COROLLARY XVII.3.2. Suppose that $\overline{A} = A$, $\overline{B} = B$, $A \cap B = \emptyset$. Then the function $f(x) := 1$, $x \in A$, $f(y) = 0$, $y \in B$, defined on the closed set $M := A \cup B$ is continuous, and hence by the theorem above it has a continuous extension F on X; the function F satisfies the hypothesis of the Urysohn's Lemma, and thus X is a normal space. □

PROOF of THEOREM XVII. 3.1. First of all let us prove (ii).

Note that on setting $h(x) := a+(b-a)f(x)$ in the Urysohn's Lemma, we obtain a "more general" formulation: suppose that A_0 and A_1 are disjoint closed sets. In order for X to be a normal space it is necessary and sufficient that for all real numbers $a < b$ there exists a continuous function $h\colon X \to [a, b]$ such that $f(A_0) = \{a\}$, $f(A_1) = \{b\}$. Now, suppose that $|f(x)| \leqslant c$ for $x \in M$; the subsets $M_1 = f^{-1}([-c, -\frac{1}{3}c])$ and $M_2 = f^{-1}([\frac{1}{3}c, c])$ are closed and disjoint, owing to the continuity of f. By Urysohn's Lemma there exists a continuous function $g_1\colon X \to [-\frac{1}{3}c, +\frac{1}{3}c]$, such that $g_1(x) = -\frac{1}{3}c$ on M_1 and $g_1(y) = \frac{1}{3}c$ on M_2.

Now, suppose that $h_1 := f-g$, where $g := g_1|_M$; then we have $|h_1(x)| \leqslant \frac{2}{3}c$ for $x \in M$; clearly, the function h_1 is continuous on M. Application of the reasoning above to h_1 (instead of f) yields (Urysohn's Lemma) a continuous function $g_2\colon X \to [-\frac{1}{3}(\frac{2}{3})c, \frac{1}{3}(\frac{2}{3})c]$ and a function h_2, continuous on M, such that $h_2 = h_1 - g_2|_M$, $|h_2(x)| \leqslant (\frac{2}{3})^2 c$. Proceeding further this way, we obtain a sequence (g_n) of functions continuous on X such that

$$|g_n(x) \leqslant \tfrac{1}{3}(\tfrac{2}{3})^{n-1}c$$

and a sequence (h_n) of continuous functions on M such that

$$h_n = h_{n-1} - g_n|_M \quad \text{and} \quad |h_n(x)| \leqslant (\tfrac{2}{3})^n c.$$

Since the series $\sum g_n$ converges uniformly on X, its sum

$$F(x) := \sum_{n=1}^{\infty} g_n(x)$$

is a continuous function, and $|F(x)| \leqslant \frac{1}{3}\sum_{n=0}^{\infty} (\frac{2}{3})^n c = c$. But the relation

$$F(x) = f(x)-h_1(x)+\sum_{n=0}^{\infty}(h_n(x)-h_{n+1}(x))$$
$$= \lim_{n\to\infty}(f(x)-h_{n+1}(x))$$

holds on M. Since $|h_{n+1}| \leqslant (\frac{2}{3})^{n+1}c \searrow 0$ so $F(x) = f(x)$ for $x \in M$. Hence F is the sought-for extension of the function f.

Now let us prove (i). The mapping $i\colon \boldsymbol{R} \to]-1, 1[$, $i(x) := x/(1+ +|x|)$ is a homeomorphism. Consider the composition $i \circ f$, where f is a given continuous function on M.

We can apply (ii) to the function $i \circ f\colon M \to [-1, +1]$; suppose that F_1 is a continuous extension of $i \circ f$ to the whole space X constructed above. Since the two-point set $\{-1\}\cup\{+1\}$ is closed and disjoint from the interval $]-1, +1[$, we have that $A := F_1^{-1}(\{-1\}\cup\{+1\})$ is a closed set in X disjoint from M. Suppose that $l\colon X \to [-1, +1]$ is a continuous function such that $l(x) = 1$ for $x \in M$ and $l(x) = 0$ for $x \in A$. We verify that $F_2(x) := F_1(x)l(x)$, $x \in X$, is an extension of the function $i \circ f\colon M \to [-1, +1]$ and that $F_2(X) \subset i(\boldsymbol{R})$. The function $F := i^{-1} \circ \circ F_2\colon X \to \boldsymbol{R}$ is continuous and $F|_M = f$, and thus is the extension sought. □

We now prove an important property of normal spaces.

THEOREM XVII.3.3. Every normal space $(X, \mathscr{T})$ can be uniformized.

PROOF. As we know from Chapter XII, there exists in the space X a weakest uniform structure $\mathfrak{U}$ such that all continuous functions $f \in C(X)$ are uniformly continuous. The sets $\mathscr{N}_{f,\varepsilon} = \{(x, y) \in X\times X\colon |f(x)- -f(y)| < \varepsilon\}$ are entourages of this structure. We are to prove that the topology $\mathscr{T}(\mathfrak{U})$ specified by $\{\mathscr{N}_{f,\varepsilon}\}$ is identical with the original topology $\mathscr{T}$. Since f is continuous, the set $\mathcal{O}_{N_{f,\varepsilon}}(x_0) = \{x : |f(x_0)-f(x)| < \varepsilon\} \in \mathscr{T}$ is open. Suppose that, conversely, $\mathcal{O}$ is some $\mathscr{T}$-open neighbourhood of the point x_0; then, as a consequence of Urysohn's Lemma there is a function g continuous on X such that $g(x_0) = 0$ and $g(y) = 1$ for $y \notin \mathfrak{U}$. We have

$$\mathcal{O}_{\mathscr{N}_{g,1}}(x_0) = \{x\colon g(x) < 1\} \subset \mathcal{O},$$

therefore,

$$\mathscr{T} = \mathscr{T}(\mathfrak{U}). \quad \square$$

COROLLARY XVII.3.4. Every compact space has a unique uniform structure.

Thus, the property of compact spaces which had been announced in Chapter XII has been demonstrated.

4. TYCHONOFF SPACES. UNIFORMIZABILITY. COMPACTIFICATION

Upon examining the proof of the last theorem of Section 3, we note that in proving the uniformizability of the normal space X we made use only of the following property of Hausdorff spaces: For every point $x_0 \in X$ and every open set $\mathcal{O} \ni x_0$ there exists a continuous function $f: X \to [0, 1]$ such that $f(x_0) = 0$ and $f(x) = 1$ in $X - \mathcal{O}$.

Such spaces are called *Tychonoff spaces* or *completely regular spaces*. We have thus proved

Theorem XVII.4.1. Every Tychonoff space can be uniformized.

It may be shown that, conversely:

Every uniform space is a Tychonoff space.

The proof is omitted here.

COROLLARY XVII.4.2. Every topological vector space is a Tychonoff space.

Let us recall that a topological space $(X, \mathscr{T})$ is said to be regular if for every point $x_0 \in X$ and every open neighbourhood $\mathcal{O} \ni x_0$ there exists a closed neighbourhood of x_0 contained in $\mathcal{O}$.

PROPOSITION XVII.4.3. Every Tychonoff space is regular.

The proof is immediate: suppose that $\mathcal{O}(x_0)$ is an open neighbourhood of the point x_0 and suppose that f is a continuous function $f: X \to [0, 1]$ such that $f(x_0) = 0$ and $f(x) = 1$ for $x \in X - \mathcal{O}$. It exists by the definition of a Tychonoff space. Then $\{x : f(x) \leqslant \frac{1}{2}\}$ is a closed subset contained in $\mathcal{O}(x_0)$, and is a neighbourhood of x_0. $\square$

This proposition "justifies" the term completely regular.

Clearly, every subspace of a Tychonoff space is itself a Tychonoff space. In particular, every subspace of a normal space, hence every subspace of a compact space is a Tychonoff space. However, as we know (Alexandrov Theorem), every locally compact space is formed by separating of one point from a compact space. The following corollary has thus been proved:

COROLLARY XVII.4.4. Every locally compact space is a Tychonoff space, and thus can be uniformized.

In addition, we give a straightforward

LEMMA XVII.4.5 (A Hausdorff space $(X, \mathcal{T})$ is a Tychonoff space) $\Leftrightarrow$ (for any point $x_0 \in X$ and neighbourhood $\mathcal{U} \ni x_0$ belonging to a fixed subbasis $\mathcal{B}$ there exists a continuous function $f\colon X \to [0, 1]$ such that $f(x_0) = 0$ and $f(x) = 1$ for $x \notin \mathcal{U}$).

PROOF. $\Rightarrow$: obvious.

$\Leftarrow$: Suppose that $\mathcal{O}$ is an open subset and suppose that $x_0 \in \mathcal{O}$. From the definition of a subbasis (the intersections of a finite number of sets belonging to $\mathcal{B}$ form a basis) it follows that there exist sets $A_1, \ldots, A_n \in \mathcal{B}$ such that $x_0 \in \bigcap A_i \subset \mathcal{O}$. Suppose that f_i is a function (by assumption) such that $f_i\colon X \to [0, 1]$, $f_i(x_0) = 0$, $f_i(x) = 1$ for $x \notin A_i$, $i = 1, \ldots, n$. Clearly, the function $f = f_1 \cup \ldots \cup f_n$ has the required properties. □

On the basis of the foregoing lemma we immediately prove

THEOREM XVII.4.6. Let $(X_i, \mathcal{T}_i)$ be Tychonoff spaces; then the product $\prod_{i \in I} X_i$ is a Tychonoff space.

PROOF. As we know, sets of the form $\mathcal{U} = P_i^{-1}(\mathcal{O}_i)$, where $\mathcal{O}_i \in \mathcal{T}_i$, constitute the subbasis of the product $\prod X_i$. Suppose that $x = (x_i)$ and suppose that $f_i\colon X_i \to [0, 1]$ is a continuous function (on X_i) such that $f_i(x_i) = 0$ and $f_i(y) = 1$ for $y \in X_i - \mathcal{O}_i$; then $f_i \circ P_i\colon X \to [0, 1]$ is continuous, $(f_i \circ P_i)(x) = f_i(x_i) = 0$, $(f_i \circ P_i)(y) = f_i(y_i) = 1$ for $y = (y_i) \notin P^{-1}(\mathcal{O}_i)$, which completes the proof as a consequence of the lemma. □

It is thus seen that just as in the case of Hausdorff spaces and regular spaces, "to be a Tychonoff space" is a hereditary property (i.e., every subset has the property of the superset) and the multiplicative property (the Cartesian product of a space with a given property also has that property). Normal spaces can be shown not to possess these properties.

Compactification (*Compact Extension*). As Alexandrov's Theorem shows, for every locally compact space X there exists a compact space $\tilde{X}$ and a homeomorphic mapping $\mathscr{C}: X \to \tilde{X}$ such that $\overline{\mathscr{C}(X)} = \tilde{X}$.

DEFINITION. Every pair $(\mathscr{C}, \tilde{X})$ with the foregoing property is called a *compactification* (or *compact extension*) *of the Hausdorff space X*.

Since a closed subset of a compact space K is itself a compact space so only Hausdorff spaces which can be homeomorphically embedded $\mathscr{C}: X \to K$ in a compact space have compactifications. Hence it follows that the set $\overline{\mathscr{C}(X)}$ is compact and thus $(\mathscr{C}, \overline{\mathscr{C}(X)})$ is a compact extension of the space X.

As we know, a multitude of compact spaces is at our disposal: all products of compact spaces, hence spaces of the form $\overline{\boldsymbol{R}}^X$, $\overline{\boldsymbol{C}}^X$, $[0, 1]^X$, etc., where X are arbitrary sets. The foregoing spaces, however, are families of functions on the space X, and thus the mapping $F: X \to [0, 1]^X$ is nothing else than a family $\mathscr{F}$ of functions on X with values from $[0, 1]$. The mapping $F: X \to [0, 1]^X$ is injective iff $\mathscr{F}$ separates the points of the set X.

As we have shown, we had on the Tychonoff space X many continuous functions $X \to [0, 1]$ which separated points from closed sets, hence a fortiori points from points. After these remarks, we can prove the following interesting

THEOREM XVII.4.7 (Tychonoff's). (i) Every Tychonoff space $(X, \mathscr{T})$ can be compactified.

(ii) If a Hausdorff space X has a compactification then it is a Tychonoff space.

PROOF. (i) Let (f_i), $i \in I$, be the family of all continuous functions on X with values in $[0, 1]$. As we know (since the space X is completely regular), the family $\mathcal{O}_i = \{x: f_i(x) < 1\}$, $i \in I$ constitutes the basis of open neighbourhoods of the topology $\mathscr{T}$ for X. Denote $F(x) := (f_i(x))_{i \in I}$. Clearly, $F: X \to [0, 1]^I$. As noted earlier, F is a continuous embedding. To prove that the mapping F^{-1} is continuous, it is sufficient to show that F is open, i.e., that the set $F(\mathcal{O}_i)$ is open in $F(X)$ for every set $\mathcal{O}_i \in \mathscr{T}$. However, $F(\mathcal{O}_i) = F(X) - \{(y_j)_{j \in I}: y_i = 1\}$ and $\{(y_j)_{j \in I}: y_i = 1\} = P_i^{-1}(1)$ is closed, as the inverse image of the closed set $\{1\}$ of the continuous mapping P_i. Thus, $(F, \overline{F(X)})$ is a compactification of X.

(ii) Suppose that $(\mathscr{C}, \overline{\mathscr{C}(X)})$ is a compactification of the Hausdorff space X; then, as a subset of a compact space, $\mathscr{C}(X)$ is a Tychonoff space, and hence, as a topological image of a Tychonoff space, X is itself a Tychonoff space. □

Hereafter we shall use the abbreviated notation rX to indicate the compactification $(r, \overline{r(X)})$ as well as the space $\overline{r(X)}$ itself. In the set $\mathscr{R}(X)$ of all compactifications of a given space X we can introduce (i) an equivalence relation and (ii) a partial order relation.

Definition. $r_1X \sim r_2X$, if there exists a homeomorphism $T\colon r_1X \to r_2X$ such that $T \circ r_1 = r_2$, i.e., if the space X is embedded in r_iX, $i = 1, 2$, and the following diagram is commutative:

$$\begin{array}{ccc} r_1X & \xrightarrow{T} & r_2X \\ \uparrow & & \uparrow \\ X & \xrightarrow{\mathrm{id}_X} & X \end{array}$$

Definition. The compactification r_1X is *greater* than r_2X if there exists a continuous mapping $F\colon r_1X \to r_2X$ such that $F \circ r_1 = r_2$. Then we write $r_1X \succ r_2X$.

The following corollary is not difficult to prove.

Corollary XVII.4.8 $(r_1X \sim r_2X) \Leftrightarrow (r_1X \succ r_2X$ and $r_2X \succ r_1X)$.

It can be demonstrated that $\mathscr{R}(X)$ has the least upper bound βX; this is called the *Stone–Čech compactification*. It is characterized by the following theorem:

Theorem XVII.4.9. (i) Any continuous mapping $t\colon X \to K$ of a Tychonoff space X into a compact space K can be extended to a continuous mapping $T\colon \beta X \to K$.

(ii) Every compactification possessing this property is a Stone–Čech compactification (i.e., is equivalent to it).

Proof. (i) Take the mapping $r\colon X \to \beta X \times K$, $r(x) := (\beta(x), t(x))$. Reasoning as in the proof of Theorem XVII.4.7, we can show that r is a homeomorphic embedding and hence $rX = \overline{rX} \subset \beta X \times K$ is a compactification of the space X. However, the compactification βX is maximal, that is to say, there exists a continuous mapping $F\colon \beta X \to rX$, such that $F \circ \beta = r$. Let $T := P_K \circ F$, where P_K is a projection onto K. The formula

$$T \circ \beta = P_K \circ F \circ \beta = P_K \circ r = t \quad \text{and hence } T \supset t,$$

clearly holds.

(ii) Suppose that the compactification αX has the property mentioned in (i); then there exists a compactification $B: \alpha X \to \beta X$ of the mapping $\beta: X \to \beta X$, that is,

$$B \circ \alpha = \beta, \text{ that is } \alpha X \succ \beta X \text{ and hence } \alpha X \sim \beta X. \ \square$$

COROLLARY XVII.4.10. Every continuous function on X can be extended to a continuous function on βX. Every compactification with this property is equivalent to βX.

Tychonoff spaces are characterized by existence of many continuous functions, or, more precisely, they are spaces in which points can be separated from closed sets by means of continuous function: if $B = \overline{B}$ and $x_0 \notin B$, then there exists a continuous function $f: X \to [0, 1]$ such that $f(x_0) = 1, f(B) = \{0\}$. Henceforth we shall consider only Tychonoff spaces of continuous functions.

DEFINITION. A family of functions separating points from closed sets (points) is said to be *(pointwise) full.*

In the proof of Tychonoff's Theorem on compactification we see that to every full family of functions $\mathscr{F}_r$ on X there corresponds a compactification rX. From the construction of the space rX as the closure of the set $\mathscr{F}_r(X)$ in the product $[0, 1]^{\mathscr{F}_r}$ we see that every function $f \in \mathscr{F}_r$ may be extended to a (continuous) function on rX: if $y \in rX$, then $\tilde{f}(y) := P_f(y)$, where P_f is the projection onto the f-th coordinate. Since all the projections are continuous, the extension $\tilde{f}: rX \to [0, 1]$ is continuous. Thus if we take the full family $\mathscr{C}(X)$ of all continuous functions on X we obtain a compactification rX of X such that every function $f \in \mathscr{C}(X)$ is extendable to a continuous function on rX, or, by Corollary XVII.4.10, we obtain a Stone–Čech compactification. In this way we have proved

THEOREM XVII.4.11. To every full family of functions $\mathscr{F}_r$ on a Tychonoff space X there corresponds a compactification rX. The elements of the family $\mathscr{F}_r$ can be extended to rX. The greatest extension, βX, that is the Stone–Čech compactification, corresponds to the family $\mathscr{F}_\beta = \mathscr{C}(X)$ of all continuous functions.

Plainly, the same compactifications may correspond to different full families. Note that if we take the family $\mathscr{C}_r(X)$ of all functions having extensions on rX (clearly, $\mathscr{C}_r(X) \supset \mathscr{F}_r$), then $\mathscr{C}_r(X)$ is an algebra (under the usual operations) and the set of continuous extensions on rX constitutes an algebra of functions which separates the points of the compact space rX, hence is the full algebra $\mathscr{C}(rX)$ of all functions continuous on rX. Conversely, given the compactification rX, we have the algebra $\mathscr{C}(rX)$; the restriction $\mathscr{C}(rX)|X =: \mathscr{C}_r(X)$ is the full algebra of functions on X. Thus, different compactifications correspond to different full algebras of functions on X. Hence, different compactifications correspond to different full algebras of functions on X. Thus we have eliminated the "inconvenience" connected with the full families $\mathscr{F}_r$.

THEOREM XVII.4.12. There exists a bijective mapping from full function algebras to compactifications rX of a Tychonoff space. The Stone–Čech compactification βX is associated with the algebra $\mathscr{C}(X)$.

The question now is whether rX can be characterized algebraically in terms of the algebra $\mathscr{C}_r(X)$? Positive answer gives the theory of maximal ideals.

5. THE THEORY OF MAXIMAL IDEALS

To begin with, we give a definition of a commutative function algebra (over the field $\boldsymbol{R}$ or $\boldsymbol{C}$).

DEFINITION. An *algebra* $\mathscr{A}$ *(of continuous functions on* X*)* is a set which is closed under addition, subtraction, and multiplication of functions, and under multiplication by numbers. The function $1(x) \equiv 1$ is the identity element of the algebra.

An *ideal* in the algebra $\mathscr{A}$ is a subalgebra $\mathfrak{i} \subset \mathscr{A}$ such that $(f \in \mathfrak{i}$ and $g \in \mathscr{A}) \Rightarrow (f \cdot g \in \mathfrak{i})$. A *maximal ideal* is an ideal $\mathfrak{m} \neq \mathscr{A}$ which is not contained in any ideal distinct from $\mathscr{A}$.

Next we consider an example of fundamental importance.

EXAMPLE. Suppose that $\mathscr{A}$ denotes an algebra of functions on the space X. Let $\mathfrak{m}(x) := \{f \in \mathscr{A} : f(x) = 0\}$; plainly this is an ideal. We shall demonstrate that $\mathfrak{m}(x)$ is a maximal ideal. Let $g \in \mathscr{A}$ and let $g \notin \mathfrak{m}(x)$, that is, $g(x) \neq 0$; let $\mathfrak{M}$ be an ideal containing both $\mathfrak{m}(x)$ and the function g. Then every function $a \in \mathscr{A}$ can be written in the form

$$a(\cdot) = \frac{a(x)}{g(x)} g(\cdot) + \left(a(\cdot) - \frac{g(\cdot)}{g(x)} a(x) \right).$$

It is seen that the function in parentheses belongs to $\mathfrak{m}(x)$, whereas the expression $\frac{a(x)}{g(x)}\, g(\cdot)$ is proportional to $g(\cdot)$; hence $a(\cdot)$, as their sum, belongs to $\mathfrak{M}$, which means that $\mathfrak{M} = \mathscr{A}$, and thus the ideal $\mathfrak{m}(x)$ is maximal. □

The following proposition can be easily proved on the basis of the Kuratowski–Zorn Lemma.

PROPOSITION XVII. 5.1. Every ideal is contained in some maximal ideal.[1]

The special type of a functional algebras, so-called Hausdorff algebras, are of great importance; this is because on the space of their maximal ideals there exists a natural Hausdorff topology.

DEFINITION. An algebra $\mathscr{H}$ is called a *Hausdorff algebra* if for any two maximal ideals $\mathfrak{m}_1 \neq \mathfrak{m}_2$ there exist elements $f_i \notin \mathfrak{m}_i$, $i = 1, 2$, such that $f_1 \cdot f_2 = 0$.

THEOREM XVII.5.2 (Stone, Gel'fand, Kolmogorov). (i) In the set $\mathscr{M}$ of maximal ideals of a Hausdorff algebra $\mathscr{H}$ can be introduced a Hausdorff topology $\mathscr{T}$, by taking sets of the form $\mathscr{O}_f = \{\mathfrak{m}: f \notin \mathfrak{m}\}, f \in \mathscr{H}$ for the basis.

(ii) $(\mathscr{M}, \mathscr{T})$ is a compact space.

PROOF. (i) The set $\mathscr{O}_0 = \emptyset$ corresponds to the zero function and to the identity element of the algebra corresponds $\mathscr{O}_1 = \mathscr{M}$, since no maximal ideal contains the identity element 1, for if it were that $1 \in \mathfrak{m}$, then for every $g \in \mathscr{H}$ we would have $g = 1 \cdot g \in \mathfrak{m}$. Therefore, $\emptyset, \mathscr{M} \in \mathscr{T}$, $\mathfrak{m} \in \mathscr{O}_f$, if $f \notin \mathfrak{m}$; $\mathfrak{m}_1 \notin \mathscr{O}_g$, if $g \in \mathfrak{m}_1$; $\mathscr{O}_{f \cdot g} = \{\mathfrak{m} \in \mathscr{M}: f \cdot g \notin \mathfrak{m}\}$, $\mathfrak{m} \notin \mathscr{O}_{f \cdot g}$, if $f \cdot g \in \mathfrak{m}$, hence $(\mathfrak{m} \notin \mathscr{O}_{f \cdot g}) \Leftrightarrow (f \cdot g \in \mathfrak{m}) \Leftrightarrow (\mathfrak{m} \ni f$ or $\mathfrak{m} \ni g)$. The last statement means that $\mathscr{M} - \mathscr{O}_{f \cdot g} = (\mathscr{M} - \mathscr{O}_f) \cup (\mathscr{M} - \mathscr{O}_g) = \mathscr{M} - (\mathscr{O}_f \cap \mathscr{O}_g)$, that is, $\mathscr{O}_{f \cdot g} = \mathscr{O}_f \cap \mathscr{O}_g$, so $\{\mathscr{O}_f: f \in \mathscr{H}\}$ satisfies the axioms for the basis of topology.

We shall now prove that $\mathscr{T}$ is a Hausdorff topology: suppose that $\mathfrak{m}_1 \neq \mathfrak{m}_2$; since $\mathscr{H}$ is a Hausdorff algebra, there exist elements $f_1 \notin \mathfrak{m}_1$ and $f_2 \notin \mathfrak{m}_2$, such that $f_1 \cdot f = 0$, that is,

$$\mathfrak{m}_1 \in \mathscr{O}_{f_1}, \quad \mathfrak{m}_2 \in \mathscr{O}_{f_2}, \quad \text{but} \quad \mathscr{O}_{f_1 \cdot f_2} = \emptyset.$$

[1] The proof can be found in the author's monograph *Methods of Hilbert Spaces*.

(ii) It remains to show that a finite covering of the space $\mathscr{M}$ can be selected from every open covering $\{\mathcal{O}_f\}_{f\in F}$. Note that the set $\mathfrak{i}$ of all (finite) sums of the form

$$f_{i_1} \cdot g_1 + \ldots + f_{i_n} \cdot g_n, \quad \text{where } f_{i_k} \in F,\ g_k \in \mathscr{H},$$

constitutes an ideal. This set is clearly an algebra and for any $g \in \mathscr{H}$ the element $(f_{i_1} \cdot g_1 + \ldots + f_{i_n} \cdot g_n)g = f_i \cdot (g_1 \cdot g) + \ldots + f_{i_n} \cdot (g_n \cdot g)$ is once again an element of the set $\mathfrak{i}$. Let us verify that $\mathfrak{i} = \mathscr{H}$. Otherwise $\mathfrak{i}$ would belong to some maximal ideal $\mathfrak{m}$, that is we would have $f \in \mathfrak{m}$, or $\mathfrak{m} \notin \mathcal{O}_f$ for every $f \in F$, which would be in contradiction with the fact that $\{\mathcal{O}_f\}$, $f \in F$, is a covering of the space $\mathscr{M}$. Hence $\mathfrak{i} = \mathscr{H}$ and thus $f_{i_1} g_1 + \ldots + f_{i_l} g_l = 1$ for some indices $i_1, \ldots, i_l$. We shall show that $\bigcup_{k=1}^{l} \mathcal{O}_{f_{i_k}} = \mathscr{M}$.

PROOF (indirect): Suppose that some $\mathfrak{m} \notin \bigcup_{k=1}^{l} \mathcal{O}_{f_{i_k}}$, that is, $f_{i_1}, \ldots$ $\ldots, f_{i_k} \in \mathfrak{m}$; accordingly $f_{i_1} \cdot g_1 + \ldots + f_{i_l} \cdot g_l = 1 \in \mathfrak{m}$, and thus $\mathfrak{m} = \mathscr{M}$, so $\mathfrak{m}$ is not a maximal ideal. □

We use this abstract theorem to examine the full algebra $\mathscr{C}_r(X)$ of continuous functions on a Tychonoff space X. We have the following theorem:

THEOREM XVII.5.3. (i) Every full algebra $\mathscr{C}_r(X)$ is a Hausdorff algebra.

(ii) Its space of maximal ideals $\mathscr{M}_r$ is a compactification equivalent to the compactification rX.

A set of maximal ideals $\mathscr{M}_\beta$ of the algebra $\mathscr{C}(X)$ is a Stone–Čech compactification βX.

The proof is preceded by a straightforward lemma:

LEMMA XVII.5.4. Suppose that $\mathscr{A}$ is a full algebra of functions on the Tychonoff space X and suppose that $\mathfrak{i}$ is an ideal in $\mathscr{A}$. If $\frac{1}{2} \leqslant f(x) \leqslant \frac{3}{2}$, $x \in X$, and $f \in \mathfrak{i}$, then $\mathfrak{i} = \mathscr{A}$.

PROOF OF THE LEMMA. As we know, no ideal different from $\mathscr{A}$ can contain the identity element; hence if $f \in \mathfrak{i}$ and $1/f \in \mathscr{A}$, then $1 = f \cdot (1/f) \in \mathfrak{i}$, or $\mathfrak{i} = \mathscr{A}$. But if $g := 1 - f \in \mathscr{A}$, then $|g| \leqslant \frac{1}{2}$, whereby $1/f = 1/(1-g) = \sum_{n=0}^{\infty} g^n \in \mathscr{A}$, and thus $\mathfrak{i} = \mathscr{A}$. □

Proof of Theorem XVII.5.3. Let $\mathfrak{m}_1 \neq \mathfrak{m}_2$ be maximal ideals. Thus, there exists an $f \in \mathfrak{m}_1 - \mathfrak{m}_2$. It is seen that $\mathfrak{i} = \{af+g: a \in \mathscr{A},\ g \in \mathfrak{m}_2\}$ is an ideal since $af \in \mathfrak{m}_1, g \in \mathfrak{m}_2$, and that $\mathfrak{i} \supset \mathfrak{m}_2$; now, $\mathfrak{m}_2$ is maximal, and thus we have $\mathfrak{i} = \mathscr{C}_r(X)$. Since $1 \in \mathscr{C}_r(X)$, there exist $h_i \in \mathfrak{m}_i$, $i = 1, 2$, such that $h_1 + h_2 = 1$. The functions $h_i' := h_i^2/(h_1^2+h_2^2) \in \mathfrak{m}_i$, $i = 1, 2$, since $h_i \in \mathfrak{m}_i$, and $\mathfrak{m}_i$ is an ideal. However, $h_1'+h_2' = 1$, $h_i' \geqslant 0$; accordingly, let

$$f_1 := (h_2'-\tfrac{1}{2}) \cup 0, \qquad f_2 := (h_1'-\tfrac{1}{2}) \cup 0,$$

now $h_2'-\frac{1}{2} = \frac{1}{2}-h_1'$, and thus if $h_2'(x) \leqslant \frac{1}{2}$, then $1 \geqslant h_1'(x) \geqslant \frac{1}{2}$, whence $h_2'(x)-\frac{1}{2} \leqslant 0$, so $f_1(x) = 0$ and, therefore, in this case $f_1 \cdot f_2(x) = 0$.

On the other hand, if $1 \geqslant h_2'(x) \geqslant \frac{1}{2}$, then $h_1'(x) = 1-h_2'(x) \leqslant \frac{1}{2}$, whence $f_2(x) = 0$, and thus in this case as well $f_1(x) \cdot f_2(x) = 0$. Therefore, we always have $f_1 \cdot f_2 = 0$.

In order to show that $\mathscr{C}_r(X)$ is a Hausdorff algebra, it is sufficient to prove that $f_i \notin \mathfrak{m}_i$, $i = 1, 2$. If it were that $f_1 \in \mathfrak{m}_1$, then also $f_1 + h_1'$ $\in \mathfrak{m}_1$; but $0 \leqslant f_1 \leqslant \frac{1}{2}$, $0 \leqslant h_i' \leqslant 1$, hence $f_1 + h_1' = (h_2'+h_1'-\frac{1}{2}) \cup h_1'$ $= \frac{1}{2} \cup h_1' \geqslant \frac{1}{2}$, or $\frac{1}{2} \leqslant f_1 + h_1' \leqslant \frac{3}{2}$; similarly, $\frac{1}{2} \leqslant f_2 + h_2' \leqslant \frac{3}{2}$.

By Lemma XVII.5.4 it follows that if $f_i \in \mathfrak{m}_i$, then $\mathfrak{m}_i = \mathscr{C}_r(X)$, which contradicts the fact that the ideal $\mathfrak{m}_i$ is maximal. Thus, $f_i \notin \mathfrak{m}_i$, $i = 1, 2$. It remains to construct a topological injection $T: X \to \mathscr{M}_r$ such that we would have $\overline{T(X)} = \mathscr{M}_r$. To this end we take $\mathfrak{m}(x) = \{f: f(x) = 0\}$, as in the example. This, as we know, is a maximal ideal. Let $T(x) := \mathfrak{m}(x)$. Then $T: X \to \mathscr{M}_r$. The mapping T is an injection, for if $x_1 \neq x_2$, then we have $\mathfrak{m}(x_1) \neq \mathfrak{m}(x_2)$ since the functions separate points.

Sets of the form $\mathscr{U}_f = \{x: f(x) \neq 0\}$, where $f \in \mathscr{C}_r(X)$, constitute a subbasis of neighbourhoods of the space X. However, by the definition of the topology $\mathscr{T}$ we have $\mathscr{O}_f = \{\mathfrak{m}: f \notin \mathfrak{m}\}$. Now, $(f \notin \mathfrak{m}(x)) \Leftrightarrow (f(x)$ $\neq 0)$, and thus $(\mathfrak{m}(x) \in \mathscr{O}_f) \Leftrightarrow (f(x) \neq 0$, or if $x \in \mathscr{U}_f)$. Hence $T^{-1}(\mathscr{O}_f)$ $= \mathscr{U}_f$, $T(\mathscr{U}_f) = \mathscr{O}_f \cap T(X)$, that is, T is a continuous, open injection. The set $T(X)$ is dense in $\mathscr{M}_r$; we have to show that $\mathscr{O}_f \cap T(X) \neq \emptyset$ for $\mathscr{O}_f \neq \emptyset$, or $f \neq 0$. Suppose that x is such that $f(x) \neq 0$; then $f \notin \mathfrak{m}(x)$, or $\mathfrak{m}(x) \in \mathscr{O}_f$, which implies that $\mathfrak{m}(x) \in \mathscr{O}_f \cap T(X)$. The proof has thus been completed. □

If the space X is compact, then the set $T(X)$ is compact and dense in $\mathscr{M}_r$ and, therefore, $T(X) = \mathscr{M}_r$, and X is homeomorphic with $\mathscr{M}_r$, but rX is homeomorphic with $\mathscr{M}_r$ and, thus, X. Accordingly, in constructing a space of maximal ideals for X and rX, we obtain the same space. However, as we know from the Stone–Weierstrass Theorem, $\mathscr{C}_r(X) = \mathscr{C}(X)$ and thus all algebras $\mathscr{C}_r(X)$ are identical. But the algebra $\mathscr{C}(rX)$, as an algebra of extensions of all functions, is isomorphic with $\mathscr{C}_r(X)$. In this way we have arrived at

THEOREM XVII.5.5. For compact spaces X, Y the following conditions are equivalent.:

(i) X and Y are homeomorphic;
(ii) some full algebras $\mathscr{C}_r(X)$ and $\mathscr{C}_r(Y)$ are isomorphic;
(iii) all full algebras of functions on X and Y are isomorphic;
(iv) the algebras $\mathscr{C}(X)$ and $\mathscr{C}(Y)$ are isomorphic.

This theorem demonstfates forcibly that examination of the algebras of continuous functions gives a deep insight into the structure of topological (compact) spaces.

Connection with the Theory of Almost Periodic Functions. The Bohr–Weil Compactification. In Chapter XI, Section 3 we gave a definition of an almost periodic function. This definition can be carried over (as Bochner and von Neumann did) to any topological groups, in the following manner:

DEFINITION. Let G be a topological group, i.e., the group operations are continuous in it (cf. Chapter XIV). A (bounded) function $f \in \mathscr{C}(G, C)$ is said to be *left* (*right*) *almost periodic* if the set $\{\tau_g f \colon g \in G\}$ is precompact in $B(G, C)$, where $\tau_g f(x) := f(g^{-1}x)$ (taking the right translation $f(x) \to f(xg)$, we obtain a definition of a right almost periodic function).

It can be shown that every left almost periodic function is right almost periodic, and vice versa, so that henceforth we shall write simply "almost periodic". Clearly, if G is a compact group then every function continuous on G is almost periodic (the proof is identical with that for Lemma XI.3.1 in which use was made only of the fact that the unit circle S^1 is compact. This observation gave André Weil the splendid idea of treating almost periodic functions as functions which are continuous on some compactification of the group G. The group structure, however, breaks down under the compactification rG, and thus a somewhat different notion of compactification needs to be introduced.

DEFINITION. A continuous compactification of a topological space X is a pair $(\varphi, \tilde{X})$, where $\varphi: X \to \tilde{X}$ is a continuous (hence, in general, non-topological) mapping of the space X into compact space $\tilde{X}$, such that the image $\varphi(X)$ is dense in $\tilde{X}$.

The following theorem, due to A. Weil, "trivializes" the theory of almost periodic functions, i.e., reduces it to a theory of continuous functions on a compact group:

Theorem XVII.5.6 (A. Weil). For any topological group G there is a continuous compactification $(\varphi, \tilde{G})$ such that almost periodic functions on G, and only these functions, are of the form $h \circ \varphi$, where h is an arbitrary function continuous on $\tilde{G}$.

Construction of the Group $\tilde{G}$. Suppose that $\mathscr{P}$ is the set of all almost periodic functions on G. Since the set $A_f := f(G)$ is relatively compact in $\boldsymbol{C}$ for every function $f \in \mathscr{P}$, so the space $\bigtimes_{f \in P} \bar{A}_f$ is compact. The mapping $\varphi: G \to \bigtimes_{f \in P} \bar{A}_f$ plainly is defined as follows:

$$\varphi(g) = \big(f(g)\big), \quad f \in \mathscr{P}.$$

Now we define the compact set $\tilde{G} := \overline{\varphi(G)}$. As a closed subset of a compact space, $\tilde{G}$ is compact. The point now is to introduce a group structure in the space $\tilde{G}$ in a manner compatible with the topology on $\tilde{G}$.

Clearly, multiplication of the elements $\varphi(g_1)$ and $\varphi(g_2)$ must be prescribed as:

$$K\big(\varphi(g_1), \varphi(g_2)\big) := \varphi(g_1 g_2).$$

The mapping $K: \varphi(G) \times \varphi(G) \to \tilde{G}$ must now be extended to the continuous mapping $\tilde{K}: \tilde{G} \times \tilde{G} \to \tilde{G}$.

Inverse elements must be defined in similar fashion in $\tilde{G}$. For this purpose, an extension $\tilde{L}$ is constructed for the mapping $L: \varphi(G) \to \varphi(G)$, where $L\big(\varphi(g)\big) := \varphi(g^{-1})$. On the basis of the density in $\tilde{G}$ of the group $\varphi(G)$ so constructed, $\tilde{G}$ is proven to be a (compact) group. Denoting the projection $\tilde{G} \to A_f$ by P_f, we see that all of the almost periodic functions on G are of the form $P_f \circ \varphi$. It can be shown that the function $h \circ \varphi$, where $h \in \mathscr{C}(\tilde{G})$, is almost periodic.

6. THE GEL'FAND THEORY OF MAXIMAL IDEALS[1]

An algebra of continuous functions on a compact space X is a Banach space in a natural fashion, $||f|| := \sup|f(X)|$. This leads to the general concept of a Banach algebra.

A *Banach algebra* is an algebra $\mathscr{A}$ which is at the same time a Banach space (over the complex field), and the inequality $||AB|| \leqslant ||A|| \cdot ||B||$, $A, B \in \mathscr{A}$ is satisfied.

Hereafter we shall concern ourselves with commutative Banach algebras; the principal results of this theory are due to I. M. Gel'fand.

Examples. 1. The algebra $\mathscr{C}(X)$ (of complex functions on X).

2. Let E be a Banach space; the set $L(E)$ of bounded linear operators is in a natural way a Banach algebra, and $A \cdot B := A \circ B$.

3. Suppose that G is a locally compact group. A convolution (cf. Chapter XVIII)

$$(f * h)(x) := \int_G f(xg^{-1})h(g)\,d\mu(g)$$

can be associated with every pair of functions, $f, h \in L^1(G, \mu)$, where μ is a (left-invariant) Haar measure.

The space $L^1(G, \mu) = L^1(G)$ is an algebra under convolution. This is a Banach algebra. This convolution algebra will be considered in more detail in Chapter XVIII. If G is an Abelian group, then the algebra $L^1(G)$ is commutative. The algebra $L^1(G)$ does not, in general, have an identity element. This is only the case when the group G is discrete, then the function $\chi_{\{e\}}$ is its identity element. The theory of the convolution algebra $L^1(G)$ is an exceptionally beautiful theory in the case of commutative algebras; this is perhaps the most impressive application of Gel'fand's theory.

4. If H is a Hilbert space, the algebra $\mathscr{L}(H)$ has an involution, i.e., the mapping $*: \mathscr{L}(H) \to \mathscr{L}(H)$, $*(A) := A^*$ of taking the adjoint operator

$$(Ah|u) = (h|A^*u) \quad \text{for all } h, u \in H.$$

DEFINITION. An *involution in a Banach algebra* is an (antilinear) homomorphism $*$ such that $(AB)^* = B^*A^*$, $(\lambda A)^* = \bar{\lambda}A^*$ for $\lambda \in \mathbf{C}$ and

[1] The proofs of the theorems in Section 6 can be found in *Methods of Hilbert Spaces*.

$(A^*)^* = A$. An involution is usually assumed to possess the property that

$$\text{(I)} \qquad \|A\| = \|A^*\|, \qquad \|A^*A\| = \|A^*\| \cdot \|A\|.$$

A Banach algebra with involution possessing the property (I) is a *C^*-algebra.*

The algebras of Examples 4 and 1 are plainly algebras with involution if we define $f^*(x) := \overline{f(x)}$.

Gel'fand and Naimark have proved that every commutative algebra $\mathscr{A}$ with involution and identity element is isometrically isomorphic with the algebra $\mathscr{C}(X)$ of continuous functions on the compact set X. The reader no doubt surmises that the space X is the set of maximal ideals of the algebra $\mathscr{A}$.

The road to the Gel'fand–Naimark Theorem leads through the Gel'fand Theorem (which we shall formulate further on); the latter in turn is obtained via the straightforward Mazur–Gel'fand Theorem which states that with every maximal ideal $\mathfrak{m}$ in a commutative Banach algebra there is associated a continuous homomorphism in $\boldsymbol{C}$ (i.e., a continuous linear and multiplicative functional $\mathscr{A} \ni A \to \hat{A}(\mathfrak{m}) \in \boldsymbol{C}$, $\mathfrak{m} \in \mathscr{M}$, since the quotient algebra $\mathscr{A}/\mathfrak{m} = \boldsymbol{C}$. Let $\hat{\mathscr{A}}$ denote the family of functions, on the set $\mathscr{M}$ of maximal ideals of the commutative Banach algebra $\mathscr{A}$, which are defined as

$$\hat{A}(\mathfrak{m}) = [A]_{\mathfrak{m}},$$

where $[\cdot]_{\mathfrak{m}}$ denotes the canonical mapping $\mathscr{A} \to \mathscr{A}/\mathfrak{m}$.

It is not difficult to notice that the mapping $\mathscr{A} \to \hat{\mathscr{A}}$ defined above is an injection if $\bigcap_{\mathfrak{m} \in \mathscr{M}} \mathfrak{m}$, called the *radical of the algebra* $\mathscr{A}$, is the zero element of the algebra. An algebra whose radical is zero is called a *semisimple algebra.* We can now give the celebrated

THEOREM XVII.6.1 (Gel'fand). Suppose that $\mathscr{A}$ is a commutative Banach algebra with identity. Then the set $\mathscr{M}$ of maximal ideals of the algebra $\mathscr{A}$ equipped by the weakest topology induced by the family $\hat{\mathscr{A}}$ (i.e., the weakest topology such that all functions $\hat{A}(\mathfrak{i})$, $\hat{A} \in \hat{\mathscr{A}}$, are continuous) is a compact space. The Gel'fand homomorphism

$$\mathscr{A} \ni A \to \hat{A} \in \hat{\mathscr{A}} = \mathscr{C}(\mathscr{M})$$

is a (continuous) isomorphism if the algebra $\mathscr{A}$ is semisimple.

Remark. It $\mathscr{A}$ does not have an identity element, then the space $\mathscr{M}$ is (clearly) locally compact, and $\hat{\mathscr{A}}$ is a space $\mathscr{C}_\infty(\mathscr{M})$ of continuous functions vanishing at infinity (a continuous function on a locally compact space vanishes at infinity if the set $\{x: |f(x)| \geqslant \varepsilon\}$ is compact for all $\varepsilon > 0$.

DEFINITION. The *spectrum of the element* $A \in \mathscr{A}$ is the set

$$\operatorname{Sp} A = \{\lambda \in \boldsymbol{C}: A - \lambda 1 \text{ has no inverse}\}.$$

We have a fact of fundamental importance for spectral theory:

THEOREM XVII.6.2 (Gel'fand's Spectral Theorem). $\operatorname{Sp} A = \hat{A}(\mathscr{M})$, that is, if a Banach algebra is generated by one element, i.e., if we consider an algebra spanned by 1 and A (the norm limits of all polynomials $\sum a_0 1 + \sum c_k A^k$), then the set of maximal ideals $\mathscr{M}$ may be identified with the spectrum $\operatorname{Sp} A$ of the element A, since these are homeomorphic sets.

Consequently, $\mathscr{M}$ is often called the *spectrum of the algebra* $\mathscr{A}$.

In the spectral theory of operators in Hilbert space an important role is clearly played by

THEOREM XVII.6.3 (Gel'fand–Naimark Theorem). Every commutative Banach algebra with involution is semisimple and is isometrically isomorphic with the algebra $\mathscr{C}(\mathscr{M})$; this isomorphism also preserves involution.

In particular, we can take a commutative C^*-algebra $\mathscr{A}$ of operators in Hilbert space $(H, (\cdot|\cdot))$. If

$$\mathscr{C}(\mathscr{M}) \ni f \to T_f \in \mathscr{L}(H)$$

is a Gel'fand–Naimark isomorphism, it can be extended to Borel-measurable functions as follows:

$$F(f; u, v) := (T_f u|v), \quad f \in \mathscr{C}(\mathscr{M}),\ u, v \in H,$$

F is a trilinear function. If u and v are fixed we have

$$F(f; u, v)| \leqslant ||T_f|| \cdot ||u|| \cdot ||v|| = ||f|| \cdot ||u|| \cdot ||v||,$$

whereby $F(\mu; u, v)$ is a complex Radon integral which we denote by $\mu_{u,v}$ or $\mu(\cdot\,; u, v)$. We write this as the formula

$$\text{(S)} \qquad \int_{\mathscr{M}} f(m)\, d\mu(m; u, v) = (T_f u|v).$$

The family $(\mu(\cdot\,; u, v))$, $u, v \in H$, is often referred to as a *spectral measure*. The identity (S) can easily be extended to all functions bounded and measurable with respect all measures $\mu(\cdot\,; u, v)$. It is more difficult to prove the existence of a basic spectral measure, i.e., a Radon measure μ on $\mathscr{M}$ such that for all u, v we have $\mu(\cdot\,; u, v) \ll \mu$, that is, (every set $\mathscr{N} \subset \mathscr{M}$ is of μ-measure zero) $\Leftrightarrow (\mu(\{\mathscr{N}\}; u, v) = 0, u, v \in H)$. A bounded operator T_g can now be associated with every function $g \in L^\infty(\mu)$; it turns out that the Gel'fand–Naimark isomorphism ceases to be (in general) an isometry but an inequality "in the proper direction" holds,

$$\|T_g\| \leqslant \|g\|_\infty.$$

We recall that $\|g\|_\infty$ is a norm in the space $L^\infty(\mu)$ and denotes the essential supremum of the function $|g|$.

To what do the characteristic functions of μ-measurable sets correspond? Since, $\chi_\Lambda = \chi_\Lambda^2 = \bar{\chi}_\Lambda$; hence $E(\Lambda) := T\chi_\Lambda$ is an idempotent operator, $E(\Lambda)^2 = E(\Lambda)$, and is Hermitian, $E(\Lambda)^* = E(\Lambda)$. If we denote $H(\Lambda) := E(\Lambda)H$, it is not difficult to see that $E(\Lambda)$ is the operator of an orthogonal projection onto the subspace $H(\Lambda)$.

Since every function f continuous on $\mathscr{M}$ can be uniformly approximated by step functions belonging to $S(\mu)$, we obtain

$$\left\|T_f - \sum f(\lambda_i)E(\Lambda_i)\right\| \leqslant \left\|f - \sum f(\lambda_i)\chi_{\Lambda_i}\right\| < \varepsilon.$$

In particular, if our algebra is generated by the Hermitian operator $L = L^*$, to which the function $f(\lambda) = \lambda$, $\mathscr{M} = \operatorname{Sp} L$, corresponds in the Gel'fand–Naimark isomorphism, we obtain the famous spectral theorem.

Theorem XVII.6.4 (Spectral Theorem). For every $f \in \mathscr{C}(\operatorname{Sp} L)$ the operator $f(L)$ admits uniform approximation by a combination of projection operators,

$$f(L) = \lim \sum f(\lambda_i)E(\Lambda_i);$$

in particular, if $f(L) = L$, then

$$(*) \qquad L = \lim \sum \lambda_i E(\Lambda_i).$$

The family $\{E(\Lambda_i)\}$ is called a *spectral family* (or *partition of unity*) specified by the algebra $\mathscr{A}$, in particular by the operator L. Formula (*) is written suggestively as

$$L = \int_{\mathrm{Sp}L} \lambda \, dE_\lambda .$$

Remark. Proofs of the Gel'fand theory are given in Chapter XX.

7. CONNECTION WITH QUANTUM MECHANICS

Commutative algebras of operators in a Hilbert space play a fundamental role in quantum mechanics. According to Dirac and von Neumann, some self-adjoint operator L corresponds to every observable physical quantity (observables). At the same time, only those physical quantities which are simultaneously measurable are commutative as operators (in the Hilbert space of physical states). However, as we know, such operators generate a commutative C^*-algebra $\mathscr{A}$ with a locally compact space $\mathscr{M}$ as its spectrum. If we have n commutative operators $L_1, \ldots, L_n$, then $\mathscr{A}(L_1, \ldots, L_n)$ has for its space of maximal ideals the set $\mathscr{M}$ which is a subset of the space R^n. Suppose that μ is the basic spectral measure of the operators $L_1, \ldots, L_n$, that is, of the algebra $\mathscr{A}(L_1, \ldots, L_n)$ which is generated by $L_1, \ldots, L_n$. Let us normalize it so that

$$\mu(\mathscr{M}) = \mu(R^n) = 1.$$

The physical states are represented by unit vectors $u \in H(\|u\| = 1)$. Accordingly, the Hilbert space H is called the *space of (physical) states*. If we have a system of commutative observables $\mathscr{A}$, then, as we know, with every state u we have associated a measure $\mu_u := \mu(\cdot\,; u, u)$. This is a probability measure $\mu_u(\mathscr{M}) = 1$; indeed,

$$(1u|u) = 1 = \int 1 d\mu_u = \mu_u(\mathscr{M}).$$

Hence, probability measures on the spectrum of the algebra $\mathscr{A}$ generated by the commutative system of observables correspond to physical states. Let us return to the algebra $\mathscr{A} = \mathscr{A}(L_1, \ldots, L_n)$ generated by n observables $L_1, \ldots, L_n$. As we know, the spectrum of the algebra $\mathscr{A}$ can be identified with the Cartesian product of the spectrum of the operators $L_1, \ldots, L_n$:

$$\mathscr{M} = \mathrm{Sp}L_1 \times \ldots \times \mathrm{Sp}L_n \subset R^n.$$

To every Borel set $\Lambda \subset \mathcal{M}$, for instance the set $\Lambda = \bigtimes_{i=1}^{n} [a_i, b_i]$, we can associate a number $\mu(\Lambda)$ (μ-basic measure), which is interpreted as the probability that a simultaneous measure of the physical quantities $L_1, \ldots, L_n$ will yield a result lying in the set Λ, e.g., that the observable L_i has the value $a_i \leqslant \lambda_i \leqslant b_i$. The coordinates of the point $\lambda = (\lambda_1, \ldots \ldots, \lambda_n) \subset \mathcal{M}$ are called *quantum numbers*. In the first stage in the development of quantum theory it was suspected that the quantum numbers run over a discrete set. This obviously is not so in the general case. This does occur only if the operators $L_1, \ldots, L_n$ representing observables have a purely discrete (point) spectrum, i.e., if $\mathcal{M}$ is a discrete subspace of the space $\boldsymbol{R}^n$. For a moment we end this short discussion. At the end of Chapter XX we shall come back to further relations with quantum mechanics and direct integrals.

8. LOCALLY FINITE FAMILIES

Let X be a topological space.

DEFINITION. The family of sets $\{A_i\}_{i \in I}$, $A_i \subset X$, is said to be *locally finite* if every point $x \in X$ has a neighbourhood $\mathscr{V}$ which intersects with, at most, a finite number of sets A_i.

THEOREM XVII.8.1. If a family $\{A_i\}_{i \in I}$ of subsets of a topological space X is locally finite, then

$$\overline{\bigcup_{i \in I} A_i} = \bigcup_{i \in I} \overline{A_i}.$$

PROOF. For every $j \in I$ we have $A_j \subset \bigcup_{i \in I} A_i$ and hence

$$\overline{A_j} \subset \overline{\bigcup_{i \in I} A_i}, \quad \text{that is} \quad \bigcup_{j \in I} \overline{A_j} \subset \overline{\bigcup_{i \in I} A_i}.$$

We shall prove that the inverse inclusion holds. Let $x \in \overline{\bigcup_{i \in I} A_i}$, that is, points of the set $\bigcup_{i \in I} A_i$ lie in every neighbourhood of the point x.

Suppose that $\mathscr{V}$ is the neighbourhood of the point x as in the definition of local finiteness (i.e., is a neighbourhood intersecting with a finite number of sets A_i). If $\mathscr{U} \subset \mathscr{V}$, then let $n(\mathscr{U})$ denote the number of sets from the family $\{A_i\}_{i \in I}$ which intersect with the set $\mathscr{U}$. Plainly,

$0 \leqslant n(\mathscr{U}) \leqslant n(\mathscr{V})$. Let us denote:

$n_0 := \min\{n(\mathscr{U}): \mathscr{U} \subset \mathscr{V}, \mathscr{U}$ is a neighbourhood of the point $x\}$.

Clearly, $n_0 \geqslant 1$, since, as we have said, every neighbourhood of x contains points from the set $\bigcup_{i\in I} A_i$. Suppose that $\mathscr{U}_0$ is a neighbourhood such that $n(\mathscr{U}_0) = n_0$. Then every neighbourhood of x contained in $\mathscr{U}_0$ meets the same sets from the family $\{A_i\}_{i\in I}$ as $\mathscr{U}_0$ does. This is so because $\mathscr{U}$ (a neighbourhood contained in $\mathscr{U}_0$) could meet fewer sets A_i, as a smaller set; however, $n(\mathscr{U}) = n(\mathscr{U}_0)$, since $n(\mathscr{U}_0) = \min n(\mathscr{U})$.

Let $A_{i_1}, \ldots, A_{i_{n_0}}$ be sets from the family $\{A_i\}_{i\in I}$ which intersect with $\mathscr{U}_0$. Let $\mathscr{O}$ be an arbitrary neighbourhood of the point x. The set $\mathscr{O}$ contains points of the sets $A_{i_1}, \ldots, A_{i_{n_0}}$ since the non-empty set $\mathscr{O} \cap \mathscr{U}_0 \subset \mathscr{U}_0$ contained in it intersects those sets.

Since an arbitrary neighbourhood of x contains elements of the sets $A_{i_1}, \ldots, A_{i_{n_0}}$, we have $x \in \bar{A}_{i_1}, \ldots, x \in \bar{A}_{i_{n_0}}$, or $x \in \bigcup_{i\in I} \bar{A}_i$. Accordingly, $\overline{\bigcup_{i\in I} A_i} \subset \bigcup_{i\in I} \bar{A}_i$, whence the theorem follows. □

Examples. 1. Suppose that $X = \boldsymbol{R}^1$ is a space with an usual topology. Take the family $A_\varepsilon := [\varepsilon, \infty[$ for $\varepsilon > 0$. This family is not locally finite since every neighbourhood of the point $x = 0$ meets an infinite number of elements of the family. As could be expected, the hypothesis of Theorem XVII.3.1 does not hold since $\bigcup_{\varepsilon>0} \bar{A}_\varepsilon = \bigcup_{\varepsilon>0} A_\varepsilon =]0, \infty[$, whereas $\overline{\bigcup_{\varepsilon>0} A_\varepsilon} = \overline{]0, \infty[} = [0, \infty[$.

2. The family $A_{n\infty} = [n, \infty[, n \geqslant 0$, is locally finite in the space $\boldsymbol{R}^1$. And indeed $\bigcup_{n=0}^{\infty} \bar{A}_n = [0, \infty[= \overline{\bigcup_{n=0}^{\infty} A_n}$.

Now we give without proof an important property of locally finite families in normal space.

LEMMA XVII.8.2 (Dieudonné). Suppose that $(X, \mathscr{T})$ is a normal space. If $X = \bigcup_{i\in I} \mathscr{O}_i$, $\mathscr{O}_i \in \mathscr{T}$ ($\mathscr{O}_i$ are open sets), and the family $\{\mathscr{O}_i\}_{i\in I}$ is locally finite, then for every $i \in I$ there exists an open set $\mathscr{V}_i$ such that the following conditions are satisfied: (i) $\bigcup_{i\in I} \mathscr{V}_i = X$, (ii) $\bar{\mathscr{V}}_i \subset \mathscr{O}_i$.

9. PARACOMPACT SPACES. PARTITION OF UNITY. METRIC SPACES ARE PARACOMPACT

DEFINITION. Let $(X, \mathcal{T})$ be a topological space, and let $X = \bigcup_{i \in I} \mathcal{O}_i$, $X = \bigcup_{j \in J} \mathcal{U}_j$. A covering $\{\mathcal{U}_j\}_{j \in J}$ is said to be a *refinement* of the covering $\{\mathcal{O}_i\}_{i \in I}$, if for every $j \in J$ there exists an $i \in I$ such that $\mathcal{U}_j \subset \mathcal{O}_i$ or, in other words, if every set $\mathcal{U}_j$ is contained in one of the sets $\mathcal{O}_i$.

Example. A subcovering is a refinement.

This statement should be understood as follows: If $I' \subset I$ and if $\bigcup_{i \in I} \mathcal{O}_i = X = \bigcup_{i \in I'} \mathcal{O}_i$, then the covering $\{\mathcal{O}_i\}_{i \in I'}$ is a refinement of the covering $\{\mathcal{O}_i\}_{i \in I}$.

Hereafter, to the end of this chapter, we shall be concerned mainly with open coverings (i.e., such that all sets $\mathcal{O}_i$, $\mathcal{U}_j$ are open). For brevity we shall write "covering" in the sense "open covering", unless some other meaning is clearly indicated.

DEFINITION. A Hausdorff space $(X, \mathcal{T})$ is said to be *paracompact* if for each covering we can find a locally finite refinement.

Example. A compact space is paracompact since a finite (and hence locally finite) subcovering (which, as we know, is a refinement) can be selected from each of its coverings.

THEOREM XVII.9.1 (Dieudonné). A paracompact space is normal.

PROOF. Let A_1 and A_2 be closed, disjoint sets in a paracompact space $(X, \mathcal{T})$. Let $x \in A_1$. Suppose that $\mathcal{O}_{xy}$ and $\mathcal{O}_y$ are disjoint neighbourhoods of the points x and y for $y \in A_2$.

Clearly, $A_2 \subset \bigcup_{y = A_2} \mathcal{O}_y$ and hence the family consisting of the set $X - A_2$ and the sets $\mathcal{O}_y$ $(y \in A_2)$ is a covering of the space X. Let $\{\mathcal{U}_i\}_{i \in I}$ be a locally finite refinement of that covering. Let $I' \subset I$ be a set of indices such that for every $i \in I'$ there exists an $y \in A_2$ such that $\mathcal{U}_i \subset \mathcal{O}_y$.

Take $\mathcal{O}_2^x := \bigcup_{i \in I'} \mathcal{U}_i$. Clearly, $A_2 \subset \mathcal{O}_2^x$ since the sets $\mathcal{U}_i$ for $i \notin I'$ lie in $X - A_2$ and hence can cover at most the set $X - A_2$. Thus, in order that $\{\mathcal{U}_i\}_{i \in I}$ be a covering, $\{\mathcal{U}_i\}_{i \in I'}$ must cover A_2.

Since the family $\{\mathcal{U}_i\}_{i \in I'}$ is locally finite, the relation

$$\overline{\mathcal{O}_2^x} = \overline{\bigcup_{i \in I'} \mathcal{U}_i} = \bigcup_{i \in I'} \overline{\mathcal{U}_i}$$

holds by virtue of Theorem XVII.3.1. However, $x \notin \overline{\mathscr{U}_i}$ for $i \in I'$ since $\mathscr{U}_i \subset \mathscr{O}_y$ and the point x has a neighbourhood $\mathscr{O}_{xy}$ which does not contain points from $\mathscr{O}_y$ (hence from $\mathscr{U}_i$ as well). Therefore, $x \notin \overline{\mathscr{O}_2^x}$. Accordingly, the open set $\mathscr{O}_x' := X - \overline{\mathscr{O}_2^x}$ is a neighbourhood of the point x, disjoint from $\mathscr{O}_2^x$.

Thus, as in the proof of the normality of a compact space, we have shown that for closed A_2 and $x \in A_1$ there exist disjoint open neighbourhoods: $\mathscr{O}_x'$ of the point x and $\mathscr{O}_2^x$ of the set A_2.

But $X = (X - A_1) \cup \{\bigcup_{x \in A_1} \mathscr{O}_x'\}$, since $A_1 \subset \bigcup_{x \in A_1} \mathscr{O}_x'$.

Take a locally finite refinement of the covering above. Denote it by $\{\mathscr{V}_j\}_{j \in J}$. Let J' denote the set of those indices for which there exists an x such that $\mathscr{V}_j \subset \mathscr{O}_x'$.

As before, it is seen that $A_1 \subset \bigcup_{j \in J'} \mathscr{V}_j =: \mathscr{O}_1$. The set $\mathscr{O}_1$ is open Moreover, by the local finiteness of the covering $\{\mathscr{V}_j\}$, we have

$$\overline{\mathscr{O}_1} = \overline{\bigcup_{j \in J'} \mathscr{V}_j} = \bigcup_{j \in J'} \overline{\mathscr{V}_j}.$$

However, $A_2 \cap \overline{\mathscr{V}_j} = \emptyset$ for $j \in J'$, since $\mathscr{V}_j \subset \mathscr{O}_x'$, whereas the set A_2 has a neighbourhood $\mathscr{O}_2^x$ disjoint from $\mathscr{O}_x'$ (and thus disjoint from $\mathscr{V}_j$). Hence $A_2 \cap \overline{\mathscr{O}_1} = \emptyset$. Therefore, the open set $\mathscr{O}_2 := X - \overline{\mathscr{O}_1}$ contains the set A_2 and is disjoint from $\mathscr{O}_1$. □

DEFINITION. The *partition of unity subordinate to the covering* $\{\mathscr{O}_i\}_{i \in I}$ is called a family of continuous functions $\{f_j\}_{j \in J}$, defined on X, such that:

(i) $0 \leqslant f_j(x) \leqslant 1$ for every $j \in J$,

(ii) the family $\{f_j^{-1}(]0, 1])\}_{j \in J}$ is locally finite (i.e., every point $x \in X$ has a neighbourhood on which only a finite number of functions f_j do not vanish),

(iii) $\sum_{j \in J} f_j(x) \underset{x}{\equiv} 1$ (the sum at every point is finite by (ii) and hence is well defined).

By (iii) the family $\{\underline{f_j}\}_{j \in J}$ (of supports of the functions f_j) is a closed covering of the space X since at every point $x \in X$ at least one of the functions f_j does not vanish.

(iv) The covering with closed sets $\{\underline{f_j}\}_{j \in J}$ is a refinement of the covering $\{\mathscr{O}_i\}_{i \in I}$.

Now we give the fundamental property of paracompact spaces, the proof of which has been the principal purpose of this chapter.

THEOREM XVII.9.2. Suppose that $(X, \mathscr{T})$ is a Hausdorff space.

(The space $(X, \mathscr{T})$ is paracompact) $\Leftrightarrow$ (For every open covering of $(X, \mathscr{T})$ there exists a partition of unity subordinate to it).

PROOF. $\Leftarrow$: If to every covering $\{\mathcal{O}_i\}_{i\in I}$ corresponds a partition of unity $\{f_j\}_{j\in J}$ then the covering $\{f_j^{-1}(]0, 1])\}_{j\in J}$ is the desired locally finite refinement of $\{\mathcal{O}_i\}_{i\in I}$ and hence the space is paracompact.

$\Rightarrow$: Let the space be paracompact and let $\{\mathcal{O}_i\}_{i\in I}$ be its covering. We shall construct a partition of unity which is subordinate to it.

Suppose that $\{\mathcal{O}'_j\}_{j\in J}$ is a locally finite refinement of $\{\mathcal{O}_i\}_{i\in I}$. Applying Lemma XVII.8.2 to the space X which is paracompact (and hence is normal by Theorem XVII.9.2), we choose a covering $\{\mathscr{V}_j\}_{j\in J}$ such that $\overline{\mathscr{V}}_j \subset \mathcal{O}'_j$. Clearly, $\{\mathscr{V}_j\}_{j\in J}$ is locally finite. Thus, Lemma XVII.8.2 can be used once again and a covering $\{\mathscr{U}_j\}_{j\in J}$ chosen so that $\overline{\mathscr{U}}_j \subset \mathscr{V}_j$.

Since the space $(X, \mathscr{T})$ is normal, by virtue of Urysohn's Lemma (Theorem XVII.2.6) for every $j \in J$ there exists a continuous function $\tilde{f}_j$ such that $0 \leqslant \tilde{f}_j \leqslant 1$, $\tilde{f}_j(x) = 1$ for $x \in \overline{\mathscr{U}}_j$ and $\tilde{f}_j(x) = 0$ for $x \notin \mathscr{V}_j$. It is seen that $\overline{\tilde{f}_j} \subset \overline{\mathscr{V}}_j$.

Take the function

$$f(x) := \sum_{j\in J} \tilde{f}_j(x).$$

The sum is well defined since in the neighbourhood of every point only a finite number of functions f_j do not vanish (because the covering $\{\mathscr{U}_j\}_{j\in J}$ is locally finite). The function f is continuous as the sum of a finite number of continuous functions (in the neighbourhood of every point).

Since $\{\mathscr{U}_j\}_{j\in J}$ was a covering, then for every $x \in X$ there exists $j \in J$ such that $x \in \mathscr{U}_j$, that is $\tilde{f}_j(x) = 1$. Therefore, $f(x) \geqslant 1$. We define the functions

$$0 \leqslant f_j(x) := \frac{\tilde{f}_j(x)}{f(x)} \leqslant 1.$$

The functions f_j so defined are continuous and have supports in $\overline{\mathscr{V}}_j \subset \mathcal{O}'_j$. The covering $\{f_j^{-1}(]0, 1])\}_{j\in J}$ thus is a locally finite refinement of $\{\mathcal{O}_i\}_{i\in I}$ since $f_j^{-1}(]0, 1]) \subset \mathscr{V}_j \subset \mathcal{O}'_j$, whereas $\{\mathcal{O}'_j\}_{i\in I}$ is a locally finite refinement of $\{\mathcal{O}_i\}_{j\in I}$. Moreover

$$\sum_{j\in J} f_j(x) = \frac{\sum \tilde{f}_j(x)}{f(x)} = \frac{f(x)}{f(x)} = 1.$$

Hence, $\{f_j\}_{j\in J}$ is the desired partition of unity subordinate to $\{\mathcal{O}_i\}_{i\in I}$. □

We conclude this chapter with a theorem which is extremely important for further applications:

THEOREM XVII.9.3. A locally compact space countable at infinity is paracompact.

PROOF. Let $\{\mathscr{U}_i\}_{i\in I}$ be an open covering of the space X and let $\{\mathcal{O}_n\}_{n\in N}$ be a sequence of relatively compact sets, of which Theorem XVII.1.1 tells. The sets $K_n := \bar{\mathcal{O}}_n - \mathcal{O}_{n-1}$ will then be compact (for convenience, we take $K_n = \emptyset$ for $n \leqslant 0$).

Clearly, the set $\mathcal{O}_{n+1} - \bar{\mathcal{O}}_{n-2}$ is an open neighbourhood of the set K_n. Let us label this set $\mathscr{V}_n$. The family $\{\mathscr{V}_n \cap \mathscr{U}_i\}_{i\in I}$ is an open covering of the compact set K_n. From this covering we take a finite subcovering corresponding to the indices $i_1, i_2, \ldots, i_{p_n}$, $n \geqslant 1$. We denote $H_{nj} := \mathscr{V}_n \cap \cap \mathscr{U}_{i_j}$, $1 \leqslant j \leqslant p_n$. Since $\{K_n\}_{n\in N}$ is a covering of X, the family $\mathscr{R}$ of all sets H_{nj} is also a covering. It is a refinement of the covering $\{\mathscr{U}_i\}_{i\in I}$.

We shall demonstrate that $\mathscr{R}$ is locally finite. Let $x \in X$ and let n be the least of the numbers for which $x \in \mathcal{O}_n$. Since $x \notin \mathcal{O}_{n-1}$, we have $x \notin \mathcal{O}_{n-2}$, whereby $x \in \mathscr{V}_n$. But $\mathscr{V}_n \cap \mathscr{V}_m \neq \emptyset$ only if $n-3 \leqslant m \leqslant n+3$; thus $\mathscr{V}_n$ intersects at most those sets H_{mj} for which $n-3 \leqslant m \leqslant n+3$. There is a finite number of such sets and hence $\mathscr{V}_n$ is the desired neighbourhood of the point x, intersecting at most a finite number of sets from the family $\mathscr{R}$.

COROLLARY XVII.9.4. A locally compact space having a countable basis of a point at infinity is paracompact.

The reader will no doubt find it easy to prove the following theorem.

THEOREM XVII.9.5. The topological union of the family $\{X_i\}_{i\in I}$ of paracompact spaces is paracompact.

Associating this theorem with the previous one, we have

COROLLARY XVII.9.6. The union (of arbitrary cardinality) of locally compact spaces countable at infinity is paracompact.

It is much more difficult (cf. Bourbaki, *Topologie Générale*) to prove that the condition above is also necessary for a locally compact space to be paracompact.

Metric spaces are paracompact. One of the most important properties of metrizable spaces is their paracompactness. This fundamental fact was proved in 1948 by A. H. Stone. The beautiful proof reproduced below was given by Mrs. M. E. Rudin in 1969. In order to simplify the wording we give the following

DEFINITION. A family $\{A_s\}_{s \in S}$ of subsets of X is *discrete* if every $x \in X$ has a neighbourhood $\mathcal{U}$ such that $\{s \in S\colon A_s \cap \mathcal{U} \neq \emptyset\}$ contains at most one element.

A family is *σ-discrete* (*σ-locally finite*) if it is a sum of countable many discrete (locally finite) families.

Example. Every discrete family is locally finite.

THEOREM XVII.9.7 (A. H. Stone). Every open covering of a metrizable space has a refinement which is both: locally finite and σ-discrete. Hence every metrizable space is paracompact.

PROOF. Let d be a distance on X inducing the same topology and let $\{\mathcal{U}_s\}_{s \in S}$ be an open covering of (X, d). We can assume that S is well ordered by the relation $\prec$. Let us define inductively a family $\mathfrak{V}_n = \{\mathcal{V}_{s,n}\}_{s \in S}$ for $n = 1, 2, \ldots$ as follows:

$$(*) \qquad \mathcal{V}_{s,n} = \bigcup K(c, 1/2^n),$$

where the summation extends over all $c \in X$ such that

(1) s is the first element of S, such that $c \in \mathcal{U}_s$,

(2) $c \notin \mathcal{V}_{t,m}$ for $m < n$ and $t \in S$,

(3) $K(c, 3/2^n) \subset \mathcal{U}_s$.

From $(*)$ it follows that $\mathcal{V}_{s,n}$ are open and by (3) $\mathcal{V}_{s,n} \subset \mathcal{U}_s$. Let $x \in X$ and let $s \in S$ be the first element of S such that $x \in \mathcal{U}_s$; choose an n such that $K(c, 3/2^n) \subset \mathcal{U}_s$. Then, either $x \in \mathcal{V}_{t,m}$ for $m < n$ and a $t \in S$, or $x \in \mathcal{V}_{s,n}$. Hence $\mathfrak{V} := \bigcup_{n=1}^{\infty} \mathfrak{V}_n$ is an open covering of X which is a refinement of $\{\mathcal{U}_s\}_{s \in S}$. Let us prove that for each $n \in N$

$$(4) \qquad (x_1 \in \mathcal{V}_{s_1,n},\ x_2 \in \mathcal{V}_{s_2,n} \text{ and } s_1 \prec s_2) \Rightarrow \left(d(x_1, x_2) > \frac{1}{2^n}\right).$$

This implies that the family $\mathfrak{V}_n$ is discrete, since each ball $K(y, 1/2^{n+1})$ intersects at most one element of $\mathfrak{V}_n$.

But from the definition of $\mathcal{V}_{s_1,n}$ and $\mathcal{V}_{s_2,n}$ follows the existence of $c_1, c_2 \in X$ satisfying (1)–(3), such that $x_i \in K(c_i, 1/2^n) \subset \mathcal{V}_{s_i,n}$ for

$i = 1, 2, \ldots$ By virtue of (3) $K(c_1, 3/2^n) \subset U_{s_1}$, and (1) $\Rightarrow (c_2 \notin \mathscr{U}_{s_1})$, whence $d(c_1, c_2) \geqslant 3/2^n$ and $d(x_1, x_2) \geqslant d(c_1, c_2) - d(c_1, x_1) - d(c_2, x_2) > 1/2^n$, whereby (4) is proved. We have only to check that for each $t \in S$ and $k, m \in N$ we have

$$(5) \qquad (K(x, 1/2^k) \subset \mathscr{V}_{t,m}) \Rightarrow (K(x, 1/2^{m+k}) \cap \mathscr{V}_{s,n} = \varnothing$$
$$\text{for } n \geqslant m+k \text{ and } s \in S,$$

since for $x \in X$ there exist $k, m \in N$ and $t \in S$ such that $K(x, 2^{-k}) \subset \mathscr{V}_{t,m}$ and, therefore, $K(x, 2^{-(m+k)})$ intersects at most $m+k-1$ elements of the covering $\mathfrak{B}$. Condition (2) implies that points c, from the definition of the set $\mathscr{V}_{s,n}$, are not in $\mathscr{V}_{t,m}$ for $n \geqslant m+k$ since $K(x, 1/2^k) \subset \mathscr{V}_{t,m}$; therefore, $d(x, c) \geqslant 1/2^k$ for every such c. But from $m+k \geqslant k+1$ and $n \geqslant k+1$ it follows that $K(x, 1/2^{m+k}) \cap K(c, 1/2^n) = \varnothing$, and (5) is proved. □

THEOREM XVII.9.8. Every metric (metrizable) space (X, d) has a σ-discrete basis.

PROOF. Let $\{\mathscr{B}_n\}$ be an open σ-discrete refinement of a covering with balls $K(y, 1/n)$. We verify that the σ-discrete family $\mathscr{B} := \bigcup \mathscr{B}_n$ is a basis of X.

COROLLARY XVII.9.9. Every metrizable space has a σ-locally finite basis.

As another corollary we obtain the famous

THEOREM XVII.9.10 (Alexandrov). Let X be a metric (metrizable) space which is locally separable, i.e. every point has a separable neighbourhood; then

1° $X = \coprod_{s \in S} X_s$, where each X_s is separable, i.e., X is a disjoint sum of separable (sub)spaces.

2° If X is connected then X is separable and hence satisfies the second axiom of countability.

From 2° there immediately follows (since a metric compact space is separable) a very important

COROLLARY XVII.9.11 (Theorem of Alexandrov). Every metrizable locally compact connected space is separable and hence is second countable.

PROOF OF THE THEOREM XVII.9.10. 1° Let $\{\mathcal{O}_i\}_{i\in I}$ be a covering of X by separable neighbourhoods $\mathcal{O}_i$ and let $\{\mathcal{U}_j\}_{j\in J}$ be a locally finite refinement of $\{\mathcal{O}_i\}_{i\in I}$. We introduce in the covering $\mathfrak{U} = \{\mathcal{U}_j\}_{j\in J}$ the following equivalence relation $\sim$: $\mathcal{U} \sim \mathcal{U}'$ if there exists a sequence $\mathcal{U}_0, \mathcal{U}_1, \ldots$ $\ldots, \mathcal{U}_n$ of elements of $\mathfrak{U}$ such that $\mathcal{U}_0 = \mathcal{U}$, $\mathcal{U}_n = \mathcal{U}'$ and $\mathcal{U}_k \cap \mathcal{U}_{k+1} \neq \emptyset$ for $k = 0, 1, \ldots, n-1$. Denote by $\{Y_s\}_{s\in S}$ the partition of $\mathfrak{U}$ into equivalence classes. Each Y_s is finite (since $\mathfrak{U}$ is local'y finite). Denote by X_s the sum of these subsets of X which are elements of Y_s. Since each $\mathcal{U}_j$, $j \in J$, is separable (as a subset of a separable space), every X_s is separable as a finite sum of separable subspaces. It is obvious that $X = \coprod_{s\in S} X_s$ (disjoint sum). Since $X_s \cap X_{s'} = \emptyset$ for $s \neq s'$, the topologies of X and $\coprod X_s$ are the same.

2° Since a subset $\mathcal{O}$ of X is open (closed) iff $\mathcal{O} \cap X_s$ is open (closed) in X_s, each X_s is open and closed in X. Let us assume that X is connected; then the index set S consists of one element and, therefore, X is separable.

Remark. Corollary XVII.9.11 was proved by P. S. Alexandrov back in 1924, twenty years before paracompact spaces were discovered. The reader may try to give an elementary proof of Theorem XVII. 9.1 (cf. pages 514–515).

XVIII. MEASURABLE MAPPINGS THE TRANSPORT OF A MEASURE CONVOLUTIONS OF MEASURES AND FUNCTIONS

In Chapter XIII, Section 20, we referred briefly to mappings of Radon integrals. Now we shall treat this subject more systematically and in more general terms. Since this concept is built after the pattern of continuous mapping, let us recall the latter.

Suppose that $(X_i, \mathscr{T}_i)$, $i = 1, 2$, is a topological space, i.e., suppose that $\mathscr{T}_i$ is a family of (open) subsets of a space, satisfying the following axioms:

(i) $\emptyset, X_i \in \mathscr{T}_i$,

(ii) $(O_\alpha \in \mathscr{T}_i) \Rightarrow \left(\bigcup_{\alpha \in A} O_\alpha \in \mathscr{T}_i\right)$, i.e., the union of any number of open sets is an open set,

(iii) $(O_1, O_2 \in \mathscr{T}_i) \Rightarrow (O_1 \cap O_2 \in \mathscr{T}_i)$, i.e., the intersection of a finite number of open sets is an open set.

The mapping $T: X_1 \to X_2$ is continuous if $T^{-1}(O_2) \in \mathscr{T}_1$ for every $O_2 \in \mathscr{T}_2$.

If $S: X_2 \to X_3$ is a continuous mapping, the composition $S \circ T: X_1 \to X_3$ is a continuous mapping.

In measure theory, as we know, a role analogous to that of topology is played by the σ-algebras of measurable subsets. We recall that:

DEFINITION. A family $\mathscr{A}$ of subsets of the space X is called a *σ-algebra* if

(i) $X \in \mathscr{A}$,

(ii) $(A_j \in \mathscr{A}, j = 1, 2, \ldots) \Rightarrow \left(\bigcup_{j=1}^{\infty} A_j \in \mathscr{A}\right)$,

(iii) $(A \in \mathscr{A}) \Rightarrow (X - A \in \mathscr{A})$.

It follows immediately from these axioms that $\emptyset \in \mathscr{A}$ and that $(A_j \in \mathscr{A}, j = 1, 2, \ldots) \Rightarrow \left(\bigcap_{j=1}^{\infty} A_j \in \mathscr{A}\right)$.

COROLLARY XVIII.0.1. Suppose that $\mathscr{A}_2$ is a σ-algebra in X_2 and suppose that the mapping $T: X_1 \to X_2$ is given; then $T^{-1}(\mathscr{A}_2) := \{T^{-1}(A): A \in \mathscr{A}_2\}$ is a σ-algebra in X_1.

PROOF—immediate. □

If we have some family $\mathscr{F} = \{A\}$ of subsets of the space X_1, the topology $\mathscr{T}(\mathscr{F})$ generated by the family $\mathscr{F}$ is the weakest topology containing $\mathscr{F}$. It is seen immediately that $\mathscr{T}(\mathscr{F})$ is formed from $\{A_j\}$ by taking finite intersections $\bigcap A_{\alpha_i}$ and arbitrary set-theoretic sums of them. It is sometimes said that the family $\mathscr{F}$ is a *subbasis* of the topology $\mathscr{T}_1$ if $\mathscr{T}_1 = \mathscr{T}(\mathscr{F})$. We recall that a family $\mathscr{B}$ is called a *basis of the topology* $\mathscr{T}_1$ if every set $O \in \mathscr{T}_1$ is a union of sets from the family $\mathscr{B}$.

In similar fashion we arrive at the concept of a generator of a σ-algebra. Every family $\mathscr{F}$ of subsets of a space X_1 generates a least σ-algebra $\mathscr{A}(\mathscr{F}) \subset \mathscr{F}$. Belonging to it are the sets X_1 and $\varnothing$, elements of the family $\mathscr{F}$ and their countable unions, and the countable intersections and complements of those sets.

DEFINITION. Let $(X, \mathscr{T})$ be a topological space. Elements of the σ-algebra $\mathscr{A}(\mathscr{T})$ are called *Borel sets* on $(X, \mathscr{T})$. The measure on $\mathscr{A}(\mathscr{T})$ is called the *Borel measure* on $(X, \mathscr{T})$.

Accordingly, closed sets, open sets, and sets which can be obtained from them by the operation of taking countable unions, countable intersections, and complements are all Borel sets.

1. MEASURABLE MAPPINGS

DEFINITION. Let $\mathscr{A}_i$ be a σ-algebra in X_i, $i = 1, 2$. The mapping $T: X_1 \to X_2$ is said to be *$(\mathscr{A}_1 - \mathscr{A}_2)$-measurable* if $T^{-1}(\mathscr{A}_2) \subset \mathscr{A}_1$, that is, if the inverse images of elements of $\mathscr{A}_2$ are elements of $\mathscr{A}_1$.

This definition leads immediately to the following

THEOREM XVIII.1.1. If a mapping $T: X_1 \to X_2$ is $(\mathscr{A}_1 - \mathscr{A}_2)$-measurable and $S: X_2 \to X_3$ is $(\mathscr{A}_2 - \mathscr{A}_3)$-measurable, the mapping $S \circ T$: $X_1 \to X_3$ is $(\mathscr{A}_1 - \mathscr{A}_3)$-measurable.

PROOF. The hypothesis follows from the formula $(S \circ T)^{-1}(A) = T^{-1}(S^{-1}(A))$, which holds for every $A \subset X_3$, whence in particular for $A \in \mathscr{A}_3$. $\square$

Remark. The measurability proof for the composition is seen to be identical with the continuity proof for the composition of mappings.

A useful theorem is the straightforward

THEOREM XVIII.1.2. Let $T: (X_1, \mathscr{A}_1) \to (X_2, \mathscr{A}_2)$ and let $\mathscr{E}_2$ generate an algebra $\mathscr{A}_2$. Then (the mapping T is $(\mathscr{A}_1 - \mathscr{A}_2)$-measurable) $\Leftrightarrow (T^{-1}(A_2) \in \mathscr{A}_1$ for every $A_2 \in \mathscr{E}_2)$.

PROOF. The family $\mathscr{F}_2 = \{A_2 \subset X_2 \colon T^{-1}(A_2) \in \mathscr{A}_1\}$ is a σ-algebra in X_2, whence $(\mathscr{A}_2 \subset \mathscr{F}_2) \Leftrightarrow (\mathscr{E}_2 \subset \mathscr{F}_2)$. But $\mathscr{A}_2 \subset \mathscr{F}_2$ is equivalent to $(\mathscr{A}_1 - \mathscr{A}_2)$-measurability of the mapping T, since

$$(\mathscr{A}_2 \subset \mathscr{F}_2) \Leftrightarrow (T^{-1}(A_2) \in \mathscr{A}_1 \text{ for } A_2 \in \mathscr{A}_2). \ \square$$

Examples. 1. Every constant mapping $T\colon X_1 \to X_2$ is $(\mathscr{A}_1 - \mathscr{A}_2)$-measurable.

2. Every continuous mapping $T\colon (X_1, \mathscr{T}_1) \to (X_2, \mathscr{T}_2)$ is Borel-measurable, i.e. is $(\mathscr{A}(\mathscr{T}_1) - \mathscr{A}(\mathscr{T}_2))$-measurable: the inverse image of a Borel set is a Borel set. This follows immediately from Theorem XVIII.1.2 and the definition of Borel sets.

The reader may wonder at this point whether a μ-measurable numerical function $T\colon X \to \bar{\boldsymbol{R}}$ can be considered as a measurable mapping, To answer this question, let us define Borel sets in $\bar{\boldsymbol{R}} = \boldsymbol{R} \cup \{+\infty\} \cup \cup \{-\infty\}$.

DEFINITION. *Borel sets* in $\bar{\boldsymbol{R}}$ are sets of the form B, $B \cup \{+\infty\}$, $B \cup \cup \{-\infty\}$, $B \cup \{+\infty\} \cup \{-\infty\}$, where B is a Borel set in the space $\boldsymbol{R}$, which is endowed with the ordinary topology of the real axis.

Remark. Introduced Borel sets on $\bar{\boldsymbol{R}}$ are nothing else than Borel sets of two-point (i.e., Čech–Stone) compactification of the real line (see Chapter XVII).

A family of Borel sets in $\boldsymbol{R}$ (or $\bar{\boldsymbol{R}}$) will be denoted by $\mathscr{B}^1$ (or $\bar{\mathscr{B}}^1$). Correspondingly, we define $\mathscr{B}^p$ and $\bar{\mathscr{B}}^p$ in the spaces $\boldsymbol{R}^p$ and $\bar{\boldsymbol{R}}^p$.

COROLLARY XVIII.1.3. We have

$$\bar{\mathscr{B}}^1 = \{B \subset \bar{\boldsymbol{R}}\colon B \cap \boldsymbol{R} \in \mathscr{B}^1\}.$$

Since the system of all intervals $]a, +\infty[$, $a \in \boldsymbol{R}$ generates the entire family $\bar{\mathscr{B}}^1$, the answer to the question posed above is provided by the following

THEOREM XVIII.1.4. A numerical function $f\colon X \to \bar{\boldsymbol{R}}$ is $(\mathscr{A} - \bar{\mathscr{B}}^1)$-measurable iff for every $a \in \boldsymbol{R}$ the set $\{x \in X\colon f(x) > a\}$ is an element of the family $\mathscr{A}$.

As we know from Chapter XIII, Sections 12–13, if the Stone axiom holds, then μ-measurable sets constitute a σ-algebra in X for any Daniell–Stieltjes integral μ on X. And Theorem XIII.13.1 states that a numerical function is μ-measurable iff it is $(\mathscr{A} - \bar{\mathscr{B}}^1)$-measurable, where $\mathscr{A}$ denotes the family of μ-measurable sets.

2. TOPOLOGIES DETERMINED BY FAMILIES OF MAPPINGS

DEFINITION. Suppose that $\{(X_i, \mathscr{T}_i)\}_{i\in I}$ is a family of topological space, S is a set, and $\{T_i\}_{i\in I}$ is a family of mappings $T_i\colon X \to X_i$. The weakest (i.e., coarsest) topology on X, under which all mappings T_i, $i \in I$, are continuous is called the *projective topology*, determined by the family of mappings $\{T_i\}_{i\in I}$.

THEOREM XVIII.2.1. Let $\mathscr{F}$ be a family of sets of the form $T_i^{-1}(O_i)$, where $O_i \in \mathscr{T}_i$, $i \in I$. Then $\mathscr{F}$ is a subbasis of $\mathscr{T}$.

Examples. 1. Suppose that $I = \{1\}$. We then have one space $(X_1, \mathscr{T}_1)$ and the mapping $T_1\colon X \to X_1$. The projective topology on X, specified by the mapping $\mathscr{T}_1$, is often called the *inverse image* of the topology $\mathscr{T}_1$ with respect to T_1.

2. The Cartesian product $X = \bigtimes_{i\in I} (X_i, \mathscr{T}_i)$ is endowed with a projective topology in the following way: By p_i we denote the projection of X onto X_i, i.e., a mapping which associates the i-th coordinate of every point $x \in X$ with that point: $p_i(x) = x_i \in X_i$. A topology specified by a family of projections $\{p_i\}_{i\in I}$ is called a *topological product* or *Tychonoff topology*.

3. Projective limit of topological spaces. The definition will be given in Section 4.

The latter example explains the origin of the term "projective topology".

Frequently we have a dual situation:

Suppose that $\{(X_i, \mathscr{T}_i)\}_{i\in I}$ are topological spaces and suppose that we are given a family of mappings $C_i\colon X_i \to X$, $i \in I$, in a set X. The strongest (i.e., finest) topology on X under which all mappings G_i are continuous is called the *inductive topology* generated by $\{G_i\}_{i\in I}$. Plainly, the basis of this topology is formed by sets $U \subset X$ such that for any $i \in I$ the inverse image $G_i^{-1}(U)$ is open in X_i.

In the case of a single space $(X_1, \mathscr{T}_1)$, the inductive topology is called the *simple image of the topology $\mathscr{T}_1$ under the mapping G_1*.

Examples. 1. The quotient topology is the most important example of an inductive topology. Let $(X_1, \mathscr{T}_1)$ be a topological space, and let $\sim$ be the equivalence relation in X_1. Let G_1 be a canonical surjection of X_1 onto the quotient space $X/\sim$.

The inductive topology, induced on $X/\sim$ by G_1 (simple image of the topology $\mathscr{T}_1$) is called a *quotient topology*.

2. Given a space X and ist subsets $X_i = (X_i, \mathscr{T}_i)$, $i \in I$, being topological spaces such that $X_i \neq X_j$, where $i \neq j$ and $X = \bigcup_i X_i$.

Let G_i be a canonical injection G_i: $X_i \to X$. A space X endowed with an inductive topology (i.e., the strongest topology under which every mapping G_i is still continuous) is called the *inductive limit of the space* X_i and is denoted by $\lim_{i \in I} \operatorname{ind} X_i$ or also $\varinjlim X_i$.

THEOREM XVIII.2.2. An inductive topology on X, generated by the family of mappings $\{G_i: X_i \to X\}_{i \in I}$, is characterized by the following property:

Let $(X_0, \mathscr{T}_0)$ be a topological space. The mapping T: $X \to X_0$ is continuous iff for every $i \in I$ the mapping $T \circ G_i$: $X_i \to X_0$ is continuous.

PROOF. The necessity of the condition given above is obvious, since a composition of continuous mappings is itself a continuous mapping. We shall show that this is also a sufficient condition.

Suppose that $O \subset X_0$ is an open set. The set $T^{-1}(O)$ is open in the inductive topology in X since for any $i \in I$ the set $G_i^{-1}(T^{-1}(O)) = (T \circ G\,)^{-1}(O)$ is open in X_i, inasmuch as $T \circ G_s$ is continuous. □

The reader will have no difficulty in proving the dual of the above theorem in a similar way:

THEOREM XVIII.2.3. Suppose that $\{T_i\}_{i \in I}$ is a family of mappings $T_i: X \to X_i$ of a set X into topological spaces $(X_i, \mathscr{T}_i)$. Given a mapping T: $X_0 \to X$, where $(X_0, \mathscr{T}_0)$ is a topological space. The mapping T is continuous under the projective topology on X iff the mapping $T_i T$: $X_0 \to X_i$ is continuous for any $i \in I$.

We can now similarly specify σ-algebras of sets "projectively" in terms of families of mappings.

To this end, the phrase "topology determined by the family of mappings $\{T_i\}_{i \in I}$" in the definition of projective topology should be replaced by "σ-algebra determined by the family of mappings $\{T_i\}_{i \in I}$", and the word "continuous" should be replaced by "$(\mathscr{A} - \mathscr{A}_i)$-measurable".

With this nomenclature, Theorems XVIII.2.1′ and XVIII.2.3′ analogous to the theorems given for projective topology are also valid.

By way of example we give the proof of Theorem XVIII.2.3′.

PROOF. The necessity of the given condition is obvious; it follows from Theorem XVIII.1.1 on composition. We shall prove the sufficiency of this condition.

The family $\mathscr{E} := \bigcup_{i\in I} T_i^{-1}(\mathscr{A}_i)$ is a generator of a σ-algebra $\mathscr{A}$ in X. Every set $E \in \mathscr{E}$ is of the form $E = T_i^{-1}(A_i)$, $A_i \in \mathscr{A}$, whereby $T_0^1(E) = (T_i \circ T_0)^{-1}(A_i)$ is an element of the algebra $\mathscr{A}_0$, by the assumed measurability of the mapping $T_i \circ T_0$. Therefore, by Theorem XVIII.1.2 it follows that T_0 is $(\mathscr{A}_0 - \mathscr{A})$-measurable. □

3. THE TRANSPORT OF A MEASURE

Suppose that $T\colon (X, \mathscr{A}) \to (X_1, \mathscr{A}_1)$ is an $(\mathscr{A} - \mathscr{A}_1)$-measurable mapping. Then every measure μ on $\mathscr{A}$ defines a measure on $\mathscr{A}_1$, which is given by the following formula:

$$\mathscr{A}_1 \ni A_1 \to \mu\big((T^{-1}(A_1)\big).$$

This measure is customarily denoted by $T\mu$ and is called the *image of the measure* μ under the mapping T or the measure obtained by the *transport* T *of the measure* μ onto the space $(X_1, \mathscr{A}_1)$. Thus we have the defining equality:

$$(T\mu)(A_1) := \mu\big(T^{-1}(A_1)\big), \quad A_1 \in \mathscr{A}_1.$$

COROLLARY XVIII.3.1. The transport of a measure is transitive, i.e.,

$$(T_2 \circ T_1)(\mu) = T_2\big(T_1(\mu)\big).$$

The proof follows immediately from the definition and from Theorem XVIII.1.1.

We can now prove a theorem whose hypothesis was taken for the definition of the image $T\mu$ of a Radon integral (cf. Chapter XIII, Section 20).

THEOREM XVIII.3.2. Let $\mathscr{A}$ be a σ-algebra of μ-measurable sets in a space X and let $\mathscr{A}_1$ be a σ-algebra in X_1. If the mapping $T\colon X \to X_1$ is $(\mathscr{A} - \mathscr{A}_1)$-measurable, then

(i) for every $(\mathscr{A}_1 - \overline{\mathscr{B}}^1)$-measurable, nonnegative function f, numerical on X_1, the function $f \circ T$ is μ-measurable and

$$(1) \qquad \int_{X_1} f\, d(T\mu) = \int_X (f \circ T)\, d\mu;$$

(ii) if f is $(\mathscr{A}_1 - \overline{\mathscr{B}}^1)$-measurable function on X_1, then

$$(f \in \mathscr{L}^1(T\mu)) \Leftrightarrow (f \circ T \in \mathscr{L}^1(\mu))$$

and formula (1) holds.

PROOF. (ii) follows from (i) and from Theorem XIII.11.10 through the decomposition $f = f^+ - f^-$. Of course we make use of the fact that $(f \circ T)^+ = f^+ \circ T$, $(fT)^- = f^- \circ T$.

Let us prove (i). Theorems XVIII.1.1 and XVIII.1.3 imply that the function $f \circ T$ is μ-measurable. However, $f \circ T \geqslant 0$ and hence the integral on the right-hand side of formula (1) is definite (perhaps equal to infinity).

First of all, for our f let us take the step function $g = \sum a_i \chi_{A_i'}$, where the sets $A_i' \in A_1$ are disjoint. In that event, $g \circ T = \sum a_i \chi_{A_i}$, where $A_i := T^{-1}(A_i') \in \mathscr{A}$. Thus

$$(T\mu)(g) = \sum a_i \cdot (T\mu)(A_i') = \sum a_i \cdot \mu(A_i) = \mu(g \circ T).$$

Now, let the function $f \geqslant 0$ be arbitrary, as in the hypothesis. A monotonic sequence of step functions $g_n \leqslant f$ (for instance, $g_n := f_{\varepsilon_n}$, where $\varepsilon_n := 1/2^n$) may be taken. Then $\sup_n g_n \circ T = f \circ T$. By the theorem of B. Levi we have

$$\mu(f \circ T) = \sup \mu(g_n \circ T) = \sup (T\mu)(g_n) = (T\mu)(f). \quad \square$$

4. THE PROJECTIVE LIMITS OF HAUSDORFF SPACES. INFINITE TENSOR PRODUCTS AND THE PROJECTIVE LIMITS OF MEASURES

This discussion of Section 2 will now be supplemented in that we shall assume that the spaces X_i are measure spaces, i.e., that $X_i = (X_i, \mu_i)$ or, more precisely, (X_i, A_i, μ_i), where in the case of an infinite set of indices we shall assume that $\mu_i(X_i) = 1$, $i \in I$. A concept of projective limit mentioned in Section 2 is often required in applications. To arrive at this concept, let us begin with the most important model, that of product. Given a nonempty set of indices I and a family of sets X_i, $i \in I$.

For every set $K \subset I$ we can define

$$X_K := \bigtimes_{i \in K} X_i, \quad \text{in particular} \quad X = X_I = \bigtimes_{i \in I} X_i.$$

We recall that X_K is the totality of mappings $\Phi_K : K \to \bigcup_{i \in K} X_i$ such that $\Phi_K(i) \in X_i$. If with each such mapping we associate its restriction to a subset $J \subset K$, we obtain the projection

(1) $\quad p_J^K : X_K \to X_J \quad$ for $J \subset K$.

We write briefly $p_J := p_J^I$, and $p_i^K = p_{(i)}^K$ for $J = \{i\}$; in particular, as in "the old notation" $p_i := p_i^I$ will be canonical projections $X \to X_i$. Clearly, the compatibility relations

(2) $\quad p_J^L = p_J^K \circ p_K^L \quad$ hold for any $J \subset K \subset L \subset I$.

In particular,

(3) $\quad p_J = p_J^K \circ p_K \quad$ for $J \subset K \subset I$.

Now let us generalize this concept by going over to the concept of projective limit.

Let $(A, \prec)$ be a directed set, let $(X_\alpha)_{\alpha \in A}$ be a family of topological spaces, and let

(1′) $\quad p_{\alpha\beta} : X_\beta \to X_\alpha \quad$ for $\alpha \prec \beta$

be continuous mappings such that compatibility relations of the type of (2) hold:

(2′) $\quad p_{\alpha\alpha} = 1_{X_\alpha}, \quad p_{\alpha\gamma} = p_{\alpha\beta} \circ p_{\beta\gamma} \quad$ for $\alpha \prec \beta \prec \gamma$.

The *projective limit* $\varprojlim X_\alpha$ ($\varprojlim p_{\alpha\beta} X_\beta$) of the (inverse) family $\{X_\alpha, p_{\alpha\beta}, \alpha, \beta \in A\}$ is a subset of the product $\bigtimes_{\alpha \in A} X_\alpha x = (x_\alpha)$ such that $x_\alpha = p_{\alpha\beta} x$ for $\alpha \prec \beta$. A mapping $p_\alpha : (\varprojlim p_{\alpha\beta} X_\beta) \to X_\alpha$ exists for every $\alpha \in A$. Plainly,

(3′) $\quad p_\alpha = p_{\alpha\beta} \circ p_\beta \quad$ for $\alpha \prec \beta$.

The topology of the projective limit X is (by definition) the projective topology induced by the family of mappings $p_\alpha : X \to X_i$. The product $\bigtimes_{i \in I} X_i$ is clearly a special case of projective limit. By $\{H\}$ let us denote a family of nonempty finite subsets $H \subset I$, directed by inclusion $H_1 \prec H_2$, if $H_1 \subset H_2$. When we denote $p_J^K = p_{KJ}$ we have $\bigtimes_{i \in I} X_i = \varprojlim p_L^K X_K$. If $(X_i, \mathscr{T}_i)$ is a Hausdorff space, the measure of the σ-algebra generated by open sets is called the *Borel measure* on X_i. The space of Borel measures on $(X_i, \mathcal{T}_i)$ will be denoted by $\mathcal{M}_b^+(X_i)$.

DEFINITION. A measure of a projective limit $X = \varprojlim X_\alpha$ is called a *projective limit of measures* μ_α *on* X_α if $p_\alpha(\mu) = \mu_\alpha$.

A necessary and sufficient condition for a measure which is the projective limit of the measure μ_i to exist at the projective limit $X = \varprojlim p_{\alpha\beta} X_\beta$ is the (ε, K)-condition given by Prokhorov for Tychonoff spaces J. Kisyński has generalized Prokhorov's Theorem to arbitrary Hausdorff spaces (cf. the author's monograph *General Eigenfunction Expansions and Unitary Representations of Topological Groups*).

THEOREM XVIII.4.1 (Prokhorov, Kisyński). Let $(X_\alpha, p_{\alpha\beta}, \alpha \prec \beta)$ be a projective system of Hausdorff spaces, where $p_{\alpha\beta}\colon X_\beta \to X_\alpha$ is a continuous surjection for $\alpha \prec \beta$. Given for every space X_α a Borel measure $\mu_\alpha \in \mathcal{M}_b^+(X_\alpha)$, such that $\mu_\alpha(X_\alpha) = 1$ (probability measure), where the family $(\mu_\alpha)_{\alpha\in A}$ obeys the compatibility condition

$$(4) \qquad p_{\alpha\beta}(\mu_\beta) = \mu_\alpha \quad \text{for } \alpha \prec \beta$$

(that is, $\mu_\beta\big(p_{\alpha\beta}^{-1}(B)\big) = \mu_\alpha(B)$ for every Borel set $B \subset X_\alpha$). Let $X = \varprojlim p_{\alpha\beta} X_\beta$.

In order that a unique measure $\mu \in \mathcal{M}_b^+(X)$ such that $p_\alpha(\mu) = \mu_\alpha$ for every $\alpha \in A$ exists on X it is necessary and sufficient that the (ε, K)-condition be satisfied:

$$(\varepsilon, K) \qquad \bigwedge_{\varepsilon>0} \; \bigvee_{K\subset X,\, K-\text{compact}} \; \bigwedge_{\alpha\in A} \mu_\alpha\big(X_\alpha - p_\alpha(K)\big) < \varepsilon .$$

The Prokhorov Theorem plays a vital role in the proof of the Minlos Theorem which is given in Chapter XIX.

The measure space $(X, \mathcal{A}, \mu)$ with measure μ such that $\mu(X) = 1$ is called a *probability space*, and the measure μ, a *probability measure*; μ-measurable subsets are called *events*, and the number $\mu(A)$ is called the *probability of occurrence of event A*.

We shall conclude this section with a formulation of a measure analogue of the Prokhorov Theorem concerning the (probability) measure on any product $\bigtimes_{i\in I} (X_i, \mathcal{A}_i)$ of σ-algebras. In other words, it is our intention to define the tensor product $\bigotimes_{i\in I} \mu_i$ of an arbitrary cardinality of normed measures. We already know the meaning of a finite tensor product $\bigotimes_{i\in H} \mu_i$ of measures, where $H \in \{H\}$ is a finite set $H \subset I$. Since this product does not depend on the order, we can define

$$(5) \qquad \mu_H := \bigotimes_{i\in H} \mu_i, \quad \mathscr{A}_H = \bigotimes_{i\in H} A_i,$$

where $\mathscr{A}_H$ is a σ-algebra on X_H spanned by sets of the form $\bigtimes_{i\in H} A_i$, $A_i \in \mathscr{A}_i$.

In keeping with the prescription we have already used on more than one occasion, the term "continuous" is replaced by "measurable" and the following natural definition can be adopted:

Definition. Let $\mathscr{A}_i$ be a σ-algebra of subsets X_i, $i \in I$. The *tensor product* $\bigotimes_{i\in I} \mathscr{A}_i =: \mathscr{A}(p_i; i \in I)$ is called the least σ-algebra $\mathscr{A}$ in $\bigtimes_{i\in I} X_i$ such that every projection $p_i: X \to X_i$ is $(\mathscr{A} - \mathscr{A}_i)$-measurable.

Then, for every $J \in \{H\}$ the mapping p_J is $(\mathscr{A} - \mathscr{A}_J)$-measurable, inasmuch as

$$(6) \qquad p_i = p_i^J \circ p_J \quad \text{for } i \in J$$

(cf. Theorem XVIII.2.3) and hence

$$(7) \qquad \mathscr{A}(p_i; i \in I) = \mathscr{A}(p_J; J \in \{H\}).$$

We seek on $\mathscr{A}$ a measure μ such that

$$(8) \qquad \mu\Big(p_J^{-1}\big(\bigtimes_{i\in J} A_i\big)\Big) = \prod_{i\in J} \mu_i(A_i)$$

for every $J \in \{H\}$ and any events $A_i \in \mathscr{A}_i$, $i \in J$. By the definition of measure image, this means that $p_J(\mu)$ associates a number $\prod_{i\in J} \mu_i(A_i)$ with every such product set $\bigtimes_{i\in J} A_i$; this is just the measure $\mu_J = \bigotimes_{i\in J} \mu_i$. It turns out that the following general theorem holds:

Theorem XVIII.4.2 (Generalization of Kolmogorov's Theorem). On the σ-algebra $\mathscr{A} = \bigotimes_{i\in I} \mathscr{A}_i$ there exists exactly one normed measure μ such that

$$(9) \qquad p_J(\mu) = \mu_J, \quad J \in \{H\}$$

for every (finite) set $J \subset I$.

DEFINITION. The measure μ, referred to in the Kolmogorov Theorem, is called the *product of (probability) measures* $(\mu_i)_{i\in I}$; it is denoted by $\bigotimes_{i\in I} \mu_i$. The probability space $\left(\bigtimes_{i\in I} X_i, \bigotimes_{i\in I} \mathscr{A}_i, \bigotimes_{i\in I} \mu_i\right)$ is called the *product of probability spaces* $((X_i, \mathscr{A}_i, \mu_i))_{i\in I}$ and is denoted by $\bigotimes_{i\in I} (X_i, \mathscr{A}_i, \mu_i)$.

Kolmogorov, the founder of modern probability theory, proved a somewhat less general theorem: viz. he considered "only" the case of the product $\boldsymbol{R}^I$. The full proof of this splendid theorem is not given here (the reader can find it, for instance, in the book by H. Bauer); we shall content ourselves merely with indicating the idea of the proof. The main steps in it are:

(i) Note that the mapping $p_J^K \colon X_K \to X_J$ is $(\mathscr{A}_K - \mathscr{A}_J)$-measurable for all $J, K \in \{H\}$ such that $J \subset K$. Indeed, the sets $\bigtimes_{i\in J} A_i$, where $A_i \in \mathscr{A}_i$, generate $\mathscr{A}_J$ and we have the formula

$$(p_J^K)^{-1}\left(\bigtimes_{i\in I} A_i\right) = \bigtimes_{i\in K} A_i',$$

where $A_i' = A_i$ for $i \in I$ and $A_i' = X_i$ for $i \in K - J$. Now, $\mu_i(X_i) = 1, i \in I$, and thus

$$\prod_{i\in K} \mu_i(A_i') = \prod_{i\in J} \mu_i(A_i). \tag{10}$$

Hence, by the Fubini–Stone Theorem,

$$p_J^K(\mu_K) = \mu_J, \quad J \subset K, \ J, K \in \{H\}. \tag{11}$$

Now we introduce the important concept of cylinder.

DEFINITION. *Cylinders* are elements of the system $\mathscr{Z} := \bigcup_{J\in\{H\}} \mathscr{Z}_J$, where

$$\mathscr{Z}_J := p_J^{-1}(\mathscr{A}_J)$$

is called the *σ-algebra of J-cylinders*.

The measurability of p_J^K shown above states that $(p_J^K)^{-1}(\mathscr{A}_J) \subset \mathscr{A}_K$ and hence by (2) we obtain

$$\mathscr{Z}_J \subset \mathscr{Z}_K \quad J \subset K, \ J, K \in \{H\}. \tag{12}$$

Any two cylinders $Z_1, Z_2 \in \mathscr{Z}$ are contained in some σ-algebra $\mathscr{Z}_J$ for an appropriate $J \in \{H\}$; indeed, if $Z_i \in \mathscr{Z}_{J_i}$, $i = 1, 2$, then $J_1 \cup J_2$

is such a set J. It thus follows that $\mathscr{Z}$ is an algebra (finite additivity), but unfortunately is not a σ-algebra (countable additivity). Owing to (7) the important equality

$$(13) \qquad \mathscr{A} = \mathscr{A}(\mathscr{Z})$$

holds, which means that the cylinders generate the algebra $\mathscr{A} = \bigotimes_{i \in I} A_i$, and thus justifies the role of cylinders.

(ii) If the problem is solvable, the measure sought after must (because of (7)) associate a number $\mu_J(A)$ with every J-cylinder $Z = p_J^{-1}(A)$. This value can, therefore, depend only on Z, and not on the special representation $Z = p_J^{-1}(\mathscr{A})$. This is indeed the case.

(iii) The function

$$\mu_0\big(p_J^{-1}(A)\big) := \mu_J(A), \qquad J \in \{H\},\ A \in \mathscr{A}_J,$$

defined (correctly) on the set $\mathscr{Z}$ (of cylinders) is additive or, as it is said, "μ_0 is a volume on $\mathscr{Z}$".

(iv) Finally, it must be shown that μ_0 is σ-countable. For then μ_0 can be uniquely extended to be measure μ (on the algebra $\mathscr{A}$); μ is then even a probability measure ($||\mu|| = 1$), because, after all, $X = p_J^{-1}(X_J)$ is a J-cylinder for every $J \in \{H\}$ and hence $\mu(X) = \mu_0(X) = \mu_J(X_J) = 1$. This part of the proof is clearly the most difficult.

The algebra of cylinders plays a particularly important role in the construction of measures on infinite-dimensional vector spaces.

5. CONVOLUTIONS OF MEASURES AND FUNCTIONS

The discussion of the present section could be conducted for finite Radon measures on $\boldsymbol{R}^n$. However, we give it in a more general form, replacing the space $\boldsymbol{R}^n$ by any locally compact commutative group and the Lebesgue integral by a Haar integral, i.e., invariant Radon integral (cf. Chapter XIX, Section 21).

As we know, a Radon measure determines a Borel measure (it can be shown that, conversely, every regular Borel measure on a locally compact space determines a Radon measure). Hence, in this section we shall concern ourselves with Borel measures.

Definition. A Borel measure on a Hausdorff space $(X, \mathscr{T})$ is *finite* if $\mu(X) < \infty$. Then we denote

$$|\mu| := \mu(X).$$

5. CONVOLUTIONS OF MEASURES AND FUNCTIONS

The set of finite Borel measures μ on the Hausdorff space $(X, \mathscr{T})$ will be denoted by $\mathscr{M}(X)$ or $\mathscr{M}(X, \mathscr{T})$.

Let $(X, \mathscr{T})$ be a topological group. Bear in mind that the group operations $X \times X \ni (x, y) \to x \cdot y \in X$ and $X \ni x \to x^{-1} \in X$ are continuous in the topological group.

The reader is familiar with examples of commutative topological groups: for instance every topological vector space is a commutative group under addition of vectors. Another example is that of the n-dimensional torus, i.e., $T^n := \boldsymbol{R}^n/\boldsymbol{Z}^n$ where $\boldsymbol{Z}^n$ is a subgroup of points from $\boldsymbol{R}^n$ with integer coordinates.

DEFINITION. Let $\mu_1, \ldots, \mu_n \in \mathscr{M}(X, \mathscr{T})$. Since the mapping

$$X^n := X \times \ldots \times X \ni (x_1, \ldots, x_n) \to T_n(x_1, \ldots, x_n) := x_1 \cdot \ldots \cdot x_n \in X$$

is continuous it is Borel-measurable. Since $\bigotimes_{j=1}^{n} \mu_j \in \mathscr{M}(X^n)$, the measure

$$\mu_1 * \ldots * \mu_n := T_n(\mu_1 \otimes \ldots \otimes \mu_n) \in \mathscr{M}(X)$$

is well defined.

This measure is called the *convolution of the measures* $\mu_1, \ldots, \mu_n$.

COROLLARY XVIII.5.1. The equality

$$|\mu_1 * \ldots * \mu_n| = |\mu_1| \ldots |\mu_n|$$

holds.

PROOF. By definition we immediately have

$$\mu_1 * \ldots * \mu_n(X) = \mu_1 \otimes \ldots \otimes \mu_n(X^n) = \prod_{j=1}^{n} \mu_j(X) = \prod_{j=1}^{n} |\mu_j|. \quad \square$$

The associativity of the convolution

$$\mu_1 * \ldots * \mu_n * \mu_{n+1} = (\mu_1 * \ldots * \mu_n) * \mu_{n+1}$$

emerges from the associativity of the tensor multiplication and the group multiplication.

In what follows we shall confine ourselves to the case of two measures. Measurability will now be constructed as Borel measurability. Since we shall concern ourselves only with commutative groups, it will be more convenient to employ additive notation:

$$x + y := x \cdot y.$$

Suppose that $(X, \mathscr{T})$ is a commutative topological group. For $\mu, \nu \in \mathscr{M}(X)$ and a measurable function $f \geqslant 0$ on X the function $f \circ T_2$: $X^2 \to \overline{\boldsymbol{R}}$ is measurable and nonnegative. Thus, on invoking Theorem XVIII.3.2 and Fubini's Theorem, we obtain

$$\int f\, d(\mu * \nu) = \int f \circ T_2\, d(u \otimes \nu)$$
$$= \iint f(x+y)\, d\mu(x)\, d\nu(y) = \iint f(x+y)\, d\nu(y)\, d\mu(x).$$

On setting $f = \chi_B$, for $B \in \mathscr{A}(\mathscr{T})$, and denoting $B-y := \{b-y\colon b \in B\}$, we arrive at an extremely important formula:

$$(\mu * \nu)(B) = \int \mu(B-y)\, d\nu(y) = \int \nu(B-x)\, d\mu(x) = (\nu * \mu)(B).$$

In this way we have proven

THEOREM XVIII.5.2. If $(X, \mathscr{T})$ is a commutative topological group, the convolution of measures from $\mathscr{M}(X)$ is commutative and associative.

Let τ_a denote a translation in a group by an element a:

$$X \ni x \to \tau_a(x) := x+a.$$

A mapping so defined is clearly a homeomorphism of the group onto itself (i.e., is bijective and continuous in both directions).

If 0 is used to denote the neutral element (or identity) of the group $(x+0 = x)$, then plainly τ_0 is an identity mapping.

DEFINITION. The measure δ_a given by

$$\mathscr{A}(\mathscr{T}) \ni B \to \delta_a(B) := \begin{cases} 1, & \text{if } a \in B, \\ 0, & \text{if } a \notin B, \end{cases}$$

is called the *Dirac measure* at the point $a \in X$.

It is easily verified that a function so defined on a σ-algebra of Borel measures is indeed a measure (i.e., is σ-additive). Clearly, $\delta_a \in \mathscr{M}(X)$ and $|\delta_a| = 1$.

The formula proved previously implies that

$$(\delta_a * \mu)(B) = \int \mu(B-y)\, d\delta_a(y) = \mu(B-a) = \mu\big(\tau_a^{-1}(B)\big)$$
$$= (\tau_a \mu)(B),$$

in accordance with the definition of the transport of a measure.

The reader will easily observe that $\tau_a \circ \tau_b = \tau_{a+b}$. Accordingly, the theorem below has been proved.

THEOREM XVIII.5.3. The following relations hold for the Dirac measure δ:

(i) $\delta_a * \mu = \tau_a \mu$,
(ii) $\delta_0 * \mu = \mu * \delta_0 = \mu$,
(iii) $\delta_a * \delta_b = \delta_{a+b}$

for $\mu \in \mathcal{M}(X)$, $a, b \in X$.

Thus, δ_0 is an identity element relative to convolution. To *translate a measure* is the same as to *convolve* it with an appropriate Dirac measure.

6. CONVOLUTIONS OF FUNCTIONS AND MEASURES ON R

In this section we assume X to be the space R^p. This entire discussion could be pursued in more general terms, for a locally compact group with a bi-invariant measure λ.

Let $\lambda = \lambda^p$ be a Lebesgue measure on R^p, and let $\mu := f\lambda$ $(0 \leqslant f \in \mathcal{L}^1(\lambda))$ be a measure with density. Since

$$|\mu| = \int f d\lambda < \infty,$$

we find that $\mu \in \mathcal{M}(R^p)$. Let us find $\mu * \nu$ for any measure $\nu \in \mathcal{M}(R^p)$.

By the translation invariance of λ and by the definition of a measure with density,

$$\int \varphi d(f\lambda) = \int (\varphi \cdot f) d\lambda \quad \text{for } 0 \leqslant \varphi\text{—measurable}$$

(cf. Chapter XIII, Section 21), for any Borel set $B \in \mathcal{B}^p$ we have

$$\mu * \nu(B) = \iint \chi_B(x+y) f(x)\, d\lambda(x)\, d\nu(y)$$
$$= \iint \chi_B(x) f(x-y)\, d\lambda(x)\, d\nu(y).$$

Now, suppose that $g(x) := \int f(x-y)\, d\nu(y)$. Clearly the function g may have an infinite value at some points, for we did not assume the function f to be integrable with respect to the measure ν. In any event, this function is well defined, since $f \geqslant 0$.

Invoking the formula derived and Tonelli's Theorem, we obtain

$$\mu*\nu(B) = \int \chi_B(x)g(x)\,d\lambda(x) = \int_B g\,d\lambda.$$

We have thus shown that the measure $\mu*\nu$ has a density with respect to the measure λ^p. Plainly, $\int g\,d\lambda = |\mu*\nu| < \infty$. Let us thus take these facts for a definition:

DEFINITION. If $0 \leqslant f \in \mathscr{L}^1(\boldsymbol{R}^p)$, then

$$(f*\nu)(x) := \int f(x-y)\,d\nu(y).$$

The function $f*\nu$ is called the *convolution of the function f with the measure* $\nu \in \mathscr{M}(\boldsymbol{R}^p)$.

We have proved the following corollary:

COROLLARY XVIII.6.1. The equality

$$(f\lambda^p)*\nu = (f*\nu)\lambda^p$$

holds.

7. CONVOLUTIONS OF INTEGRABLE FUNCTIONS

Suppose now that the measures μ and ν both have a density with respect to $\lambda = \lambda^p$: $\mu = f\lambda$, $\nu = g\lambda$, where $0 \leqslant f$, $g \in \mathscr{L}^1(\boldsymbol{R}^p)$. It follows from the preceding section that the function

$$\boldsymbol{R}^p \ni x \to (f*g\lambda)(x) = \int f(x-y)g(y)\,d\lambda(y)$$

is a density of the measure $\mu*\nu$ with respect to λ. This function will be denoted by $f*g$ (that is, $(f*g)(x) := \int f(x-y)g(y)\,d\lambda(y)$) and is called the *convolution of the functions f* and g.

THEOREM XVIII.7.1. If $0 \leqslant f$, $g \in \mathscr{L}^1(\boldsymbol{R}^p)$, then

(i) $0 \leqslant f*g \in \mathscr{L}^1(\boldsymbol{R}^p)$ (i.e., the convolution of nonnegative integrable functions is itself a nonnegative integrable function),

(ii) $f*(g+h) = f*g+f*h$, $f*(\alpha g) = (\alpha f)*g = \alpha(f*g)$ for $\alpha \in \boldsymbol{R}_+$,

(iii) $(f\lambda)*(g\lambda) = (f*g)\lambda$,

(iv) $f*g = g*f$ (commutativity of convolution),

(v) $(f*g)*h = f*(g*h)$ (associativity of convolution).

PROOF. Item (i) has already been proved. Formulae (ii) and (iii) follow from the definition of a convolution. For (iv) we have

$$(f*g)(x) = \int f(x-y)g(y)\,d\lambda(y) = \int f(x+y)g(-y)\,d\lambda(y)$$
$$= \int f(t)g(x-t)\,d\lambda(t) = (g*f)(x).$$

The convolution is proved to be associative just as it was shown to be commutative. □

The convolution of a function of any sign is reduced to the case already considered by decomposing the function into positive and negative parts.

We reemphasize that the function $x \to \int f(x-y)g(y)\,d\lambda(y)$ need not be well defined for every x. However, since it is integrable it must be well defined and finite λ-almost everywhere. In the general case, therefore, we take

$$(f*g)(x) := \int f(x-y)g(y)\,d\lambda(y)$$

only for those $x \in \boldsymbol{R}^p$, for which the right-hand side is defined.

XIX. THE THEORY OF DISTRIBUTIONS HARMONIC ANALYSIS

1. THE SPACE $C_0^\infty(\Omega)$

A very important role in modern analysis is played by the set of functions which are infinitely differentiable on an open set $\Omega \subset \boldsymbol{R}^n$ and which have compact supports.

This set will be denoted $C_0^\infty(\Omega)$ (as usual, $C_0(\Omega)$ denotes the set of continuous functions with compact supports).

The construction of such functions was given at a relatively late date (1938, Friedrichs, Sobolev, Bochner). We shall now present it.

Let

$$f(t) := \begin{cases} e^{1/t} & \text{for } t < 0, \\ 0 & \text{for } t \geqslant 0. \end{cases}$$

From Part I we know that $f \in C^\infty(\boldsymbol{R}^1)$. Now take

$$\boldsymbol{R}^n \ni x \to \varphi(x) := f(||x||^2 - 1)$$
$$= \begin{cases} \exp\left(-\dfrac{1}{1-||x||^2}\right) & \text{for } ||x|| < 1, \\ 0 & \text{for } ||x|| \geqslant 1, \end{cases}$$

where $||x||^2 = x_1^2 + x_2^2 + \ldots + x_n^2$. Since the function $||x||^2 - 1$ belongs to the set $C^\infty(\boldsymbol{R}^n)$, its composition with the function f is also infinitely differentiable. Clearly, $\underline{\varphi} = \overline{K(0, 1)} \subset \boldsymbol{R}^n$ ($\underline{\varphi}$ is the support of the function φ) and thus $\varphi \in C_0^\infty(\boldsymbol{R}^n)$.

We shall show how other C_0^∞-functions can be formed by means of φ.

We write $\alpha := \int \varphi \, d\lambda^n$; then the function $\psi(x) := \dfrac{1}{\alpha} \cdot \varphi(x)$ has the following properties:

(i) $\psi \in C_0^\infty(\boldsymbol{R}^n)$,

(ii) $\underline{\psi} = \overline{K(0, 1)} \subset \boldsymbol{R}^n$,

(iii) $\int \psi \, d\lambda^n = 1$.

Now let $u \in \mathscr{L}^1(\boldsymbol{R}^n)$ and let $\underline{u} \subset K \subset \Omega$, where K is an open set. We denote

$$\delta := d(K, \complement\Omega) = \inf_{x \in K}\left(\inf_{y \notin \Omega} d(x, y)\right)$$

(here, as usually, $d(x, y)$ denotes the distance of these points in the sense of the Pythagorean metric in $\boldsymbol{R}^n$, whereas $\complement\Omega := \boldsymbol{R}^n - \Omega$).

Take the function

$$u_\varepsilon(x) := \int u(x-\varepsilon y)\,\psi(y)\,d\lambda^n(y)$$

for $\varepsilon < \frac{1}{2}\,\delta$.

Recall that ψ is measurable as a continuous function and the product of measurable functions is a measurable function.

Since ψ is a bounded function, this product belongs to $\mathscr{F}^1(\boldsymbol{R}^n)$ and hence by Theorem XIII.11.10 is an integrable function, which means that the formula above holds for all x. Applying the change-of-variables theorem to the mapping $z = \Phi(y) := x-\varepsilon y$, we find that

$$u_\varepsilon(x) = \frac{1}{\varepsilon^n}\int u(z)\,\psi\left(\frac{x-z}{\varepsilon}\right)d\lambda^n(z),$$

since $|\det \Phi'| = \varepsilon^n$. Hence $u_\varepsilon = u * \psi_\varepsilon$, where $\psi_\varepsilon(y) := \varepsilon^{-n}\psi(y/\varepsilon)$.

We make use of properties of the convolution to prove the following

THEOREM XIX.1.1 (On Regularization). Let Ω be an open set in $\boldsymbol{R}^n$. If $u \in \mathscr{L}^p(\lambda^n)$ and $\underline{u} \subset K \subset \Omega$ (K—an open set), then for $2\varepsilon < \delta = d(K, \complement\Omega)$ we have

(i) $u_\varepsilon \in C_0^\infty(\Omega)$,

(ii) $u_\varepsilon \xrightarrow[\varepsilon\to 0]{} u$ in the sense of $\mathscr{L}^p$ and $N_p(u_\varepsilon) \leqslant N_p(u)$,

(iii) if $u \in C_0(\Omega)$, then $u_\varepsilon \to u$ uniformly.

PROOF. (i) Let $a := \sup_x \left|\frac{\partial}{\partial x_i}\,\psi\left(\frac{x-z}{\varepsilon}\right)\right|$. Then

$$\left|\frac{\partial}{\partial x_i}u(z)\,\psi\left(\frac{x-z}{\varepsilon}\right)\right| = |u(z)|\left|\frac{\partial}{\partial x_i}\,\psi\left(\frac{x-z}{\varepsilon}\right)\right| \leqslant a\cdot|u(z)|.$$

Since $a\cdot|u| \in \mathscr{L}^1$, the hypotheses of Theorem XIII.10.3 on the differentiation of integrals with a parameter are thus satisfied. By this theorem the function u_ε is continuously differentiable.

On applying similar reasoning to higher-order derivatives, we deduce that the function u_ε has as many continuous derivatives as ψ does. Thus $u_\varepsilon \in C^\infty$.

The formula for u_ε implies that

$$\underline{u_\varepsilon} \subset K_\varepsilon := \{x \in \Omega\colon d(x, K) = \inf_{y\in K} d(x, y) \leqslant \varepsilon\},$$

and thus $\underline{u_\varepsilon}$ is a compact set, that is, $u_\varepsilon \in C_0^\infty(\Omega)$.

(iii) If $u \in C_0(\Omega)$, then this function is uniformly continuous as a function continuous on a compact set. Thus, for every number $\eta > 0$ there exists an $\varepsilon > 0$ such that $(d(x, x') \leqslant \varepsilon) \Rightarrow (|u(x) - u(x')| < \eta)$. Accordingly

$$|u_\varepsilon(x) - u(x)| = \left| \int u(x-\varepsilon y)\psi(y)\,dy - u(x)\int \psi(y)\,dy \right|$$
$$\leqslant \int |u(x-\varepsilon y) - u(x)| \cdot \psi(y)\,dy \leqslant \eta \cdot \int \psi = \eta.$$

The last inequality follows from the fact that we integrate only over the ball $\|y\| \leqslant 1$, and $d((x-\varepsilon y), x) \leqslant \varepsilon$. Thus, by appropriately taking $\varepsilon > 0$, we can make the quantity $|u_\varepsilon(x) - u(x)|$ arbitrarily small, regardless of x.

(ii) Let $p > 1$, and let $q > 1$ be a real number such that $1/p + 1/q = 1$. Since $\psi \geqslant 0$, we have $\psi = \psi^{1/p}\psi^{1/q}$. Hence, invoking the Hölder Theorem, we have

$$|u_\varepsilon(x)| = \left| \int u(x-\varepsilon y)\psi^{1/p}(y)\psi^{1/q}(y)\,dy \right|$$
$$\leqslant \left\{ \int |u(x-\varepsilon y)\psi^{1/p}(y)|^p dy \right\}^{1/p} \left\{ \int |\psi(y)|\,dy \right\}^{1/q}$$
$$= \left\{ \int |u(x-y)|^p \psi(y)\,dy \right\}^{1/p}.$$

Thus

$$N_p(u_\varepsilon) = \left\{ \int |u_\varepsilon(x)|^p \right\}^{1/p} \leqslant \left\{ \int \left(\int |u(x-\varepsilon y)|^p \psi(y)\,dy \right) dx \right\}^{1/p}$$
$$= \left\{ \frac{1}{\varepsilon^n} \int \left(\int |u(z)|^p \psi\left(\frac{x-z}{\varepsilon} \right) dz \right) dx \right\}^{1/p}.$$

Application of the Fubini Theorem leads to

$$N_p(u_\varepsilon) \leqslant \frac{1}{\varepsilon^{n/p}} \left\{ \int |u(z)|^p \left(\int \psi\left(\frac{x-z}{\varepsilon} \right) dx \right) dz \right\}^{1/p}.$$

However

$$\int \psi\left(\frac{x-z}{\varepsilon} \right) dx = \varepsilon^n \int \psi(x)\,dx = \varepsilon^n,$$

and thus

$$N_p(u_\varepsilon) \leqslant \left\{ \int |u(z)^p|\,dz \right\}^{1/p} = N_p(u),$$

which is to say that we have proved the predicted inequality for $p > 1$.

For $p = 1$ the inequality follows from the fact that the norm of the convolution is not greater than the product of the norm of the functions convoluted ($N_1(\psi) = 1$).

We shall now prove that $u_\varepsilon \xrightarrow[\varepsilon \to 0]{} u$ in the sense of the norm N_p.

Since $\mathscr{L}^p = \overline{C_0(\mathbf{R}^n)}$ (closure in the sense of the norm N_p), for some $\eta > 0$ we choose a function $v \in C_0(\mathbf{R}^n)$ such that $N_p(u-v) < \frac{1}{3}\eta$. Suppose, moreover, that we are given a number $0 < \alpha < \frac{1}{2}d(K, \complement \Omega)$.

A metric space is normal (Lemma XVII.2.1) and thus by Urysohn's Lemma there is a function $\varphi \in C(\Omega)$, $0 \leqslant \varphi(x) \leqslant 1$, equal to unity on K and equal to zero on $\complement K_\alpha$. Since K_α is a compact set, we have $\underline{\varphi} \subset \complement K_\alpha \subset \Omega$, or $\varphi \in C_0(\Omega)$.

Take the function $w := \varphi \cdot v$. Now,

$$|u(x) - w(x)| = |u(x) - v(x)| \quad \text{for } x \in \underline{u} \subset K$$

and

$$|u(x) - w(x)| = |w(x)| \leqslant |v(x)| = |u(x) - v(x)| \quad \text{for } x \notin \underline{u},$$

whence

$$|u(x) - w(x)| \leqslant |u(x) - v(x)|,$$

or

$$N_p(u - w) \leqslant N_p(u - v) < \tfrac{1}{3}\eta.$$

Take an $\varepsilon < \frac{1}{2}\alpha$ so that

$$\sup_x |w_\varepsilon(x) - w(x)| \leqslant \tfrac{1}{3}\eta \cdot |K_{3\alpha/2}|^{1/p}$$

(as usual, $|K_{3\alpha/2}|$ denotes the Lebesgue measure of the set $K_{3\alpha/2}$). Such a number ε does exist by virtue of point (iii) which has already been proved for this theorem, since $w \in C_0(\Omega)$.

Bearing in mind that $\underline{w} \subset K_\alpha \subset K_{3\alpha/2}$, $\underline{w_\varepsilon} \in K_{3\alpha/2}$, by Minkowski's inequality we have

$$\begin{aligned} N_p(u_\varepsilon - u) &\leqslant N_p(u_\varepsilon - w_\varepsilon) + N_p(w_\varepsilon - w) + N_p(w - u) \\ &\leqslant 2N_p(u - w) + \Big(\int_{K_{3\alpha/2}} |w_\varepsilon - w|^p\Big)^{1/p} \\ &\leqslant \tfrac{2}{3}\eta + \big((\tfrac{1}{3}\eta)^p |K_{3\alpha/2}|^{-1} |K_{3\alpha/2}|\big)^{1/p} = \tfrac{2}{3}\eta + \tfrac{1}{3}\eta = \eta, \end{aligned}$$

and thus for any $\eta > 0$ there is an ε such that $N_p(u_\varepsilon - u) < \eta$. □

Remark. Regularization of some function, measure, or distribution is an approximation (in the sense of the topology of the objects considered) by regular (i.e., differentiable) functions. Theorem XIX.1.1 was the first on the regularization of functions $u \in \mathscr{L}^p(\Omega)$, or possibly $u \in C_0(\Omega)$. A theorem on the regularization of distributions will be given somewhat further on. As is seen, the construction of regular functions u_ε approximating the function u consists in forming the convolution $u_\varepsilon = u * \psi_\varepsilon$, where $0 \leqslant \psi_\varepsilon \in C_0^\infty(R^n)$, $\underline{\psi_\varepsilon} \in K(0, \varepsilon)$ and $\int \psi_\varepsilon d\lambda^n = 1$. If pointwise convergence is introduced in the space $\mathscr{M}(R^n)$ of finite Radon measure on R^n, then $\psi_\varepsilon \to \delta_0$, $\varepsilon \to 0$, under this convergence. As we know, δ_0 is the convolution identity element and accordingly the family $\{\psi_\varepsilon\}$ is often referred to as "approximate identity element".

When subsequently we shall define a convolution of distributions, an analogous theorem on regularization will be obtained by convoluting the distribution T with the approximate identity element ψ_ε.

Let $\Omega \subset R^n$ be an open set. Now introduce the following notation: $\mathscr{L}^p(\lambda^n, \Omega) = \overline{C_0(\Omega)}$ (N_p-closure).

Clearly, if $u \in \mathscr{L}^p(\lambda^n)$ then $u|_\Omega \in \mathscr{L}^p(\lambda^n, \Omega)$.

A question arises whether functions from $\mathscr{L}^1(\lambda^n, \Omega)$ admit approximation by functions from $C_0^\infty(\Omega)$. The answer is contained in the following theorem:

THEOREM XIX.1.2. The set $C_0^\infty(\Omega)$ is dense in $\mathscr{L}^p(\lambda^n, \Omega)$ for every $p \geqslant 1$. To put it another way: $\overline{C_0^\infty(\Omega)} = \mathscr{L}^p(\lambda^n, \Omega)$ (N_p-closure).

PROOF. Since the set $C_0(\Omega)$ is dense in $\mathscr{L}^p(\lambda^n, \Omega)$ for every $p \geqslant 1$, it is sufficient to prove that $C_0^\infty(\Omega)$ is dense in $C_0(\Omega)$, that is, that every continuous function with compact support is arbitrarily N_p-approximable by functions from $C_0^\infty(\Omega)$.

This fact, however, follows from item (ii).

Remark. Since $C_0^\infty(\Omega)$ is dense in the set of elementary functions $E = C_0(\Omega)$ for any Radon integral, all of its properties are deducible from knowledge of the integral on $C_0^\infty(\Omega)$. However, $C_0^\infty(\Omega)$ cannot be used for E, since this set is not a lattice.

In conclusion, we give the following fact which we have already used in proving theorems on the theory of functions of complex variables:

THEOREM XIX.1.3. Let K be a compact subset of an open set $\Omega \subset R^n$. There exists a neighbourhood $D \subset \Omega$ of the set K and a function $f_K \in C_0^\infty(\Omega)$ such that $0 \leqslant f_K \leqslant 1$ and is equal to unity on D.

PROOF. Let $\delta = d(K, \complement\Omega)$. Put $\varepsilon = \frac{1}{4}\delta$, $\varepsilon' = \frac{1}{8}\delta$, and let

$$u(x) := \begin{cases} 1 & \text{for } x \in K_{\varepsilon'}, \\ 0 & \text{for } x \notin K_{\varepsilon'}. \end{cases}$$

Plainly, $u \in \mathscr{L}^1(\Omega)$. Take $D := K_{\delta/8}$. It is seen that D is a neighbourhood of the set K. Let us set $f_K := u_\varepsilon \in C_0^\infty(\Omega)$. Note that

$$0 \leqslant f_K(x) = \int u(x - \varepsilon y)\psi(y)\,dy \leqslant 1 \cdot \int \psi(y)\,dy = 1.$$

When $x \in D$ we have $(x - \varepsilon y) \in K_{\varepsilon'}$, for $\|y\| \leqslant 1$, whence $u(x - \varepsilon y) = 1$, and thus $f_K(x) = \int \psi(y)\,dy = 1$. □

2. A DIFFERENTIABLE PARTITION OF UNITY ON R^n

Let $\Omega \subset R^n$ be an open set. This set can be treated as a metric space (Ω, d_1), a subspace of the space R^n (d_1 denotes the restriction of the Pythagorean metric d in R^n to the set Ω).

As we learned in Chapter XVII, a metric space is paracompact and hence a partition of unity exists in it (cf. Theorem XVII.9.2). It turns out, however, a stronger fact can be proved, viz. the existence of an arbitrarily differentiable partition of unity, that is, one such that $f_j \in C_0^\infty(\Omega)$, $j \in J$.

THEOREM XIX.2.1. If $\Omega \subset R^n$ is an open set, then for every open covering of Ω there is a differentiable partition of unity $\{f_j\}_{j\in J}$, $f_j \in C_0^\infty(\Omega)$ subordinate to a refinement of that covering.

PROOF. This theorem differs from Theorem XVII.9.2 only in that Theorem XIX.1.3 should be used instead of Urysohn's Lemma to construct the functions f_j. The only difficulty is that the sets V_j, referred to in the proof of Theorem XVII.9.2, which are closed in Ω, need not be compact (e.g., if the covering, of which the partition of unity is a refinement, consists of one single element Ω).

To overcome these difficulties, note that the family of all open balls lying together with the closure in the set Ω, constitutes a covering of Ω. Denote this family by $\{D_\alpha\}_{\alpha\in\mathscr{A}}$, where $\alpha = (x, r)$ is the index labelling the ball.

Given now a covering $\{O_i\}_{i\in I}$, of which a differentiable partition of unity is to be a refinement.

Take a “finer” covering: $\{O_i \cap D_\alpha\}_{(i,\alpha)\in I\times\mathscr{A}}$. By the paracompactness of the space (Ω, d_1) there exists a locally finite covering $\{O'_j\}_{j\in J}$ which is a refinement of this covering and hence is also a refinement of the original $\{O_i\}_{i\in I}$.

Proceeding with the proof as in Theorem XVII.9.2, we observe that every set O'_j lies in one of the sets $O_i \cap D_\alpha$ whereby $\overline{O'_j} \subset \overline{D_\alpha}$ is a compact set. Consequently, $\overline{V_j} \subset \overline{O'_j}$ is also compact.

Now, if we invoke Theorem XIX.1.3 and set

$$\tilde{f}_j := f_{\overline{V}_j} \in C_0^\infty(V_j) \subset C_0^\infty(\Omega),$$

then also $f_j = \dfrac{\tilde{f}_j}{\sum \tilde{f}_j} \in C_0^\infty(\Omega)$. □

3. THE SPACE OF TEST FUNCTIONS. DISTRIBUTIONS

In the set $C_0^\infty(\Omega, C)$ (complex-valued functions) we define the following convergence:

DEFINITION. A sequence $\varphi_m \in C_0^\infty(\Omega, C)$ is said to *converge to zero* (*zero* or *null function*) if:

(i) there exists a compact set $K \subset \Omega$ such that the supports of all functions φ_m lie in K ($\underline{\varphi_m} \subset K$),

(ii) the functions and their partial derivatives of all orders tend uniformly to zero.

Somewhat further on, in the set $C_0^\infty(\Omega)$ we shall introduce a locally convex topology, i.e., one defined in terms of seminorms, and we shall note that this locally convex space is a Montel space.

Now we introduce the following notation: Let α be the multi-index $\alpha = (\alpha_1, \alpha_2, \ldots, \alpha_n)$ (where n is the dimension of the space). Then $|\alpha| := \alpha_1 + \alpha_2 + \ldots + \alpha_n$, whereas

$$D^\alpha \varphi := \frac{\partial^{|\alpha|}\varphi}{\partial x_1^{\alpha_1} \partial x_2^{\alpha_2} \ldots \partial x^{\alpha_n}}.$$

The second point of the definition for the convergence of the sequence φ_m to zero can now be recast as follows:

$$\bigwedge_\alpha D^\alpha \varphi_m \to 0 \text{ uniformly.}$$

The set $C_0^\infty(\Omega, \boldsymbol{C})$ with a pseudotopology so introduced is called a *space of test functions* and is denoted by $\mathscr{D}(\Omega)$.

The space $\mathscr{D}(\Omega)$ is complete under the pseudotopology above.

LEMMA XIX.3.1. Let (φ_m) be a Cauchy sequence in $\mathscr{D}$. Then there exists a $\varphi \in C_0^\infty(\Omega, \boldsymbol{C})$ such that $\varphi_m \xrightarrow[m\to\infty]{} \varphi$ in the sense of $\mathscr{D}$.

PROOF. Since (φ_m) is a Cauchy sequence in a uniform metric, there is a $\varphi \in C(\Omega)$ such that $\varphi_m \xrightarrow[m\to\infty]{} \varphi$ uniformly. Also, since all derivatives of the functions φ_m converge uniformly, we have $\varphi \in C^\infty(\Omega)$ generalized to the space $\boldsymbol{R}^n$ by Theorem V.13. But $\underline{\varphi_m} \subset K$, whence $\underline{\varphi} \subset K$, or $\varphi \in C_0^\infty(\Omega)$. □

DEFINITION. A *distribution* is a (complex-valued) linear functional continuous on the space $\mathscr{D}(\Omega)$.

The space of distributions (dual of $\mathscr{D}(\Omega)$) is denoted by $\mathscr{D}'(\Omega)$.

There is an important fact, the deeper meaning of which is given in the next section:

THEOREM XIX.3.2. A linear functional on $C_0^\infty(\Omega)$ is a distribution iff for every compact $K \subset \Omega$ there exist constants $a = a(K)$ and $p = p(K)$ such that

$$|T(\varphi)| \leqslant a \sum_{|\alpha| \leqslant p} \sup |D^\alpha \varphi(K)| \quad \text{for } \varphi \in C_0^\infty(K).$$

DEFINITION. If a constant p can be chosen independently of K, the distribution T is said to be *of finite order* (in Ω) and the least such number p is called the *order of the distribution* T (in Ω).

Examples. 1. By the formula

$$C_0^\infty(\Omega) \ni \varphi \to \mu(\varphi) \in \boldsymbol{C}$$

the Radon integral μ defines a distribution of order 0.

2. $n = 1$, $\Omega =]0, 1[$. $T(\varphi) := \sum_{j=1}^{\infty} D^j \varphi(1/j)$ is a distribution of infinite order.

PROOF OF THEOREM XIX.3.2. The sufficiency of the condition given is clear. We shall prove its necessity indirectly.

Suppose that the given evaluation does not hold, i.e. for every $a = p = m$ there exists a function $\varphi_m \in C_0^\infty(\Omega)$ such that $T(\varphi_m) = 1$, whereas

$\sup|D^\alpha\varphi_m(K)| \leqslant 1/m$ for $|\alpha| \leqslant m$. This is because the needed inequality is homogeneous in φ (i.e., if the inequality holds for φ, it is also valid for $b\varphi$ when $b \in C$).

Thus, if there is a K such that the inequality does not hold for any a and p, we obtain a sequence (φ_m) which is convergent to zero in the sense of the space $\mathscr{D}$ and for which $T(\varphi_m) \nrightarrow 0$. □

Further examples of distributions:

3. An important example of a Radon integral (cf. Example 1) is the Dirac delta $\delta_a(\varphi) = \varphi(a)$ familiar from Chapter XVIII.

4. Distributions represented by locally integrable functions (i.e., integrable on every compact subset of the space R^n).

Suppose that f is such a function. We define the distribution

$$T_f(\varphi) := \int f(x)\varphi(x)\,d\lambda^n(x) \in C.$$

The linearity of the functional T_f is trivial. Its continuity on $\mathscr{D}$ follows from the Lebesgue Theorem on dominated convergence. For, suppose that $\varphi_m \to 0$ in $\mathscr{D}$. Therefore, $\underline{\varphi_m} \subset K$, $a_m := \sup|\varphi_m(K)| \to 0$. Let $a := \max a_m$. Then

$$\int f\cdot\varphi_m\,d\lambda^n = \int \chi_K\cdot f\cdot\varphi_m\,d\lambda^n.$$

But $\chi_K\cdot f\cdot\varphi_m \to 0$ pointwise and $|\chi_K\cdot f\cdot\varphi_m| < a\cdot\chi_K\cdot f \in \mathscr{L}^1(\lambda^n)$, whence $\int \chi_K\cdot f\cdot\varphi_m\,d\lambda^n \xrightarrow[m\to\infty]{} 0$, which completes the proof. □

Distributions represented by locally integrable functions are used so often that in practice the function f is identified with the distribution T_f. Such procedure is not rigorous to the extent that an entire class of functions equal to each other λ^n-almost everywhere defines the same distribution. The reader no doubt guesses that the mapping of classes,

$$L^1_{\text{loc}} \ni \dot{f} \to T_{\dot{f}} \in \mathscr{D}', \quad \text{where } T_{\dot{f}}(\varphi) := \int f\cdot\varphi\,d\lambda^n,\ f \in \mathscr{D},$$

is already an injection and defines the embedding L^1_{loc} in $\mathscr{D}'$.

Thus, we can write $L^1_{\text{loc}}(\Omega) \subset \mathscr{D}'(\Omega)$.

By analogy with functions, a so-called "function" notation of distributions is used very frequently (especially in physics); i.e., if $T \in \mathscr{D}'$, $\varphi \in \mathscr{D}$, we denote

$$T(\varphi) = \int_{R^n} T(x)\varphi(x)\,dx.$$

It should be emphasized that in general the symbol $T(x)$ is not in itself meaningful (if T is not represented by a function). For instance, for the Dirac distribution:

$$\delta_0(\varphi) = \int \delta_0(x)\varphi(x)\,dx = \varphi(0).$$

If $\delta_0(x)$ were to be a function, then it would have to be equal to zero for $x \neq 0$ so that the integral $\int \delta_0(x)\varphi(x)\,dx$ would vanish for functions endowed with a support outside $x = 0$. Such a function would, however, define a zero distribution (being equal to zero dx-almost everywhere), even if it were that $\delta_0(0) = \infty$. Before the theory of distributions came onto the scene, δ was said to be a function which at zero took on a value equal to "infinity" and the following conditions were written:

$$\delta_0(x) = 0 \quad \text{for } x \neq 0,$$

$$\delta_0(0) = \infty,$$

$$\int \delta_0(x)\,dx = 1.$$

5. A *dipole* at a point $x_0 \in \boldsymbol{R}^n$, with a moment of $e \in \boldsymbol{R}^n$:

$$\mathscr{D}(\boldsymbol{R}^n) \ni \varphi \to T(\varphi) := (\nabla_e \varphi)(x_0).$$

This is a first-order distribution.

6. *Multipoles*: a 2^k-pole at x_0, with moments of $e_1, \ldots, e_k \in \boldsymbol{R}^n$ is a distribution of order k, given by the formula

$$\mathscr{D}(\boldsymbol{R}^n) \ni \varphi \to T(\varphi) := \Big(\nabla_{e_1}\big(\nabla_{e_2} \ldots (\nabla_{e_k}(\varphi)(x_0)) \ldots\big)\Big).$$

Linear operations in $\mathscr{D}'$ are defined in the usual way:

$$(a_1 T_1 + a_2 T_2)(\varphi) := a_1 T_1(\varphi) + a_2 T_2(\varphi).$$

In accordance with the notation of Chapter VII, we shall also write

$$\int T(x)\varphi(x)\,dx = \int \varphi(x)\,dT(x) = T(\varphi) = \langle \varphi, T \rangle.$$

Plainly, the expression $\langle \varphi, T \rangle$ is bilinear.

All of these symbols will be used interchangeably.

4. INDUCTIVE LIMITS. THE TOPOLOGY OF THE SPACE $\mathscr{D}$

In this section we give a concept of inductive limit which is sufficient for our needs. The reason is that in general the inductive limit of locally convex spaces is not even a topological vector space (algebraic operations are not continuous). We need a concept of inductive limit in the "category" of l.c.v.s.

DEFINITION. Given a linear space X and its vector subspaces $X_i = (X_i, \mathscr{T}_i)$ being locally convex topological vector spaces such that $X_i \neq X_j$, $i \neq j$, and $X = \bigcup_i X_i$. Let G_i be a canonical injection $G_i: X_i \to X$. A space X endowed with the strongest locally convex topology under which every mapping G_i is still continuous is called the *inductive limit* of l.c.v. spaces X_i and is denoted by $\operatorname{limind}_{i \in I} X_i$ or $\varinjlim X_i$.

DEFINITION. The inductive limits of locally convex metrizable spaces (hence, such that their topology is given by a sequence of seminorms) are called *Mackey* (or *bornological*) *spaces.*

Since $\Omega \subset R^n$ is a countable space at infinity, there exists a sequence of compact sets $K_j \nearrow \Omega$. To be more precise, $K_j \subset \operatorname{int} K_{j+1}$, and every compact subset $K \subset \Omega$ is contained in some K_p. We denote

$$X_i = \mathscr{D}(\Omega, K_i) = \{\varphi \in C_0^\infty(\Omega) : \underline{\varphi} \subset K_i\}.$$

The space X_i has a topology specified by a countable family of seminorms

$$||\varphi||_{p, K_i} := \sup_{|\alpha| \leqslant p} |D^\alpha \varphi(K_i)|, \quad p = 1, 2, \ldots$$

As we know, $\mathscr{D}(\Omega, K_i)$ is a Fréchet space (i.e., is metrizable and complete).

DEFINITION. $\mathscr{D}(\Omega) := \operatorname{limind} \mathscr{D}(\Omega, K_i)$.

Hence, $\mathscr{D}(\Omega)$ is a Mackey space.

Theorem XIX.3.2 thus makes the statement, no more, no less, that a distribution is a continuous functional on the space $\mathscr{D}(\Omega)$.

Remark. The foregoing corollary is a special case of the theorem which says that a linear mapping A of a Mackey space E is continuous iff it is sequentially-continuous, i.e., if $A\varphi_n \to 0$ for every sequence $\varphi_n \to 0$.

The space of linear *continuous* functionals on a topological vector space X is denoted by X' (as distinct from X^*, the space of *all* linear functionals on X). This is precisely why the space of distributions is denoted by $\mathscr{D}'(\Omega)$.

The concept of inductive limit will now be used to define the following Mackey spaces:

DEFINITION. Let m be a natural number. We define

$$\mathscr{D}^m(\Omega) := \varinjlim \mathscr{D}^m(\Omega, K_i),$$

where $\mathscr{D}^m(\Omega, K_i)$ are spaces of functions from $C_0^m(\Omega)$, whose support is contained in K_i and whose topology is given by means of a single seminorm:

$$||\varphi||_{K_i} := \sup_{|\alpha| \leqslant m} |D^\alpha \varphi(K_i)|.$$

Hence, it is seen that $\mathscr{D}^m(\Omega)$ is a space $C_0^m(\Omega)$ endowed with a topology of the inductive limit of the Banach spaces $\mathscr{D}^m(\Omega, K_i)$.

In mathematics there is at times need of a somewhat more general concept of inductive limit, one in which the spaces X_i are no longer subsets of the space X, and hence G_i are not canonical embeddings, but only $X = \bigcup_{i \in I} G_i(X_i)$. In that case an inductive topology is also introduced in X. The previous definition is obtained by taking canonical injections for the G_i.

5. THE PASTING TOGETHER PRINCIPLE FOR DISTRIBUTIONS. THE SUPPORT OF A DISTRIBUTION

In this section Ω denotes an open, nonempty subset in $\boldsymbol{R}^n$.

First of all, we carry out the *localization of a distribution.* Note that if we have some distribution T on Ω, that is, $T \in \mathscr{D}'(\Omega)$, and if O is an open subset in Ω, then T defines the (unique) distribution $r_{O\Omega} T \in \mathscr{D}'(O)$, given by the formula $r_{O\Omega}(\varphi) = T(\varphi)$, $\varphi \in C_0(O)$. Plainly, the mapping $r_{O\Omega}: \mathscr{D}'(\Omega) \to \mathscr{D}'(O)$ is linear. For every pair of open sets $U \subset V \subset \Omega$ we thus have by analogy the linear mappings (homomorphism)

$$r_{UV}: \mathscr{D}'(V) \to \mathscr{D}'(U),$$

these mappings satisfying the relations: r_{UU} is an identity and $r_{WV} = r_{WU} \circ r_{UV}$ for all $W \subset U \subset V$.

5. PASTING TOGETHER PRINCIPLE FOR DISTRIBUTIONS

We shall now prove a fundamental theorem of distribution theory:

THEOREM XIX.5.1 (Pasting Together Principle). Let $\{O_i\}_{i\in I}$ be a family of sets open in Ω and let $O = \bigcup_{i\in I} O_i$. If $T_i \in \mathscr{D}'(O_i)$ and if $T_i = T_j$ on $O_i \cap O_j = O_{ij}$ (that is, $r_{O_{ij}O_i}(T_i) = r_{O_{ij}O_j}(T_j)$) for any $i, j \in I$, then there exists one, and only one, distribution $T \in \mathscr{D}'(O)$ such that $r_{O_iO}(T) = T$, $i \in I$. (Hence $\mathscr{D}'(\Omega)$ is a sheaf of v.s. on Ω.)

PROOF. First of all, we shall show that if such a distribution does exist, it is unique. Let $T, S \in \mathscr{D}'(O)$ and let $r_{O_iO}(T) = r_{O_iO}(S)$ for all $i \in I$. Let us take the differentiable partition of unity $\{f_j\}_{j\in J}$ associated with the covering $\{O_i\}_{i\in I}$ of the set O (cf. Theorem XIX.2.1). If $\varphi \in \mathscr{D}(O)$, it follows that $\varphi = \sum \varphi \cdot f_j$, but $\varphi \cdot f_j \in \mathscr{D}(O_i)$ for some $i \in I$, whereby

$$T(\varphi) = \sum T(\varphi \cdot f_j) = \sum S(\varphi \cdot f_j) = S(\varphi).$$

Now we shall show that the functional T does indeed exist.

Let us define it as $T(\varphi) := \sum_j T_{i(j)}(\varphi \cdot f_j)$, where i is an index such that $f_j \subset O_{i(j)}$.

Next we verify that this formula really does define a functional, i.e., does not depend on the choice of the partition of unity. For this it is sufficient to prove that if

$$\psi_i \in \mathscr{D}(O_i), \sum \psi_i = 0, \text{ then } \sum T_i(\psi_i) = 0.$$

However, since $f_k \cdot \psi_i \in C_0^\infty(O_i \cap O_k)$, we have

$$\sum_i T_i(\psi_i) = \sum_i \sum_k T_i(f_k \cdot \psi_i)$$

$$= \sum_k \sum_i T_k(f_k \cdot \psi_i) = \sum_k T_k\Big(f_k \cdot \sum_i \psi_i\Big) = 0.$$

The functional T is clearly linear, and thus it is sufficient to prove that it is also continuous. Let K be a subset compact in O, and let $C_0^\infty(K) \ni \varphi_k \to 0$, if $k \to \infty$ ($\mathscr{D}(O)$-convergence of the sequence φ_k). Accordingly, for every j we have $\varphi_k \cdot f_j \to 0$ in the sense of $\mathscr{D}(O_i)$, where $f_j \subset O_i$ and thus

$$T(\varphi_k) = \sum T_{i(j)}(\varphi_k \cdot f_j) \to 0, \quad \text{as } k \to \infty. \ \square$$

Remark. The theorem above states, among other things, that a distribution is completely defined by its local behaviour. For instance, if a distribution vanishes in the neighbourhood of every point, it is equal to zero (equal to a zero or null distribution).

These facts may be used in order to introduce the concept of the support of a distribution:

DEFINITION. The *support of a distribution* $T \in \mathscr{D}'(\Omega)$, denoted by $\underline{T}$ or suppt T, is a set of points $x \in \Omega$ which have no open neighbourhood $O \ni x$, such that $r_O(T) = 0$.

It may also be said that $\underline{T} = \Omega - N$, where N is a set on which T vanishes. Since N is an open set as a sum of open sets, T is closed in Ω. Note that if T is represented by a continuous function, then the foregoing definition of a support is consistent with that given earlier: "the closure of a set on which T does not vanish".

Example. The support of the delta: suppt $\delta_a = \{a\}$, for if $\varphi \in C_0^\infty(\Omega - - \{a\})$, then $\delta_a(\varphi) = \varphi(a) = 0$ and hence δ_a vanishes on $\Omega - \{a\}$.

6. THE SPACE $\mathscr{E}(\Omega)$. DISTRIBUTIONS WITH COMPACT SUPPORTS

In the vector space $C^\infty(\Omega)$, of indefinitely differentiable functions, a topology is introduced by means of a countable family of seminorms, which were mentioned at the beginning of the previous section:

$$f \to ||f||_{K_j,r} := \sup_{|\alpha| \leqslant r} |D^\alpha f(K_j)|,$$

where $K_j \nearrow \Omega$ is an increasing sequence of compact subsets, exhausting the set Ω. A space $C^\infty(\Omega)$ endowed with the topology above (i.e., topology of compact convergence with derivatives of all orders) will be denoted by $\mathscr{E}(\Omega)$. This is a metrizable complete space and hence is a Fréchet space.

Plainly, $\mathscr{D}(\Omega) \subset \mathscr{E}(\Omega)$, and the identical embedding is continuous ($\mathscr{D}(\Omega)$-convergence is stronger).

As an example of convergence in $\mathscr{E}(\Omega)$ we give the following

LEMMA XIX.6.1. Let $f_j \in \mathscr{E}(\Omega)$ be a sequence of functions such that any compact set $K \subset \Omega$ intersects only a finite number of supports of the functions f_j. Then the sequence

$$\varphi_n := \sum_{j=1}^{n} f_i$$

converges in the topology of the space $\mathscr{E}(\Omega)$ to $\varphi := \sum_{j=1}^{\infty} f_j$.

PROOF. For every compact set K the sum $\sum f_j(x)$ has only a finite number of non-zero terms for $x \in K$. □

Next we prove the following interesting

THEOREM XIX.6.2. $\mathscr{E}'(\Omega) \subset \mathscr{D}'(\Omega)$ (i.e., linear functionals continuous on $\mathscr{E}$ are distributions) and $\mathscr{E}'(\Omega)$ is exactly the set of distributions with compact supports.

PROOF. A functional linear and continuous on $\mathscr{E}$ is also defined and linear on a smaller set $\mathscr{D}(\Omega) \subset \mathscr{E}(\Omega)$. It is continuous on $\mathscr{D}(\Omega)$ because the convergence of a sequence of functions in $\mathscr{D}(\Omega)$ implies their convergence in $\mathscr{E}(\Omega)$. This functional is thus a distribution.

We shall show that this distribution must have a compact support.

Suppose that suppt T is not a compact set. Then there exists a sequence $\{x_n\} \subset \Omega$ such that $x_n \in \text{suppt}\, T$ and $\{x_n\}$ has no limit point in Ω. Of course $r_n := \inf_{n \neq m} d(x_n, x_m)$ is greater than zero. It follows from the definition of a support that there are functions $\varphi_n \in D(\Omega)$ such that $\underline{\varphi_n} \subset K(x_n, \frac{1}{2} r_n)$ and such that $T(\varphi_n) \neq 0$ for them.

By a choice of suitable multipliers a situation may be produced so that, for instance, $T(\varphi_n) = 1$ for every n.

Since $\underline{\varphi_n} \cap \underline{\varphi_m} = \emptyset$ for $n \neq m$ and each compact set meets only a finite number of x_n, we find that $\mathscr{D}(\Omega) \ni \psi_k := \sum_{n=1}^{k} \varphi_n$ satisfies the condition $\psi_k \to \sum_{n=1}^{\infty} \varphi_n = \psi \in \mathscr{E}$ (cf. Lemma XIX.6.1) and this is an $\mathscr{E}$-convergence. If it were that $T \in \mathscr{E}'$, we would have arrived at a contradiction

$$T(\psi_k) = \sum_{n=1}^{k} T(\varphi_n) = k \to \infty,$$

whereas

$$T(\psi_k) \to T(\psi) < \infty.$$

Consequently, $T \in \mathscr{E}'$, or every distribution $T \in \mathscr{E}'$ has a compact support.

Now we shall prove that the converse is true, i.e., that every distribution with a compact support belongs to $\mathscr{E}'$.

Let $T \in \mathscr{D}'$ and let $\operatorname{suppt} T = K$ (compact set). Take a function $0 \leqslant f \leqslant 1$ such that $f \in \mathscr{D}$ and $f(x) = 1$ for x lying in some neighbourhood $O = O(K)$ of the set K (cf. Theorem XIX.1.3).

Now, suppose that $\varphi \in \mathscr{E}$. Then $\varphi \cdot f \in \mathscr{D}$. Let us define the distribution $\tilde{T}(\varphi) := T(\varphi \cdot f)$. The value of $\tilde{T}(\varphi)$ does not depend on the choice of the function f

$$T(\varphi \cdot f_1) = T(\varphi \cdot f_2), \quad \text{or} \quad T(\varphi \cdot (f_1 - f_2)) = 0,$$

inasmuch as for every $x \in O_1 \cap O_2$ we have $f_1(x) = f_2(x)$, that is,

$$\underline{\varphi \cdot (f_1 - f_2)} \subset \{\Omega - O_1 \cap O_2\} \subset \{\Omega - \operatorname{suppt} T\}.$$

It will be shown that $\tilde{T} \in \mathscr{E}'$. The linearity of the distribution is obvious and hence it is sufficient to prove its continuity.

Let $\varphi_n \to 0$ in $\mathscr{E}$, that is, $D^\alpha \varphi_n \to 0$ for every α, and let this convergence be almost uniform, i.e. uniform, say, on the support $\underline{f}$. Consequently, as is easily seen, $D^\alpha(f \cdot \varphi_n) \to 0$ uniformly. Moreover, the supports of all functions $f \cdot \varphi_n$ lie in the compact set $\underline{f}$ and hence $f \cdot \varphi_n \to 0$ in $\mathscr{D}$. Therefore, $\tilde{T}(\varphi_n) = T(f \cdot \varphi_n) \to 0$, which proves the continuity.

Now we shall demonstrate that $\tilde{T}|\mathscr{D} = T$, which means that $T \subset \mathscr{E}'$.

If $\varphi \in \mathscr{D}$, then $\tilde{T}(\varphi) - T(\varphi) = T(\varphi \cdot (f-1)) = 0$, since $f(x) - 1 = 0$ for x lying in a neighbourhood of the set $\operatorname{suppt} T$. □

7. OPERATIONS ON DISTRIBUTIONS

As we already know, the concept of distribution is at the same time a generalization of the differentiable function and the Radon integral. One would expect that some operations defined on objects from these classes are transferable to distributions. We have already defined linear combinations of distributions:

$$(aT_1 + bT_2)(\varphi) := aT_1(\varphi) + bT_2(\varphi).$$

The general prescription for defining operations on distributions reads: "Treat a distribution as a (regular) density in Lebesgue measure and rewrite the formula so that it be meaningful for distributions. Take this identity (in the case of functions or measures) for the definition of operations on distributions."

Now we shall show how to apply this rule.

1. *Differentiation of Distributions.* If $T \in C^\infty(\Omega)$, we use the formula for integration by parts $|\alpha|$ times; the integrals around the boundary vanish since $T \cdot \varphi$ vanishes everywhere beyond a particular ball:

$$D^\alpha T(\varphi) = \int_{R^n} D^\alpha T(x) \varphi(x) \, d\lambda^n(x)$$

$$= (-1)^{|\alpha|} \int_{R^n} T(x) (D^\alpha \varphi)(x) \, d\lambda^n(x) = (-1)^{|\alpha|} T(D^\alpha \varphi).$$

Hence we adopt the following definition:

DEFINITION. $D^\alpha T$ is a *distribution* given by the formula

$$\mathscr{D}(\Omega) \ni \varphi \to (D^\alpha T)(\varphi) := (-1)^{|\alpha|} T(D^\alpha \varphi) \in \boldsymbol{C}.$$

In a different notation:

$$\langle \varphi, D^\alpha T \rangle := (-1)^{|\alpha|} \langle D^\alpha \varphi, T \rangle.$$

The foregoing definition does indeed define a linear and continuous functional on $\mathscr{D}(\Omega)$, since $D^\alpha \colon \mathscr{D} \to \mathscr{D}$ is a continuous linear mapping, for if $\varphi_j \to 0$ in $\mathscr{D}$, then $D^\alpha \varphi_j \to 0$ in $\mathscr{D}$ (differentiation does not increase the support and φ_j would converge together with all the derivatives).

Distributions thus can always be differentiated. It was this very desire to define a class of objects which would comprise functions and to which the differentiation operation could be extended that was one of the principal impulses for the formulation of this beautiful theory. Hence, every locally integrable function (be it even noncontinuous) can be differentiated in the sense of a distribution. The resultant object will in general no longer be a locally integrable function but will be a distribution of order greater than zero. A number of examples to illustrate the differentiation of distributions will be given in a subsequent section.

2. *Multiplication by Functions.* $f \in C^\infty(\Omega)$. If $T \in \mathscr{L}^1_{\text{loc}}(\Omega)$, then for $f \in C^\infty(\Omega)$ we have

$$(fT)(\varphi) = \int (fT) \cdot \varphi \, d\lambda^n = \int T \cdot (f \cdot \varphi) \, d\lambda^n = T(f \cdot \varphi).$$

Thus, let us adopt the following definition:

DEFINITION. $\mathscr{D} \ni \varphi \to (f \cdot T)(\varphi) := T(f \cdot \varphi)$ or to put it another way, $\langle \varphi, f \cdot T \rangle := \langle f \cdot \varphi, T \rangle$.

Again, we verify that if $\varphi_j \to 0$ in $\mathscr{D}$, then $(f \cdot \varphi_j) \to 0$ in $\mathscr{D}$ (continuity) and the mapping $\mathscr{D} \ni \varphi \to f \cdot \varphi \in \mathscr{D}(\Omega)$ is linear. Hence, the formula above does indeed define a distribution.

Remark. L. Schwartz showed that multiplication of a distribution by a distribution cannot be defined so that the operation be associative. This has resulted in some physicists becoming disenchanted with distributions; thus even L. Schwartz has not managed to satisfy everybody!

3. *The Tensor Product of Distributions.* The tensor product of any integrals (measures) was defined in Chapter XIII. If Ω_1, Ω_2 were arbitrary locally compact spaces (e.g., open sets in the spaces $\boldsymbol{R}^n$, $\boldsymbol{R}^p$) and if T_1 and T_2 were Radon measures on these spaces, the tensor product $T_1 \otimes T_2$ was defined by

$$(T_1 \otimes T_2)(\varphi_1 \otimes \varphi_2) := T_1(\varphi_1) \cdot T_2(\varphi_2), \qquad \varphi_i \in C_0(\Omega_i).$$

By linearity this formula carried over to any linear combination $\sum_j \varphi_1^j \otimes \varphi_2^j$, where $\varphi_i^j \in C_0(\Omega_i)$, $j = 1, 2, 3, \ldots$ and $i = 1, 2$. Since such linear combinations were dense in $C_0(\Omega_1 \times \Omega_2)$, in a topology of uniform convergence under a stationary support, $T_1 \otimes T_2$ were extended to ntegrals on $\Omega_1 \times \Omega_2$ by virtue of continuity (cf. Chapter XIII, Section 18).

This formula will be taken here as the definition of the tensor product of distributions. For the following theorem is true:

THEOREM XIX.7.1. Let $T_i \in \mathscr{D}'(\Omega_i)$, $i = 1, 2$. Then there exists a unique distribution $T_1 \otimes T_2 \in \mathscr{D}'(\Omega_1 \times \Omega_2)$ such that

$$(T_1 \otimes T_2)(\varphi_1 \otimes \varphi_2) = T_1(\varphi_1) \cdot T_2(\varphi_2), \qquad \varphi_i \in (\Omega_i).$$

Moreover, the "Fubini identity" also holds:

$$\int \varphi(x, y)\, d(T_1 \otimes T_2)(x, y) = \int \left(\int \varphi(x, y)\, dT_2(y) \right) dT_1(x)$$
$$= \int \left(\int \varphi(x, y)\, dT_1(x) \right) dT_2(y).$$

PROOF. The proof will be carried out in two stages.

The proof of uniqueness will be analogous to the proof of uniqueness for the tensor product of Radon measures. Namely, we shall show in

Lemma XIX.7.3 that the set of linear combinations

$$\mathscr{D}(\Omega_1)\otimes\mathscr{D}(\Omega_2) = \left\{\sum \varphi_1^j\otimes\varphi_2^j : \varphi_i^j \in \mathscr{D}(\Omega_i)\right\}, \qquad i = 1, 2$$

(the *algebraic tensor product* of the spaces $\mathscr{D}(\Omega_1)$ and $\mathscr{D}(\Omega_2)$) is dense in the space $\mathscr{D}(\Omega_1\times\Omega_2)$. Therefore, two such distributions would coincide on dense sets and hence would be equal to each other.

Again, the existence of the distributions will be demonstrated as in the case of Radon measures: the right-hand side of the "Fubini equality" is taken as the definition of the tensor product of distributions. Upon verification that it does indeed define a continuous linear functional on $\mathscr{D}(\Omega_1\times\Omega_2)$, it remains to prove that this definition does not depend on the "order of integration". This can be demonstrated as follows: If $\varphi = \varphi_1\otimes\varphi_2$, then

$$\int\left(\int\varphi(x, y)\,dT_2(y)\right)dT_1(x) = T_1(\varphi_1)\cdot T_2(\varphi_2)$$
$$= \int\left(\int\varphi(x, y)\,dT_1(x)\right)dT_2(y).$$

Both sides are linear, so that this equality carries over to the whole space $\mathscr{D}(\Omega_1)\otimes\mathscr{D}(\Omega_2)$ and, since the latter is dense in $\mathscr{D}(\Omega_1\times\Omega_2)$, the equality holds everywhere.

What remains to be proven is that the right-hand side of the Fubini equality defines a distribution.

Suppose that $\varphi\in\mathscr{D}(\Omega_1\times\Omega_2)$, and denote $\psi^x(y) := \varphi(x, y)$. Clearly, $\psi^x\in\mathscr{D}(\Omega_2)$ for $x\in\Omega_1$. By K_i we denote the compact projection of the support of the function φ onto Ω_i. Now take the function

$$f(x) := T_2(\psi^x) = \int\varphi(x, y)\,dT_2(y) = T_2(\varphi(x, \cdot)).$$

The uniform continuity of the function φ and its derivatives (continuous on a compact set) and the fact that $\underline{\psi^x}\subset K_2$ imply that $\psi^x \xrightarrow[x\to x_0]{} \psi^{x_0}$ in the sense of $\mathscr{D}(\Omega_1)$ and hence

$$f(x) = T_2(\psi^x)\to T_2(\psi^{x_0}) = f(x_0),$$

which means that the function f is continuous.

In similar fashion we shall show f to be differentiable, and continuously differentiable at that:

$$\frac{f(x+e_i\cdot h)-f(x)}{h} = T_2\left(\frac{\psi^{x+e_ih}-\psi^x}{h}\right).$$

By the mean-value theorem

$$\left(\frac{\psi^{x+e_ih}-\psi^x}{h}\right)(y) = \frac{\varphi(x+e_ih, y)-\varphi(x, y)}{h}$$

$$= \frac{\partial\varphi(x+e_ih\theta, y)}{\partial x_i} \in \mathscr{D}(\Omega_2)$$

and thus it also follows from the uniform continuity of the derivatives of the function φ that

$$\frac{\psi^{x+e_ih}-\psi^x}{h} \underset{h\to 0}{\rightarrow} \frac{\partial\varphi}{\partial x_i}(x, \cdot),$$

the convergence being in the sense of the space $\mathscr{D}(\Omega_2)$. This means that

$$\frac{f(x+e_ih)-f(x)}{h} \to T_2\left(\frac{\partial}{\partial x_i}\varphi(x_i, \cdot)\right) = \frac{\partial}{\partial x_i}f(x).$$

Since $\dfrac{\partial}{\partial x_i}\varphi \in \mathscr{D}(\Omega_1\times\Omega_2)$, just as we did for φ we infer that $\dfrac{\partial}{\partial x_i}f$ is a continuous function.

Reasoning similarly for any α ($|\alpha|$ times), we arrive at the formula $D^\alpha f(x) = T_2(D_x^\alpha\varphi(x, \cdot))$, where D_x^α denotes differentiation with respect to the variable x. As before, we deduce that these derivatives are continuous, and thus the function f is infinitely differentiable. Since $\psi^x \equiv 0$ for $x \notin K$, we have $\underline{\psi^x} \subset K_1$, or $f \in \mathscr{D}(\Omega_1)$.

Hence the right-hand side of the "Fubini equality" exists:

$$T_1(f) = \int\left(\int\varphi(x, y)\,dT_2(y)\right)dT_1(x).$$

As in the case of measures, it may be assumed that:

$$(T_1\otimes T_2)(\varphi) := T_1(f) = T_1(T_2(\varphi)).$$

It is an obvious fact that this expression is linear in the argument φ. The continuity of the expression remains to be proved.

Thus, suppose that $\varphi_n \to 0$ in $\mathscr{D}(\Omega_1\times\Omega_2)$. Then, among other things, the inclusion $\underline{\varphi_n} \subset K \subset K_1\times K_2$ holds (K_i are the corresponding projections of K onto Ω_i).

Now, T_2 is a distribution (continuous functional) on $\mathscr{D}(\Omega_2, K_2)$ so that there exists a neighbourhood W of zero in $\mathscr{D}(\Omega_2, K_2)$ such that $|T_2(\psi)| \leqslant 1$ for $\psi \in W$. The topology in $\mathscr{D}(\Omega_2, K_2)$ is prescribed by means of seminorms, hence in particular there exists a neighbourhood W of the form

$$\{\psi \in \mathscr{D}(\Omega_2, K_2): ||\psi||_p \leqslant 1/M\}$$

for some finite p.

The linearity of the distribution T_2 implies that $|T_2(\psi)| < M \cdot ||\psi||_p$ for any $\psi \in \mathscr{D}(\Omega_2, K_2)$. Thus

$$\begin{aligned}|f_n(x)| &= |T_2(\psi_n^x)| \leqslant M \cdot ||\psi_n^x||_p = M \cdot \sup_{|\alpha| \leqslant p} |D^\alpha \psi_n^x(K_2)| \\ &= M \cdot \sup_{|\alpha| \leqslant p} |D_y^\alpha \varphi_n(x, K_2)| \\ &\leqslant M \cdot \sup_{|\alpha| \leqslant p} |D^\alpha \varphi_n(K)| = M \cdot ||\varphi_n||_p \underset{n \to \infty}{\to} 0,\end{aligned}$$

since $\varphi_n \to 0$ in $\mathscr{D}$.

We have thus proved that $f_n \to 0$ uniformly.

Identical reasoning is applicable to the derivatives of the function f, because $D^\alpha f_n(x) = T_2(D_x^\alpha \varphi_n(x, \cdot))$ and $D_x^\alpha \varphi_n \to 0$ in $\mathscr{D}$.

Moreover, $f_n \subset K_1$ for every n, and thus $f_n \to 0$ in $\mathscr{D}(\Omega_1)$. Consequently, $(T_1 \otimes T_2)(\varphi_n) = T_1(f_n) \to 0$, which completes the proof of continuity. □

Before proceeding to prove that $\mathscr{D}(\Omega_1) \otimes \mathscr{D}(\Omega_2)$ is dense in $\mathscr{D}(\Omega_1 \times \times \Omega_2)$, we shall define a convolution of distributions.

4. *Convolutions of Distributions.* Let $\Omega = R^n$. Let us recall how the convolution $T * S$ of arbitrary finite Borel measures $T, S \in \mathscr{M}(R^n)$ was defined (cf. Chapter XVIII, Section 5).

Suppose that t denotes the mapping $t\colon R^n \times R^n \to R^n$, given by the group operation (vector addition)

$$R^n \times R^n \ni (x, y) \to t(x, y) = x + y \in R^n.$$

This operation is Borel-measurable (i.e., is a morphism of a Borel structure into a Borel structure), and thus enables the measure $T_1 \otimes T_2 \in \mathscr{M}(R^n \times R^n)$ to be transported onto R^n. The image $t(T_1 \otimes T_2) =: T_1 * T_2 \in \mathscr{M}(R^n)$ satisfies the identity

$$\int \varphi \, d(T_1 * T_2) = \int (\varphi \circ t) \, d(T_1 \otimes T_2) = \iint \varphi(x+y) \, dT_1(x) \, dT_2(y).$$

If the notation $(t^*\varphi)(x, y) := \varphi(t(x, y))$ is introduced, the foregoing formula can be recast in the form

$$(T_1 * T_2)(\varphi) := (T_1 \otimes T_2)(t^*\varphi), \quad \varphi \in \mathscr{D}(R^n).$$

This formula will be adopted as the definition of a convolution of distributions, when at least one of the distributions has a compact support. This precaution is called for by the following considerations:

To begin with, note that the support of the tensor product of distributions is the Cartesian product of the supports: $\underline{T_1 \otimes T_2} = \underline{T_1} \times \underline{T_2}$. On the other hand, the support ot the function $t^*\varphi = \tilde{\varphi}$ is the union of subsets which are parallel of the diagonal $x+y = 0$ and is compact only if $\varphi \equiv 0$. The right-hand side of the definition of convolution is meaningful only when the part common to the support of the function $t^*\varphi$ and the suppoıt of the distribution $T_1 \otimes T_2$ is bounded in $\boldsymbol{R}^{2n}$. This is so, for instance, if $\underline{T_1} = K$ is a compact set. Indeed, suppose that φ has a compact support $\underline{\varphi} = A$:

$$\operatorname{suppt} {}^*\varphi = \{(x, y) : x+y \in \operatorname{suppt} \varphi\}.$$

Since $\underline{T_1} \otimes \underline{T_2} \subset K \times \boldsymbol{R}^n$, any point of the common part B of the two supports obeys the relationship: $x \in K$, $x+y \in A$, whence $y \in A \dot{-} K$, and thus $B \subset K \times (A \dot{-} K)$, or B is bounded, and is even compact (we have denoted $A \dot{-} K := \{a-k \in \boldsymbol{R}^n : a \in A, k \in K\}$).

Thus, we are to prove the following

THEOREM XIX.7.2. Let $T_1 \in \mathscr{E}'(\boldsymbol{R}^n)$, $T_2 \in \mathscr{D}'(\boldsymbol{R}^n)$. Then there exists a distribution $T_1 * T_2$ given by

$$\mathscr{D}(\boldsymbol{R}^n) \ni \varphi \to (T_1 \otimes T_2)(t^*\varphi) =: (T_1 * T_2)(\varphi) \in \boldsymbol{C}.$$

The distribution $T_1 * T_2 \in \mathscr{D}'(\boldsymbol{R}^n)$ is called the *convolution of the distributions* T_1 and T_2. The convolution is commutative, i.e., $T_1 * T_2 = T_2 * T_1$.

PROOF. Let $f \in C_0^\infty(\boldsymbol{R}^n)$ and let $f = 1$ on some neighbourhood U of the set $K = \underline{T_1}$ (cf. Theorem XIX.1.3). Then

$$\begin{aligned} \int \varphi \, d(T_1 * T_2) &= \iint \varphi(x+y) \, dT_1(x) \, dT_2(y) \\ &= \iint f(x) \varphi(x+y) \, dT_1(x) \, dT_2(y). \end{aligned}$$

However, $(x, y) \to f(x)\varphi(x+y)$ has a compact support and is equal to $\varphi(x+y)$ on a neighbourhood of the set $B = \underline{T_1 \otimes T_2} \cap \underline{t^*\varphi}$. Thus, if $\mathscr{D}(\boldsymbol{R}^n) \ni \varphi_j \to 0$, $\underline{\varphi_j} \subset A$, $j = 1, 2, \ldots$, the supports of the functions $(x, y) \to f(x)\varphi_j(x+y)$ are contained in the prescribed compact set

$K\times(A \dot{-} K) \subset R^n \times R^n$, and the functions themselves are uniformly convergent to zero along with derivatives of any order.

Since $T_1 \otimes T_2 \in \mathscr{D}'(R^n \times R^n)$, we have

$$(T_1 * T_2)(\varphi_j) = \iint f(x)\varphi_j(x+y)\,dT_1(x)\,dT_2(y) \to 0,$$

which means that $T_1 * T_2$ is a continuous functional on $\mathscr{D}(R^n)$, and thus is a distribution (the linearity is obvious).

The commutativity and associativity of the convolution follow from the corresponding properties of the tensor product. □

A lemma of fundamental importance will be proved next.

LEMMA XIX.7.3. The set $\mathscr{D}(R^n) \otimes \mathscr{D}(R^m)$ is dense in $\mathscr{D}(R^n \times R^m)$.

PROOF. Let $f \in \mathscr{D}(R^n \times R^m)$. Choose a number $a > 0$ such that $((x, y) \in \underline{f}) \Rightarrow (|x_i| < a, |y_j| < a$ for all $i = 1, \ldots, n$; $j = 1, \ldots, m)$.

Let $\varrho(t)$ be a function of a real variable $t \in R^1$ such that $\varrho(t) = 1$ if $|t| \leqslant a$ and $\varrho(t) = 0$, if $|t| > 2a$. Let $\varrho \in C_0^\infty(R^1)$ (cf. Theorem XIX.1.3). We define the functions $g \in \mathscr{D}(R^n)$ and $h \in \mathscr{D}(R^m)$ as follows:

$$g(x) := \varrho(x_1) \cdot \varrho(x_2) \cdot \ldots \cdot \varrho(x_n),$$
$$h(y) := \varrho(y_1) \cdot \varrho(y_2) \cdot \ldots \cdot \varrho(y_m).$$

Let $\alpha = (\alpha_1, \ldots, \alpha_{n+m})$. The function $D^\alpha f$ on the compact set

$$K := \{(x, y): |x_i| \leqslant 2a, |y_i| \leqslant 2a\}$$

can be uniformly approximated, to within ε, by means of polynomials and hence by functions of the form

$$u(x, y) = \sum a_k x^{\beta^k} y^{\gamma^k} = \sum a_k x^{\beta^k} \otimes y^{\gamma^k},$$

where we have written

$$x^\beta := x_1^{\beta_1} \cdot x_2^{\beta_2} \ldots x_n^{\beta_n}$$

for the index $\beta = (\beta_1, \ldots, \beta_n)$. Correspondingly, y^γ denotes a monomial of the variables $(y_1, \ldots, y_m)$. We have

$$\frac{\partial^{|\alpha|-1}}{\partial x_1^{\alpha_1} \ldots \partial x_{n+m}^{\alpha_{n+m}-1}} f(x, y)$$
$$= \int_{-2a}^{y_m} D^\alpha f(x_1, \ldots, x_n, y_1, \ldots, y_{m-1}, t)\,dt.$$

Thus, if we take

$$u_1(x, y) := \int_{-2a}^{y_m} u(x_1, \ldots, x_n, y_1, \ldots, y_{m-1}, t)\,dt,$$

then for $(x, y) \in K$:

$$\left|\left(\frac{\partial^{|\alpha|-1}}{\partial x_1^{\alpha_1} \ldots \partial x_{n+m}^{\alpha_{n+m}-1}} f - u_1\right)(x, y)\right| \leqslant \int_{-2a}^{+2a} \varepsilon\,dt = 4a\varepsilon.$$

Integration in this way $|\alpha|$ times yields a polynomial v such that $|v-f| \leqslant (4a)^{|\alpha|} \cdot \varepsilon$ on K. If previously we took $a \geqslant \frac{1}{4}$, all derivatives up to and including D^α are evaluated similarly:

$$|D^{\alpha'}v - D^{\alpha'}f| \leqslant (4a)^{|\alpha|-|\alpha'|} \cdot \varepsilon \leqslant (4a)^{|\alpha|} \cdot \varepsilon$$

for $\alpha' = (\alpha_1', \ldots, \alpha_{n+m}')$ such that $\alpha_i' \leqslant \alpha_i$ (such a situation will be denoted by $\alpha' \leqslant \alpha$). Now, the polynomial v is multiplied by the function $g \otimes h$, and we then have

(i) $w := (g \otimes h) \cdot v \in \mathscr{D}(R^n) \otimes \mathscr{D}(R^m)$, $\underline{w} \subset K$,

(ii) $|D^{\alpha'}w - D^{\alpha'}f| \leqslant |D^{\alpha'}w - D^{\alpha'}v| + |D^{\alpha'}v - D^{\alpha'}f|$
$\leqslant |D^{\alpha'}v(g \otimes h - 1)| + (4a)^{|\alpha|}\varepsilon.$

However, the function $g \otimes h - 1$ does not vanish on such a subset of K that the function f in turn vanishes on it; hence

$$|D^{\alpha''}v| = |D^{\alpha''}v - D^{\alpha''}f| \leqslant (4a)^{|\alpha|}\varepsilon$$

for $\alpha'' \leqslant \alpha' \leqslant \alpha$, on the set $\underline{(g \otimes h - 1)} \cap K$.

Differentiation of the product of the functions $v \cdot (g \otimes h - 1)$ yields $2^{|\alpha'|}$ terms and thus, finally, we have

$$|D^{\alpha'}w - D^{\alpha'}f| \leqslant 2^{|\alpha'|} \cdot (4a)^{|\alpha|}\varepsilon \sup_{\alpha'' \leqslant \alpha'} |D^{\alpha''}(g \otimes h - 1)| + (4a)^{|\alpha|}\varepsilon$$

$$\leqslant (4a)^{|\alpha|}\varepsilon \cdot \big(1 + 2^{|\alpha'|} \sup_{\alpha'' \leqslant \alpha} |D^{\alpha''}(g \otimes h - 1)|\big).$$

On taking

$$\varepsilon \leqslant \frac{1}{k} \cdot [(4a)^{|\alpha|} \cdot \big(1 + 2^{|\alpha'|} \sup_{\alpha'' \leqslant \alpha} |D^{\alpha''}(g \otimes h - 1)|\big)]^{-1},$$

we obtain a function $w_{k,\alpha}$ such that

(i) $w_{k,\alpha} \in \mathscr{D}(R^n) \otimes \mathscr{D}(R^m)$,

(ii) $|D^{\alpha'}w_{k,\alpha} - D^{\alpha'}f| \leqslant 1/k$ for $\alpha' \leqslant \alpha$.

Now, let $\alpha^k = (k, k, \ldots, k)$ ($n+m$ times). Take the sequence $w_k := w_{k,\alpha^k}$. It is seen that $w_n \xrightarrow[n\to\infty]{} f$ in the sense of $\mathscr{D}(R^n \times R^m)$. □

5. *Translation of a Distribution.* Let $\Omega = R^n$. As we know, the translation τ_n: $x \to x+h$ defines a translation of the function $(\tau_h\varphi)(x) := \varphi(x-h) = \varphi(\tau_{-h}x)$ and a translation of Radon measures on R^n: $(\tau_h\mu)(\varphi) := \mu(\varphi \circ \tau_h) = \mu(\tau_{-h}\varphi)$.

This formula will be adopted as a definition:

DEFINITION. A distribution given by the formula

$$(\tau_h T)(\varphi) := T(\tau_{-h}\varphi)$$

is called a *translation of a distribution* $T \in \mathscr{D}'(R^n)$. This definition is correct since τ_{-h}: $\mathscr{D} \to \mathscr{D}$ is a continuous linear mapping and hence the composition $\tau_h T - T \circ \tau_{-h}$ is a distribution.

Remark. A translation in functional notation is written $\langle\varphi, \tau_h T\rangle = \langle\tau_{-h}\varphi, T\rangle$. In function notation

$$\tau_h T(\varphi) = \int \varphi(x)(\tau_h T)(x)\,dx = \int \varphi(x)T(x-h)\,dx.$$

Example. $\delta_a = \tau_a\delta_0$. In function notation: $\delta_a(x) = \delta_0(x-a) =: \delta(x-a)$.

8. THE CONVOLUTION ALGEBRA $\mathscr{E}'(R^n)$

In a set of distributions with compact supports $\mathscr{E}'(R^n)$, as we know, a convolution is always defined, is commutative, and associative. It also follows from the properties of the tensor product that the convolution is also linear. These and other facts, not yet proved, will be grouped together in the following

THEOREM XIX.8.1. (i) Under convolution the vector space $\mathscr{E}'(R^n)$ is a commutative algebra: $S*(aT_1+bT_2) = a(S*T_1)+b(S*T_2)$.

(ii) The distribution δ_0 is the neutral element of the algebra $\mathscr{E}'(R^n)$: $\delta_0 * T = T*\delta_0 = T$.

(iii) $\delta_h * T = \tau_h T$.

(iv) $\tau_h(T*S) = \tau_h T*S = T*\tau_h S$.

PROOF. Hypothesis (i) has already been proved in Theorem XIX. 7.2 and

$$(\delta_h * T)(\varphi) = \iint \varphi(x+y)\,d\delta_h(x)\,dT(y)$$

$$= \int \varphi(h+y)\,dT(y) = T(\tau_{-h}\varphi) = (\tau_h T)(\varphi),$$

which means that we have proved (ii) and (iii).

Hypothesis (iv) follows from (iii) and from the properties of the convolution (commutativity and associativity):

$$\tau_h(T*S) = \delta_h*(T*S) = (\delta_h*T)*S = (\tau_h T)*S. \ \square$$

Note that translation and differentiation commute for $\varphi \in \mathscr{D}(\boldsymbol{R})$:

$$(D^\alpha\varphi)(x+y) = \tau_{-y}(D^\alpha\varphi)(x) = D_x^\alpha\tau_{-y}\varphi(x).$$

The commutation of convolution and differentiation follows from:

THEOREM XIX.8.2. The relation

$$D^\alpha(T*S) = T*(D^\alpha S) = (D^\alpha T)*S$$

holds.

PROOF. By the definition of the derivative of a distribution

$$T*(D^\alpha S)(\varphi) = \int\left(\int \varphi(x+y)(D^\alpha S)(x)\,dx\right)dT(y)$$

$$= (-1)^{|\alpha|}\int\left(\int (D^\alpha\varphi)(x+y)\,S(x)\,dx\right)dT(y)$$

$$= (-1)^{|\alpha|}\cdot(T*S)(D^\alpha\varphi) = D^\alpha(T*S)(\varphi). \ \square$$

Setting $T = \delta_0$ and bearing in mind that δ_0 is the identity element in the convolution algebra, we obtain

COROLLARY XIX.8.3. We have $D^\alpha S = (D^\alpha\delta_0)*S$, that is to say, differentiation can be replaced by convolution with a derivative of the Dirac measure.

This corollary can be restated in a somewhat more general form: Action of a differential operator is equivalent to convolution with a distribution having a support at zero:

$$\sum a_\alpha D^\alpha S = \left(\sum a_\alpha D^\alpha\delta_0\right)*S.$$

Remark. It was recently shown (Bruhat, Kac, Maurin) how to define distributions on arbitrary locally compact groups (not necessarily Lie groups). Since no difficulty is encountered in defining convolution, a device which suggests itself is that of defining "differential operators", invariant under right translations, as a (left) convolution with a distribution having a support in the identity of the group.

9. THE IMAGE OF A DISTRIBUTION

The translation of a distribution was the simplest mapping induced by the mapping of the space R^n. The question is whether, just as for measures, one can define a broader class of mappings of distributions which, in a sense, play the role of measurable mappings.

DEFINITION. A continuous mapping $\Phi\colon X \to Y$ of a locally compact space X into a locally compact space Y is said to be *proper* if the inverse image of every compact set in Y is compact in X, that is, if $K \subset Y$ is compact, then $\Phi^{-1}(K) \subset X$ is compact.

Proper differentiable mappings define mappings of the space of distributions, for the following straightforward theorem holds:

THEOREM XIX.9.1. Let $\Omega_1 \subset R^n$, $\Omega_2 \subset R^p$ and let $\Phi\colon \Omega_1 \to \Omega_2$ be a proper, differentiable mapping. Then the mapping $\Phi^*\colon \mathscr{D}(\Omega_2) \to \mathscr{D}(\Omega_1)$, defined as in Chapter XVIII, i.e., $(\Phi^* f)(x) = (f \circ \Phi)(x)$, $f \in \mathscr{D}(\Omega_2)$, is linear and continuous.

Moreover, the mapping $\Phi := (\Phi^*)^*$, transposed to Φ^* and given by the formula $(\Phi T)(f) := T(\Phi^* f)$, $T \in \mathscr{D}'(\Omega_1)$, is continuous linear mapping $\Phi\colon \mathscr{D}'(\Omega_1) \to \mathscr{D}'(\Omega_2)$.

Remark. The distribution ΦT is called the *image of the distribution T* (with respect to the map Φ).

PROOF. If $f \in \mathscr{D}(\Omega_2)$, then $f \circ \Phi$ is an differentiable function as a composition of differentiable mappings. This function has a compact support: $\Phi^{-1}(f)$ and hence belongs to $\mathscr{D}(\Omega_1)$. Since the linearity of the mapping Φ^* is obvious, its continuity remains to be proved. However, if $f_j \to 0$ in the sense of $\mathscr{D}(\Omega_2)$, then $f_j \subset K$ and $\sup|D^\alpha f_j(K)| \to 0$, whereby it follows that $\Phi^* f_j \subset \Phi^{-1}(K)$ and that $\sup|\Phi^* f_j(\Omega_1)| \leqslant \sup|f_j(\Omega_2)|$.

The derivatives of the function $\Phi^* f$ also uniformly converge to zero. This follows from the fact that on differentiating the function $\Phi^* f = f \circ \Phi$, we obtain a combination of derivatives of the function f_j, multiplied by the corresponding partial derivatives of the mapping Φ. However, being continuous, the derivatives of Φ are bounded on the compact set $\Phi^{-1}(K)$ and thus their products with the derivatives of the function f_j, which are uniformly convergent to zero, also converge uniformly to zero.

Consequently, the mapping Φ^* is continuous.

It thus follows that ΦT, as the composition $T \circ \Phi^*$, is also linear and continuous and hence is a distribution.

As it is the transposed to the linear continuous mapping Φ^*, the mapping Φ is linear and continuous. □

Example. Every diffeomorphism Φ: $\Omega_1 \to \Omega_2$ (i.e., differentiable one-to-one mapping whose inverse is also differentiable) defines a mapping of the space of distributions. If T has a compact support, then so does T.

10. REMARKS ON THE TENSOR PRODUCTS $E\overline{\otimes}F$ AND $E\hat{\otimes}F$. THE KERNEL THEOREM

In this book we have already used the term "tensor product" on several occasions. In the present section we shall cursorily introduce the concept of the tensor product of vector spaces and we shall tell of various locally convex topologies which play an important role in analysis.

The *tensor (algebraic) product* $E\otimes F$ of vector spaces E and F is the linear space of all formal (finite) sums $g = \sum e_r \otimes f_r$, where $e_r \in E, f_r \in F$, the following expressions being identified with each other:

$$(e_1+e_2)\otimes f = e_1\otimes f + e_2\otimes f,$$

$$e\otimes(f_1+f_2) = e\otimes f_1 + e\otimes f_2,$$

$$\alpha(e\otimes f) = (\alpha e)\otimes f = e\otimes(\alpha f), \quad \alpha \in \boldsymbol{C} \text{ (or } \boldsymbol{R}).$$

The bilinear mapping B: $E\times F \to E\otimes F$, given by

$$B(e,f) := e\otimes f,$$

will be said to be *canonical*.

10. TENSOR PRODUCTS $E\overline{\otimes}F$ AND $E\hat{\otimes}F$

In analysis, the spaces E and F usually have locally convex topologies, i.e., topologies induced by families of semi-norms which separate points. Back before the war J. von Neumann introduced various locally convex topologies in the tensor products of Banach spaces. Incidentally, von Neumann's investigations were aimed at laying the mathematical foundations for quantum theory.

A new chapter in the history of locally convex tensor products was opened by the Ph.D. thesis (1955) of the then young genius, Arthur Grothendieck. Grothendieck, a pupil of L. Schwartz, wanted to understand "the Kernel Theorem" (which will be discussed a little further on) and for this purpose he erected the edifice of the theory of locally convex tensor products. These considerations put him on the road to the concept of extremely important locally convex spaces which, constituting an abstract background for the kernel theorem, have been called nuclear spaces. Grothendieck's thesis, a very difficult work, has undoubtedly been the most important development in functional analysis since the war. The studies on nuclear spaces have not been completed. Interesting applications of these spaces can be found in the author's monograph *General Eigenfunction Expansions and Unitary Representations of Topological Groups*.

Let E and F be locally convex spaces (e.g., $E = \mathscr{D}(\Omega_1)$, $F = \mathscr{D}(\Omega_2)$). Their product, $E \times F$, has a natural locally convex topology. The tensor product $E \otimes F$ admits introduction of a finest locally convex topology such that

(i) the canonical mapping $B\colon E \times F \to E \otimes F$ be (still) continuous,
(ii) the canonical mapping B be (still) partially continuous.

DEFINITION. A space $E \otimes F$ endowed with a locally convex finest topology such that B is partially continuous, is denoted by $E \otimes_i F$ and is called a *tensor product with inductive topology*. The completion of the space $E \otimes_i F$ is denoted by $E \overline{\otimes} F$.

DEFINITION. The (finest) topology under which the mapping B is continuous is called *projective*. A space $E \otimes F$ endowed with this topology is designated by $E \otimes_p F$, and its completion by $E \hat{\otimes} F$.

Remark. Completion of these spaces is necessary for they can be shown to be complete only if the dimension of at least one of spaces E or F is finite.

The two topologies are in general different (they are identical, say, if both E and F are Banach spaces or, more generally, Fréchet spaces; this fact follows from the Mazur–Orlicz–Bourbaki Theorem which can be found, for instance, in my *Methods of Hilbert Spaces*). An inductive topology is in general finer (stronger).

This leads up to the interesting Grothendieck Theorem which is given here without proof:

THEOREM XIX.10.1. The following identities hold:
(i) $\mathscr{D}(\Omega_1 \times \Omega_2) = \mathscr{D}(\Omega_1)\overline{\otimes}\mathscr{D}(\Omega_2)$,
(ii) $\mathscr{D}'(\Omega_1 \times \Omega_2) = \mathscr{D}'(\Omega_1)\hat{\otimes}\mathscr{D}'(\Omega_2)$.

Now for some historical comments on the kernel theorem.

In his classical monograph, Dirac often writes various linear operators in "integral" form, e.g.,

$$f(x) = \int \delta(x-y) f(y)\,dy \quad \text{or} \quad (Af)(x) = \int G(x, y) f(y)\,dy,$$

treating the "kernel" $G(\cdot\,, \cdot)$ as a function of two variables. Of course, in general there are no such functions (cf. remarks on the subject of δ in Section 3).

Two things led Laurent Schwartz, the founder of the theory of distributions, to formulate the celebrated "Kernel Theorem" in 1950. On the one hand, there were the successes of the Dirac notation, which has been adopted by physicists. On the other hand, there was the enormous role of integral operators in the theory of boundary-value problems for differential operators where the inverse operator A^{-1} is often an integral operator whose kernel, the Green's function of the boundary-value problem, is a true function of two variables. The full proof for the kernel theorem was not given until several years later (Grothendieck, Schwartz). It turns out that Dirac's intuitive conclusions could be endowed with a precise meaning by construing the "integral kernel" G as a distribution on $\Omega_1 \times \Omega_2$, that is, $G \in \mathscr{D}'(\Omega_1 \times \Omega_2)$.

THEOREM XIX.10.2 (First Version of the "Kernel Theorem"). Any continuous linear mapping $L\colon \mathscr{D}(\Omega_1) \to \mathscr{D}'(\Omega_2)$, where $\mathscr{D}'(\Omega_2)$ has a weak topology, can be assigned exactly one distribution $G \in \mathscr{D}'(\Omega_1 \times \Omega_2)$, called the *kernel of the mapping L*, such that

$$\langle \psi, L\varphi \rangle = \langle \psi \otimes \varphi, G \rangle$$

for $\varphi \in \mathscr{D}(\Omega_1)$, $\psi \in \mathscr{D}(\Omega_2)$.

In the Dirac notation this equality takes on the form

$$\iint \psi(x)(L\varphi)(x)\,dx = \iint \psi(x)\varphi(y)G(x, y)\,dy\,dx,$$

that is,

$$(L\varphi)(x) = \int \varphi(y)g(x, y)\,dy.$$

Since to operators L there correspond in one-to-one fashion bilinear forms which are partially continuous, i.e., continuous in each variable when the other is arbitrarily fixed, the kernel theorem can be reformulated in a different version:

THEOREM XIX.10.3 (Second Version of the "Kernel Theorem"). To every bilinear, partially continuous form K, defined on $\mathscr{D}(\Omega_1)\times\mathscr{D}(\Omega_2)$, there corresponds a unique distribution $G \in \mathscr{D}'(\Omega_1 \times \Omega_2)$ such that $K(\varphi, \psi) = \langle\varphi\otimes\psi, G\rangle$ for any functions $\varphi \in \mathscr{D}(\Omega_1)$, $\psi \in \mathscr{D}(\Omega_2)$. The reader will notice the close connection of the "Kernel Theorem" with Theorem XIX.10.1 (i).

A relatively simple proof can be found in Chapter III of *Methods of Hilbert Spaces* by the present author.

11. THE TENSOR PRODUCT $E\underline{\otimes}F$ OF HILBERT SPACES

If $E = (E, (\ |\ )_E)$, $F = (F, (\ |\ )_F)$ are Hilbert spaces, then the scalar product

$$(e_1\otimes f_1 \mid e_2\otimes f_2) := (e_1 \mid e_2)_E\cdot(f_1 \mid f_2)_F$$

can be introduced in a natural manner on the tensor product $E\otimes F$.

Since any element of the space $E\otimes F$ can be written in the form $\sum e_i\otimes f_i$, the definition above can be extended by linearity to the whole space $E\otimes F$.

It is seen without difficulty that if $\{e_i\}_{i\in I}$ is an orthonormal basis in E and $\{f_j\}_{j\in J}$ is an orthonormal basis in F, then $\{e_i\otimes f_j\}_{(i,j)\in I\times J}$ is an orthonormal basis in $E\otimes F$.

The completion of the space $E\otimes F$ in the norm given by the scalar product above is already a Hilbert space which we shall denote by $E\underline{\otimes}F$. This space plays a great role in quantum mechanics.

The following general example is of interest.

Example. Let $E = L^2(X, \mu)$, $F = L^2(Y, \nu)$. Bear in mind that

$$(e_1 \mid e_2)_E = \int e_1(x)\cdot\overline{e_2(x)}\,d\mu(x)$$

and, similarly, in F. It can be shown that

$$L^2(X\times Y, \mu\otimes\nu) = L^2(X, \mu)\overline{\otimes} L^2(Y, \nu).$$

It is a general fact that if E and F are infinite-dimensional Hilbert spaces, then $E\hat{\otimes}F = E\overline{\otimes}F \neq E\otimes F$.

Linear mappings $F' \to E$ defined by

$$(e\otimes f)(f') := \langle f, f'\rangle\cdot e$$

can be associated with the elements of the space $E\otimes F$.

In the case of a Hilbert space, the space F' is anti-isomorphic with F. The anti-isomorphism $A\colon F' \to F$ is given by the Fréchet–Riesz Theorem (Theorem XIII.14.4):

$$(g\,|\,Af) := \langle g, f'\rangle.$$

Definition. The elements of the space $E\overline{\otimes}F$, where we have identified F' with F by the Fréchet–Riesz anti-isomorphism, are called a *Hilbert–Schmidt mappings*. Let $\{e_i\}$, $\{f_j\}$ be orthonormal bases in E and F, respectively. To the tensor $T = \sum_{j\in J} e_j\otimes f_j$ we associate the operator $F \to E$ which sends $f\in F$ onto $Tf := \sum_{j\in J}(f|f_j)\,e_j$. Then we have

$$\sum \|Tf_j\|^2 = \sum \|e_j\otimes f_j\|^2 =: |||T|||^2 < \infty.$$

The norm $|||\ |||$ in $E\overline{\otimes}F$ is often referred to as the *Hilbert–Schmidt norm* of the mapping $T\colon F \to E$. The following theorem has thus been proved:

Theorem XIX.11.1. Let E and F be Hilbert spaces and let $\{f_j\}_{j\in J}$ be an orthonormal basis in F. Then

(i) The mapping $T \in \mathscr{L}(F, E)$ is a Hilbert–Schmidt mapping if and only if $|||T|||^2 = \sum \|Tf_j\|_E^2 < \infty$.

(ii) Every Hilbert–Schmidt mapping $T\colon F \to E$ is a $|||\ |||$-limit of finite-dimensional mappings T_n (i.e., such that their range $T_n(F)$ is finite-dimensional):

$$T_n = \sum_{j=1}^{n} e_j\otimes f_j.$$

Remark. The Hilbert–Schmidt norm is finer than the norm in $\mathscr{L}(F, E)$: $\|T\| \leqslant |||T|||$.

It can be shown (cf. *Methods of Hilbert Spaces*) that the completion of the space $E\otimes F$ in the sense of the norm $\|\ \|$ yields compact operators, i.e., operators such that they map a ball in the space F into a precompact set in the space E. Thus, every Hilbert–Schmidt operator is compact.

Example. As we know from the last example, every Hilbert–Schmidt mapping $T: L^2(Y, \nu) \to L^2(X, \mu)$, is an integral operator

$$(Tf)(x) = \int_Y K(x, y) f(y) d\nu(y),$$

where the kernel K is an element of the space $L^2(X \times Y, \mu \times \nu)$ and hence

$$\iint |K(x, y)|^2 d\mu(x) d\nu(y) < \infty .$$

The foregoing condition is called the *Hilbert–Schmidt condition.* It was precisely Hilbert and Schmidt who dealt with the particular case when $E = F = L^2(\Omega, \lambda^n)$.

We shall end this section by giving a definition of nuclear spaces. Many equivalent definitions for this important class of spaces exist, but the one in terms of Hilbert–Schmidt mappings is perhaps the simplest and the most operative.

Let $E = (E, \|\ \|_i, i \in I)$ be a locally convex space. With each seminorm $\|\ \|_i$ we can associate a normed space $E_i := E/N_i$, where $N_i = \{e \in E: \|e\|_i = 0\}$ (cf. Chapter XIII, Section 5).

The norm in E_i induced by the seminorm $\|\ \|_i$ will be denoted by

$$|\ |_i: |e|_i := \|e\|_i.$$

The completion of the space $(E_i, \|\ \|_i)$ will be designated by $\overline{E}_i$.

The family of seminorms $(\|\ \|_j, j \in J)$ on E is said to be *equivalent to the family* $(\|\ \|_i, i \in I)$, if both induced the same topology. This means that for every $i \in I$ there exists a $j \in J$ and a constant $c > 0$ such that $\|\ \|_i \leqslant c\|\ \|_j$, and for every $j \in J$ there exists an $i \in I$ and a constant $c > 0$ such that $\|\ \|_j \leqslant c\|\ \|_i$.

DEFINITION. A locally convex space $(E, \|\ \|_j, j \in J)$ is *nuclear* if there exists an equivalent system of seminorms $(\|\ \|_i, i \in I)$ such that

(i) the spaces $\overline{E}_i$ are Hilbert spaces for $i \in I$,

(ii) for every $i \in I$ there exists a $k \in I$ such that $\|\ \|_i \leqslant \|\ \|_k$ and such that the natural mapping $A_{ik}: \overline{E}_k \to \overline{E}_i$ is a Hilbert–Schmidt mapping.

The following theorem is a straightforward consequence.

THEOREM XIX.11.2. A Hilbert space is nuclear if and only if it is finite-dimensional.

PROOF. The sufficiency of the condition given above is plain since in a finite-dimensional space every operator is a Hilbert–Schmidt operator.

We shall demonstrate that the condition is necessary. Since the family of norms consists of one seminorm, which is even a norm, we have $N = \{e: ||e|| = 0\} = \{0\}$, that is, $E/N = E$. Therefore, A is an identity mapping.

But A is a Hilbert–Schmidt mapping, and thus if $\{e_i\}_{i\in I}$ is an orthonormal basis in E, then

$$\infty > |||A|||^2 = \sum ||e_i||^2 = \dim E. \ \square$$

The theory of distributions provides us with basic examples of nuclear spaces. One proves that the spaces $\mathscr{D}(\Omega)$, $\mathscr{D}'(\Omega)$, $\mathscr{E}(\Omega)$, $\mathscr{E}'(\Omega)$ are nuclear. For the proof the reader is referred to the authors monograph *General Eigenfunctions Expansions and Unitary Representations of Topological Groups*.

12. REGULARIZATION OF DISTRIBUTIONS

As the beginning of this chapter we learned a method of regularizing functions from $\mathscr{L}^p$, that is, approximating them in the sense of the norm N_p by infinitely differentiable functions. It turns out that distributions, objects which are more general than functions from $\mathscr{L}^p$, can also be regularized (in a somewhat different sense, of course). One of the simpler theorems on this subject follows:

THEOREM XIX.12.1. Given a function $f \in \mathscr{L}^1(\lambda^n)$ such that $\int f = 1$. Let $g_\varepsilon(x) := \frac{1}{\varepsilon^n} f\left(\frac{x}{\varepsilon}\right)$. Then $g_\varepsilon \underset{\varepsilon\to 0}{\to} \delta_0$ in the sense of the weak topology in $\mathscr{D}'(\boldsymbol{R}^n)$, which is to say that $\langle \varphi, g_\varepsilon\rangle \underset{\varepsilon\to 0}{\to} \langle \varphi, \delta_0\rangle = \varphi(0)$ for every $\varphi \in \mathscr{D}(\boldsymbol{R}^n)$.

PROOF. Let $\varphi \in \mathscr{D}(\boldsymbol{R}^n)$. Then

$$\langle \varphi, g_\varepsilon\rangle = \int g_\varepsilon(x)\varphi(x)\,dx = \frac{1}{\varepsilon^n}\int f\left(\frac{x}{\varepsilon}\right)\varphi(x)\,dx$$
$$= \int f(y)\varphi(\varepsilon y)\,dy.$$

However, $f(y)\cdot\varphi(\varepsilon y) \underset{\varepsilon\to 0}{\to} f(y)\cdot\varphi(0)$ for every y and

$$|f(y)\cdot\varphi(\varepsilon y)| \leqslant |f(y)|\sup_x|\varphi(x)|$$

and hence, by the Lebesgue Theorem on dominated convergence,

$$\langle\varphi, g_\varepsilon\rangle = \int f(y)\cdot\varphi(\varepsilon y)\,dy \to \int f(y)\cdot\varphi(0)\,dy = \varphi(0)\cdot 1$$
$$= \delta_0(\varphi). \ \square$$

Example. As is known, $\int e^{-\pi\|x\|^2}dx = 1$ (cf. Chapter XIII, Section 22). If we set $\varepsilon = a^{-1/2}$, we have $a^{n/2}e^{-\pi\|x\sqrt{a}\|^2} \underset{a\to\infty}{\to} \delta_0$.

Putting $\pi a = c$ gives us

$$\left(\sqrt{\frac{c}{\pi}}\right)^n \cdot e^{-c\|x\|^2} \underset{c\to\infty}{\to} \delta_0.$$

Before going on to the general theorem on regularization, we shall make several comments on convolution,

Every function from $\mathscr{D}(R^n)$ is locally integrable and is thereby a distribution: $\mathscr{D} \subset \mathscr{L}^1_{\text{loc}} \subset \mathscr{D}'$. Since it has a compact support, it is a distribution from $\mathscr{E}'$: $\mathscr{D} \subset \mathscr{E}'$. Thus, if $\varphi \in \mathscr{D}$, $T \in \mathscr{D}$, it is meaningful to speak of the convolution $T * \varphi \in \mathscr{D}'$.

As for the convolution of a measure with an integrable function (Chapter XVIII, Section 5), we note that the distribution $T*\varphi$ is represented by a function.

To this end, invoking Fubini Theorem for distributions, we write:

$$(T*\varphi)(f) = \iint T(x)\varphi(y)f(x+y)\,dx\,dy$$
$$= \int T(x)\int\varphi(y)f(x+y)\,dy\,dx$$
$$= \int T(x)\int\varphi(y-x)f(y)\,dy\,dx$$
$$= \iint T(x)\varphi(y-x)f(y)\,dx\,dy$$
$$= \int\left(\int T(x)\varphi(y-x)\,dx\right)f(y)\,dy.$$

Thus the distribution $T*\varphi$ is represented by the function

$$(T*\varphi)(x) := \int T(x)\varphi(y-x)\,dx = \langle T, \tau_x\check{\varphi}\rangle.$$

Here $\check{\varphi}$ denotes the function $x \to \varphi(-x)$ on R^n. The function $T*\varphi$ is easily seen to be continuous. The following stronger theorem is valid.

THEOREM XIX.12.2. The relation

$$T*\varphi \in C^\infty(\boldsymbol{R}^n) \quad \text{holds for } \varphi \in C_0^\infty(\boldsymbol{R}^n).$$

PROOF. This theorem is proved in the same way as the fact that $T(\varphi) \in \mathscr{D}(\Omega_1)$ for $\varphi \in \mathscr{D}(\Omega_1 \times \Omega_2)$. The only difference is that the function $\tilde{\varphi}$ does not have a compact support and hence $T*\varphi$ need not have one either. The fact that the function φ and all its derivatives are uniformly continuous, which constituted the basis of the previous proof (cf. Theorem XIX.7.1), remains unaltered. □

Remark. We already know that $T*\varphi$ is infinitely differentiable and the derivative of a differentiable function taken in the usual sense is equal to its derivative taken in the sense of a distribution; therefore, on making use of the convolution property, we have:

$$D^\alpha(T*\varphi) = T*(D^\alpha\varphi) = (D^\alpha T)*\varphi.$$

If we associate Theorem XIX.12.1 with the theorem above, we obtain the following in quite a straightforward manner.

THEOREM XIX.12.3 (On the Regularization of Distributions). The set $C^\infty(\boldsymbol{R}^n)$ is dense in $\mathscr{D}'(\boldsymbol{R}^n)$ in the weak topology.

To put it more exactly: if $\varphi \in \mathscr{D}(\boldsymbol{R}^n)$ and $\int \varphi\, d\lambda^n = 1$, then writing

$$\psi_\varepsilon(x) := \frac{1}{\varepsilon^n}\varphi\left(\frac{x}{\varepsilon}\right)$$

we obtain $C^\infty(\boldsymbol{R}^n) \ni \psi_\varepsilon * T \xrightarrow[\varepsilon\to 0]{} T$ in the sense of the weak topology for any $T \in \mathscr{D}'(\boldsymbol{R}^n)$.

PROOF. Let $f \in \mathscr{D}(\boldsymbol{R}^n)$. Since $(\check{f})^\vee(x) = \check{f}(-x) = f(x)$, we have

$$\begin{aligned}\langle f, \psi_\varepsilon * T\rangle &= [(\psi_\varepsilon * T)*\check{f}](0)\\ &= [\psi_\varepsilon*(T*\check{f})](0) \to \langle (T*\check{f})^\vee, \delta_0\rangle\\ &= (T*\check{f})^\vee(0) = (T*\check{f})(0) = \langle f, T\rangle,\end{aligned}$$

and just this means that the distribution $\psi_\varepsilon * T$ converges weakly to T.

In the proof we have made use of the fact that $\psi_\varepsilon \to \delta_0$. □

13. EXAMPLES OF DISTRIBUTIONS IMPORTANT FOR APPLICATIONS

1. We shall solve the equation $x \cdot T(x) = 0$ on $\boldsymbol{R}$ or, in other words, we shall find a distribution which, when multiplied by the function

$f(x) = x$ yields a zero distribution (the function notation of distributions is employed very often in the present section).

Let $\varphi \in \mathcal{D}$ and let $\varphi(0) = 0$. Then the function $\psi(x) := \varphi(x)/x$ does exist and $\psi \in \mathcal{D}(\psi(0)) = \lim_{x\to 0} (\varphi(x)/x) = \varphi'(0)$. Therefore we have

$$\langle \varphi, T\rangle = \langle \psi\cdot x, T\rangle = \langle \psi, x\cdot T\rangle = \langle \varphi, 0\rangle = 0,$$

which means that T takes value zero on every function from $\mathcal{D}$ which vanishes at zero.

Now take a fixed function $\psi_0 \in \mathcal{D}$ such that $\psi_0(0) = 1$ and put $T(\psi_0) = c$. For any function $\varphi \in \mathcal{D}$ we have

$$\begin{aligned} T(\varphi) &= T(\varphi - \varphi(0)\cdot\psi_0 + \varphi(0)\cdot\psi_0) \\ &= T(\varphi - \varphi(0)\cdot\psi_0) + \varphi(0)\cdot T(\psi_0). \end{aligned}$$

But $(\varphi - \varphi(0)\cdot\psi_0)(0) = \varphi(0) - \varphi(0)\cdot\psi_0(0) = 0$ and thus the first summand vanishes. Hence we have obtained the result that $T(\varphi) = c\cdot\varphi(0)$, which means that the class of distributions $T = c\cdot\delta_0$ is the only solution of our equation.

2. For our next example, let us solve the equation $T' = 0$ on $\boldsymbol{R}$, where $T' = dT/dx$. As we know, in the set of differentiable functions a constant function is the only solution of this equation. Let us see whether we increase the number of solutions by allowing a largeı of number of objects, all distributions, "to enter the competition".

Suppose that $\varphi \in \mathcal{D}$ and $\int \varphi(x)dx = 0$. Then the function $\psi(x) := \int_{-\infty}^{x} \varphi(t)dt$ is infinitely differentiable and has a compact support: if $\underline{\varphi} \in [a, b]$, then for $x \geqslant b$ we have

$$\psi(x) = \int_{-\infty}^{x} \varphi(t)dt = \int_{-\infty}^{\infty} \varphi(t)dt = 0.$$

Accordingly $\langle \varphi, T\rangle = \langle \psi', T\rangle = -\langle \psi, T'\rangle = -\langle \psi, 0\rangle = 0$.

It has thus been shown that T vanishes on all functions from $\mathcal{D}$ whose integral over the whole of $\boldsymbol{R}$ is zero.

Now take a fixed function $\Psi_0 \in \mathcal{D}$ such that $\int \Psi_0 = 1$, and put $T(\Psi_0) = c$. For any function $\varphi \in \mathcal{D}$ we have

$$\begin{aligned} T(\varphi) &= T\Big(\varphi - \Psi_0\cdot\int\varphi + \Psi_0\cdot\int\varphi\Big) \\ &= T\Big(\varphi - \Psi_0\cdot\int\varphi\Big) + T(\Psi_0)\cdot\int\varphi. \end{aligned}$$

However,

$$\int\left(\varphi-\Psi_0\cdot\int\varphi\right)=\int\varphi-\int\Psi_0\cdot\int\varphi=0,$$

and thus the first summand vanishes. In view of this,

$$T(\varphi)=T(\Psi_0)\int\varphi=c\int\varphi=\int c\cdot\varphi(x)\,dx,$$

which means that $T=c$ (a distribution represented by a constant function).

3. In the space $\boldsymbol{R}$ the function $f(x):=|x|$ is locally integrable and hence is a distribution. Let us evaluate its derivative:

$$\langle |x|',\varphi\rangle=-\langle |x|,\varphi'\rangle=-\int_{-\infty}^{\infty}|x|\cdot\varphi'(x)\,dx$$

$$=\int_{-\infty}^{0}x\cdot\varphi'(x)\,dx-\int_{0}^{\infty}x\cdot\varphi'(x)\,dx=-\int_{-\infty}^{0}\varphi(x)\,dx+\int_{0}^{\infty}\varphi(x)\,dx$$

$$=\int_{-\infty}^{\infty}\theta(x)\varphi(x)\,dx=\langle\varphi,\theta\rangle,$$

where

$$\theta(x):=\begin{cases}-1 & \text{for } x<0,\\ 1 & \text{for } x\geqslant 0.\end{cases}$$

Therefore, $|x|'=\theta(x)$.

Now let us evaluate the derivative of the last distribution:

$$\langle\theta',\varphi\rangle=-\langle\theta,\varphi'\rangle=-\int_{-\infty}^{\infty}\theta(x)\varphi'(x)\,dx=\int_{-\infty}^{0}\varphi'(x)\,dx-$$

$$-\int_{0}^{\infty}\varphi'(x)\,dx=\varphi(x)\Big|_{-\infty}^{0}-\varphi(x)\Big|_{0}^{\infty}=2\varphi(0)=2\langle\delta_0,\varphi\rangle,$$

and thus $\theta'=2\delta_0$. Furthermore, $\langle\delta_0',\varphi\rangle=-\langle\delta_0,\varphi'\rangle=-\varphi'(0)$, which means that the derivative of the delta function is a dipole placed at zero, with moment opposite to the direction of the coordinate axis.

It is easily seen that in similar fashion we obtain a formula for the m-th derivative of the delta function:

$$\langle\delta_0^{(m)},\varphi\rangle=(-1)^m\varphi^{(m)}(0).$$

4. As is known, the function $\log|x|$ is locally integrable on $\boldsymbol{R}$ and is thus a distribution. Let us evaluate its derivative:

$$\langle(\log|x|)', \varphi\rangle = -\langle\log|x|, \varphi'\rangle = -\int_{-\infty}^{\infty} \log|x|\,\varphi'(x)\,dx$$

$$= \lim_{\varepsilon\to 0}\left(-\int_{-\infty}^{-\varepsilon} \log(-x)\cdot\varphi'(x)\,dx - \int_{\varepsilon}^{\infty} \log x\cdot\varphi'(x)\,dx\right)$$

$$= \lim_{\varepsilon\to 0}\left(\int_{-\infty}^{-\varepsilon} \frac{\varphi(x)}{x}\,dx - \log(-x)\cdot\varphi(x)\Big|_{-\infty}^{-\varepsilon} + \right.$$

$$\left. + \int_{\varepsilon}^{\infty} \frac{\varphi(x)}{x}\,dx - \log x\cdot\varphi(x)\Big|_{\varepsilon}^{\infty}\right)$$

$$= \lim_{\varepsilon\to 0}\left(\int_{-\infty}^{-\varepsilon} \frac{\varphi(x)}{x}\,dx + \int_{\varepsilon}^{\infty} \frac{\varphi(x)}{x}\,dx + (\log\varepsilon)[\varphi(\varepsilon)-\varphi(-\varepsilon)]\right).$$

But φ is a differentiable function, whereby

$$\varphi(\varepsilon) = \varphi(0)+\varphi'(0)\cdot\varepsilon+r(\varepsilon),$$
$$\varphi(-\varepsilon) = \varphi(0)-\varphi'(0)\cdot\varepsilon+r(-\varepsilon),$$

where $r(\varepsilon)\varepsilon \underset{\varepsilon\to 0}{\to} 0$. Therefore

$$(\log\varepsilon)\,[\varphi(\varepsilon)-\varphi(-\varepsilon)] = (\log\varepsilon)\cdot[2\varphi'(0)\cdot\varepsilon+r(\varepsilon)+r(-\varepsilon)] \underset{\varepsilon\to 0}{\to} 0,$$

since $\log\varepsilon$ tends to infinity more slowly than $\varepsilon\to 0$ and $r(\varepsilon)$ and $r(-\varepsilon)$ approach zero more quickly than ε does.

We introduce the notation

$$\lim_{\varepsilon\to 0}\left(\int_{-\infty}^{-\varepsilon} \frac{\varphi(x)}{x}\,dx + \int_{\varepsilon}^{\infty} \frac{\varphi(x)}{x}\,dx\right) =: P\int_{-\infty}^{\infty} \frac{\varphi(x)}{x}\,dx.$$

The number so defined is called the *principal value of the integral* $\int_{-\infty}^{\infty} \frac{\varphi(x)}{x}\,dx$. It should be emphasized that the integral $\int_{-\infty}^{\infty} \frac{\varphi(x)}{x}\,dx$

itself does not in general exist (unless φ vanishes at a sufficiently rapid rate at zero). It may happen, for instance, that

$$\int_{-\infty}^{\varepsilon} \frac{\varphi(x)}{x} dx \underset{\varepsilon\to 0}{\to} -\infty, \quad \int_{\varepsilon}^{\infty} \frac{\varphi(x)}{x} dx \underset{\varepsilon\to 0}{\to} +\infty .$$

Approaching zero simultaneously from both sides ensures that the sum of these expressions always has a finite limit.

Use is often made of the following notation, which is compatible with the function notation for distributions:

$$P \int_{-\infty}^{\infty} \frac{\varphi(x)}{x} dx = \int_{-\infty}^{\infty} P\left(\frac{1}{x}\right) \varphi(x)\, dx.$$

Thus

$$\langle (\log|x|'), \varphi\rangle = \langle P(1/x), \varphi\rangle, \quad \text{or} \quad \log|x|' = P(1/x).$$

Remark. The distribution $P(1/x)$ is not a function, that is to say, there is no locally integrable function f which would obey the equality

$$\int P\left(\frac{1}{x}\right) \varphi(x)\, dx \underset{\varphi}{\equiv} \int f(x)\varphi(x)\, dx.$$

5. Now we consider the example of a distribution on R^3 frequently used in physics. Let us denote: $r(x, y, z) = \sqrt{x^2+y^2+z^2}$ (x, y, z are the coordinates of a point in R^3).

The function $1/r$ is locally integrable and hence is a distribution. Let us calculate $\Delta \dfrac{1}{r}$, where Δ is the familiar Laplacian operator:

$$\Delta f = \operatorname{div} \operatorname{grad} f = \frac{\partial^2 f}{\partial x^2} + \frac{\partial^2 f}{\partial y^2} + \frac{\partial^2 f}{\partial z^2}.$$

We have

$$\left\langle \Delta \frac{1}{r}, \varphi \right\rangle = (-1)^2 \left\langle \frac{1}{r}, \Delta\varphi \right\rangle$$

$$= \int \frac{\Delta\varphi}{r} dx\, dy\, dz = \lim_{\varepsilon\to 0} \int_{\varepsilon \leqslant r \leqslant a} \frac{\Delta\varphi}{r} dx\, dy\, dz,$$

where a is the radius of a ball containing the support of the function φ. Invoking Green's formula (Theorem XV.5.4), we have

$$\int\limits_{\varepsilon\leqslant r\leqslant a} \frac{\Delta\varphi}{r}\,dx\,dy\,dz$$

$$= \int\limits_{\varepsilon\leqslant r\leqslant a} \varphi\cdot\left(\Delta\frac{1}{r}\right)dx\,dy\,dz - \int\limits_{r=\varepsilon}\left(\frac{1}{r}\cdot\frac{d\varphi}{dn} - \varphi\frac{d}{dn}\frac{1}{r}\right)d\sigma$$

(the integral over the surface $r = a$ vanishes because it is beyond the support of the function φ). However, $\operatorname{grad}\frac{1}{r} = (-x/r^3, -y/r^3, -z/r^3)$. The reader can easily calculate that $\Delta\frac{1}{r} = \operatorname{div}\operatorname{grad}\frac{1}{r} = 0$ for $r \neq 0$. Moreover,

$$\frac{d}{dn}\frac{1}{r} = \left(\operatorname{grad}\frac{1}{r}\middle|\, n\right) = \left(\operatorname{grad}\frac{1}{r}\middle|\left(\frac{x}{r}, \frac{y}{r}, \frac{z}{r}\right)\right) = -\frac{1}{r^2},$$

and thus

$$\left\langle \Delta\frac{1}{r}, \varphi\right\rangle = \lim_{\varepsilon\to 0}\left(-\int\limits_{r=\varepsilon}\frac{1}{r}\cdot\frac{d\varphi}{dn}d\sigma - \int\limits_{r=\varepsilon}\varphi\cdot\frac{1}{r^2}d\sigma\right).$$

But $\left|\frac{d\varphi}{dn}\right| = |(\operatorname{grad}\varphi|n)| \leqslant \|\operatorname{grad}\varphi\| < c$, since $\varphi \in C_0^\infty(R^n)$ and thus

$$\left|\int\limits_{r=\varepsilon}\frac{1}{r}\cdot\frac{d\varphi}{dn}d\sigma\right| \leqslant c\cdot\int\limits_{r=\varepsilon}\frac{1}{r}d\sigma = c\frac{1}{\varepsilon}\cdot\int\limits_{r=\varepsilon}d\sigma = c\frac{1}{\varepsilon}4\pi\varepsilon^2 \underset{\varepsilon\to 0}{\to} 0.$$

On the other hand,

$$\int\limits_{r=\varepsilon}\varphi\cdot\frac{1}{r^2}d\sigma = \frac{1}{\varepsilon^2}\int\limits_{r=\varepsilon}\varphi\,d\sigma = \frac{1}{\varepsilon^2}4\pi\varepsilon^2 S_\varepsilon(\varphi),$$

where $S_\varepsilon(\varphi)$ denotes the mean value of the function φ on the sphere $\{\|x\| = \varepsilon\}$. Obviously, $\lim\limits_{\varepsilon\to 0} S_\varepsilon(\varphi) = \varphi(0)$ whence

$$\left\langle \Delta\frac{1}{r}, \varphi\right\rangle = -4\pi\varphi(0) = \langle -4\pi\delta_0, \varphi\rangle,$$

or

$$\Delta\frac{1}{r} = -4\pi\delta_0.$$

14. THE FOURIER TRANSFORMATION. THE SPACE $\mathscr{S}$

The role of Fourier transformations in modern mathematics and theoretical physics can scarcely be overrated. It renders invaluable service in both computations and theoretical considerations.

One of the most beautiful areas of modern mathematics is that of "abstract harmonic analysis", i.e., the transposition of classical harmonic analysis (Fourier transformations on $\boldsymbol{R}^n$ and Fourier series) to arbitrary locally compact Abelian groups and to some noncommutative groups.

This theory has come into being thanks to mathematicians of the stature of Hermann Weil, von Neumann, Stone, Gel'fand and his school, André Weil and the Bourbaki school, Beurling, Selberg, and Harish Chandra.

It has turned out that classical harmonic analysis is a special chapter of the theory of group representations, but the grandest sections of this continually growing theory would not have come into being without a knowledge of the theorems of classical harmonic analysis. Today, the Fourier transformation provides the readiest access to the beautiful theorems which are fundamental to the whole of topology and the duality theory.

Harmonic analysis originated as vibration theory and an attempt to decompose periodic functions into sums and series of the simplest vibrations, i.e., harmonic vibrations. In speaking of a function which is a Fourier transformation, physicists customarily say that it has been "decomposed into plane waves".

The enormous role of Fourier transforms in the theory of differential equations is explained by the fact that the Fourier transformation diagonalizes all partial derivatives simultaneously.

It is no accident that Laurent Schwartz, an eminent "harmonic analyst", was the founder of the theory of distributions. When he introduced the space $\mathscr{S}$ which is fundamental to the theory of distributions he was guided precisely by the needs of Fourier transformations.

All of this induced one eminent mathematician to say: "people have devised many transformations, but the Fourier transformation was created by God" (this mathematician is an "atheist").

Now let us give a definition of the Fourier transformation for a function $f \in \mathscr{L}^1(\lambda^n)$.

DEFINITION. The function

$$R^n \ni \xi \to \hat{f}(\xi) := \int f(x)\cdot e^{-2\pi i\langle x,\xi\rangle}\, d\lambda^n(x),$$

where $x \in R^n$, $\xi \in (R^n)' = R^n$, $\langle x, \xi\rangle = x_1\xi_1 + \ldots + x_n\xi_n$, is called the *Fourier transform* of an integrable function f.

The mapping $\mathscr{F}: f \to \hat{f}$ is called the *Fourier transformation*:

$$\hat{f} = \mathscr{F}f.$$

If $f \in C_0^\infty(R^n)$, then on integrating by parts $|\alpha|$ times, we arrive at the formula

$$(\mathscr{F}D^\alpha f)(\xi) = \int e^{-2\pi i\langle x,\xi\rangle} D^\alpha f(x)\, d\lambda^n(x) = (2\pi i)^{|\alpha|}\xi^\alpha(\mathscr{F}f)(\xi),$$

where $\xi^\alpha := \xi_1^{\alpha_1} \ldots \xi_n^{\alpha_n}$.

Differentiating the Fourier-transform formula and invoking the theorem on the integration of integrals with a parameter, we obtain

$$(D^\beta\mathscr{F}f)(\xi) = (-2\pi i)^{|\beta|}\cdot[\mathscr{F}(x^\beta\cdot f)](\xi).$$

To be able to differentiate and to perform Fourier transformations ("trans-fourierize", as Schwartz puts it) an arbitrary number of times, one must have a linear function space of arbitrarily differentiable functions which would be integrable upon multiplication by a polynomial of any order. The space $C_0^\infty(R^n)$, for example, is such a space, but it has two inconvenient features: it is too small and, moreover, the transform of a function from C_0^∞ is no longer a function from C_0^∞.

The space $\mathscr{S}$ introduced by Schwartz is a more extensive set of infinitely differentiable functions, and it does not have these drawbacks:

DEFINITION. The space $\mathscr{S}$ is the linear set of all functions $f \in C^\infty(R^n)$ such that

$$\|f\|_{\alpha,\beta} := \sup_{x\in R^n} |x^\beta\cdot D^\alpha f(x)| < \infty$$

for all multi-indices $\alpha = (\alpha_1, \ldots, \alpha_n)$, $\beta = (\beta_1, \ldots, \beta_n)$.

A locally convex topology of the space $\mathscr{S}$ is given by a countable family of seminorms $\|\ \|_{\alpha,\beta}$.

COROLLARY XIX.14.1. The inclusion

$$\mathscr{S} \subset \mathscr{L}^1(\lambda^n)$$

holds.

PROOF. The function $f \in \mathscr{S}$ is measurable as a continuous function and

$$|f| \leqslant c \cdot (1+||x||)^{-2n};$$

the majorant is an integrable function. □

LEMMA XIX.14.2. The Fourier transformation maps $\mathscr{S} \to \mathscr{S}$ in a continuous manner.

PROOF. If the formulae given on the preceding page for the derivative of a transform and the transform of a derivative are combined, the result is:

$$\xi^\beta D^\alpha \hat{f}(\xi) = \xi^\beta(-2\pi i)^{|\alpha|}(x^\alpha f)^\wedge(\xi) = \frac{(-2\pi i)^{|\alpha|}}{(2\pi i)^{|\beta|}}(D^\beta x^\alpha f)^\wedge(\xi).$$

The definition of the space $\mathscr{S}$ implies that $D^\beta x^\alpha f$ belongs to $\mathscr{S}$ and hence is an integrable function. From this fact and from the theorems on the continuity and differentiability of integrals with a parameter it follows that it was possible to integrate under the integral sign (i.e., the formulae which we used were valid) and that $\xi^\alpha D^\alpha \hat{f}$ is a continuous function for any α and β, and thus $\hat{f} \in C^\infty$.

The definition of the space $\mathscr{S}$ also implies that there exist multi-indices α' and β' such that

$$\sup_x (1+||x||)^{2n}|(D^\beta x^\alpha f(x))| \leqslant c(a, \beta)||f||_{\alpha'\beta'}.$$

Setting $\int(1+||x||)^{-2n} =: C$, we have

$$\begin{aligned}\xi^\beta D^\alpha \hat{f}(\xi)| &= \left|\int e^{-2\pi i\langle x,\xi\rangle}\frac{(-2\pi i)^{|\alpha|}}{(2\pi i)^{|\beta|}}D^\beta(x^\alpha f(x))d\lambda^n(x)\right| \\ &\leqslant \left|\frac{(-2\pi i)^{|\alpha|}}{(2\pi i)^{|\beta|}}\right|\int c(\alpha,\beta)||f||_{\alpha'\beta'}(1+||x||)^{-2n}dx \\ &\leqslant (2\pi)^{|\alpha|-|\beta|}C\cdot c(\alpha,\beta)||f||_{\alpha'\beta'}.\end{aligned}$$

If the constant obtained is denoted by $C(\alpha, \beta)$, the inequality which we have proved can be read as follows:

For any multi-indices α and β there exist multi-indices α' and β' and a constant $C(\alpha, \beta)$ such that

$$||\mathscr{F}f||_{\alpha\beta} = ||\hat{f}||_{\alpha\beta} \leqslant C(\alpha, \beta)||f||_{\alpha'\beta'}.$$

This means that:

(i) the seminorms $\|\hat{f}\|_{\alpha\beta}$ of $\hat{f}$ have finite values and hence belong to the space $\mathscr{S}$;

(ii) the mapping $\mathscr{F}$ is continuous since, if $f_n \to 0$, then for all α and β: $\|f_n\|_{\alpha'\beta'} \to 0$ and thus also

$$\|\hat{f}_n\|_{\alpha\beta} \to 0, \quad \text{or} \quad \hat{f} \to 0. \ \square$$

LEMMA XIX.14.3. If $f \in \mathscr{S}$, then as $\varepsilon \to 0$ the functions $x \to g_\varepsilon(x) := f(\varepsilon x)$ tend to a constant function, equal to $f(0)$. This convergence is almost uniform.

PROOF. Since a compact set is a subset of some ball, it is sufficient to prove that the convergence is uniform on every ball.

Making use of the continuity of the function f at the point $x = 0$, we can establish that for every number $\varepsilon > 0$ there exists an r such that $|f(x)-f(0)| < \varepsilon$ if $\|x\| < r$. Accordingly, $|g_\varepsilon(x)-f(0)| = |f(\varepsilon x)-f(0)| < \varepsilon$ for $\|x\| < R$ if $0 < \varepsilon < r/R$. $\square$

As an example of the considerations given above, let us evaluate the Fourier transform of the following function from the space $\mathscr{S}(\boldsymbol{R}^1)$:

$$h(x) := e^{-ax^2} \quad (a > 0).$$

We have

$$\hat{h}(\xi) = \int_{-\infty}^{\infty} e^{-ax^2} e^{-2\pi i\langle x,\xi\rangle} dx$$

$$= \int_{-\infty}^{\infty} e^{-a\left(x+\frac{i\pi}{a}\xi\right)^2} e^{-\frac{\pi^2}{a}\xi^2} dz = e^{-\frac{\pi^2\xi^2}{a}} \int_{-\infty}^{\infty} e^{-a\left(x+i\frac{\pi}{a}\xi\right)^2} dx.$$

Let Γ_b denote the following closed contour in the complex plane $\boldsymbol{C}$:

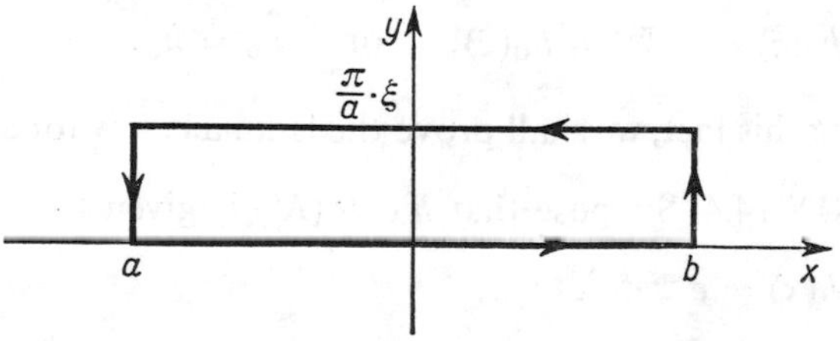

Fig. 27

Since the function e^{-az^2} is analytic throughout the entire plane, we have

$$0 = \int_{\Gamma_b} e^{-az^2} dz = \int_{-b}^{b} e^{-ax^2} dx + \int_{0}^{\frac{\pi}{a}\xi} e^{-a(b+iy)^2} dy -$$

$$- \int_{-b}^{b} e^{-a\left(x+i\frac{\pi}{a}\xi\right)^2} dx + \int_{\frac{\pi}{a}\xi}^{0} e^{-a(-b+iy)^2} dy.$$

However

$$\left| \int_{0}^{\frac{\pi}{a}\xi} e^{-a(\pm b+iy)^2} dy \right| \leqslant \frac{\pi}{a}\xi \sup_{0 \leqslant y \leqslant \frac{\pi}{a}\xi} |e^{-a(\pm b+iy)^2}|$$

$$= \frac{\pi}{a}\xi e^{-a\left(b^2 - \frac{\pi^2}{a^2}\xi^2\right)} \underset{b\to\infty}{\to} 0,$$

whence

$$0 = \lim_{b\to\infty} \int_{\Gamma_b} e^{-az^2} dz = \int_{-\infty}^{\infty} e^{-ax^2} dx - \int_{-\infty}^{\infty} e^{-a\left(x+i\frac{\pi}{a}\xi\right)^2} dx.$$

Accordingly

$$\hat{h}(\xi) = e^{-\frac{\pi^2\xi^2}{a}} \int_{-\infty}^{\infty} e^{-ax^2} dx = \sqrt{\frac{\pi}{a}}\, e^{-\frac{\pi^2\xi^2}{a}}.$$

In particular, if for $a = \pi$ we take $h_0(x) := e^{-\pi x^2}$ we obtain

$$\hat{h}_0(\xi) = e^{-\pi\xi^2} = h_0(\xi), \quad \text{or} \quad \hat{h}_0 = h_0.$$

Employing this fact, we shall prove the lemma below for any space $\boldsymbol{R}^n$:

LEMMA XIX.14.4. Suppose that $h \in \mathscr{S}(\boldsymbol{R}^n)$ is given by

$$h(x) = e^{-\pi\|x\|^2}.$$

Then $\hat{h} = h$.

PROOF. We have

$$\hat{h}(\xi) = \int_{R^n} \exp\Big(-\pi \sum_{k=1}^{n} x_k^2\Big) \cdot \exp\Big(-2\pi i \sum_{k=1}^{n} x_k \cdot \xi_k\Big) dx_1 \ldots dx_n$$

$$= \prod_{k=1}^{n} \int e^{-\pi x_k^2} \cdot e^{-2\pi i x_k \xi_k} dx_k = \prod_{k=1}^{n} \exp(-\pi \xi_k^2)$$

$$= \exp\Big(-\pi \sum_{k=1}^{n} \xi_k^2\Big) = e^{-\pi \|\xi\|^2}. \ \square$$

The next theorem is fundamental to further considerations.

THEOREM XIX.14.5. Let f and g be functions from $\mathscr{S}(R^n)$. Then

(i) $(g \cdot \hat{f})\hat{}\,(x) = \int \hat{g}(x+y) f(y)\, dy$,
(ii) $\int g \cdot \hat{f} = \int \hat{g} \cdot f$,
(iii) $\hat{\hat{f}} = \check{f}$, where $\check{f}(x) := f(-x)$,
(iv) $f(x) = (\mathscr{F}^{-1}\hat{f})(x) = \int e^{2\pi i \langle x, \xi \rangle} \hat{f}(\xi)\, d\xi$,
(v) $(\hat{f})\check{} = (\check{f})\hat{}$,
(vi) $\mathscr{F}\mathscr{S} = \mathscr{S}$,
(vii) $\mathscr{F}$ is a topological isomorphism of $\mathscr{S}$.

PROOF. (i) We have

$$(g\hat{f})\hat{}\,(x) = \int e^{-2\pi i \langle x, \xi \rangle} \cdot g(\xi) \Big(\int f(y) e^{-2\pi i \langle y, \xi \rangle} dy\Big) d\xi$$

$$= \int f(y) \cdot \Big(\int g(\xi) e^{-2\pi i \langle x+y, \xi \rangle} d\xi\Big) dy = \int f(y) \hat{g}(x+y)\, dy$$

(the second equality follows from Fubini's Theorem).

(ii) Note that $\hat{\varphi}(0) = \int \varphi(x) \cdot e^{-2\pi i \langle x, 0 \rangle} dx = \int \varphi(x) dx$. Hence

$$\int g \cdot \hat{f} = (g \cdot \hat{f})\hat{}\,(0) = \int \hat{g}(y+0) f(y)\, dy = \int \hat{g} \cdot f.$$

(iii) Let $g_\varepsilon(x) := h(\varepsilon x)$, where $h(x) = \exp(-\pi \|x\|^2)$.
Observe that $|g_\varepsilon(x)| = \exp(-\pi \varepsilon^2 \|x\|^2) \leqslant 1$ and $g_\varepsilon(x) \underset{\varepsilon \to 0}{\to} 1$ pointwise (and even almost uniformly; cf. Lemma XIX.14.3). Therefore

$$|g_\varepsilon(\xi) \hat{f}(\xi) \exp(-2\pi i \langle x, \xi \rangle)| \leqslant |\hat{f}(\xi)| \quad (\hat{f} \in \mathscr{S} \subset \mathscr{L}^1)$$

and

$$g_\varepsilon(\xi)\hat{f}(\xi)\exp(-2\pi i\langle x, \xi\rangle) \underset{\varepsilon\to 0}{\to} \hat{f}(\xi)\exp(-2\pi i\langle x, \xi\rangle)$$

pointwise. By Lebesgue's Theorem on dominated convergence

$$\begin{aligned}&(g_\varepsilon\cdot\hat{f})\hat{}\,(x)\\ &= \int g_\varepsilon(\xi)\hat{f}(\xi)e^{-2\pi i\langle x,\xi\rangle}\,d\xi \underset{\varepsilon\to 0}{\to} \int \hat{f}(\xi)e^{-2\pi i\langle x,\xi\rangle}d\xi = \hat{\hat{f}}(x).\end{aligned}$$

However

$$\begin{aligned}\hat{g}_\varepsilon(x) &= \int e^{-2\pi i\langle x,\xi\rangle}h(\varepsilon\xi)\,d\xi\\ &= \frac{1}{\varepsilon^n}\int e^{-2\pi i\langle x, t/\varepsilon\rangle}h(t)\,dt = \frac{1}{\varepsilon^n}\hat{h}\left(\frac{x}{\varepsilon}\right).\end{aligned}$$

Consequently, (i) of the hypothesis gives us

$$\begin{aligned}\hat{\hat{f}}(x) &= \lim_{\varepsilon\to 0}(g_\varepsilon\cdot\hat{f})\hat{}\,(x) = \lim_{\varepsilon\to 0}\int \hat{g}_\varepsilon(x+y)f(y)\,dy\\ &= \lim_{\varepsilon\to 0}\int \hat{g}_\varepsilon(s)f(s-x)\,ds = \lim_{\varepsilon\to 0}\frac{1}{\varepsilon^n}\int \hat{h}\left(\frac{s}{\varepsilon}\right)f(s-x)\,ds.\end{aligned}$$

Putting $z = s/\varepsilon$ and invoking Lemma XIX.14.4, we obtain

$$\hat{\hat{f}}(x) = \lim_{\varepsilon\to 0}\int h(z)f(\varepsilon z-x)\,dz.$$

However

$$|h(z)\cdot f(\varepsilon z-x)| \leqslant |h(x)|\cdot\{\sup_{x\in R^n}|f(x)|\}$$

and

$$h(z)\cdot f(\varepsilon z-x) \underset{\varepsilon\to 0}{\to} h(z)\cdot f(-x) \quad \text{pointwise.}$$

When we invoke Lebesgue's Theorem a second time, we obtain

$$\hat{\hat{f}}(x) = \int h(z)f(-x)\,dz = f(-x)\int_{R^n} e^{-\pi\|z\|^2}dz = f(-x) = \check{f}(x).$$

(iv) follows immediately from (iii).

(v) $(\hat{f})\check{} = (\hat{\hat{\hat{f}}}) = (\hat{\hat{f}})\hat{} = (\check{f})\hat{}$.

(vi) Every function $f \in \mathscr{S}$ is a transform of some function, viz. $(h = \hat{f})^{\vee} = \hat{\hat{\hat{f}}}$: $\hat{h} = (\hat{f})^{\hat{}} = (\check{f})^{\vee} = f$.

(vii) The mapping $\mathscr{F}$ is a surjection (vi), injection ((iv), if the transforms are equal, so are the functions), and is continuous (Lemma XIX.14.2).

The inverse $\mathscr{F}^{-1}$ is also continuous since, as we have shown in (vi), it is a composition of continuous mappings $\mathscr{F}^{-1} = \mathscr{F} \circ \mathscr{F} \circ \mathscr{F}$). □

COROLLARY XIX.14.6 (Riemann–Lebesgue Lemma). If $f \in \mathscr{L}^1(\lambda^n)$, then $\hat{f}(\xi) \to 0$, as $\|\xi\| \to \infty$.

PROOF. If $f_n \in \mathscr{S}$, then $\hat{f}_n \in \mathscr{S}$, and thus $\hat{f}_n$ vanishes rapidly at ∞. Since $C_0^\infty(R^n) \subset \mathscr{S}$, and $C_0^\infty(R^n)$ is dense in $\mathscr{L}^1(\lambda^n)$ (Theorem XIX.1.2), $\mathscr{S}$ is dense in $\mathscr{L}^1(\lambda^n)$. Take the sequence $\mathscr{S} \ni f_n \underset{n\to\infty}{\to} f$ in the sense of $\mathscr{L}^1(\lambda^n)$. We then have

$$|\hat{f}(\xi) - \hat{f}_n(\xi)| \leqslant \int |e^{-2\pi i \langle x, \xi\rangle}| \, |f(x) - f_n(x)| dx$$
$$= \|f - f_n\|_{\mathscr{L}^1} \to 0,$$

as $n \to \infty$, and thus $\hat{f}_n \to \hat{f}$ uniformly. But $\hat{f}_n$ vanishes at ∞, whence the hypothesis. □

THEOREM XIX.14.7. Let $f, g \in \mathscr{S}$. Then

(i) $(f \cdot g)^{\hat{}} = \hat{f} * \hat{g}$,

(ii) $(f * g)^{\hat{}} = \hat{f} \cdot \hat{g}$.

PROOF. (i) When the function f is replaced by $\hat{\hat{\hat{f}}} = \check{\hat{f}}$ in (i) of Theorem XIX.14.5 we obtain:

$$(g \cdot f)^{\hat{}}(x) = \int \hat{g}(x+y) \check{\hat{f}}(y) dy = \int \hat{g}(x+y) \hat{f}(-y) dy$$
$$= \int \hat{g}(x-z) \hat{f}(z) dz = (\hat{g} * \hat{f})(x).$$

(ii) The substitution of $\hat{g}$ for g in (i) of Theorem XIX.14.5 yields

$$(\hat{g} \cdot \hat{f})^{\hat{}}(x) = \int \hat{\hat{g}}(x+y) f(y) dy = \int g(-x-y) f(y) dy$$
$$= (g * f)(-x) = (g * f)^{\vee}(x).$$

Hence $(\hat{g} \cdot \hat{f})^{\hat{\hat{}}} = (g * f)^{\hat{\vee}} = (g * f)^{\check{\hat{}}}$, or $(\hat{g} \cdot \hat{f}) = (g * f)^{\hat{}}$ □.

15. THE FOURIER TRANSFORMATION AS A UNITARY OPERATOR ON THE SPACE $\mathscr{L}^2(\mathbf{R}^n)$

DEFINITION. A linear operator $V: H \to H$, preserving the inner product, i.e., $(Vu|Vv) = (u|v)$, $u, v \in H$, is called an *isometric operator* (*isometry*) in the Hilbert space H.

Remark. We have $||Vu|| = (Vu|Vu)^{1/2} = (u|u)^{1/2} = ||u||$, and thus the isometric operator preserves distance.

DEFINITION. A bijective isometry is called a *unitary operator*.

LEMMA XIX.15.1. An isometric mapping U_0 of a set D dense in a Hilbert space H can be extended by continuity to an isometric mapping defined throughout the whole space H.

Moreover, if U_0 was a bijection of D onto D, then the extended operator U is unitary.

PROOF. The first part of the hypothesis follows from the theorem on the extension of a continuous mapping. Thus, suppose that $U_0: D \to D$ is a bijection. We shall show that $U: H \to H$ is also a bijection.

Let $u \in H$. We shall prove the existence of a $v \in H$ such that $Uv = u$, that is, such that U is a surjection. There exists a sequence $D \ni u_n \to u$. As U_0 is injective, it follows that there is a sequence $v_n \in D$ such that $U_0 v_n = Uv_n = u_n$. However

$$||v_n - v_m|| = ||U_0(v_n - v_m)|| = ||u_n - u_m|| \to 0.$$

Hence, there exists an element $v = \lim_{n\to\infty} v$ and

$$Uv = \lim_{n\to\infty} U_0 v_n = \lim_{n\to\infty} u_n = u.$$

The mapping U is also an injection, for if $Uv = Uu$, then

$$||v-u|| = ||U(v-u)|| = 0, \quad \text{or } v = u. \; \square$$

We recall that the space $L^2(\lambda^n, \mathbf{C})$ (of classes of square-integrable complex functions) is a Hilbert space over $\mathbf{C}$ with the scalar product

$$(\dot{f}|\dot{g}) = \int f \cdot \bar{g}, \quad \text{where } \bar{g}(x) := \overline{g(x)}.$$

Clearly, $\mathscr{S} \subset L^2$ if the function is identified with the class it represents and if $\mathscr{S}$ is dense in L^2.

This brings us to the following important

THEOREM XIX.15.2 (Plancherel). A Fourier transformation is isometric on $\mathscr{S}$, which is to say that by Lemma XIX.15.1 and item (vii) of Theorem XIX.14.5 it is extendible by continuity to a unitary operator in the space L^2.

PROOF. Let $f \in \mathscr{S}$. Then

$$\overline{\hat{f}(\xi)} = \overline{\int f(x)\exp(-2\pi i\langle x, \xi\rangle)dx} = \int \bar{f}(x)\exp(2\pi i\langle x, \xi\rangle)dx$$
$$= \int \bar{f}(x)\exp(-2\pi i\langle x, -\xi\rangle)dx = \hat{\bar{f}}(-\xi) = (\hat{\bar{f}})^{\vee}(\xi).$$

Accordingly,

$$\hat{\bar{\hat{f}}} = [(\hat{\bar{f}})^{\vee}]^{\wedge} = (\hat{\hat{\bar{f}}})^{\vee} = \check{\bar{f}} = \bar{f}.$$

If in item (ii) of Theorem XIX.14.5 we put the function $\bar{\hat{f}}$ instead of f, then for $f, g \in \mathscr{S}$ we obtain

$$(\mathscr{F}g|\mathscr{F}f) = (\hat{g}|\hat{f}) = \int \hat{g}\cdot\bar{\hat{f}} = \int g\cdot\hat{\bar{\hat{f}}} = \int g\cdot\bar{f} = (g|f). \ \square$$

16. TEMPERED DISTRIBUTIONS. THE FOURIER TRANSFORMATION IN $\mathscr{S}'$

To begin with, we shall prove the following straightforward

LEMMA XIX.16.1. (i) The set $C_0^\infty(R^n)$ is dense in the space $\mathscr{S}$.
(ii) The embedding $\mathscr{D} \to \mathscr{S}$ is continuous.

PROOF. (i) Let $f \in \mathscr{S}$. Take a function $g \in C_0^\infty(R^n)$ such that $g(x) = 1$ for $||x|| \leqslant 1$, and take $g_\varepsilon(x) := g(\varepsilon x)$. Now, let $f_\varepsilon(x) := g_\varepsilon(x)\cdot f(x)$. Clearly, $f_\varepsilon \in C_0^\infty(R)^n$. Since $f \in \mathscr{S}$, we have $x^\alpha D^\beta f(x) \underset{x\to\infty}{\to} 0$. Given a number $\eta < 0$. Since, for example, for $0 \leqslant \varepsilon \leqslant 1$ all functions $D^\alpha(g_\varepsilon - 1)$ are uniformly bounded in common by constants which depend only on the order of the derivatives, it is possible to choose an r such that $x^\alpha D^\beta[(g_\varepsilon(x)-1)\cdot f(x)]| < \eta$ for $||x|| > r$. However, we have $f_\varepsilon(x) - f(x) = \big(g(\varepsilon x)-1\big)\cdot f(x) = 0$, for $||x|| \leqslant 1/\varepsilon$, and thus if $\varepsilon \leqslant 1/r$, then

$$||f_\varepsilon - f||_{\alpha,\beta} = \sup_{||x||\geqslant r} |x^\alpha D^\alpha[\big(g_\varepsilon(x)-1\big)f(x)]| < \eta.$$

It has thus been shown that $C_0^\infty(R^n) \ni f_\varepsilon \underset{\varepsilon\to 0}{\to} f$ in the sense of $\mathscr{S}$.

Item (ii) of the hypothesis is obvious. $\square$

COROLLARY XIX.16.2. If $T \in \mathscr{S}'$ (i.e., T is a continuous linear functional on $\mathscr{S}$) and $T(\mathscr{D}(R^n)) = 0$, then $T = 0$.

COROLLARY XIX.16.3. $(T \in \mathscr{S}') \Rightarrow (T|\mathscr{D}(R^n)) \in \mathscr{D}'(R^n)$.

These two corollaries imply that $\mathscr{S}'$ can be identified with (embedded in) a subset of the space $\mathscr{D}'$. This justifies the following definition:

DEFINITION. The space $\mathscr{S}'$ is called the *space of tempered distributions*.

Since $\mathscr{S} \subset \mathscr{E}$ and the embedding $\mathscr{S} \to \mathscr{E}$ is continuous, we have the natural injections:

$$\mathscr{E}' \subset \mathscr{S}' \subset \mathscr{D}'.$$

Now every locally integrable function, which is a distribution, is a tempered distribution. It is easily seen, however, that integrable functions are tempered distributions: $\mathscr{L}^1 \subset \mathscr{S}'$. An interesting property of these distributions is given by

LEMMA XIX.16.4. If $f \in \mathscr{L}^1$, then for any $\varphi \in \mathscr{S}$ we have

$$\langle \varphi, \hat{f} \rangle = \langle \hat{\varphi}, f \rangle.$$

PROOF. Fubini's Theorem implies that

$$\begin{aligned}\langle \varphi, \hat{f} \rangle &= \int \varphi(x) \left(\int f(\xi) \exp(-2\pi i \langle x, \xi \rangle) d\xi \right) dx \\ &= \int f(\xi) \left(\int \varphi(x) \exp(-2\pi i \langle x, \xi \rangle) dx \right) d\xi \\ &= \int f(\xi) \hat{\varphi}(\xi) d\xi = \langle \hat{\varphi}, f \rangle. \ \square\end{aligned}$$

In accordance with what was said at the beginning of Section 8 on extending the definition of function operations to distributions, we adopt the following definition:

DEFINITION. The *Fourier transform of a tempered distribution* T is given by the formula

$$\langle \varphi, \hat{T} \rangle := \langle \hat{\varphi}, T \rangle \quad \text{for } \varphi \in \mathscr{S}.$$

The foregoing definition is consistent, i.e., T is a tempered distribution as the composition of a Fourier transformation on $\mathscr{S}$, which is linear and continuous, with the distribution T.

Using a different notation, we write: $(\mathscr{F}T)(\varphi) := T(\mathscr{F}\varphi)$ or $\hat{T}(\varphi) := T(\hat{\varphi})$. The Fourier transform of a tempered distribution is often written in function notation, especially in papers by physicists:

$$\hat{T}(\xi) = \int T(x) \cdot e^{-2\pi i \langle x, \xi \rangle} dx.$$

THEOREM XIX.16.5. (i) The Fourier transformation $\mathscr{F}$ is a bijection $\mathscr{S}' \to \mathscr{S}'$.

(ii) If $\mathscr{S}'$ is endowed with a weak topology (i.e., pointwise convergence), then $\mathscr{F}$ is a topological isomorphism of the space $\mathscr{S}'$.

(iii) $\hat{\hat{T}} = \check{T}$ for every $T \in \mathscr{S}'$, where $\check{T}(\varphi) := T(\check{\varphi})$.

PROOF. Item (i) follows from the fact that $F: \mathscr{S} \to \mathscr{S}$ is the topological isomorphism; (ii) and (iii) are obvious. □

Examples. 1. $\hat{\delta}_0 = 1$ (a distribution represented by a constant function) since

$$\hat{\delta}_0(\varphi) = \delta_0(\hat{\varphi}) = \hat{\varphi}(0) = \int 1 \cdot \varphi(x) dx = 1(\varphi).$$

2. $\hat{1} = \delta_0$, since $\hat{1} = \hat{\hat{\delta}}_0 = \check{\delta}_0 = \delta_0$.
3. $(D^\alpha T)^\wedge = (2\pi i\xi)^\alpha \hat{T}$ for $T \in \mathscr{S}'$.
4. $((-2\pi i x)^\alpha T)^\wedge = D^\alpha \hat{T}$ for $T \in \mathscr{S}'$.
5. It follows from 2 and 4 that $((-2\pi i x)^\alpha)^\wedge = D^\alpha \delta_0$.
6. $(\tau_h T)^\wedge = e^{-2\pi i \langle h, \xi \rangle} \hat{T}$.
7. $(e^{2\pi i \langle x, h \rangle} T)^\wedge = \tau_h T$.
8. Let $\mathscr{S}'(R^1) \ni T := \sum_{n=-\infty}^{\infty} \delta_n$. Since $e^{2\pi i n} = 1$, we have $e^{2\pi i x} T = T$.

Moreover, clearly $\tau_1 T = T$. Therefore, from Examples 6 and 7 we have (one-dimensional case):

$$\tau_1 \hat{T} = \hat{T}, \quad e^{2\pi i \xi} \hat{T} = \hat{T}, \quad \text{or} \quad (e^{2\pi i \xi} - 1)\hat{T} = 0.$$

This last fact implies that the support of the distribution $\hat{T}$ is contained in a set on which the function $e^{2\pi i \xi} - 1$ vanishes, i.e., is contained in the set of integers. In other words: $\hat{T} = \sum S_n$, where suppt $S_n \subset \{n\}$. As in the example given at the beginning of Section 14, it can be shown that $S_n = c_n \delta_n$, $c_n \in R$.

The invariance of the distribution T under translations by integers implies that $c_n = \text{const} = c$, that is, $\hat{T} = cT$.

Take $h(x) = e^{-\pi x^2}$. As we know, $\hat{h} = h$. Accordingly

$$0 \neq T(h) = T(\hat{h}) = \hat{T}(h) = cT(h),$$

whence $c = 1$, or $\hat{T} = T$. This formula yields the famous *Poisson summation formula*:

$$\sum_{n=-\infty}^{\infty} \varphi(n) = \sum_{n=-\infty}^{\infty} \delta_n(\varphi) = T(\varphi) = \hat{T}(\varphi) = T(\hat{\varphi})$$

$$= \sum_{n=-\infty}^{\infty} \delta_n(\hat{\varphi}) = \sum_{n=-\infty}^{\infty} \hat{\varphi}(n).$$

Thus, we have proved

THEOREM XIX.16.6 (Poisson Summation Formula). Every function $\varphi \in \mathscr{S}(\boldsymbol{R})$ obeys the formula

$$\sum_{n=-\infty}^{\infty} \varphi(n) = \sum_{n=-\infty}^{\infty} \hat{\varphi}(n).$$

Examples. 9. Let us evaluate the transform of the distribution given by the function

$$\operatorname{sgn} x = \begin{cases} +1 & \text{for } x > 0, \\ -1 & \text{for } x < 0. \end{cases}$$

We have

$$\langle \varphi, \hat{\operatorname{sgn}} \rangle = \langle \hat{\varphi}, \operatorname{sgn} \rangle$$

$$= -\int_{-\infty}^{0} \left(\int_{-\infty}^{0} \varphi(x) e^{-2\pi i \langle x, \xi \rangle} dx \right) d\xi + \int_{0}^{\infty} \left(\int_{-\infty}^{\infty} \varphi(x) e^{-2\pi i \langle x, \xi \rangle} dx \right) d\xi$$

$$= \lim_{a \to \infty} \lim_{\varepsilon \to 0} \left[-\int_{-a}^{0} \int_{-\infty}^{-\varepsilon} - \int_{-a}^{0} \int_{\varepsilon}^{\infty} + \int_{0}^{a} \int_{-\infty}^{-\varepsilon} + \int_{0}^{a} \int_{\varepsilon}^{\infty} \right] \times$$

$$\times [\varphi(x) e^{-2\pi i \langle x, \xi \rangle}] dx \, d\xi$$

$$= \lim_{a \to \infty} \lim_{\varepsilon \to 0} \left[\frac{1}{\pi i} \left(\int_{-\infty}^{-\varepsilon} \frac{\varphi(x)}{x} dx + \int_{\varepsilon}^{\infty} \frac{\varphi(x)}{x} dx \right) - \right.$$

$$-\frac{1}{2\pi i}\left(\int_{-\infty}^{-\varepsilon}\frac{\varphi(x)}{x}e^{2\pi iax}dx+\int_{\varepsilon}^{\infty}\frac{\varphi(x)}{x}e^{2\pi iax}dx+\right.$$

$$\left.+\int_{-\infty}^{-\varepsilon}\frac{\varphi(x)}{x}e^{-2\pi iax}dx+\int_{\varepsilon}^{\infty}\frac{\varphi(x)}{x}e^{-2\pi iax}dx\right)\Bigg]$$

$$=\frac{1}{\pi i}\left\langle\varphi, P\frac{1}{x}\right\rangle-\lim_{a\to\infty}\lim_{\varepsilon\to 0}\frac{1}{2\pi i}\left[\int_{-\infty}^{-\varepsilon}\frac{e^{2\pi iax}}{x}(\varphi(x)-\varphi(-x))\,dx+\right.$$

$$\left.+\int_{\varepsilon}^{\infty}\frac{e^{2\pi iax}}{x}(\varphi(x)-\varphi(-x))\,dx\right].$$

L'Hospital's rule, however, implies that

$$\lim_{x\to 0}\frac{\varphi(x)-\varphi(-x)}{x}=2\varphi'(0)<\infty,$$

where from the function $g(x):=(\varphi(x)-\varphi(-x))/x$ is continuous, vanishes rapidly at infinity, and hence is integrable. Accordingly, we have

$$\langle\varphi, \hat{\mathrm{sgn}}\rangle=\frac{1}{\pi i}\left\langle\varphi, P\frac{1}{x}\right\rangle-\frac{1}{2\pi i}\lim_{a\to\infty}\int_{-\infty}^{\infty}g(x)e^{2\pi iax}\,dx$$

$$=\frac{1}{\pi i}\left\langle\varphi, P\frac{1}{x}\right\rangle-\frac{1}{2\pi i}\lim_{a\to\infty}\hat{g}(-a).$$

By the Riemann–Lebesgue Lemma (Corollary XIX.14.6) we know that $\lim_{a\to\infty}\hat{g}(-a)=0$ and hence finally

$$\hat{\mathrm{sgn}}=\frac{1}{\pi i}P\frac{1}{x}.$$

It follows immediately from this that $\hat{P\frac{1}{x}}=\pi i\check{\mathrm{sgn}}=-\pi i\,\mathrm{sgn}$.

In function notation we often write:

$$P\frac{1}{x}=\pi i\int_{-\infty}^{\infty}\sin(\xi)e^{-2\pi ix\xi}\,d\xi,$$

$$P\int_{-\infty}^{\infty}\frac{1}{\xi}e^{-2\pi ix\xi}d\xi=-\pi i\,\mathrm{sgn}(x).$$

10. Calculate the transform of the distribution given by the function

$$\theta(x) = \begin{cases} 1 & \text{for } x > 0, \\ 0 & \text{for } x < 0. \end{cases}$$

Since $\theta + \check{\theta} = 1$, whereas $\theta - \check{\theta} = \operatorname{sgn}$, we have

$$\hat{\theta} + \hat{\check{\theta}} = \delta_0 \quad \text{and} \quad \hat{\theta} - \hat{\check{\theta}} = \widehat{\operatorname{sgn}} = \frac{1}{\pi i} P \frac{1}{x}.$$

Adding sides, we obtain

$$\hat{\theta} = \tfrac{1}{2}\delta_0 + \frac{1}{2\pi i} P \frac{1}{x}.$$

Remark. In function notation the results of Examples 2 and 10 are written as:

$$\delta(x) = \int_{-\infty}^{\infty} e^{-2\pi i x\xi} d\xi, \qquad \int_{0}^{\infty} e^{-2\pi i x\xi} d\xi = \tfrac{1}{2}\delta(x) + \frac{1}{2\pi i} P \frac{1}{x}.$$

The distribution $\hat{\theta}$ is frequently denoted in physics by δ_+ and

$$\hat{\check{\theta}}(x) = \int_{-\infty}^{0} e^{-2\pi i x\xi} d\xi = \tfrac{1}{2}\delta(x) - \frac{1}{2\pi i} P \frac{1}{x}$$

is denoted by δ_-. The equality $\delta = \delta_+ + \delta_-$ holds.

We know that the Fourier transformation on $\mathscr{S}'$ is a generalization of the classical transformation of a function from $\mathscr{L}^1$. We shall show how it is related to the Plancherel transformation on L^2:

LEMMA XIX.16.7. The inclusion

$$\mathscr{L}^2(\lambda^n) \supset \mathscr{S}'(\boldsymbol{R}^n)$$

holds.

PROOF. Clearly, $\mathscr{S} \subset \mathscr{L}^2$, and hence by the Schwarz inequality we have

$$\left| \int \varphi(x) f(x) d\lambda^n(x) \right| \leqslant ||\varphi||_{L^2} \cdot ||f||_{L^2} < \infty$$

for $\varphi \in \mathscr{S}$, $f \in L^2$. From the Lebesgue Theorem on dominated convergence in the space L^2 it follows that this expression is continuous with respect to φ. □

THEOREM XIX.16.8. If $f \in \mathscr{L}^2(R^n)$, the transformation of this distribution-function is representable by a function from $\mathscr{L}^2$(i.e., $\hat{f} \in L^2$).

This function is equal to the L^2-transform of f (the Fourier–Plancherel transform $\mathscr{F}f$).

PROOF. Let $f \in \mathscr{L}^2$, $\varphi \in \mathscr{S}$. We have

$$\left|\int f(x)\hat{\varphi}(x)dx\right| \leqslant \|f\|\cdot\|\hat{\bar{\varphi}}\| = \|f\|\cdot\|\hat{\varphi}\| = \|f\|\cdot\|\varphi\|,$$

where the last equality follows from the Plancherel Theorem (the norms are taken in the sense of L^2, that is, $\|\ \| = N_2$). Hence, the functional $L_f(\varphi) := \int f(x)\hat{\varphi}(x)\,dx$ is N_2-continuous in the argument φ, that is, can be extended to a functional continuous on L^2. The Fréchet–Riesz Theorem (XIII.14.4) implies the existence of a function $g \in \mathscr{L}^2$ such that

$$L_f(\varphi) = (\varphi|g) = \int \bar{g}(x)\varphi(x)\,dx = \langle\varphi, \bar{g}\rangle.$$

However $L_f(\varphi) = \langle\hat{\varphi}, f\rangle = \langle\varphi, \hat{f}\rangle$, where $\hat{f}$ here denotes the $\mathscr{S}'$-transform. This equality holds for any $\varphi \in \mathscr{S}$, whereby $\hat{f} = \bar{g} \in \mathscr{L}^2$. However,

$$(\hat{f}|\varphi) = \langle\bar{\varphi}, \hat{f}\rangle = \langle\hat{\bar{\varphi}}, f\rangle = (f|\bar{\hat{\bar{\varphi}}}) = (f|(\check{\varphi})^\wedge)$$
$$= \left(\mathscr{F}f|(\mathscr{F}(\check{\varphi})^\wedge)\right) = (\mathscr{F}f|(\check{\varphi})^{\hat{}}) = (\mathscr{F}f|\varphi).$$

Since $\mathscr{S}$ is a set dense in L^2, it follows from this equality that $\hat{f} = \mathscr{F}f$. □

17. THE LAPLACE–FOURIER TRANSFORMATION FOR FUNCTIONS AND DISTRIBUTIONS. THE PALEY–WIENER–SCHWARTZ THEOREM

It is a great temptation to cite the beautiful proofs of the theorems of Paley, Wiener, and Schwartz which rank among the most splendid developments in harmonic analysis. The original proofs of these profound theorems were difficult, but Hörmander has come up with amazingly straightforward proofs which are unquestionably classical. These theorems concern Laplace tıansforms, or Fourier–Laplace or Laplace–Fourier transforms as they are often called, and for this reason I will preface the formulation of these theorems with remarks of a terminological nature.

Let $\zeta = (\zeta_1, \ldots, \zeta_n) \in C^n$. Since R^n can be considered a subset of the subspace C^n, we shall use the notation

$$\langle \zeta, x \rangle := \sum_{j=1}^{n} \zeta_j x_j \quad \text{for } x \in R^n.$$

DEFINITION. Let $T \in \mathscr{D}'(R^n)$; if for some $\eta \in R^n$ the distribution $e^{-2\pi\langle \eta, \cdot \rangle} T \in \mathscr{S}'(R^n)$, then

(1) $\qquad \mathscr{L}(T) := \mathscr{F}(e^{-2\pi\langle \eta, \cdot \rangle} T)$

is called the *Laplace transform of the distribution T.*

Remark. Strictly speaking, the notation $\mathscr{L}_\eta(T)$ should be used because the transform given by (1) depends on the parameter η. However, the index η will be dropped in the present discussion.

If T is represented by a function $f \in C_0^\infty(R^n)$, then a Laplace transform exists for every η. Setting $\zeta = \xi + i\eta$, we write $\mathscr{L}(f)(\zeta) = \int_{R^n} e^{-2\pi i \langle \zeta, x \rangle} f(x)\, dx$ and we adopt the following definition:

DEFINITION. The mapping

$$\mathscr{E}(R^n) \ni f \to \mathscr{L}(f) \in C^n,$$

where

$$\mathscr{L}(f)(\zeta) := \int_{R^n} e^{-2\pi i \langle \zeta, x \rangle} f(x) dx = \mathscr{F}(e^{-2\pi\langle \eta, \cdot \rangle} f),$$

is called the *Laplace–Fourier transformation.*

It is left to the reader to prove the following interesting fact:

PROPOSITION XIX.17.1. Let $T \in \mathscr{D}'(R^n)$; then the set of $\eta \in R^n$, for which $e^{-2\pi\langle \eta, \cdot \rangle} T \in \mathscr{S}'$, is convex.

This proposition shows that the definition of the Laplace transform associates with every distribution $T \in \mathscr{D}'(R^n)$ some convex, perhaps empty, set $W(T) \subset R^n$, of those η for which $\mathscr{L}(T)$ is defined.

The Laplace transformation plays an important role in quantum field theory (in the Wightman approach). It associates functions of n complex variables with functions on R^n (more generally, with distributions). This is why physicists are interested in the theory of functions of many complex variables which, unfortunately, cannot be said of most mathematicians. This theory is indeed difficult, as mentioned before, but it is extremely elegant. It will be discussed in Part III of the book

The Fourier–Laplace transforming of functions of the class $\mathscr{D}$ are described in the celebrated

THEOREM XIX.17.2. (Paley–Wiener Theorem). An entire function F is the Fourier–Laplace transform of a function $f \in \mathscr{D}$ supported on the sphere $K(r) = \{x: ||x|| < r\}$ iff

$$(2) \qquad \bigwedge_{p \in N} \bigvee_{c(p)>0} \bigwedge_{\zeta \in C^n} |F(\zeta)| \leqslant c(p)(1+|\zeta|)^{-p} e^{r|\mathrm{Im}\,\zeta|}.$$

Laurent Schwartz has generalized this theorem to distributions with a compact support in the following way:

THEOREM XIX.17.3 (Schwartz–Paley–Wiener Theorem). An entire function F is the Fourier–Laplace transform of a distribution with support in $K(r)$ iff for some constants c, p

$$(3) \qquad |F(\zeta)| \leqslant c(1+|\zeta|)^{p} e^{r|\mathrm{Im}\,\zeta|}.$$

PROOF OF THE PALEY–WIENER THEOREM. $\Rightarrow$: As we know, the equality

$$\prod_{j=1}^{n} (2\pi i \zeta_j)^{\alpha_j} F(\zeta) = \int_{||x|| \leqslant r} e^{2\pi i \langle \zeta, x\rangle} D^\alpha f(x)\, dx$$

holds for $f \in \mathscr{E}(R^n)$, whence the assertion.

$\Leftarrow$: Let

$$f(x) := \int_{R^n} e^{2\pi i \langle x, \xi\rangle} F(\xi) d\xi.$$

Then, as in the proof of Theorem XIX.17.3, we demonstrate that $\hat{f} = F$ and that $f \in C^\infty(R^n)$. To show that $\operatorname{suppt} f \subset K(r)$, we observe that (2) allows the integration to be moved into the complex domain (Cauchy's Theorem); we have

$$(4) \qquad f(x) = \int_{R^n} e^{2\pi i \langle x, \xi + i\eta\rangle} F(\xi + i\eta)\, d\xi,$$

for any complex $\eta \in R^n$. Let us evaluate the integral in (4) by means of (2): setting $p = n+1$, we obtain

$$|f(x)| \leqslant c(p) e^{r|\eta| - 2\pi \langle x, \eta\rangle} \int_{R^n} (1+|\xi|)^{-(n+1)} d\xi < \infty.$$

On taking $\eta = tx$ and passing to the limit $t \to +\infty$, we find that for $|x| > r$ we have $f(x) = 0$, or $\operatorname{suppt} f \subset K(r)$. □

PROOF OF THE SCHWARTZ–PALEY–WIERER THEOREM. $\Rightarrow$: As we know from the definition of the space $\mathscr{E}'(\boldsymbol{R}^n)$, if $T \in \mathscr{E}'(\boldsymbol{R}^n)$, then there exist positive constants c, p, r such that

$$(5) \qquad |T(\varphi)| \leqslant c \sum_{|\alpha| \leqslant p} \sup |D^\alpha \varphi(K(r))|, \quad \varphi \in \mathscr{E}(\boldsymbol{R}^n).$$

If in (5) we take $\varphi_\xi(x) := \exp(-2\pi i \langle x, \xi \rangle)$, we have

$$(6) \qquad T(e^{-2\pi i\langle \cdot, \zeta \rangle}) \leqslant c \sum_{|\alpha| \leqslant p} |2\pi\zeta|^\alpha e^{r|\mathrm{Im}\,\zeta|} \leqslant c_1(1+|\zeta|)^p e^{r|\mathrm{Im}\,\zeta|}.$$

The hypothesis thus follows from (6) and this important

PROPOSITION XIX.17.4. The Fourier–Laplace transform of a distribution $T \in \mathscr{E}'$ is given by the formula

$$(7) \qquad \mathscr{L}(T)(\zeta) = T(e^{-2\pi i\langle \cdot, \zeta \rangle}), \quad \mathscr{L}(T) \in \mathscr{A}(\boldsymbol{C}^n).$$

(On setting $\zeta \in \boldsymbol{R}^n$, we obtain the Fourier transform.)

The proof of the assertion will be given further on. Let us now go on to the second part of the proof of the Schwartz–Paley–Wiener Theorem.

$\Leftarrow$: Let (φ_ε) be an approximation of the distribution δ: $\varphi_\varepsilon \to \delta_0$, $\operatorname{suppt} \varphi_\varepsilon \subset K(\varepsilon)$. Since $F \in \mathscr{S}'$, and $\mathscr{F}(\mathscr{S}') = \mathscr{S}'$, we have $F = \hat{f}$ for some $f \in \mathscr{S}'$. Let $f_\varepsilon := f * \varphi_\varepsilon$; hence $\hat{f}_\varepsilon = \hat{f} \cdot \hat{\varphi}_\varepsilon = F \cdot \hat{\varphi}_\varepsilon$. Now, $\operatorname{suppt} \varphi \subset K(\varepsilon)$, so that on applying (2) we have for an arbitrary q the constant c_1 such that

$$|\hat{\varphi}_\varepsilon(\zeta)| \leqslant c_1(1+|\zeta|)^{-p} e^{\varepsilon|\mathrm{Im}\,\zeta|},$$
$$|\hat{f}_\varepsilon(\zeta)| = |F \cdot \hat{\varphi}_\varepsilon(\zeta)| \leqslant c_2(1+|\zeta|)^{p-q} e^{(r+\varepsilon)|\mathrm{Im}\,\zeta|}.$$

When we invoke (2) $\Leftarrow$ we see that $\operatorname{suppt}(f * \varphi_\varepsilon) \subset K(r+\varepsilon)$. If $\varepsilon \searrow 0$, then $f * \varphi_\varepsilon \to f$. Thus $\operatorname{suppt} f \subset K(r)$. $\square$

PROOF OF PROPOSITION XIX.17.4. If $T \in C_0^\infty(\boldsymbol{R}^n)$, formula (7) is obvious. Again take $\varphi_\varepsilon \to \delta_0$. As we know, regular functions (with compact supports) $T * \varphi_\varepsilon \to T$ in the weak topology of the space $\mathscr{E}'$, hence in $\mathscr{S}'$ as well. Thus $\mathscr{F}(T * \varphi_\varepsilon) \to \hat{T}$ in $\mathscr{S}'$ as $\varepsilon \to 0$. However,

$$\mathscr{A}(\boldsymbol{C}^n) \ni \mathscr{L}(T * \varphi_\varepsilon) = (T * \varphi_\varepsilon)(e^{-2\pi i\langle x, \xi \rangle})$$
$$= T(\check{\varphi}_\varepsilon * e^{-2\pi i\langle x, \xi \rangle}) = \hat{\varphi}_\varepsilon(\xi) T(e^{-2\pi i\langle x, \xi \rangle}).$$

Since $\hat{\varphi}_\varepsilon(\xi) \to 1$ almost uniformly on $\boldsymbol{C}^n$ as $\varepsilon \to 0$, the function $T(e^{-2\pi i\langle x, \xi \rangle})$ is holomorphic on $\boldsymbol{C}^n$ as the compact convergence limit of holomorphic functions on $\boldsymbol{C}^n$. Its restriction to $\boldsymbol{R}^n$ is the Fourier transform of the distribution T. □

Concluding Remarks. Every mapping $T \in \mathscr{D}'(\boldsymbol{R}^n)$, as we know, defines the convex set $W(T)$ of the $\eta \in \boldsymbol{R}^n$ for which T has Laplace transforms.

A question which interests us is: Given a convex set $\Gamma \subset \boldsymbol{R}^n$, what distributions have a Laplace transform defined for $\eta \in \Gamma$? This question is of particular interest to quantum physicists. The answer is provided by the following

THEOREM XIX.17.5. Let the set Γ be open and convex in $\boldsymbol{R}^n$ and let $e^{-\langle x, \eta\rangle} T \in \mathscr{S}'$ for $\eta \in \Gamma$. Then $\mathscr{L}(T) \in \mathscr{A}(\boldsymbol{R}^n + i\Gamma)$ and for any compact $K \subset \Gamma$ there exists a polynomial P_k such that

$$|\mathscr{L}(T)(\xi - i\eta)| \leqslant |P_K(\xi)| \tag{8}$$

for $\eta \in K$.

Conversely, if $F \in \mathscr{A}(\boldsymbol{R}^n + i\Gamma)$ satisfies (8) for every compact subset $K \subset \Gamma$, then $F = \mathscr{L}(T)$, where T is unique and $\Gamma = W(T)$.

The proof of this theorem will not be given here. The reader will find it along with many theorems of a similar nature in the excellent slim volume by Streater and Wightman, *PCT, Spin Statistics, and All That*, New York, 1964.

18. FUNDAMENTAL SOLUTIONS OF DIFFERENTIAL OPERATORS

Let A be a differential operator of order k, with constant coefficients, on the space $\boldsymbol{R}^n$: $A = \sum_{|\alpha| \leqslant k} a_\alpha D^\alpha$, and let φ be a function from $\mathscr{D}(\boldsymbol{R}^n)$. Consider the differential equation

$$Au = \varphi.$$

DEFINITION. A distribution $T \in \mathscr{D}'(\boldsymbol{R}^n)$ which satisfies the equation $AT = \delta_0$ is called the *fundamental solution of the operator A.*

Of course, in general the operator A has an infinite number of fundamental solutions: if u is an arbitrary solution of the homogeneous equation $Au = 0$, then $A(T+u) = \delta_0$, and thus $T+u$ is also the fundamental solution.

THEOREM XIX.18.1. If T is a fundamental solution of the operator A, then for any $\varphi \in \mathscr{D}(R^n)$ the function $u := T*\varphi$ is a solution of the equation $Au = \varphi$.

PROOF. We have

$$Au = A(T*\varphi) = (AT)*\varphi = \delta_0*\varphi = \varphi. \quad \Box$$

Thus, knowledge of the fundamental solution allows the equation to be solved "with arbitrary right-hand side".

In order to find a fundamental solution of a given operator, note that $(AT)^\wedge = \hat{\delta}_0 = 1$. However,

$$(AT)^\wedge = \left(\sum a_\alpha D^\alpha T\right)^\wedge = \sum a_\alpha (D^\alpha T)^\wedge = \sum a_\alpha (2\pi i x)^\alpha \hat{T}$$

(cf. Example 3, Section 17). Thus, the Fourier transform of a fundamental solution obeys the equation

$$\left(\sum a_\alpha (2\pi i x)^\alpha\right) \cdot \hat{T} = 1.$$

The notation $\hat{T} = [\sum a_\alpha (2\pi i x)^\alpha]^{-1}$ is sometimes used. This notation is not meaningful, however, since the right-hand side of this equation is not a locally integrable function; its integral is divergent at points at which the polynomial $\sum a_\alpha (2\pi i x)^\alpha$ vanishes. Nevertheless, this expression can often be imparted a distributional sense in the same way as, for example, we formed the distribution $P\dfrac{1}{x}$ from the function $\dfrac{1}{x}$ which is not locally integrable.

A general method has been given by L. Hörmander (cf. K. Maurin, *Methods of Hilbert Spaces*, and Chapter XXI).

Examples. 1. Let $n = 1$. Take the equation $\dfrac{d}{dx} u = \varphi (\varphi \in \mathscr{D}(R^1))$.

The equation for $\hat{T}$ here is of the form: $\hat{T} \cdot 2\pi i x = 1$.

Let us, for example, write $\hat{T} = \dfrac{1}{2\pi i} P \dfrac{1}{x}$. Then

$$T(x) = \frac{1}{2\pi i}\left(P\frac{1}{x}\right)^{\wedge\vee} = \frac{1}{2\pi i}(-\pi i \operatorname{sgn} x)^\vee = -\tfrac{1}{2}\operatorname{sgn}(-x) = \tfrac{1}{2}\operatorname{sgn} x.$$

Hence

$$u(x) = (T*\varphi)(x) = \tfrac{1}{2}\int_{-\infty}^{\infty} \varphi(x-y)\operatorname{sgn} y\, dy$$

$$= \tfrac{1}{2}\Big(-\int_{-\infty}^{\infty} \varphi(x-y)\,dy + \int_{0}^{\infty} \varphi(x-y)\,dy\Big)$$

$$= \tfrac{1}{2}\Big(\int_{\infty}^{x} \varphi(z)\,dz - \int_{x}^{-\infty} \varphi(z)\,dz\Big) = \int_{-\infty}^{x} \varphi(z)\,dz - \tfrac{1}{2}\int_{-\infty}^{\infty} \varphi(z)\,dz.$$

From another place we know that in general $u(x) = \int_{-\infty}^{x} \varphi(z)\,dz + C$. It is seen that a particular way of imparting meaning to the expression $1/x$ resulted in a particular choice of the constant $C\,(= -\tfrac{1}{2}\int \varphi(z)\,dz)$.

Observe, moreover, that if T is an arbitrary solution of the equation $\hat{T}\cdot 2\pi i x = 1$, then

$$\left(\hat{T} - \frac{1}{2\pi i} P\frac{1}{x}\right)\cdot x = 0,$$

and thus

$$\hat{T} - \frac{1}{2\pi i} P\frac{1}{x} = c\cdot\delta_0, \quad \text{or} \quad \hat{T} = \frac{1}{2\pi i} P\frac{1}{x} + c\cdot\delta_0.$$

In this case, therefore, we have succeeded in finding all the fundamental solutions and all the solutions of the given equation:

$$T = \left(\frac{1}{2\pi i} P\frac{1}{x} + c\delta_0\right)^{\check{}} = \tfrac{1}{2}\operatorname{sgn} + c,$$

whence

$$u(x) = \int_{-\infty}^{x} \varphi(z)\,dz - \tfrac{1}{2}\int_{-\infty}^{\infty} \varphi(z)\,dz + c\int_{-\infty}^{\infty} \varphi(z)\,dz.$$

2. Poisson's equation (cf. Theorem XV.6.1)

$$\Delta u = 4\pi\varrho$$

appears in electrostatics. This is an equation which relates the electric potential $u(x)$ and the charge density ϱ.

In the simplest case, $T(x) = 1/(-4\pi||x||)$ is taken for a fundamental solution. As we know, $\Delta \dfrac{1}{||x||} = -4\pi\delta_0$. Thus

$$u(x) = \frac{1}{-4\pi}\left(\frac{1}{r} * \varrho\right)(x) = \frac{1}{-4\pi}\int \frac{\varrho(y)}{||x-y||}dy.$$

This law is frequently enunciated in physics as follows: "The charge $Q = \varrho(y)\,dy$ sets up a potential $\dfrac{Q}{-4\pi r} = \dfrac{Q(y)\,dy}{-4\pi||x-y||}$ at a point x. To obtain the total potential at x, take the sum of the contributions from all charged points, i.e. integrate over the variable y."

19. POSITIVE-DEFINITE FUNCTIONS. POSITIVE DISTRIBUTIONS. THE THEOREMS OF BOCHNER AND MINLOS

Positive-definite functions play a significant role in many areas of analysis, probability theory, and, as of late, in quantum theory as well. Their enormous role in group theory will be mentioned in the next section.

DEFINITION. Let X be an Abelian group (additive notation will be used in this section). A function $f\colon X \to \boldsymbol{C}$, which does not vanish identically, is said to be *positive-definite* if

$$\sum_{j,k=1}^{n} f(x_j - x_k)c_j\bar{c}_k \geqslant 0$$

for any $x_1, \ldots, x_n$ and $c_1, \ldots, c_n \in \boldsymbol{C}$.

A straightforward theorem follows:

THEOREM XIX.19.1. If a function f is positive-definite, then

(i) $f(-x) = \overline{f(x)}$,

(ii) $f(0) > 0$,

(iii) $|f(x)| \leqslant f(0)$, hence f is bounded,

(iv) $|f(x)-f(y)| \leqslant 2f(0)\cdot \mathrm{Re}\big(f(0)-f(x-y)\big)$.

PROOF. (i) If we set $n = 2$, $x_1 = 0$, $x_2 = x$, $c_1 = 1$, $c_2 = c$, then by the definition of a positive-definite function we have

(1) $\quad (1+|c|^2)f(0)+cf(x)+\bar{c}f(-x) \geqslant 0.$

Putting $c = 1$ in this inequality, we see that $f(x)+f(-x)$ is a real number. When we set $c = i$, we see that $i\big(f(x)-f(-x)\big)$ is a real number. Consequently,

$$\overline{f(x)}+\overline{f(-x)} = f(x)+f(-x),$$
$$-\overline{(f(x)-f(-x))} = f(x)-f(-x).$$

When we add by sides we have $2\overline{f(-x)} = 2f(x)$.

(ii) Suppose, first of all, that $f(0) = 0$. Then on substituting in (1) $c = -\overline{f(x)}$ and taking account of (i), we have $-2|f(x)|^2 \geqslant 0$, and thus $f(x) = 0$. Since this case was excluded in the definition, $f(0) \neq 0$. However, if we take $n = 1$, $x_1 = 0$, $c_1 = 1$, by the definition we get $f(0)\cdot 1 \geqslant 0$, and thus $f(0) > 0$.

(iii) We make the substitution $c = -\overline{f(x)}/f(0)$ in equation (1). Division of both sides of the equation by $f(0)$ yields

$$1+\frac{|f(x)|^2}{|f(0)|^2}-2\left|\frac{f(x)}{f(0)}\right|^2 \geqslant 0.$$

It thus follows that $|f(x)| \leqslant |f(0)| = f(0)$.

(iv) Let us take $n = 3$, $x_1 = 0$, $x_2 = x$, $x_3 = y$, $c_1 = 1$, and $c_2 = \dfrac{t|f(x)-f(y)|}{f(x)-f(y)}$, where t is a real number $c_3 = -c_2$. Then from the definition of a positive-definite function it follows that

$$f(0)\,(1+2t^2)+2t|f(x)-f(y)|-2t^2\,\mathrm{Re}f(x-y) \geqslant 0,$$

the discriminant of the second-degree trinomial in the variable t must therefore be negative. However, this discriminant is of the form

$$|f(x)-f(y)|^2-2f(0)\,\mathrm{Re}\big(f(0)-f(x-y)\big) \leqslant 0,$$

and thus (iv) has been proved. □

COROLLARY XIX.19.2. Let X be a commutative topological group. Then every positive-definite function which is continuous at zero is uniformly continuous.

The proof follows from (iv).

Examples. 1. Let $X = \boldsymbol{R}^1 \ni \xi$. The function $f(x) = e^{-2\pi i x\xi}$ is continuous and positive-definite, since:

$$\sum_{j,k} c_j\bar{c}_k e^{-2\pi i\xi(x_j-x_k)} = \Big(\sum_j c_j e^{-2\pi i\xi x_j}\Big)\Big(\sum_k \bar{c}_k e^{2\pi i\xi x_k}\Big)$$
$$= \Big|\sum_j c_j e^{-2\pi i\xi x_j}\Big|^2 \geqslant 0.$$

2. $X = \boldsymbol{R}^n$, $\boldsymbol{R}^{n\prime} = \boldsymbol{R}^n$; the function $f(x) = e^{-2\pi i\langle x, \xi\rangle}$ is a positive-definite continuous function. Proof, as above.

3. Any linear combination, with positive coefficients, of exponential functions is a continuous positive-definite function:

$$\boldsymbol{R}^n \ni x \to f(x) := \sum_{r=1}^{M} a_r e^{-2\pi i\langle x, \xi_r\rangle}, \qquad \xi_r \in \boldsymbol{R}^n,\ a_r > 0.$$

4. Let μ be an arbitrary finite Radon measure on $\boldsymbol{R}^n$; then

$$\boldsymbol{R}^n \ni x \to f(x) := \int_{\boldsymbol{R}^n} e^{-2\pi i\langle x, \xi\rangle} d\mu(\xi)$$

is a continuous function (by the theorem on integrals with a parameter) and positive-definite since

$$\sum_{j,k} c_j \cdot \overline{c_k} \cdot \int e^{-2\pi i\langle x_j - x_k, \xi\rangle} d\mu(\xi)$$
$$= \int \Big| \sum_j c_j \cdot e^{-2\pi i\langle x_j, \xi\rangle} \Big|^2 d\mu(\xi) \geqslant 0$$

(cf. Example 1).

Samuel Bochner observed that Example 4 is general:

THEOREM XIX.19.3 (Bochner's Theorem). For every continuous function f which is positive-definite on $\boldsymbol{R}^n$ there exists a (unique) finite Radon measure μ on $\boldsymbol{R}^n$ such that

$$f(x) = \int_{\boldsymbol{R}^n} e^{-2\pi i\langle x, \xi\rangle} d\mu(\xi), \qquad \hat{f} = \check{\mu}.$$

The proof will be preceded by an important theorem on positive distributions.

THEOREM XIX.19.4. Let the distribution $T \in \mathscr{D}'(\boldsymbol{R}^n)$ be positive, i.e., $T(\varphi) \geqslant 0$ if $\mathscr{D}(\boldsymbol{R}^n) \ni \varphi \geqslant 0$. Then there exists a Radon measure μ such that $T(\varphi) = \mu(\varphi)$, $\varphi \in \mathscr{D}(\boldsymbol{R}^n)$. To put it briefly, positive distributions are measures.

PROOF. Proceeding as in the proof of Theorem XIII.17.3. let us assume that $\mathscr{D} \ni \varphi_n \to 0$ uniformly and let $\varphi_n \subset K$, where K is a fixed compact set. Take the function $0 \leqslant g \in C_0^\infty(\overline{\boldsymbol{R}^n})$ such that $g(x) = 1$ for $x \in K$. Thus we have $|\varphi_n(x)| \leqslant \varepsilon_n \cdot g(x)$, where $\varepsilon_n \searrow 0$. Let $\varphi_n = u_n^{(1)} + iu^{(2)}$, where $u_n, v_n \in \mathscr{D}$ are real.

Since $-\varepsilon_n g \leqslant u_n^{(j)} \leqslant \varepsilon_n g$, $j = 1, 2$, the fact that distribution T is positive implies that $-\varepsilon_n T(g) \leqslant T(u_n^{(j)}) \leqslant \varepsilon_n T(g)$, whence $|T(u_n^{(j)})| \leqslant \varepsilon_n T(g)$, $j = 1, 2$, or $|T(\varphi_n)| \leqslant 2\varepsilon_n T(g)$.

Thus we have shown that $T(\varphi_n) \to 0$, if $\varphi_n \to 0$, in the sense of $C_0(\boldsymbol{R}^n)$. However, $C_0^\infty(\boldsymbol{R}^n)$ is dense in $C_0(\boldsymbol{R}^n)$ endowed with a locally convex topology of the inductive limit of the space $C_0(K_r)$, $K_r \nearrow \boldsymbol{R}^n$.

Therefore, the functional T is uniquely extendible to a linear functional continuous on $C_0(\boldsymbol{R}^n)$, that is, to a Radon integral μ, one which perhaps is non-positive. It remains to verify that $\mu \geqslant 0$. Let $0 \leqslant \varphi \in C_0(\boldsymbol{R}^n)$. We regularize φ: $0 \leqslant \varphi * \overset{\varepsilon}{\Psi} \to \varphi$, as $\varepsilon \to 0$ in the sense of convergence in $C_0(\boldsymbol{R}^n)$ (cf. Theorem XIX.1.1, item (iii)). Consequently

$$0 \leqslant \mu(\varphi * \overset{\varepsilon}{\Psi}) \to \mu(\varphi),$$

and hence $\mu(\varphi) \geqslant 0$, which is to say that μ is a positive Radon integral. □

In the course of the proof we have demonstrated

COROLLARY XIX.19.5. For a distribution $T \in \mathscr{D}'(\Omega)$ to be determined by a Radon integral, it is necessary and sufficient that T be continuous on the set $C_0^\infty(\Omega)$ endowed with a topology $C_0(\Omega)$.

Bochner's Theorem will be proved in two stages. First, we shall show that $\hat{f}$ is a positive distribution, i.e., that there exists a Radon integral μ such that $\hat{f}(\varphi) = \check{\mu}(\varphi)$. Second, we shall demonstrate that the integral is finite: $\|\mu\| < \infty$.

(i) Let us denote $\varphi^*(x) := \overline{\varphi(-x)}$. Note first of all that if f is a continuous positive-definite function on $\boldsymbol{R}_n$, then

$$\iint f(x-y)\varphi(x)\overline{\varphi(y)}\,d\lambda^n(x)\,d\lambda^n(y) \geqslant 0, \qquad \varphi \in \mathscr{S}.$$

This inequality is obtained from the definition of a positive-definite function by approximating the complex measure $\varphi\lambda^n$ by the measures $\sum c_j \delta_{x_j}$ and taking account of the fact that f is bounded (Theorem XIX.19.1, item (iii)). To put it in less "learned" words, we approximate the integral by finite Riemann sums, which are positive by virtue of the definition of a positive-definite function. However,

$$\iint f(x-y)\varphi(x)\overline{\varphi(y)}\,dx\,dy = \iint f(x)\varphi(x)\overline{\varphi(x-z)}\,dx\,dz$$
$$= \int f(z)\cdot(\varphi * \varphi^*)(z)\,dz = \int f(\varphi * \varphi^*) \geqslant 0 \quad \text{for } \varphi \in \mathscr{S}.$$

Since $f \in \mathscr{S}'$, we can apply a Fourier transformation to f. Bear in mind that $(g \cdot \bar{h})\hat{} = \hat{g} * \hat{h}^*$ and thus $\hat{f}(|u|^2) = f((u \cdot \bar{u})\hat{}\,) = f(\hat{u} * \hat{u}^*) \geqslant 0$.

Now, let $0 \leqslant \varphi \in \mathscr{D}$. We shall approximate this function in $\mathscr{D}$ by functions of the form $u \cdot \bar{u}$, $u \in \mathscr{D}$. Let $0 \leqslant g \in \mathscr{D}$, $g(x) = 1$ for $x \in \overline{\varphi}$; then $\varphi = \lim_{n \to 0} g^2(\varphi + 1/n)$ in the sense of convergence in $\mathscr{D}$. Thus, on setting $u_n := g\sqrt{\varphi + n^{-1}}$, we have

$$\varphi = \lim_{n \to \infty} u_n \cdot \bar{u}_n.$$

Therefore $0 \leqslant \hat{f}(u_n \bar{u}_n) \xrightarrow[n \to \infty]{} \hat{f}(\varphi)$, whereby $\hat{f}(\varphi) \geqslant 0$. By virtue of Theorem XIX.19.4 $\hat{f}(\varphi) = \check{\mu}(\varphi)$, $\check{f} = \hat{\hat{f}} = \check{\hat{\mu}}$, or $f = \hat{\mu}$.

(ii) Let $h \in \mathscr{D}$ and let $\varphi := \hat{h} \geqslant 0$, $h(0) = 1$ (for h we can take the function $h = c \cdot u * u^*$ for any $u \in \mathscr{D}$, where $c^{-1} = (u * u^*)(0)$, for then $\hat{h} = c|\hat{u}|^2 > 0$). Making use of the equalities just proved and of inequality (iii) of Theorem XIX.19.1, we have

$$\hat{f}(|h|^2) = \int h^2 d\check{\mu} = f(\varphi * \varphi^*) = \iint f(x)\varphi(x-y)\overline{\varphi(-y)}\,dx\,dy$$
$$\leqslant f(0)\left(\int \varphi\, d\lambda^n\right)^2 = f(0)(\hat{\varphi}(0))^2 = f(0)h(0)^2 = f(0).$$

When in this inequality we replace h by $h(\varepsilon\xi)$ and make use of the fact that $h(\varepsilon\xi) \xrightarrow[\varepsilon \to 0]{} 1$ uniformly on every ball (Lemma XIX.14.3) we obtain

$$f(0) \geqslant \int |h(\varepsilon\xi)|^2 d\mu(\xi) \geqslant \int_{K(0,r)} |h(\varepsilon\xi)|^2 d\mu(\xi) \to \int_{K(0,r)} d\mu$$

for any $r > 0$. It thus follows that $\|\mu\| \leqslant f(0)$. □

Minlos generalized the Bochner Theorem to nuclear spaces. The following theorem holds:

THEOREM XIX.19.6 (Minlos Theorem). Let E be a complete real nuclear space, and let f be a continuous positive-definite function on E. Then a finite Borel measure μ such that

$$f(x) = \int_{E'} e^{-2\pi i\langle x, \xi\rangle} d\mu(\xi)$$

exists on E'.

The proof can be found in the present author's monograph, *General Eingenfunction Expansions and Unitary Representations of Topological Groups*.

20. REPRESENTATIONS OF LOCALLY COMPACT GROUPS. THE RELATION BETWEEN UNITARY REPRESENTATIONS AND POSITIVE-DEFINITE FUNCTIONS

In Part I of the book we mentioned one-parameter groups of mappings (dynamic systems, Stone's Theorem, etc.). These were homomorphisms of the (additive) group R to the group of operators (or, more generally, the transformations) of a space H, i.e., the representation of the group R to the group of automorphisms of the space H. The theory of group representations has a long history: as every great mathematical theory it has grown out of concrete needs. It is surprising that there is probably no major area of mathematics and physics in which the concept of the representation of a group has not intervened in an essential manner. One can scarcely imagine modern geometry, mechanics (analytical dynamics, quantum mechanics, and in recent years, even hydrodynamics), many areas of physics, number theory, analysis (modular functions, automorphic forms, theta functions, the theory of so-called special functions) without the concepts and methods of the theory of group representation. Accordingly, we are now giving several general concepts of the theory of topological group representations; later we shall go on to the most important representations, unitary representations. Further on in the book we shall continually come across these relations. The literature of this subject is enormous; the books devoted to the theory of representations of groups and its applications are legion (cf. for instance, the author's monographs *Methods of Hilbert Spaces* and *General Eigenfunction Expansions and Unitary Representations of Topological Groups*). The reader who would like quickly to get an idea of the great variety of applications of this theory is referred to the many lecture notes of one of the fathers of the modern theory of group representation, George W. Mackey.

Basic Concepts

DEFINITION. Let G be a (topological) group, and let H be a Hausdorff space. The pair (U, H), where $G \ni x \to U(x) \in \operatorname{Aut} A$ is a homomorphism of the group G into the group of automorphisms of the space H is called the *representation of G on H*. Thus we have $U(g_1 g_2) = U(g_1) \circ U(g_2)$, $U(e) = \mathrm{id}_H$, $U(g^{-1}) = U(g)^{-1}$. The representation (U, H) is continuous if the mapping $G \times H \ni (g, h) \to U(g)h \in H$ is continuous. Henceforth we shall assume that G is locally compact and that H is a complete, locally convex space over $\boldsymbol{C}$ (or $\boldsymbol{R}$), whose topology is given by the family of continuous subsets $(\| \|_p, p \in \Pi)$.

LEMMA XIX.20.1. The continuity of (U, H) is equivalent to the set of the following two conditions:

(i) For every $h \in H$, the mapping $G \ni g \to U(g)h \in H$ is continuous (this condition is called the *partial continuity of the representation*).

(ii) For every compact $K \subset G$ the set $\{U(g): g \in K\}$ is equicontinuous, (i.e., for every neighbourhood $\mathcal{O}_1$ of zero in H there exists a neighbourhood $\mathcal{O}_2$ of zero such that $U(K)(\mathcal{O}_2) \subset \mathcal{O}_1$).

PROOF. The continuity of the representation (U, H) clearly implies (i) and (ii). Suppose that (i) and (ii) are satisfied; we then prove the continuity of (U, H). By (i) the continuity of (U, H) will be proved if for any net (h_n) in H tending to 0 we have $U(g_n)h_n \to 0$ as $g_n \to g$ in G.

Take a compact neighbourhood K of the point $g \in G$ such that $g_n \in K$ for $n > n_0$; then (by (ii)) the set $\{U(g_n): n > n_0\}$ is equicontinuous. Let $\mathcal{O}_1$ be a neighbourhood of 0 in H, and take a neighbourhood $\mathcal{O}_2$ of zero in H such that $U(g_n)(\mathcal{O}_2) \subset \mathcal{O}_1$ for $n > n_0$. Changing the notation (i.e., taking a later $\bar{n}_0 > n_0$), we can assume that $h_n \in \mathcal{O}_2$ for $n > n_0$; therefore, $U(g_n)h_n \in \mathcal{O}_1$ for $n > n_0$.

COROLLARY XIX.20.2. If H is a Banach space or, more generally, a barrelled space (hence, say, a Fréchet, Montel, complete nuclear, or sequentially complete Mackey space), then (i) $\Rightarrow$ (ii). That is to say, every partially continuous representation in a barrelled space is continuous.

Indeed, the Banach–Steinhaus Theorem holds: Let H be barrelled and let F be locally convex; then any set $A \subset \mathscr{L}(H, F)$, which is bounded at every point is equicontinuous (the proof of the Banach–Steinhaus Theorem is given in Chapter XX).

In the case of a Banach space H (it is sufficient to assume that H is normed, not necessarily continuous) examination of partial continuity can be facilitated substantially by the following

PROPOSITION XIX.20.3. Let G be a locally compact group and let (U, H) be a representation in the normed space H. Let F be a dense subset in H. If

(iii) the mapping $G \ni g \to U(g)f \in H$ is continuous for every $f \in F$, then the representation (U, H) is continuous.

PROOF. We are to show that for every compact $K \subset G$ the family $\{U(g): g \in K\}$ is equicontinuous, i.e., that $\sup\|U(K)\| < \infty$. However, we know that the upper envelope of functions continuous on G is lower semicontinuous and that lower semicontinuous functions on the locally compact space G can be majorized on a non-empty, open subset $A \subset G$. Let $r(x) = \|U(x)\| = \sup_f \|U(x)f\|$, $f \in \mathscr{F}$, $\|f\| \leqslant 1$. Thus r is lower semicontinuous on G and $\|U(x)\| \leqslant M$, $x \in A$, A being open and nonempty. The compact set K can be covered by a finite number of translations of the set A: let $K \subset \bigcup_{i \in 1}^{n} Ax_i$. Then for $x \in K$ we have

$$\|U(x)\| = \|U(a)U(x_i)\| \leqslant \|U(a)\| \, \|U(x_i)\| \leqslant M \sup_{1 \leqslant i \leqslant n} \|U(x_i)\|.$$

Hence, condition (ii) is satisfied.

Let $\varepsilon > 0$, $h_0 \in H$, $g_0 \in G$; there exists a compact neighbourhood K of the point g_0 such that $\sup\|U(K)\| = M < \infty$. Thus

$$\|U(g)h - U(g_0)h_0\| \leqslant \|U(g)h - U(g)h_0\| + \|U(g)h_0 - U(g)f\| + \|U(g)f - U(g_0)f\| + \|U(g_0)f - U(g_0)h_0\|.$$

Take $f \in F$ such that $\|h_0 - f\| < \varepsilon/4M$; then there exists a neighbourhood $\mathcal{O}$ of the element h_0 and a neighbourhood $\mathscr{V}$ of the point g_0 such that $\|h - h_0\| < \varepsilon/4M$, $h \in \mathcal{O}$; $\|U(g)f - U(g_0)f\| < \varepsilon/4$, $g \in \mathscr{V}$. That is, for $(g, h) \in \mathscr{V} \times \mathcal{O}$ we have $\|U(g)h - U(g_0)h_0\| < \varepsilon$. □

Examples of Representations

A. *Bounded and Unitary Representations.* If H is a normed space when there exists an $M > 0$ such that $\|U(g)\| \leqslant M$ for all $g \in G$, the representation U is called *bounded.*

Assume H to be a Hilbert space. *Unitary representation* is a homomorphism of G into the group of unitary operators in H.

B. *Regular Representations.* Suppose that (G, X) is a left G-space, i.e., G is a group of transformations of the space X:

$$G \times X \to X, \quad G \times X \in (g, x) \to g \cdot x \in X,$$
$$g_1 \cdot (g_2 \cdot x) = (g_1 g_2) \cdot x, \quad e \cdot x = x, \; x \in X.$$

Then G is said to *operate on X from the left.*

Suppose that $\mathscr{H}$ is a linear space of functions on the G-space X; then $U(g)f(x) := f(g^{-1}x)$ is called the (*left*) *regular representation of G in H.*

Examples. 1. Let $H = L^p(G, dg)$, where G is locally compact, $1 \leqslant p < \infty$, where dg is the (left) Haar measure on G. Clearly, G is a left (right) G-space: $g_1 \cdot g_2 := g_1 g_2$. Since dg is a left Haar measure, we have $\|U(g)f\|_p = \|f\|_p$ for $f \in L^p(G)$, $g \in G$. Hence, the representation is bounded. Take $C_0(G)$ as a set dense in $\mathscr{L}^p(G)$. The mapping $g \to U(g)f \in L^p(G, dg)$ is continuous for every $f \in C_0(G)$. Indeed, $U(g)f$ tends uniformly to $U(g_0)f$ as g approaches g_0 because the supports of the function $U(g)f$ are contained in a fixed compact set. Thus we have proved the following

PROPOSITION XIX.20.4. A left (right) regular representation in $L^p(G, dg)$, $1 \leqslant p < \infty$, is continuous.

2. A particularly important example is that of $p = 2$. We then immediately have

PROPOSITION XIX.20.5. A left regular representation in a Hilbert space $L^2(G, dg)$ (with scalar product $(f_1 | f_2) = \int f_1 \bar{f}_2 dg$) is a continuous unitary representation.

3. We now give an example of a discontinuous representation. Let $H = L^\infty(G, dg)$ and let $U(g)f(x) := f(g^{-1}x)$. Plainly, the operator $U(g)$ is an isometry with norm 1, but the set $C_0(G)$ is no longer dense in $L^\infty(G, dg)$ and the reasoning of Example 1 no longer applies. Moreover, a regular representation in $L^\infty(G, dg)$ is not continuous. Indeed, the continuity of $g \to U(g)f$ at the point e would mean that $f(g^{-1}x) \to f(x)$ uniformly on G as $g \to e$, that is, f would have to be uniformly continuous (which is absurd, because f is an arbitrary measurable and essentially bounded function).

Extension of the Representation $(U, \mathscr{H})$ *to* $M_0(G)$. Let $M_0(G)$ be a vector space of complex Radon measures with compact supports on G. Let $(U, \mathscr{H})$ be a continuous representation of a group G in a Banach space $\mathscr{H}$. We set

$$U(\mu)h := \int_G U(x)h\,d\mu(x), \qquad \mu \in M_0(G),\ h \in H.$$

(The reader should recall the definition of the integral of a function with values in a Banach space). If we put $\mu = \delta_g$ (Dirac measure concentrated at the point g), we obtain $U(\delta_g)h = U(g)h$. Thus $(U(\mu))$ is indeed an extension of $(U(g))$ if the group G is embedded in $M_0(G)$:

$G \ni g \to \delta_g \in M_0(G)$. We recall that $M_0(G)$ is a convolution algebra, where the convolution $\mu * \nu$ is defined as

$$\mu * \nu(f) := \iint f(xy)\, d\mu(x)\, d\nu(y) = \int d\mu(x) \int f(xy)\, dy$$
$$= \int d\nu(y) \int f(xy)\, d\mu(y)$$

for $f \in C_0(G)$. If G is Abelian, then of course $\mu * \nu = \nu * \mu$. The Banach space of measure with finite mass, the completion of the space $M_0(G)$, will be denoted by $M^1(G)$. This, too, is a convolution algebra which in a natural manner contains the space $L^1(G, dg)$: we assign the measure $\mu_f := f dg$ to the function $f \in L^1(G, dg)$. We can now prove a useful

THEOREM XIX.20.6. Let G be a locally compact group and let (U, H) be a continuous representation of G in the Banach space H. Then

(1) $U(\mu) \in \mathscr{L}(H)$ for $\mu \in M_0(G)$,

(2) $\mu \to U(\mu)$ is a representation of the algebra $M_0(G)$ in E,

(3) if (U, H) is bounded (e.g. if (U, H) is unitary), then U is extendible to a continuous representation of the (convolution) Banach algebra $M^1(G)$ into the Banach algebra $\mathrm{Aut}\, H$.

PROOF. (1) If $h \to 0$, then $U(x)h \to 0$ uniformly on a compact support and hence $U(\mu)h \to 0$. The linearity of $U(\mu)$ is obvious.

(2) The linearity of the mapping $\mu \to U(\mu)$ is obvious. We have

$$U(\mu * \nu)h = \int U(x) h\, d(\mu * \nu)(x) = \iint U(xy) h\, d\mu(x)\, d\nu(y)$$
$$= \iint U(x)U(y) h\, d\mu(x)\, d\nu(y) = \int U(\mu)U(y) h\, d\nu(y)$$
$$= U(\mu) \int U(y) h\, d\nu(y) = U(\mu)U(\nu).$$

(3) If $\mu \in M_0(G)$ then

$$\|U(\mu)h\| = \left\| \int U(x) h\, d\mu(x) \right\|$$
$$\leqslant \int \|U(x)h\|\, d|\mu|(x) \leqslant M \|h\|\, \|\mu\|.$$

That is to say, the mapping $H \times M_0(G) \ni (h, \mu) \to U(\mu)h \in H$ is continuous. Hence, this mapping admits extension to the continuous mapping $H \times M^1(G) \to H$. □

COROLLARY XIX.20.7. Since $U(\delta_g) = U(g)$, and $U(f) = U(f\,dg)$ for $f \in L^1(G, dg)$, we have

$$U(g)U(\mu)h = U(\delta_g)U(\mu)h = U(\delta_g * \mu)h.$$

Let us now consider an extremely important

Case of Unitary Representation. Suppose that (U, H) is a unitary representation of the group G. Now, $U(x)^* = U(x^{-1})$, so that

$$(U(\mu)^*h|a) = (h|U(\mu)a) = \int (h|U(x)a)\, d\bar{\mu}(x)$$

$$= \int (U(x^{-1})|h\ a)\, d\bar{\mu}(x) = (U(\mu^*)h|a)$$

(where we have set $\mu^* := \bar{\check{\mu}} = \check{\bar{\mu}}$ and $\check{\mu}$ is the image of the measure μ under the mapping $x \to x^{-1}$). Thus we have the formula

$$U(\mu)^* = U(\mu^*).$$

Since $(\mu * \nu)^* = \nu^* * \mu^*$, we see that the unitary representation (U, H) is extendible to a $*$-representation of the algebra with involution $M^1(G)$ into the $*$-algebra $\mathscr{L}(H)$ of operators in the Hilbert space H.

Representations of Lie Groups. Regularization. The Gårding Theorem. Let us now consider the case when G is a Lie group, i.e., when it is a differentiable manifold (and hence, is analytic), on which the group operation is differentiable. Now let us define a procedure which is analogous to the regularization of functions (cf. Section 1 of this chapter) and which in actual fact is a natural generalization of the latter.

DEFINITION. Let G be a Lie group, let H be a locally convex complete space, and let (U, H) be a continuous representation. The vector $h \in H$ is differentiable (for U) if the mapping $\tilde{h}: G \to H$ defined by $\tilde{h}(x) := U(x)h$ is an element of the space $C^\infty(G, H)$ of differentiable mappings of G into H. The space of differentiable vectors will be denoted by H_∞.

Thus we have a theorem of fundamental importance for the theory of representations of Lie groups:

THEOREM XIX.20.8 (Gårding's). Let (U, H) be a continuous representation in a locally convex complete space H. Then

(i) $U(f)h \in H_\infty$ for $f \in C_0^\infty(G)$, $h \in H$.

(ii) The space H_0 of finite linear combinations of vectors of the form $U(f)h, f \in C_0^\infty(G), h \in H$, the Gårding space of the representation (U, H)—is dense in H (a fortiori H_∞ is dense in H).

(iii) The space H_0 (likewise H_∞) is U-invariant, that is, $U(g)H_0 \subset H_0$ for $g \in G$.

PROOF. (i) We are to show that the mapping $x \to U(x)U(f)h$ is differentiable. However, $U(x)U(f)h = U(\delta_x * f)h$, and the mapping $G \ni x \to U(\delta_x * f)h \in H$ is arbitrarily differentiable for every $h \in H$.

(ii) Relatively compact neighbourhoods $\mathcal{O}$ of a point e constitute a directed set under inclusion. For every $\mathcal{O}$ we take $f_\mathcal{O} \in C_0^\infty(G)$ such that suppt $f_\mathcal{O} \subset \mathcal{O}, f_\mathcal{O} \geqslant 0, \int f_\mathcal{O} dg = 1$. Then the net $(f_\mathcal{O})$ approximates δ_e ("approximate identity" or "Dirac sequence") in the sense that

(a) $f_\mathcal{O} * \varphi \to \delta_e^* \varphi = \varphi$ for every $\varphi \in C_0^\infty(G)$,

(b) $U(f_\mathcal{O})h \to U(\delta_e)h = U(e)h = h$ for every $h \in H$.

Indeed, suppose that $\|\ \|_{i \in I}$ is a family of seminorms determining the topology of the space H. We fix h and $i \in I$, $\varepsilon > 0$; owing to the continuity of (U, H) we can find a relatively compact neighbourhood $\mathcal{O}_i$ of the identity element of G such that $\|U(x)h - h\|_i < \varepsilon$ for $x \in \mathcal{O}_i$. Thus we have

$$\|U(f_{\mathcal{O}_i})h - h\|_i = \left\|\int U(x) f_{\mathcal{O}_i}(x)dx - h\int f_{\mathcal{O}_i}(x)dx\right\|_i$$
$$= \left\|\int \big(U(x)h - h\big) f_{\mathcal{O}_i}(x)dx\right\|_i$$
$$\leqslant \|U(x)h - h\|_i f_{\mathcal{O}_i}(x)dx < \varepsilon\int f_{\mathcal{O}_i} dx = \varepsilon.$$

(iii) $U(g)U(f)h = U(\delta_g * f)$, but $\delta_g * f \in C_0^\infty(G)$, which completes the proof. □

The space H_∞ of differentiable vectors was introduced and investigated by F. Bruhat in his celebrated doctoral thesis *Sur les representations induites des groupes de Lie*, Bull. Math. France 84 (1956). After Bruhat, we endow H_∞ with a topology which in general is finer than the relative topology from H; we do this in a natural manner as follows:

Consider the mapping $H_\infty \ni h \to \tilde{h} \in C^\infty(G, E)$. This mapping is an injection and, hence, it identifies H_∞ with a closed subspace of $C^\infty(G, E)$. Indeed, suppose that $\tilde{h}_n$ converges to $a \in C^\infty(G, E)$, $h_n \in H_\infty$, and then we have $a(x) = U(x)a(e)$ for every $x \in G$, because $\tilde{h}_n(x) = U(x)\tilde{h}_n(e)$, that is, a is a function of the form $\tilde{h}$ associated with the element h

$= a(e) \in H_\infty$. Now we endow H_∞ with the relative topology induced by $C^\infty(G, H)$ via the identification $h \to \tilde{h}$. Since the space $C^\infty(G, H)$ is complete, H_∞ is complete and (as can be shown) is a Fréchet space if H is a Banach space, in particular if H is a Hilbert space. (The construction of a countable family of seminorms requiring a topology of the space H_∞, which we owe to Goodman, will be given in Part III.)

Stone Theorem.

LEMMA XIX.20.9. Let (U, H) be the unitary representation of an Abelian topological group X. Then for every $h \in H$ the function $x \to (U_x h|h)$ is a continuous positive-definite function.

PROOF. We have

$$\sum_{j,k=1}^{n} c_j\overline{c_k}(U_{x_j - x_k} h|h) = \sum_{j,k=1}^{n} c_j\overline{c_k}(U_{x_k}^* h|U_{x_j}^* h)$$

$$= \Big(\sum_k \overline{c_k} U_{x_k}^* h \Big| \sum_k \overline{c_k} U_{x_k}^* h\Big) \geqslant 0. \ \square$$

THEOREM XIX.20.10 (Stone). Let (U_t) be a one-parameter group of unitary operators in a Hilbert space H (that is, $t \to U_t$ is a unitary representation of the Abelian group $\boldsymbol{R}^1$). Then there exists a spectral family $\{E(\lambda)\}_{\lambda \in \boldsymbol{R}^1}$ such that

$$(U_t f|g) = \int e^{2\pi it\lambda} d(E(\lambda)f|g), \quad f, g \in H.$$

On denoting $A := \int \lambda dE(\lambda)$, that is, $(Af|g) := \int \lambda d\big(E(\lambda)f|g\big)$, we can recast the right-hand side of this equality in the form $(e^{2\pi itA}f|g)$ and thus the hypothesis can be written briefly as $U_t = e^{2\pi itA}$, $t \in \boldsymbol{R}^1$.

Remark. The definition of the spectral family $\big(E(\lambda)\big)$ is recalled at the end of the preceding section. The reader who wishes to become more closely acquainted with the spectral theory of operators in Hilbert space is referred to the present author's monograph, *Methods of Hilbert Spaces*.

PROOF. By Lemma XIX.20.9 and Bochner's Theorem we know that there exists a finite measure μ_h on $\boldsymbol{R}^1$ such that

$$(U_t h|h) = \int e^{2\pi it\xi} d\mu_h(\xi), \quad \int d\mu_h = (h|h).$$

The "polarization formula"

$$4(U_t h|g) = \big(U_t(h+g)|(h+g)\big) - \big(U_t(h-g)|(h-g)\big) + $$
$$+ i\big(U_t(h+ig)|(h+ig)\big) - i\big(U_t(h-ig)|(h-ig)\big)$$

holds for any $h, g \in H$. Thus every pair $h, g \in H$ can be assigned a complex finite measure $\mu_{h,g}$ on $\boldsymbol{R}^1$, known as spectral measure, such that

$$(U_t h|g) = \int e^{2\pi i t \xi} d\mu_{h,g}(\xi).$$

Since $(U_t h|g) = \overline{(U_{-t} g|h)}$, we have $\mu_{g,h} = \overline{\mu_{h,g}}$. Therefore, for every $\lambda \in \boldsymbol{R}^1$ the bilinear form

$$(h, g) \to B_\lambda(h, g) := \int_{-\infty}^{\lambda - 0} d\mu_{h,g}(s)$$

is Hermitian: $B_\lambda(h, g) = \overline{B_\lambda(g, h)}$. Accordingly, a Hermitian operator E^λ such that $B_\lambda(h, g) = (E_\lambda h|g)$ corresponds to it. We can thus write

$$(E_\lambda h|g) = \int_{-\infty}^{\lambda - 0} d\mu_{h,g}(s),$$

or

$$(U_t h|g) = \int e^{2\pi i t \lambda} d(E_\lambda h|g).$$

To put it in somewhat more general terms: A Hermitian operator $E(M)$, given by the identity

$$(E(M)h|g) = \int \chi_M \, d\mu_{h,g},$$

can be associated with every Borel set M on $\boldsymbol{R}^1$. Hence we have $E_\lambda = E(]-\infty, \lambda])$.

The equality

$$E(M \cap S) = E(M) \cdot E(S)$$

will be shown to hold for $M, S \in \mathscr{B}^1$. If we set $S = M$, we obtain $E(M) = E(M)E(M)$ and, hence, the operators $E(M)$ are Hermitian and idempotent; similarly, for the operators E_λ we have: $E_\lambda = E_\lambda^*$, $E_\lambda = E_\lambda^2$. Clearly, $E(\boldsymbol{R}) = I$.

Note that the family $\{E(M)\colon M \in \mathscr{B}^1\}$ is an increasing family in the sense that for every $h \in H$ and for $M \supset S$ we have

$$(E(M)h|h) = \|E(M)h\|^2 \geqslant \|E(S)h\|^2.$$

It remains to prove formula (1):

$$(E(M)E(S)h|g) = \int_{\lambda} \chi_M(\lambda)\, d(E_\lambda E(S)h|g) =$$

$$= \int \chi_M(\lambda)\, d_\lambda \int_{-\infty}^{\lambda} \chi_S(t)\, d(E_t h|g) = \int \chi_M(\lambda)\chi_S(\lambda)\, d(E_\lambda h|g)$$

$$= \int \chi_{M\cap S}(\lambda)\, d(E_\lambda h|g) = (E(M\cap S)h|g)$$

for any $h, g \in H$.

Accordingly, $E(M)E(S)h = E(M\cap S)h$ for any h, that is, $E(M)E(S) = E(M\cap S)$. □

DEFINITION. Let Λ be any topological space, and let $\mathscr{B}(\Lambda)$ be its σ-algebra of Borel sets. The mapping $\mathscr{B}(\Lambda) \supset M \to E(M)$, where $E(M)$ are Hermitian idempotent operators with the properties

(i) $E(\Lambda) = I$,

(ii) $E(\emptyset) = 0$,

(iii) $E(M)E(S) = E(M\cap S)$,

is called the *spectral family on* $\mathscr{B}(\lambda)$.

21. THE HAAR INTEGRAL

One of the greatest achievements of the theory of integration in the 20th century has undoubtedly been the proof given by Alfred Haar (Hungary) in 1930 for the existence of a left-invariant Radon integral on a locally compact group. Haar himself presented the proof for groups which satisfy the second axiom of countability. The existence proof for an invariant Radon integral was provided for any locally compact group by A. Weil, S. Banach, and others.

For Lie groups, the construction of an invariant integral was given as long ago as 1897 by Adolf Hurwitz (a friend of Hilbert and Minkowski). The Hurwitz formula was given in Chapter XIV, Section 15; cf. also the end of the present section. The Hurwitz integral proved to be the principal instrument in the global theory of Lie groups and in the theory of the representation of these groups. Perhaps the loveliest application is to be found in the classical works of Hermann Weyl on the theory of representations of compact groups (cf., for example, *General Eigenfunction Expansions and Unitary Representations of Topol-*

ogical Groups) and the theory of almost-periodic functions on groups (cf. *Methods of Hilbert Spaces*). However, the proofs of the existence of the Haar integral are not based on the existence of the Hurwitz integral. This fact could, of course, be shown on the basis of Yamabe's Theorem which states that every locally compact connected group is the projective limit of a Lie group, making use of the Hurwith integral, and treating the Haar integral as a projective limit of Hurwitz integrals. However, such a procedure would be extremely involved and, moreover, the proof of Yamabe's Theorem is in fact based on the existence of the Haar integral. The invariant integral on uniform spaces was introduced by Loomis and, as we know, locally compact groups are uniform spaces. That construction is an adaptation of A. Weil's method for proving the existence of the Haar integral. This construction will be prefaced by some straightforward remarks.

PROPOSITION XIX.21.1. On a locally compact group G every function $f \in C_0(G)$ is uniformly left-(right-)continuous, i.e., is uniformly continuous with respect to the left (right) uniformity of B (B^1) given by a family of subsets $\beta \subset G \times G$ of the form $(x, y) \in \beta$, if $xy^{-1} \in V$, where V runs over the neighbourhood basis of the neutral element $e \in G$. In other words,

$$\bigwedge_{\varepsilon>0} \bigvee_{V} \bigwedge_{x,y} (xy^{-1} \in V) \Rightarrow |f(x)-f(y)| < \varepsilon,$$

$$\text{or} \quad |f(gx)-f(x)| < \varepsilon$$

for $g \in V$, $x \in G$.

In the case of the right uniform structure B^1:

$$\bigwedge_{\varepsilon>0} \bigvee_{\beta^1 \in B^1} \bigwedge_{(x,y)\in\beta^1} |f(x)-f(y)| < \varepsilon,$$

where $(x, y) \in \beta^1$, if $x^{-1}y \in V$.

PROOF. Let K denote a compact subset in G, let

$$C_K = \{f \in C_0(G): \underline{f} \subset K\},$$

and let U be a compact symmetric neighbourhood of the element e, that is, $U = U^{-1} = \{g \in G: g^{-1} \in U\}$. Since the set

$$W = \{g \in G: |f(gx)-f(x)| < \varepsilon, x \in UK\}$$

is open and since $W \ni e$ (but $f(x) = 0 = f(gx)$ for $g \in U$, $x \notin UK$), the inequality $|f(gx)-f(x)| < \varepsilon$ is satisfied for any x if $g \in V := W \cap \cap U$. □

We now go on to the fundamental theorem:

FUNDAMENTAL THEOREM XIX.21.2 (Haar, A. Weil). A left (right) Haar integral μ exists, i.e.,

$$\int f(gx)\,d\mu(x) = \int f(x)\,d\mu(x) \quad \text{for } f \in C_0(G),\ g \in G.$$

The proof is carried out in four stages.

(i) *Haar quotient* $(u:v)$, where $u, v \in C_0^+(G)$, $v > 0$, is the term used to denote a nonnegative number defined as follows:

Let $c_i \in \boldsymbol{R}_+$, $g_i \in G$ be such that

$$u(x) \leqslant \sum_{n=1}^{n} c_i v(g_i x)$$

for any $x \in G$.

Such c_i clearly do exist because there exists a nonempty open set $U \subset G$ such that $\inf\limits_{y \in U} v(y) > 0$ and the support of the function v can be covered with a finite number of left translations of U. Let $C(u, v)$ denote the totality of sequences (c_i) for which suitable sequence (g_i) exists. Suppose that

$$(u:v) = \inf \sum c_i,$$

where (c_i) run over $C(u, v)$.

The Haar quotient possesses the following properties:

(a) $(L_g u:v) = (u:v)$ (*left invariance*),

(b) $(u_1+u_2:v) \leqslant (u_1:v)+(u_2:v)$ (*subadditivity*),

(c) $(\alpha u:v) = \alpha(u:v)$, $\alpha > 0$ (*positive homogeneity*),

(d) $(u_1 \leqslant u_2) \Rightarrow \big((u_1:v) \leqslant (u_2:v)\big)$ (*isotonicity*),

(e) $(u:w) \leqslant (u:v)\,(v:w)$,

(f) $(u:v) \geqslant (\sup u)/(\sup v)$.

PROOF. (f) Take x such that $u(x) = \sup u(G)$; then $\sup u \leqslant \sum c_i v(g_i x) \leqslant (\sum c_i) \sup v$, and hence $\sum c_i \geqslant (\sup u)/(\sup v)$, whereby the hypothesis follows.

Points (a)–(d) emerge immediately from the definition.

(e) If $(c_i) \in C(u, v)$, $(d_j) \in C(v, w)$, or $u(x) \leqslant \sum c_i v(g_i x)$, $v(x) \leqslant \sum d_j w(s_j x)$, then $u(x) \leqslant \sum_{i,j} c_i d_j w(s_j g_i x)$, and thus

$$(u:v) \leqslant \inf \Big(\sum_{i,j} c_i d_j\Big) \leqslant \inf\Big(\sum_i c_i\Big) \inf \sum_j d_j = (u:v)\,(v:w). \square$$

(ii) The semi-additive nonnegative functional I_φ,

$$C_0^+(G) \ni u \to I_\varphi(u) \in \mathbf{R}^+,$$

is defined as follows: fix $0 \neq u_0 \in C_0^+(G)$ and set

$$I_\varphi(u) := \frac{(u:\varphi)}{(u_0:\varphi)};$$

by (e), it follows that

$$\text{(g)} \qquad \frac{1}{(u_0:u)} \leqslant I_\varphi(u) \leqslant (u:u_0).$$

This means that the functional I_φ is uniformly bounded with respect to φ.

(iii) The functional I_φ has all the prescribed properties, with the exception of additivity. We shall show that I_φ is "a fortiori additive", the smaller the support of the function. To be precise, the following lemma is true:

LEMMA XIX.21.3. The relation

$$\bigwedge_{u_1 u_2 \subset C_0^+(G)} \ \bigwedge_{\varepsilon > 0} \ \bigvee_{\substack{V \ni e \\ V\text{—open}}} \quad I_\varphi(u_1) + I_\varphi(u_2) \leqslant I_\varphi(u_1 + u_2) + \varepsilon$$

is satisfied for any function $\varphi \in C_V^+$.

The Haar integral is a generalized limit of the functional I_φ if $\underline{\varphi} \to \delta_e$, that is, if the supports $\underline{\varphi}$ are close to $\{e\}$.

PROOF OF THE LEMMA. Take a function $v \in C_0^+$ such that $v(x) = 1$ for $x \in \operatorname{suppt}(u_1 + u_2)$. Also take arbitrary $\delta, \eta \in \mathbf{R}_+$, and let

$$u := u_1 + u_2 + \delta v,$$

$$w_i(x) = \begin{cases} u_i/u(x) & \text{for } u(x) > 0, \qquad i = 1, 2. \\ 0 & \text{for } u(x) = 0. \end{cases}$$

Since $w_i \in C_0(G)$, by Proposition XIX.21.1 there exists a set V such that $|w_i(x) - w_i(y)| < \eta$ for $y^{-1} x \in V$. Now, let $\varphi \in C_V^+$ and $u(x)$

$\leqslant \sum_j c_j \varphi(g_j x)$; hence if $\varphi(g_j x) \neq 0$, that is, if $g_i x \in V$, then $|w_i(x) - w_i(g_j^{-1})| < \eta$. Hence we have

$$u_i(x) = u(x) w_i(x) \leqslant \sum_j c_j \varphi(g_j x) w_i(x)$$
$$\leqslant \sum_j c_j \varphi(g_j x) \left(w_i(g_j^{-1}) + \eta\right),$$

or $(u_i : \varphi) \leqslant \sum c_j \left(w_i(g)_j^{-1} + \eta\right)$. Since $w_1 + w_2 = (u_1 + u_2)/u$, we have $u \geqslant u_1 + u_2$, whereby $w_1 + w_2 \leqslant 1$, and we obtain $(u_1 + u_2 : \varphi) \leqslant \sum c_j (1 + 2\eta)$. Dividing both sides by $(u_0 : \varphi)$, taking $\sum c_j$ sufficiently close to $(u : \varphi)$, and bearing in mind that $I_\varphi(f) = \dfrac{(f : \varphi)}{(u_0 : \varphi)}$, we get the inequality

$$I_\varphi(u_1) + I_\varphi(u_2) \leqslant I_\varphi(u)\,(1 + 2\eta)$$
$$\leqslant \left(I_\varphi(u_1 + u_2) + \delta I_\varphi(v)\right)(1 + 2\eta)$$

(since $u = u_1 + u_2 + \delta v$).

However, $I_\varphi(f) \leqslant (f : \varphi_0)$, and so we obtain the hypothesis of the lemma by taking δ and η so small as to have

$$2\eta(u_1 + u_2 : u_0) + \left(\delta(1 + 2\eta)\,(v : u_0)\right) < \varepsilon. \;\square$$

(iv) *Construction of the Haar Functional μ as the Generalized Limit of I_φ.* Let

$$P := \bigtimes_u \left[\frac{1}{(u_0 : u)}, (u : u_0)\right];$$

P is a compact space (Tychonoff's Theorem!). Since

$$\frac{1}{(u_0 : u)} \leqslant I_\varphi(u) \leqslant (u : u_0),$$

the functional I_φ belongs to P. Let $S(V) := \overline{\{I_\varphi \in P : \varphi \in C_V^+\}}$. As a closed subset of the compact space P, the set $S(V)$ is compact. Since $S(V_1) \cap \ldots \cap S(V_n) = S\left(\bigcap_{i=1}^n V_i\right) \neq \emptyset$, the family $\{S(V)\}$ possesses the infinite intersection property and thus $\bigcap_S S(V) \neq \emptyset$, where V runs over the compact neighbourhood of the neutral element. Let $\mu \in \bigcap_S S(V)$. Now, $\mu \in \overline{\{I_\varphi : \varphi \in C_V^+\}}$ for any V, so that for any $u_1, \ldots, u_k \in C_0^+(G)$,

$\varepsilon > 0$, there exists a $\varphi \in C_V^+$ such that $|I_\varphi(u_j) - \mu(u_j)| < \varepsilon$, $i = 1, \ldots, k$. Since I_φ is left-invariant, subadditive, positive, and positive-homogeneous, μ has the same properties. That μ is additive follows from the lemma:

$$\mu(u_1 + u_2) \leqslant \mu(u_1) + \mu(u_2) \leqslant \mu(u_1 + \mu_2) + 2\varepsilon,$$

because $\varepsilon > 0$ is arbitrary. We now extend μ to the whole space $C_0(G)$, setting $\mu(u_1 - u_2) := \mu(u_1) - \mu(u_2)$, $u_j \in C_0^+(G)$. Since the inequality $\mu(u) \geqslant 1/(u_0 : u) > 0$ guarantees that the Radon functional μ is not trivial, μ is a (left-invariant) Haar integral. □

Uniqueness of the Haar Integral. It did not follow from its construction that the integral μ is unique up to a factor; this is the subject of the following fundamental

THEOREM XIX.21.4 (von Neumann, A. Weil). The Haar integral is unique to within a multiplicative constant.

The proof follows from the next lemma:

LEMMA XIX.21.5 Let μ and ν be a left-invariant and a right-invariant Haar integral, respectively; than there exists a real-valued positive function k on G such that

$$(1) \qquad \int k u \, d\nu = \int u \, d\mu, \qquad u \in C_0(G),$$

$$(2) \qquad k(y) \int v \, d\nu = \int v(yx^{-1}) \, d\mu(x), \qquad y \in G,\ v \in C_0(G).$$

PROOF OF THE LEMMA. By Fubini's Theorem and by the invariance of the integrals μ and ν we have

$$\begin{aligned}(3) \qquad \int u \, d\mu \int v \, d\nu &= \iint u(x) v(y) d(\mu \otimes \nu)(x, y) \\ &= \int d\nu(y) \int u(yx) v(y) d\mu(x) = \int d\mu(x) \int u(yx) v(y) d\nu(y) \\ &= \int d\mu(x) \int u(y) v(yx^{-1}) d\nu(y) = \int u(y) \left(\int v(yx^{-1}) \right) d\mu(x) d\nu(y).\end{aligned}$$

Note that, on taking any function v, $0 \neq v \in C_0^+(G)$, we obtain a continuous function

$$(4) \qquad y \to k(y) := \left(\int v \, d\nu \right)^{-1} \int v(yx^{-1}) \, d\mu(x)$$

on the right-hand side of (2). Thus (4) yields (2) for any $u \in C_0(G)$. Now, v does not appear on the right-hand side of (1), so that $\int (k_1 -$

$-k_2)u\,dv = 0$ identically for $u \in C_0(G)$, and thus $k_1 = k_2$, that is, (1) determines k, which is to say k does not depend on the choice of v. If we take v so that $\int v\,dv = 1$, then (3) gives us

$$\int u\,d\mu = \int u(y)k(y)dv(y), \quad u \in C_0(G),$$

or equality (1). □

PROOF OF THEOREM XIX.21.4. Let μ_1 and μ_2 be left-invariant Haar integrals, and let ν be a right-invariant Haar integral. By Lemma XIX.21.5 we know that there exist positive continuous functions k_1 and k_2 on G such that

$$\int k_i u\,dv = \int u\,d\mu_i, \quad u \in C_0(G),\ i = 1, 2.$$

In that case, $f := k_1/k_2$ is a positive continuous function such that $\int u\,d\mu_1 = \int f k_2 u\,dv = \int f u\,d\mu_2$ for any $u \in C_0(G)$. By the left-invariance of the integrals μ_1 and μ_2, for any $g \in G$ we have

$$\int f u\,d\mu_2 = \int u\,d\mu_1 = \int u(gx)\,d\mu_1(x)$$
$$= \int f(x)u(gx)\,d\mu_2(x) = \int f(g^{-1}x)u(x)\,d\mu_2(x)$$

identically for $u \in C_0(G)$, whence $f(x) = f(g^{-1}x)$ identically for x, or $f =$ const, and thus $\int u\,d\mu_1 =$ const $\int u\,d\mu_2$, $u \in C_0(G)$. □

The Invariant Integral on a Uniform Space. Let X be a locally compact space on which a locally compact group G actstr ansitively, i.e., we have a continuous mapping $A: G \times X \to X$ such that

$$A(g_1 g_2, x) = A(g_1, A(g_2, x)), \quad g_1, g_2 \in G,\ x \in X.$$

Hereafter we shall write $gx := A(g, x)$. If (X, B) is a uniform space with uniformity B (inducing the topology of X), we say that G acts *uniformly* on (X, B), if B is G-invariant, i.e.,

$$((x, y) \in \beta) \Leftrightarrow ((gx, gy) \in \beta), \quad \beta \in B,\ g \in G.$$

It is not difficult to verify that: (G acts uniformly on X) $\Leftrightarrow$ (for every neighbourhood U of a point $x \in X$ there exists a neighbourhood $W \ni x$ such that

$$(*) \qquad (W \cap (gW) \neq \varnothing) \Rightarrow (gW \subset U)).$$

If condition $(*)$ is satisfied, the family $\bigcup_{g \in G} gU \times gU$, where U runs over the topology of X may be taken for B. (Of course, for every g the mapping $X \ni x \to gx \in X$ is uniformly continuous with respect to the uniformity B.)

Examples. 1. Let $X = G$ and let B a family of sets of the form $\{(x, y) \in G \times G: x^{-1}y \in V\}$, where V runs over the filter of neighbourhoods of the element ε. Then the group G acts uniformly on itself.

2. Taking the family $\{(x, y): xy^{-1} \in V\}$, we obtain another (right) uniformity under which G also acts uniformly.

The reader should verify that the following generalization of the fundamental theorem can be proved by using the method employed to prove the Haar Theorem:

THEOREM XIX.21.6. If G acts uniformly on (X, B), then there exists a G-invariant Radon integral μ on X,

$$\int u(gx)\, d\mu(x) = \int u(x)\, d\mu(x), \qquad u \in C_0(X),\ G \in g.$$

Remark. While using the method of proof employed for the Haar Theorem, we nevertheless somewhat "modify" point (iv) of that proof. Namely, the role of the sets $S(V)$ is played now by the sets

$$S_1(\beta) := \overline{\{p_\varphi \in P: \underline{u} \times \underline{u} \subset \beta\}}, \qquad \beta \in B,$$

where $p_\varphi \in P$, $p_\varphi(f) := (f:\varphi)/(u_0:\varphi)$. The projection p_φ was previously denoted by l_φ.

If (X, G) is a uniform space, it is advisable to introduce a somewhat more general concept of the invariance of the integral.

DEFINITION. A Radon integral on X is said to be *χ-quasi-invariant* if there exists a function $\chi: F \to \boldsymbol{R}$ such that

$$\int u(gx)\, d\mu(x) = \chi(g) \int u(x)\, d\mu(x), \qquad u \in C_0(G).$$

Clearly, the function χ is a homomorphism of the group G into the multiplicative group $\boldsymbol{R} - \{0\}$.

Examples. 1. Let $X = \boldsymbol{R}^n$, $G = \mathrm{GL}_n(\boldsymbol{R})$ the group of real $n \times n$ matrices. By the change-of-variables theorem for the Lebesgue integral λ^n on $\boldsymbol{R}^n$ we know that λ^n is $|\det|^{-1}$-quasi-invariant with respect to $\mathrm{GL}_n(\boldsymbol{R})$.

Remark. Clearly, $\mathrm{GL}_n(\boldsymbol{R})$ does not act transitively on $\boldsymbol{R}^n$: the point $O = (0, \ldots, 0)$ is invariant. The formulation above should thus be interpreted as: $X = \boldsymbol{R}^n - \{0\}$. On this space $\mathrm{GL}_n(\boldsymbol{R})$ now acts transitively: $\lambda^n(O) = 0$.

2. Let E and F be finite-dimensional vector spaces (to focus our attention, let $E = \boldsymbol{R}^m$ and let $F = \boldsymbol{R}^n$, where $m \leqslant n < \infty$). Let $X = L(E, F)$. If σ is a Haar measure on the Abelian group X, then σ is $|\det|^{-1}$-quasi-invariant with respect to the linear group $G = L(F)$.

PROOF. Let $e_1, \ldots, e_m$ be a basis of the space F. Construct the operation of the group G preserving the isomorphism $X \cong F^m$: $X \ni f \to (f(e_1), \ldots, f(e_m)) \in F^m$. Observe that f is not unique only if $|f(e_1) \wedge \ldots \ldots \wedge f(e_n)|^2 = 0$; hence, monomorphisms constitute an open subset $O \subset X$ such that $\sigma(X - O) = 0$. If $E = F$, then σ is a left $|\det|^{-1}$-quasi-invariant measure on G. Thus, we obtain a left-invariant Haar integral (actually speaking, this is a Hurwitz integral, see Example 3) given by

$$\mu(u) = \int_G u\, d\mu = \int u |\det|^n d\sigma.$$

This example will enable us to give a general formula for the Hurwitz integral.

3. *The Hurwitz Integral.* Let G be an n-dimensional Lie group. We know (Chapter XIV, Section 15) that there exists a left-invariant n-form, unique within a multiplicative constant. It gives us the left-invariant Radon integral μ, i.e., Haar (Hurwitz) integral. Suppose that $g = (g_1, \ldots \ldots, g_n)$ and $h = (h_1, \ldots, h_n)$, where g_i and h_j are local coordinates in the neighbourhoods of the points $e, h \in G$; $(h \cdot g)_i = \varphi_i(h_1, \ldots, h_n, g_1, \ldots, g_n)$, $i = 1, \ldots, n$ are functions of class C^1. It is easily verified that $\left[\det\left(\frac{\partial \varphi_i}{\partial g_k}\right)\right]^{-1}(h, e)\, dh^1 \ldots dh^n$ is left-invariant n-form. Hence

$$\mu(u) := \int u(h)\left[\det\left(\frac{\partial \varphi_i}{\partial g_k}\right)\right]^{-1}(h, e)\, d\lambda^n(h)$$

(where we have assumed for simplicity that the support of the function u is "contained in the local chart or map") is a left-invariant integral.

INDEX OF SYMBOLS

INDEX OF SYMBOLS

INDEX OF SYMBOLS

INDEX OF SYMBOLS

INDEX OF SYMBOLS

INDEX OF SYMBOLS

INDEX OF SYMBOLS

$\left.\begin{matrix}SL_2(R)\\SL_2(Z)\end{matrix}\right\}$—the group 496

$ds^2_{\mathfrak{H}}$—the hermitian metric (of Poincaré–Klein) on $\mathfrak{H}$ 496

$K_{\mathfrak{H}}$—the Gaussian curvature 496

$\overline{\mathscr{H}}(M)$—the superharmonic functions 505

$\underline{\mathscr{H}}(M)$—the subharmonic functions 505

$P(z, \zeta) \equiv P_z(\zeta)$—the Poisson kernel 506

P_z—the Poisson integral 506

$\mathfrak{P}$—the Perron family 509

H_f— 510

$b = b^a$—the barrier at a 510

μ_x^B—the harmonic measure of x with respect to B 515

H_f^B—the generalize dsolution of the Dirichtet prcblεm (in the sense of Perron–Wiener–Brelot) 518

$H(x, E, B)$—the harmonic measure of the set E at a point x 520

S_P^M—the set of positive superharmonic functions 524

g_P^M—the Green's function of M with pole p 524

$A^*(\Omega)$—the space of nonvanishing holomorphic functions in Ω 541

$\sin \pi z$— 551

$\cos \pi z$— 551

$\Gamma(z)$—the gamma function 551

γ—the Euler's constant 552

(ω_1, ω_2)—the fundamental pair 554

$[\omega_1, \omega_2]$, L, $L(\omega_1, \omega_2) := Z\omega_1 + Z\omega_2$—the lattice (of periods) generated by ω_1 and ω_2 554

$x+\omega_1$, $x+\omega_1+\omega_2$, $x+\omega_2$—the period parallelogram 555

$\wp(z)$—the Weierstrass pe-function 558

G_n—the Eisenstein series of weight n 559

$D(x_1, \ldots, x_n) = \prod_{i<j} (x_i - x_j)^2$—the discriminant of the polynomial 562

$\sigma(z, L) \equiv \sigma(z)$—the Weierstrass sigma function 564

$Z(z, L) = Z(z)$—the Weierstrass zeta function 565

$\zeta(z)$—the Riemann zeta function 565

$\theta(z)$—the theta function 566

g—the Möbius transformation 566

$\left.\begin{matrix}\Gamma := SL_2(Z)\\ \Gamma_+ := SL_2(Z)/\{\pm 1\}\end{matrix}\right\}$—the modular groups 566

$M(k, \Gamma)$—the space of modular (cusp) forms of weight $2k$ 572

Δ—the discriminant of the Weierstrass equation 573

J—the Dedekind-Klein modular function 574

j—the modular invariant (elliptic modular function) 574

B_k—the Bernoulli numbers 575

$\tau(\cdot)$—the Ramanujan tau function 577

$C^{\#} := \pi^{-1}C$—the inverse immage of C 579

$\text{ord}_x f$—the order of f at x 584
$M := \bigoplus_{k=0}^{\infty} M(k, \Gamma)$—the graded algebra 588
$C(G_4, G_6)$—the polynomial algebra 588
h—the composite function 589
j_A—the j-invariant of A 591
$E(P)$—the elliptic-integral 591
$\text{Per}(\varphi)$—the lattice of periods of the holomorphic 1-form 593
$\text{Jac}(X)$—the Jacobi variety (or jacobian) of X 594
$\text{Div}(Y)$—the Abelian group of divisors on Y pointwise addition 595
deg—the degree 595
$\text{Div}_0(X)$—the subgroup of divisors of degree 0 596
$\text{Div}_p(X)$—the subgroup of principal divisors of meromorphic functions 596
$\text{Pic}(X) := \text{Div}_0(X)/\text{Div}_p(X)$—the Picard group of the compact Riemann surface X 596
F_Γ—the fundamental domain 603
$\{i_\infty\}$—the cusp of Γ (or F_Γ) 603
$\Gamma(N)$—the principal congruence subgroup of level N 606
$f_a(\omega_1, \omega_2)$—the first Weber function 607
$J_\gamma(x)$—the Jacobi determinant of the map γ at the point x 608
$i(D) := \dim M^1_{-D}$—the speciality index of the divisor D 611
$b(f, x) := m(f, x)$—the index of ramification of f at x 615
$b(f)$—the total index of ramification of f 615
θ—the elliptic integral 620
$\text{Rh}^1_A(X) := A^1(X)/dA(X)$—the holomorphic de Rham cohomology space 636
$d\mu$—the $SL_2(\mathbf{R})$-invariant Radon measure μ on the upper half plane $\mathfrak{H}$ 640
$m(A)$— 640
$\text{sn}(w, k)$—the Jacobi elliptic function 649
$\mathscr{A}$—the algebra of continuous functions 667
$\mathfrak{m}(x)$—the ideal in the algebra $\mathscr{A}$ 667
$\mathfrak{M}$ —the ideal 667
$\mathscr{H}$—the Hausdorff algebra 668
$L^1(G, \mu) = L^1(G)$—the algebra under convolution 673
C^*—the algebra 674
$[\cdot]_{\mathfrak{m}}$—the canonical mapping $\mathscr{A} \to \mathscr{A}/\mathfrak{m}$ 674
$\text{Sp}\, A$—the spectrum of the element A 675
$(\mu(\cdot\,; u, v)), u, v \in H$—the spectral measure 676
$\|g\|_\infty$—the norm in the space $L^\infty(\mu)$ 676
$|A(\Lambda_i)$—the spectral family (or partition of unity) 677
$\lim_{i \in I} \text{ind}\, X_i (\varinjlim X_i)$—the inductive limit of the space X_i 693
$T\mu$—the image of the measure μ under the mapping T 694
$\varprojlim X_\alpha$—the projective limit 696
$(X, \mathscr{A}, \mu)$—the probability space 697
μ—the probability measure 697
$\mu(A)$—the probability of occurrence of event A 697

INDEX OF SYMBOLS

SUBJECT INDEX

(Numerals in italics regard to pages of Part I of "Analysis")

SUBJECT INDEX

NAME INDEX